Pictorial Atlas of
Soil and Seed Fungi

Morphologies of Cultured Fungi and Key to Species

Second Edition

Pictorial Atlas of Soil and Seed Fungi

Morphologies of Cultured Fungi and Key to Species

Second Edition

Tsuneo Watanabe

CRC PRESS

Boca Raton   London   New York   Washington, D.C.

Library of Congress Cataloging-in-Publication Data

Watanabe, Tsuneo, 1937–
  Pictorial atlas of soil and seed fungi : morphologies of cultured fungi and key to species / Tsuneo Watanabe.—2nd ed.
     p.   cm.
  Includes bibliographical references (p.).
  ISBN 0-8493-1118-7 (alk. paper)
  1. Soil fungi—Atlases. 2. Seeds—Microbiology—Atlases. I. Title.

QR111 .W267 2002
632′.4—dc21   2002017443

This book contains information obtained from authentic and highly regarded sources. Reprinted material is quoted with permission, and sources are indicated. A wide variety of references are listed. Reasonable efforts have been made to publish reliable data and information, but the author and the publisher cannot assume responsibility for the validity of all materials or for the consequences of their use.

Neither this book nor any part may be reproduced or transmitted in any form or by any means, electronic or mechanical, including photocopying, microfilming, and recording, or by any information storage or retrieval system, without prior permission in writing from the publisher.

The consent of CRC Press LLC does not extend to copying for general distribution, for promotion, for creating new works, or for resale. Specific permission must be obtained in writing from CRC Press LLC for such copying.

Direct all inquiries to CRC Press LLC, 2000 N.W. Corporate Blvd., Boca Raton, Florida 33431.

Trademark Notice: Product or corporate names may be trademarks or registered trademarks, and are used only for identification and explanation, without intent to infringe.

Visit the CRC Press Web site at www.crcpress.com

© 2002 by CRC Press LLC

No claim to original U.S. Government works
International Standard Book Number 0-8493-1118-7
Library of Congress Card Number 2002017443
Printed in the United States of America 1 2 3 4 5 6 7 8 9 0
Printed on acid-free paper

Preface to the Second Edition

Since the publication of the English edition of this book in 1994, the classification system of organisms has changed dramatically, following the eight-kingdom concept based on the knowledge obtained with molecular techniques. All soil fungi now belong to the kingdom of Fungi, Chromista, or Protozoa; therefore, soil fungi described in this book are based on the traditional broad concept of fungi belonging to one of these kingdoms. Although some fungal species have been identified with the molecular techniques without observing the morphological characteristics, the fungi have been originally named on morphology, and new fungi will be described and accepted scientifically on the basis of morphological descriptions following the traditional classification system. Therefore, compiled knowledge on these fungi is still essential and needed for any workers on this line.

In this revised second edition, the original text is partially revised by rearranging, correcting, rewording, redrawing electronically and adding the recent knowledge, together with descriptions of 45 additional fungal species, their pictures, illustrations, and references.

I would like to thank John Sulzycki, Senior Editor; Samar Haddad, Project Editor; Erika Dery, Project Coordinator; and Barbara E. Norwitz, Publisher, at CRC Press, for their help, advice, and considerations in publishing this edition.

Tsuneo Watanabe

Preface

Biological elements related to higher plants and animals have been studied in relation to environmental problems, but the significance of microorganisms, another biological element, has been neglected. Microbial activities include nitrogen circulation and fixation, and disintegration of organic matter. Most living entities may not exist on Earth without the activities of microorganisms.

Some edible mushrooms are cooked and eaten directly. Bread, cheese, syouyu (soy sauce), miso, and various other foods, as well as alcoholic beverages, chemicals, medicines, including acids and penicillin, and other agricultural chemicals such as gibberellin, are products of microbial activities of yeasts, *Aspergillus*, *Penicillium*, and other fungi. Some fungi, however, attack directly plants, animals, and human beings, causing diseases. Also, some fungi are often poisonous and carcinogenic, excrete toxins contaminating foods. Dusts and fungi inhaled occasionally inside houses and buildings, or in the fields, cause asthma and allergenic troubles. Houses, foods, industrial products, books, and arts and craft materials are often damaged due to fungal activities. Some fungi are useful, however, influencing our lives directly or indirectly, but most fungi live under soil conditions separate from our lives and we do not understand their usability.

In this book, fungi detected or isolated from soil or plant seeds are provisionally termed as "soil and seed fungi." All these fungi influence one another, with antagonism or mutual symbiotic or cooperative relationships. Some of them also influence plant growth, and may degrade lignocellulose or deleterious atmospheric materials, including useless plastics and dioxin and its related compounds. By studying the individual fungi mycologically and physiologically, and increasing our knowledge of them, we may understand more about plant pathogenic and usable fungi, concomitantly from ecological views with plant protection and further improvement of human life.

Many books on mushrooms and fungi and the diseases they cause have been published. Drawings and pictures of these fungi and diseases are helpful for identification and disease diagnosis; however, only a few books are available on soil and seed fungi and their descriptions are very limited. In the plant pathological fields, mycological studies are prerequisite to determining the causal agents of the diseases, including isolation, preservation, classification, and identification.

The fungi in this text were detected, isolated, and stocked during such etiological and ecological studies. Most of them are not pathogenic to plants, and have been neglected for now. However, biocontrol practices have been emphasized recently to control plant diseases, and various saprophytic fungi have been studied to find effective biocontrol agents among them. In addition, physiologically active substances from fungal products may be screened and found by studying soil fungi together with gene bank practices.

In this text, 308 species of fungi that have been studied pathologically and mycologically during the course of soilborne plant disease studies, were selected among 10,000 living stock cultures of the author's own, and were described with pictures and illustrations.

Although purposes, materials, and methods of research are different, knowledge of soil fungi itself may be helpful for those interested in mycological work. Furthermore, only a few references and monographs are available on soil and seed fungi, and this book may therefore foster interest in these fungi.

For the publication of this book, the author would like to thank Susumu Yoshida, President, and Muneo Okazaki, Editor-in-Chief of Soft Science Publications, Tokyo, Japan, for their help and advice.

Tsuneo Watanabe

Preface to the English Edition

Soil fungal floras and fungi associated with plant seeds have been studied all over the world, and some of them have been published. However, publications treating morphologies of these fungi identified are rather limited, except those done by Gilman (1945), Barron (1968), Domsch and Gams (1972), and Domsch et al. (1980a,b).

In this book, morphologies of 308 species of fungi, mainly collected in Japan, are shown based on the author's own photomicrographs and illustrations.

This book was originally written in Japanese (original title: *Photomicrographs and Illustrations of Soil Fungi*), and published by Soft Science Publications, Tokyo, Japan, in May 1993. In the English edition, the original Japanese edition was partially revised, with four newly described fungi, together with 30 additional references.

The English edition was issued by invitation of Jon R. Lewis, Editor-in-Chief, Lewis Publishers, Inc., who asked me to write a book on "ecology and fungi" or a related topic in May 1991, when I was in the process of writing this book in Japanese, and by the acceptance of its translation into English.

I hope the English edition may contribute more easily and widely to the understanding of the ecology and distribution of fungi.

The author would like to thank Jon R. Lewis, Editor-in-Chief; Ed Norman and Jennifer Pate, Project Editors, Lewis Publishers, Inc.; and the persons concerned with the publication of the English edition for their help and advice; and S. Yoshida, President, Soft Science Publications, Tokyo, Japan, who kindly agreed to publish the English edition of this book.

Tsuneo Watanabe

Manual Design and Usage

More than 350 fungal species are described, including 46 Mastigomycetous species belonging to seven genera, 33 Zygomycetous species (12 genera), 36 Ascomycetous species (16 genera), 9 Basidiomycetous species (more than 2 genera), and 240 mitosporic fungal (Deuteromycetous) species (116 genera).

Latin binominals are adopted, mainly following publications by Farr et al. (1989), Hawksworth et al. (1995), and Jong and Gantt (1987). Also, major synonyms that appeared in recent literature are frequently consulted.

All of the fungi in this text are described alphabetically in order of Mastigomycotina, Zygomycotina, Ascomycotina, Basidiomycotina, and Deuteromycotina. For each fungus, the literature, morphology, and dimensions of each organ and material are described in order, occasionally with the remarks. Colony characteristics are not generally included because of too much variation in media used, and time of observation. Keys are prepared for all of the fungi to the genus level, and to the species level for the genus with more than three species.

All illustrations and pictures are based on the author's own work using the materials described. The fungi examined are isolated from soil, plant roots, and seeds. Most of the samples were collected in Japan, but some samples are from the Dominican Republic, Paraguay, Switzerland, and Taiwan, the Republic of China (ROC). Details of isolation methods are described elsewhere in the text. Scientific terms are mostly adopted following Hawksworth et al. (1995).

Living cultures of some fungi are deposited at the Gene bank, Ministry of Agriculture, Forestry and Fishery (MAFF), National Institute of Agrobiological Sciences in Tsukuba, at Fermentation Institute, Osaka (IFO), at American Type Culture Collection (ATCC), and at Centraalbureau Voor Schimmelcultures (CBS), and most of them are listed in the appendix of this book.

Acknowledgments

We wish to acknowledge the following journals for granting us copyright permissions: *Ann. Phytopath. Soc. Jpn.* (the Phytopathological Society of Japan), *Mycologia* (the Mycological Society of America, the New York Botanical Garden) *Mycologia Helvetica* (the Swiss Mycological Society), *Phytopathology* (the American Phytopathological Society), *Trans. Br. Mycol. Soc.* (the British Mycological Society), and *Trans. Mycol. Soc. Jpn. and Mycoscience* (the Mycological Society of Japan).

The Author

Tsuneo Watanabe earned his Ph.D. degree in plant pathology from the University of California, Berkeley in 1967. He has been active as a plant pathologist and mycologist in the field of soilborne diseases, their pathogens, and soil fungi for more than 30 years at the National Institute of Agroenvironmental Sciences, Forestry and Forest Products Research Institute, and National Institute of Advanced Industrial Science and Technology of Japan.

He is the author of more than 150 scientific papers on the subject, and the book titled *Dictionary of Soilborne Plant Diseases* (Asakura Publishing Company, Ltd., Tokyo, 1998).

Dr. Watanabe is currently studying lignocellulose and dioxin-decomposing fungi as a part of the national Bioconsortia Project of Japan.

Table of Contents

1. Study on Soil and Seed Fungi 1
   1.1 Fungi in Soil and Fungus Flora 1
   1.2 Relationships between Soilborne Plant Pathogenic Fungi and
       Other Soil Microorganisms 3
   1.3 Research Problems on Soil Fungus Floras 4
   1.4 Problems on Classification of Root-Inhabiting Fungi 5
   1.5 Study of Soil Fungi in Relation to Seed Fungi 6

2. Materials and Methodology 13
   2.1 Collection Sites and Samples 13
   2.2 Principles of Isolation Method 13
   2.3 Isolations and Cultures of Soilborne Fungi from Plants 14
   2.4 Isolation of Fungi from Soil 14
       2.4.1 Direct Inoculation Method 15
       2.4.2 Dilution (Plate) Method and Isolation Media 15
       2.4.3 Trapping (Bait) Method and the Substrates 15
   2.5 Preservation of Cultures 15
   2.6 Morphogenesis on Agar Cultures and Their Observations 16

3. Identification of Fungi 17
   3.1 Basal Knowledge for Identification 17
   3.2 Necessity of Experimentation 18
   3.3 Selection of Appropriate Binomials 18
   3.4 Morphologies to be Observed for Identification 18
   3.5 References for Fungal Taxonomy and Identification 20

4. Key to Classes of Soil Fungi 21
   Key to Mastigomycetes, Kingdom of Chromista 22
   Key to Zygomycetes, Kingdom of Fungi 23
   Key to Ascomycetes, Kingdom of Fungi 23
   Key to Basidiomycetes, Kingdom of Fungi 24
   Key to Deuteromycetes (Mitosporic Fungi), Kingdom of Fungi 25
   A. Pycnidium-Forming Fungi 25
   B. Sporodochium-Forming Fungi 26
   C. Synnema-Forming Fungi 27
   D. Aleuriosporae 27
   E. Arthrosporae 28
   F. Blastosporae 28
   G. Phialosporae 30
   H. Porosporae 31
   I. Sympodulosporae 32
   J. Annelosporae and Others 32
   K. Sterile (Nonspore-Forming) Fungi 33

5	List of Fungus Genera	35
	Mastigomycotina	35
	Zygomycotina	35
	Ascomycotina	35
	Basidiomycotina	35
	Deuteromycotina (Mitosporic Fungi)	36
6	Morphology of Soil Fungi	39
	Mastigomycotina	39
	Zygomycotina	93
	Ascomycotina	133
	Basidiomycotina	177
	Deuteromycotina (Mitosporic Fungi)	183

References 457

Appendix: List of Living Cultures of Soil Fungi Deposited and Publicized 471

Index 479

Afterword to the Second Edition 485

Afterword 486

1

Study on Soil and Seed Fungi

1.1 FUNGI IN SOIL AND FUNGUS FLORA

In soil, there live numerous kinds of bacteria, actinomycetes, fungi, algae and various other plants, and various protozoa including amoebas, nematodes, earthworms, mites, and other soil animals that keep the individual units and populations in balance. Some of them disintegrate dead plants, carcasses of dead animals, and organic matter, forming humus and maintaining soil fertility.

Some bacteria, including ammonifying bacteria, *Nitrosomonas* bacteria, and nitrifying bacteria fix free nitrogen and are involved with the nitrogen cycle on Earth. In addition, there are some soilborne plant pathogens and parasitic, harmful animals, although these organisms are few in number and small in population.

In this text, fungi that live in soil and are detected or isolated from soil are tentatively termed as soil fungi. Among such soil fungi, some are typical which have been isolated only from soil, and some are atypical, which are readily and frequently isolated from other atmospheres including various kinds of organic matter found around us.

Fungi from underground parts, and especially associated with soilborne diseases, appear to be typical "soil fungi," but some of them are also often isolated from seed, and thus named as "seed fungi."

Some fungi may live in soil associated with the organic matter derived from the aboveground living entities after their deaths.

Some airborne fungi may be contaminated with soil and may be casually isolated from soil. In addition, some fungi may be isolated from soil animals, and they may also be classified as one of the soil fungi. For example, 63 fungal species are isolated from cysts of soybean cyst nematode, *Heterodera schachili* (Carris et al., 1989).

In this text, no rigid definition of soil fungi is given, but fungi detected or isolated from soil, seeds, and roots of plants are tentatively termed as soil fungi. Isolates from plant residues in soil, soilborne animals, or mushrooms are also included as soil fungi.

The science of soil microbes has begun to study mechanisms of nitrogen fixation and other biochemical reactions in soil. During the course of study, various bacteria, yeasts, and fungi were isolated and described.

Adametz (1886) was the pioneer of soil fungus study, isolating and naming four species of yeasts, and 11 fungal species including *Aspergillus glaucus*, *Penicillium glaucus*, and *Mucor stolonifer*.

Jensen (1912) isolated 35 species of soil fungi during the course of study on classification of soil microbes, their morphology, distribution, and ecology, including fungus survival in winter.

In Holland, Oudemans and Koning isolated 45 fungal species including *Trichoderma koningi* and *Mucor racemosus*.

Butler (1907) isolated six *Pythium* species and some pythiaceous fungi in India, and Hagem (1907) isolated 16 species belonging to Mucorales, including eight new species.

In Japan, Takahashi (1919) isolated some fungi from field soils at Komaba in Tokyo, identified 25 species belonging to 13 genera, and reported in the first volume of *Ann. Phytopathol. Soc. Jpn.* It must be the pioneer soil fungal study in Japan, although some other early works on this line may have been conducted (note the early issues of *J. Agrochem., Jpn.*).

Waksman is famous for his discovery of streptomycin, and published a series of papers on soil fungi since 1916 and advocated a theory that common fungi live in any soil. To demonstrate this, he isolated more than 200 fungal species belonging to 42 genera from 25 locations in the U.S. Among these fungi, four genera, viz. *Aspergillus, Mucor, Penicillium,* and *Trichoderma*, live commonly in soil of any location. Especially, he noted that *Mucor* and *Penicillium* commonly occur in soil of temperate or cool-climate areas, *Aspergillus* in tropical soil, and *Trichoderma* occurs frequently in wet or acidic soils. In addition to these four genera, the following eight genera were pointed out to be dominant: *Acrostalagmus* (syn. *Verticillium*), *Alternaria, Cephalosporium* (syn. *Acremonium*), *Cladosporium, Fusarium, Rhizopus, Verticillium,* and *Zygorhynchus*. He further found that soil fungi tend to colonize on moribund plant residues, many kinds and numerous populations of bacteria and Actinomycetes live in soil, and the more fertile the soils, the more numerous and massive are the fungi.

Although there are differences in fungal species among soil isolates by different isolation methods, 17 common soil fungi were nominated including 12, listed by Waksman, and the other five, namely *Absidia, Botrytis, Chaetomium, Cylindrocarpon,* and *Stemphylium*, by Burges (1965).

Some fungi, such as *Pythium, Mortierella, Rhizoctonia,* and Basidiomycetous fungi, had not been listed or were very rarely listed as members of soil fungi before 1949 (Chesters, 1949), but by the hyphal isolation method devised by Warcup (1959), these fungi became common as soil fungus members. Especially, about 40 species of Basidiomycetous fungi have been isolated from soil and were identified in his work.

According to Burges (1965), over 600 fungus species, including 200 Phycomycetous species (*Mastigomycotina* and *Zygomycotina*), 32 Ascomycetous species, and 385 Deuteromycetous (Mitosporic fungal) species are recorded as soil fungi in the first edition of "A Manual of Soil Fungi" written by Gilman in 1945. Since then, soil fungus study has been serious and active, but nearly 1800 to 2000 species may be considered reasonable as the total number of soil fungi. This is calculated on the basis of Gilman's figure (600) in 1945 plus 1200, the figure increased since 1945. Its nearly 800 species of fungi are now being catalogued in the *Index of Fungi* every year (Hawksworth et al., 1995); if 20 of them (2.5%) are assumed to be soilborne, more than 1120 species must be added to the number of soil fungi for the past 56 years.

These figures are less than 2.7% of 72,065 species of total fungi recorded (Hawksworth et al., 1995). These small figures must be due to the lack of attention to individual organisms, because soil microorganisms have been treated previously en masse in the categories of bacteria, actinomycetes, fungi, algae, and so on.

Researchers want to identify soil fungi and know their genus and species names, their numbers, and the common and dominant species in soil surrounding us, although research purposes may vary considerably. Concomitantly, it is helpful to compare fungi or fungus floras in different soils, paying attention to the characteristics of the floras, and their similarity rates. To get this information, various isolation methods and media have been used, including soil dilution and soil plate methods, often with rose bengal streptomycin agar medium.

For particular genera or species, we like to compare their occurrence or detection frequency (isolation frequency) and coefficient of similarity. Values of coefficient of similarity are obtained, following the equation $S = 2W/(a + b)$, where "w" is the number of common species in two fungal populations and "a" and "b" are the numbers of species in both populations; namely, as the similarity between two populations increases, the value of the coefficient of similarity approaches one.

By using these approaches, interrelationships among habitats of higher plants, soil fungus floras, many different soils including forest, grassland, uncultivated and cultivated soils, and comparison of soil fungus floras in diseased vs. healthy soils, various soil atmospheres including soil pH, soil type, organic contents, soil depth in different seasons, and factors influencing soil fungus floras become subjects of study; through the knowledge of soil fungus floras we are able to understand the various habitats of higher plants and their activities, soil fertility, and disease occurrence areas and some environmental problems.

1.2 RELATIONSHIPS BETWEEN SOILBORNE PLANT PATHOGENIC FUNGI AND OTHER SOIL MICROORGANISMS

Although fungi resting or dormant in soil may be affected by soil temperatures, water content, pH, and physical or chemical soil elements, no competitive relationships may exist among these organisms. However, they are active and become cooperative with or antagonistic against other organisms with complicated relationships. For example, on water agar plates with a part of the washed roots of any plants without sterilization, some fungus organs such as conidia, chlamydospores, and sporangia or pycnidia, perithecia and other fruiting bodies may be readily observed in an elapse of time. Concomitantly, bacteria and actinomycetes growing around the hyphae penetrate, parasitize, and disintegrate the hyphae, or the hyphae themselves intermingle with each other.

Antagonism among amoeba, bacteria, or fungi under the microscope, or in nematodes escaped from the root tissues, may be observed on agar plates, trapped by a kind of nematode trapper, i.e., *Arthrobotrys* sp., and wriggling. All these phenomena may always happen, especially in the vicinities of plant roots. For example, *Rhizoctonia solani,* the notorious soilborne plant pathogen, may be parasitized by some typical soil fungi including *Trichoderma viride, Aspergillus,* and *Penicillium* spp., whereas *R. solani* itself may parasitize *Pythium debaryanum,* the damping-off fungus (Butler, 1957).

Verticillium species isolated from strawberry roots was endoparasitic to nematode and antagonistic against many soilborne plant pathogenic fungi (Watanabe, 1980). It is generally easy to detect and isolate *in vitro* soil fungi which are parasitic on any soilborne plant pathogens or inhibitory against them. However, their antagonistic activities appear to be very much limited in natural soil. For example, about 40% of 3500 isolates of bacteria and actinomycetes obtained from soils of 60 locations in Australia were shown to be antagonistic against one or more species of nine soilborne plant pathogens *in vitro,* but only 4% were antagonistic in soil (Broadbent et al., 1971). Antibiotic substances have been extracted from these antagonistic organisms, since Weindling in 1932 obtained viridin and gliotoxin from *T. viride* (if possible, from *Gliocladium virense*; Webster and Lamos, 1964) antagonistic against *R. solani*.

On one hand, there are some organisms that are antagonistic, but on the other hand, some organisms are known to influence the morphogenesis of other organisms. For example, *Phytophthora cinnamomi* are induced and stimulated for the formation of sporangia by some bacteria such as *Chromobacterium violaceum, Pseudomonas* spp. (Zentmyer, 1965; Marx and Haasis, 1965); its oospores — which are not formed in single cultures because of their heterothallic nature with different mating types — are induced to form singly by the influence of volatile substances excreted by *T. viride* and *Trichoderma* spp. This phenonomenon is called the "Trichoderma effect" (Brasier, 1971). Furthermore, chlamydospores of *Fusarium solani* f. *phaseoli* and sclerotia of *R. solani* and *Sclerotium rolfsii* are induced or stimulated to form by some bacteria and actinomycetes including *Arthrobacter* sp., *Bacillus subtilis, B. lichenoformis,* and *Bacillus* spp., *Protaminobacter* sp., and *Streptomyces griseus* (Ford et al., 1970; Henis and Inbar, 1968).

Differentiation of the rhizomorph and its growth of *Armillaria mellea* are stimulated by the activity of *Aureobasidium pullulans,* and its activity is due to the ether effect (Pentland, 1967). A similar stimulative effect was shown by the culture broth of *Macrophoma* sp. (Watanabe, 1986).

Sphaerostilbe repens, the root rot pathogen of tea and various plants in the tropics, is induced to form hyphal bundles and synnema by the influence of *Aspergillus* spp., *Penicillium* spp., and *Verticillium lamelicola* (Botton and El-Khouri, 1978). All these are examples of the morphogenesis of certain fungi under the influence of activities of other organisms, and these activities are related to alcohols including hexanal, ethyl and methyl alcohol; fatty acids including linoleic acid; oils and fats; and unidentified substances excreted by microorganisms.

1.3 RESEARCH PROBLEMS ON SOIL FUNGUS FLORAS

Fungus floras in soils or associated with plant roots have been studied worldwide by compiling fungal members together with their isolation or detection frequencies. As one of the problems related to these studies, mycologists tend to specialize in particular fungi, paying attention to the morphologies, whereas pathologists are interested in the pathogenicity of any of the organisms, and soil microbiologists pick up microbial problems en masse and ecologically. Whatever approaches are taken, the study of fungus floras is most important for the etiological study of plant diseases, and to understand environmental factors for soilborne disease occurrence.

With different purposes in research, particular groups of fungi or all fungi may be studied. Kinds of isolated fungi and their frequencies of isolation (or detection) may be studied at first, but treatment of each fungus may be different; for example, some of them may be described together with the process of identification, but others may just be listed without any detailed descriptions on identification procedures. Therefore, it is almost impossible to refer to publications equally.

A list of fungi in fungus floras may be of no use for some mycologists because of incomplete descriptions of the fungi in the list, for plant pathologists because of the lack of pathogenicity experiments, and for ecologists because of too many superfluous mathematical treatments of individual populations. However, by accumulating data on this line, knowledge may be increased, and concomitantly, natural phenomena may be understood and clarified gradually.

Technical problems such as isolation methods, media, and differences in isolation and incubation temperatures may occur for any particular fungus floras. For example, *Mortierella* spp. and *Pythium* spp., a kind of Zygomycetous and Mastigomycetous fungi, respectively, may be detected from almost any soil by a trapping method using cucumber seeds as a trapping substrate, but they were not listed in some literature using other isolation methods (see the Supplement at the end of this chapter). We often find that some 10 genera may be isolated from any plant root tissue plated on any nutrient-rich media after surface sterilization with chemicals, but more than 30 genera may be isolated by single hyphal tippings from the same sample plated on water agar after just washing without sterilization.

Many fungi remain unidentified because of the lack of the technology to induce sporulation.

Synonyms may be another troublesome problem for the study of fungal floras. For example, *Acrostalagmus, Cephalosporium, Hormodendrum,* and *Papularia* recorded as members of fungus floras in old literature are now believed to be synonyms of *Verticillium, Acremonium, Cladosporium,* and *Arthrinium,* respectively. These problems often bring some contradictions and disorder to the study of soil fungi.

In Japan, some works have been published on soil fungus floras, but in the U.S., Holland, and the U.K., abundant data and knowledge have been accumulated. For example, using different research methods or technologies, and in different times and locations, it was clarified that some soil fungi live in sand and frozen soil, and that some common fungi are also present in such soils. To my experience, the common fungi happened to be isolated in soils of the Dominican Republic, Japan, Taiwan (ROC), Switzerland, and South and North America.

Generally speaking, with the increase of research on soil fungus floras, soil fungi may be observed more individually, rather than en masse, and new fungi may be discovered and knowledge on their classification will be more and more increased.

1.4 PROBLEMS ON CLASSIFICATION OF ROOT-INHABITING FUNGI

Among various fungi observed on plant roots, there are pathogens inhibiting plant growth, damaging or collapsing them, but some fungi are mycorrhizal, which may stimulate plant growth. Most of these fungi, after penetrating through natural openings and wounds, colonizing the roots, take nutrients from substances excreted by the roots. However, the activities of many fungi are unknown.

Numerous fungal genera are associated with plant roots. For example, 58, 46, and 38 genera were found to be associated with roots of strawberry, sugarcane, and paulownia, respectively (Watanabe, 1977; Watanabe et al., 1974; 1987).

The number of fungi isolated is different on the basis of sample site, time and size of sample collection, isolation methods, and media used, and may be limited in a given site. With the increase of the number of isolation techniques and isolation media, however, the number may be generally increased, or may be influenced by disorder or the development of taxonomy itself or recognition of study of the fungus floras. Namely, when we pay attention to particular groups of fungi, we tend to neglect other individual fungi, without further identification.

In a given site, there may be the cosmopolitan and indigenous soil fungi. For studying these fungi qualitatively, their versatilities may be increased with the following study attitudes.

1. Changing experimental purposes, or increasing the number of approaches to clarify.
2. Using more versatile isolation methods and media.
3. Increasing the sample quantity.
4. Changing isolation atmospheres including temperatures, pH, moisture content, and timing.
5. Including plant residues, seeds, remnants, and carcasses of soil animals.
6. Treating the samples with heat or chemicals.
7. Including nonculturable fungi, such as rhizomorphic fungi.
8. Improving the techniques for sporulation and identification.

For Mastigomycetous fungi, *Pythium* spp. are generally isolated from old or declining roots, occasionally with *Aphanomyces*, and *Phytophthora*, and aqueous fungi such as *Dictyuchus, Saprolegnia,* and *Pythiogeton*. Among Zygomycetous fungi, the genus *Mortierella* is the most common, followed by *Mucor* and *Rhizopus*. In addition, *Absidia, Cunninghamella, Gongronella*, and *Syncephalastrum* may often be isolated. *Saksenaea* is not so often recorded as a member of soil fungus floras, but it was isolated from sugarcane roots in Okinawa. *Helicocephalum*, one of the Zygomycetous fungi which are not cultured singly and purely *in vitro*, was detected on agar cultures together with various organisms, and lived for more than 6 months by the mixed culture. These are rather rare fungi, because for over 30 years, the author has only once isolated them. For the ascomyceteous fungi, *Chaetomium* is most commonly isolated, and the number of its species isolated from soil is numerous. In addition, *Thielavia* and seven others are often isolated from soil. Basidiomycetous fungi have not been commonly isolated from soil, but *Coprinus* and *Thanatephorus* are examples of fruiting *in vitro,* but most of the Basidiomycetous isolates are not successful for fruiting *in vitro*. They are just judged to be Basidiomycetous because of their having clamp connections.

Mitosporic (Deuteromycetous) fungi are most frequently isolated, and they are rich in species. *Alternaria* and *Penicillium* are always listed as the members of fungus floras. Generally speaking, a total of 9 to 74 mitosporic fungus genera have been isolated from the soil of each location (see the Supplement at the end of this chapter).

Unsporulated, sterile, and unidentified fungi are different in their treatment in the literature, and they are classified as independent items. It is not clear if these fungi do not sporulate because of their innate nature, or if we could not induce their sporulation because of lack of technique. In addition, there are a few unidentified fungi, although they sporulated.

In the fungal taxonomy, fungi are classified, identified, and described mainly based on morphology observed in nature. For example, fungi belonging to the genus *Pestalotia* (syn., *Pestalotiopsis*) form sporodochia with morphologically characteristic conidia, but they do not usually form acervuli on agar cultures, the morphology of which has often been used as criteria for taxonomy. It appears to be difficult to induce morphogenesis by inoculation in nature, because of the excessive labor and time required. However, with the increase of knowledge on agar cultures of various fungi, unknown fungi may be identified more readily.

1.5 STUDY OF SOIL FUNGI IN RELATION TO SEED FUNGI

Unemerged seeds in soil or pre-emergence damping-off may often be caused by seedborne plant pathogens, and even after emergence, young seedlings often become collapsed by damping off or root rots. Most fungi associated with seeds, including plant pathogens, are similarly inhabitants of soil, and therefore the study of seed fungi is "the study of soil fungi." Among the fungi associated with commercial kidney bean seeds, *Colletotrichum lindemuthianum, Macrophomina phaseolina,* and *Rhizoctonia solani* were isolated from the seeds at the rates of 1 per 22 to 41 seeds (Watanabe, 1972b).

Christensen and Kaufmann (1965), working with the fungi associated with seeds, classified them into two groups, i.e., "field fungi," which must be contaminated with seeds in the field during harvest, and "storage fungi," which must be contaminated during transit and storage. We cannot specify when and how the contamination occurred on these seeds, but most of these fungi can live under both seed and soil conditions. There must be a few fungi only limited to living in seed, which infect only seeds and complete their life cycle on the seeds. Many organisms may be introduced into soil by sowing, but on the other hand, some organisms penetrate, contaminate, and colonize plant tissues directly or indirectly in various growth stages, repeatedly. Among these fungi, some influence seed quality and reduce germinability. Aflatoxin and toxic substances are produced from toxicogenic fungi and affect the health of animals and human beings; therefore, the study of seed fungi is very important as it concerns our health. In addition, there are many mycologically interesting fungi to which we must pay attention.

Supplement: Examples of Studies of Fungus Floras

Sample	Isolation (method and media)	Fungi (no. of genera, species, and remark)[a]	Ref.
		North America	
Canada Manitoba, 75 samples	Dilution plate with two media	Total: 64 g, 178 spp. (*Zygomycota* 7 g, 22 spp.; *Ascomycota* 9 g, 13 spp.; *Basidiomycota* 2 g, 2 spp.; *Deuteromycotina* 46 g, 141 spp.)	Bisby et al. (1933, 1935)
Canada Ottawa, Rhizosphere soil of wheat, oats, alfalfa, and pea	Dilution plate	Total: 17 g, 38 spp. (*Zygomycota* 3 g, 3 spp.; *Deuteromycotina* 14 g, 35 spp.) Floras compared between rhizo- and nonrhizosphere	Timonin (1940)
Canada 10 forest soils	Dilution with soil extract agar	Total: 22 g, over 56 spp. (*Zygomycota* 2 g, 12 spp.; *Ascomycota* 1 g, 1 spp.; *Deuteromycotina* 19 g, 43 spp.) Dominant fungi: *Mortierella*, *Pullularia*, *Trichoderma*, and *Penicillium*	Morral and Vanterpool (1968)
Canada Ontario, four conifer forest soils	Soil washing	Total: 41 g, 75 spp. (*Zygomycota* 5 g, 10 spp.; *Ascomycota* 5 g, 5 spp.; *Deuteromycotina* 31 g, 60 spp.) Dominant fungi: *Mortierella*, *Penicillium*, and *Trichoderma*	Widden and Parkinson (1973)
Canada Tundra at the North pole	Soil washing and plate	Total: 26 g, 46 spp. (*Zygomycota* 1 g, 1 sp.; *Ascomycota* 2 g, 2 spp.; *Deuteromycotina* 23 g, 43 spp.) Dominant fungi: Sterile fungi, *Penicillium*, *Chrysosporium*, and *Cylindrocarpon*; *Trichoderma* not detected. 27 species from leaves also studied	Widden and Parkinson (1979)
U.S. New Jersey, grassland and others, 8 samples	Dilution and direct inoculation with four media	Total: over 29 g, 106 spp. (*Zygomycota* 4 g, 19 spp., *Ascomycota* 2 g, 5 spp. *Deuteromycotina* 23 g, 82 spp.) Dominant genera: *Penicillium*, *Mucor*, *Aspergillus*, *Trichoderma*, and *Cladosporium*	Waksman (1916)
U.S. New Jersey, grassland and others, 25 samples	Dilution and direct inoculation with four media	Total: 42 g, 137 spp. (*Zygomycota* 4 g, 18 spp.; *Ascomycota* 5 g, 7 spp.; *Deuteromycotina* 33 g, 112 spp.)	Waksman (1917)

Supplement: Examples of Studies of Fungus Floras (continued)

Sample	Isolation (method and media)	Fungi (no. of genera, species, and remark)[a]	Ref.
U.S. Texas, forest soils, 4 positions	Dilution with Waksman's medium	Total: 13 g, 32 spp. (*Zygomycota* 3 g, 5 spp.; *Ascomycota* 1 g, 1 sp.; *Deuteromycotina* 9 g, 26 spp.) Dominant genera: *Penicillium, Aspergillus*	Morrow (1940)
U.S. Wisconsin southern hardwood forest soils, 13 locations	Dilution with soil extract agar	Total: 20 g, 50 spp. Dominant fungi: (*Zygomycota* 4 g, 9 spp.; *Ascomycota* 1 g, 1 sp.; *Deuteromycotina* 15 g, 40 spp.) Dominant genera: *Absidia, Mucor, Mortierella,* and *Zygorhynchus* Fungus floras reflected in higher plant vegetation	Tresner et al. (1954)
U.S. Georgia, forest and cultivated soils, 45 samples	Dilution, direct inoculation and others with rose bengal streptomycin agar and various media	Total: 63 g, 165 spp. (*Zygomycota* 11 g, 22 spp. *Ascomycota* 8 g, 10 spp. *Basidiomycota* 2 g, 2 spp; *Deuteromycotina* 42 g, 131 spp.) Dominant genera: *Penicillium, Aspergillus, Cunninghamella, Trichoderma,* and *Rhizopus* Fungus floras compared between summer and winter The floras became poorer with soil depth	Miller et al. (1957)
U.S. Southern Wisconsin forest soils, five locations	Dilution	Total: >36 g, 199 spp. (*Zygomycota* >4 g, >9 spp.; *Ascomycota* >2 g, >2 spp.; *Deuteromycotina* >30 g, >83 spp.) Dominant genera: *Penicillium, Paecilomyces, Gliocladium,* and *Humicola* Similarity rates among the respective samples, 18.9–40.7	Christensen et al. (1962)
U.S. Southern 10 states, 30 forest nurseries	Dilution and soil plate with rose bengal streptomycin agar	Total: 45 g, 121 spp. (*Zygomycota* 8 g, 16 spp.; *Ascomycota* 4 g, 11 spp.; *Deuteromycotina* 33 g, 94 spp.) Dominant fungi: *Aspergillus, Penicillium, Trichoderma,* and *Fusarium*; *Pythium* and *Rhizoctonia* not isolated	Hodges (1962)
U.S. Northern Wisconsin bogs and swamps, 15 locations	Dilution	Total: 57 g, 130 spp. (*Zygomycota* 2 g, 11 spp.; *Ascomycota* 8 g, 14 spp.; *Deuteromycotina* 47 g, 105 spp.)	Christensen and Whitlingham (1965)

Supplement: Examples of Studies of Fungus Floras (continued)

Sample	Isolation (method and media)	Fungi (no. of genera, species, and remark)[a]	Ref.
U.S. Alaska, uncultivated soils, 19 samples	Dilution with ether treatment	Total: 14 g, 23 spp. (*Zygomycota* 3 g, 3 spp.; *Ascomycota* 2 g, 3 spp.; *Deuteromycotina* 9 g, 17 spp.) Dominant fungi: *Mortierella*, *Penicillium*	Yokoyama et al. (1979)
U.S. South Dakota, grassland 10 samples	Dilution	Total: 13 g, 62 spp. (*Zygomycota* 1 g, 1 sp.; *Deuteromycotina* 12 g, 40 spp.) Dominant fungi: *Penicillium*, *Acremonium*, *Aspergillus*, *Chrysosporium*, and *Fusarium* Fungus floras compared domestically and internationally	Clarke and Christensen (1981)
U.S. and Mexico Arizona-Mexico Sonoran desert, 24 locations, 30 samples	Soil dilution and soil plate with 5 media	Total: 104 g, 230 spp. (*Mastigomycota* 1 g, 1 sp.; *Zygomycota* 7 g, 15 spp.; *Ascomycota* 21 g, 77 spp.; *Basidiomycota* 1 g, 4 spp.; *Deuteromycotina* 74 g, 133 spp.) No specific fungus flora in desert. *Curvularia*, colored fungi, and pycnidium-forming fungi frequently isolated	Ranzoni (1968)
Central and South America			
Honduras Banana rhizosphere soils	Dilution with rose bengal streptomycin agar	Total: 37 g, 52 spp.: (*Zygomycota* 2g, 2 spp.: *Ascomycota* 4 g, 4 spp.; *Basidiomycota* 1 g, 1 sp.; *Deuteromycotina* 30 g, 45 spp.) Fungus floras on roots, rhizoplanes, and rhizospheres compared	Goos and Timonin (1962)
Honduras Banana plantation soils	Dilution with rose bengal streptomycin agar	Total: 48 g, 64 spp.; (*Zygomycota* 6 g, 6 spp.; *Ascomycota* 8 g, 10 spp; *Basidiomycota* 1 g, 1 sp.; *Deuteromycotina* 33 g, 47 spp.) *Pythium* and *Mortierella* not isolated	Goos (1963)
Jamaica Sugarcane rhizosphere	Dilution and soil plate	Total: 67 g, 91 spp. (*Mastigomycota* 1 g, 1 spp.; *Zygomycota* 6 g, 6 spp.; *Ascomycota* 10 g, 10 spp.; *Deuteromycotina* 50 g, 50 spp.) Dominant fungi: *Aspergillus*, *Penicillium*, *Paecilomyces*, and *Cephalosporium*	Robison (1970)

Supplement: Examples of Studies of Fungus Floras (continued)

Sample	Isolation (method and media)	Fungi (no. of genera, species, and remark)[a]	Ref.
Panama and Costa Rica			
Panama and Costa Rica 31 samples	Direct inoculation with 17 media	Total: 73 g, 135 spp. (*Mastigomycota* 1 g, 1 sp.; *Zygomycota* 8 g, 11 spp.; *Ascomycota* 9 g, 20 spp. *Deuteromycotina* 39 g, 100 spp. *Myxomycetes* 3 spp.) Dominant fungi: *Penicillium, Aspergillus, Fusarium,* and *Cunninghamella*; *Mortierella* and Basidiomycetous fungi not isolated	Farrow (1954)
Panama and Costa Rica Banana field soils, 30 locations	Dilution with rose bengal streptomycin agar	Total: 33 g, 47 spp. (*Zygomycota* 4 g, 5 spp.; *Ascomycota* 3 g, 6 spp.; *Deuteromycotina* 21 g, 35 spp.) *Mortierella, Pythium,* and Basidiomycetous fungi not isolated	Goos (1960)
Trinidad Fallow and sugarcane field soils	Dilution and direct inoculation	Total: 40 g, 44 spp. (*Mastigomycota* 1 g, 1 sp.; *Zygomycota* 2 g, 2 spp.; *Ascomycota* 4 g, 8 spp.; *Deuteromycotina* 33 g, 33 spp.)	Mills and Vlitos (1967)
Australia			
Australia Southern Australia, wheat field soils	Dilution, soil plate, and hyphal isolation	Total: 57 g, 94 spp. (*Mastigomycota* 3 g, 4 spp.; *Zygomycota* 9 g, 19 spp.; *Ascomycota* 11 g, 13 spp.; *Basidiomycota* 3 g, 3 spp.; *Deuteromycotina* 31 g, 53 spp.) Isolation methods compared Basidiomycetous fungi isolated only by hyphal isolation method	Warcup (1957)
Australia Wheat field soils	Dilution, soil plate, and plant residue	Total: 54 g, 74 spp. (*Mastigomycota* 1 g, 8 spp.; *Zygomycota* 10 g, 11 spp.; *Ascomycota* 6 g, 8 spp.; *Basidiomycota* 5 g, 5 spp.; *Deuteromycotina* 32 g, 42 spp.) Fumigation effect on fungus floras studied	Warcup (1976)
Asia			
India Tea rhizosphere soils	Dilution with rose bengal streptomycin agar	Total: 26 g, 50 spp. (*Zygomycota* 4 g, 5 spp.; *Ascomycota* 4 g, 5 spp.; *Deuteromycotina* 18 g, 40 spp.) More numerous species isolated from roots infected by *Usulina zonata* than from healthy roots	Agnihothrudu (1960)

Supplement: Examples of Studies of Fungus Floras (continued)

Sample	Isolation (method and media)	Fungi (no. of genera, species, and remark)[a]	Ref.
Iraq Liwaas, Central Iraq, 5 locations	Not recorded	Total: 41 g, 150 spp. (*Mastigomycota* 1 g, 8 spp.; *Zygomycota* 3 g, 18 spp.; *Ascomycota* 3 g, 5 spp.; *Deuteromycotina* 34 g, 119 spp.) Dominant fungi: *Cladosporium, Aspergillus, Fusarium, Alternaria, Penicillium, Humicola, Mucor,* and *Pythium*	Al-Doory et al. (1959)
Israel Peanut fields, 12 locations	Dilution with two media	Total: 42 g, 95 spp. (*Zygomycota* 8 g, 14 spp.; *Ascomycota* 6 g, 30 spp.; *Deuteromycotina* 28 g, 51 spp.) Dominant fungi: *Mucor, Rhizopus, Aspergillus, Trichoderma, Cephalosporium,* and *Fusarium*	Joffe and Borut (1966)
Israel Mikve, cultivated soils	Dilution and direct inoculation with 13 media	Total: 46 g, 147 spp. (*Zygomycota* 7 g, 13 spp.; *Ascomycota* 7 g, 60 spp.; *Deuteromycotina* 32 g, 74 spp.) Dominant fungi: *Rhizopus, Aspergillus, Penicillium, Alternaria,* and *Fusarium* Effect of fertilization and vegetation on fungal floras studied	Joffe (1963)
Japan Osaka, paddy field soils, 4 locations	Dilution plate, isolated at 42°C	Total: 21 g, 37 spp.: (*Zygomycota* 2 g, 3 spp.; *Ascomycota* 9 g, 16 spp.; *Deuteromycotina* 10 g, 18 spp.) Fungus floras in both meso- and thermophilic fungi	Ito et al. (1981)
Japan Ogasawara (the Bonin Islands), forest and uncultivated soil, 11 locations	Direct inoculation, baiting with cucumber seeds and toothpicks	Total: 47 g, 81 spp.: (*Mastigomycota* 1g, 10 spp. *Zygomycota* 3 g, 9 spp. *Ascomycota* 1 g, 2 spp. *Basidiomycota* 1g, 1 sp. *Deuteromycotina* 41g. 59 spp.) the Bonin Islands: known as the Ocean Islands	Watanabe et al. (2001a)
Malaysia Forest and cultivated soils in the west	Dilution plate	Total: 26 g, 54 spp. (*Zygomycota* 3 g, 4 spp.; *Ascomycota* 4 g, 5 spp.; *Basidiomycota* 1 g, 7 spp,; *Deuteromycotina* 18 g, 38 spp.) *Pythium* and *Mortierella* not isolated	Varghese (1972)

Supplement: Examples of Studies of Fungus Floras (continued)

Sample	Isolation (method and media)	Fungi (no. of genera, species, and remark)[a]	Ref.
Europe			
England Brown forest and podozol soils	Sieve soaking plate and hyphal isolation	Total: 22 g, 50 spp. (*Zygomycota* 3 g, 16 spp.; *Deuteromycotina* 19 g, 33 spp.) Dominant fungi: *Trichoderma, Mucor, Penicillium,* and *Mortierella*	Thornton (1956)
England Suffolk grassland, five locations	Soil plate	Total: 48 g, 148 spp. (*Mastigomycota* 1 g, 3 spp.; *Zygomycota* 10 g, 30 spp.; *Ascomycota* 9 g, 24 spp.; *Deuteromycotina* 28 g, 91 spp.) Dominant fungi: *Penicillium, Mortierella, Absidia, Cephalosporium,* and *Fusarium*	Warcup (1951)
England Kidney bean root and its rhizosphere soils	Soil plate	Total: 17 g, 52 spp. (*Mastigomycota* 1 g, 1 sp.; *Zygomycota* 3 g, 5 spp.; *Ascomycota* 1 g, 1 sp.; *Deuteromycotina* 12 g, 19 spp.)	Dix (1964)
Sweden Forest soils in the south	Soil washing	Total: 21 g, 90 spp. (*Zygomycota* 3 g, 18 spp.; *Deuteromycotina* 18 g, 36 spp.) *Mortierella, Penicillium,* and *Trichoderma* occupied more than 71% of the total isolates	Söderström (1975)
Africa			
Nyasaland Coffee field	Soil plate	Total: 39 g, 81 spp. (*Zygomycota* 7 g, 8 spp.: *Ascomycota* 3 g, 4 spp.; *Basidiomycota* 3 g, 3 spp.; *Deuteromycotina* 26 g, 34 spp.) Dominant fungi: *Aspergillus, Trichoderma, Cephalosporium,* and *Fusarium*	Siddiqi (1964)

[a] *Aspergillus, Penicillium,* and *Rhizoctonia* are variously treated for their ana- or teleomorph in the original literature. The total number of genera and species of the respective classes in each work do not always coincide because figures of unknown and sterile fungi may be included in one work, but not in another.

2
Materials and Methodology

All fungi in the text are isolated, identified, and described on the basis of the following experimental methods and samples.

2.1 COLLECTION SITES AND SAMPLES

Soil samples were collected from cultivated and uncultivated soils (sand) with various habitats including forests, grasslands, paddy fields, and mountains in Japan, sugarcane fields in Taiwan, paulownia plantations in Paraguay, and black pepper plantations in the Dominican Republic.

As plant samples, more than 100 plants were assayed including various crops, flower plants, fruit and forest trees, and seeds of various crops including pea, radish, and black pepper, and forest seeds including Japanese black pine and flowering cherry seeds.

2.2 PRINCIPLES OF ISOLATION METHOD

After plating samples on agar cultures and incubating under certain conditions for a given period, samples may elongate hyphae and finally result in sporulation or fruit body formation.

By single hyphal tippings or isolation of spores directly at the tip of a transfer needle, or directly from fruiting bodies, pure cultures are established. For single hyphal tip isolation, a single hyphal tip elongated from substrates on water agar, less than 2 mm long in one piece, is selected and cut with agar blocks to establish pure cultures under the dissecting microscope at 30×.

At least 10 transfer needles may be prepared for fungal isolation, making them ready to use one by one after heat sterilization and cooling. As isolation media, 2% water agar has been frequently used. The agar plates should be thin, containing 7-ml/9-cm petri dishes. This is to avoid extra hyphal growth in thick agar and to isolate single hyphal tips readily. Bacteria are rather difficult to grow on plain water agar. Therefore, it is not necessary to prepare antibiotic- or acetic acid-containing agar in most cases.

In the single spore isolation method, spore suspensions appropriately diluted are poured onto plain water agar, and left for over 20 min for spore sedimentation and subsequently to remove extra water. After incubation for more than 2 h, single spores often germinated are dissected together with tiny agar blocks under a dissecting microscope. Spores with germ tubes are readily distinguished on agar. However, it is better to check and confirm single spores on each agar block under a compound microscope at 100× or more.

The best isolation techniques may be performed to use clean petri dishes without scratches or pits, and to prepare thin, solid plain water agar plates, and appropriate spore suspensions.

Using glass needles with capillary tips, single spores may be separated from a spore mass on water agar under a dissecting microscope. In this technique, thick plates (5 mm thick) of 3 to 4% plain water agar may be prepared. Isolations may be practiced with aseptic glass needles sterilized by soaking in boiled water for each isolation procedure.

2.3 ISOLATIONS AND CULTURES OF SOILBORNE FUNGI FROM PLANTS

Methods of isolation of soil fungi are different, according to the research purposes and samples used. The general procedures for isolation of any fungi from plant materials are as follows:

1. Wash plant materials under running tap water for at least 30 min.

2. For isolation from diseased plants, freshly infected parts may be selected, and they are cut into tissue segments of less than 5 mm. From the aged, infected tissues, the more numerous saprophytic fungi may be isolated.

3. Sterilized plant tissues may be prepared with antiformin (a strongly alkaline solution of sodium hypochlorite) or ethanol, together with unsterilized ones. Concentrations of chemicals may be different, according to the samples used, but generally they may be soaked in 1 to 5% antiformin or 70% ethanol for 30 sec to 5 min. Unsterilized samples may be included to reduce failure of isolation of the fungi susceptible to such chemicals. Part of the samples used for microscopic examination may also be used for samples for fungal isolation.

4. Plant tissue segments placed on isolation media are incubated at the appropriate temperature for 1 to 7 days. The isolation media, specific for individual fungi, may be used, but for general purposes, plain water agar may be one of the best media because bacteria and some contaminated fungi may be suppressed for lack of nutrition, and individual hyphae can be observed readily during isolation procedures. The treated plates may be incubated, under lower temperatures below 15°C, at 20 to 25°C, the optimum for most fungi, and at higher temperatures above 34°C. On the plates incubated under variable temperatures, morphogenesis may be stimulated, resulting in swift sporulation. Identification practices may become easy.

5. Hyphae may be elongated from the tissue segments plated within a few days, and single hyphal tippings may be practiced as soon as possible to get rid of extra contamination. The plates used may be further incubated for continuous observations and finally, ascocarps and other fruiting structures may be formed on such plates.

6. For isolation of Pythiaceous fungi, including *Aphanomyces* spp., the fungi may be trapped initially by susceptible plants or some other trapping substrates, and these materials may be soaked further under water for observation and isolation.

2.4 ISOLATION OF FUNGI FROM SOIL

For ecological studies of soil fungi such as distribution, populations, and activities, and their subsequent use for prediction of soilborne disease occurrence, fungi have been isolated from soil.

There are many isolation methods including dilution plate, soil plate, immersion tube, plant debris, hyphal isolation, flotation, and trapping (bait) methods. In this study, direct inoculation, trapping, soil dilution, and flotation methods have been used for soil fungus isolations.

The kinds of soil fungi and their isolation frequencies are different according to the isolation methods, the media, and the temperatures during isolation procedures, and satisfactory results may be obtained by a combination of a few of these methods.

2.4.1 Direct Inoculation Method

The direct inoculation method may be best for isolating various and general soil fungi simply, readily, and economically.

The method is to isolate fungi in pure cultures by single hyphal tippings from hyphae grown out of soil particles sprinkled over agar media (Waksman, 1916). Czapek agar was originally used as an isolation medium by Waksman, but plain water agar is equally well suited and recommendable after comparisons of different kinds of fungus genera and isolation frequencies in potato-dextrose agar (PDA), Czapek, and various agar media. No addition of streptomycin and other antibiotics or chemicals is necessary to avoid bacterial contamination. The drawback of this method is more frequent selection of the fast-growing fungi, while neglecting more slowly growing fungi.

2.4.2 Dilution (Plate) Method and Isolation Media

The diluted soil samples are plated onto isolation media, and pure cultures are obtained from colonies appearing on the selective media after 2 to 3 days' incubation at the appropriate temperatures.

For the selective isolation of *Fusarium* spp., pepton-pentachloronitrobenzene (PCNB) agar (Nash and Snyder, 1965), V-8 juice-dextrose-yeast extract agar (VDYA)-PCNB (Papavizas, 1967), and Komada's synthetic agar (Komada, 1972) have often been used. For isolation of *Phytophthora*, the pimaricin-vancomycin-PCNB ($P_{10}VP$) (Tsao and Occana, 1969) and hymexazol (3-hydroxy-5-methylisoxazole, HMI) containing $P_{10}VP$ or PDA at concentrations of 25 to 50 µm/ml, together with various antibiotics, were devised by Masago et al. (1977) and Tsao and Guy (1977).

The selective medium for *Pythium aphanidermatum* was also devised by Burr and Stanghellini (1973). A monograph of chemicals and selective media for various fungi was summarized by Tsao (1970).

2.4.3 Trapping (Bait) Method and the Substrates

This method is often used for isolation of Mastigomycetous fungi. The substrates, mixed with wet soil samples and incubated below 10°C or at 25°C for 1 to 7 days, are removed from soils and subsequently washed under running tap water. These substrates are then placed on plain water agar and incubated for more than 1 day. Pure isolates are obtained by cutting single hyphal tips grown out of the substrate together with agar blocks. The substrates used are various, according to the individual research purposes, but potato tubers, apple fruit, roots of sweet potato and carrots, and seeds of cucumber, corn, and lupine are often used.

2.5 PRESERVATION OF CULTURES

The preservation of cultures is basic and important for mycological work. Over 10,000 cultures have been stocked for more than 30 years in the author's collection by pouring potato dextrose (PD) broth in test tubes with steel caps (7 ml per tube) into old and shrunken cultures in tubes whenever necessary. The medium is economically prepared because it does not contain agar. Without any particular experience, it is possible to pour the medium into nearly 150 cultures in tubes within an hour, and it is not necessary to paste new labels for each test tube. However, it is rather difficult to observe colony characteristics in tubes with PD broth as compared with slant agar cultures.

Dried specimens are prepared by drying agar cultures in plastic petri dishes without lids within one day after treatment in paper bags or envelopes at 60°C. If time allows, the cultures in plates are dried with lids, changing the surface and reversing alternately every day for a few days until dried. The specimens are helpful for further studies on identification, and may be used as type specimen for description of the new species.

2.6 MORPHOGENESIS ON AGAR CULTURES AND THEIR OBSERVATIONS

For particular fungi, sporulation and formation of sclerotia may occur on hosts or in soil, and it may be possible to identify them on the basis of such morphologies. However, no signs of morphologies may be observed in most of the etiological work of plant diseases and, therefore, cultures obtained from plant tissues or soils may be directly observed on agar cultures for identification. Therefore, it is prerequisite to induce sporulation for the cultures on agar.

Most of Mitosporic (Deuteromycetous) fungi may sporulate on rich agar media such as PDA, but others are difficult to sporulate. The following trials are recommended to obtain successful sporulation.

1. Use some rich agar media, including cornmeal agar, Czapek (Dox) agar, malt agar, oatmeal agar, PDA, and V-8 juice agar. Some Ascomycetous fungi may often form fruiting structures in agar after a long incubation period, and therefore the quantity of media in the plates should be increased.
2. Alter the cultural environments, including light conditions and temperatures, for successful sporulation.
3. Change the balance of nutrition in agar. The cultures may be drastically changed from nutrient-rich to poor nutrient cultures, including plain water agar culture.
4. Use natural media prepared by mixing dried and propylene oxide-treated plant tissues into agar (Hansen and Snyder, 1947).
5. Soak a bit of culture, infected plant tissue, or substrate in pond water, well water, or petri salts solution to induce zoospore discharge for some Mastigomycetous fungi.
6. Case (cover) cultures with soils, or grow in wood chip medium, including rice bran, to obtain successful fruiting for some Basidiomycetous fungi.

3

Identification of Fungi

Fungus species, or the taxa (singular: taxon) are usually named on the basis of morphological characteristics. The names of fungi may be determined by comparing the already known species in morphologies. Generally speaking, any fungi may be identified if they are known species, and their morphologies are observed clearly.

Observations may be conducted at various levels from the naked eye through stereomicroscope or compound microscope to electron microscope. However, on the basis of satisfactory observations by compound microscope, identification may be possible, and the fungi may be correctly named due to individual abilities of observation of morphologies, or technical abilities for inducing sporulation in agar cultures.

3.1 BASAL KNOWLEDGE FOR IDENTIFICATION

Spores may be one of the most important morphological characteristics for identification. There are various types of spores including oospores, zygospores, ascospores, basidiospores, conidia, sporangiospores, and chlamydospores, and based on spore morphology, fungi are easily classified into Mastigomycetous, Zygomycetous, Ascomycetous, Basidiomycetous, and Mitosporic (Deuteromycetous) fungi.

Some fungi may be identified on the basis of other morphological characteristics besides spores. For example, Mastigomycetous or Zygomycetous fungi may be differentiated from other fungi because of the lack of hyphal septum. Furthermore, after soaking cultures in water, some Mastigomycetous fungi may emit zoospores, whereas most Zygomycetous fungi may form sporangiospores. Some Basidiomycetous fungi may be readily differentiated on the basis of clamped hyphae.

Particular representative isolates are selected at first for identification and observed for their morphological characteristics. Fruiting structures, spores, mycelia, growing habits, and morphologies of various organs in nature may be observed initially and in cultures subsequently.

Observations under a dissecting and a compound microscope are conventionally conducted. Without placing a cover slip, the habit of sporulation, spores in chains, or the spore head may be readily observed under a microscope. Observations with an oil lens are also essential.

Specialists may identify some fungi by the partial observation of those fungi on the basis of accumulated experience and knowledge, but observations must be repeated to know morphological characteristics in detail, and to access the most suitable taxon.

Keys must be prepared at various levels, including division, class, order, family, and genus. After trial and error, the most suitable taxon must be accessed, hopefully to the species level. After consulting the literature related to the expected taxon and rechecking morphologies, we may finish the identification.

Keys are based on the standard and general characteristics, but some keys are too artificial by nature, including some exceptions. Therefore, we just refer to such selections after keying out.

Some fungi may be named without detailed studies and, therefore, with further studies, other significant characteristics may be found and added; old literature may be consulted later, resulting in reclassification or the production of synonyms or new combinations.

All these things may occur routinely, and we understand this readily because there are numerous synonyms. Therefore, without completing identification just by following the keys, we have to check and consult fungi on the basis of the original descriptions, often comparing them with the type of specimen. At any rate, overall judgment is essential for identification.

3.2 NECESSITY OF EXPERIMENTATION

All descriptions, classifications, and naming may be based on the morphologies formed in nature. However, some fungi may exceptionally form morphological characteristics just by culturing. There are many fungi whose names are coined on the basis of the names of host plants, and such customs may be still present. Several synonyms are combined into one species on the basis of the results of inoculation experiments (i.e., in *Exobasidium* or rust).

Numerous heterothallic fungi may be classified based on the anamorphs, but by observing the teleomorphs artificially produced by fertilization experiments, more scientifically reliable identification may be achieved. Therefore, we have to get accustomed to observing morphologies formed on the host by inoculation tests or culturing.

3.3 SELECTION OF APPROPRIATE BINOMIALS

It is quite difficult to select the best binomial among various synonyms. For example, the binomial of the rice blast fungus, *Pyricularia oryzae* Cavara, has been traditionally used, but we are now at a loss as to whether we should use *P. grisea* (Cooke) Sacc. which has priority mycologically, or *Magnaporthe grisea* (Hebert) Barr, based on the recent work of the teleomorph. However, this fungus generally does not form the teleomorph in nature and *in vitro* and, therefore, on the basis of the rules of nomenclature, we can use either name scientifically (Rossman et al., 1990).

For the fungus with a few synonyms currently used, the best binomial must be selected on the basis of the individual scientific sense with future prospect. If possible, synonyms may be included in any scientific descriptions for such fungal names, but to adopt the most suitable name, recent literature should always be consulted. For example, there are three currently used binomials for the charcoal rot fungus, i.e., *Macrophomina phaseolina* (Tassi) Goid., *Sclerotium bataticola* Taubenhouse, and *Rhizoctonia bataticola* (Taub.) Butl., but the first name is now most commonly used.

Verticillium albo-atrum sensu lato had been commonly used as the wilt pathogen of various plants, but after controversy for recognition of *V. dahliae* for a long time, the latter name has been used, together with *V. albo-atrum* sensu strict.

3.4 MORPHOLOGIES TO BE OBSERVED FOR IDENTIFICATION

Although observations of the morphologies are most important for identification, cultural characteristics are similarly emphasized for some fungi and should not be neglected.

Identification of Fungi

The following points should be paid attention for observation, although there are some differences in individual fungi. In addition, physiological characteristics such as temperature responses and host ranges are included in the keys, and should be similarly studied, together with morphological characteristics.

A. **Cultural Characteristics**
   1. Color and tint in colony surface and reverse
   2. Smell or fragrance
   3. Quantity of aerial hyphae
   4. Colony surface texture: cottony, shrunken, sloppy, resupinate, velvety, powdery (floury), crustaceous, water soaked, embedded, yeast-like, sticky, homogeneous or heterogeneous, presence or absence of elevation
   5. Colony margin: smooth, irregular, restricted, spreading
   6. Pattern: zonate, radiate, flowery, arachnoid
   7. Pigment exuded: color, watery
   8. Organs formed: fruiting structures, sclerotia, rhizomorphs, synnema, sporodochia, stroma, setae

B. **Morphology**
   1. Size: length, width, thickness, etc.
   2. Color: refer to standard color charts (i.e., Rayner (1970), Ridgway (1912))
   3. Shape
      a. General characteristics for all fungi

         Hyphae (septate, aseptate, septum location, clamp connection, hyphopodia), appressoria, chlamydospores, rhizomorphs, synnema (pl. -ta), and others

      b. Differences in the respective classes

         Mastigomycetous fungi: oogonium (pl.-a), antheridium (pl. -a), oospores (plerotic, aplerotic), sporangium (pl. -a), or hyphal swellings (hypha-like sporangia, sphaerosporangia, lobate sporangia)

         Zygomycetous fungi: sporangium (columella, apophysis), sporangiola, sporangiophores, sporangiospores, zygospores, rhizoid, creeping hyphae, zygospores

         Ascomycetous fungi: ascocarp (nakid ascocarp, perithecium, apothecium, etc.), appendages (hairs, setae), ascoma wall [peridium, tissue (textura) type], stroma, ascus (pl. asci) (disposition, apical structure, evanescent or nonevanescent), paraphysis (pl. paraphyses), ascospores

         Basidiomycetous fungi: basidiocarp (volva, stipe, annulus, umbrella, hymenium (location, formation pattern, shape, gill, pore, needle shape, cystidia), basidia, spore print, basidiospores

         Mitosporic fungi (*Deuteromycotina*): apothecium (pl. -a), pycnidium (pl. -a), sporodochium (pl. -a), conidiophore (erect, creeping, resupinate, simple, or branched; branching pattern), conidial types (aleuriosporae, annellosporae, arthrosporae, blastosporae, phialosporae, porosporae, radulasporae, sympodulosporae), papulaspore

   4. Number: number of spore septum, zoospore flagellum, oospores per oogonium, antheridium per oogonium, oil globules per spore, ascospores per ascus, and basidiospores per basidium, etc.
   5. External and internal structures of tissues: smooth, echinulate, warty, presence or absence of hair, texture of peridium, component tissues of ascocarp or pycnidia (conidiocarp), component hyphae in Basidiomycetes (presence or absence of primary hyphae, skeleton hyphae, and/or uniting hyphae)
   6. Presence or absence of protuberances: number and shape

7. Positions of organs: monoclinous or diclinous, paragynous, amphigynous, hypogynous, etc., in positions of sexual organs in Mastigomycetous fungi
8. Relations to other organs: sexual patterns (conjugations of two aplanogametangia, of an aplanogamentangium and a gametangium, and of two gametangia), location
9. Proliferation: internal proliferation, external proliferation
10. Germination pattern: direct germination (by germ tubes), indirect germination (by zoospores)
11. Swimming pattern: monoplanetism, diplanetism
12. Connection conditions: presence or absence of catenulation in conidia and chlamydospores, and number and origin of catenulate spores
13. Shapes, formation order, and arrangement of appendaged hairs, setae, and others and/or supplementary organs
14. Formation pattern: formation of fruiting structures (discrete, aggregate, caespitose)
15. Positions of occurrence (aerial hyphae, embedded, erumpent)

C. **Physiological Characteristics**
   1. Temperatures: growth temperatures, optimum temperature for growth, cardinal temperatures for growth, growth rate
   2. Growth media and nutritional requirements: suitable media for sporulation and growth
   3. Reaction for reagent and staining: lactophenol, cotton blue, acid fuchsin, KOH, $FeSO_4$, Melzer reagent
   4. Resistance against chemicals
   5. Hyphal anastomosis
   6. Co-culture reaction
   7. Parasitic nature, pathogenicity

3.5 REFERENCES FOR FUNGAL TAXONOMY AND IDENTIFICATION

General: Ainsworth et al. (1973), Alexopoulos and Mims (1979), Aoshima et al. (1987), Arx (1981), Bessey (1961), Domsch and Gams (1972), Domsch et al. (1980), Farr et al. (1989), Hasegawa (1989), Hawksworth et al. (1995), Hiura (1967), Kobayashi et al. (1992), Miyaji and Nishimura (1991), Udagawa et al. (1973), Ulloa and Hanlin (2000), Webster (1980).

Zygomycetes: O'Donnell (1979), Zycha et al. (1969).

Mastigomycetes: Erwin et al. (1980), Ito (1936), Katsura (1971), Middleton (1943), Plaats-Niterink (1981).

Ascomycetes: Dennis (1978), Hanlin (1989, 1998, 1999).

Basidiomycetes: Ito (1955), Imazeki and Hongo (1987, 1989), Singer (1986).

Deuteromycetes: Barnett and Hunter (1987), Barron (1958), Carmichael et al. (1980), Ellis (1971, 1976), Kiffer and Morelet (1999), Matsushima (1975, 1995), Sutton (1980).

4
Key to Classes of Soil Fungi

Although soil fungi now belong to three kingdoms, viz. Kingdom of Protozoa including *Plasmodiophora,* Kingdom of Chromista including *Phytophthora* and *Pythium,* and Kingdom of Fungi including the rest of fungi according to the recent classification system (Hawksworth et al., 1995), they are treated as they were without the special separation.

Key words: ascospore(s), basidiospore(s), clamp connection, conidium (pl. -a), hypha (pl. -e), oospore(s), rhizomorph, sclerotium (pl. -a), sporangiospore(s), zoospore(s), zygospore(s)

1. Hyphae   aseptate 2
        septate 6

2. Sporangiospores   formed .... Zygomycetes
        none 3

3. Oospores   formed .... Mastigomycetes
        none 4

4. Zoospores   formed .... Mastigomycetes
        none 5

5. Zygospores   formed .... Zygomycetes
        none .... Mastigomycetes or Zygomycetes

6. Hyphae   with clamp connection .... Basidiomycetes
        without clamp connection 7

7. Spores   formed 8
        none 9

8. Ascospores   formed .... Ascomycetes
   Basidiospores   formed .... Basidiomycetes
   Conidia   formed .... Deuteromycetes

9. Sclerotia and   formed 10
   other organs   not formed .... Deuteromycetes and others

10. Sclerotia	well differentiated with rind and medulla	*Sclerotium*
	not well differentiated	11
11. Hyphae	constricted near branching area	*Rhizoctonia*
	not so	12
12. Papulaspores	formed	*Papulaspora*
	none	13
13. Rhizomorphs	formed	*Armillaria* and others
	none	Unidentifiable fungi

KEY TO MASTIGOMYCETES, KINGDOM OF CHROMISTA

Key words: antheridium (pl. -a), chlamydospore(s), oogonium (pl. -a), oospore(s), sporangiospore(s), sporangium (pl. -a), vesicle, zoospore(s), zoospore differentiation, zoosporangium (pl. -a)

1. Sexual organs	formed	2
	none	10
2. Oogonia	amphigynous	*Phytophthora*
	not so	3
3. Sporangia	globose	4
	not so	5
4. Zoospores	differentiated inside sporangia	*Phytophthora*
	differentiated in vesicles formed outside sporangia	*Pythium*
5. Zoosporangia	both lobate and hypha-like	*Plectospira*
	either lobate or hypha-like	6
6. Zoosporangia	hypha-like	7
	not so	8
7. Zoospores	encysted in a mass at the tip of hypha-like sporangia	*Aphanomyces*
	not so	*Pythium*
8. Zoosporangia	lobate	*Pythium*
	cylindrical	9
9. Zoospores	liberated outside sporangia, and oogonia bearing more than three oospores per oogonium	*Saprolegnia*
	encysted, forming a network within sporangia	*Dictyuchus*
10. Sporangia	large, unbalanced, mostly ellipsoidal	*Pythiogeton*
	not so	Heterothallic isolates (*Phytophthora* and others)

KEY TO ZYGOMYCETES, KINGDOM OF FUNGI

Key words: apophysis (pl. -ses), columella, homothallic, heterothallic, hyphal swellings (vesicles), merosporangia, rhizoid(s), sporangiophore(s), sporangium (pl. -a), zygospores

1. Vesicles	formed between sporangiophore and sporangia	2
	not formed	4
2. Sporangia	formed directly on vesicle	3
	at apexes of branches developed on vesicle	*Umbelopsis*
3. Sporangiospores	one spore per sporangium (conidia)	*Cunninghamella*
	more than two spores per cylindrical sporangium (merosporangium)	*Syncphalastrum*
4. Sporangia	globose	5
	flask-shaped	*Saksenaea*
5. Sporangia	with apophysis	6
	without apophysis	7
6. Apophysis	globose	*Gongronella*
	not so	*Absidia*
7. Sporangia	columellate	8
	without columella or poorly columellate	*Mortierella*
8. Columella	twisted or coiled	*Helicocephalum*
	not so	9
9. Rhizoid	formed just below sporangiophore	*Rhizopus*
	not formed	10
10. Sporangiophores	partially twisted	11
	not so	*Mucor*
11. Zygospores	formed in single culture (homothallic), zygospores with unequal suspensors	*Zygorhnchus*
	not formed (heterothallic)	*Circinella*

KEY TO ASCOMYCETES, KINGDOM OF FUNGI

Key words: ascocarp, ascospores, ascus (pl. asci), bitunicate, evanescent, gelatinous sheath, nonevanescent, ostiole, papilla, seta (pl. -e), unitunicate

1. Ascocarp	ostiolate or papillate	2
	not so	11

2. Ascus	1-spored .... *Monosporascus*
	4 to 8-spored 3

3. Ascospores	1-celled 4
	over 2-celled 9

4. Ascocarps	hairy or setose 5
	not so 7

5. Ascocarps	hairy 6
	setose, and cylindrical asci and pigmented ascospores ... *Phaeotrichosphaeria*

6. Ascocarps	with well-developed hairs .... *Chaetomium*
	not so, ascospores dark .... *Achaetomium*

7. Ascospore	with gelatinous sheath .... *Sordaria*
	not so 8

8. Asci	globose .... *Microascus*
	cylindrical .... *Glomerella*

9. Ascospores	composed of dark and hyaline unequal cells .... *Apiosordaria*
	with two homogenously colored cells 10

10. Asci	bitunicate .... *Didymella*
	unitunicate .... *Nectria*

11. Ascospores	1-celled 12
	over 2-celled 13

12. Asci	evanescent .... *Thielavia*
	nonevanescent .... *Anixiella*

13. Asci	bitunicate 14
	unitunicate 15

14. Ascospores	lunar-shaped .... *Massarina*
	cylindrical .... *Preussia*

15. Ascospores	2-celled, hyaline .... *Eudarluca*
	over 2-celled, composed of hyaline and dark cells .... *Zopfiella*

KEY TO BASIDIOMYCETES, KINGDOM OF FUNGI

Key words: basidiocarp (annulus [ring], hymenium (pl. -a) (arrangement, echinulate, lamellae (gill), pileus (cup), pore, shape, stipe [stalk], volva), basidiospore(s), basidium (pl. -a), cystidium (pl. -a)

Key: none

KEY TO DEUTEROMYCETES (Mitosporic Fungi), KINGDOM OF FUNGI

Key words: acervulus (pl. -li), conidioma (pl. –ta), pycnidium (pl. -a), spore type (aleuriospore, annellospore, arthrospore, blastospore, phialospore, porospore, sympodulospore), sporodochium (pl. -a), synnema (pl. -mata)

1. Conidiomata   formed 2
                    not formed 3

2. Conidiomata   Pycnidia formed .... Pycnidium-former
                    Sporodochia (or acervuli) formed .... Sporodochium-former
                    Synnemata formed .... Synnema-former

3. Conidia   formed 4
            not formed .... Sterile fungi

4. Conidia   Aleurospore-type .... Aleuriosporae
            Arthrospore-type .... Arthrosporae
            Blastospore-type .... Blastosporae
            Phialospore-type .... Phialosporae
            Porospore-type .... Porosporae
            Sympodurospore-type .... Sympodurosporae
            Others .... Annellosporae and others

A. Pycnidium-Forming Fungi

Key words: aggregate, appendage, conidia, discrete, filiform, ostiole, pycnidia, setae

1. Pycnidia   globose 2
             occasionally cup-shaped .... *Hainesia*

2. Conidia   1-celled 3
            over 2-celled 8

3. Conidia   hyaline 4
            pigmented 12

4. Conidia   two types .... *Phomopsis*
            one type 5

5. Pycnidia   with well-developed papillae .... *Cytospora*
             not so 6

6. Pycnidia   hairy or setose 7
             not so .... *Phoma, Macrophomina*

7. Pycnidia   with setae around ostioles .... *Pyrenochaeta*
             covered with setae .... *Chaetomella*

8. Conidia	2-celled	9
	over 3-celled	*Stagonospora*
9. Conidia	hyaline	10
	pigmented	11
10. Conidia	with filiform appendages	*Robillarda*
	not so, 2-celled with unequal size	*Apiocarpella*
11. Pycnidia	aggregate	*Botryodiplodia*
	discrete	*Diplodia*
12. Conidia	6 to 7-angled	*Microsphaeropsis*
	ovate or ellipsoidal	13
13. Conidia	curved significantly	*Selenophoma*
	not so	*Coniothyrium*

B. Sporodochium-Forming Fungi

Key words: appendage, conidia, conidiophores, filiform, setae, sporodochia, vesicle

1. Setae	formed	2
	not formed	6
2. Setae	curved	*Sarcopodium*
	not curved	3
3. Conidia	1-celled	4
	over 2-celled	*Wiesneriomyces*
4. Conidia	with filiform appendages	*Mycoleptodiscus*
	without appendages	5
5. Sporodochia	well differentiated	*Volutella*
	not so	*Colletotrichum, Myrothecium*
6. Conidia	simple	7
	complicated	13
7. Conidia	with filiform appendages	8
	not so	9
8. Conidia	cylindrical, concolor	*Hyphodiscosia*
	ellipsoidal with central dark cells and hyaline end cells	*Pestalotia*
9. Conidia	1-celled	*Myrothecium*
	over 2-celled	10
10. Conidia	lunar-shaped, with foot cell	*Fusarium*
	cylindrical	11

Key to Classes of Soil Fungi 27

11.	Conidiophores	penicillate with stipe and terminal vesicle .... *Cylindrocladium*
		simple, nonpenicillate 12
12.	Conidia	clavate or ellipsoidal .... *Cylindrocarpon*
		long-cylindrical or thread-like........................ *Gloeocercorpora*
13.	Conidia	of one kind, branched .... *Tetracladium*
		more than two kinds, smooth or echinulate .... *Spegazzinia*

C. Synnema-Forming Fungi

Key words: conidia, setae, synnema

1.	Synnema	formed *in vitro*.. 2
		not formed, with pear-shaped hyphal swellings .... *Dematophora*
2.	Setae	formed among synnema, conidia 1-celled..................... *Trichurus*
		not so, conidia 2-celled .... *Didymostilbe*

D. Aleuriosporae

Key words: clamp connection, conidia, conidiophore, septa, setae, sterigma (pl. -ta)

1.	Conidia	1-celled... 2
		over 2-celled... 8
2.	Hyphae	with clamp connection................................ *Sporotrichum*
		not so 3
3.	Conidia	hyaline .... *Sepedonium*
		pigmented.. 4
4.	Conidiophores	well developed 5
		not so 6
5.	Synnemata and setae	formed .... *Botryotrichum*
		not formed .... *Staphylotrichum*
6.	Conidiophores	globose, hyaline...................................... *Nigrospora*
		not so 7
7.	Conidia	globose... *Humicola*
		ellipsoidal... *Mammaria*
8.	Conidia	hyaline 9
		pigmented.. 13
9.	Conidia	with appendages at both ends .... *Hyphodiscosia*
		without appendages 10

10. Conidia	allantoid, ellipsoidal, or spindle-shaped 11
	cylindrical, often radiately branched with two to three arms ..... *Trinacrium*

11. Conidia	allantoid .... *Neta*
	ellipsoidal or spindle-shaped 12

12. Conidia	spindle-shaped, with broad enlarged middle cell .... *Monacrosporium*
	not so .... *Dactylella, Vermispora*

13. Conidia	longitudinally and transversely septate (muriform) 14
	mainly transversely septate 16

14. Conidia	borne directly on hyphae .... *Fumago*
	not so 15

15. Conidia	globose .... *Epicoccum*
	widely ellipsoidal .... *Pithomyces*

16. Conidia	branched .... *Tetracladium*
	not so 17

17. Conidiophores	developed 18
	lacking or poorly developed .... *Trichocladium*

18. Conidiophores	with sterigmata .... *Camposporium*
	not so .... *Sporidesmium*

E. Arthrosporae

Key words: catenulation, conidia, conidiophore(s)

1. Conidiophores	well developed 2
	poorly developed 3

2. Conidia	1-celled .... *Oidiodendron*
	2-celled .... *Trichothecium*

3. Conidia	mainly cylindrical, truncate at both ends .... *Geotrichum*
	globose 4

4. Conidia	mainly catenulate .... *Basipetospora*
	not catenulate .... *Chrysosporium*

F. Blastosporae

Key words: conidiophore(s), conidia, conidiogenous cells, muriform

1. Conidiophores	poorly developed 2
	well developed 8

Key to Classes of Soil Fungi

2. Conidiogenous cells	globose, simple	3
	not so	4
3. Conidia	1-celled, often with germ slit	*Arthrinium*
	muriform	*Naranus*
4. Hyphae and conidia	both hyaline	5
	not so	6
5. Sterigmata	developed on conidia	*Sporobolomyces*
	undeveloped	*Candida*
6. Conidia	1-celled	*Aureobasidium*
	over 2-celled	7
7. Conidia	branched to 3–4 directions	*Tetraploa*
	branched to 4–5 directions	*Tripospermum*
8. Conidiogenous cells	differentiated, cylindrical, or globose	9
	undifferentiated	12
9. Conidia	1-celled	10
	over 2-celled	*Cephaliophora*
10. Conidiogenous cells	formed limitedly at the apex	11
	formed everywhere	*Gonatobotrys*
11. Conidia	formed at the apex of conidiogenous cell, 1–16 per cell	*Oedocephalum*
	formed apically and laterally, numerous	*Chromelosporium*
12. Conidia	1-celled	13
	over 2-celled	16
13. Conidia	hyaline	14
	pigmented	15
14. Conidia	homogeneous in size, catenulate in a long chain	*Monilia*
	heterogeneous in size, catenulate in a short chain	*Hyalodendron*
15. Conidia	globose	*Periconia*
	ellipsoidal and others	*Cladosporium*
16. Conidia	2-celled	17
	more than 2-celled	19
17. Conidia	hyaline	*Trichothecium*
	pigmented	18
18. Conidiophores	branched	*Cladosporium*
	almost unbranched	*Bispora*

19. Conidia	globose..*Torula*
	not so 20

20. Conidia	cylindrical.. *Septonema*
	taenia-like.. *Taeniolella*

G. Phialosporae

Key words: appendage, catenulation, conidia, conidiophores, fertile (conidium-forming) area, foot cell, penicillate, phialide(s), spore mass

1. Conidia	more than 2-celled... 2
	1-celled.. 5

2. Conidiophores	penicillate with stipe and terminal vesicles, conidia cylindrical......................................*Cylindrocladium*
	not so 3

3. Conidia	formed in long chains, dark............................. *Sporoschisma*
	not so 4

4. Conidia	lunate with a foot cell .... *Fusarium*
	cylindrical, without a foot cell........................ *Cylindrocarpon*

5. Conidiophores	with inflated apical cells bearing numerous phialides .... *Aspergillus*
	without an inflated apical cell...................................... 6

6. Conidia	pigmented.. 7
	hyaline 11

7. Conidiophores	poorly developed 8
	well developed 9

8. Conidia	aggregate in a mass .... *Phialophora*
	catenulate .... *Torulomyces*

9. Conidia	branched... 10
	almost unbranched..................................... *Stachybotrys*

10. Conidia	catenulate .... *Phialomyces*
	aggregate in a mass .... *Myrothecium*

11. Conidia	globose... *Cladorhinum*
	not so 12

12. Conidia	boat-shaped or lunate, with or without appendages............ *Codinaea*
	not so 13

13. Conidia	clavate... 14
	not so 16

Key to Classes of Soil Fungi

14. Conidia	catenulate	15
	aggregate in a spore mass	*Stachybotryna*
15. Chlamydospores	solitary	*Chalara*
	catenulate	*Thielaviopsis*
16. Conidiophores	poorly developed	17
	well developed	18
17. Phialides and conidia	both hyaline	*Acremonium*
	only phialides pigmented	*Chloridium*
18. Conidia	dry	19
	wet	22
19. Conidiophores	pigmented, conidia catenulate	*Thysanophora*
	hyaline, spore aggregate in a row	20
20. Conidia	cylindrical	*Metarhizium*
	not so	21
21. Conidia	globose, conidiophores densely penicillate	*Penicillium*
	limoniform, conidiophores poorly penicillate	*Paecilomyces*
22. Spore mass	only at the apex of conidiophores	*Gliocladium*
	formed at apical parts of conidiophores	23
23. Conidiophores	hyaline	24
	pigmented	*Gonytrichum*
24. Conidiophores	verticillate	*Verticillium*
	irregularly branched	*Trichoderma*

H. Porosporae

Key words: beak, conidia, conidiophore(s), longitudinally or transversely septate, sympodulate

1. Conidia	transversely and longitudinally septate (muriform)	2
	mainly transversely septate	4
2. Conidia	ovate or clavate, often well beaked	*Alternaria*
	elipsoidal, or round without a beak	3
3. Conidiophores	proliferated apically	*Stemphylium*
	developed sympodially	*Ulocladium*
4. Conidia	catenulate	5
	not so	6
5. Conidia	globose	*Torula*
	long ellipsoidal	*Corynespora*

6. Conidia	mostly curved.. *Curvularia*
	not so 7

7. Conidia	obclavate, germinated from any cell................. *Helminthosporium*
	broadly fusiform, germinated from end cell.................. *Biporalis*

I. Sympodulosporae

Key words: appendages, biconical, conidia, conidiophore(s), filiform, hilum, sporogeneous (fertile) area

1. Conidia	coiled .... *Helicomyces*
	not so 2

2. Conidia	1-celled.. 3
	over 2-celled... 7

3. Conidia	biconical.. *Beltrania*
	not so 4

4. Conidiophores	with significant zigzag fertile area .... *Tritirachium*
	not so 5

5. Conidiophores	almost simple..................................... *Ramichloridium*
	branched... 6

6. Conidia	apiculate at one end................................. *Hansfordia*
	with a hilum....................................... *Nodulisporium*

7. Conidia	mostly 2-celled.. 8
	over 3-celled.. 11

8. Conidia	pigmented.. *Scolecobasidium*
	hyaline 9

9. Conidia	ovate, 2-celled with unequal size .... *Arthrobotrys*
	cylindrical, or clavate... 10

10. Conidia	long and cylindrical................................... *Dactylaria*
	clavate, narrowed toward one end....................... *Candelabrella*

11. Conidia	pluriseptate, clavate; conidiophores, simple and short ..... *Mycocentrospora*
	allantoid, aleurioconidia also formed .... *Neta*

J. Annelosporae and Others

Key words: annellation, conidiogenous cells, conidiophores

1. Conidiophores	with annellated conidiogenous cells .... *Scopulariopsis*
	not so .... others to be inspected

K. Sterile (Nonspore-Forming) Fungi

Key words: medulla, papulaspore, rind, sclerotia

1. Sclerotia   well developed, composed of rind and medulla .... *Sclerotium*

   not so 2

2. Hyphae   constricted near branching junction .... *Rhizoctonia*

   not so 3

3. Cellular bodies   papulaspores formed .... *Papulaspora*

   microsclerotia (and pycnidia) formed .... *Macrophomina*

5

List of Fungus Genera

MASTIGOMYCOTINA

1. *Aphanomyces*
2. *Dictyuchus*
3. *Phytophthora*
4. *Plectospira*
5. *Pythiogeton*
6. *Pythium*
7. *Saprolegnia*

ZYGOMYCOTINA

1. *Absidia*
2. *Circinella*
3. *Cunninghamella*
4. *Gongronella*
5. *Helicocephalum*
6. *Mortierella*
7. *Mucor*
8. *Rhizopus*
9. *Saksenaea*
10. *Syncephalastrum*
11. *Umbelopsis*
12. *Zygorhynchus*

ASCOMYCOTINA

1. *Achaetomium*
2. *Anixiella*
3. *Apiosordaria*
4. *Chaetomium*
5. *Didymella*
6. *Eudarluca*
7. *Glomerella*
8. *Massarina*
9. *Microascus*
10. *Monosporascus*
11. *Nectria*
12. *Phaeotrichosphaeria*
13. *Preussia*
14. *Sordaria*
15. *Thielavia*
16. *Zopfiella*

BASIDIOMYCOTINA

Armillaria, *Coprinus*, and six taxa

DEUTEROMYCOTINA (Mitosporic Fungi)

1. *Acremonium*
2. *Alternaria*
3. *Apiocarpella*
4. *Arthrinium*
5. *Arthrobotrys*
6. *Aspergillus*
7. *Aureobasidium*
8. *Basipetospora*
9. *Beltrania*
10. *Biporalis*
11. *Botyodiplodia*
12. *Botryotrichum*
13. *Camposporium*
14. *Candelabrella*
15. *Candida*
16. *Cephaliophora*
17. *Chaetomella*
18. *Chalara*
19. *Chloridium*
20. *Chromelosporium*
21. *Chrysosporium*
22. *Cladorrhinum*
23. *Cladosporium*
24. *Chloridium*
25. *Codinea*
26. *Colletotrichum*
27. *Coniothyrium*
28. *Corynespora*
29. *Curvularia*
30. *Cylindrocarpon*
31. *Cylindrocladium*
32. *Cytospora*
33. *Dactylaria*
34. *Dactylella*
35. *Dematophora*
36. *Didymostilbe*
37. *Diplodia*
38. *Epicoccum*
39. *Fumago*
40. *Fusarium*
41. *Geotrichum*
42. *Gliocladium*
43. *Gloeocercospora*
44. *Gonatobotrys*
45. *Gonytrichum*
46. *Hainesia*
47. *Hansfordia*
48. *Helicomyces*
49. *Helminthosporium*
50. *Humicola*
51. *Hyalodendron*
52. *Hyphodiscosia*
53. *Macrophomina*
54. *Mammaria*
55. *Metarhizium*
56. *Microsphaeropsis*
57. *Monacrosporium*
58. *Monilia*
59. *Mycocentrospora*
60. *Mycoleptodiscus*
61. *Myrioconium*
62. *Myrothecium*
63. *Naranus*
64. *Neta*
65. *Nigrospora*
66. *Nodulisporium*
67. *Oedocephalum*
68. *Oidiodendron*
69. *Paecilomyces*
70. *Papulaspora*
71. *Penicillium*
72. *Periconia*
73. *Pestalotia*
74. *Phialomyces*
75. *Phialophora*
76. *Phoma*
77. *Phomopsis*
78. *Pithomyces*
79. *Pyrenochaeta*
80. *Ramichloridium*
81. *Rhizoctonia*
82. *Robillarda*
83. *Sarcopodium*
84. *Sclerotium*

DEUTEROMYCOTINA (continued)

85. *Scolecobasidium*
86. *Selenophoma*
87. *Sepedonium*
88. *Septonema*
89. *Spegazzinia*
90. *Sporidesmium*
91. *Sporobolomyces*
92. *Sporoschisma*
93. *Sporotrichum*
94. *Stachybotrys*
95. *Stagonospora*
96. *Staphylotrichum*
97. *Stemphylium*
98. *Taeniolella*
99. *Tetracladium*
100. *Tetraploa*
101. *Thielaviopsis*
102. *Thysanophora*
103. *Torula*
104. *Torulomyces*
105. *Trichocladium*
106. *Trichoderma*
107. *Trichothecium*
108. *Trichurus*
109. *Trinacrium*
110. *Tripospermum*
111. *Tritirachium*
112. *Ulocladium*
113. *Vermispora*
114. *Verticillium*
115. *Volutella*
116. *Wiesneriomyces*

6
MORPHOLOGY OF SOIL FUNGI
Mastigomycotina

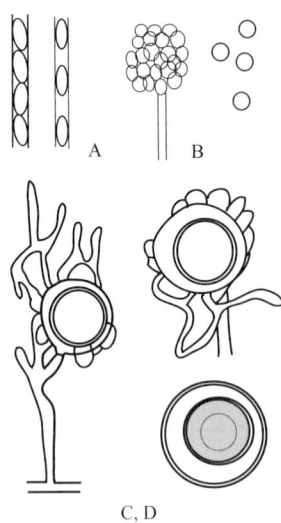

Aphanomyces de Bary

Jahrb. Wiss. Bot. 2:178, 1860.
Type species: *A. stellatus* de Bary

Aphanomyces cladogamus Drechsler

References: Drechsler 1929; McKeen 1952b; Scott 1961.

Morphology: Zoospores encysted at the tips of hypha-like sporangia, forming a mass of zoospores encysted after moving inside the sporangium in a row. Oogonia mostly terminal, often aggregated, covered with characteristically branched hypha-like antheridia, monoclinous. Oospores aplerotic, with single oil globules.

Dimensions: Zoospores encysted ca. 7–8 µm in diameter. Oogonia 18.7–35 µm in diameter. Oospores 15–25 µm in diameter: oil globule 12.5–16.3 µm in diameter.

Material: 78-2252, 78-2273 (Tomato field soil, Morioka, Japan); 83-388 (Forest soil, Shizuoka, Japan).

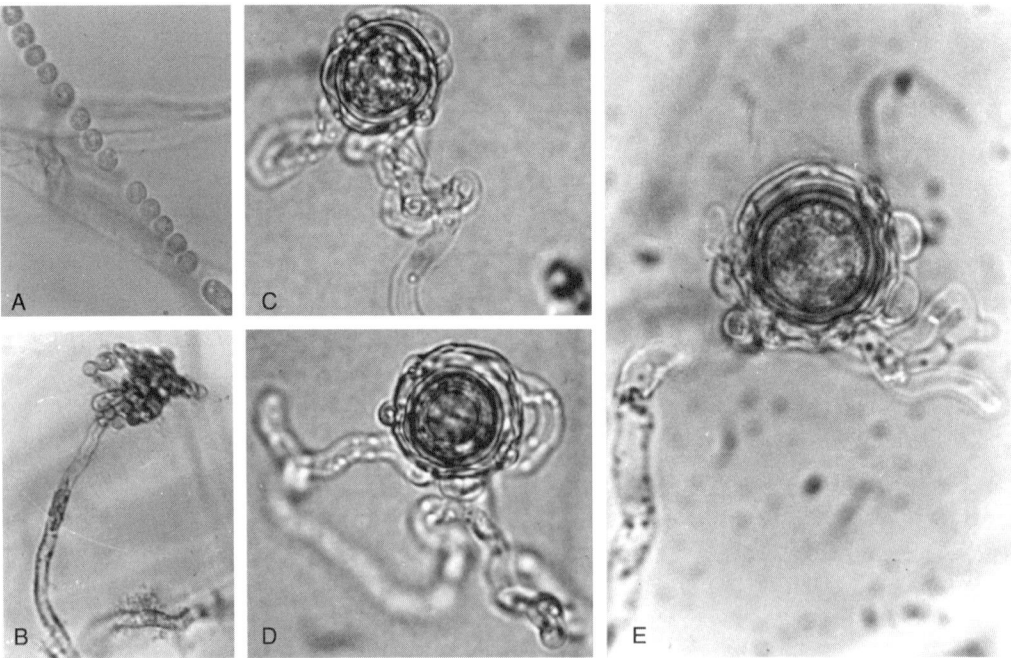

Aphanomyces cladogamus. A,B: Hypha-like sporangia containing encysted zoospores in a row (A) and bearing a mass of encysted zoospores at the tip (B). C–E: Oogonia, antheridia, and oospores.

Dictyuchus Leitg.

Bot. Ztg. 26:503, 1868.
Type species: *D. monosporus* Leitg.

Dictyuchus sp.

References: Johnson 1951; Nagai 1931, 1933; Padgett and Seymour 1974; Rattan et al., 1979.

Morphology: Zoosporangia inflated, terminal, cylindrical with the primary zoospores developed, and readily encysted inside the sporangia, thus showing a net-like appearance, each zoospore releasing a secondary zoospore. Immature sexual organs with irregularly distributed spines are formed. Chlamydospores (gemmae) globose or subglobose.

Dimensions: Zoosporangia 125–250 × 20-30 µm. Oogonia 27–28 µm in diameter. Chlamydospores 22–23 µm in diameter.

Material: 81-787, 81-789 (Lettuce field soil, Katsuura, Chiba, Japan).

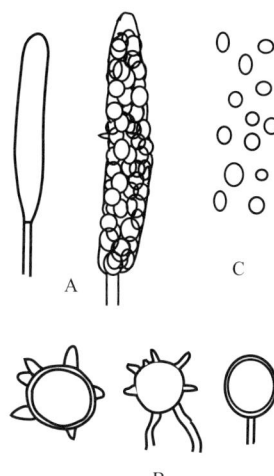

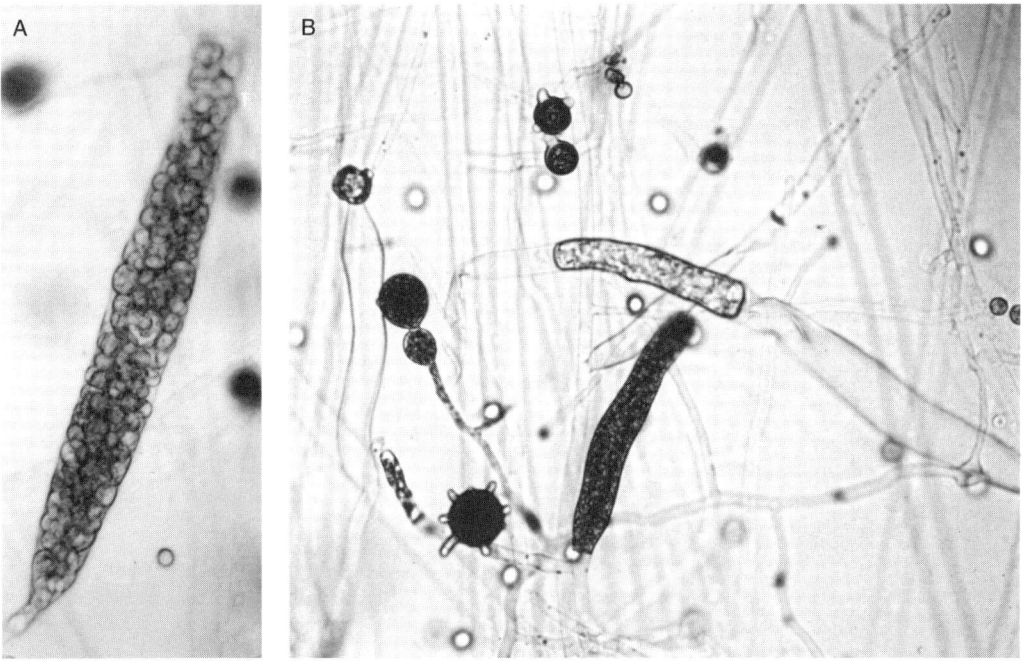

Dictyuchus sp.   A: Zoosporangium with encysted zoospores. B: Sexual organs, chlamydospores, and vacant and intact zoosporangia.

Phytophthora de Bary

J. R. Agric. Soc. 2, 12:240, 1876.
Type species: *P. infestans* (Mont.) de Bary

Key to Species

1.	Sexual organs	formed in single culture..2
		not formed 5
2.	Antheridia	amphigynous ....*P. melonis*
		amphigynous and paragynous 3
3.	Sporangia	significantly papillate...........*P. nicotianae* var. *parasitica* (homothallic)
		not so 4
4.	Antheridiumphores	sessile, oospores aplerotic .... *P. erythroseptica*
		not so, oospores plerotic ....*P. megasperma*
5.	Chlamydospores	single 6
		catenulate ....*P. cyptogea*
6.	Sporangiophores	deciduous .... *P. capsici*
		not so .... *P. nicotianae* var. *parasitica* (heterothallic)

Phytophthora capsici Leonian

References: Aragaki and Uchida 2001; Erwin et al. 1983; Tucker 1931; Waterhouse 1963, 1970.

Morphology: Sporangia long ellipsoidal, often triangular, or irregular, distinctly papillate, often 2 to 3 papillae per sporangium. Zoospores developed inside sporangia. Sporangiophores readily broken and detached. Sexual organs not formed. Chlamydospores rarely formed, globose, thick-walled. Heterothallic.

Dimensions: Sporangia 20–50 (-55) × 15–42.5 (-50) µm: papilla 6.2–7.5 (-12.5) µm wide, 1.2–6 µm deep. Chlamydospores ca. 35 µm in diameter.

Material: 71-317 (Eggplant root, Ishikawa, Japan); 82-534 (Tomato root, Osaka, Japan); 82-690 (Chrysanthemum field soil, Mie, Japan).

Remarks: Sexual organs and oospores may be formed in dual cultures. A group of *Phytophthora* isolates previously referred to this species were newly named as *Phytophthora tropicalis* Aragaki and Uchida (2001).

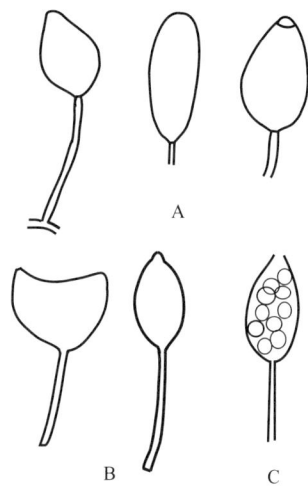

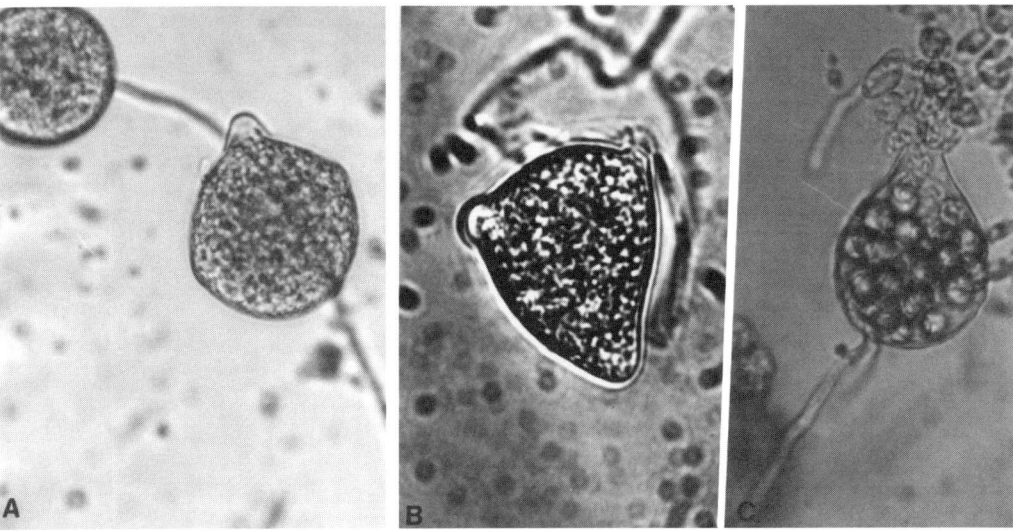

Phytophthora capsici.   A,B: Sporangia. C: Differentiated zoospores inside the sporangium and their release.

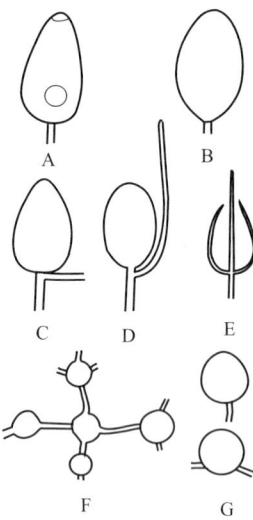

Phytophthora cryptogea Pethybridge & Lafferty

References: Erwin et al. 1983; Tucker 1931; Waterhouse 1963, 1970.

Morphology: Sporangia obovate or ellipsoidal, internally proliferated, slightly or faintly papillate or nonpapillate. Zoospores developed inside sporangia. Sexual organs not formed. Hyphal swellings globose, clustered. Heterothallic.

Dimensions: Sporangia 22.5–57.5 × 17.5–42.5 µm: exit tubes 1–20 µm wide. Hyphal swellings 12.5–20 µm in diameter.

Material: 75-261 (*Lagenaria* root, Fukuoka, Japan).

Remarks: Sexual organs may be formed in dual cultures. This fungus was pathogenic to bottle gourd seedlings.

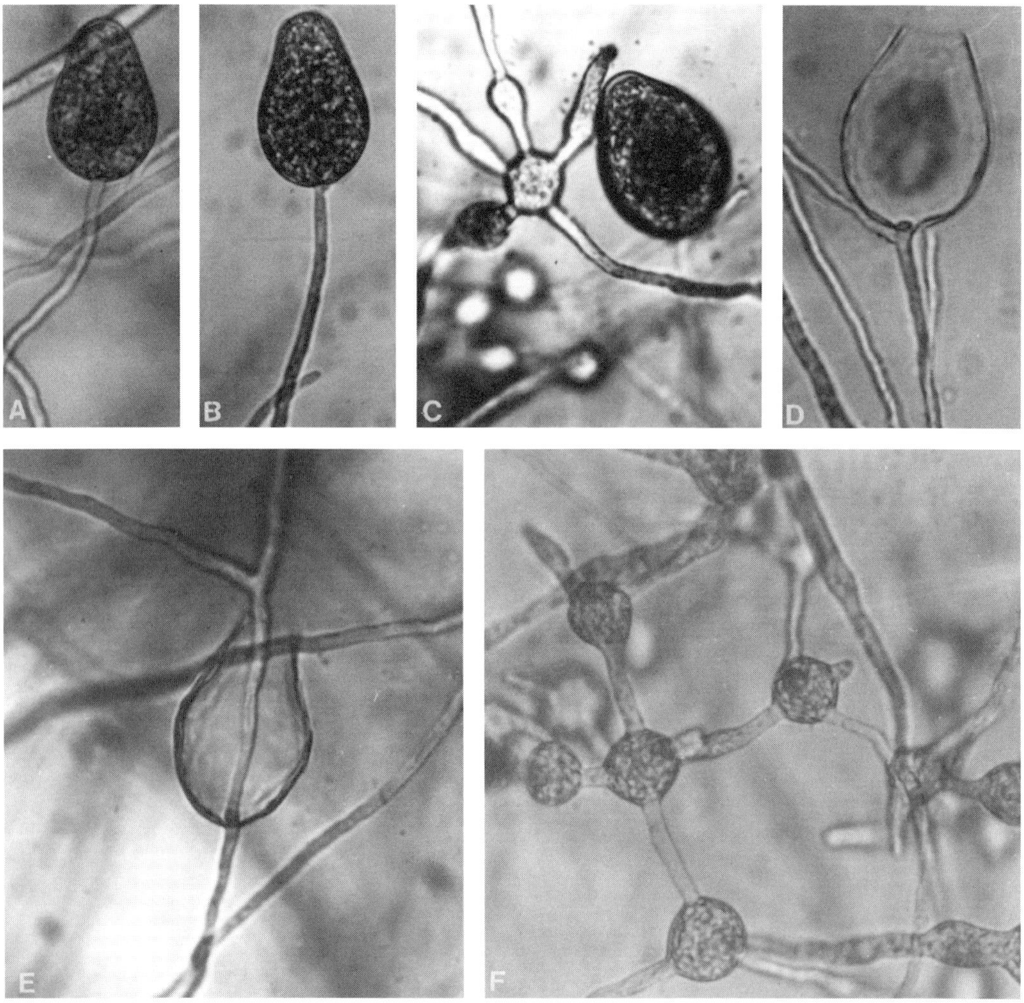

Phytophthora cryptogea. A–E: Sporangia with external and internal proliferation (E). F,G: Clustered (F) or simple hyphal swellings.

Phytophthora erythroseptica Pethybridge

References: Tucker 1931; Waterhouse 1963, 1970.

Morphology: Sporangia ellipsoidal, nonpapillate. Zoospores developed inside sporangia. Sporangiophores often 1-septate near sporangia. Oogonia globose, with single paragynous or amphigynous antheridia, which are sessile, and cylindrical in shape. Oospores mostly aplerotic. Homothallic.

Dimensions: Sporangia 30–45 × 22.5–42.5 µm: exit tubes 7.5–17.5 µm wide. Zoospores encysted 9–11.3 µm in diameter. Oogonia 25–40 µm in diameter. Antheridia 12.5–22.5 × 15–22.5 µm. Oospores 15–35 µm in diameter: oospore wall ca. 0.5 µm thick.

Material: 71-428 (Cultivated soil, Fukui, Japan).

Remarks: This fungus was trapped from soil with potato cubes as the substrate.

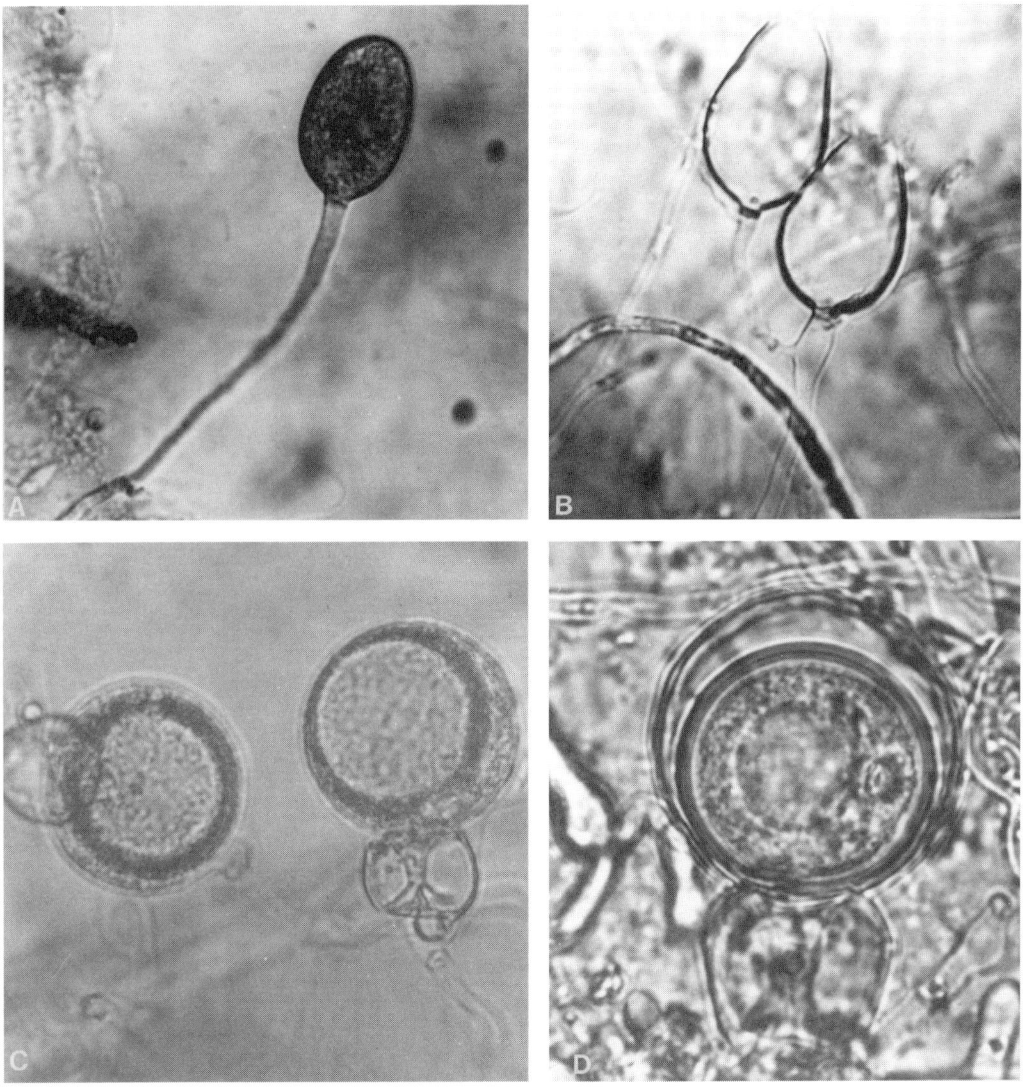

Phytophthora erythroseptica. A,B: Sporangia. C,D: Oogonia, antheridia, and oospores.

Phytophthora megasperma Drechsler

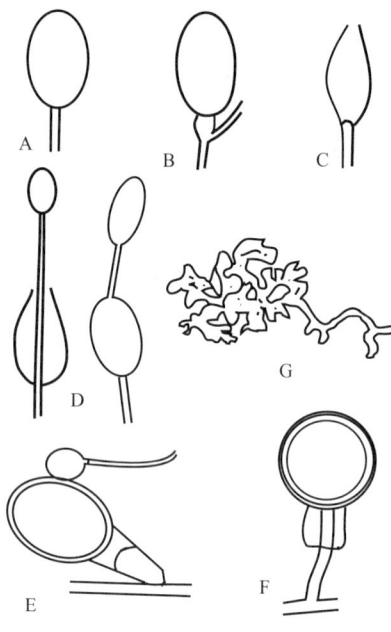

References: Drechsler 1931; Erwin 1965; Nagai et al. 1978; Waterhouse 1963, 1970.

Morphology: Sporangia ellipsoidal or ovate, internally and externally proliferated, nonpapillate. Zoospores developed inside sporangia. Sporangiophores often inflated near sporangia. Oogonia globose or ovate, funnel-shaped with an attenuated end, with paragynous, rarely amphigynous antheridia. Oospores plerotic, often with single globules. Antheridia globose, stalked. Hyphae often well twisted and knobby. Homothallic.

Dimensions: Sporangia 23–70 × 18–42.5 μm. Sporangiophores 150–350 × 2.5–10 μm. Zoospores encysted 8.7–12.5 μm in diameter. Oogonia 35–55 μm in diameter. Oogoniumphores 7.5–120 × 10–25 μm. Antheridia 7.5–25 × 10–20 μm. Oospores 27.5–49 μm in diameter: oospore wall 1.2–3.8 μm thick: oil globules 13.7–15 μm in diameter.

Material: 77-360, 77-361 (Rose roots, Chiba, Japan).

Remarks: Nonpapillate proliferating sporangia, large oogonia and oospores, predominance of paragynous antheridia, and the ready formation of sexual organs characterize this species (Erwin, 1965).

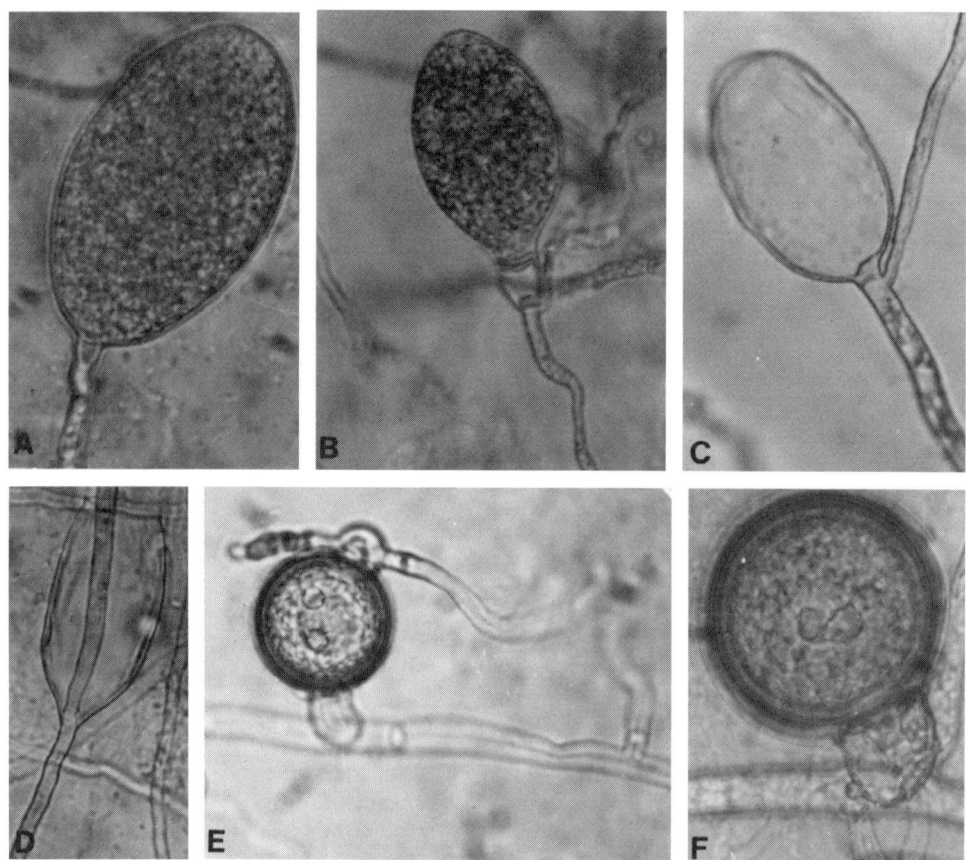

Phytophthora megasperma. A–D: Sporangia and external (B,C) and internal proliferation (D). E,F: Oogonia, antheridia, and oospores (E: paragynous; F: amphigynous). G: Twisted and knobby hypha.

Phytophthora melonis Katsura

References: Erwin et al. 1983; Katsura 1976.

Morphology: Sporangia ellipsoidal, internally proliferated, slightly papillate or nonpapillate. Zoospores developed inside sporangia. Sporangiophores often inflated and septate near the sporangia. Oogonia globose, with single amphigynous, antheridia. Antheridia ellipsoidal or rectangular, sessile. Oospores yellowish brown, aplerotic. Homothallic.

Dimensions: Sporangia (32.5-) 55–75 × (17.5-) 35–45 µm: subsporangial branches 6–12.5 µm wide. Oogonia 22.5–35 µm in diameter. Oospores 20–32.5 µm in diameter: wall ca. 2 µm thick: oil globules ca. 24 µm in diameter. Antheridia 17.5–30 µm wide, 17.5–25 µm deep.

Material: 68-103 (Cucumber root, Shizuoka, Japan); 80-201 (Cucumber field soil, Fukuoka, Japan).

Remarks: This fungus was pathogenic to cucumber and tomato seedlings.

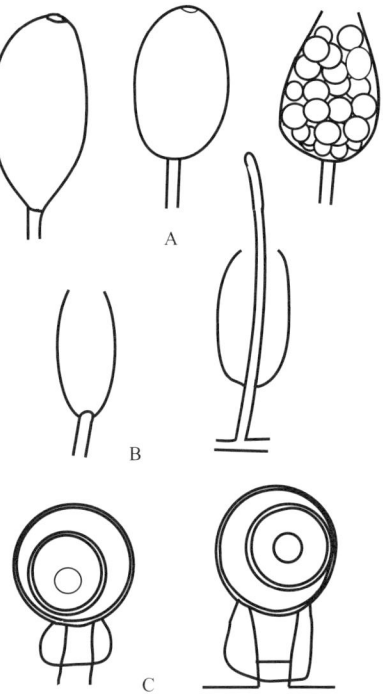

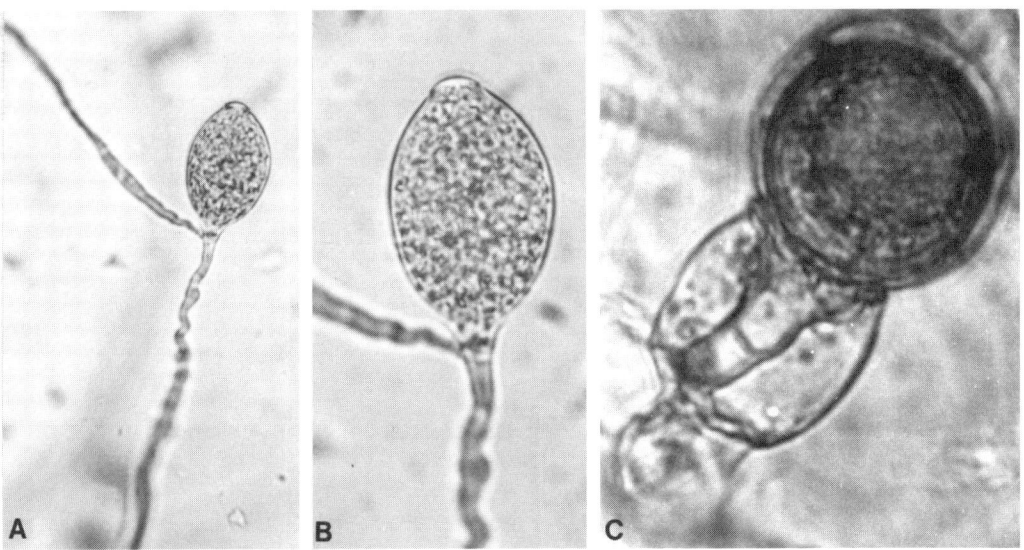

Phytophthora melonis.   A,B: Sporangia. C: Oogonium with amphigynous antheridium, and oospore.

Phytophthora nicotianae van Breda de Haan var. *parasitica* (Dastur) Waterhouse (Homothallic)

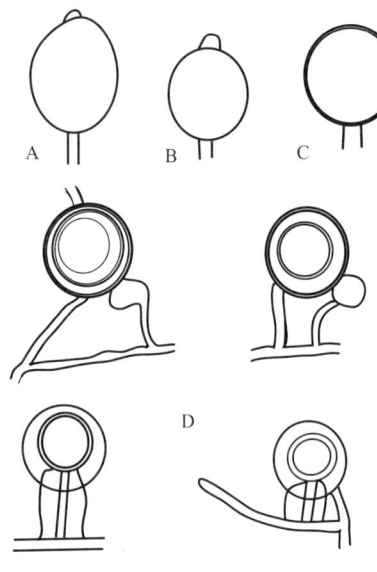

Synonym: *P. parasitica* Dastur

References: Erwin et al. 1983; Tucker 1931; Newhook et al. 1978; Waterhouse 1963, 1970.

Morphology: Sporangia ellipsoidal, subglobose, often triangular or irregular, conspicuously papillate, often with two papillae per sporangium, well recognizable even after release of zoospores. Zoospores developed inside sporangia. Oogonia globose, mainly terminal, with single paragynous or amphigynous antheridia. Oospores golden yellow, aplerotic. Antheridia globose or rectangular, mono- or diclinous. Chlamydospores globose, thick-walled. Homothallic.

Dimensions: Sporangia 20–35 (-40) × 20–30 μm: papilla 6–7.5 μm wide, 2.5–3 μm deep. Oogonia 17.5–32.5 μm in diameter. Oospores 14–30 μm in diameter: oospore wall 2.5–3 μm wide. Antheridia 12.5–15 × 10–12.5(-13.8) μm. Chlamydospores 22.5–30 μm in diameter.

Material: 75-410 (Eggplant root, Nagasaki, Japan).

Remarks: Homothallic isolates may not be so common in Japan.

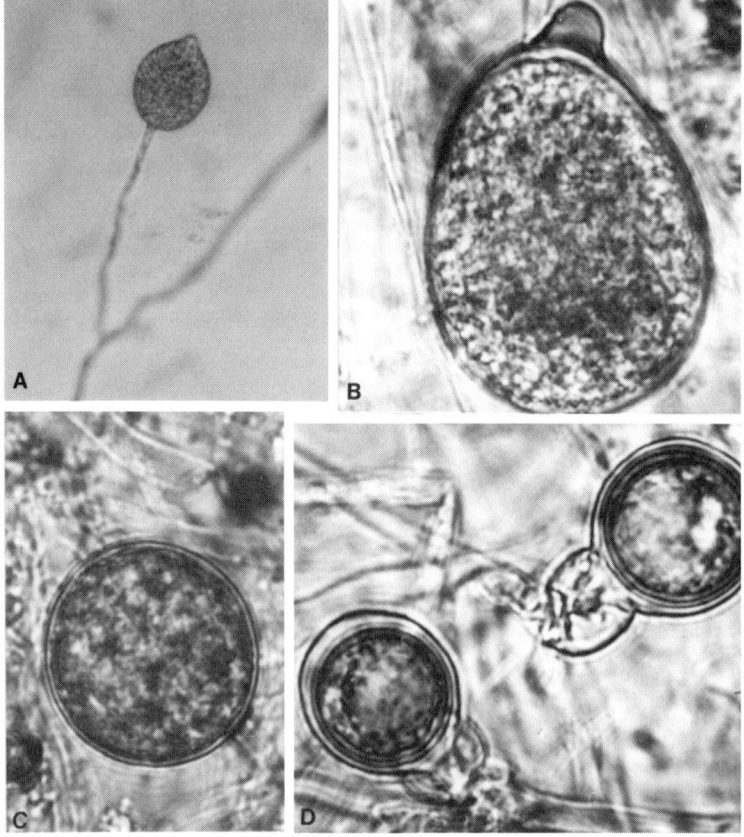

Phytophthora nicotianae var. *parasitica* (homothallic)   A,B: Sporangia. C: Chlamydospore. D: Oogonia, antheridia, and oospores.

Phytophthora nicotianae van Breda de Haan var. *parasitica* (Dastur) Waterhouse (Heterothallic)

Synonym: *P. parasitica* Dastur

References: Erwin et al. 1983; Tucker 1931; Newhook et al. 1978; Waterhouse 1963, 1970.

Morphology: Sporangia ellipsoidal, conspicuously papillate. Papilla well recognizable even after release of zoospores. Zoospores developed inside sporangia. Sexual organs not formed in single cultures. Chlamydospores well developed, globose, thick-walled. Heterothallic.

Dimensions: Sporangia 30–65 × (15-) 30–45 µm: papilla 5–7.5 × 2.5–5 (-7.5) µm. Chlamydospores 22.5–40 µm in diameter.

Material: 68-101 (Fishdragon root, Shizuoka, Japan); 69-529, 69-530 (Pineapple rhizome, Okinawa, Japan); 79-197 (Passionflower field soil, Amami-Oshima, Kagoshima, Japan); 82-618, 82-619 (Zanthoxylum root, Nara, Japan); 82-644, 82-645 (Strawberry root, Nara, Japan); 82-736, 82-737 (Bell pepper field soil, Wakayama, Japan); 82-793, 82-794 (Zanthoxylum field soil, Nara, Japan).

Remarks: Sexual organs may be formed in dual cultures.

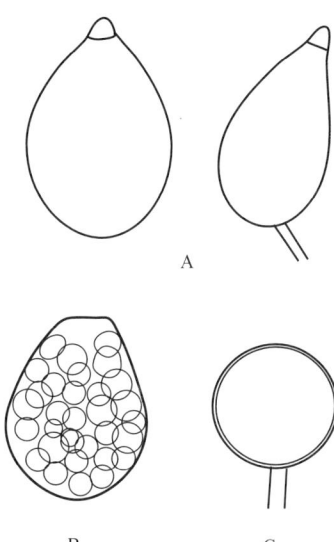

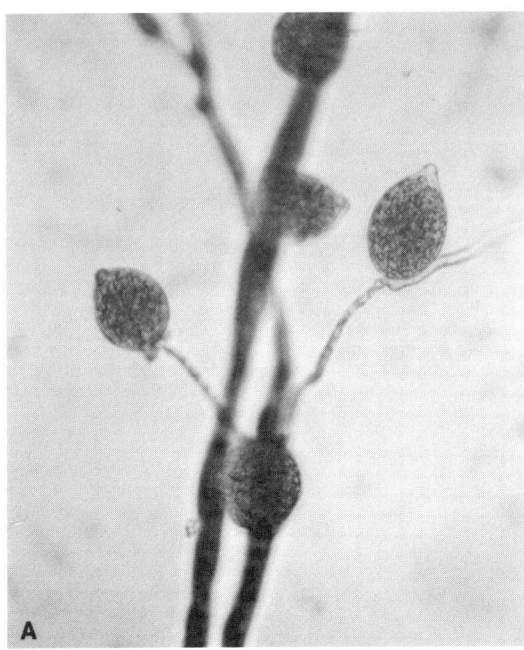

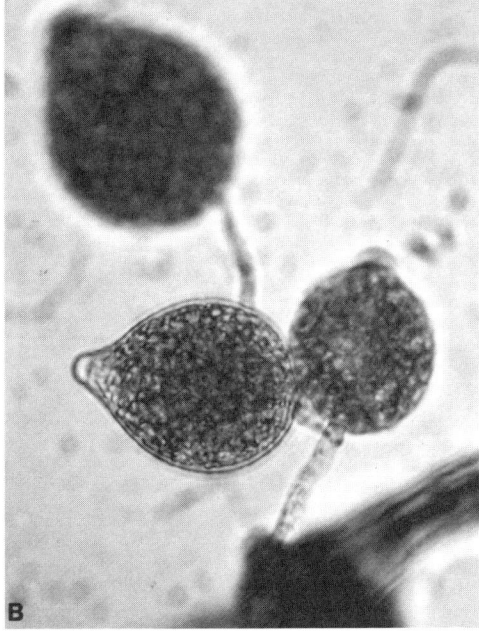

Phytophthora nicotianae var. *parasitica* (heterothallic).   A,B: Sporangia. D: Chlamydospore.

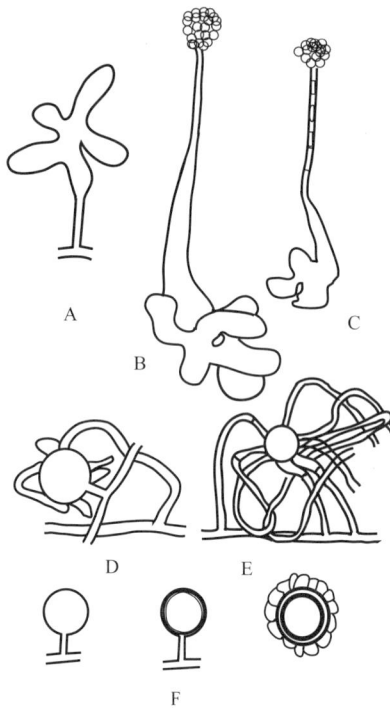

Plectospira Drechsler

J. Agric. Res. 34:288, 1927.
Type species: *P. myriandra* Drechsler

Plectospira myriandra Drechsler

References: Drechsler 1927, 1929, 1953; Watanabe 1987.

Morphology: Sporangia simple, curved, inflated, branched, lobate, often aggregated into complex masses usually with one to three tapering evacuation tubes. Zoospores encysted at the exit of long cylindrical tubes extended from lobate sporangia, forming encysted zoospore masses after moving inside the tube in a row just like *Aphanomyces*. Oogonia terminal. Oospores plerotic, formed sexually or parthenogenetically. Antheridia hypha-like, mostly diclinous, characteristically aggregated surrounding oogonia.

Dimensions: Lobate sporangia up to 400 μm long, up to 25 μm wide: evacuation tubes 95–275 μm long, 5–12.5 μm wide at base: 2.5–4.5 μm wide at apex. Zoospores encysted 5.5–15 μm in diameter. Oogonia including parthenospores 17.5–27.5 μm in diameter. Oospores 15–23.8 μm in diameter: oospore wall 0.6–1.3 μm thick. Antheridia 1.7–5 μm wide.

Material: 84-209 (= ATCC 64139, Bamboo field soil, Nankoku, Tosa, Japan).

Remarks: This fungus is morphologically *Pythium*- and *Aphanomyces*-like because of formation of lobate sporangia, filamentous tubes with zoospore movement in a row, and encysted zoospores at the exit. The isolate 84-209 was reisolated 59 years after the original descriptions.

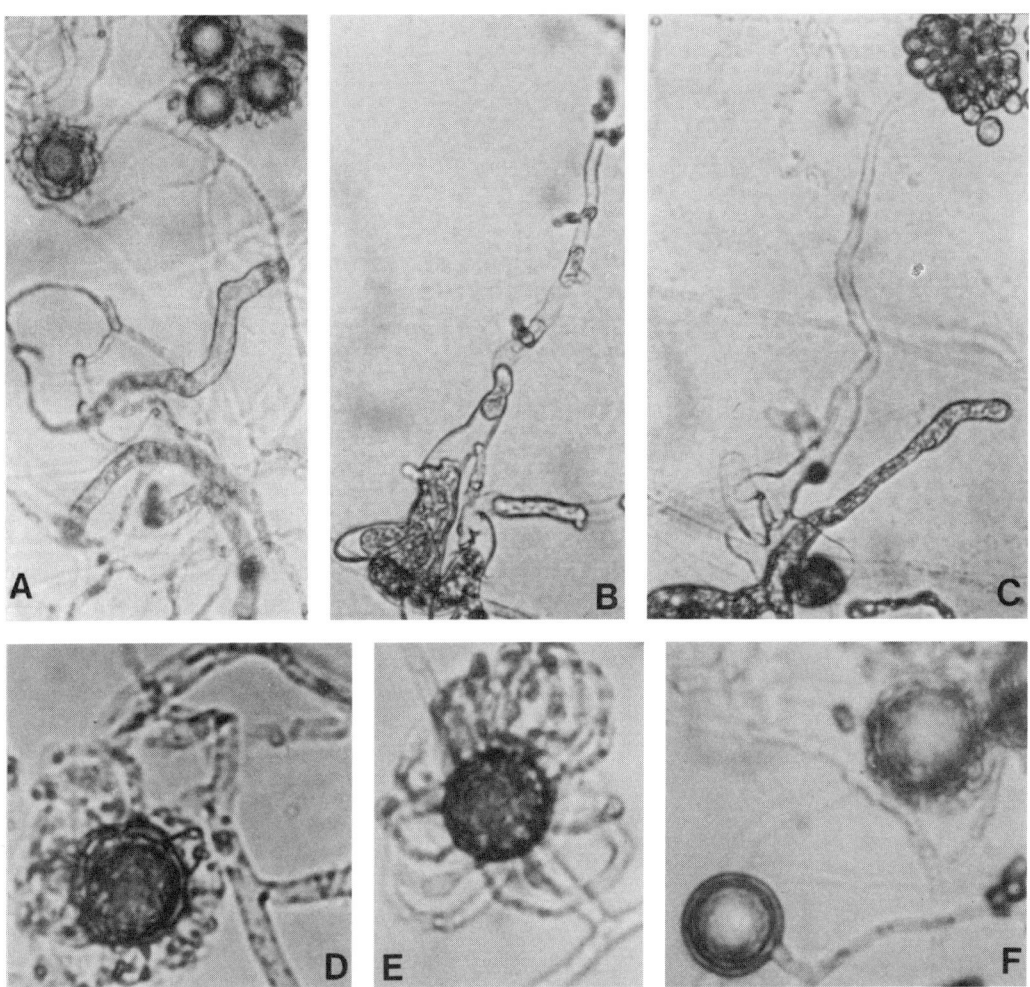

Plectospira myriandra. A: Sporangia. B,C: Sporangia and encysted zoospores at the exit (C). D,E: Oogonia and antheridia. F: Oospores formed parthenogenetically (left) and sexually (right). (From Watanabe, T. 1987. *Mycologia*, 79:77–81. With permission.)

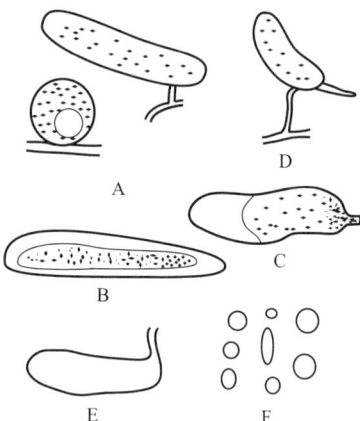

Pythiogeton Minden

Mycol. Unters. 2(2):241, 1916.
Type species: *P. ramosum* Minden

Pythiogeton ramosum Minden

References: Jee et al. 2000; Sparrow 1960; Watanabe 1974a.

Morphology: Sporangia cylindrical, globose, ellipsoidal, or irregular in shape, often curved, dense in protoplasm, mostly terminal, occasionally intercalary or tangential on the sporogenous hyphae, often sessile, often germinated directly elongating 1 to several germ tubes, zoospores developed from a part of protoplast flowed out into water through emission tubes. Sporogenous hyphae branched heavily. No sexual organs formed. Heterothallic.

Dimensions: Sporangia 48.6–185 μm long, 21.4–30 μm wide on water agar, but swelled up to 280 μm long, 50 μm wide in water; emission tubes ca. 200 μm long. Zoospores encysted 7.5–15 μm in diameter. Sporogenous hyphae 1.9–3.9 μm wide.

Material: 72-X112 (Sugarcane root, Taiwan, ROC).

Remarks: The fungus was characterized by large cylindrical sporangia obliquely attached on sporangiophores, and differentiation of zoospores from protoplast flowed out from sporangia in water. A taxonomic key of eight species is provided by Jee et al. (2000).

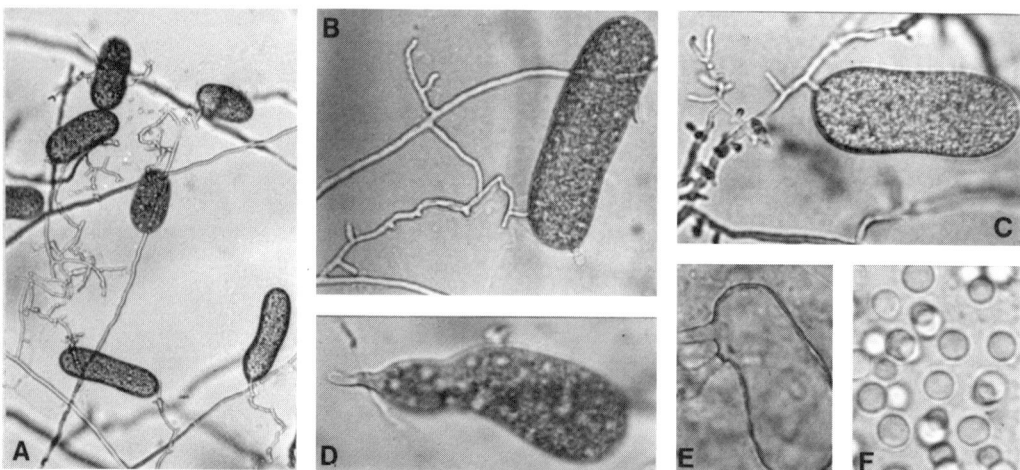

Pythiogeton ramosum. A–C: Cylindrical sporangia on slender sporogenous hyphae. D: Germinated sporangium with a single germ tube. E: Vacant sporangium with an emission tube. F: Zoospores encysted. (From Watanabe, T. 1974a. *Trans. Mycol. Soc. Jpn.*, 15:343–357. With permission.)

Pythium Pringsheim

Jahrb. Wiss. Bot. 1:304, 1858.
Type species: *P. monospermum* Pringsheim

Key to Species

1.	Sexual organs	not formed..2
		formed in a single culture 10
2.	Hypha-like sporangia and chlamydospore-like structures	formed... *P. afertile*
		not formed..3
3.	Sporangia	ellipsoidal or cylindrical..................................4
		globose 5
4.	Oospores	formed parthenogenetically .... *P. dissimile*
		not formed..................................... *P. elongatum*
5.	Oospores	formed parthenogenetically .... *P. conidiophorum*
		not formed..6
6.	Sporangia	single..7
		catenulate, deciduous .... *P. intermedium*
7.	Sporangia	under 30 μm in diameter 8
		over 40 μm in diameter .... *P. splendens*
8.	Sporangia	zoospores released 9
		only germ tubes elongated directly .... *P. sylvaticum*
9.	Sporangia	proliferated internally....................... *P. carolinianum*
		not so, with subsporangial vacant inflation .... *P. nayoroense*
10.	Oogonia	echinulate, with over 2 protuberances 11
		smooth..16
11.	Protuberances	numerous, evenly distributed 12
		2–4, irregularly distributed....................... *P. irregulare*
12.	Sporangia	not formed.................................. *P. acanthophoron*
		formed..13
13.	Sporangia	single..14
		formed in aggregates .... *P. acanthicum*
14.	Sporangia	globose or subglobose 15
		lobate... *P. periplocum*

15.	Oogonial protuberances	acute .... *P. echinulatum*
		digitate .... *P. spinosum*
16.	Sporangia	almost globose 17
		not so 23
17.	Zoospores	not formed 18
		formed 21
18.	Oogonia and hyphal swellings	intercalary, born in chains .... *P. salpingophorum*
		not so 19
19.	Oospores	plerotic, with sessile antheridia .... *P. rostratum*
		aplerotic 20
20.	Antheridia	1 per oogonium, sessile or short-stalked .... *P. ultimum*
		1–3 per oogonium, furrowed on the upper surface ...... *P. sulcatum*
21.	Sporangia	internally proliferated, antheridia lunate, with a furrow on the upper wall .... *P. oedochilum*
		not so 22
22.	Antheridia	1 per oogonium, dampbell-shaped, stalked .... *P. vexans*
		3 per oogonium, unstalked .... *P. paroecandrum*
23.	Sporangia	hypha-like 24
		lobate 27
24.	Antheridia	1 per oogonium, diclinous .... *P. apleroticum*
		over 2 per oogonium, monoclinous 25
25.	Oospores	plerotic .... *P. torulosum*
		aplerotic 26
26.	Antheridia	unstalked .... *P. dissotcum*
		stalked from a few origins .... *P. angustatum*
27.	Oospores	plerotic 28
		aplerotic 30
28.	Chlamydospore-like globose structures	formed in chains .... *P. catenulatum*
		not formed 29
29.	Antheridia	crook-necked, both monoclinous and diclinous ...... *P. graminicola*
		not so, mainly diclinous .... *P. inflatum*
30.	Antheridia	1 per oogonium 31
		over 2 per oogonium 32

31.	Oogoniumphores	bent toward antheridium.......................... *P. deliense*
		not so, antheridia often intercalary............ *P. aphanidermatum*
32.	Antheridiumphores	coiled around oogoniumphores spirally 2–3 times .... *P. zingiberum*
		not so.. 33
33.	Oogonia	over 30 μm in diameter .... *P. myriotylum*
		less than 20 μm in diameter, closely connected with hyphal swellings........................ *P. indigoferae*

Pythium acanthicum Drechsler

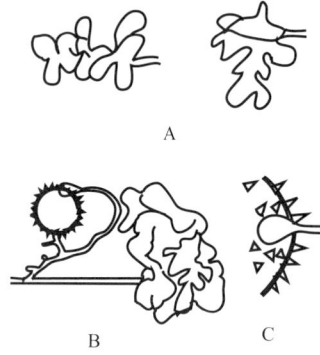

References: Middleton 1943; Plaats-Niterink 1981; Watanabe 1974a.

Morphology: Sporangia composed of aggregates of globose or lobate cells. Zoospores not formed. Oogonia terminal, with acute protuberances evenly distributed, bearing mostly single, rarely 2–3 antheridia, monoclinously. Oospores usually plerotic. Antheridia globose or clavate.

Dimensions: Sporangia ca. 92.5 μm in diameter. Oogonia excluding protuberances 21.8–30 μm in diameter: protuberances 2.5–5 × 2.2–3.6 μm. Oospores (18.2-) 20–22.5 μm in diameter.

Material: 72-X100 (Sugarcane roots, Taiwan, ROC).

Remarks: This fungus is morphologically close to *P. oligandrum* Drechsler, and interpretation of protuberances, namely acute or blunt, differentiates both fungi.

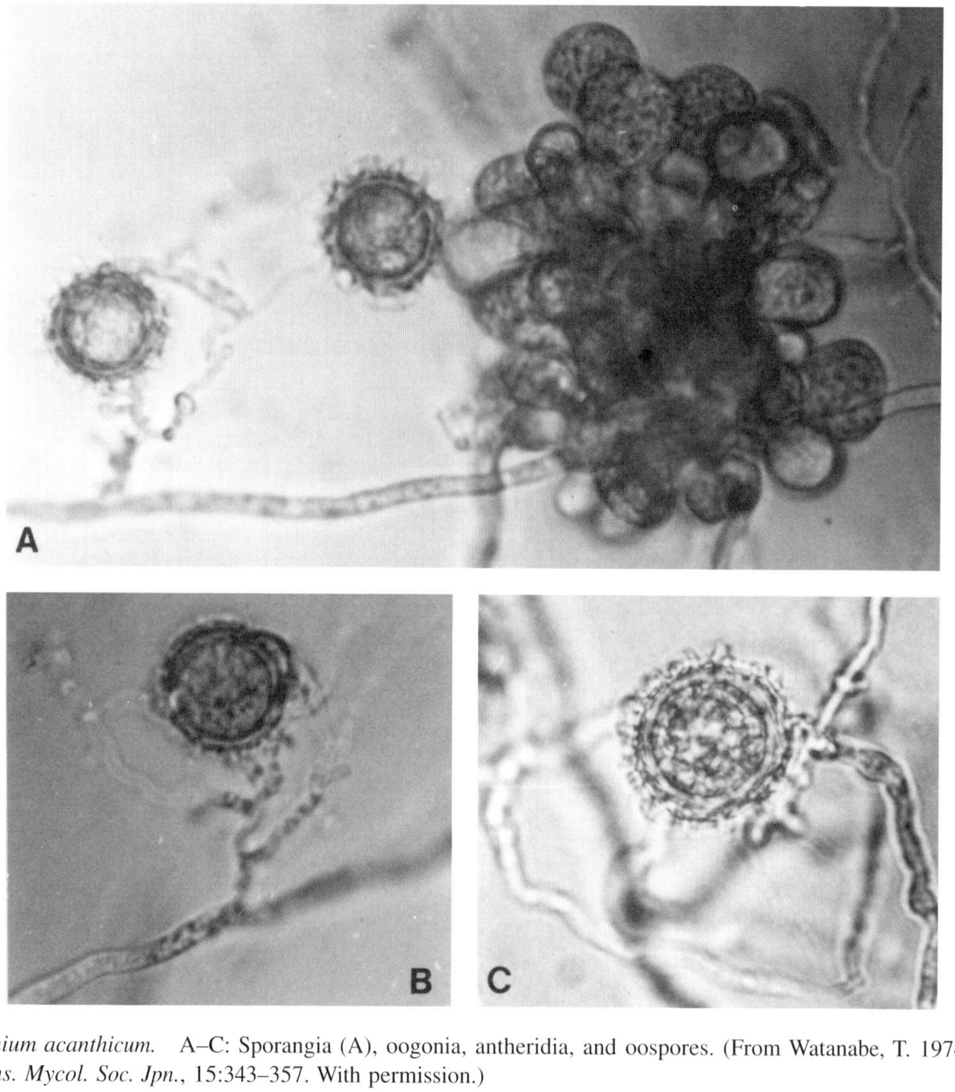

Pythium acanthicum. A–C: Sporangia (A), oogonia, antheridia, and oospores. (From Watanabe, T. 1974a. *Trans. Mycol. Soc. Jpn.*, 15:343–357. With permission.)

Pythium acanthophoron Sideris

References: Middleton 1943; Plaats-Niterink 1981; Sideris 1932; Watanabe 1990.

Morphology: Sporangia and inflated hyphae not formed. Hyphae rather knobby. Zoospores not formed. Oogonia terminal or intercalary, often papillate, with homogeneous blunt protuberances, bearing mostly single, rarely 2 antheridia in broad contact. Antheridia cylindrical or subglobose, curved, constricted at the upper middle part, mainly monoclinous. Oospores plerotic.

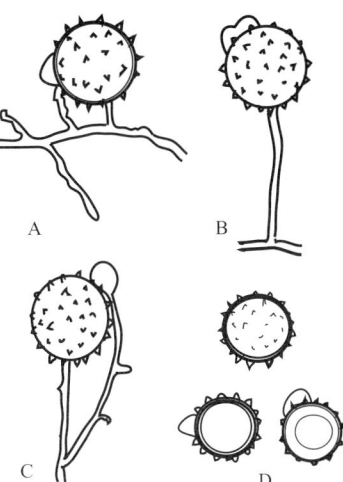

Dimensions: Oogonia (18-) 20–30 µm in diameter: protuberances under 3 µm long.

Material: 81-339 (Forest soil, Hamatonbetsu, Hokkaido, Japan); 81-393 (Forest soil, Chitose, Hokkaido, Japan); 82-744 (Radish field soil, Wakayama, Japan).

Remarks: Colonies on PDA are nonaerial, with flower petal-like pattern. This species may include the isolates with globose hyphal swellings and similar sexual organs. In some literature, oospores are aplerotic.

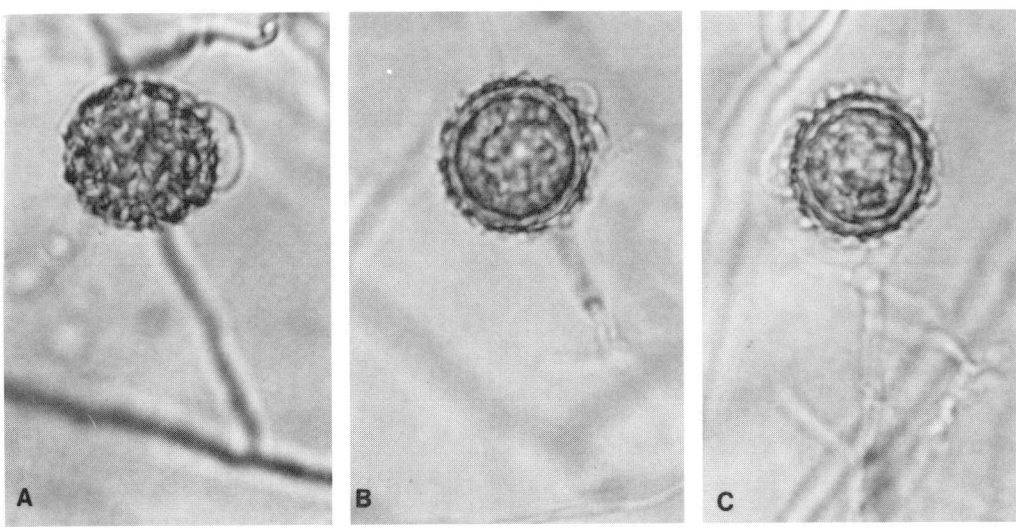

Pythium acanthophoron. A–D: Oogonia, antheridia, and oospores. (From Watanabe, T. 1990b. *Ann. Phytopathol. Soc. Jpn.*, 56:549–556. With permission.)

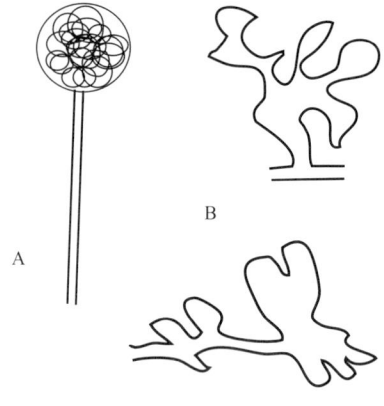

Pythium afertile Kanouse & Humphrey

References: Matthews 1931; Middleton 1943; Plaats-Niterink 1981; Watanabe et al. 1977.

Morphology: Sporangia hypha-like. No sexual organs formed. Chlamydospore-like hyphal swellings terminal or intercalary, globose, ellipsoidal, or irregular.

Dimensions: Hyphae 4.5–7.3 μm wide. Vesicles 30–35 (-65) μm: exit tubes ca. 200–500 μm long. Encysted zoospores 7.5–10 μm in diameter. Chlamydospore-like hyphal swellings ca. 6–40 (-80) μm long, 10–15 μm wide.

Material: 72-X128 (Sugarcane roots, Taiwan, ROC); 74-663 (= ATCC 36439, Strawberry roots, Shizuoka, Japan).

Remarks: Among the isolates identified as *P. afertile*, *P. flevoense* Van der Plaats-Niterink, one of heterothallic *Pythium*, forms zoospores from hypha-like sporangia. The isolates forming zoospores from hypha-like sporangia without sexual organs were treated as H-Zs as the provisional species in my work.

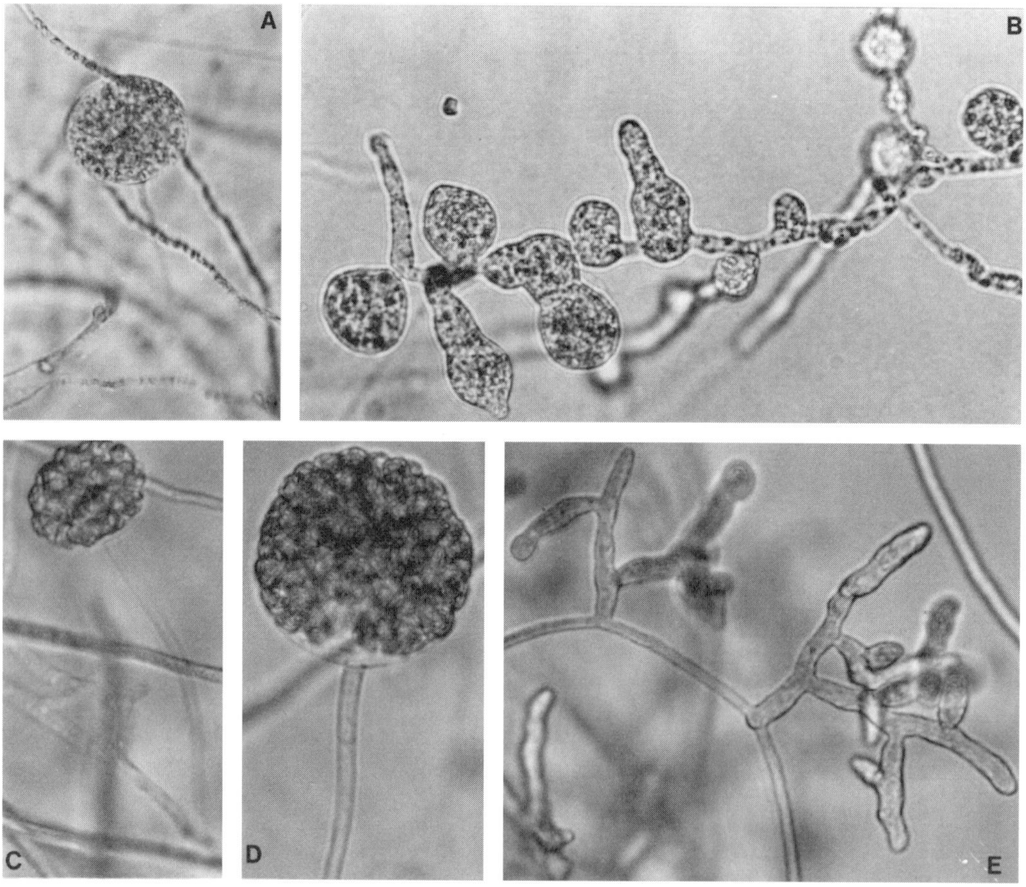

Pythium afertile. (A,B) and H-Zs (C–E). A,C,D: Vesicle formed from hypha-like sporangia. B: Chlamydospore-like structures. E: Somewhat thick-walled dendroid hyphae. (From Watanabe, T. et al. 1977. *Phytopathology*, 67:1324–1332. With permission.)

Pythium angustatum Sparrow

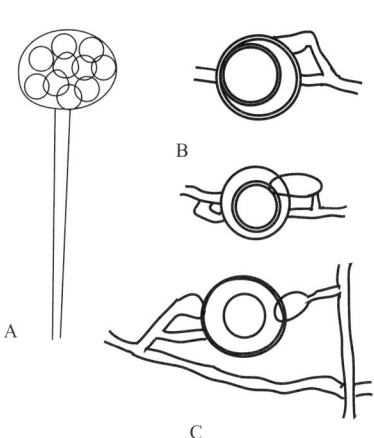

References: Middleton 1943; Plaats-Niterink 1981; Sparrow 1931; Watanabe et al. 1977.

Morphology: Sporangia hypha-like. Zoospores readily discharged after vesicle formation. Oogonia terminal or intercalary, bearing one to three antheridia per oogonium. Oospores aplerotic. Antheridia monoclinous or diclinous, clavate or crook-necked, often limited by the septum or short antheridiumphores, or nonstipitate.

Dimensions: Vesicles 25–45 µm in diameter. Zoospores encysted 8–10 µm in diameter. Oogonia 20–26.3 µm in diameter. Oospores 15–23.8 µm in diameter: oospore wall 0.2–0.8 µm thick. Antheridia 10–15 × 5–7.5 µm.

Material: 73-257 (= ATCC 36485, Strawberry roots, Shizuoka, Japan); 74-843, 74-846 (Strawberry root, Saitama, Japan).

Remarks: Oospore wall of this fungus is under 1 µm thick, whereas that of *P. dissotcum,* over 1.5 µm thick.

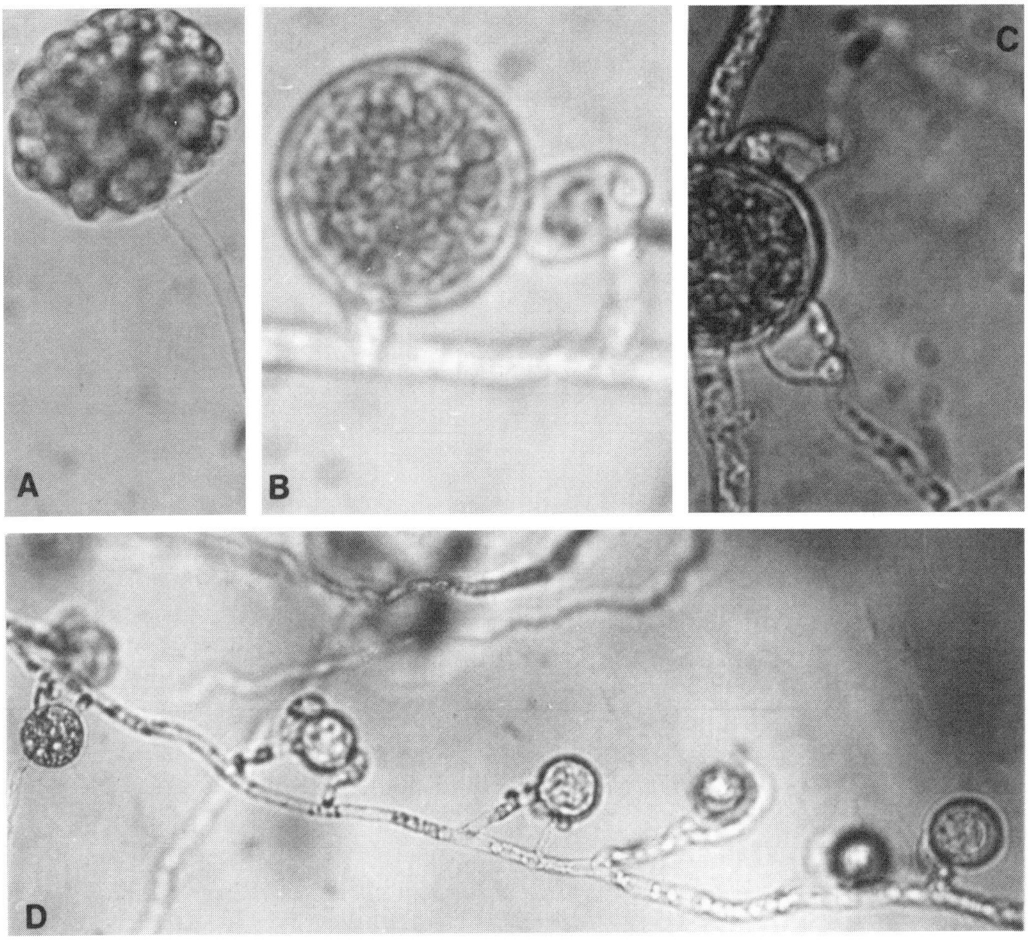

Pythium angustatum. A: Zoospore differentiation inside vesicle developed from hypha-like sporangium. B–D: Oogonia and antheridia (B,D: monoclinous; C: diclinous). (From Watanabe, T. et al. 1977. *Phytopathology*, 67:1324–1332. With permission.)

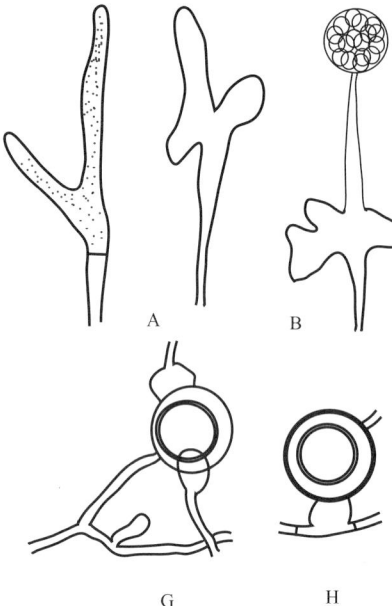

Pythium aphanidermatum (Edson) Fitz.

Synonym: *P. butleri* Subramaniam

References: Hickman 1943; Middleton 1943; Plaats-Niterink 1981; Watanabe 1974, 1983; Watanabe et al. 1977.

Morphology: Sporangia lobate, branched. Oogonia mainly terminal, bearing single (or 2) antheridia per oogonium. Oospores aplerotic. Antheridia cylindrical or barrel-shaped, often intercalary and terminal, monoclinous or diclinous.

Dimensions: Sporangia 107–200 × 7–13.4 µm: vesicles 30–50 µm in diameter. Zoospores encysted 10–12 µm in diameter. Oogonia 25–32.5 µm in diameter. Oospores 17.5–25 µm in diameter: wall 0.5–2.8 µm thick. Antheridia 10–22.5 × 10–12.5 µm.

Material: 72-X109 (Sugarcane root, Taiwan, ROC); 73-54 (= ATCC 36431, Strawberry field soil, Shizuoka, Japan); 78-2175 (Spinach root, Yamagata, Japan); 79-71 (= ATCC 64140, Cucumber roots, Naha, Okinawa, Japan); 80-73 (Cabbage root, Fukuoka, Japan).

Remarks: This fungus was often isolated from aerial substrates including stem and/or fruit of cucumber and soybean. The optimum temperatures for growth are at or near 34°C. The distribution and ecology of this fungus in Japan were discussed by Watanabe (1983b).

Mastigomycotina 61

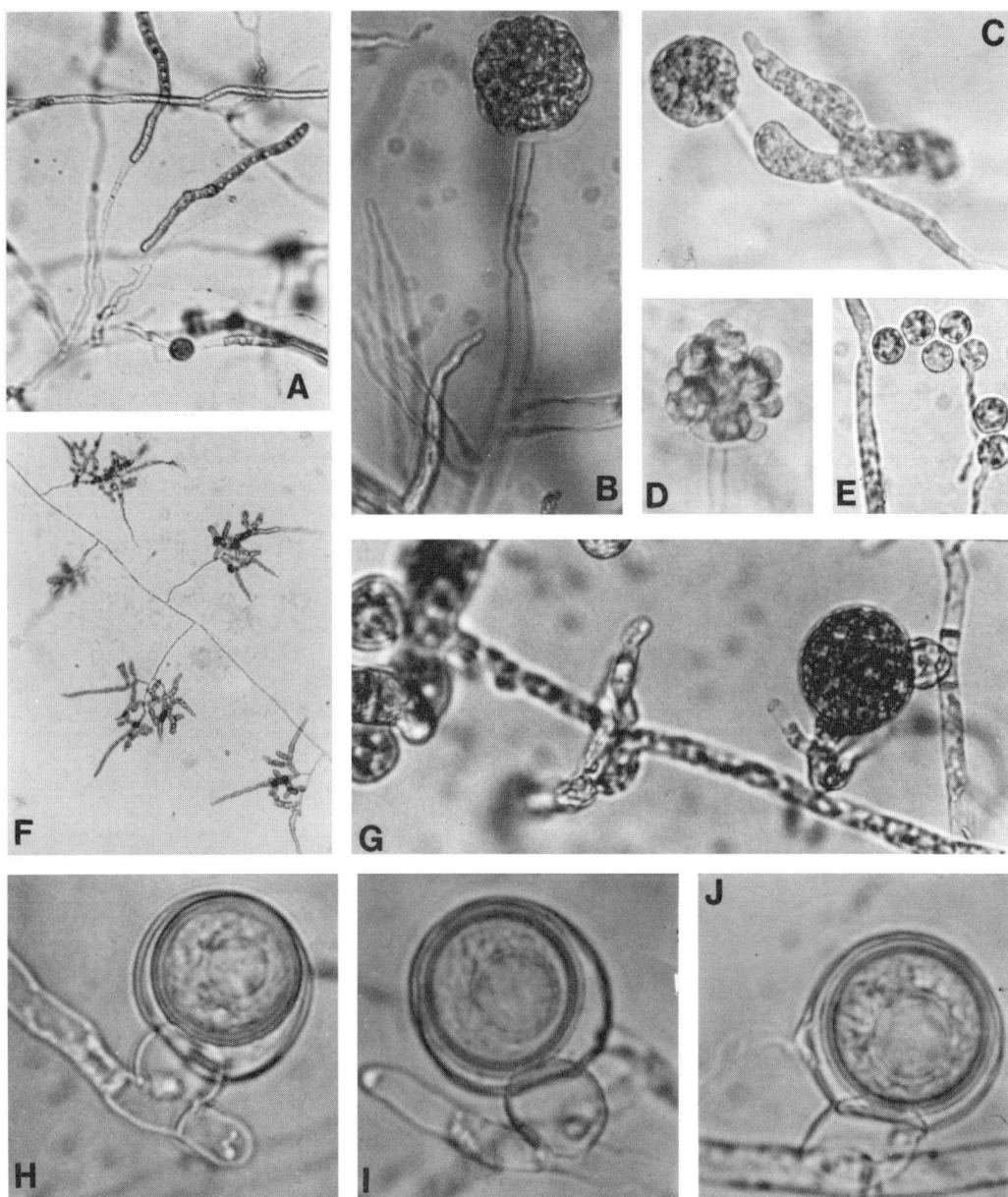

Pythium aphanidermatum. A–D,F: Sporangia and vesicle formation. E: Encysted zoospores and germination. G: Sporangia and sexual organs (diclinous). H–J: Oogonia, antheridia, and oospores. (From Watanabe, T. 1974a. *Trans. Mycol. Soc. Jpn.*, 15:343–357. With permission.)

Pythium apleroticum Tokunaga

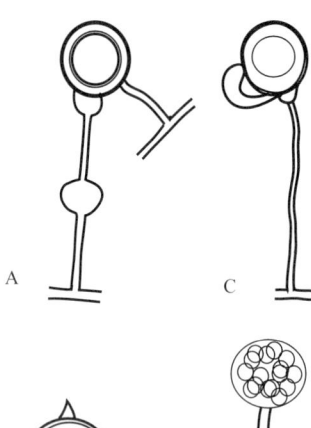

References: Ito 1936; Middleton 1943; Plaats-Niterink 1981; Watanabe et al. 1977.

Morphology: Sporangia hypha-like. Zoospores readily discharged after vesicle formation. Oogonia terminal, bearing single antheridia per oogonium. Oospores aplerotic, often formed parthenogenetically. Antheridia diclinous, subglobose, crook-necked, curved, cylindrical, various in shape. Conidial-like structures rarely formed.

Dimensions: Zoospores encysted 7.5–10.5 µm in diameter. Oogonia 21–25 µm in diameter. Oospores 17.5–22.5 µm in diameter. Antheridia ca. 15 µm long, 7 µm wide. Conidial-like structures 20–27.5 µm in diameter.

Material: 74-841 (= ATCC 36441, Strawberry roots, Saitama, Japan).

Remarks: This fungus is characterized by oogonia bearing single antheridia diclinously and hypha-like sporangia.

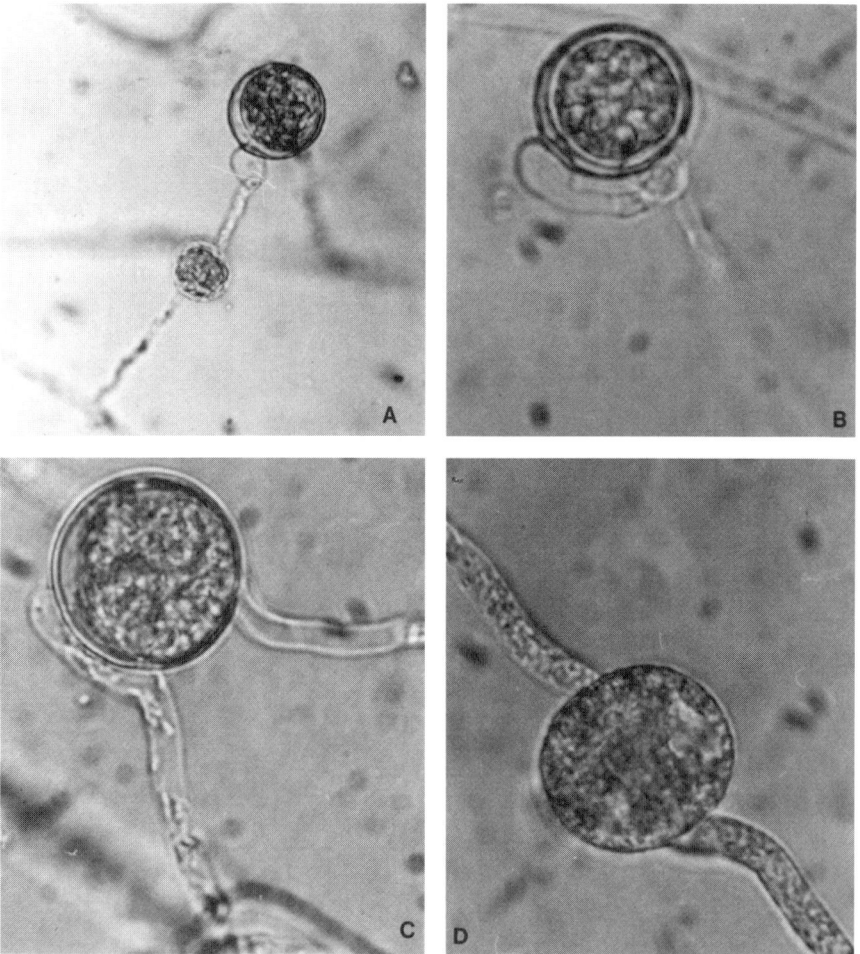

Pythium apleroticum. A–C: Oogonia, antheridia, and oospores (B). D: Conidia. E: Vesicle formation. (From Watanabe, T. 1974a. *Trans. Mycol. Soc. Jpn.*, 15:343–357. With permission.)

Pythium carolinianum Matthews

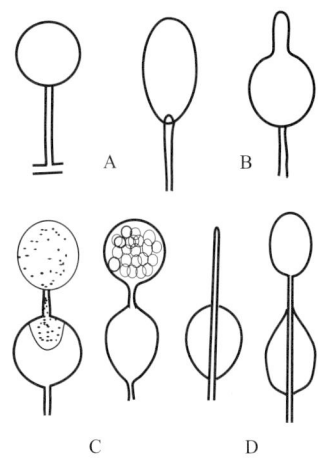

References: Matthews 1931; Middleton 1943; Plaats-Niterink 1981; Watanabe et al. 1977.

Morphology: Sporangia globose, ellipsoidal, mainly terminal, internally proliferated, significantly papillate. Long exit tubes often left after discharge of zoospores. Heterothallic.

Dimensions: Sporangia 20–35 (-37.5) µm in diameter; vesicles 30–35 µm in diameter; exit tubes 5–35 (-90) µm long. Zoospores encysted ca. 12–13 µm in diameter.

Material: 73-5 (= ATCC 36434, Strawberry field soil, Shizuoka, Japan); 79-170 (Cabbage field soil, Naha, Okinawa, Japan); 85-54 (= ATCC 66260, Flowering cherry seed, Hachioji, Tokyo, Japan).

Remarks: This fungus is characterized by well-papillated, internally proliferating (occasionally nesting) sphaerosporangia, and heterothallism.

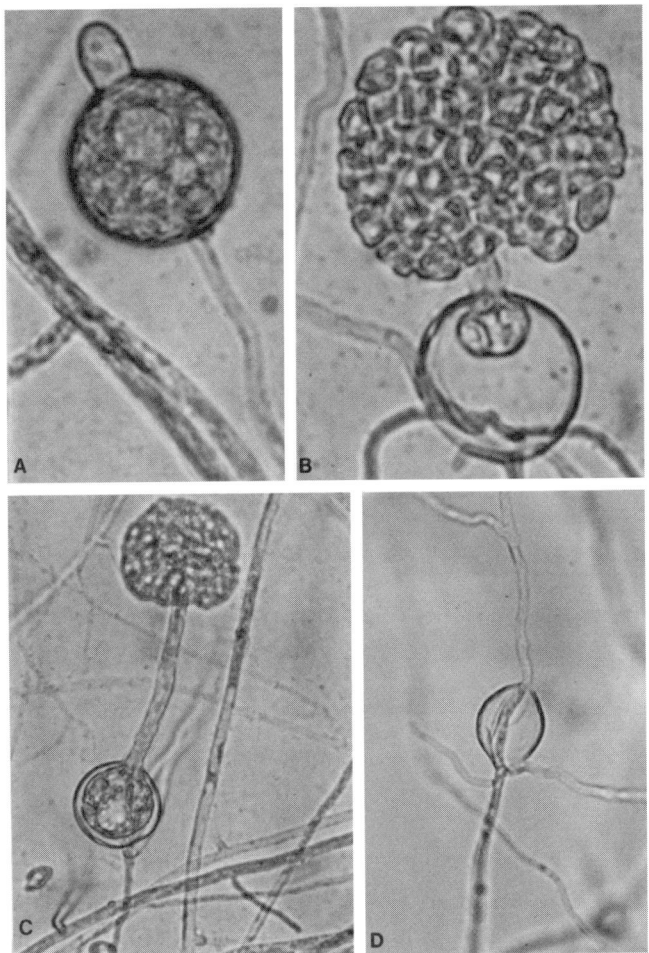

Pythium carolinianum. A–C: Sporangia and vesicle formation (B,C). D: Internally proliferated sporangium. (From Watanabe, T. et al. 1977. *Phytopathology*, 67:1324–1332. With permission.)

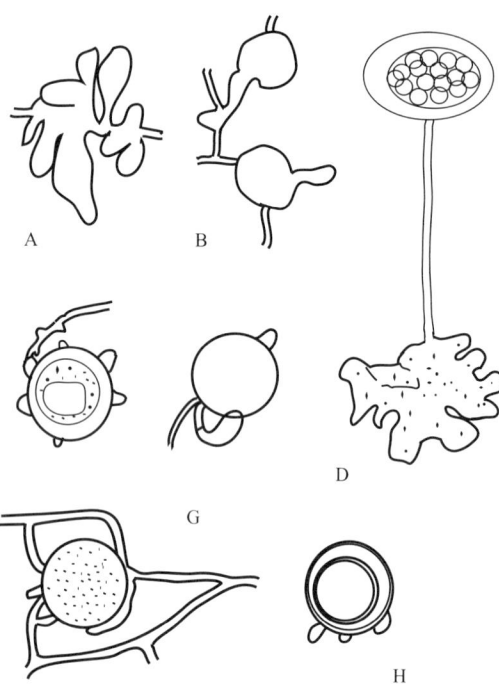

Pythium catenulatum Matthews

References: Hendrix and Campbell 1969; Matthews 1931; Middleton 1943; Plaats-Niterink 1981; Watanabe 1974; Watanabe et al. 1977.

Morphology: Sporangia lobate, aggregates of globose cells. Zoospores discharged after vesicle formation. Conidia globose, single or a few in short chains, intercalary and terminal, germinated directly by germ tubes. Oogonia mainly terminal or intercalary bearing 1–6 antheridia per oogonium, monoclinous or diclinous. Antheridia crook-necked, cylindrical, terminal. Oospores mainly plerotic, with a rather thick oospore wall.

Dimensions: Sporangia 17–20 μm wide: vesicles ca. 55 μm in diameter. Conidia 17.5–42.5 μm in diameter. Oogonia 26.4–42.5 μm in diameter. Oospores 18.7–35 μm in diameter: oospore wall 1.2–3.8 μm wide. Antheridia ca. 6.2–6.5 μm wide.

Material: 72-X99 (= ATCC 38892, Sugarcane roots, Taiwan, ROC).

Remarks: This fungus was pathogenic to roots of sugarcane seedlings.

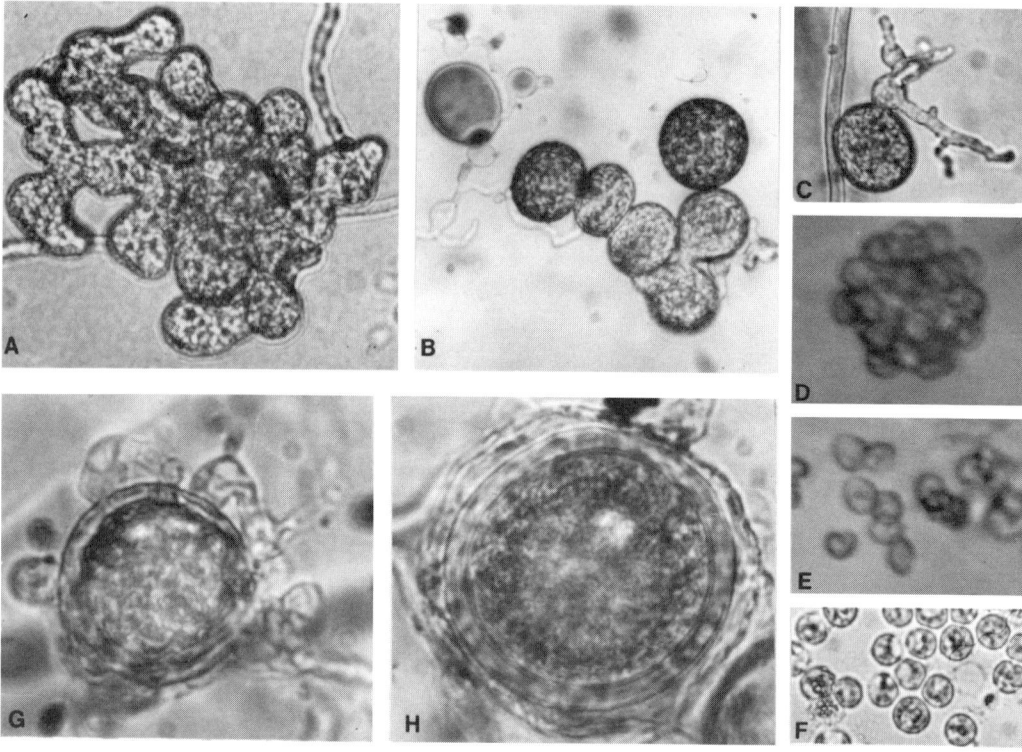

Pythium catenulatum. A,B: Lobate sporangia (A) and conidia (B). C: Direct germination of conidium. D,E: Zoospore differentiation in vesicle (D) and zoospore discharge (E). F: Encysted zoospores. G,H: Sexual organs and oospores (H). (From Watanabe, T. 1974a. *Trans. Mycol. Soc. Jpn.*, 15:343–357. With permission.)

Pythium conidiophorum Jokl

References: Ali-Shtayeh 1986; Matthews 1931; Middleton 1943; Plaats-Niterink 1981; Watanabe 1990b.

Morphology: Sporangia (conidia) globose, terminal or intercalary, only directly germinated with germ tubes. Oogonia mainly terminal, bearing no antheridia. Oospores plerotic, often formed parthenogenetically, with single large globules. Appressorium-like structures ellipsoidal, curved, spindle-shaped, formed in clusters.

Dimensions: Conidia 15–25 µm in diameter. Oogonia 16.2–22.5 µm in diameter: oospore wall ca. 1.2 µm wide. Appressoria ca. 30 µm long, 7.5–12.5 µm wide.

Material: 81-410 (Notsuke grassland soil, Hokkaido, Japan).

Remarks: Appressorium-like structures were not previously described.

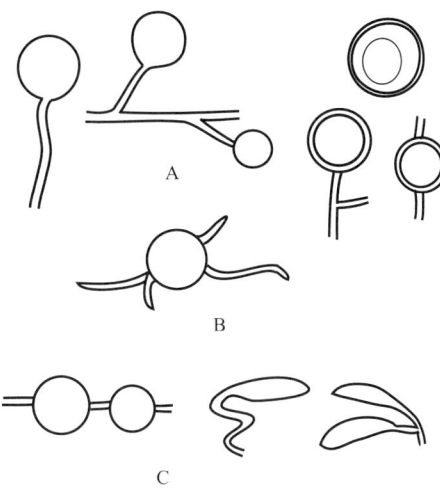

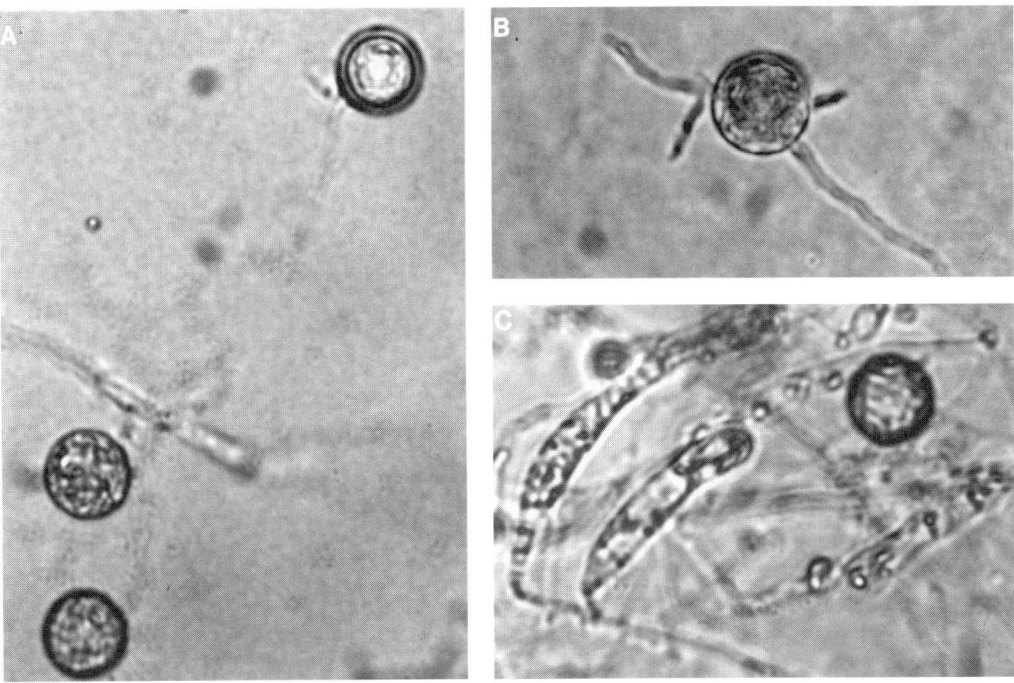

Pythium conidiophorum. A: Sporangia and parthenogenetically-formed oospores. B: Germination of conidia. C: Sporangia and appressoria. (From Watanabe, T. 1990b. *Ann. Phytopathol. Soc. Jpn.*, 56:549–556. With permission.)

Pythium deliense Meurs

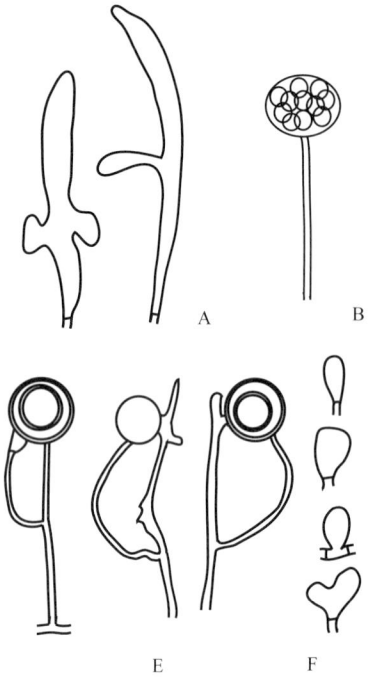

References: Meurs 1934; Middleton 1943; Plaats-Niterink 1981; Watanabe 1974a, 1981.

Morphology: Sporangia lobate, simple or branched. Zoospores discharged after vesicle formation. Oogoniumphores often curved toward antheridia, monoclinous. Oogonia terminal, often aggregated, bearing single antheridia. Oospores aplerotic. Antheridia terminal or intercalary, long-ellipsoidal, subglobose, or various in shape.

Dimension: Sporangia 30–200 × (7.2-) 10–17.5 μm. Oogonia 15–22.5 (-26.3) μm in diameter. Oospores (13.3-) 15.7–20.7 μm in diameter: oospore wall 0.3–0.5 μm thick. Antheridia 5–13.8 (-22.5) × (2.2-) 5–7.5 (-8.6) μm.

Material: 72-X108 (= ATCC 38893, Sugarcane roots, Taiwan, ROC); 79-76 (= ATCC 64141), 79-78 (= ATCC 64142) (Pumpkin roots, Naha, Okinawa, Japan); 79-226 (Adzuki bean field soil, Kagoshima, Japan).

Remarks: The fungus was intentionally isolated from the Ryukyu Islands based on the knowledge of its distribution in Taiwan, ROC, adjacent to the Ryukyu Islands. Bending of oogoniumphore to antheridia is very characteristic for this fungus.

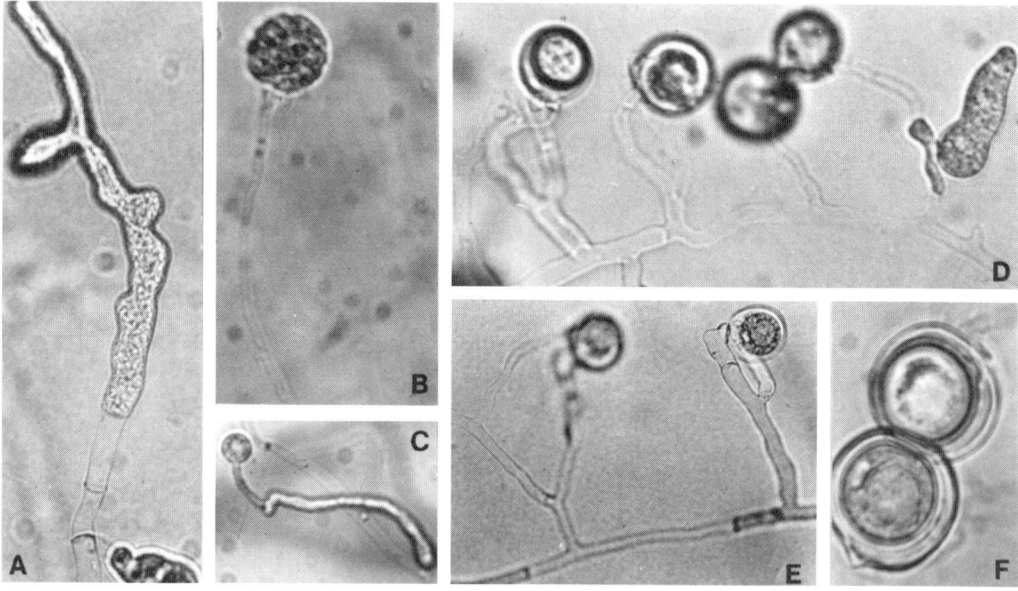

Pythium deliense. A: Lobate sporangium. B: Vesicle formation. C: Germination of encysted zoospores. D–F: Oogonia, antheridia, and oospores (D,F). (From Watanabe, T. 1974a. *Trans. Mycol. Soc. Jpn.*, 15:343–357. With permission.)

Pythium dissimile Vaartaja

References: Plaats-Niterink 1981; Vaartaja 1965; Watanabe 1990b.

Morphology: Sporangia lobate, terminal or intercalary, rarely discharging zoospores, mostly functioned as conidia germinated by germ tubes. Oogonia mainly terminal, bearing no antheridia. Antheridia, possibly hypogynous, but not recognizable. Oospores formed parthenogenetically, plerotic, with thick oogonium wall, containing single oil globules.

Dimensions: Sporangia up to 380 μm long, 15–25 μm wide. Zoospores encysted ca. 10 μm in diameter. Oogonia 21.2–25 μm in diameter; oospore wall ca. 2 μm wide.

Material: 81-584 (Corn field soil, Nakashibetsu, Hokkaido, Japan).

Remarks: Appressorium-like small hyphal swellings were catenulate or clustered.

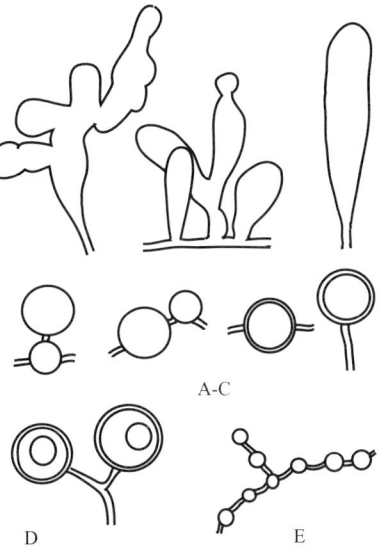

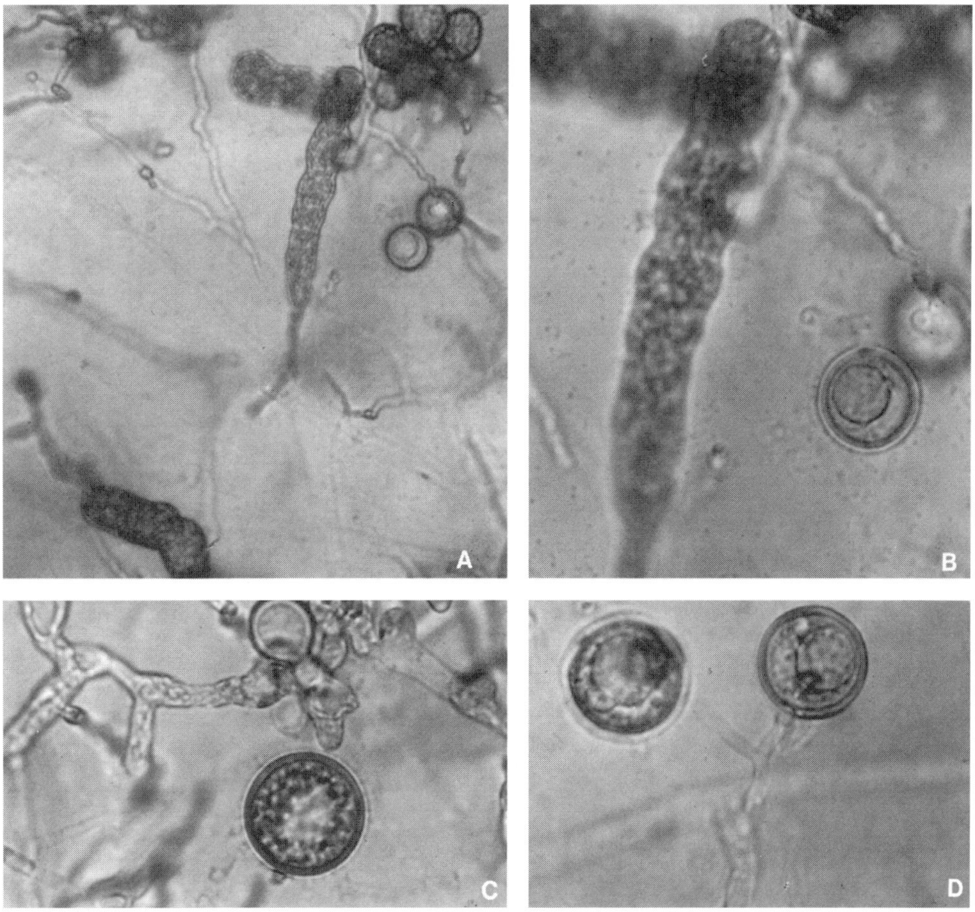

Pythium dissimile. A,B: Sporangia and oospores formed parthenogenetically (B: close-up of A). C–D: Oospores formed parthenogenetically. E: Catenulate appressorium-like hyphal swellings. (From Watanabe, T. 1990b. *Ann. Phytopathol. Soc. Jpn.*, 56:549–556. With permission.)

Pythium dissotocum Drechsler

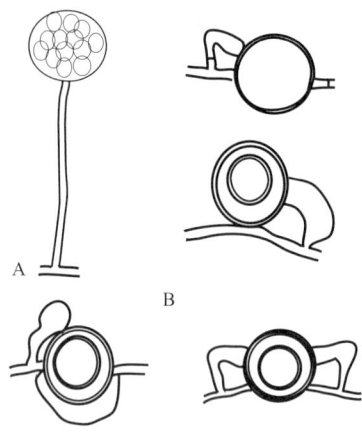

References: Drechsler 1930; Middleton 1943; Plaats-Niterink 1981.

Morphology: Sporangia hypha-like. Zoospores discharged after forming vesicles. Oogonia terminal or intercalary, bearing 1–2 antheridia per oogonium. Oospores aplerotic. Antheridia simple, cylindrical, crook-necked, or irregular-shaped, often curved, often nonstipitate, monoclinous or diclinous.

Dimensions: Vesicles ca. 30–35 (-50) µm in diameter. Zoospores encysted (5-) 7.5–10 (–12.5) µm in diameter. Oogonia 15–30 µm in diameter; oospores 14–20 µm in diameter; oil globules 12.5–15.5 µm in diameter; oospore wall 1.5–2 µm thick. Antheridia 4–5 µm wide.

Material: 80-571 (Watermelon field soil, Kagoshima, Japan), 81-708 (= ATCC 64149, Lettuce field soil, Chiba, Japan).

Remarks: *Pythium angustatum* may be better treated as a synonym of this fungus.

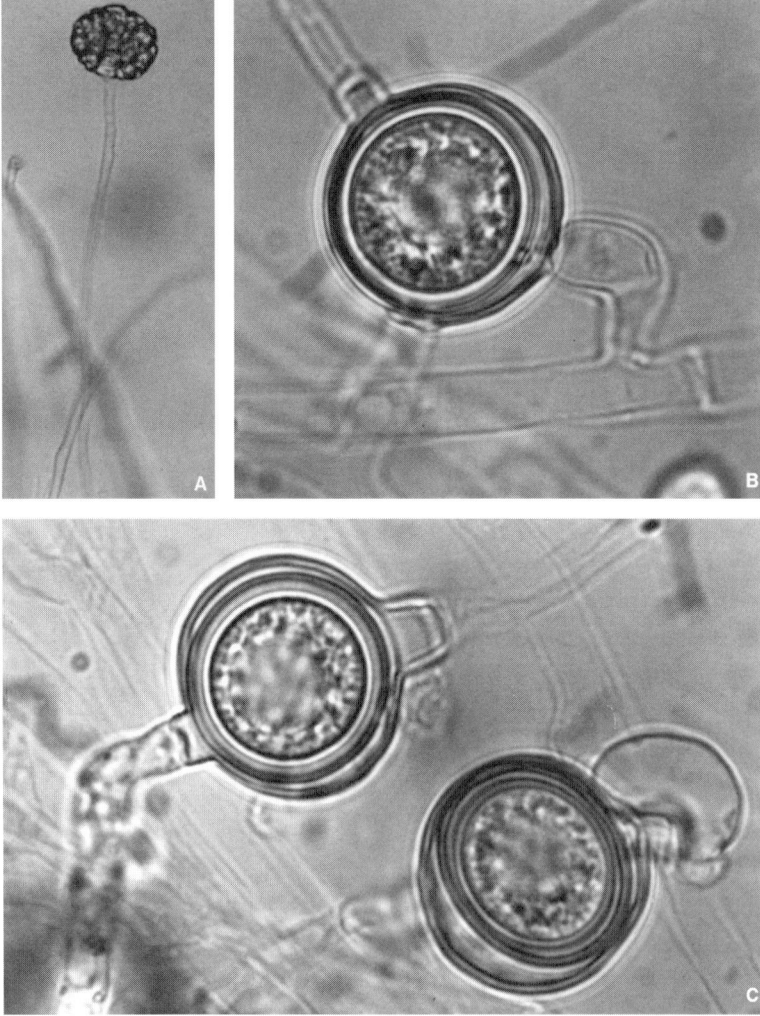

Pythium dissotocum. A: Vesicle formation from hypha-like sporangium. B,C: Oogonium, sessile antheridium, and oospore.

Pythium echinulatum Matthews

References: Matthews 1931; Middleton 1943; Milanez 1978; Plaats-Niterink 1981; Watanabe et al. 1977.

Morphology: Sporangia mainly subglobose, or cylindrical, terminal or intercalary. Zoospores rarely discharged unnoticeably. Oogonia echinulate with acute protuberances regularly distributed, terminal, bearing single antheridia per oogonium. Oospores aplerotic, hypogynous, often unstalked, monoclinous.

Dimensions: Sporangia 20–70 × (20-) 22.5–37.5 µm. Oogonia 20–30 µm in diameter: protuberances 3.7–5 × 2.2–3.7 µm. Oospores 17.5–22.5 µm in diameter: oospore wall 0.5–1.3 µm thick.

Material: 73-29 (= ATCC 38891, Strawberry roots, Shizuoka, Japan); 74-822 (= ATCC 36437, Strawberry roots, Saitama, Japan); 81-219 (= ATCC 64153, Radish field soil, Kanagawa, Japan); 84-178 (Watermelon field soil, Zentsuji, Kagawa, Japan).

Remarks: This fungus is characterized by echinulate oogonia with acute protuberances, containing aplerotic oospores and discrete spherical structures. However, it is rather difficult to interpret whether spherical hyphal structures are discrete or aggregated, and oogonial protuberances (spines) are acute or blunt.

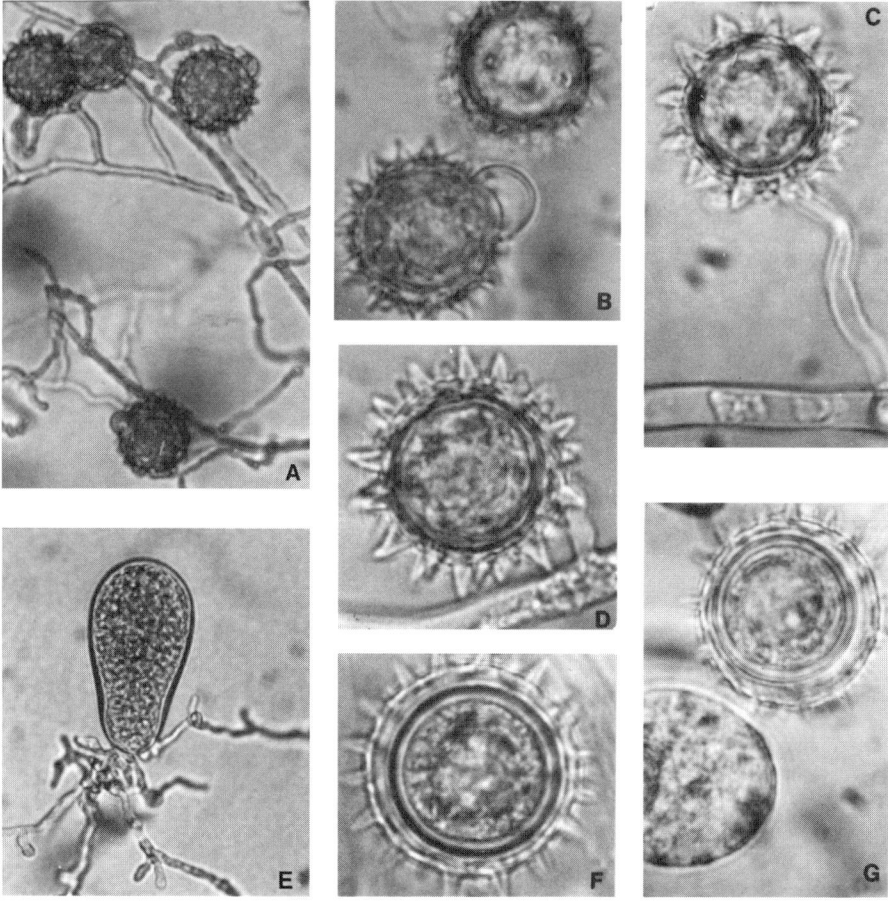

Pythium echinulatum. A: Immature oogonium. B: Oogonia and antheridium. C,D: Oogonia containing plerotic oospores. E: Direct germination of sporangium. F,G: Oogonia, aplerotic oospores, and a part of the sporangium (G). (From Watanabe, T. et al. 1977. *Phytopathology*, 67:1324–1332. With permission.)

Pythium elongatum Matthews

References: Matthews 1931; Middleton 1943; Plaats-Niterink 1981; Watanabe 1989a,b; Watanabe et al. 1977.

Morphology: Hyphal swellings (sporangia) terminal or intercalary, lobate, subglobose or irregular, with long exit tubes. Zoospores discharged after vesicle formation. No sexual organs formed.

Dimensions: Hyphal swellings (15-) 25–28 µm wide: exit tubes of lobate sporangia over 100 µm long. Vesicles 15–30 (-45) µm in diameter.

Material: 79-72 (= ATCC 64152, Tomato roots, Naha, Japan); 79-228 (Adzuki root, Naze, Kagoshima, Japan); 80-222 (Soybean field soil, Miyakonojyo, Kagoshima, Japan).

Remarks: This fungus is characterized by lobate or globose hyphal swellings, both of which elongate long exit tubes before vesicle formation.

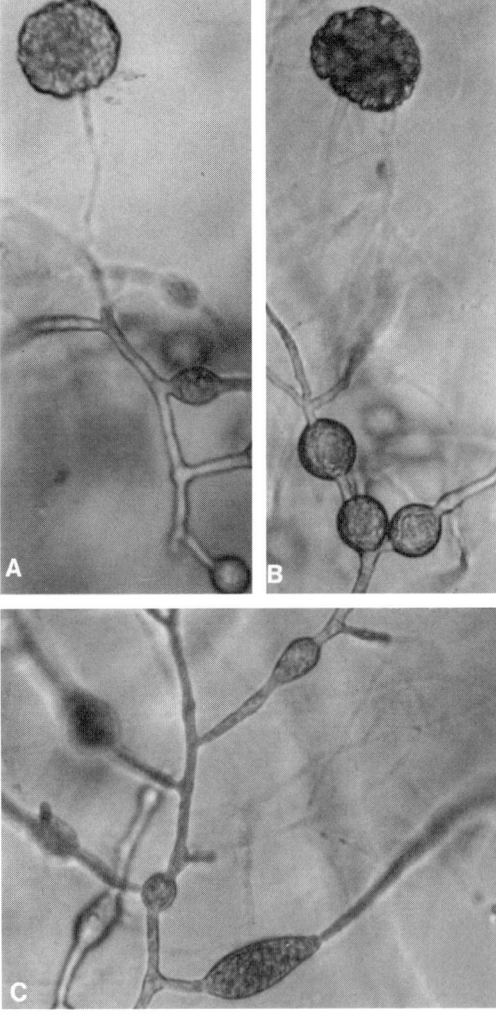

Pythium elongatum. A,B: Lobate sporangia and vesicle formation. C: Inflated hyphae. (From Watanabe, T. et al. 1977. *Phytopathology,* 67:1324–1332. With permission.)

Pythium graminicola Subramaniam

References: Drechsler 1936; Matthews 1931; Middleton 1943; Plaats-Niterink 1981; Watanabe 1974a.

Morphology: Sporangia lobate, irregular, well branched, zoospores rarely formed unnoticeably. Oogonia mainly terminal, bearing (1) 2–6 antheridia per oogonium. Oospores plerotic or aplerotic, with single oil globules. Antheridia cylindrical, hypha-like, crook-necked, intercalary or terminal, monoclinous or diclinous. Appressoria composed of ovate cells in chains.

Dimensions: Sporangia 6–10.3 μm broad. Oogonia 20–28.8 (-40) μm in diameter. Oospores 20–25 μm in diameter: oil globules 10–15 μm in diameter. Antheridia 3.6–5.5 μm wide. Appressoria 6–17.5 μm wide.

Material: 72-X115, 72-116 (Sugarcane root, Taiwan, ROC); 80-214 (= ATCC 64150, Soybean field soil, Miyakonojyo, Miyazaki, Japan); 80-302 (= ATCC 64151, Corn field soil, Kumamoto, Japan).

Remarks: Both plerotic and aplerotic oospores formed in most of the isolates studied, with mono- and diclinous antheridia, and thus *P. graminicola* is not easily differentiated from the related species.

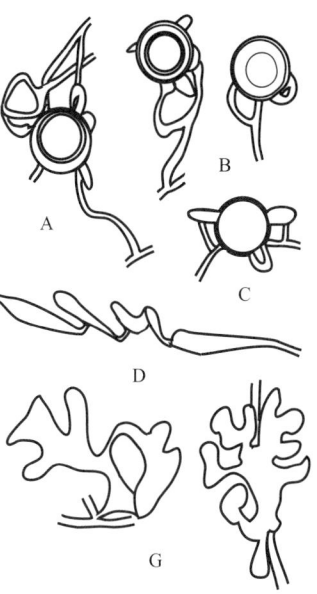

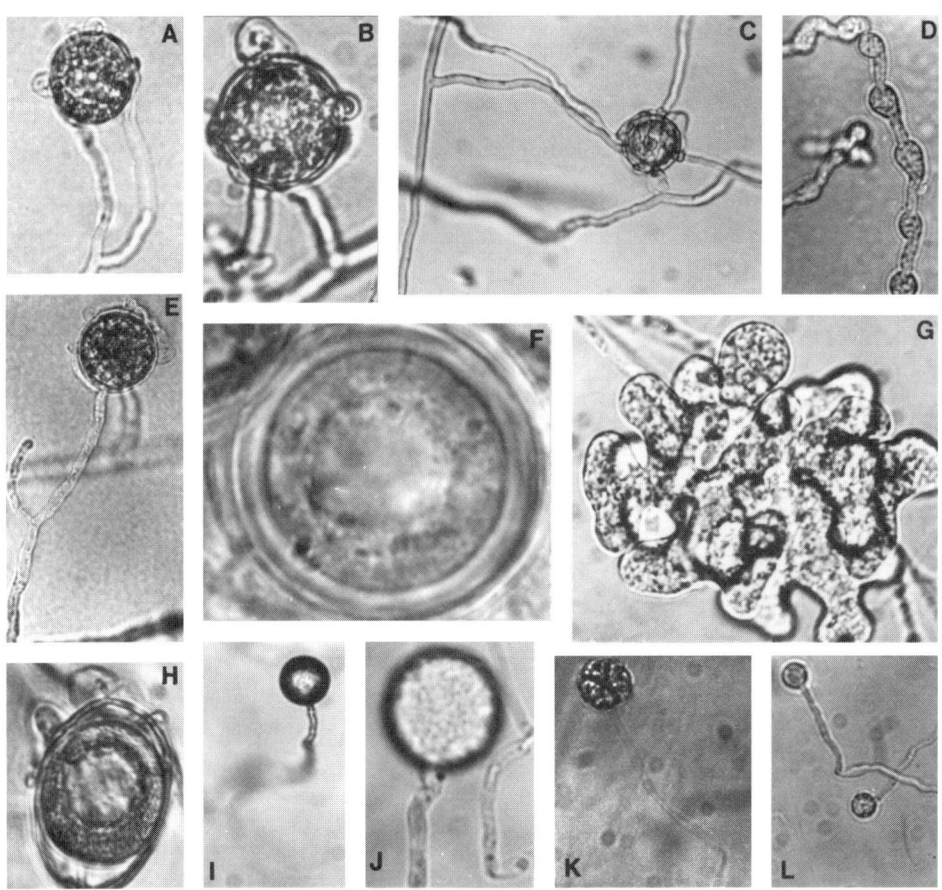

Pythium graminicola. A–C,E,H: Oogonia and antheridia. D: Appressoria. F: Oospore. G: Lobate sporangia. I,J: Immature oogonium-like structures formed aerially and on agar. K: Vesicle formation. L: Germination of encysted zoospores. (From Watanabe, T. 1974a. *Trans. Mycol. Soc. Jpn.*, 15:343–357. With permission.)

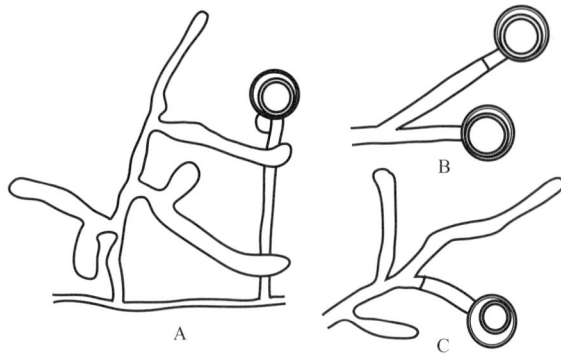

Pythium cf. *indigoferae* Butler

References: Matthews 1931; Middleton 1943; Plaats-Niterink 1981; Watanabe et al. 1998.

Morphology: Hyphae often showing dendroid branching, sometimes slightly swollen, often directly connected with sexual organs. No zoospore discharged. Oogonia terminal, borne on thick oogoniumphores, with single or occasionally double, mostly monoclinous or rarely hypogynous antheridia. Oospores aplerotic, occasionally developed parthenogenetically.

Dimensions: Hyphae mostly 7.5 μm thick. Oogonia 17.5–20 μm: oogoniumphores, mostly 5 μm wide. Oospores 12.5–15 μm in diameter: oospore wall 1.5–2 μm thick.

Material: 90-104 (= ATCC 200701, Mountain soil, Dischmatal, Switzerland); 90-113 (= ATCC 200702, Mountain soil, Malans, Switzerland).

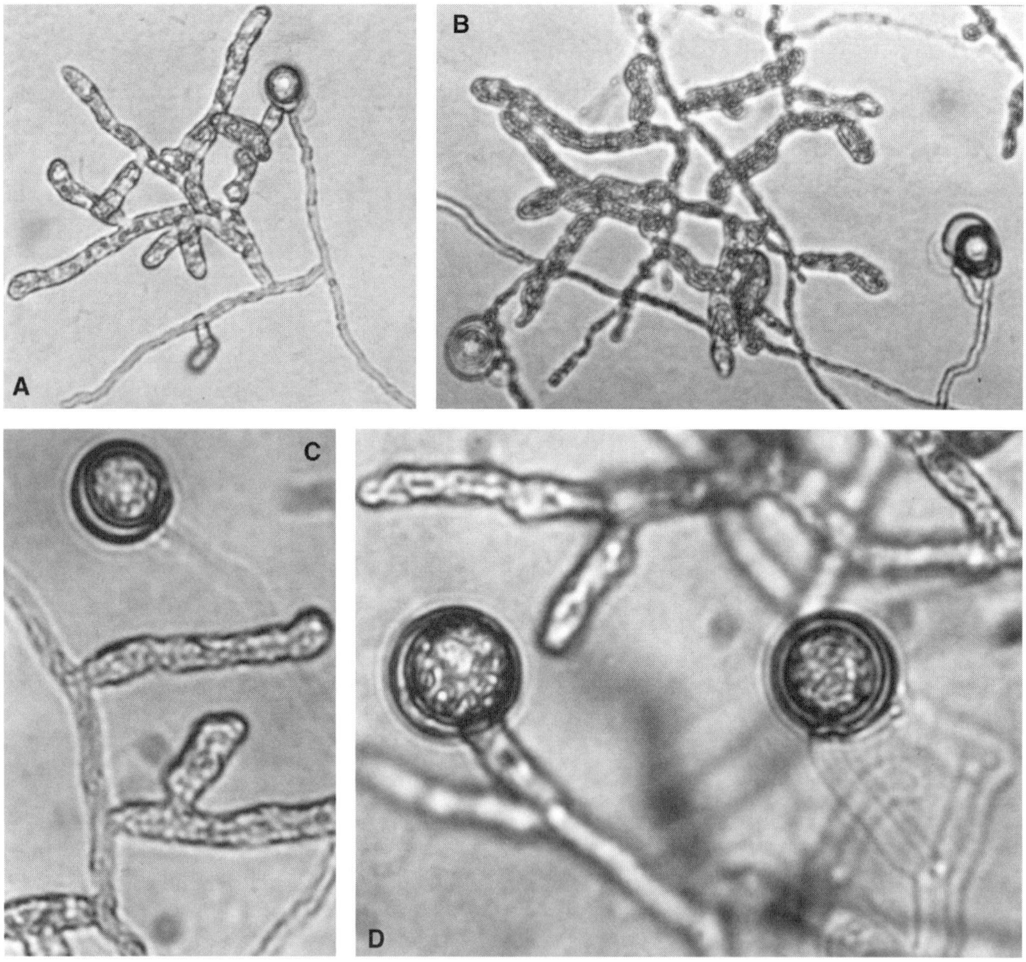

Pythium cf. *indigoferae*. A–D: Lobate hyphal swellings and sexual organs with aplerotic oospores. Note parthenogenetically developed oospores (C,D). (From Watanabe, T. et al. 1998. *Mycologia Helvetica*, 10:3–13. With permission.)

Pythium inflatum Matthews

References: Matthews 1931; Middleton 1943; Plaats-Niterink 1981; Watanabe 1974a, 1989a.

Morphology: Sporangia irregular in shape, dense in protoplasm, zoospores discharged with difficulty. Oogonia terminal or intercalary, bearing 2–3 antheridia per oogonium. Antheridia cylindrical, attaching oogonia only apically and partially, mostly diclinous. Oospores plerotic.

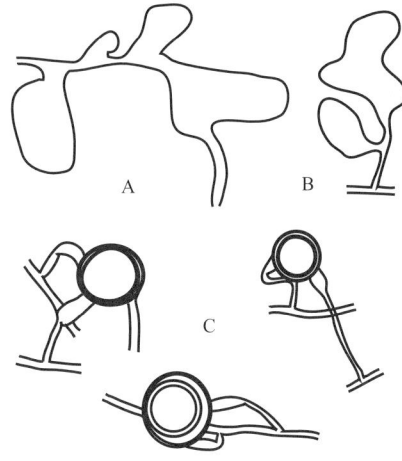

Dimensions: Lobate sporangia 75–110 × 12–22 µm. Oogonia 17.5–25 (-32.5) µm in diameter. Antheridia (7.5-) 12.5–15 × 5–7.5 µm. Oospores ca. 15–20 µm in diameter.

Material: 72-X117 (= ATCC 38894, Sugarcane roots, Taiwan, ROC); 73-51 (= ATCC 36436, Strawberry roots, Shizuoka, Japan); 80-150, 80-152 (Watermelon field soil, Fukuoka, Japan).

Remarks: This fungus is characterized by lobate sporangia and oogonia containing plerotic oospores, bearing single diclinous antheridia.

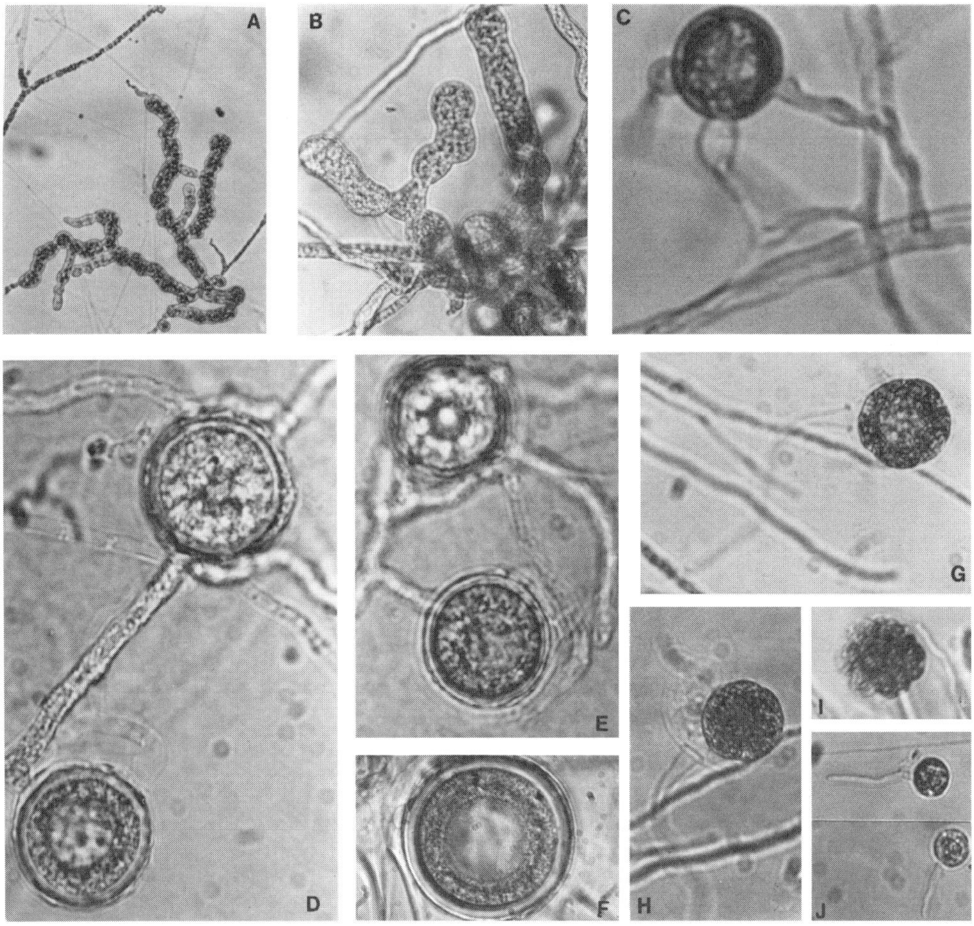

Pythium inflatum. A,B: Lobate sporangia. C–E: Oogonia and antheridia. F: Oospore. G–I: Vesicle formation. J: Germination of encysted zoospores. (From Watanabe, T. 1974a. *Trans. Mycol. Soc. Jpn.*, 15:343–357. With permission.)

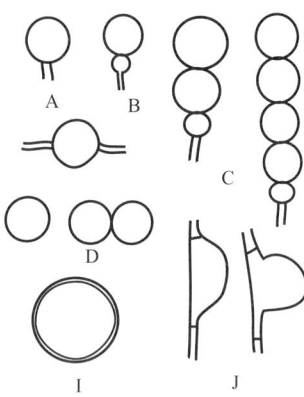

Pythium intermedium de Bary

References: Matthews 1931; Middleton 1943; Plaats-Niterink 1981; Watanabe 1983a, 1985; Watanabe et al. 1977.

Morphology: Sporangia globose, over 3 sporangia in chains developed basipetally, deciduous. No sexual organs formed. Thick-walled chlamydospores occasionally formed, solitary.

Dimensions: Conidia 15–20 µm in diameter. Chlamydospores 18–20 µm in diameter.

Material: 73-483-2 (= ATCC 36445, Strawberry field soil, Shimane, Japan); 81-395 (= ATCC 64143, Forest soil, Hokkaido, Japan); 83-398 (Piedmont forest soil of Mt. Fuji, Shizuoka, Japan).

Remarks: Encysted zoospores were rarely observed, but zoospore discharge was never observed by the author. Zoospores discharged from some 10 to 20% of sporangia (Stanghellini et al., 1988).

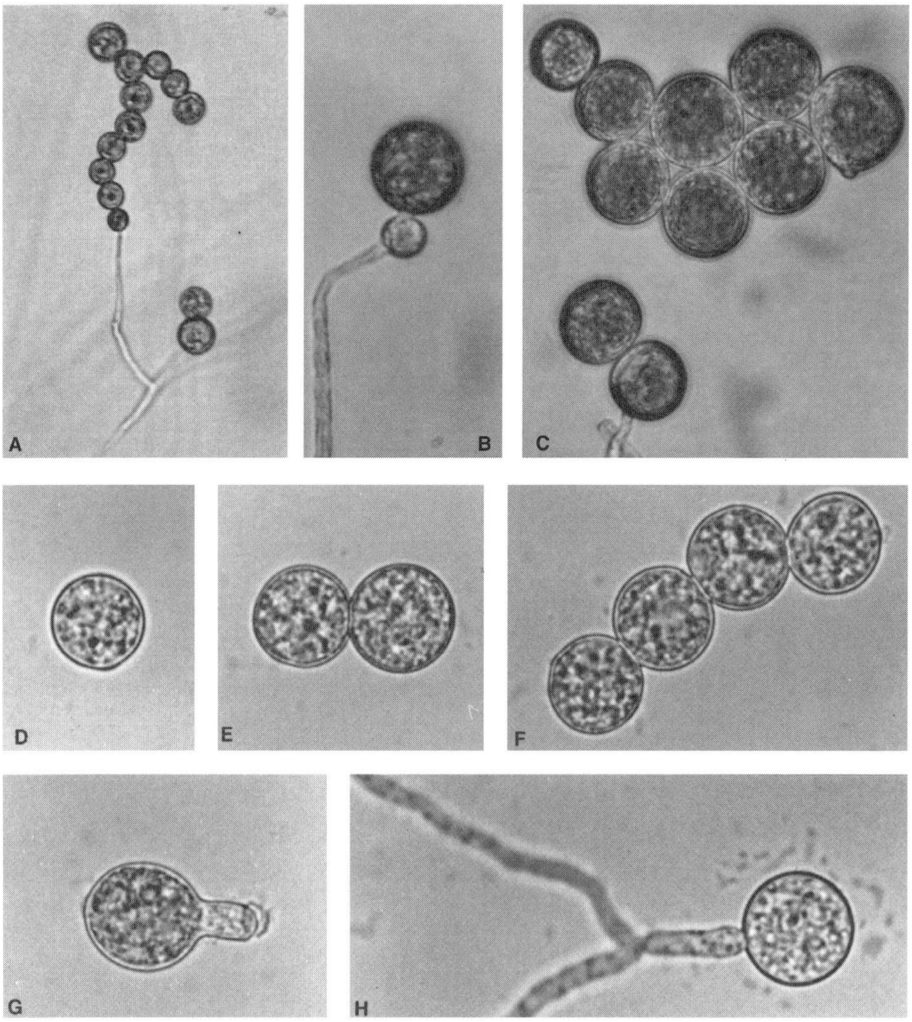

Pythium intermedium. A: Globose sporangia in a chain. B: Young sporangia basipetally developed. C–F: Detached sporangia. G,H: Direct germination of sporangia. I: Chlamydospore. J: Inflated hyphae. (From Watanabe, T. et al. 1977. *Phytopathology,* 67:1324–1332. With permission.)

Pythium irregulare Buisman

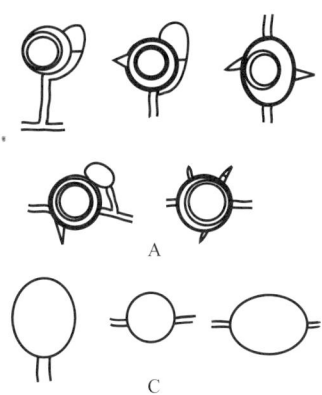

References: Biesbrock and Hendrix 1967; Matthews 1931; Middleton 1943; Plaats-Niterink 1981.

Morphology: Sporangia (conidia) mostly ellipsoidal, subglobose, cylindrical, terminal or intercalary. Oogonia terminal or intercalary, smooth or echinulate with 1–4 (8) protuberances irregularly distributed, bearing single antheridia per oogonium. Oospores aplerotic. Antheridia clavate or cylindrical, separated from antheridium branches with septum, partially covering oogonia, monoclinous.

Dimensions: Conidia 22.5–30 (-37.5) μm in diameter. Oogonia 15–25 (-30) μm in diameter: protuberances ca. 4–7 μm long, 2.2–2.3 μm wide. Oospores (12.5-) 14–20 (-22.5) μm in diameter. Antheridia 5–8 μm wide.

Material: 78-2452 (= ATCC 64144, Peanut field soil, Aomori, Japan); 81-30 (= ATCC 64145, Cabbage field, Kanagawa, Japan).

Remarks: In some isolates, antheridium branches are rather long and curved.

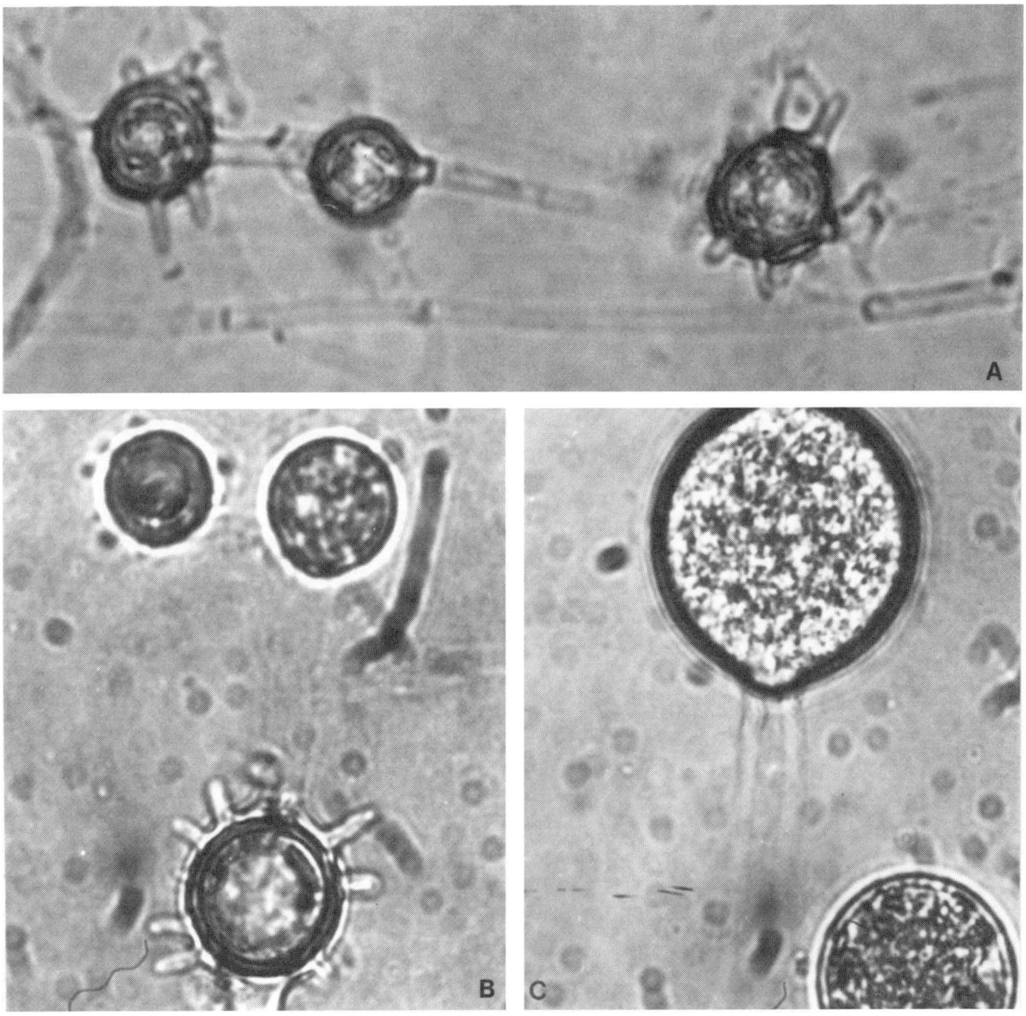

Pythium irregulare. A–C: Sporangia (B,C), oogonia, and oospores.

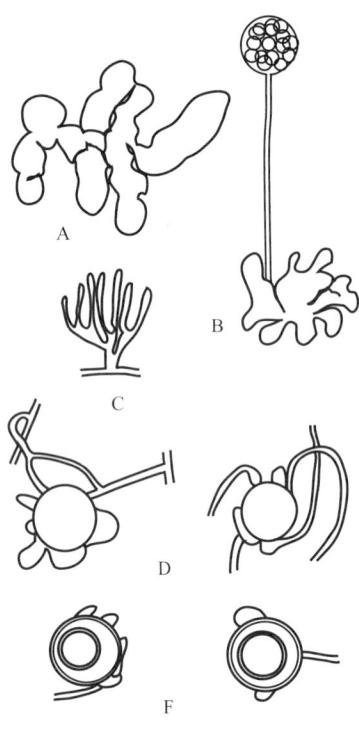

Pythium myriotylum Drechsler

References: Drechsler 1930, 1943; Matthews 1931; Middleton 1943; Nagai et al. 1988; Plaats-Niterink 1981; Watanabe 1977a; Watanabe et al. 1977.

Morphology: Sporangia lobate with long exit tubes. Zoospores discharged after vesicle formation. Oogonia mainly terminal, bearing 2–7 (mostly 3–4) antheridia per oogonium. Oospores aplerotic. Antheridia cylindrical, crook-necked, or various in shape, diclinous. Appressoria broom-like.

Dimensions: Sporangia ca. 10 μm wide; vesicles 40–55 μm in diameter: exit tubes up to 450 μm long. Zoospores encysted 7.5–11 μm in diameter. Oogonia 28.7–37.5 μm in diameter. Oospores 20–32.5 μm in diameter: oospore wall 2–3 μm thick. Antheridia ca. 15 μm long, 7.5 μm wide.

Material: 74-864 (= ATCC 36440, Strawberry roots, Saitama, Japan); 83-193 (Arrowhead roots, Saitama, Japan); 85-CH-7526 (Dasheen root, Chiba, Japan).

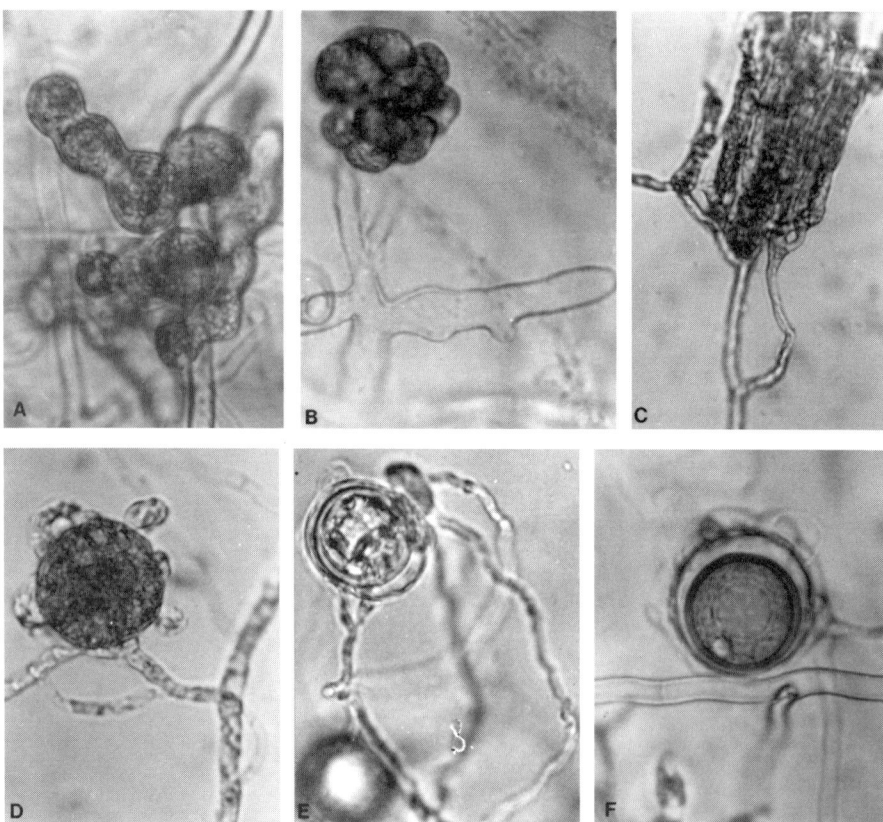

Pythium myriotylum. A,B: Lobate sporangia and vesicle formation (B). C: Appressorium. D–F: Oogonia, antheridia, and oospores (F). (From Watanabe, T. 1977a. *Ann. Phytopathol. Soc. Jpn.*, 43:306–309. With permission.)

Pythium nayoroense T. Watanabe

References: Watanabe 1990.

Morphology: Sporangia of two kinds: globose with subsporangial vacant inflation, somewhat thick-walled, with well-developed exit tubes originating from papilla, often irregularly papillate, and ellipsoidal, usually intercalary. Zoospores discharged after vesicle formation. No sexual organs formed.

Dimensions: Sporangia: globose 22.5–40 µm in diameter; ellipsoidal 35–60 × 20–22.5 (-30) µm: subsporangial swellings 12.5–13.8 (-37.5) µm in diameter: exit tubes 32.5–160 × 7.5–10 µm. Zoospores encysted 7.5–12.5 µm in diameter.

Material: 81-499, 81-500 (Potato field soil, Nayoro, Hokkaido, Japan).

Remarks: Subsporangial vacant inflation is characteristic for this fungus.

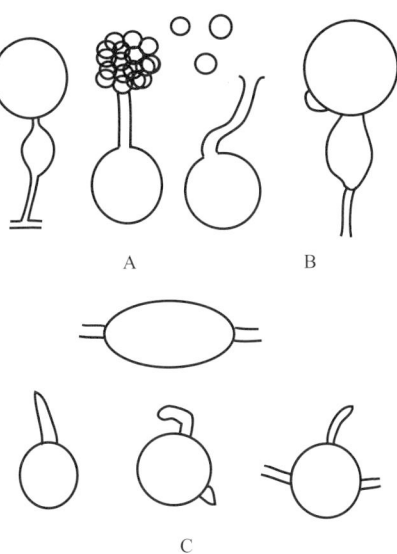

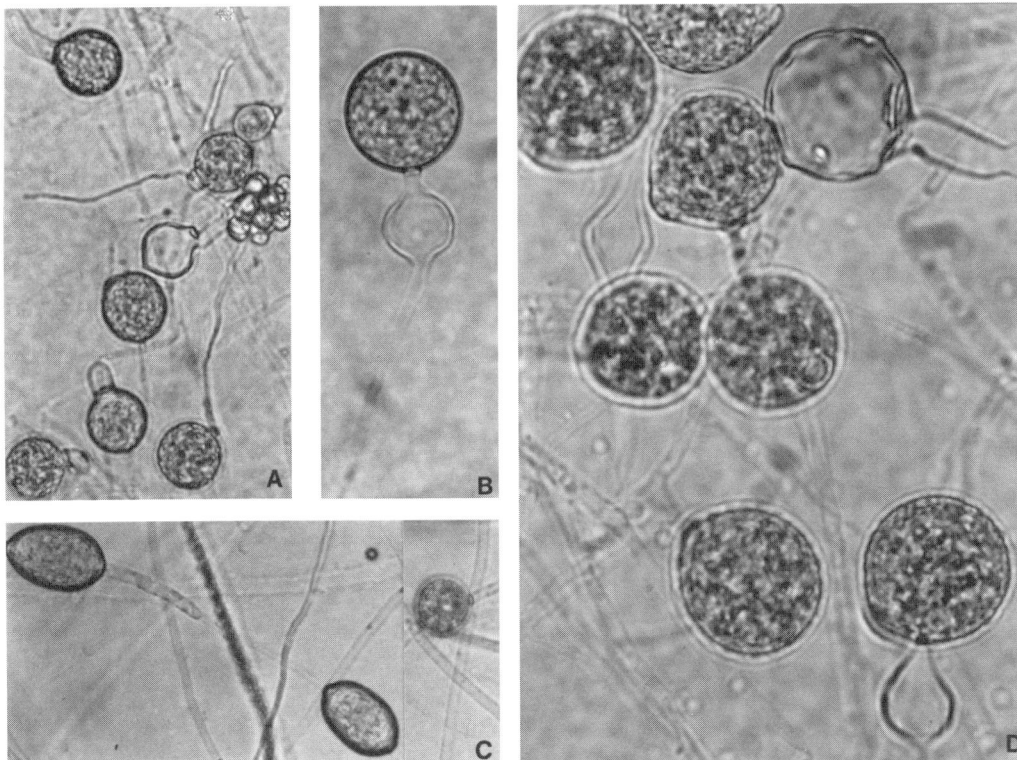

Pythium nayoroense. A,C,D: Sphaerosporangia and vesicle formation. B: Sporangia and subsporangial hyphal swellings.

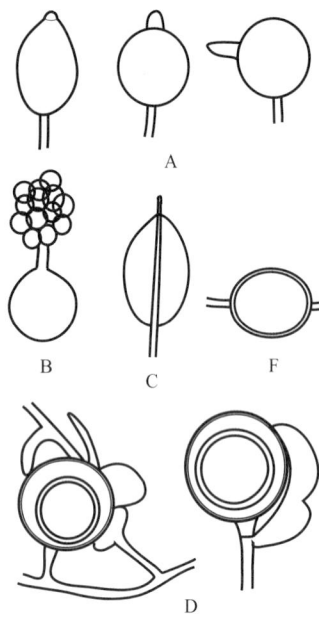

Pythium oedochilum Drechsler

References: Drechsler 1930, 1941; Middleton 1943; Plaats-Niterink 1981; Watanabe et al. 1977.

Morphology: Sporangia globose, ellipsoidal or ovate, conspicuously papillate, terminal or intercalary, internally proliferated. Zoospores discharged after vesicle formation. Oogonia terminal, occasionally papillate, bearing 1–3 antheridia. Oospores mostly aplerotic, several oil globules included. Antheridia cylindrical, lunar-shaped, curved, wavy, mostly furrowed on the upper surface, occasionally 1-septate, diclinous.

Dimensions: Sporangia 27.5–37.5 (-40) µm in diameter: papilla 5–6.5 µm wide. Zoospores encysted 10–15 µm in diameter. Oogonia 25–32.5 (-33.8) µm in diameter. Oospores 22.5–30 µm in diameter: oospore wall ca. 0.5 µm thick. Antheridia ca. 25–27.5 µm long, 5–7.5 µm wide.

Material: 73-165 (= ATCC 36433, Strawberry roots, Shizuoka, Japan); 74-888 (Strawberry root, Saitama, Japan); 82-634 (Spinach roots, Nara, Japan).

Remarks: The antheridial stalks occasionally coiled around the oogonial branches. Such coiling also occurs in *P. helicoides* Drechsler, *P. palingenes* Drechsler, *P. polytylum* Drechsler, and *P. ostracodes* Drechsler (Plaats-Niterink, 1981).

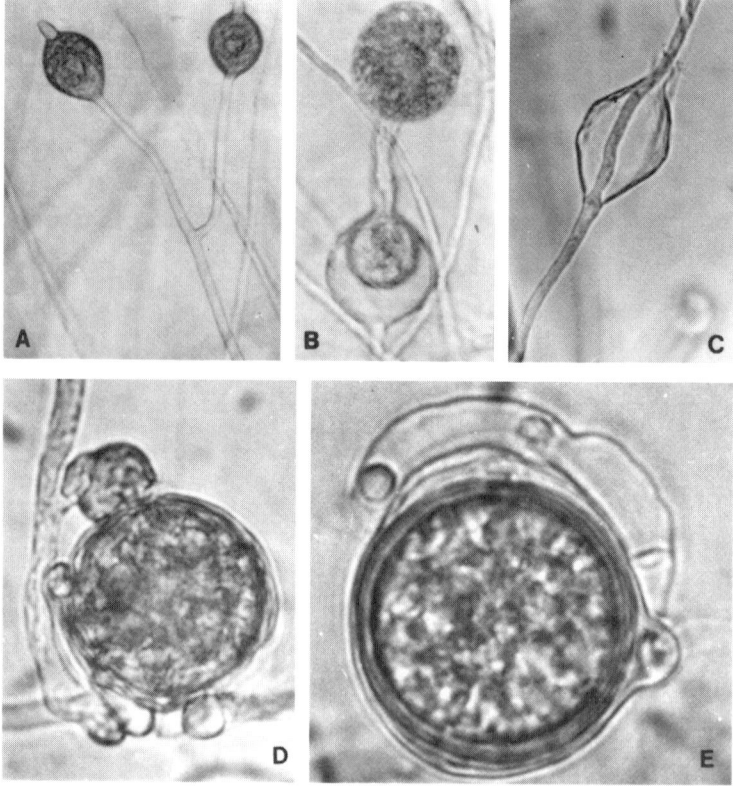

Pythium oedochilum. A–C: Sphaerosporangia with vesicle formation (B) and internally proliferated (C). D,E: Oogonia, antheridia, and oospores (E). F: Chlamydospore. (From Watanabe, T. et al. 1977. *Phytopathology*, 67:1324–1332. With permission.)

Pythium paroecandrum Drechsler

References: Drechsler 1930, 1940; Middleton 1943; Plaats-Niterink 1981; Watanabe 1985; Watanabe et al. 1977.

Morphology: Sporangia globose or subglobose, terminal or intercalary, rarely 2–3 sporangia in a chain. Zoospores discharged after vesicle formation. Oogonia terminal or intercalary, bearing single antheridia per oogonium. Antheridia cylindrical, crook-necked, typically sessile, or often unstalked, diclinous or monoclinous. Oospores aplerotic.

Dimensions: Sporangia (15-) 25–32.5 µm in diameter. Zoospores encysted 6.5–10 µm in diameter. Oogonia (15-) 20–25 µm in diameter. Oospores 11.2–20 µm in diameter: oospore wall 0.7–1.5 µm thick.

Material: 74-827 (= ATCC 36432, Strawberry roots, Saitama, Japan); 83-400 (Piedmont forest soil of Mt. Fuji, Shizuoka, Japan).

Remarks: Sessile antheridia may be one of the main characteristics of this fungus.

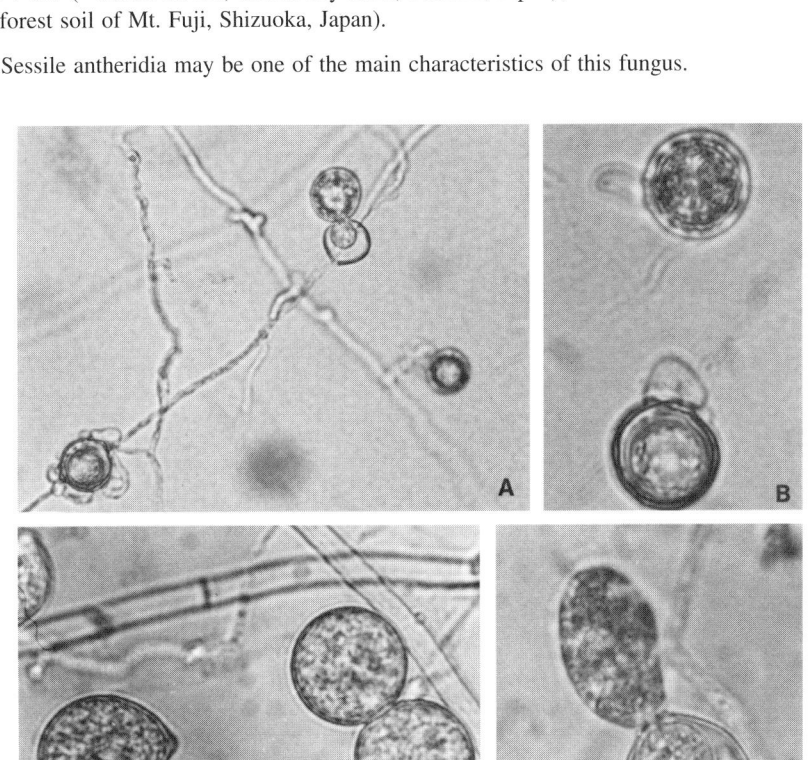

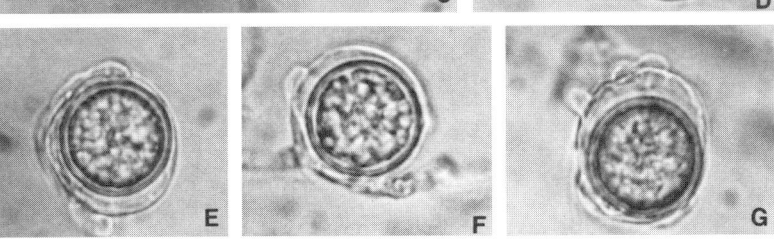

Pythium paroecandrum. A,B: Sphaerosporangia and sexual organs. C,D: Sphaerosporangia and vesicle formation (D). E–G: Oogonia, antheridia, and oospores (E,G). (From Watanabe, T. 1985. *Trans. Mycol. Soc. Jpn.*, 26:41–45. With permission.)

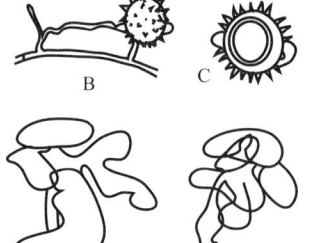

Pythium periplocum Drechsler

References: Drechsler 1930, 1939; Middleton 1943; Plaats-Niterink 1981; Watanabe 1988c; Watanabe et al. 1987b.

Morphology: Sporangia lobate, terminal or intercalary, zoospores rarely or not discharged. Oogonia echinulate, with protuberances regularly distributed, bearing 1–4 antheridia per oogonium. Antheridia cylindrical, clavate and various, furrowed partially in the upper surface, mainly diclinous. Oospores aplerotic.

Dimensions: Sporangia ca. 13 μm wide. Oogonia (excluding protuberances) 17.5–25 (-28.8) μm in diameter: protuberances 2.1–3.6 × 1.8–2 μm. Oospores 15–23.8 μm in diameter: oospore wall ca. 0.6 μm thick. Antheridia ca. 16.2 × 5 μm.

Material: 78-2243 (Watermelon field soil, Aomori, Japan); 79-71 (Cucumber roots, Naha, Okinawa, Japan); 84-375 (Poplar roots, Sakaide, Kagawa, Japan); 85-53 (= ATCC 66261, Flowering cherry seeds, Hachioji, Tokyo, Japan).

Remarks: It is sometimes difficult to observe both asexual and sexual structures together.

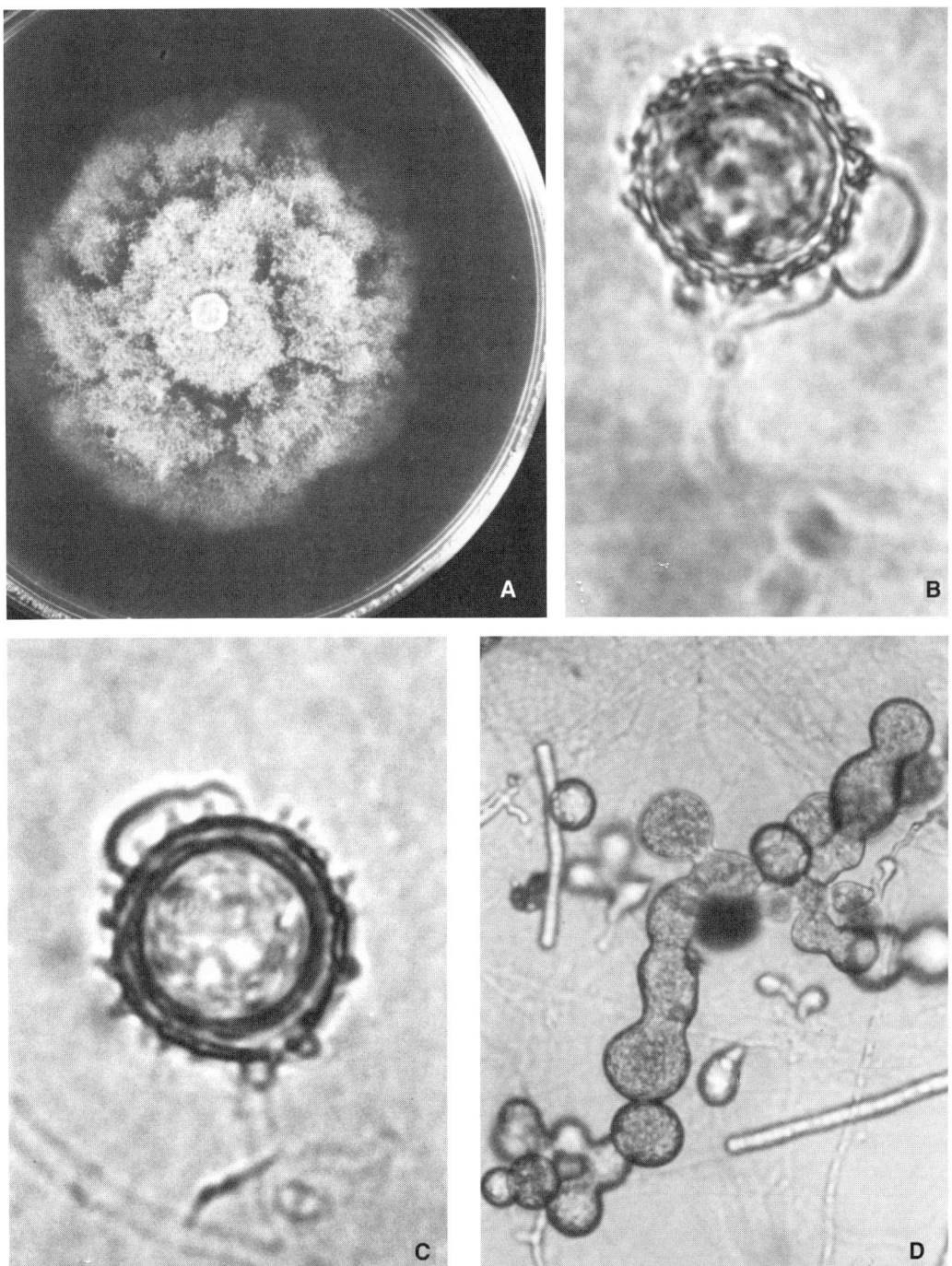

Pythium periplocum. A: Culture on PDA. B,C: Oogonia, antheridia, and oospores (C). D: Lobate sporangia. (From Watanabe, T. et al. 1987b. *Trans. Mycol. Soc. Jpn.*, 28:475–481. With permission.)

Pythium rostratum Butler

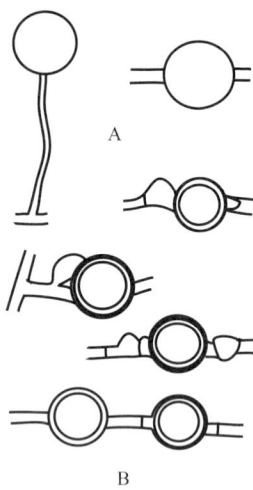

References: Matthews 1931; Middleton 1943; Plaats-Niterink 1981; Watanabe 1988b.

Morphology: Sporangia (conidia) globose, mostly terminal. No zoospores formed. Oogonia smooth, occasionally papillate, mostly intercalary, bearing 1–2 antheridia per oogonium. Antheridia sessile in a stalk or hypogynous. Oospores plerotic.

Dimensions: Conidia 12.5–30 µm in diameter. Oogonia 20–27.5 µm in diameter. Oospores 12.5–25 µm in diameter: oospore wall ca. 2.5 µm thick.

Material: 81-424 (Potato field soil, Hokkaido, Japan); 82-143 (Melon field soil, Chiba, Japan); 84-411 (Buckwheat and oat field soil, Tokushima, Japan).

Remarks: This fungus is characterized by intercalary oogonia with plerotic oospores and conidia.

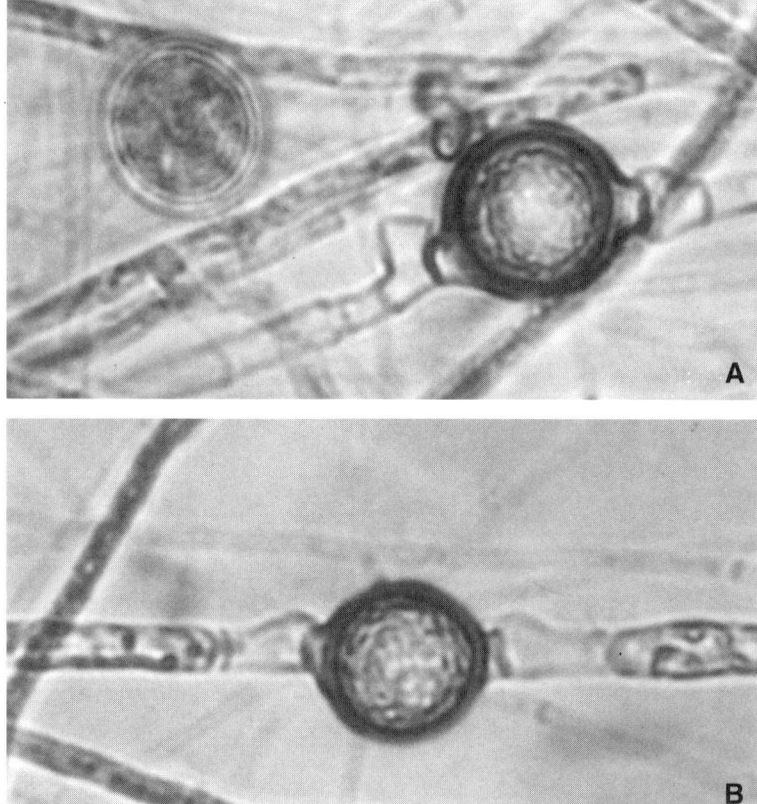

Pythium rostratum. A,B: Sporangia (A), oogonia, antheridia, and oospores (A,B). (From Watanabe, T. 1988b. *Ann. Phytopathol. Soc. Jpn.*, 54:523–528. With permission.)

Pythium salpingophorum Drechsler

References: Matthews 1931; Middleton 1943; Plaats-Niterink 1981.

Morphology: Hyphal swellings, globose, not proliferated. Oogonia usually intercalary, smooth-walled, usually formed in a series, bearing aplerotic oospores. Oospores often formed parthenogenetically, aplerotic.

Dimensions: Oogonia (17) 20–26 µm. Oospores 15–20 µm in diameter.

Material: 01-105 (Kita-kou, Hahajima, the Bonin Island, Tokyo, Japan).

Remarks: Formation of sexual organs in a series may be significant for identification of this species. Proliferation of sporangia is emphasized for identification by Middleton (1943), but not by Plaats-Niterink (1981).

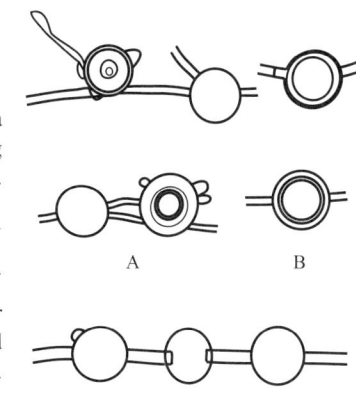

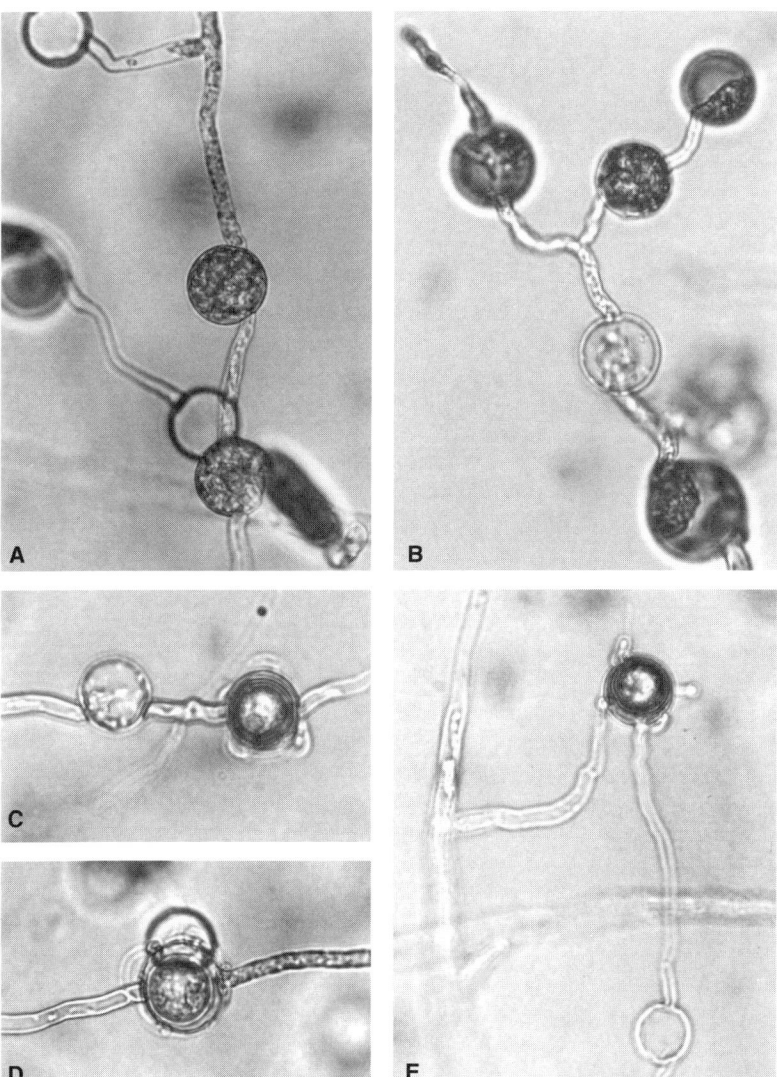

Pythium salpingophorum. A,B: Intercalary intact and vacant globose hyphal swellings. C-E: Oogonia with monoclinous (C,D) and diclinous antheridia (E).

Pythium spinosum Sawada

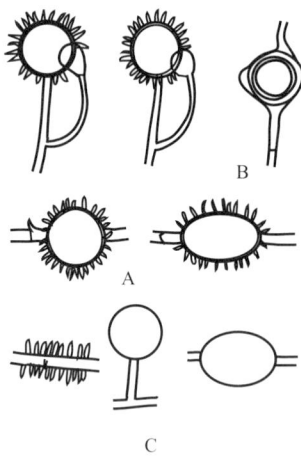

References: Ito 1936; Matthews 1931; Middleton 1943; Plaats-Niterink 1981; Watanabe et al. 1977.

Morphology: Sporangia (conidia) globose, terminal or intercalary, No zoospores formed. Oogonia smooth or echinulate with obtuse, digitate protuberances, bearing single antheridia per oogonium. Antheridia sessile or hypogynous, monoclinous or diclinous. Oospores plerotic.

Dimensions: Conidia 16.2–30 μm in diameter. Oogonia (excluding protuberances) 15–28.5 μm in diameter: protuberances 5–13.8 μm long, 1.7–3.7 μm wide. Oospores 16–17.5 μm in diameter.

Material: 73-160 (= ATCC 36438, Strawberry roots, Shizuoka, Japan); 79-468 (= ATCC 64146, Eggplant field soil, Kagoshima, Japan).

Remarks: This fungus is characterized by echinulate oogonia with digitate protuberances and plerotic oospores, and conidia.

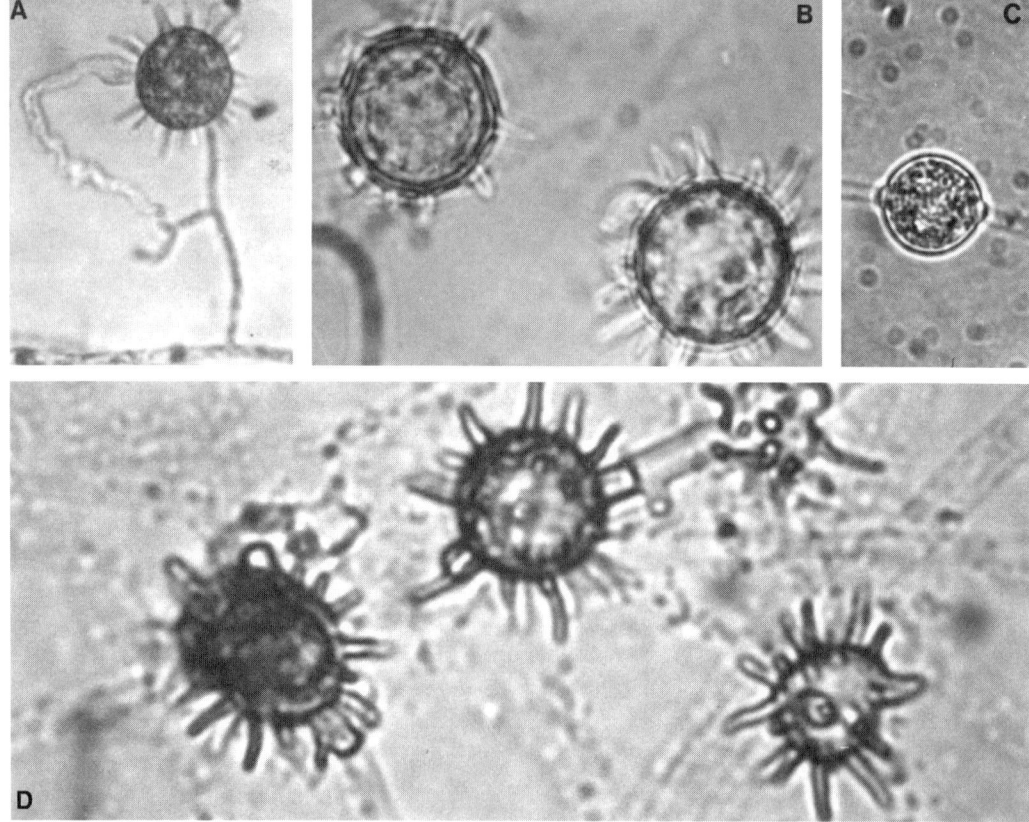

Pythium spinosum. A,B: Oogonia, antheridia, and oospores (B). C: Sporangia. D: Immature oogonia with digitate protuberances. (From Watanabe, T. et al. 1977. *Phytopathology*, 67:1324–1332. With permission.)

Pythium splendens Braun

References: Braun 1925; Matthews 1931; Middleton 1943; Plaats-Niterink 1981; Watanabe et al. 1977.

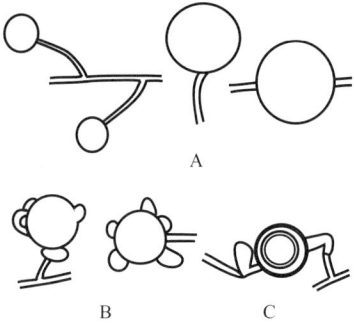

Morphology: Sporangia (conidia) globose, large, pale brown, mostly terminal, occasionally thick-walled. Zoospores not formed. Heterothallic, or rarely homothallic. Oogonia smooth, terminal, bearing 1–4 antheridia per oogonium. Antheridia mainly crook-necked, diclinous, rarely monoclinous. Oospores aplerotic.

Dimensions: Conidia (22-) 35–50 μm in diameter. Oogonia 25–37.5 μm in diameter. Oospores 20–27.5 μm in diameter: oogonial wall ca. 0.5 μm thick. Antheridia 9.5–17.5 μm long, 10–12.5 μm wide.

Material: 73-192 (= ATCC 36444, Strawberry roots, Shizuoka, Japan); 80-212 (Peanut field soil, Miyakonojyo, Miyazaki, Japan); 82-17 (= ATCC 64147, Melon roots, Chiba, Japan).

Remarks: This fungus is characterized by large terminal sporangia, over 40 μm in diameter.

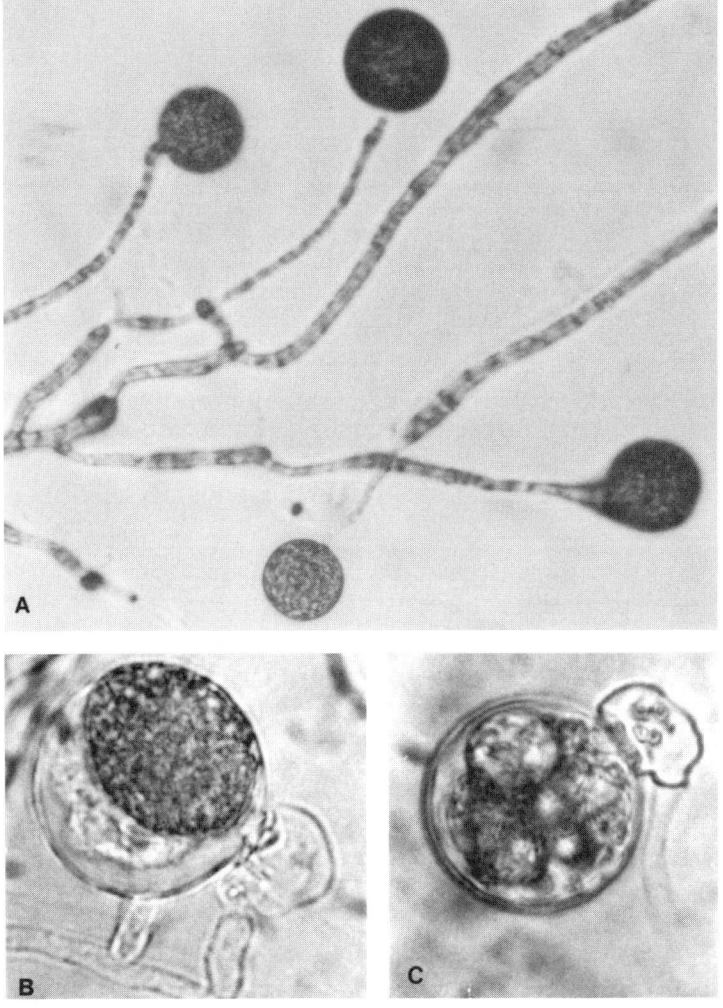

Pythium splendens. A: Sporangia. B,C: Aborted oogonia and antheridia. (From Watanabe, T. et al. 1977. *Phytopathology*, 67:1324–1332. With permission.)

Pythium sulcatum Pratt & Mitchell

References: Plaats-Niterink 1981; Pratt and Mitchell 1973; Watanabe et al. 1986b.

Morphology: Sporangia (conidia) globose, rarely peanut-shaped, terminal. Zoospores not formed. Oogonia terminal, bearing usually 1–2 (rarely 3–4) antheridia per oogonium. Antheridia clavate, crook-necked, hypha-like, globose or various in shape, stalked, furrowed partially on the upper surface, mainly monoclinous. Oospores aplerotic, with single oil globules. Appressoria often sickle- or sausage-shaped, catenulate.

Dimensions: Conidia 20–30 μm in diameter. Oogonia 15–26.3 (-28.8) μm in diameter. Oospores 13.7–22.5 μm in diameter: oospore wall 0.3–0.9 μm thick. Appressoria 2.2–6.3 μm in diameter.

Material: 83-160, 83-161 (Carrot roots, Chiba, Japan).

Remarks: Although this fungus is described as the species with hypha-like sporangium, no zoospores were observed.

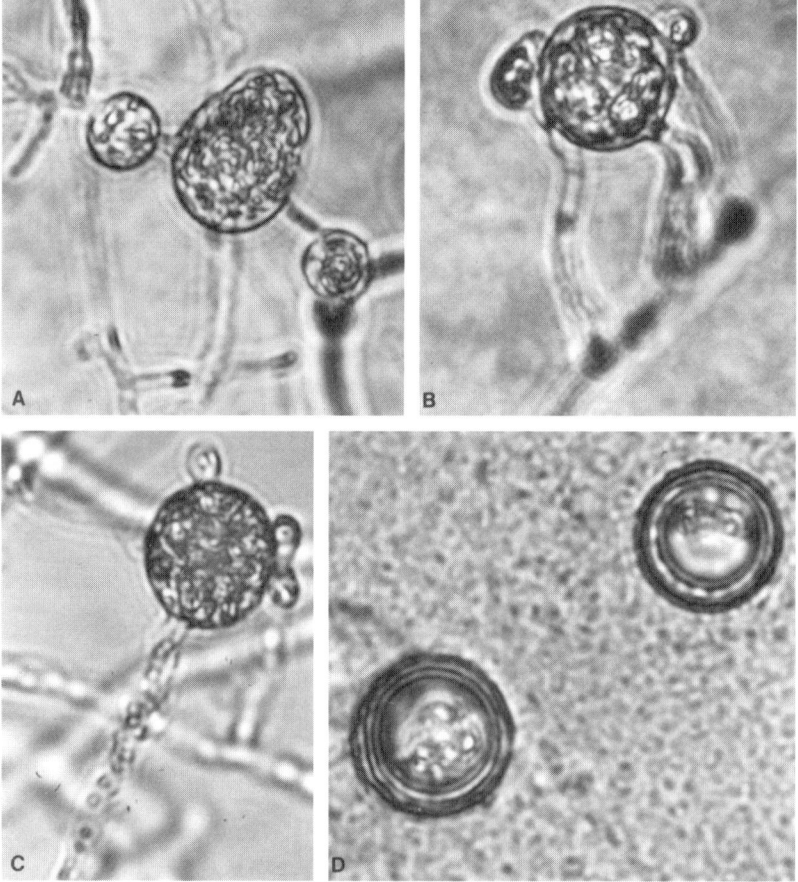

Pythium sulcatum. A: Sporangia. B,C: Oogonia and antheridia. D: Oospores. (From Watanabe, T. et al. 1986b. *Ann. Phytopathol. Soc. Jpn.*, 52:287–291. With permission.)

Pythium sylvaticum Campbell et Hendrix

References: Campbell and Hendrix 1967; Hendrix and Campbell 1974; Plaats-Niterink 1981; Watanabe 1985; Watanabe et al. 1977.

Morphology: Sporangia (conidia) terminal or intercalary, globose or ellipsoidal. Zoospores not formed. Heterothallic or homothallic. Oogonia smooth, terminal or intercalary, bearing 2–3 (4) antheridia per oogonium. Antheridia mainly terminal, globose, cylindrical, occasionally slightly constricted, hypha-like or various in shape, mostly diclinous. Oospores aplerotic. Sporangia often surrounded by hypha-like, appressorium-like structures. Appressoria long-ellipsoidal or clavate.

Dimensions: Conidia: globose 17.5–30 μm in diameter: ellipsoidal up to 45 μm long, 20–25 μm wide. Oogonia 22.5–25 μm in diameter. Oospores 12.5–20 μm in diameter: oogonial wall ca. 0.5 μm thick. Antheridia (5-) 7.5–20 × 6–10 μm. Appressoria 27.5–30 × 12.5 μm.

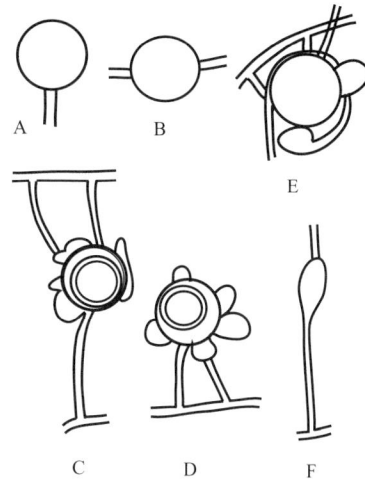

Material: 74-866 (= ATCC 36442, Strawberry roots, Shizuoka, Japan); 80-164 (Watermelon field soil, Fukuoka, Japan); 83-391 (Piedmont forest soil, Shizuoka, Japan).

Remarks: This fungus is most common in Japanese soil. Heterothallic isolates are characterized in forming globose conidia, usually 20–25 μm in diameter. Homothallic isolates form oogonia containing aplerotic oospores with mostly 3 globose, somewhat constricted antheridia surrounding diclinously, together with conidia in single cultures.

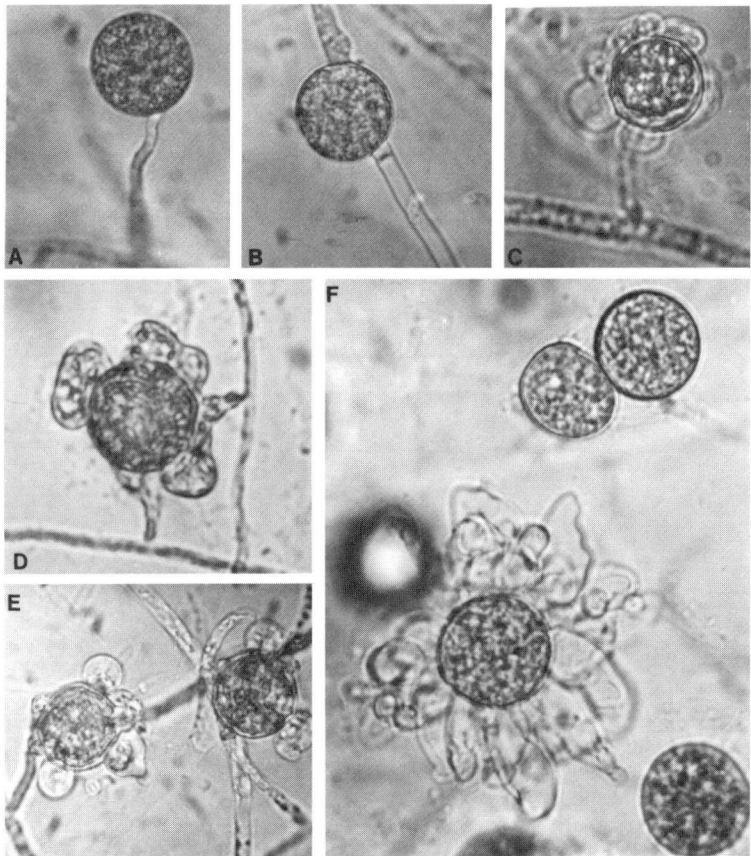

Pythium sylvaticum. A,B: Sporangia. C–E: Oogonia and antheridia. F: Sporangium surrounded by hypha-like, appressorium-like structures. (From Watanabe, T. et al. 1977. *Phytopathology*, 67:1324–1332. With permission.)

Pythium torulosum Coker et Patterson

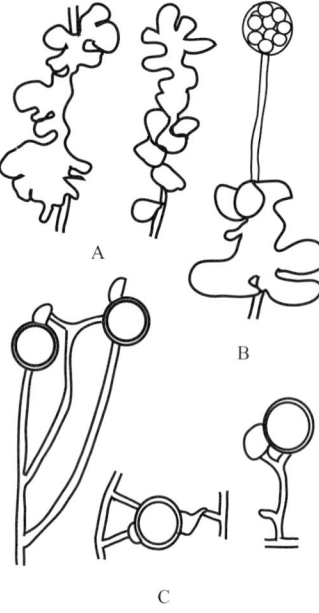

References: Matthews 1931; Middleton 1943; Plaats-Niterink 1981; Watanabe 1985; Watanabe et al. 1977.

Morphology: Sporangia lobate, with long exit tubes developed. Zoospores readily discharged after vesicle formation. Oogonia mainly terminal, bearing 1–2 antheridia per oogonium. Antheridia clavate, allantoid or various in shape, very short or sessile, mostly monoclinous. Oospores plerotic.

Dimensions: Sporangia ca. 12.5–14 µm wide: vesicles 25–40 µm in diameter: exit tubes over 150 µm long. Zoospores encysted ca. 7–7.5 µm in diameter. Oogonia 12.5–20 µm in diameter. Oospores 7.5–15 (-17.5) µm in diameter: oospore wall ca. 1.5–2.3 µm wide. Antheridia 7.5–10 × 5 µm.

Material: 73-256 (= ATCC 36484, Strawberry roots, Shizuoka, Japan); 81-327 (= ATCC 64148, Grassland, Hokkaido, Japan); 83-390 (Piedmont forest soil, Shizuoka, Japan).

Remarks: This fungus is characterized by lobate sporangia and small oogonia containing plerotic oospores, usually under 20 µm in diameter, with single monoclinous antheridia.

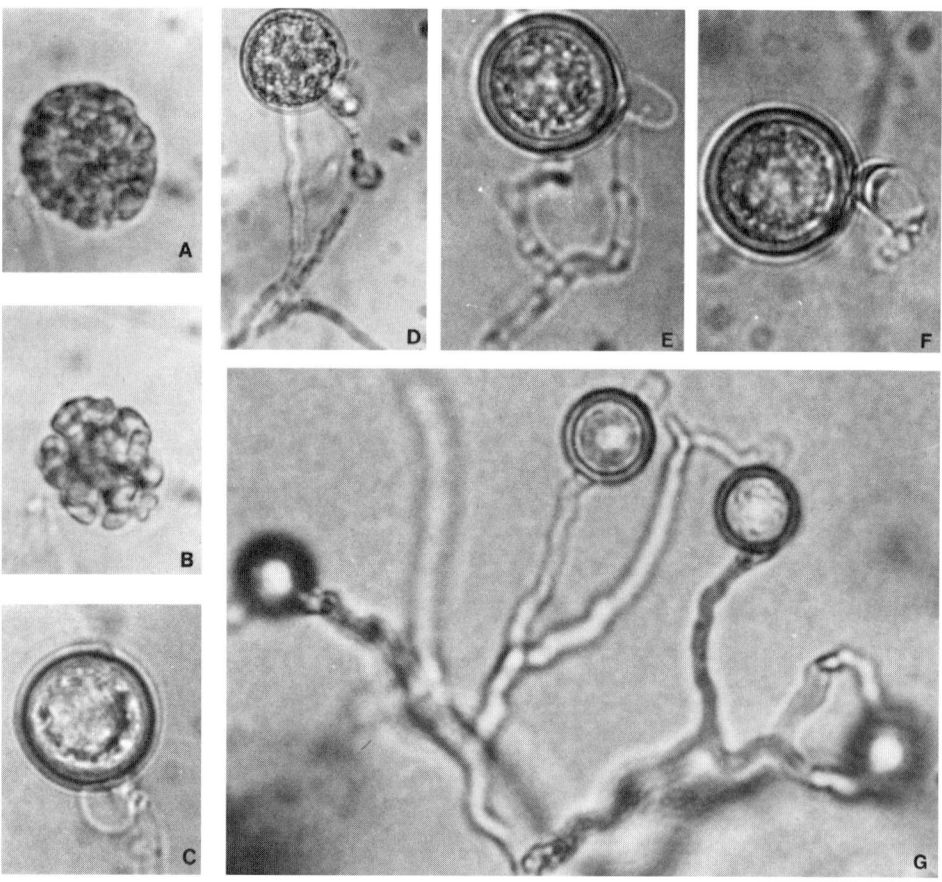

Pythium torulosum. A,B: Vesicle formation from lobate sporangia. C–F: Oogonia, antheridia, and oospores (E–G). (From Watanabe, T. et al. 1977. *Phytopathology*, 67:1324–1332. With permission.)

Pythium ultimum Trow

References: Matthews 1931; Middleton 1943; Plaats-Niterink 1981; Watanabe et al. 1977.

Morphology: Sporangia (conidia) globose, terminal or intercalary. Zoospores not formed. Oogonia smooth, terminal, bearing single antheridia per oogonium. Antheridia globose, crook-necked, lunar-shaped, curved or various in shape, sessile or short-stalked, monoclinous. Oospores aplerotic, rarely plerotic.

Dimensions: Conidia (15-) 20–30 (-35) μm in diameter. Oogonia 17.5–27.5 μm in diameter. Oospores (11.2-) 15–20 μm in diameter: oospore wall ca. 1.3–2.3 μm thick. Antheridia ca. 10–17.5 μm long, 3.7–7.5 (-15) μm wide.

Material: 74-901 (= ATCC 36443, Strawberry roots, Shizuoka, Japan); 80-225 (Sweet potato field soil, Kagoshima, Japan).

Remarks: This fungus is characterized by terminal oogonia containing aplerotic, thick-walled oospores, with single monoclinous hypogynous antheridia originating near the oogonia and conidia.

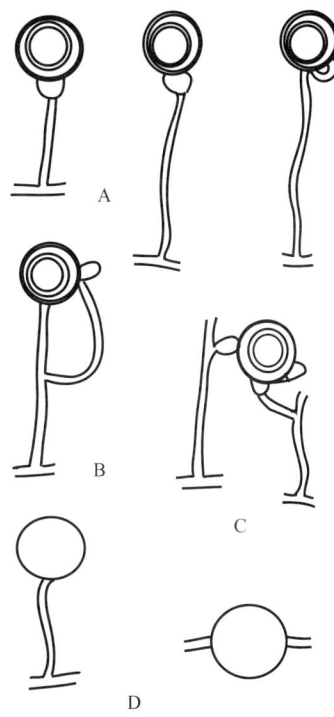

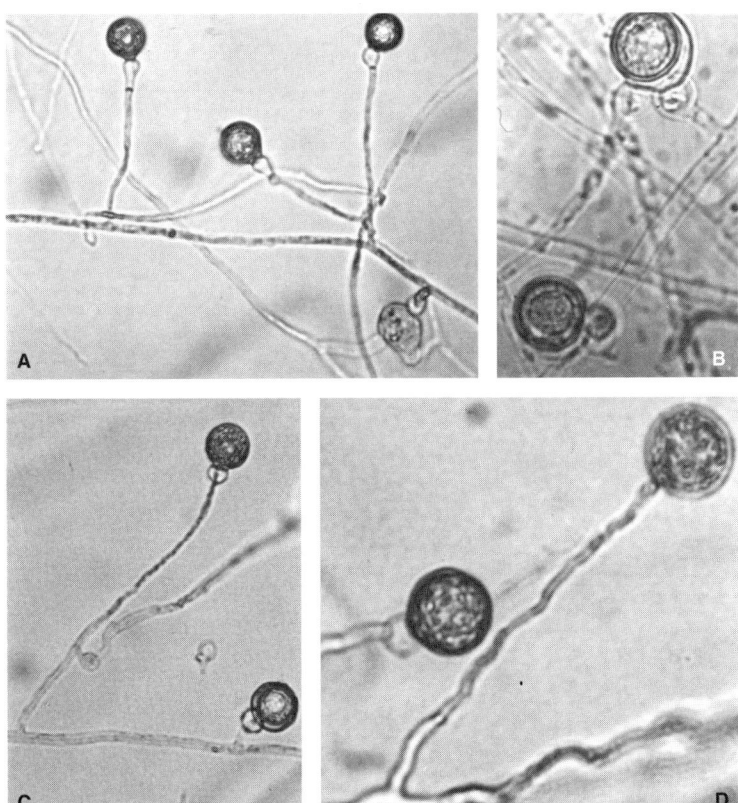

Pythium ultimum. A–C: Oogonia, antheridia, and oospores. D: Sporangia and sexual organs. (From Watanabe, T. et al. 1977. *Phytopathology*, 67:1324–1332. With permission.)

Pythium vexans de Bary

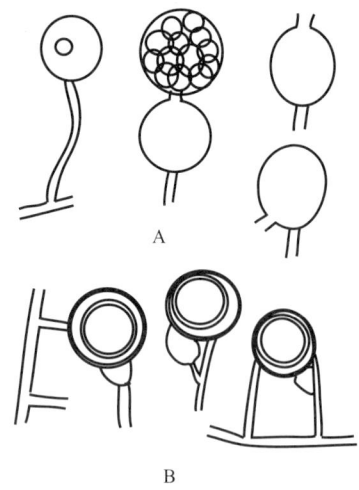

References: Matthews 1931; Middleton 1943; Plaats-Niterink 1981; Watanabe 1985.

Morphology: Sporangia globose, ovoid, terminal or intercalary. Zoospores readily discharged after vesicle formation. Oogonia terminal, occasionally papillate, bearing single antheridia per oogonium. Antheridia characteristically damp-bell-shaped, lunar-shaped, clavate or various in shape, stalked, mostly monoclinous. Oospores aplerotic.

Dimensions: Sporangia 15–25 (-30) µm in diameter: vesicles 22.5–25 µm in diameter: exit tubes ca. 35 × 5 µm. Oogonia 17.5–30 µm in diameter. Oospores 14–20 µm in diameter. Antheridia: clavate (7.5-) 10–15 × 5–7.5 µm; damp-bell-shaped 7.5–10 µm wide, 2.5–8 µm deep.

Material: 79-203 (Chinese cabbage field soil, Kagoshima, Japan); 82-763 (Forest soil, Shizuoka, Japan); 83-410 (Piedmont forest soil, Shizuoka, Japan).

Remarks: This fungus is characterized by sphaerosporangia and terminal oogonia containing single aplerotic oospores with single monoclinous damp-bell-shaped antheridia. Sexual organs are formed on well-developed stalks.

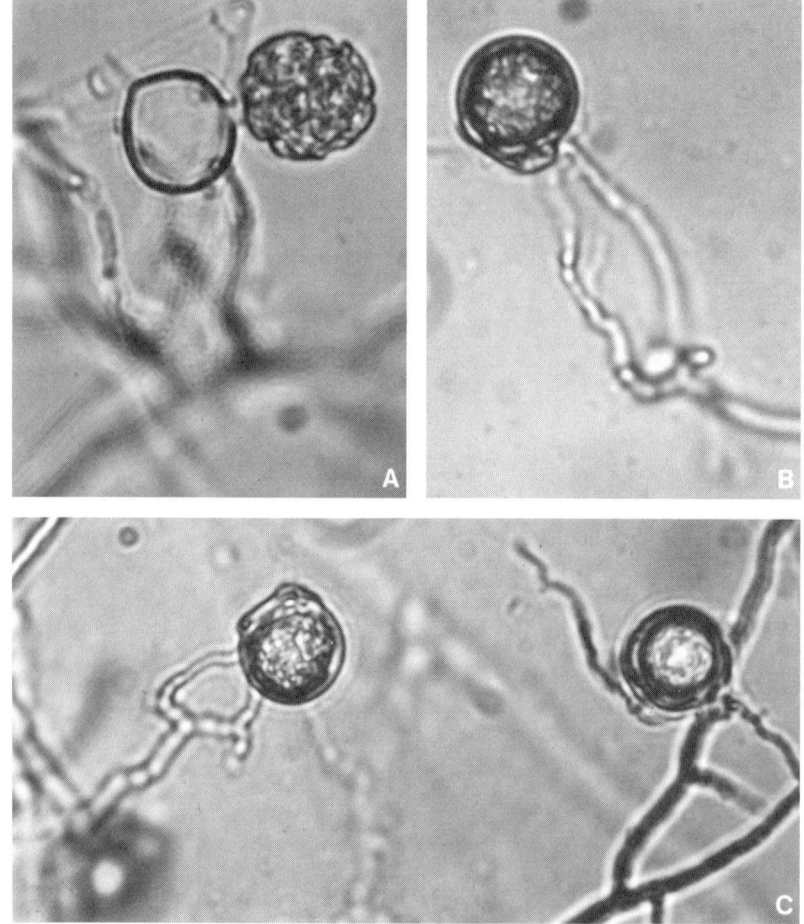

Pythium vexans. A: Vesicle formation from sphaerosporangium. B,C: Oogonia, antheridia, and oospores (C). (From Watanabe, T. 1986. *Trans. Mycol. Soc. Jpn.*, 26:41–45. With permission.)

Pythium zingiberum Takahashi

References: Takahashi 1954; Plaats-Niterink 1981; Watanabe 1984, 1992d.

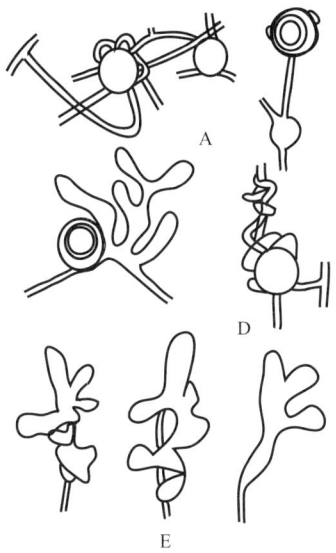

Morphology: Sporangia lobate, long exit tubes extended before vesicle development, often connected with sexual organs. Zoospores formed. Oogonia mainly terminal, often aggregated in a few, occasionally papillate at one or two positions, bearing 1–3 antheridia per oogonium. Antheridia hypha-like, clavate, often coiling around oogoniumphores, diclinous. Oospores aplerotic. Appressoria composed of globose or clavate apical cells and stalked hyphae.

Dimensions: Lobate sporangia ca. 10 µm wide: exit tubes ca. 200–220 µm long. Oogonia 20–35 µm in diameter. Oospores ca. 17.5–22.5 µm in diameter. Apical cells of appressoria 10–20 µm in diameter.

Material: 82-724, 82-725 (Ginger field soil, Wakayama, Japan).

Remarks: This fungus is morphologically close to *P. myriotylum*, but it is also treated as a synonym of *P. volutum* Vanterpool and Truscott. It is also spelled as *P. zingiberis* (Plaats-Niterink, 1981).

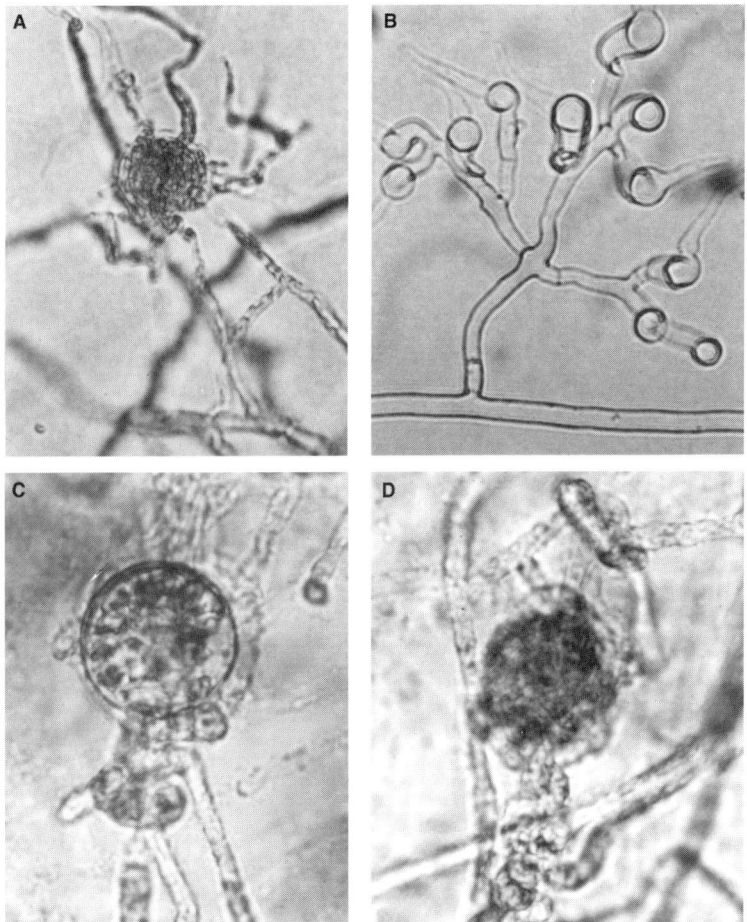

Pythium zingiberum. A: Oogonium bearing antheridium diclinously. B: Appressorium. C,D: Hypha-like antheridia coiling around oogoniumphores. E: Lobate sporangia. (From Watanabe, T. 1992b. *Ann. Phytopathol. Soc. Jpn.*, 58:65–71. With permission.)

Saprolegnia Nees

Nova Acta Leop.-Carol. 11:513, 1823.
Type species: *S. ferax* (Gruith.) Thuret

Saprolegnia anisospora de Bary

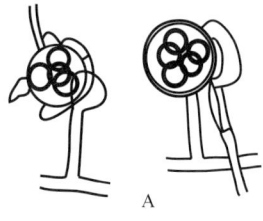

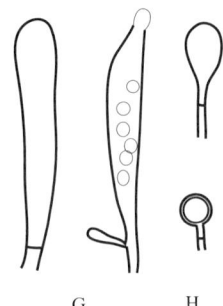

References: Seymour, 1970.

Morphology: Zoosporangia inflated, terminal, clavate or cylindrical, internally proliferated. Zoospores developed inside zoosporangia, and mostly readily discharged. Oogonia globose with diclinous antheridia, including 2–8 eccentric oospores per oogonium. Antheridia tubular, clavate, lunate or cylindrical on simple or branched antheridiumphores. Gemmae (chlamydospores) globose or subglobose.

Dimensions: Sporangia 84–152.5 × 14–25 μm. Zoospores encysted various in size, mostly 12–13 μm. Oogonia 40–45 μm.

Material: 81-97 (Forest soil, Sado, Niigata, Japan); 81-783, 81-787, 81-796 (Lettuce field soil, Kamogawa, Chiba, Japan).

Remarks: A total of 22 species are accepted in the genus. Various-sized encysted zoospores, oospore number (2–8) per oogonium, diclinous antheridia, and eccentric oospores characterize this species.

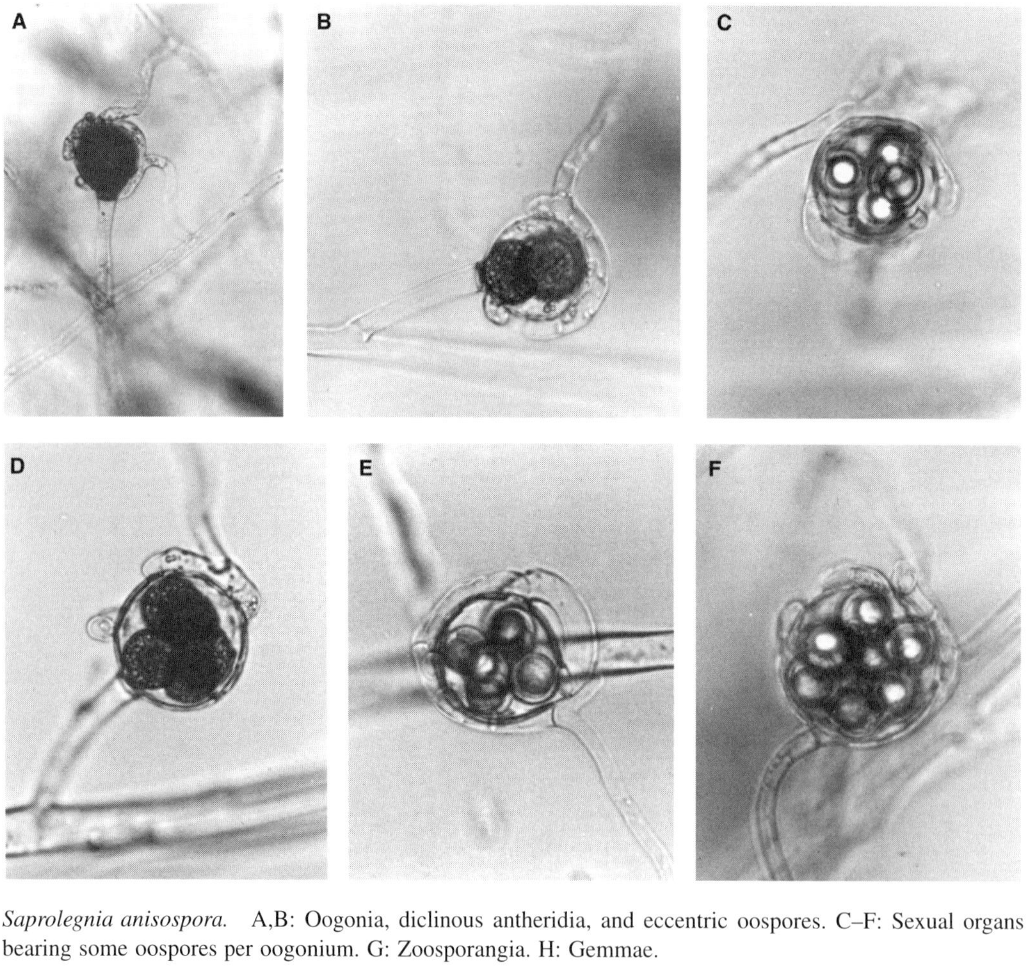

Saprolegnia anisospora. A,B: Oogonia, diclinous antheridia, and eccentric oospores. C–F: Sexual organs bearing some oospores per oogonium. G: Zoosporangia. H: Gemmae.

MORPHOLOGY OF SOIL FUNGI

Zygomycotina

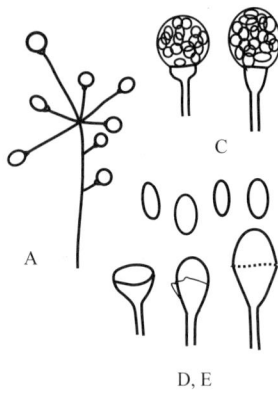

Absidia van Tieghem

Ann. Sci. Nat. Bot., Ser. 6, 4:350, 1876.

Type species: *A. reflexa* van Tieghem

Absidia repens van Tieghem

References: Hesseltine and Ellis 1966.

Morphology: Sporangiophores hyaline, erect, branched, often curved, bearing columellate and apophysate sporangia terminally. Sporangia pale brown; columellae hemispherical; apophyses flask-shaped; both columellae and apophyses subglobose or ellipsoidal. After deliquescence of sporangia, sporangial wall partially left. Sporangiospores subhyaline, ellipsoidal or ovate, 1-celled.

Dimensions: Sporangiophores over 125 µm tall; 5–6.3 µm wide at apex. Sporangia 20–25 µm in diameter; columellae + apophyses 25–40 × 15–40 µm (columellae 12.5–20 µm wide, 10–15 µm deep; apophyses 10–25 × 2.5–20 µm). Sporangiospores 3.5–3.8 × 2.5–2.8 µm.

Material: 70-1201 (Pineapple field soil, Hachijo, Tokyo, Japan).

Remarks: Apophyses of *Absidia* and *Gongronella* are funnel-shaped, and globose or subglobose, respectively.

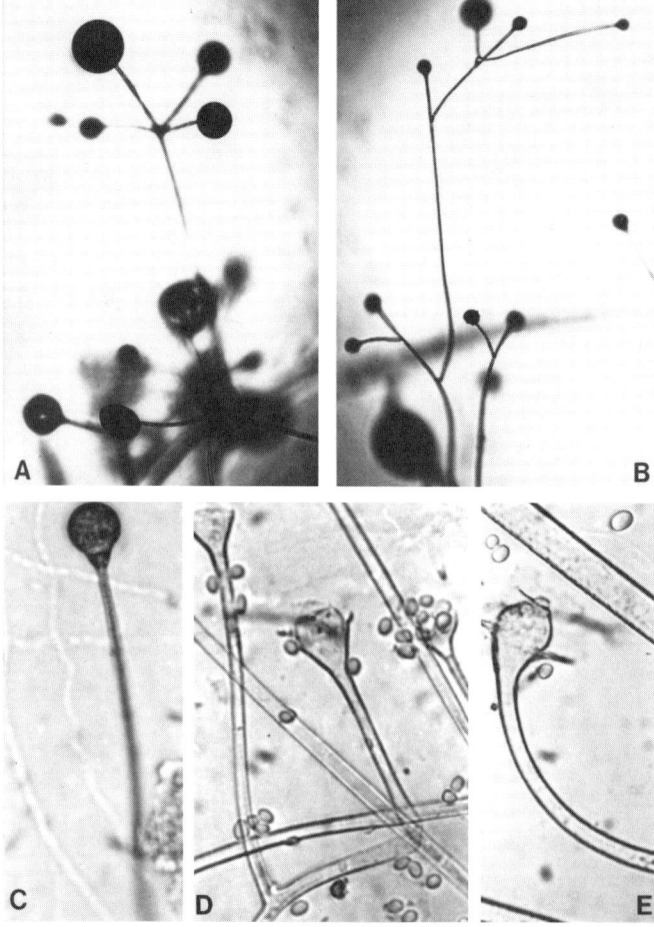

Absidia repens. A–C: Sporangiophores and sporangia. D,E: Sporangia after deliquescence showing apophyses, columellae, and sporangiospores.

Circinella van Tieghem & Le Monn

Ann. Sci. Nat. Bot., Ser. 5, 17:298, 1873.
Type species: *C. umbellata* van Tieg. & Le Monn

Circinella muscae (Sorok.) Berl. & de Toni

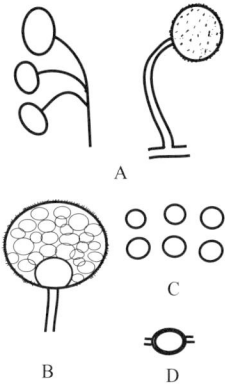

References: Hesseltine and Fennell 1955.

Morphology: Sporangiophores erect, curved, simple or branched, bearing columellate sporangia terminally. Sporangia terminal, black, globose or subglobose, finely echinulate; columellae globose. Sporangiospores globose, dark green, 1-celled, granulate. Chlamydospores ellipsoidal, black, thick-walled.

Dimensions: Sporangiophores 60–100 µm tall; ca. 10 µm wide. Sporangia 45–50 µm in diameter; columellae ca. 20 µm in diameter. Sporangiospores 5–10 µm in diameter. Chlamydospores 17–18 × 15 µm.

Material: 69-269 (Pineapple field soil, Naha, Okinawa, Japan).

Remarks: As indicated by the genus name, sporangiophores of the fungus may be more twisted.

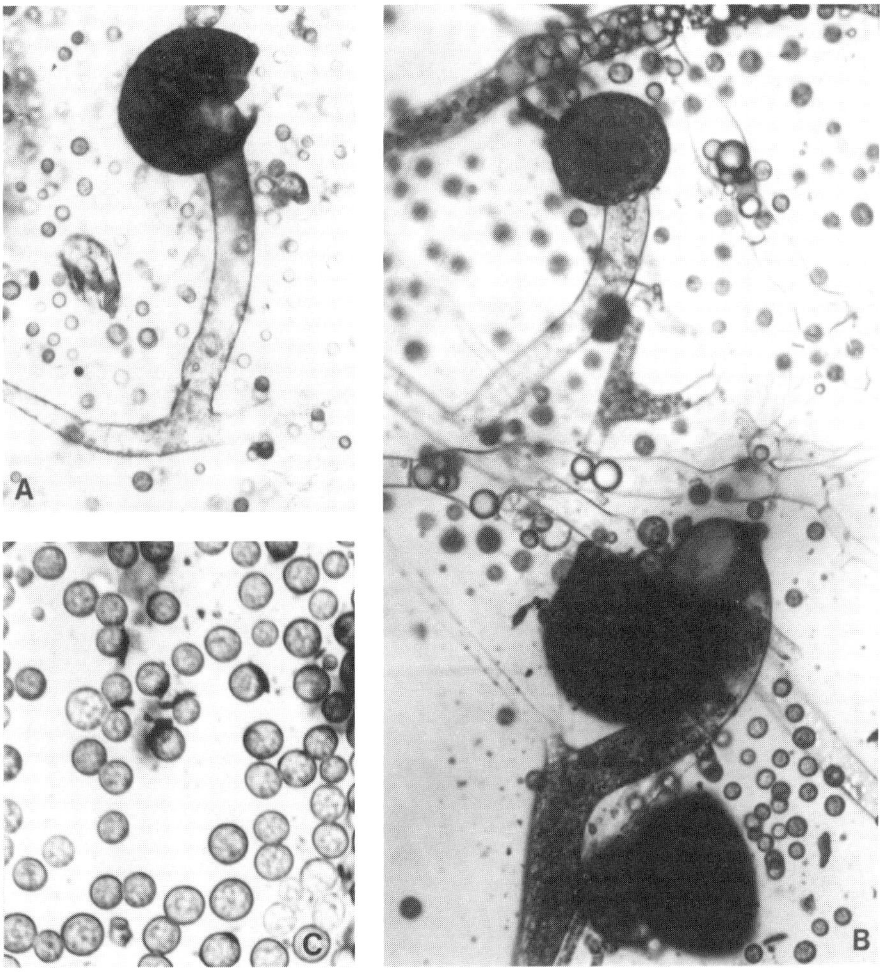

Circinella muscae. A,B: Sporangiophores, sporangia, and columella (B) and sporangiospores. C: Sporangiospores. D. Chlamydospore.

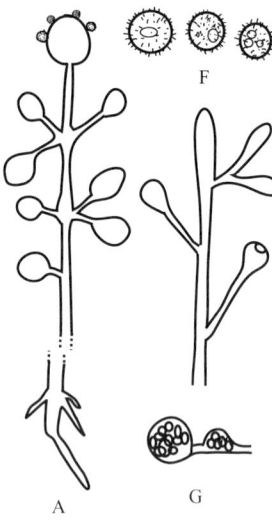

Cunninghamella Matr.

Ann. Mycol. 1:46, 1903.
Type species: *C. echinulata* (Thaxter) Thaxter

Cunninghamella echinulata (Thaxter) Thaxter

References: Alcorn and Yeager 1938; Watanabe 1975a.

Morphology: Sporangiophores erect, simple or branched with verticillate or sympodial branches in a few positions, terminated in vesicles with sterigmata and spores (sporangiospores), rhizoidal basally. Vesicles hyaline, pale brown, globose or subglobose. Spores yellowish brown, globose, oil globules included, 1-celled, echinulate conspicuously. Hyphal swellings globose, bright yellow, often granulate.

Dimensions: Sporangiophores over 800 µm tall; branches 10–91 × 5 µm; vesicles on sporangiophores 25–37.5 × 20–35 µm; vesicles on branches 13–23 × 11–20 µm. Sporangiospores 8.7–14 µm in diameter; protuberances (spines) 2.5–3.8 µm long.

Material: 69-265 (Pineapple field soil, Okinawa, Japan), 72-11-2 (= ATCC 32321, Sugarcane roots, Taiwan, ROC).

Remarks: This fungus differs from *C. elegans* in forming conspicuously echinulate sporangiospores and small vesicles, under 37.5 µm in diameter.

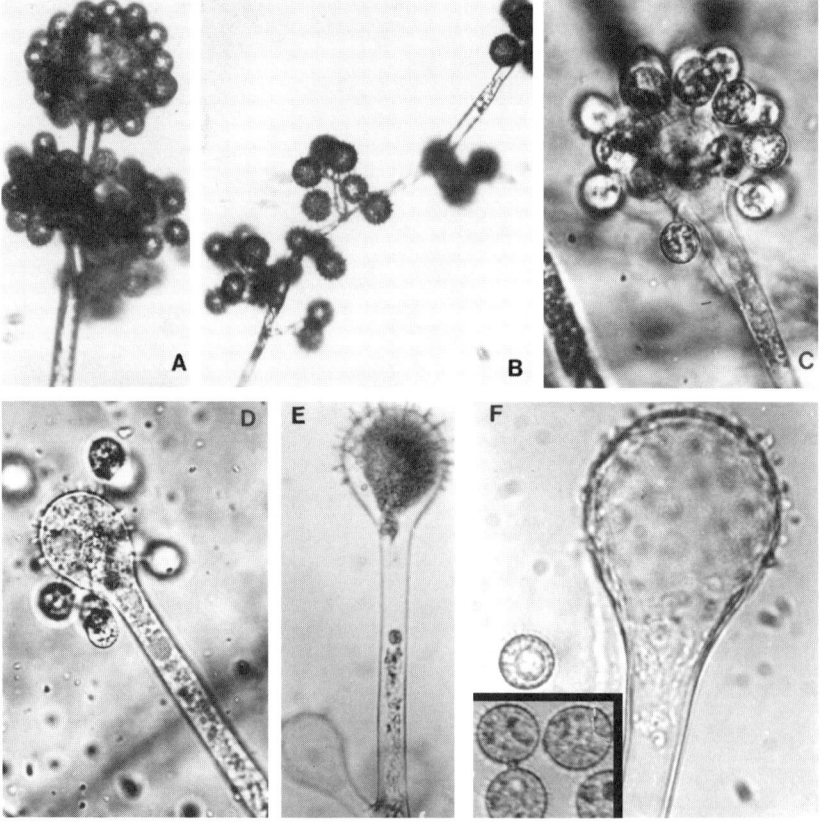

Cunninghamella echinulata. A–F: Sporangiophores, vesicles (C–F), and sporangiospores. G: Hyphal swellings. (From Watanabe, T. 1975a. *Trans. Mycol. Soc. Jpn.*, 16:18–27. With permission.)

Cunninghamella elegans Lendner

References: Domsch et al. 1980.

Morphology: Sporangiophores erect, simple or branched with verticillate or sympodial branches in a few positions, terminated in vesicles with sterigmata and spores (sporangiospores), rhizoidal basally. Vesicles hyaline, pale brown, globose or subglobose. Spores yellowish brown, globose, oil globules included, 1-celled, smooth or finely echinulate.

Dimensions: Vesicles on main sporangiophores 22.5–52.5 μm in diameter; vesicles on branches 10–25 μm in diameter. Sporangiospores 6.2–10 × 5–7.5 μm.

Material: 74-667 (Strawberry roots, Shizuoka, Japan), 87-46 (Flowering cherry tree roots, Hachioji, Tokyo, Japan).

Remarks: The fungus differs from *C. echinulata* in forming smooth or finely echinulate sporangiospores, and large vesicles on main sporangiophores, over 50 μm in diameter.

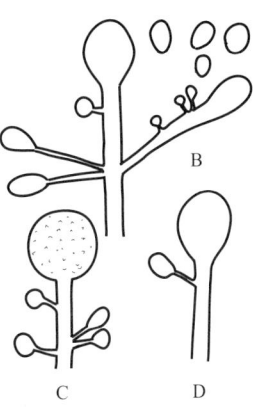

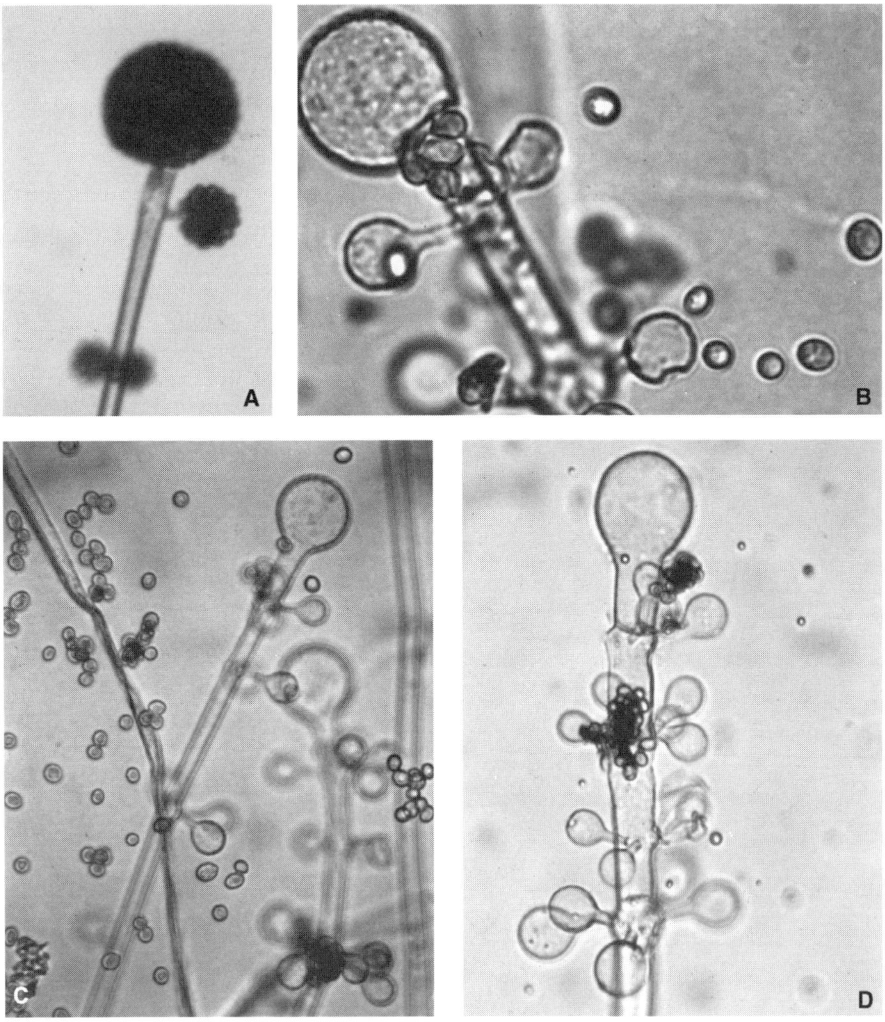

Cunninghamella elegans. A–D: Sporangia, and sporangiophores showing vesicles (B–D), and sporangiospores (B,C).

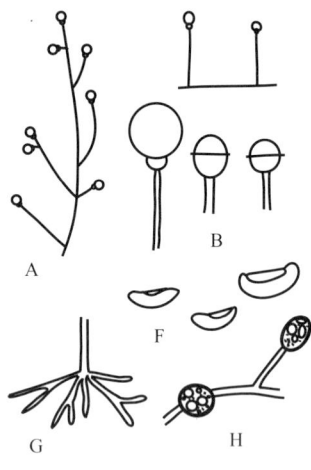

Gongronella Ribaldi

Riv. Biol. Gen., N. S. 44:164, 1952.

Type species: *G. urceolifera* Ribaldi = *G. butleri* (Lendn.) Peyronel & Dal Vesco

Gongronella butleri (Lendn.) Peyronel & Dal Vesco

Synonym: *Absidia butleri* Lendner

References: Hesseltine and Ellis 1964; Watanabe 1975a; Zycha et al. 1969.

Morphology: Sporangiophores hyaline, erect, simple or branched, septate near sporangia, bearing sporangia at the apexes of branches, and rhizoidal basally. Sporangia globose, gray, columellate and conspicuously apophysate: apophyses hemispherical and columellae half-globose. Sporangiospores hyaline, boat-shaped. Chlamydospores grayish green, globose, granulate, solitary.

Dimensions: Sporangiophores 54–340.5 × 4.8–4.9 µm. Sporangia (12.5-) 15–20.7 µm in diameter; columellae 7.5–12.5 × 6.2–10 µm; apophyses 7.5–11 × 6.5–12.5 µm. Sporangiospores 2.5–3.8 × 1.4–2 µm. Chlamydospores 4.5–7.3 µm in diameter.

Material: 69-332 (Pineapple field soil, Okinawa, Japan), 70-1107 (Pineapple field soil, Hachijo, Tokyo, Japan), 72-X98 (= ATCC 32323, Sugarcane roots, Taiwan, ROC).

Remarks: Hemispherical apophyses characterize the genus *Gongronella*.

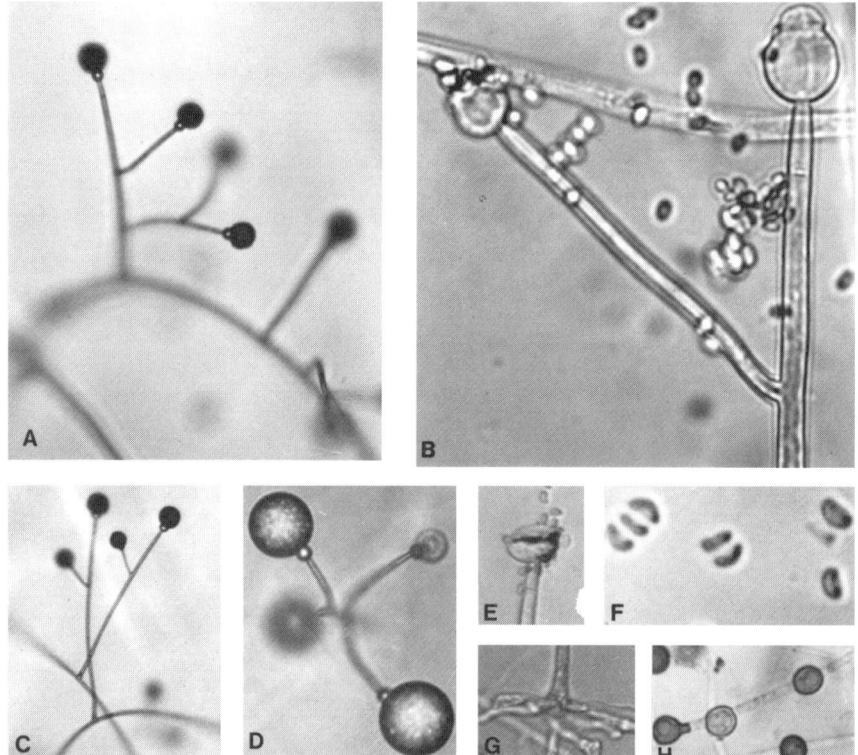

Gongronella butleri. A,C,D: Sporangiophores and sporangia. B,E: Columellae and apophyses. F: Sporangiospores. G: Rhizoid. H: Chlamydospores. (From Watanabe, T. 1975a. *Trans. Mycol. Soc. Jpn.*, 16:18–27. With permission.)

Helicocephalum Thaxter

Bot. Gaz. 16:201, 1891.
Type species: *H. sacophilum* Thaxter

Helicocephalum oligosporum Drechsler

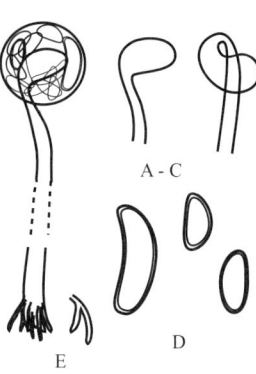

References: Barron 1975; Drechsler 1934; Watanabe and Koizumi 1976.

Morphology: Sporangiophores hyaline, erect, gradually tapering from base toward apex, spirally coiled, terminated ellipsoidal or spindle-shaped inflated columellae, bearing merosporangia, rhizoidal basally with 5–20 rhizoids. Sporangia borne in transparent mucilaginous water drops, ellipsoidal, yellowish brown, 6–13 spores included. Spores yellowish brown, ellipsoidal or ovate, slightly curved, 1-celled, thick-walled.

Dimensions: Sporangiophores 480–680 × 15–20 µm; rhizoids 25–40 × 2.2–2.5 µm. Sporangia 50–75 µm in diameter: columellae 30–40 × 15 µm. Sporangiospores 39–60 × 17–25 µm.

Material: This fungus was observed on water agar plates with root segments of gentian.

Remarks: This fungus is not culturable, and the fourth species in this genus was reported by Kitz and Embree (1989).

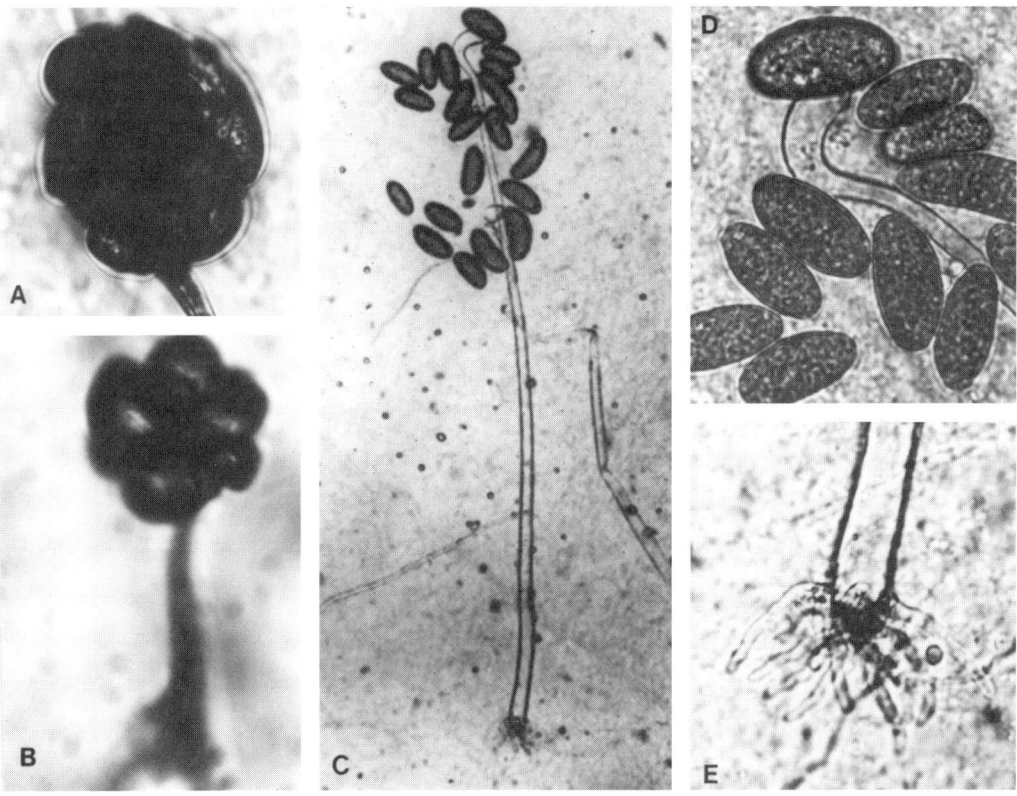

Helicocephalum oligosporum. A–D: Sporangiophores (A,B), sporangia (B,C), and sporangiospores (C,D). Note apexes of sporangiophores after deliquescence of sporangia (A,D). C,E: Rhizoids. (From Watanabe, T. and Koizumi, S. 1976. *Trans. Mycol. Soc. Jpn.*, 17:1–3. With permission.)

Mortierella Coemans

Bull. Acad. R. Sci. Belg. 2, 15:288, 1863.
Type species: *M. polycephala* Coemans

Key to Species

1.	Sporangiospores	formed 2
		not formed, chlamydospores echinulate .... *M. chlamydospora*
2.	Sporangiophores	simple 3
		branched 5
3.	Sporangiophores	under 120 µm tall .... *M. alpina*
		over 500 µm tall 4
4.	Sporangiophores	nonrhizoidal, spores 4–10 µm long .... *M. tsukubaensis*
		rhizoidal, spores 12–16 µm long .... *M. boninense*
5.	Sporangiophores	branched above the middle part 6
		below the middle part 9
6.	Subsporangial branches	present .... *M. ambigua*
		absent 7
7.	Sporangiophores	with one branch .... *M. uniramosa*
		with verticillate or sympodial branches 8
8.	Sporangiophore branches	verticillate .... *M. exigua*
		sympodial .... *M. epicladia* var. *chlamydospora*
9.	Sporangiospores	1–3 per sporangium 10
		numerous per sporangium 11
10.	Sporangiospores	1 per sporangium .... *M. humilis*
		1–3 per sporangium .... *M. verticillata*
11.	Sporangiophores	verticillate or sympodial 12
		with only 1 branch .... *M. minutissima*
12.	Sporangiophores	verticillate, spores under 2 µm in diameter .... *M. isabellina*
		sympodial, spores over 5 µm long 13
13.	Chlamydospores	solitary 14
		catenulate or in clusters .... *M. zychae*
14.	Chlamydospores	over 60 µm in diameter .... *M. gemmifera*
		under 25 µm in diameter 15

15. Sporangiophores   220–350 μm tall.................................... *M. hyalina*

                                   under 200 μm tall .... *M. elongata*

Remarks: About 90 species were reported in genus *Mortierella* (Gams, 1977).

Mortierella alpina Peyron.

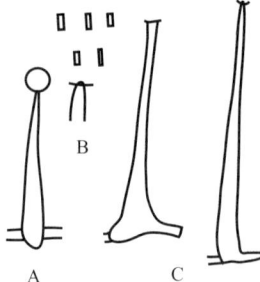

References: Gams 1977; Watanabe et al. 1995a.

Morphology: Sporangiophores hyaline, erect, simple, rarely branched at the upper part, tapering gradually from base toward apex, bearing single terminal sporangia, inflated basally. Sporangiospores hyaline, cylindrical, 1-celled.

Dimensions: Sporangiophores (17.5-) 60–110 × 5–7.5 µm. Sporangia (6-) 7.5–13.8 (-17) µm in diameter. Sporangiospores (2-) 2.5–4.3 × (0.5-) 1–1.5 (-2.5) µm.

Material: 93-318 (= MAFF 425590, Root of *Phellodendron amurense*, Tokyo, Japan), 98-67 (= MAFF 238010, Uncultivated soil, Tsukuba, Japan).

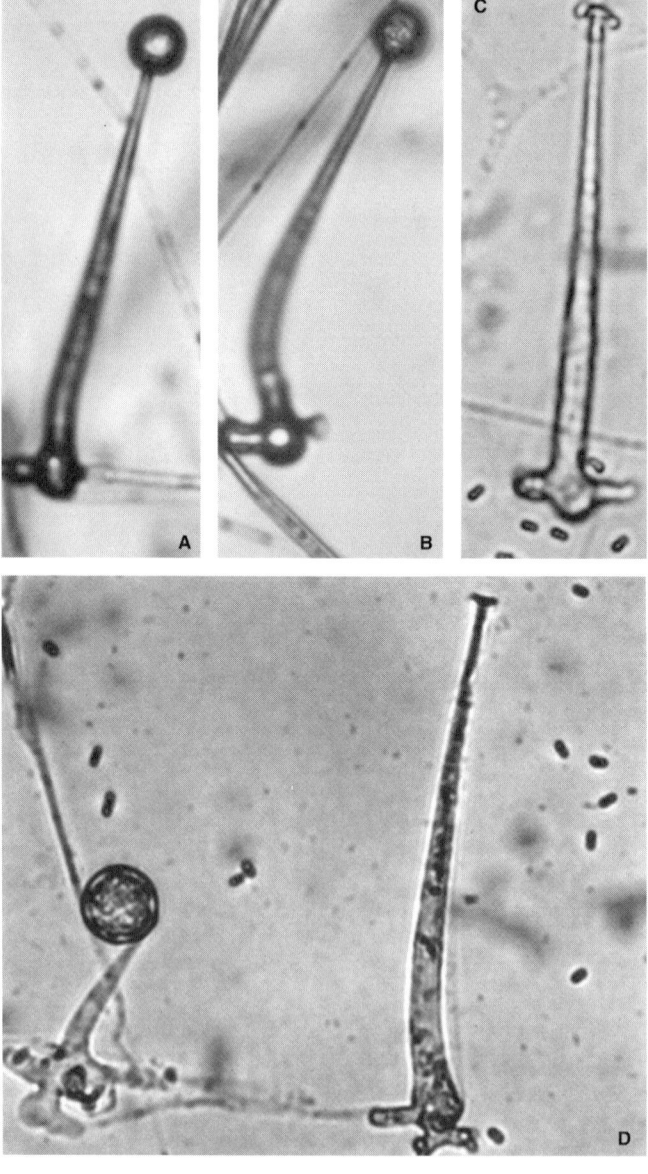

Mortierella alpina.   A,B: Sporangiophores and sporangia. C,D: Sporangiospores on deliquescens of sporangia.

Mortierella ambigua Mehrotra

References: Mehrotra et al. 1963; Watanabe 1975a.

Morphology: Sporangiophores pale brown, erect, simple or branched, gradually tapering from base toward apex, with sporangia terminally, rhizoidal basally. Sporangia noncolumellate. Secondary sporangia often developing from subsporangial swellings (or apophyses) of respective sporangia, 1–6 spores per sporangium. Sporangiospores hyaline, ellipsoidal, broadly ellipsoidal, often curved, 1-celled. Chlamydospores brown, globose, thick-walled. Heterothallic.

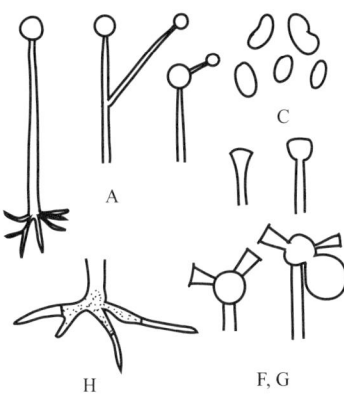

Dimensions: Sporangiophores (24.3-) 80–380 μm tall; 7.2–25 μm wide basally; 1.7–3 μm wide apically; subsporangium branches 7.5–15 μm long. Sporangia 9.7–37.5 μm in diameter; apophyses 5–12.5 μm in diameter. Rhizoids 30–75 × 5–11.3 μm. Sporangiospores 4.8–11.3 × 3–5.2 μm. Chlamydospores 20–62.5 × 16.2–29.2 μm.

Material: 72-X46 (= ATCC 32324, Sugarcane roots, Okinawa, Japan); 84-353 (Forest nursery soil, Kagawa, Japan).

Remarks: This fungus is characterized by subsporangial branching from the uppermost inflated region in clusters.

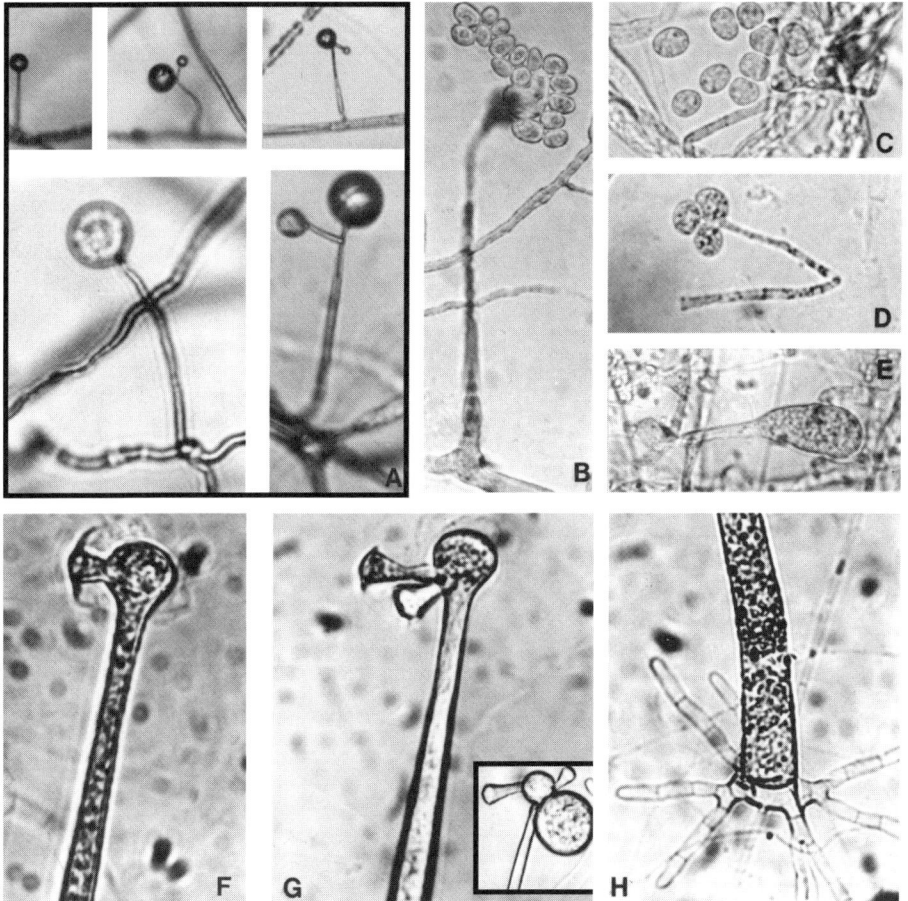

Mortierella ambigua. A: Sporangiophores and sporangia. B: Sporangiophore and sporangiospores. C,D: Sporangiospores. E: Chlamydospores. F,G: Apical parts of sporangiophores. H: Rhizoids. (From Watanabe, T. 1975a. *Trans. Mycol. Soc. Jpn.*, 16:18–27. With permission.)

Mortierella boninense T. Watanabe

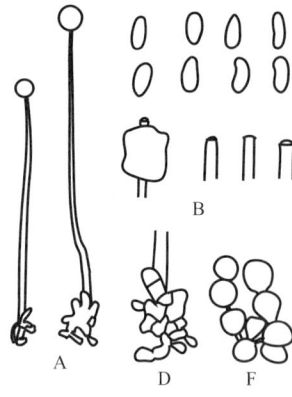

References: Watanabe 2001a.

Morphology: Sprangiophores erect, hyaline, simple, often aerial, rhizoidal, bearing sporangia terminally. Sporangia multisporous, after deliquescens, leaving a trace of columella. Sporangiospores hyaline, cylindrical or ovoid. Chlamydospores globose or subglobose, clustered.

Dimensions: Sprangiophores usually over 500 μm up to 700 μm tall, 8–12 μm wide basally, 3.6–6 μm wide apically. Sporangia 45–70 μm in diameter. Sporangiospores 12–16 × 5–8 μm. Chlamydospores 10 μm in diameter.

Material: TW 01-126 (Uncultivated soil, Higashikou, Hahajima, the Bonin Islands, Tokyo Japan).

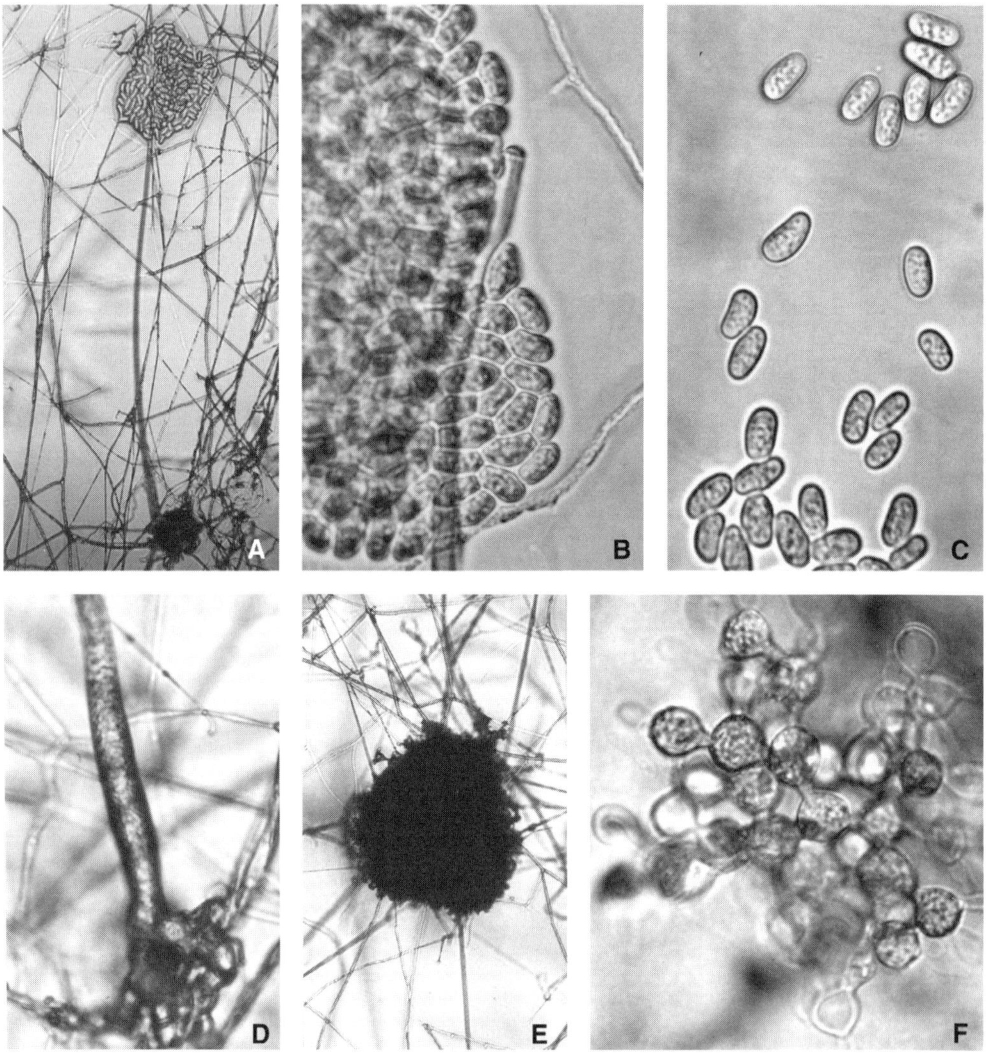

Mortierella boninense. A: Sporangiophore, sporangiospores, and rhizoid on dehiscence of sporangium. B: Sporangiospores and apex of sporangiophore. C: Sporangiospores. D: Rhizoid. E,F: Aggregates of chlamydospores. (From Watanabe, T. et al. 2001e. *Three New Mortierella from Soil in the Bonin Islands, Japan.* In preparation. With permission.)

Mortierella chlamydospora (Chesters) Plaats-Niterink

References: Plaats-Niterink et al. 1976; Watanabe 1990a.

Morphology: Sporangia not formed. Chlamydospores spiny with protuberances (stylospores), or rarely smooth, globose or subglobose, ellipsoidal, terminal or intercalary, often aerial, thick-walled; protuberances cylindrical, often curved, aerial chlamydospores apparently sunflower-like in appearance. Zygospores globose, thick-walled, with single well-developed globose or subglobose suspensors, often 1-septate, and another poorly developed suspensor (often disappeared). Homothallic.

Dimensions: Chlamydospores 15–30 μm in diameter; protuberances 1.2–2.3 μm long, 0.5–5 μm wide. Zygospores 27.5–52.5 μm in diameter; suspensors 25–42.5 μm wide.

Material: 73-34 (= IFO 32541, Strawberry field soil, Shizuoka, Japan), 74-667 (Strawberry roots, Shizuoka, Japan), 81-260 (Eggplant field soil, Fukushima, Japan).

Remarks: *M. echisphaera* is close to this fungus, but it does not form zygospores, and *M. indohii*, another morphologically similar fungus, is heterothallic, and when fertilized it forms zygospores covered by hyphae originated from suspensors.

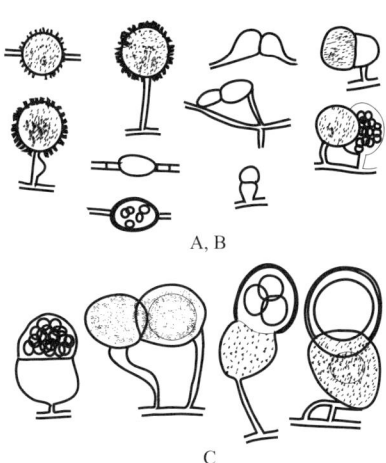

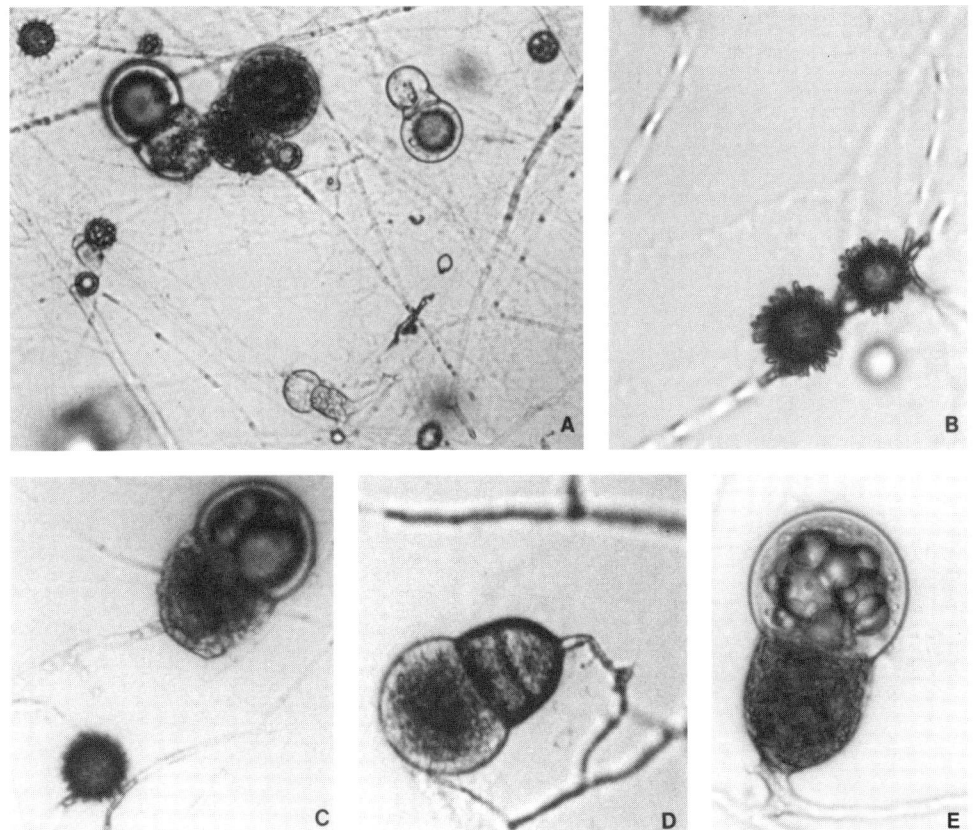

Mortierella chlamydospora. A–C: Sunflower-like chlamydospores and zygospores in various stages. D,E: Immature zygospores with well-developed suspensors. (From Watanabe, T. 1990a. *Mycologia*, 82:278–282. With permission.)

Mortierella elongata Linnemann

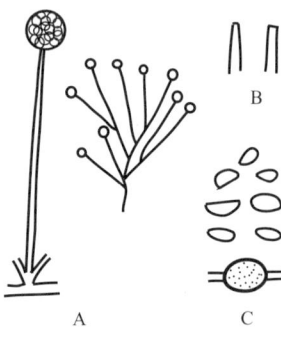

References: Watanabe 1975a; Zycha et al. 1969.

Morphology: Sporangiophores often aerial, erect., simple or branched symypodially at base, tapering gradually from base toward apex, forming sporangia terminally. Sporangiospores hyaline, ellipsoidal, cylindrical, ovate, often curved, 1-celled. Chlamydospores brown, subglobose, thick-walled. Heterothallic.

Dimensions: Sporangiophores 160–330 μm tall; 5.5–10 μm wide at base. Sporangia 11.2–30 μm in diameter. Sporangiospores 6.5–11.3 (-15) × 4.5–7.5 (-8.8) μm. Chlamydospores 10.5–25 × 6.2–15 μm.

Material: 70-1388 (Pineapple field soil, Hachijo, Tokyo, Japan), 72-X129 (= ATCC 32325, Sugarcane roots, Okinawa, Japan).

Remarks: This fungus formed aggregates of white mycelial mat.

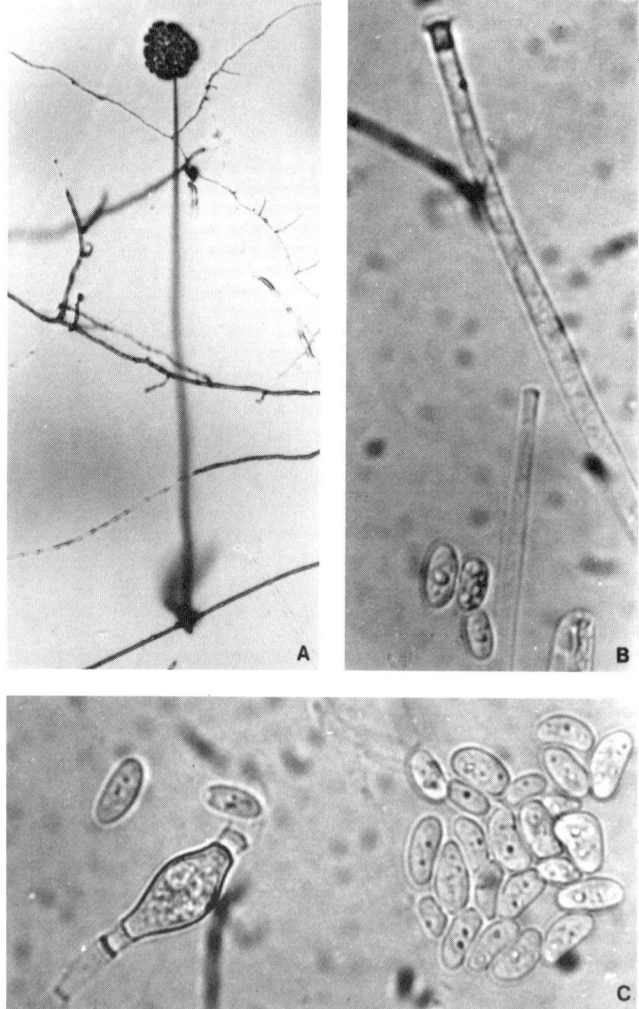

Mortierella elongata. A: Sporangiophores and sporangia. B: Apexes of sporangiophores and sporangiospores. C: Sporangiospores and chlamydospore. (From Watanabe, T. 1975a. *Trans. Mycol. Soc. Jpn.*, 16:18–27. With permission.)

Mortierella epicladia var. *chlamydospora* T. Watanabe

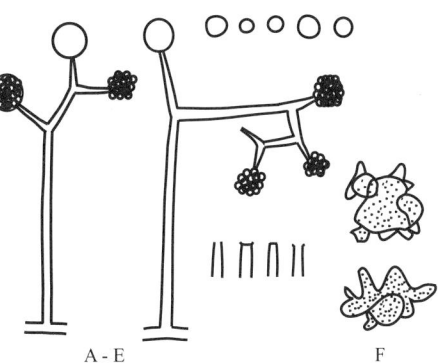

References: Gams 1977; Watanabe et al. 2001.

Morphology: Sporangiophores erect, hyaline, simple or branched sympodially 1–5 times, tapering toward apex, constricted at base, bearing sporangia terminally. Sporangia many-spored, noncolumellate on dehiscence or nonapophysate; collarettes inconspicuous. Sporangiospores hyaline, globose or subglobose. Chlamydospores irregularly lobate.

Dimensions: Sporangiophores 40–90 (-140) µm long, (2.4-) 3–6 µm wide at base, branches usually 6–36 µm long. Sporangia usually 8–15 µm in diameter. Sporangiospores 2–7 µm in diameter. Chlamydospores 18–34 µm in diameter.

Material: 00-21 (= MAFF 238166, Forest soil, Mt. Chibusa, Hahajima, the Bonin Islands, Tokyo, Japan).

Remarks: This fungus is unique in forming irregularly shaped chlamydospores.

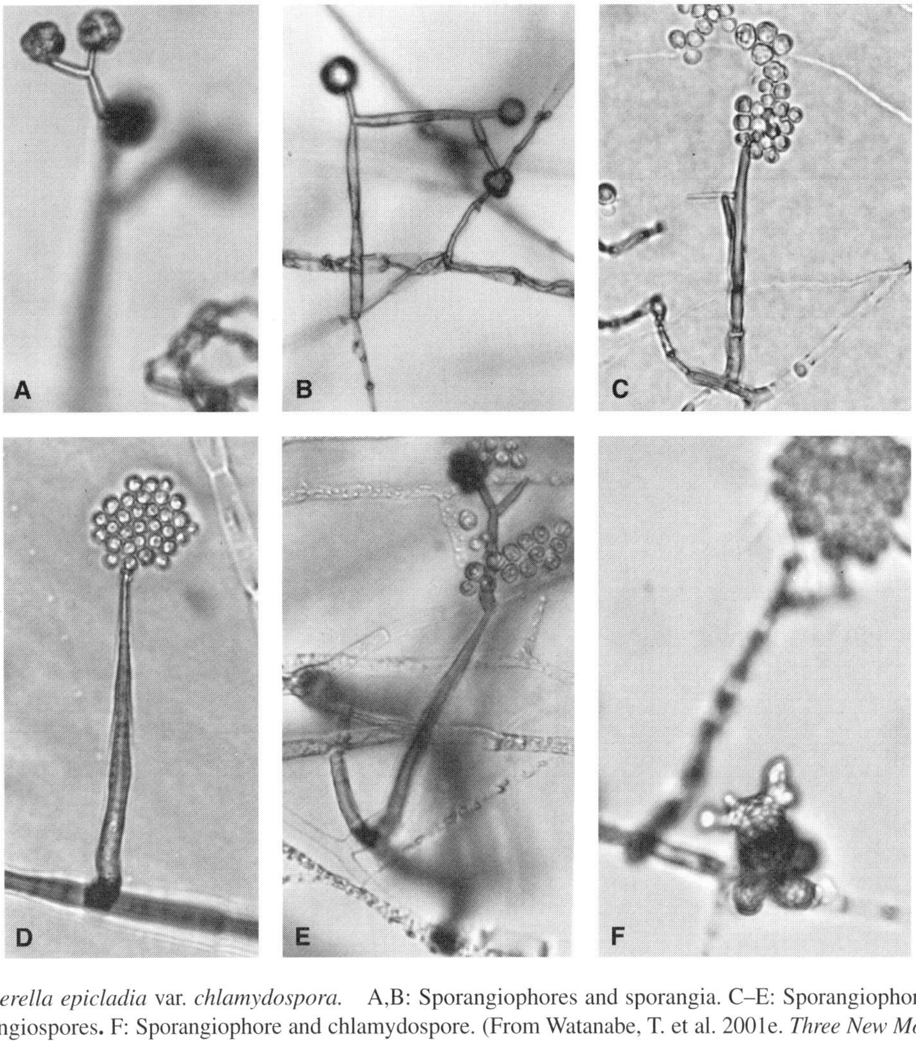

Mortierella epicladia var. *chlamydospora*. A,B: Sporangiophores and sporangia. C–E: Sporangiophore and sporangiospores. F: Sporangiophore and chlamydospore. (From Watanabe, T. et al. 2001e. *Three New Mortierella from Soil in the Bonin Islands, Japan*. In preparation. With permission.)

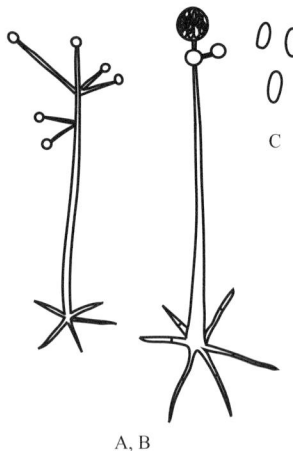

Mortierella exigua Linnemann

References: Domsch et al. 1980a,b.

Morphology: Sporangiophores erect, simple or branched verticillately or alternately 1–3 times near apex, gradually tapering from base toward apex, bearing sporangia terminally, often rhizoidal basally. Sporangiospores hyaline, ellipsoidal, 1-celled.

Dimensions: Sporangiophores ca. 330 µm tall; 5.5 µm wide at base. Sporangia ca. 20 µm in diameter. Sporangiospores ca. 7.5 × 5 µm.

Material: 70-1462 (Pineapple field soil, Hachijo, Tokyo, Japan).

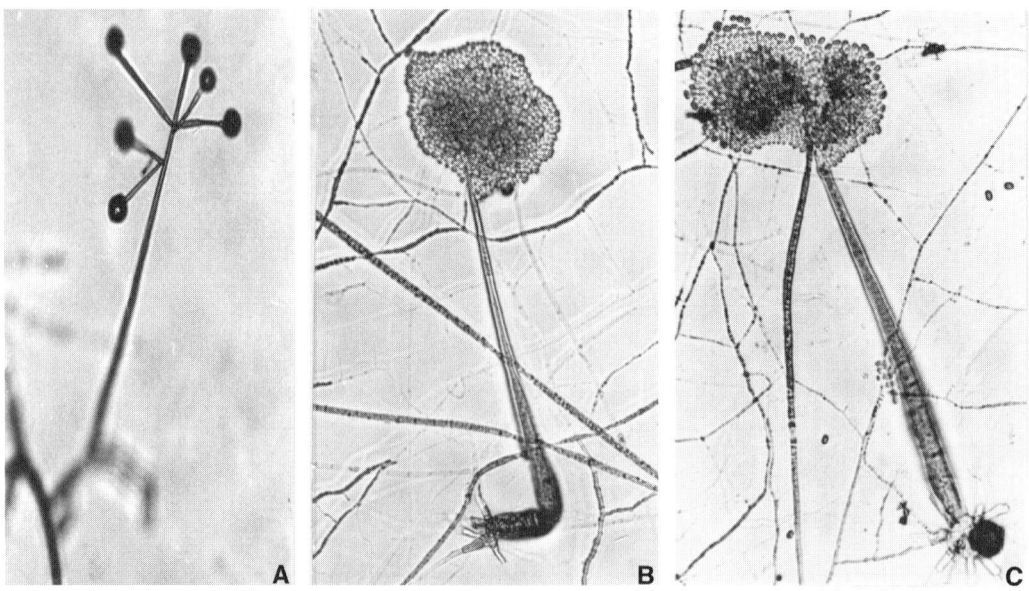

Mortierella exigua. A: Sporangiophore and sporangia. B,C: Sporangiophores and sporangiospores.

Mortierella gemmifera Ellis

References: Ellis 1940; Gams 1976, 1977.

Morphology: Sporangiophores hyaline, erect, tapering gradually from base toward apex, simple or branched at the lower half, bearing sporangia terminally. Sporangia hyaline, globose, lacking columellae, several spores included. Sporangiospores hyaline, broadly ellipsoidal or subglobose. Chlamydospores yellowish brown, subglobose, large, rich in protoplast, granulate conspicuously, terminal or intercalary. Heterothallic.

Dimensions: Sporangiophores 325–490 μm tall; 6–6.3 μm wide at base; 2–3 μm wide at apex. Sporangiospores 9.5–10 × 7.5–8.8 μm. Chlamydospores up to 95 μm in diameter.

Material: 92-125 (Forest soil, Hamamatsu, Shizuoka, Japan).

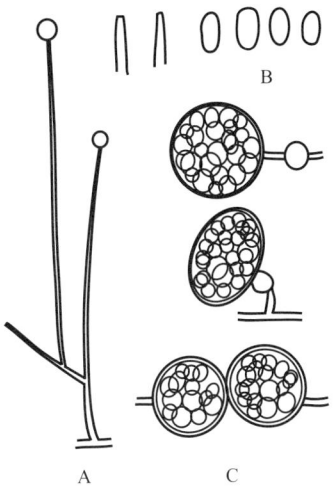

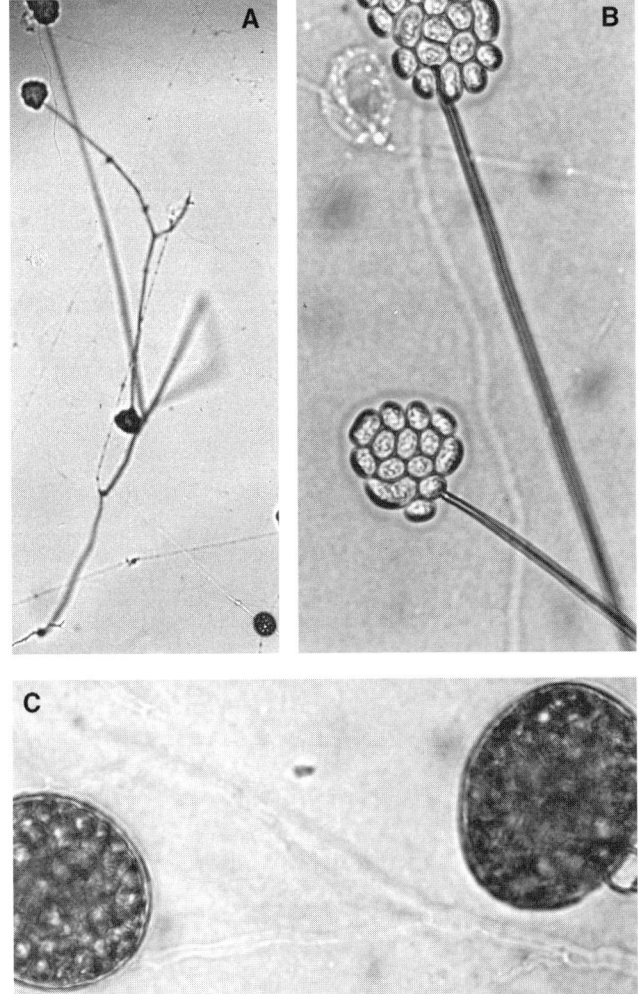

Mortierella gemmifera. A: Sporangiophore and sporangia. B: Apexes of sporangiophores and sporangiospores. C: Chlamydospores.

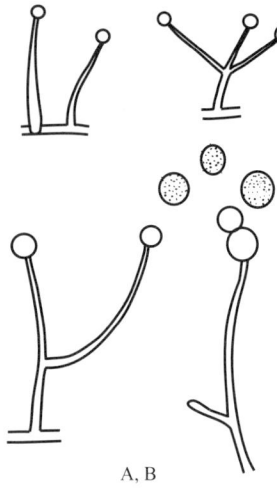

Mortierella humilis Linnem. ex Gams

References: Domsch et al. 1980a,b; Gams 1976, 1977.

Morphology: Sporangiophores hyaline, erect, simple or branched, alternately or verticillately at the lower half, tapering gradually from base toward apex, bearing sporangia terminally. Sporangiospores (conidia) 1 per sporangium, hyaline, globose, smooth or finely echinulate (spiny).

Dimensions: Sporangiophores (57.5-) 65–112.5 (-187.5) × 5–7.5 μm; 2.5–7.5 μm wide basally; 1.2–1.3 μm wide apically. Sporangia (conidia) 7.5–12.5 μm in diameter.

Material: 92-123 (Grassland soil, Hamamatsu, Shizuoka, Japan).

Remarks: Both *M. bisporalis* (Thaxt.) Bjorling and *M. lignicola* (G. W. Martin) W. Gams and R. Moreau resemble this fungus, but the former species is often curved at the apical part of sporangiophores, and the latter, extremely narrowed from the middle of sporangiophores.

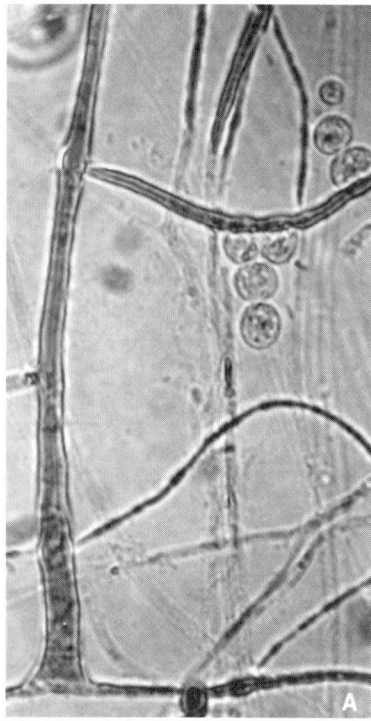 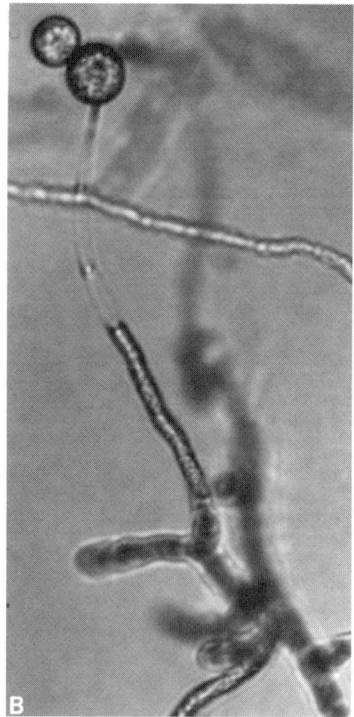 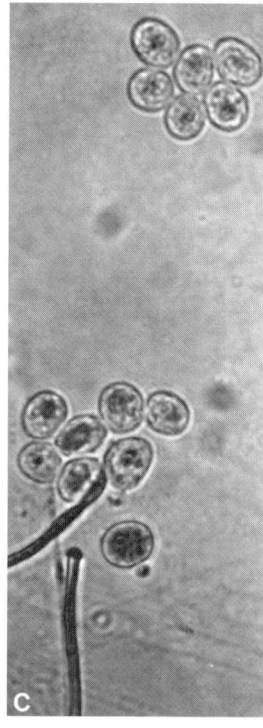

Mortierella humilis. A,B: Sporangiophore and sporangiospores. C: Apical part of sporangiophore and sporangiospores.

Mortierella hyalina (Harz.) W. Gams

Synonym: *M. hygrophia* Linnemann

References: Domsch et al. 1980a,b.

Morphology: Sporangiophores hyaline, simple or sympodially branched from lower part, elongating 3 to several branches, bearing sporangia terminally, nonrhizoidal. Sporangia noncolumellate on dehiscence. Sporangiospores hyaline, subglobose or ellipsoidal, 1-celled. Chlamydospores ellipsoidal, solitary.

Dimensions: Sporangiophores 220–350 × 10 µm; 10 µm wide basally. Sporangia 20–30 µm in diameter. Sporangiospores 7.5–15 × 5.5–11.3 µm. Chlamydospores 17.5–25 × 10–15 µm.

Material: 70-1345; -1713 (Pineapple roots, Hachijo, Tokyo, Japan).

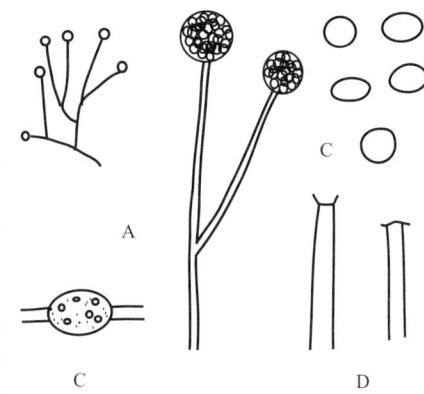

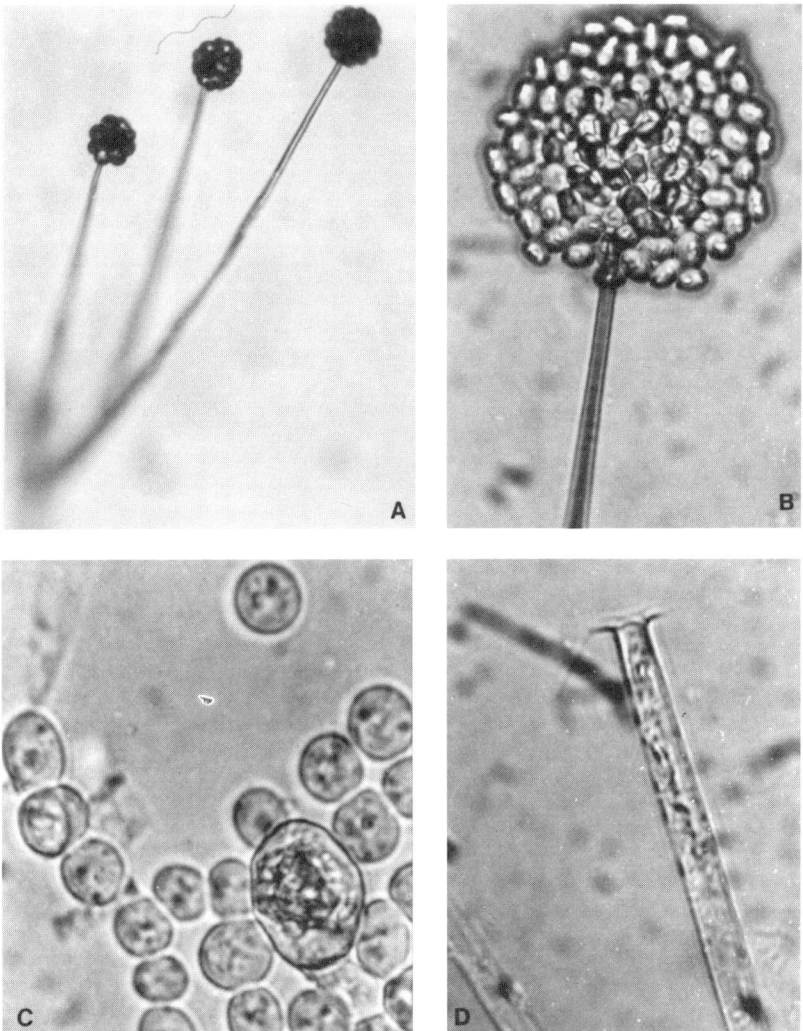

Mortierella hyalina. A: Sporangiophores and sporangia. B: Collapsed sporangium and sporangiospores. C: Sporangiospores and chlamydospore. D: Apex of sporangiophore.

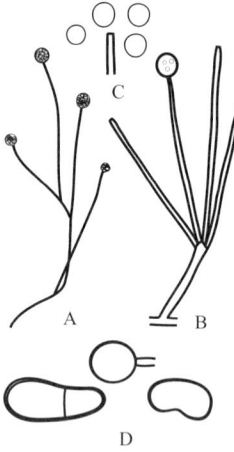

Mortierella isabellina Oudem.

References: Domsch et al. 1980a,b; Turner 1963.

Morphology: Sporangiophores hyaline, erect, branched sympodially near the middle part, apparently verticillate, tapering gradually toward apex, often septate, bearing sporangia terminally. Sporangia noncolumellate on dehiscence. Sporangiospores hyaline, globose or subglobose, often slightly angular. Chlamydospores pale yellow, globose, cylindrical, ovate, thick-walled, often septate.

Dimensions: Sporangiophores 150–370 × 2.5–3.6 µm; ca.1.8 µm wide apically. Sporangia 7.5–15 µm in diameter. Sporangiospores 1.4–2 µm in diameter. Chlamydospores: cylindrical ones 47.5–82.5 × 35–42.5 µm; globose ones 32.5–40 µm in diameter.

Material: 84-477 (Japanese black pine seed, Okawa, Kagawa, Japan).

Remarks: PDA colonies are pale yellowish brown, and zonate.

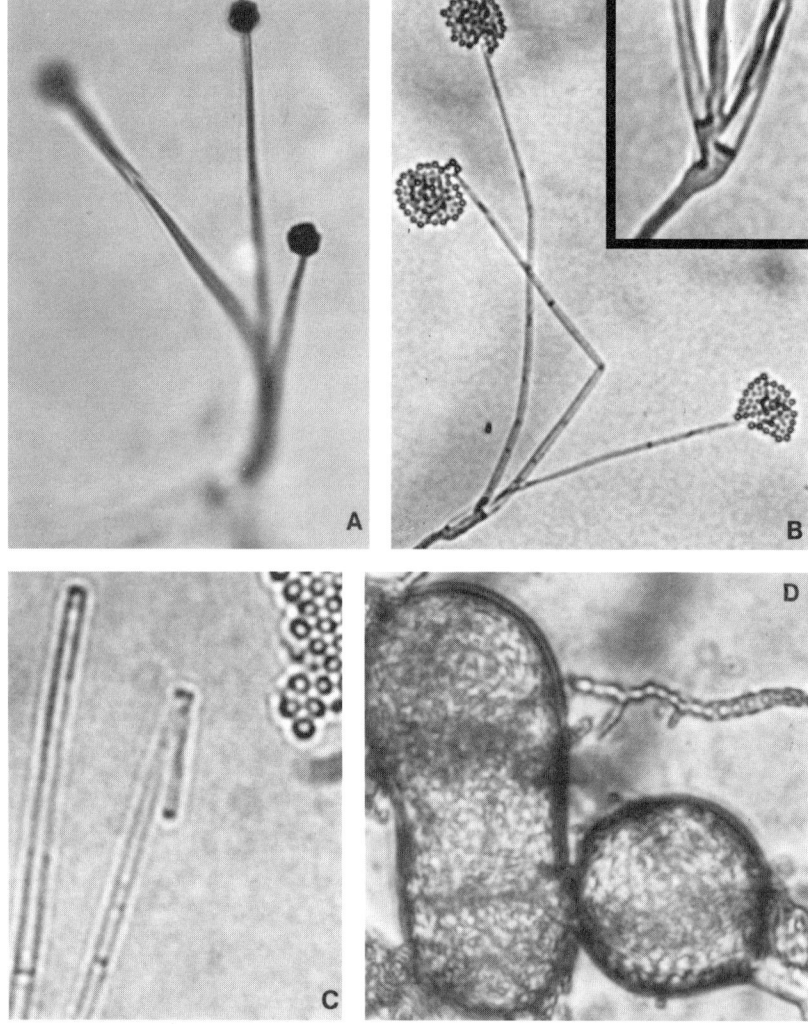

Mortierella isabellina. A: Sporangiophores and sporangia. B: Sporangiophores; the close-up of a branching part in the inset. C: Apexes of sporangiophores and sporangiospores. D: Chlamydospores.

Mortierella minutissima van Tieghem

References: Turner and Pugh 1961; Watanabe et al. 1987a.

Morphology: Sporangiophores hyaline, erect, simple or rarely branched at the middle part, gradually tapering toward apex, bearing sporangia terminally. Sporangia hyaline, globose, noncolumellate on dehiscence. Sporangiospores hyaline, globose.

Dimensions: Sporangiophores 60–175 µm tall; 5–7.5 µm wide at base; 1.7–2 µm wide at apex. Sporangia 17.5–22.5 µm in diameter. Sporangiospores 5–10 µm in diameter.

Material: 85-P55 (Paulownia roots, Itapua, Paraguay).

Remarks: This fungus occasionally forms single sporangiospore per sporangium (conidium).

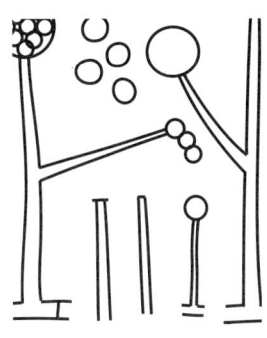

A, B

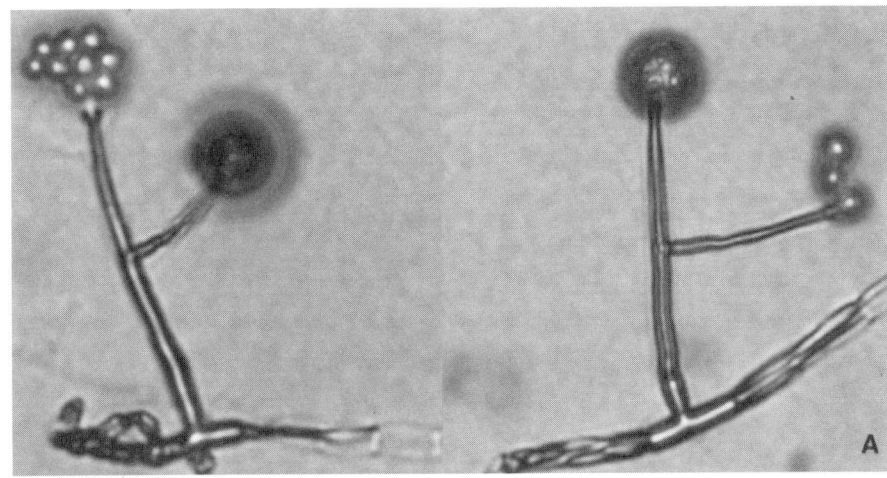

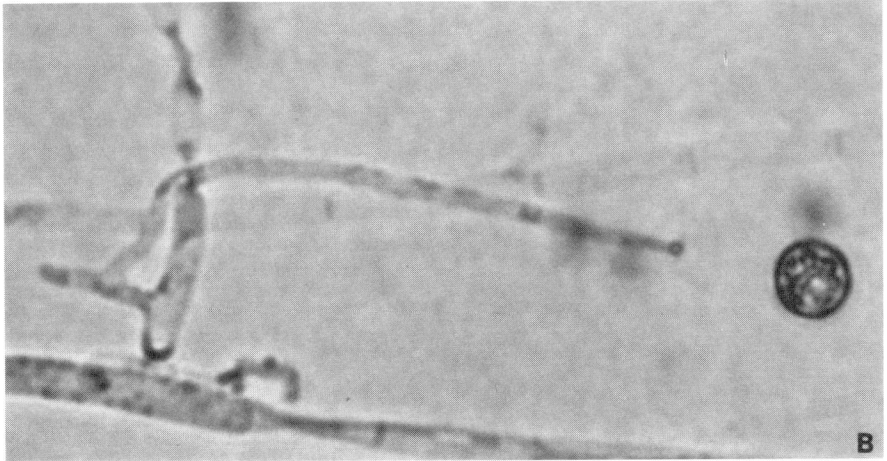

Mortierella minutissima. A,B: Sporangiophores, sporangia, and sporangiospores.

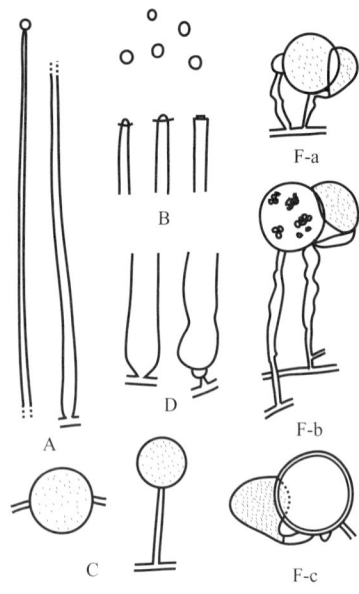

Mortierella tsukubaensis T. Watanabe

References: Watanabe et al., 2000.

Morphology: Sporangiophores erect, hyaline, simple, tapering from base toward apex, bearing sporangia terminally, leaving minute columellae and conspicuous collarettes on dehiscence, rarely rhizoidal, surrounded with empty vesicles. Sporangiospores hyaline, mostly ellipsoidal, subglobose, or angular. Chlamydospores globose or subglobose, pale brown, usually intercalary. Homothallic. Zygospores superficial, of naked type, with unequal suspensors, major suspensors globose, ovate, ellipsoidal, or club-shaped, nearly equal to zygospore size, minor suspensors globose, subglobose, cylindrical, slightly swollen hypha-like, curved, not well differentiated from vegetative hyphae, often aborted.

Dimensions: Sporangiophores mostly 600–2800 μm tall, 6–40 μm wide at base, 2–10 μm wide near apices, columellae mostly 5–7 μm, 4 μm tall. Sporangia mostly 20–50 μm in diameter. Sporangiospores 4–10 × 4–7 μm. Chlamydospores 10–54 μm in diameter.

Zygospores mostly 34–52 μm in diameter: zygospore wall 1–1.6 μm thick. Major susupensors 18–52 × 14–38 μm. Minor suspensors 10–18 × 5–7 μm.

Material: 98-120 (= MAFF 237778 = ATCC 204319, Uncultivated soil, Tsukuba, Japan).

Remarks: The simple large sporangiophores with many-spored sporangia and naked zygospores are characteristic for this species. The key of 12 homothallic *Mortierella* species was prepared by Watanabe et al. (2001).

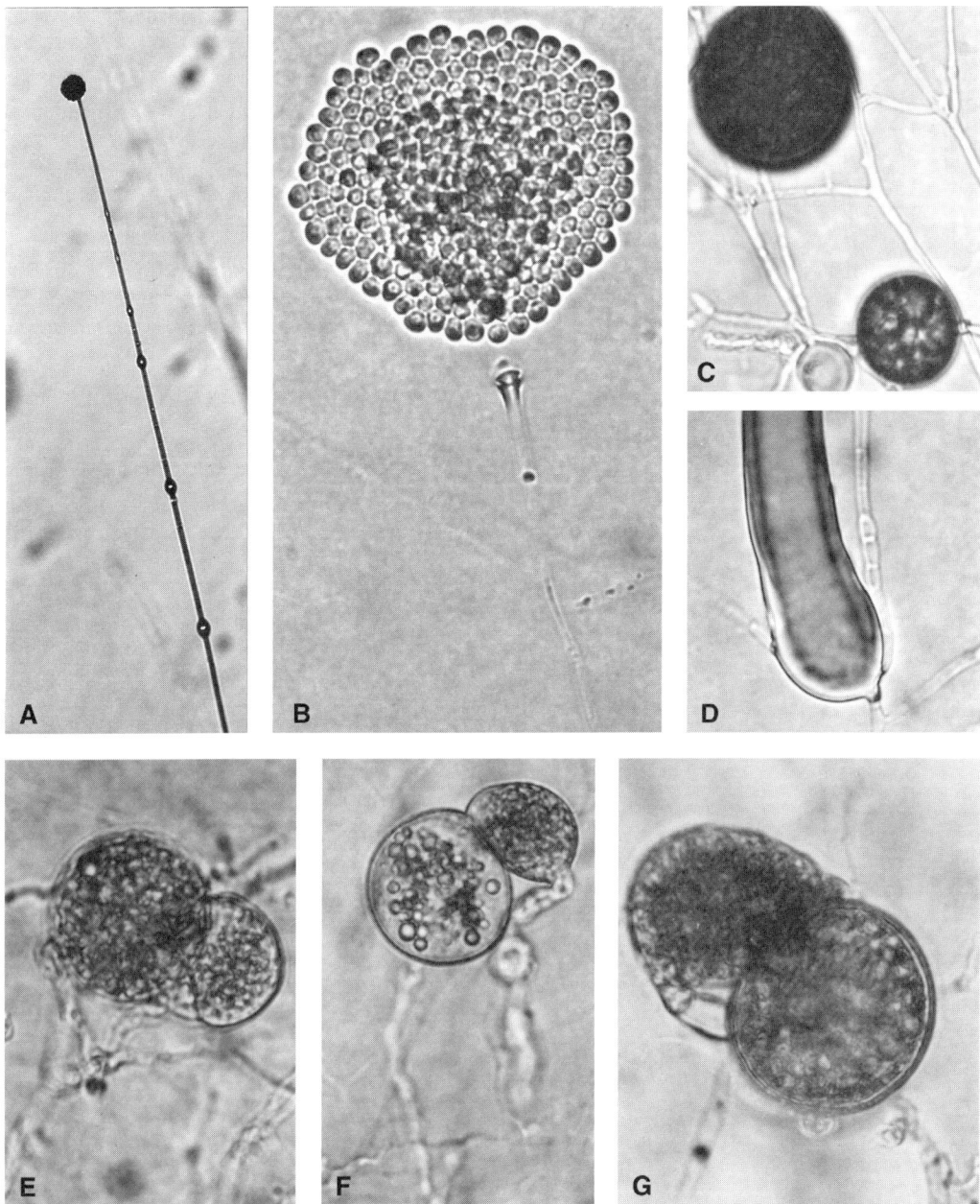

Mortierella tsukubaensis. A: Sporangiophore and sporangium. B: Sporangiospores and apex of sporangiophore. C: Chlamydospores. D: Sporangiophore base. E–G: Immature (E,F) and mature (G) zygospores. (From Watanabe, T. et al. 2001a. *Mycol. Res.*, 105:506–509. With permission.)

Mortierella uniramosa T. Watanabe

References: Gams 1977; Watanabe et al. 1995.

Morphology: Sporangiophores hyaline, erect, simple or branched once in the upper portion, tapering toward apex, bearing sporangia terminally. Sporangia many-spored, a trace of columella left on dehiscence. Sporangiospores ellipsoidal or cylindrical. Chlamydospores irregularly lobate.

Dimensions: Sporangiophores (85-) 100–250 (-275) μm long (3.6-) 5–10 μm wide at base. Sporangia 14–15 μm in diameter. Sporangiospores (5-) 6–9 (-10) × (2.4-) 3–5(-7) μm. Chlamydospores 16–30 × 10–16 μm in diameter.

Material: TW 00-33 (= MAFF 238167, Forest soil, Mt. Chibusa, Hahajima, the Bonin Islands, Tokyo, Japan).

Remarks: This fungus is unique in forming irregularly shaped chlamydospores.

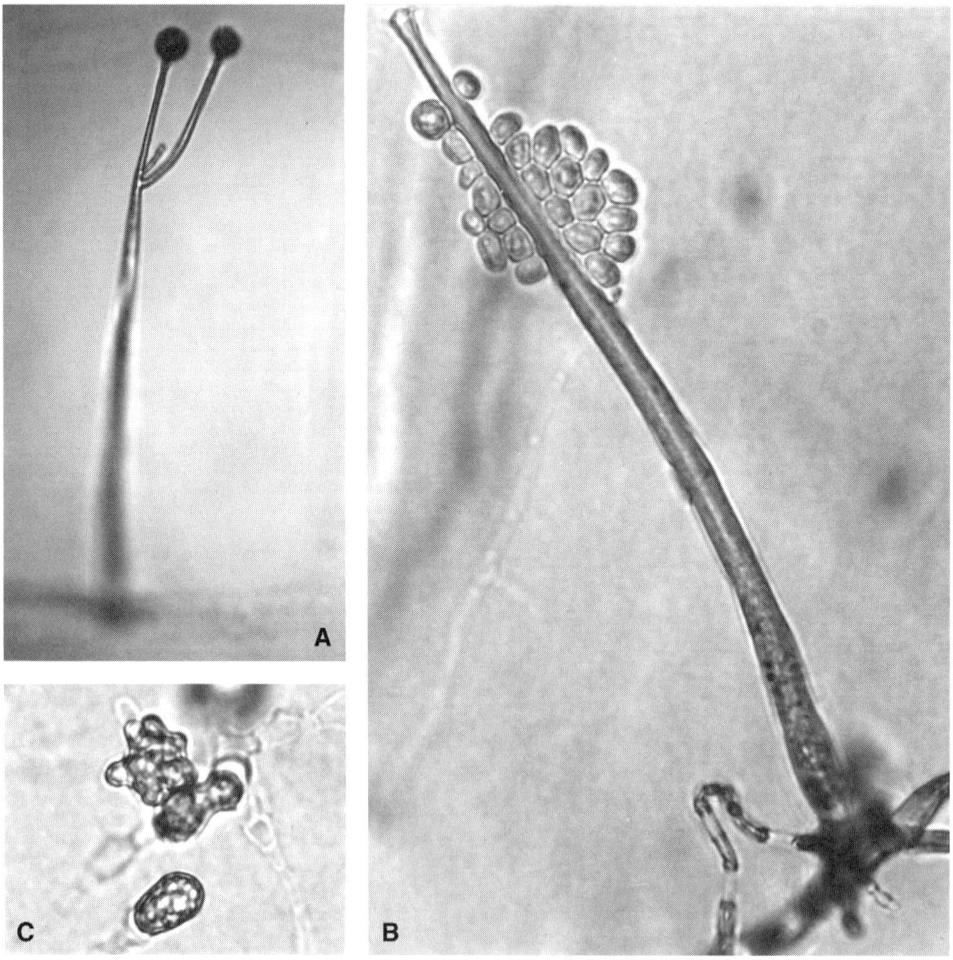

Mortierella uniramosa. A: Sporangiophore and sporangia. B: Sporangiophore and sporangiospores after deliquescence of sporangium. C: Chlamydospores. (From Watanabe, T. 2001e. *Three New Mortierella from Soil in the Bonin Islands, Japan.* In preparation. With permission.)

Mortierella verticillata Linnemann

References: Chien et al. 1974; Gams 1977.

Morphology: Sporangiophores verticillate basally, with more than 3 branches, tapering gradually toward apex. Sporangiospores (conidia) single per sporangium, terminal, globose, smooth.

Dimensions: Sporangiophores 30–120 µm tall; branches 30–42.5 µm long; ca. 5 µm wide at base. Sporangiospores 5–8 µm in diameter.

Material: 83-396; -397 (Piedmont forest soil, Shizuoka, Japan).

Remarks: *M. marburgensis* Linnemann was treated as a synonym of this fungus (Gams, 1977).

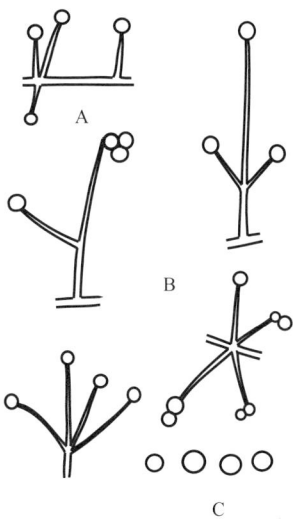

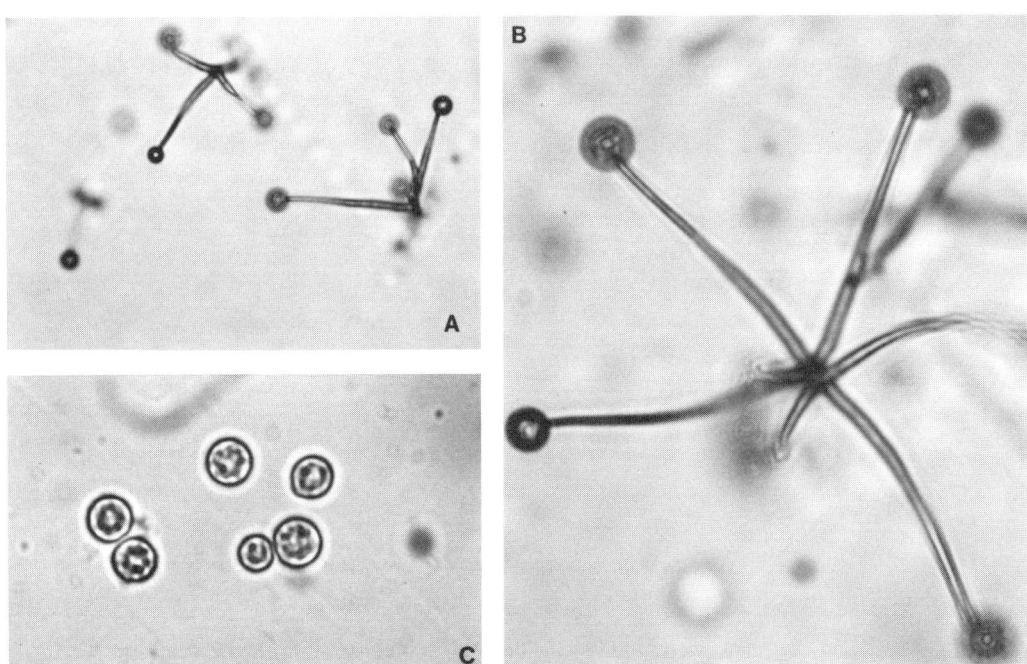

Mortierella verticillata. A,B: Sporangiophores and sporangia. C: Sporangiospores.

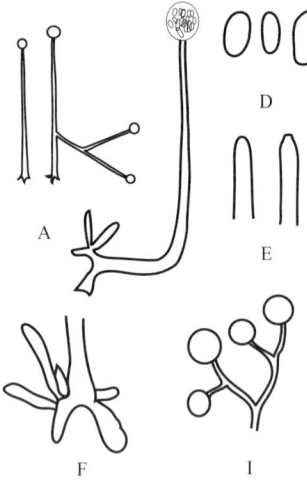

Mortierella zychae Linnemann

References: Watanabe 1975a; Zycha et al. 1969.

Morphology: Sporangiophores hyaline, erect, simple or branched sympodially near base, developed occasionally from aerial hyphae, tapering gradually from base toward apex, bearing sporangia terminally, poorly rhizoidal basally. Sporangia noncolumellate on dehiscence. Sporangiospores hyaline, ellipsoidal, 1-celled. Chlamydospores globose, catenulate, and clustered.

Dimensions: Sporangiophores 200–800 μm tall; 8–12.5 μm wide basally; 3–4.5 μm wide apically. Sporangia (30-) 37.5–45 (-75) μm in diameter. Sporangiospores 8.7–20 × 4–12 μm. Chlamydospores 14–17.5 μm in diameter.

Material: 70-1015 (Pineapple field soil, Hachijo, Tokyo, Japan); 72-X91-555 (= ATCC 32326, Sugarcane roots, Taiwan, ROC).

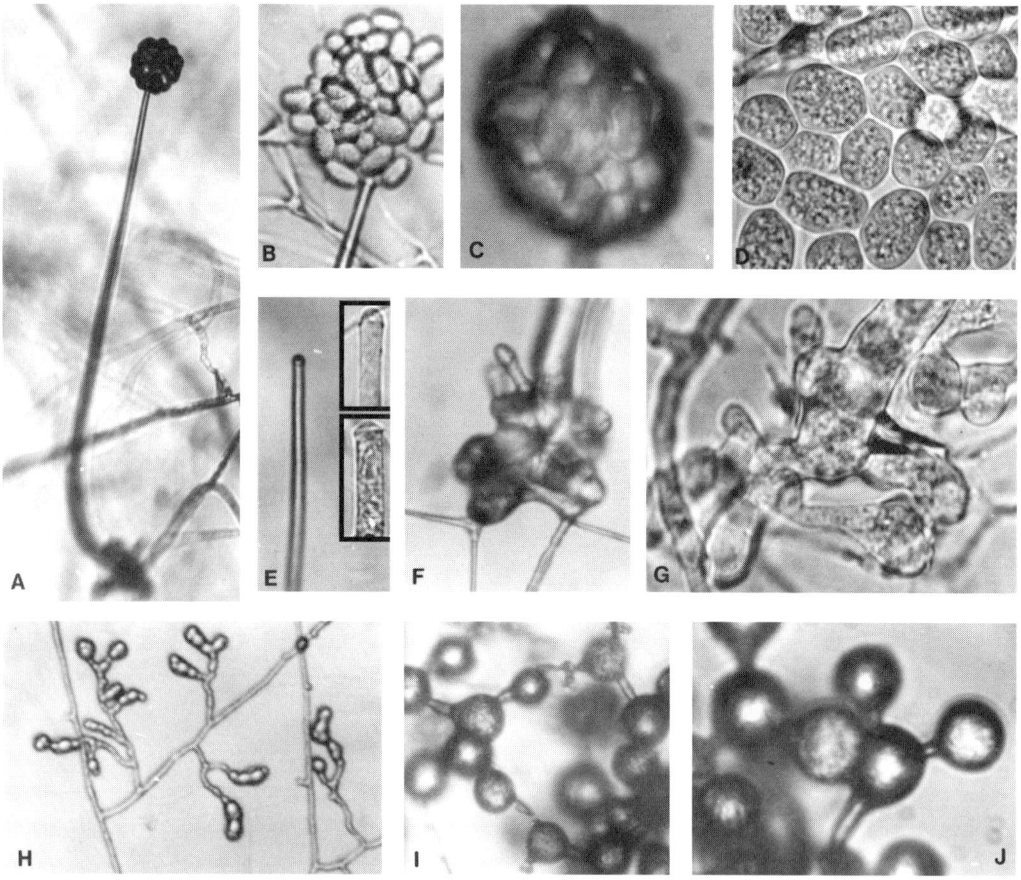

Mortierella zychae. A: Sporangiophores and sporangium. B,C: Apical parts of sporangiophores and sporangiospores. D: Sporangiospores. E: Apex of sporangiophore. F,G: Rhizoids. H–J: Chlamydospores. (From Watanabe, T. 1975a. *Trans. Mycol. Soc. Jpn.*, 16:18–27. With permission.)

Zygomycotina

Mucor Micheli : Fr.

Syst. Mycol. 3:317, 1832.
Type species: *M. mucedo* Fr.

Key to Species

1. Zygospores	formed in single culture	*M. hachijoensis*
	not formed	2
2. Columellae	protuberant	*M. plumbeus*
	not so	3
3. Sporangiospores	hyaline, ellipsoidal	4
	pigmented, subglobose	*M. circinoides*
4. Sporangiospores	up to 4.3 µm long	*M. microsporus*
	5–6.3 µm long	*M. hiemalis*

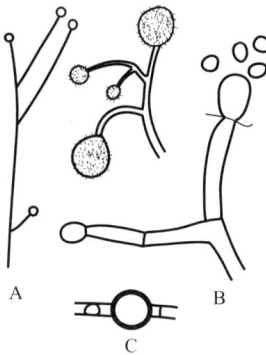

Mucor circinelloides van Tieghem

References: Domsch et al. 1980; Zycha et al. 1969.

Morphology: Sporangiophores hyaline, erect, mostly branched sympodially. Sporangia terminal, dark brown or black, finely echinulate or smooth, columellate on dehiscence; columellae hyaline, globose or subglobose with collar. Sporangiospores hyaline or pale brown especially in mass, globose, subglobose, or ovate, 1-celled. Chlamydospores pale yellow, subglobose.

Dimensions: Sporangiophores ca. 3 mm tall; 12.5–20 µm wide basally. Sporangia 22.5–70 (-95) µm in diameter; columellae 13–33 (47.5-) × 11.2–35 (37.5) µm. Sporangiospores 5–8.5 µm in diameter. Chlamydospores 15–30 µm in diameter.

Material: 85-42; -43; -44 (Flowering cherry seed, Hachioji, Tokyo, Japan); 86-41 (Japanese cedar seed, Toyama, Japan).

Remarks: This species is classified at the subspecies level: based on colony characteristics, conidial morphology, and presence or absence of collar on the vesicles. Both f. *janssenii* and f. *griseocyanus* are dark gray or grayish brown in colony colors, and their conidia are globose, ovate, or subglobose, whereas f. *circinelloides* is yellowish brown in colony color, and its conidia ovate.

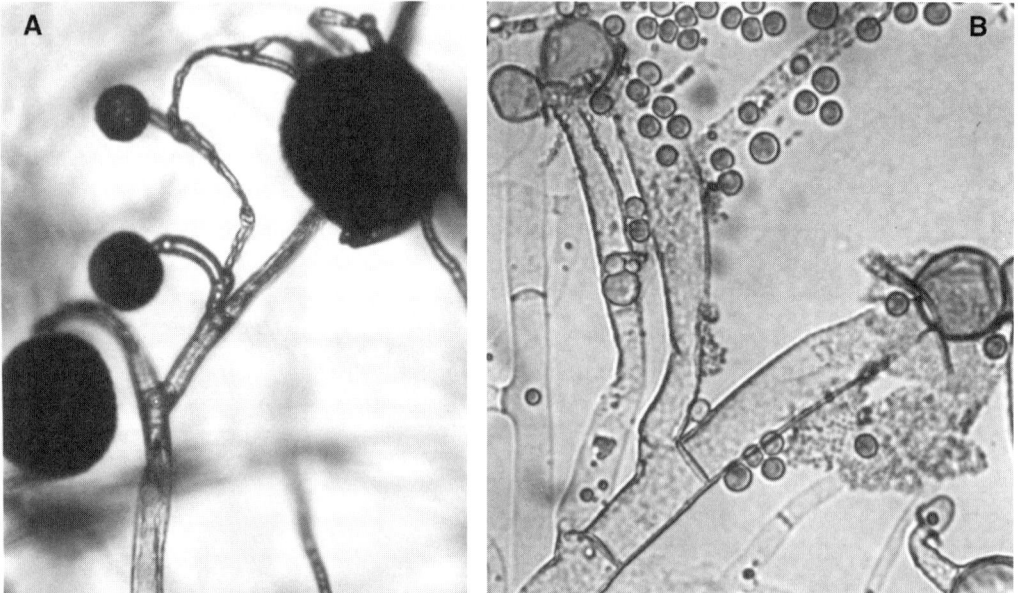

Mucor circinelloides. A: Sporangiophores and sporangia. B: Sporangiophores on dehiscence of sporangia and sporangiospores. C: Chlamydospore.

Mucor hachijoensis T. Watanabe

References: Schipper 1973, 1978; Watanabe 1994.

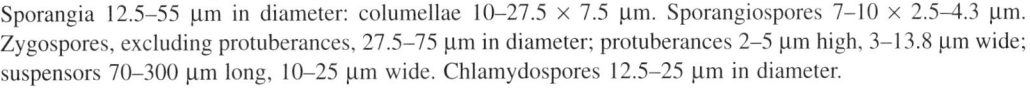

Morphology: Sporangiophores erect, mostly simple, subhyaline, occasionally branched, bearing sporangia terminally. Sporangia globose, subhyaline or brown in color, columellate on dehiscence; columellae globose, subglobose, or broadly ellipsoidal with indistinct collar. Sporangiospores hyaline to subhyaline, ellipsoidal, globulate. Homothallic. Zygospores on aerial hyphae, dark brown to black, warty, containing single large oil globules, connected with opposed suspensors in equal or unequal length and width. Chlamydospores globose, solitary, brown, thick-walled.

Dimensions: Sporangiophores over 4 mm tall, ca. 7.5–16.3 µm wide basally, 2.5–5 µm wide just below the sporangium. Sporangia 12.5–55 µm in diameter: columellae 10–27.5 × 7.5 µm. Sporangiospores 7–10 × 2.5–4.3 µm. Zygospores, excluding protuberances, 27.5–75 µm in diameter; protuberances 2–5 µm high, 3–13.8 µm wide; suspensors 70–300 µm long, 10–25 µm wide. Chlamydospores 12.5–25 µm in diameter.

Material: 70-1179 (Pineapple field soil, Hachijo, Tokyo, Japan).

Remarks: This fungus does not grow above 37°C. It is indistinctly rhizoidal, but not directly connected to sporangiophores. It differs morphologically from any of the five homothallic *Mucor* species known.

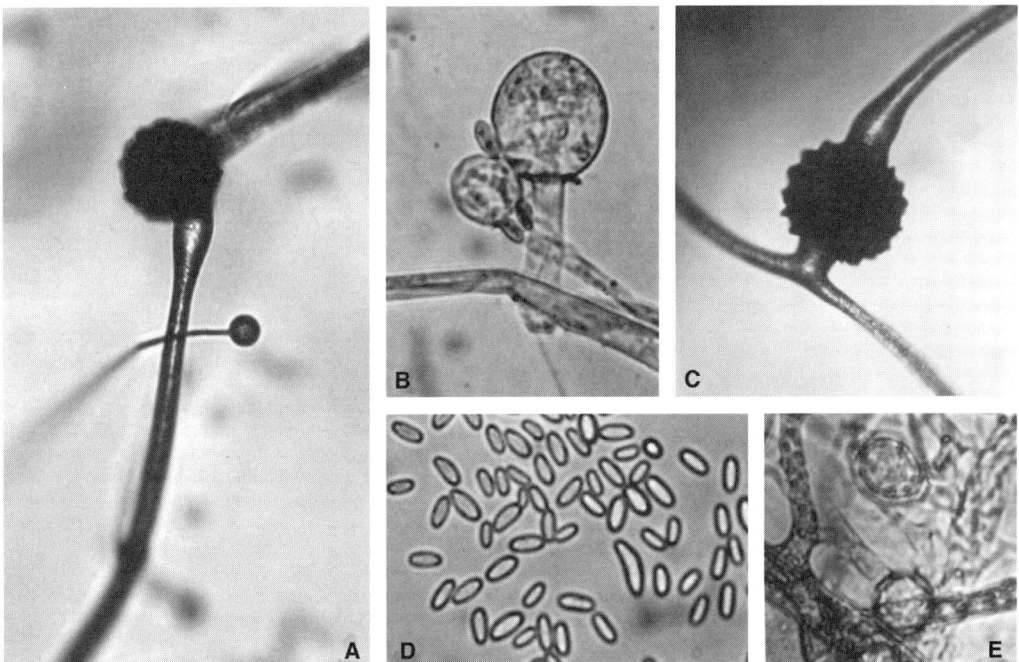

Mucor hachijoensis. A: Sporangium and zygospore. B: Columellae. C: Zygospore with opposite unequal suspensors. D: Sporangiospores. E: Chlamydospores and hyphae. (From Watanabe, T. 1994b. *Mycologia*, 86:691–695. With permission.)

Mucor hiemalis Wehmer f. *luteus* (Linnemann) Schipper

Synonym: *M. luteus* Linnemann

References: Gilman 1957; Schipper 1973, 1978; Watanabe 1971.

Morphology: Sporangiophores erect, pale reddish yellow, simple or branched, bearing sporangia terminally. Sporangia globose, yellow, smooth, columellate on dehiscence: collumellae globose. Sporangiospores long ellipsoidal, various in size, guttulate.

Dimensions: Sporangiophores over 1 mm tall; 5–12.5 µm wide basally. Sporangia 57–85 µm in diameter; columellae 22.5–28 µm in diameter. Sporangiospores (3.7-) 5–6.3 × 2–2.8 (-3.5) µm.

Material: 69-334 (Pineapple field soil, Okinawa, Japan).

Remarks: Colonies on PDA are yellow. Fertilization experiments are required to determine forma speciales (Schipper, 1973).

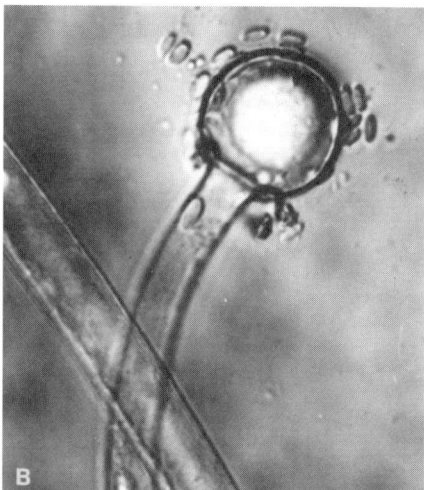
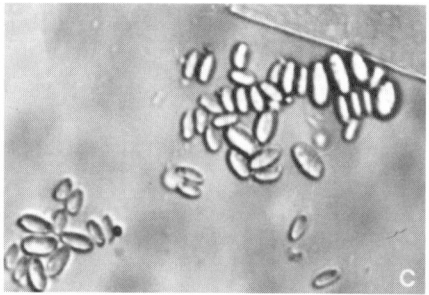

Mucor hiemalis f. *luteus*.   A: Sporangiophores and sporangia. B: Columella on the apical part of sporangiophore. C: Sporangiospores. (From Watanabe, T. 1971. *Trans. Mycol. Soc. Jpn.*, 12:35–47. With permission.)

Mucor microsporus Namyslowski

References: Zycha et al. 1969.

Morphology: Sporangiophores hyaline, erect, branched sympodially, bearing sporangia terminally, tapering gradually from base toward apex, rhizoidal, but not directly connected to sporangiophores. Sporangia terminal, black, columellate on dehiscence: columellae globose or subglobose, nonapophysate. Sporangiospores hyaline, ellipsoidal, 1-celled.

Dimensions: Sporangiophores 2–3 mm tall; 4.5–8.8 µm wide apically. Sporangia 25–60 µm in diameter; columellae 16–35 µm in diameter. Sporangiospores 3–4.3 × 2.2–2.5 µm.

Material: 70-1043; -1087 (Pineapple roots, Hachijo, Tokyo, Japan).

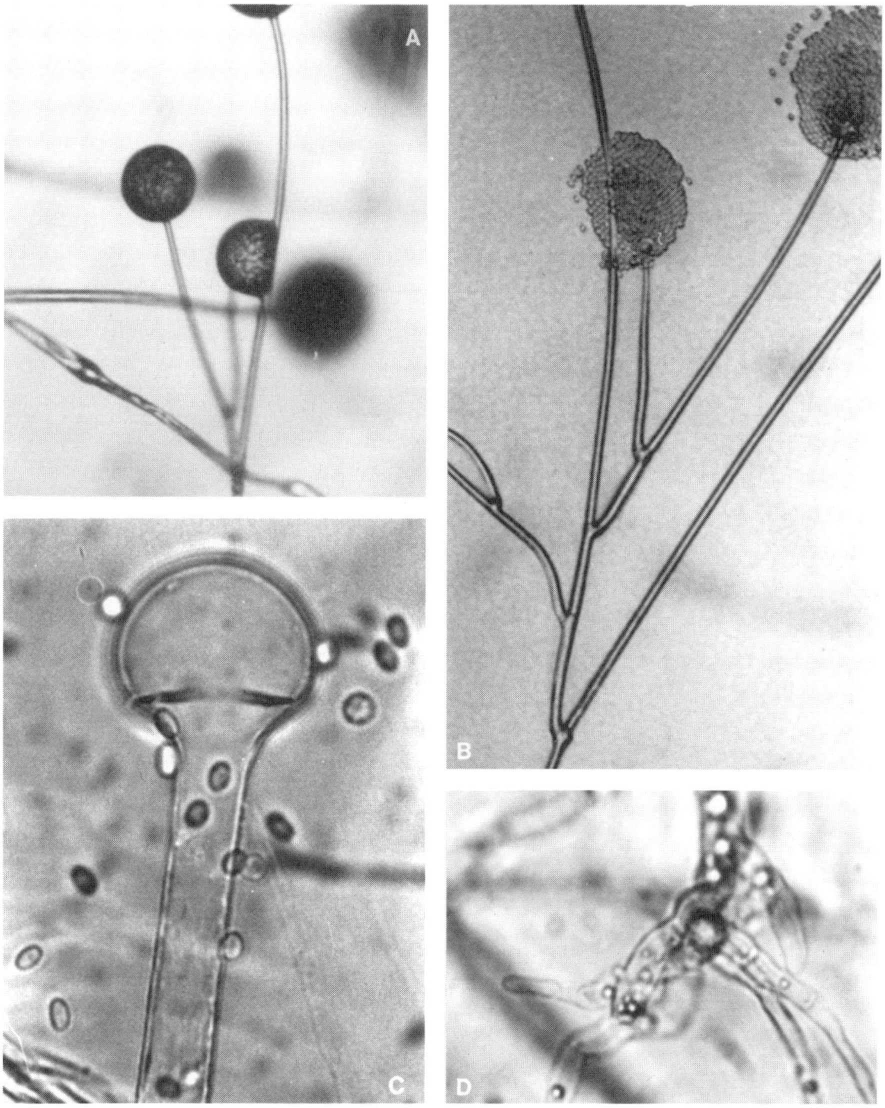

Mucor microsporus. A: Sporangiophores and sporangia. B: Collapsed sporangia and apical parts of sporangiophores. C: Columella on apical sporangiophore and sporangiospores. D: Rhizoids.

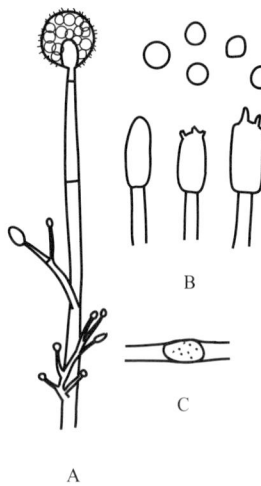

Mucor plumbeus Bonorden

References: Domsch et al. 1980; Watanabe et al. 1986a.

Morphology: Sporangiophores hyaline, erect, branched, septate rather frequently, tapering from base toward apex, bearing sporangia terminally. Sporangia brown, globose, finely echinulate, columellate on dehiscence; columellae pale brown, cylindrical, characteristically protuberant apically. Sporangiospores pale brown, globose, 1-celled. Chlamydospores pale brown, subglobose, cylindrical or broadly ellipsoidal, solitary.

Dimensions: Sporangiophores 2–3 mm tall; 15–25 µm wide basally. Sporangia 50–100 µm in diameter; columellae 30–40 × 20–22.5 µm; protuberances ca. 2.5 µm wide, 5 µm deep. Sporangiospores 5–7.5 (-11.3) µm in diameter. Chlamydospores 17.5–30 × 15–20 µm.

Material: 84-508 (Japanese red pine seed, Nagano, Japan).

Remarks: This fungus was often isolated from air, but Waksman (1916) illustrated it as one of the soil fungi.

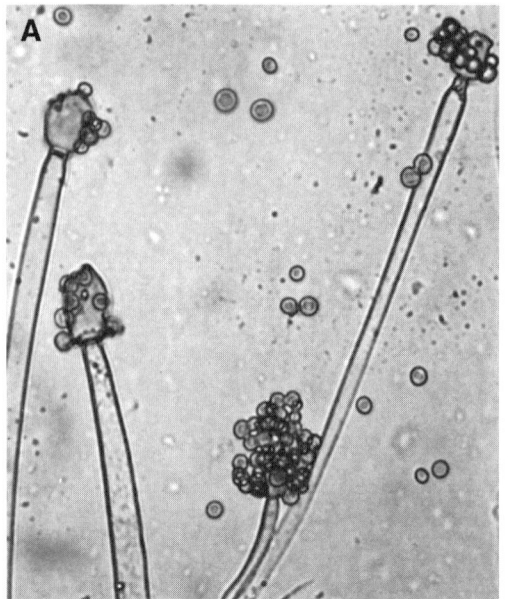
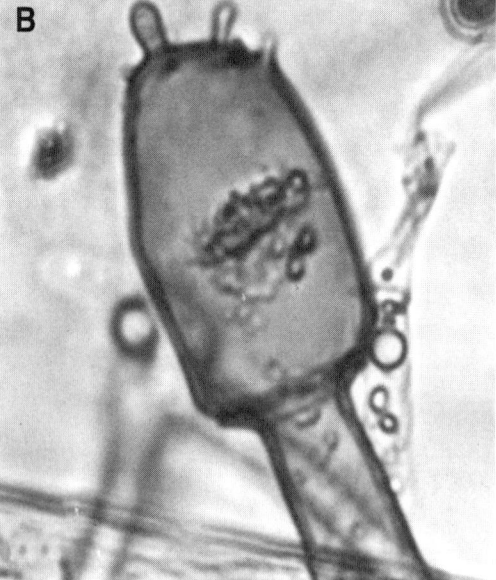

Mucor plumbeus. A,B: Sporangiophores and columellae on apical parts of sporangiophores and sporangiospores (A). C: Chlamydospore.

Rhizopus Ehrenb.

Icon. Fung. 2:20, 1838.

Type species: *R. stolonifer* (Ehrenb. : Fr.) Vuill.

Rhizopus oryzae Went & Prnisen Geerligs

Synonym: *R. arrhizus* A. Fisher

References: Domsch et al. 1980a,b; Schipper 1984.

Morphology: Sporangiophores erect, simple or branched, yellowish to dark brown, rhizoidal, connected directly to sporangiophores, bearing sporangia terminally. Sporangia globose, dark brown to black, minutely spiny, apparently subglobose after maturity, columellate on dehiscence; columellae globose, brown. Sporangiospores subglobose to subellipsoidal, pale brown, with bluish stripes (lines).

Dimensions: Sporangiophores 1.5–2.5 mm tall; 22.5–25 μm wide basally; ca. 15 μm wide apically. Sporangia 107.2–180 μm in diameter; columellae 77–125 × (72-) 90–115 μm. Sporangiospores 7.7–11.3 × 5.7–7.5 μm. Chlamydospores 105–125 × 35–50 μm.

Material: 70-90 (Snap bean seed, Hachioji, Tokyo, Japan); 85-200 (Flowering cherry seed, Hachioji, Tokyo, Japan).

Remarks: Ellipsoidal structure (Figure D) near rhizoid may be an immature zygospore.

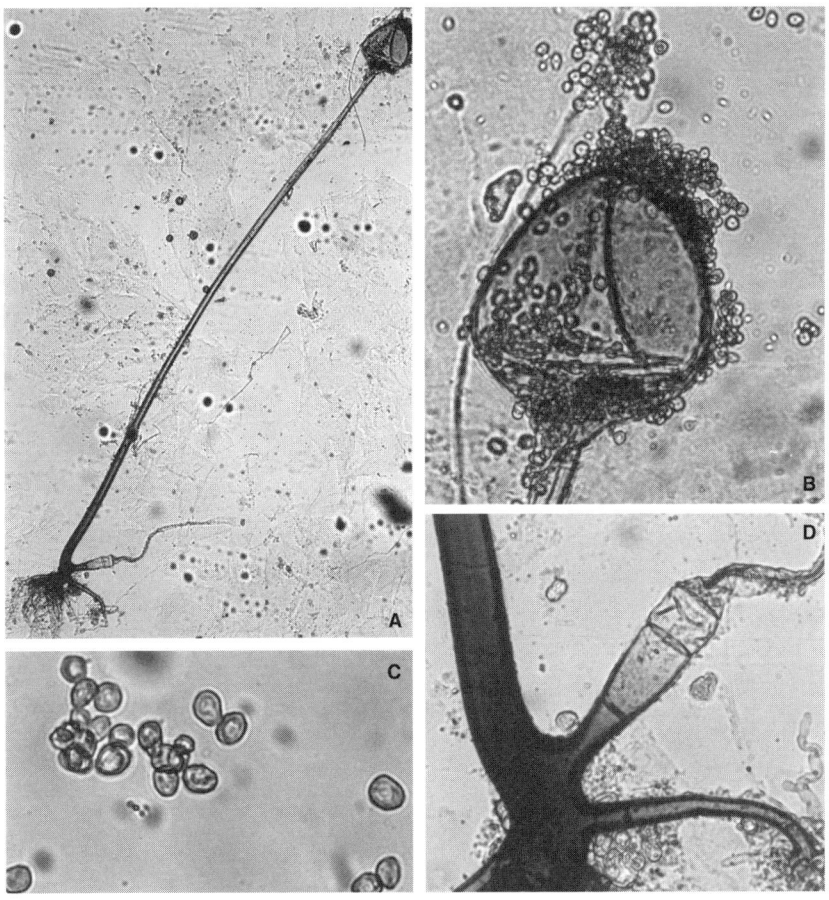

Rhizopus oryzae. A: Sporangiophore with columella on dehiscence of sporangium and rhizoid. B: Columella and sporangiospores. C: Sporangiospores. D: Chlamydospore-like structure at the basal part of sporangiophore.

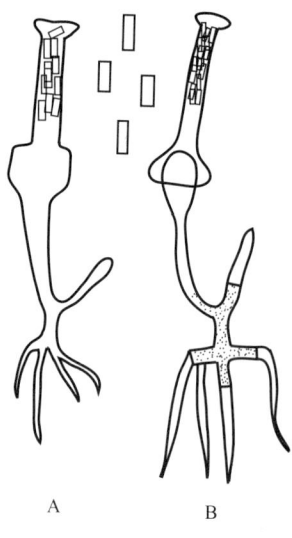

Saksenaea Saksena

Mycologia 45:434, 1953.
Type species: *S. vasiformis* Saksena.

Saksenaea Saksena

References: Saksena 1953; Watanabe 1971.

Morphology: Sporangiophores brown, branched, bearing sporangia terminally, conspicuously rhizoidal basally. Sporangia characteristically flask-shaped with long cylindrical structures ended in widened ostioles; rhizoides brown, branched into 4 directions. Sporangiospores hyaline, cylindrical, 1-celled.

Dimensions: Sporangiophores up to 60 μm tall, 7–7.5 μm wide. Sporangia ca. 32.5 μm tall, 32.5 μm wide: cylindrical part ca. 32.5 × 7.5–10 μm, ostioles ca. 12.5 μm wide; columellae 17.5 μm long. Rhizoids ca. 50 μm long, 4.5 μm wide. Sporangiospores up to 5 μm long, 2–2.5 μm wide.

Material: 69-323 (Pineapple field soil, Okinawa, Japan).

Remarks: This fungus is noteworthy as a pathogen of zygomycosis.

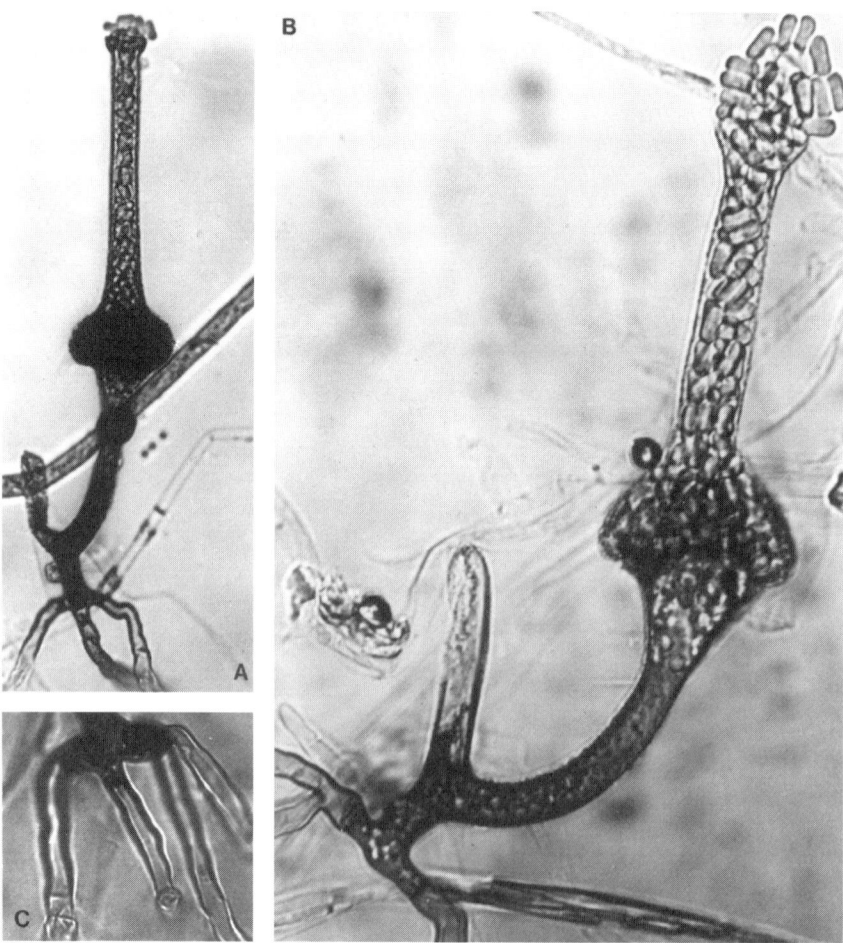

Saksenaea vasiformis. A,B: Sporangiophores, sporangia, and sporangiospores. C: Rhizoids. (From Watanabe, T. 1971. *Trans. Mycol. Soc. Jpn.*, 12:35–47. With permission.)

Syncephalastrum Schroeter

Krypt. Fl. Schles. 3(2):217, 1886.
Type species: *S. racemosum* (Cohn) Schroeter

Syncephalastrum racemosum (Cohn) Schroeter

References: Benjamin 1966; Domsch et al. 1980; Watanabe 1975a; Watanabe et al. 1986a; Zycha et al. 1969.

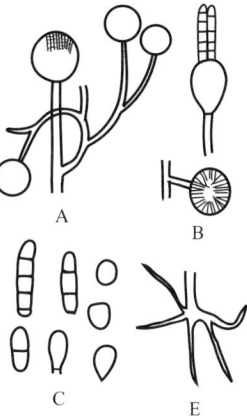

Morphology: Sporangiophores erect, simple or branched, bearing merosporangia on the terminal vesicles, rarely rhizoidal basally; vesicles hyaline or pale brown, globose or subglobose, echinulate: merosporangia pale brown, cylindrical, with 1–8 sporangiospores in a row. Sporangiospores hyaline, pale brown, ovate, globose; boat-shaped, cuneiform, or various in shape; occasionally truncate or sharpened, often with frills, only slightly spiny or verruculose. Heterothallic.

Dimensions: Sporangiophores ca. 3.5 mm tall; 6.5–12 µm wide basally. Vesicle + merosporangia 40–110 µm in diameter: vesicles 12–50 (-62.5) µm in diameter. Merosporangia up to 27.5 µm long. Sporangiospores: globose ones 2.4–4.4 (5.5) µm in diameter; ovate ones 3.6–5.1 (-6.3) × 3.4–4.2 µm.

Material: 72-X156 (= ATCC 32330, Sugarcane roots, Taiwan, ROC); 84-529 (Japanese red pine seed, Nagano, Japan); 86-62, -63 (Japanese cypress seed, Nagano, Japan).

Remarks: Another species, *S. verruculosum* Misra, is included in this genus.

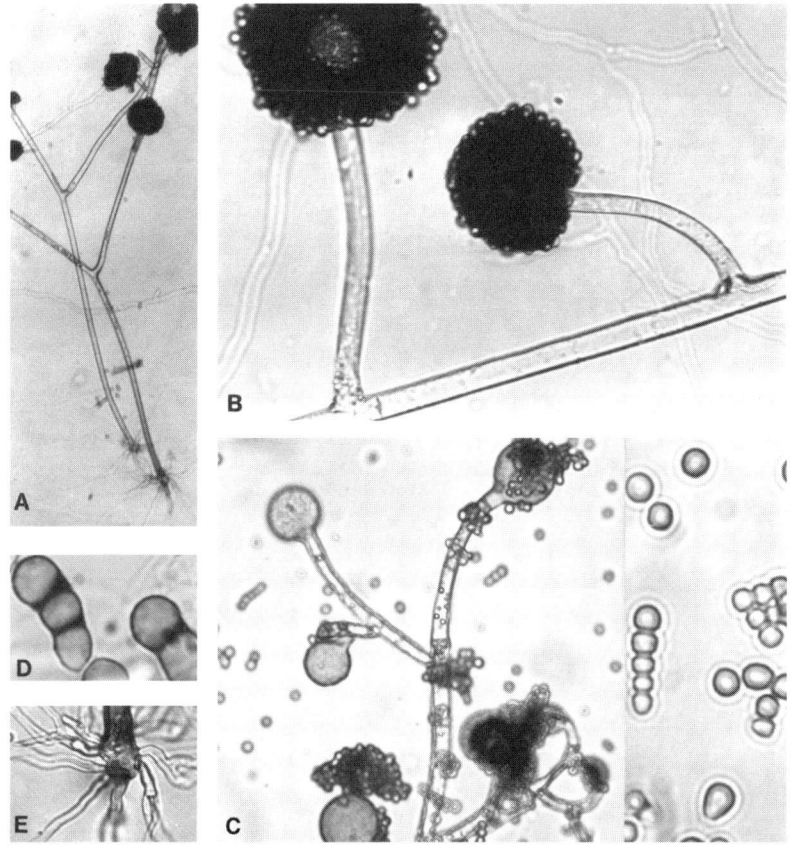

Syncephalastrum racemosum. A: Sporangiophores and sporangia. B,C: Vesicle and sporangiospores. D: Merosporangia. E: Rhizoids. (From Watanabe, T. 1975a. *Trans. Mycol. Soc. Jpn.*, 16:18–27. With permission.)

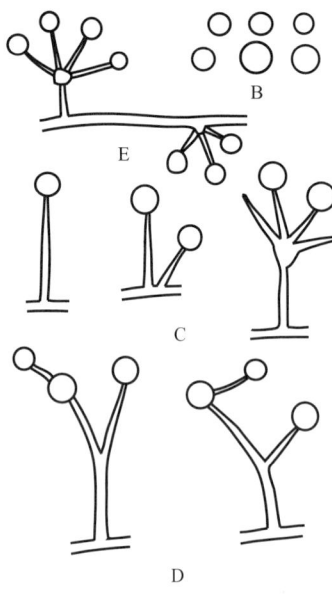

Umbelopsis Amos & Barnett

Mycologia 58:805, 1966.
Type species: *V. versiforme* Amos & Barnett

Umbelopsis nana (Linn.) v. Arx

Synonym: *Mortierella nana* Linn.; *U. versiformis* Amos & Barnett

References: Amos and Barnett 1966; Arx 1982; Kendrick et al. 1994; Watanabe et al. 2001.

Morphology: Sporangiophores hyaline, erect, short, forming vesicles (globose inflations) at the middle, elongating 2 to more than 12 branches with the terminal single sporangia, often proliferated from the sporangia. Sporangiospores hyaline, 1-celled, deciduous.

Dimensions: Sporangiophores 1–50 µm long from base to primary vesicle, 1.6–4 µm wide, 10–54 µm long from primary vesicles to sporangia. Sporangia (3-) 4–7 (-10) µm in diameter. Sporangiospores 2.5–3 (-4.5) µm in diameter.

Material: 99-454 (= MAFF 238015, Forest soil, Takabotchi, Shiojiri, Nagano, Japan), 99-481 (= MAFF 238018, Forest soil, Kataoka, Shiojiri, Nagano, Japan); 99-485 (= MAFF 238019, Forest soil, Ina, Nagano, Japan).

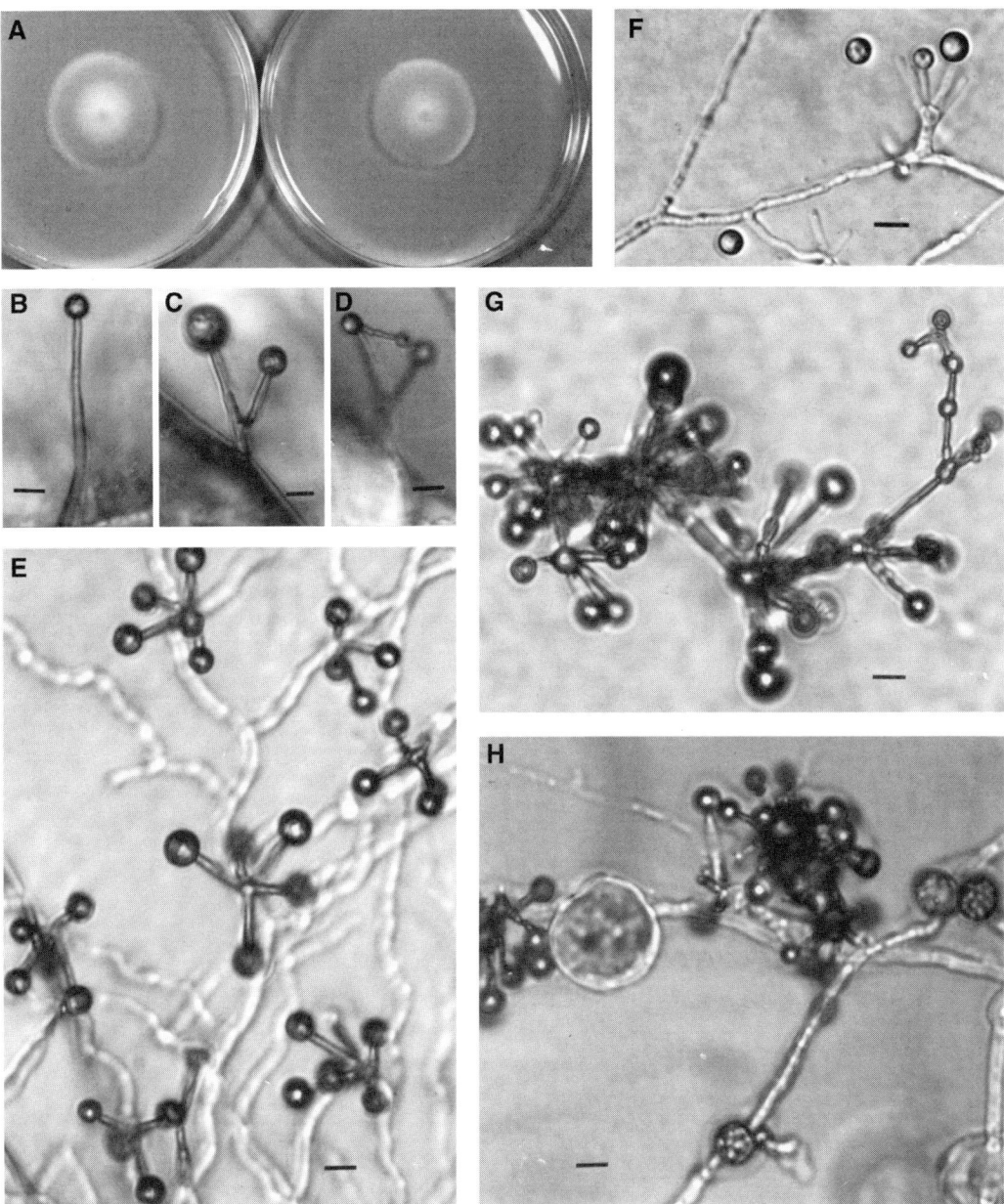

Umbelopsis nana. A: Six-day-old colonies of two isolates on PDA. B–E: Sporangiophores and sporangia. F: Sporangiophores after detachment of sporangiospores, and sporangiospores. G,H: Proliferated and aggregated sporangiophores and sporangia with intercalary chlamydospores and vacant hyphal swelling (H). (From Watanabe, T. et al. 2001b. *Mycoscience.* In Press. With permission.)

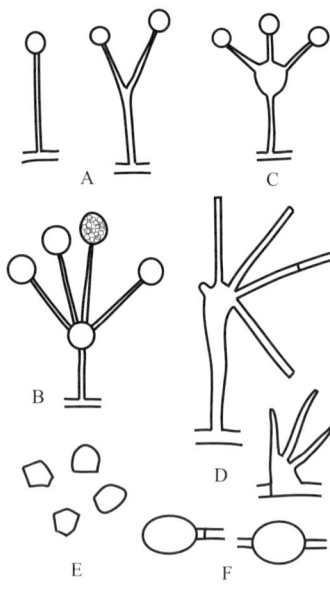

Umbelopsis vinacea (Dixon-Stewart) v. Arx

Synonym: *U. multispora* T. Watanabe

References: Arx 1982; Watanabe 1977b; Watanabe et al. 2001.

Morphology: Sporangiophores hyaline, erect, forming vesicles (globose inflations) at the middle part, and subsequently with 3–6 verticillate branches, forming sporangia terminally, occasionally developing secondary and tertiary vesicles without forming sporangia. Chlamydospores brown, globose, ellipsoidal, barrel-shaped, thick-walled, granulate. Sporangiospores hyaline, 1-celled, angular with 4–7 edges.

Dimensions: Sporangiophores 5–42.5 (-230) µm long from base to primary vesicle, 2–4.5 µm wide; 10–65 × 2–4.5 µm long from primary vesicles to sporangia. Primary vesicles 7.5–17.5 µm in diameter. Sporangia 7.5–25 µm in diameter. Sporangiospores 2.5–3 (-4.5) µm in diameter. Chlamydospores 4.5–9.5 µm in diameter.

Material: 73-723 (= S15X23 = ATCC 38089 = CBS 236.82, Strawberry roots, Yasugi, Shimane, Japan); 99-463 (= MAFF 238017, Forest soil, Takabotchi, Shiojiri, Nagano, Japan).

Remarks: Colonies on PDA are nonaerial, pinkish orange with slight radiation. The fungus was also treated as one of Mitosporic (Deuteromycetous) fungi (Barnett and Hunter, 1987).

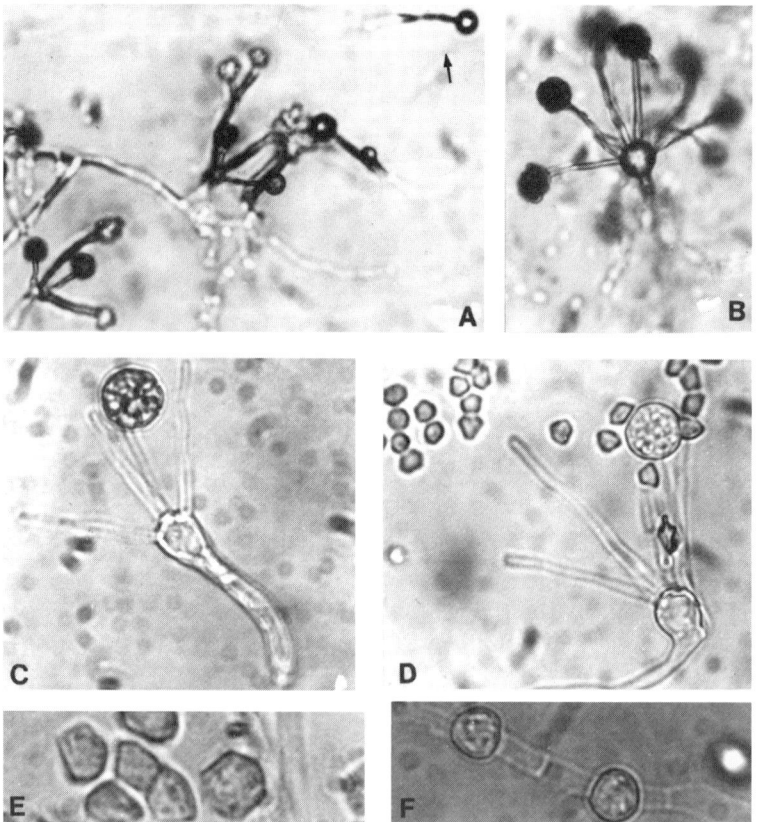

Umbelopsis vinacea. A: Sporangiophores and sporangia. B,C: Vesicles and sporangia. D: Sporangiophore and sporangiospores. E: Sporangiospores. F: Chlamydospores. (From Watanabe, T. 1977b. *Trans. Mycol. Soc. Jpn.*, 18:242–244. With permission.)

Zygorhynchus Vuill.

Bull. Soc. Mycol. Fr. 19:116, 1903.
Type species: *Z. heterogamus* Vuill.

Zygorhynchus moelleri Vuillemin

References: Hesseltine et al. 1959; Watanabe 1975a; Zycha et al. 1969.

Morphology: Sporangiophores hyaline, erect, simple or branched, twisted spirally or curved, bearing sporangia terminally. Sporangia brown, globose, minutely echinulate, columellae and remnant of sporangial wall left on dehiscence: columellae subglobose with collar. Sporangiospores hyaline, ellipsoidal, 1-celled. Sexual organs readily formed. Zygospores brown, half-globose, warty with pigmented protuberances, bearing single conspicuously developed suspensors. Chlamydospores globose.

Dimensions: Sporangiophores 75–85 × 4.3–5.4 µm: branches 20–75 µm long. Sporangia 29.1–56.4 µm in diameter: columellae 8.5–19.5 × 7.2–13.4 µm. Sporangiospores 3.4–4.4 × 2.4–2.7 µm. Zygospores (26-) 35–45 µm in diameter: suspensors 29.1–38.9 µm long, 12.1–17.1 µm wide. Chlamydospores ca. 10 µm in diameter.

Material: 72-X11 (= ATCC 32331, Sugarcane roots, Taiwan, ROC); 74-55 (Grassland soil, Sapporo, Hokkaido, Japan).

Notes: The genus name is also spelled as *Zygorrhynchus*.

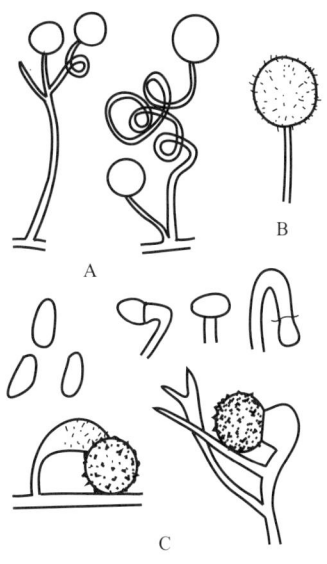

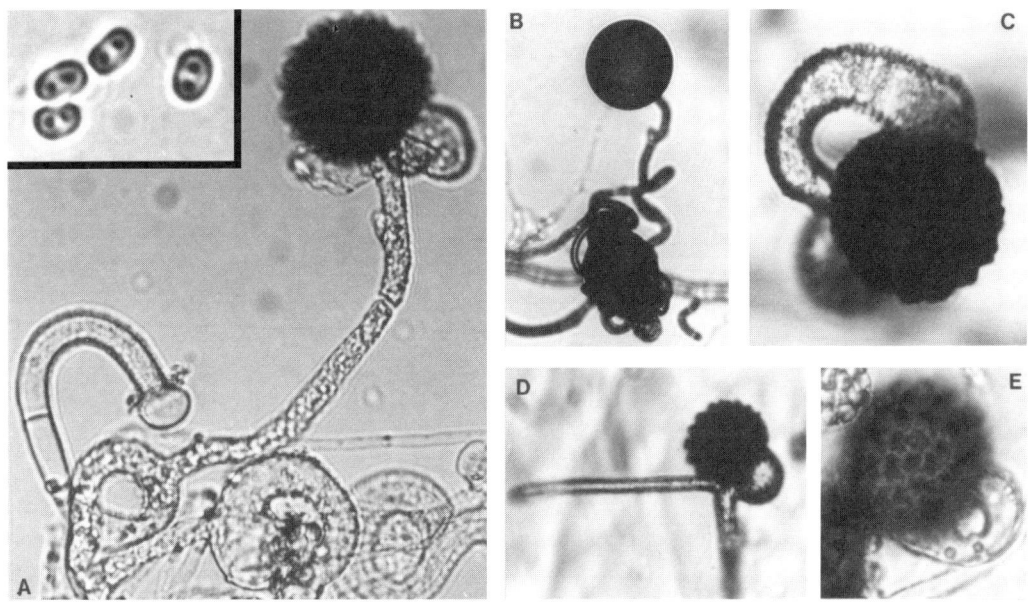

Zygorhynchus moelleri. A: Sporangiophores, sporangiospores, and zygospore. B: Sporangiophores and sporangia. C–E: Zygospores. (From Watanabe, T. 1975a. *Trans. Mycol. Soc. Jpn.*, 16:18–27. With permission.)

MORPHOLOGY OF SOIL FUNGI

Ascomycotina

Achaetomium Rai, Tewari & Mukerji

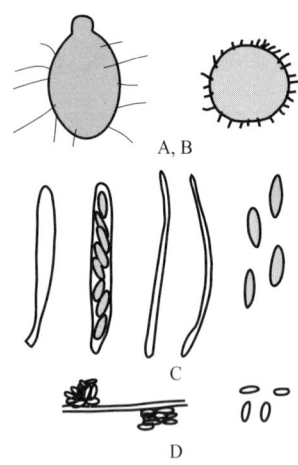

Can. J. Bot. 42:693, 1964.
Type species: *A. globosa* Rai, Tewari & Mukerji.

Achaetomium sp. 99-473

References: Cannon 1986; Rai et al. 1964.

Morphology: Perithecia black or dark brown, flask-shaped or subglobose, ostiolate, covered with soft hairs. Asci hyaline, cylindrical or clavate, 8-ascosporous in one row. Paraphyses hyaline, cylindrical, curved. Ascospores ellipsoidal or fusiform, brown or dark green. Anamorph: *Aureobasidium*-like. Conidiophores lacking or very short. Conidia hyaline, cylindrical.

Dimensions: Perithecia 150–330 × 150–200 μm; Ostiolar regions 50 μm long, 60–80 μm wide. Asci 68–84 × 5–7 μm. Paraphyses nearly 50–54 μm long, 2 μm wide. Ascospores 10–12 × 2.8–4 μm. Conidia 3–4.4 × 1.2–1.6 μm.

Material: 99-473 (Forest soil, Agematsu, Shiojiri, Nagano, Japan).

Remarks: The genus *Achaetomium* is characterized in forming tomentose perithecia, evanescent unitunicate asci, and 1-celled phaeosporus ascospores with one polar pore at each end and, therefore, this fungus partially resembles *Achaetomium*, but differs from it in forming nonevanescent asci and paraphyses together with *Aureobasidium* anamorph. This fungus also resembles *Phaeotrichosphaeria* sp. 99-467, with cylindrical unitunicate asci, phaeosporous ascospores, and paraphyses, together with *Aureobasidium* anamorph, but differs from it in forming tomentose perithecia and dimensions of ascospores.

Ascomycotina 135

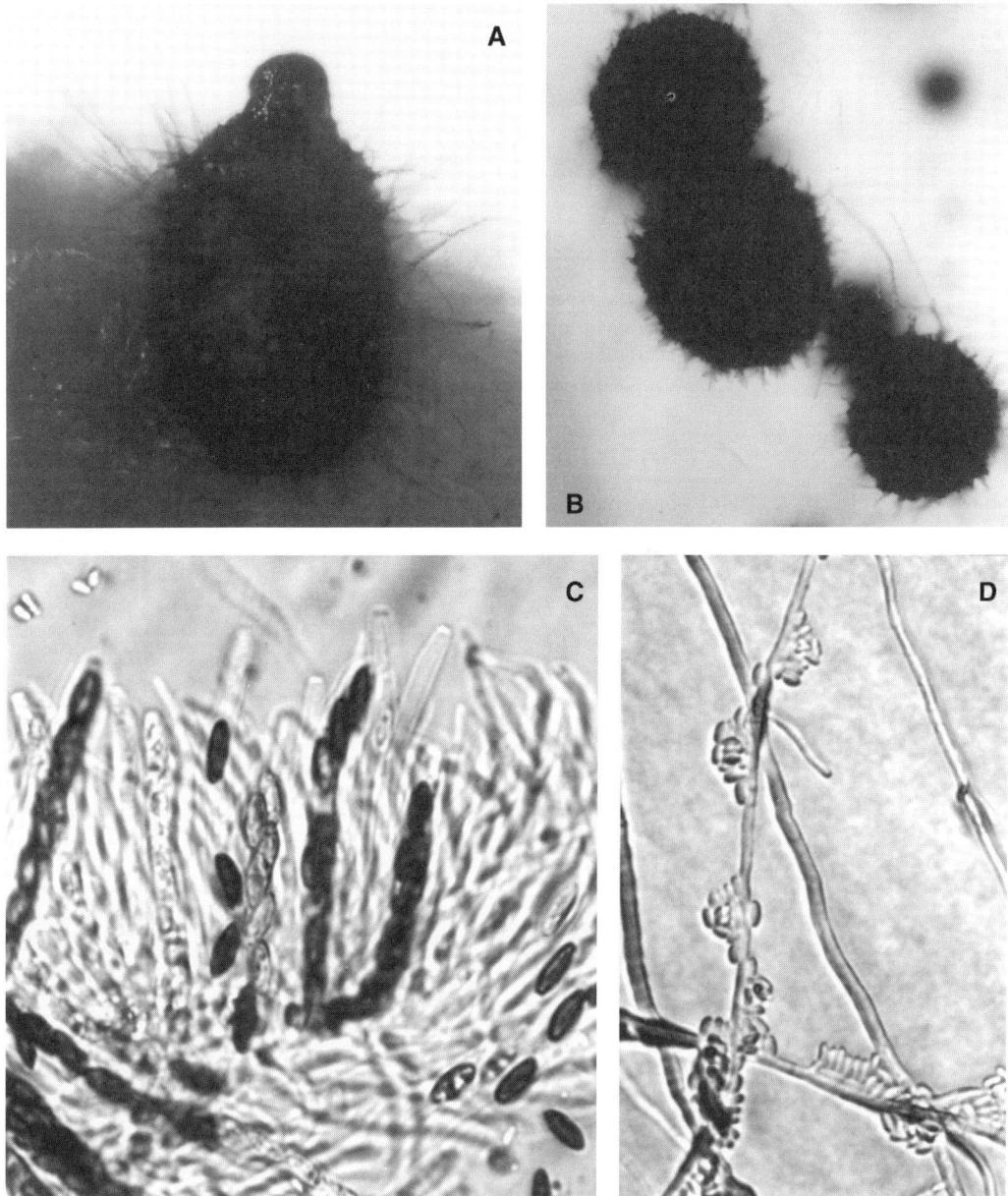

Achaetomium sp.   A,B: Erumpent (A) and embedded perithecia (B). C: Asci, paraphyses and ascospores. D: *Aureobasidium* anamorph.

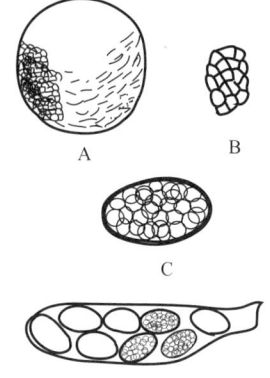

Anixiella Saito & Minoura

J. Ferment. Technol., Osaka 26:4, 1948.
Type species: *A. reticulispora* Saito & Minoura

Anixiella reticulata (Booth et Ebben) Cain

Synonym: *Gelasinospora reticulata* (Booth et Ebben) Cailleux

References: Cailleux 1971; Cain 1950, 1961; Udagawa 1965.

Morphology: Cleistothecia black or dark brown, globose, semitransparent or opaque; peridium thin, with soft skin, pseudoparenchymatous. Asci transparent, cylindrical, 8-ascosporus in 2 rows. Paraphyses unclear. Ascospores ellipsoidal, yellowish or dark brown, germ pores indistinct, rich in oil globules when immature.

Dimensions: Cleistothecia 197–300 μm in diameter. Asci ca. 150 × 30 μm. Ascospores 27.5–30 × 20–22.5 μm.

Material: 79-371, 79-372 (Fallow soil, Nishigahara, Tokyo, Japan).

Remarks: This fungus was first recorded by Saito and Minoura (1948). It was readily isolated from the nursery stock soil over 30 days after autoclaving.

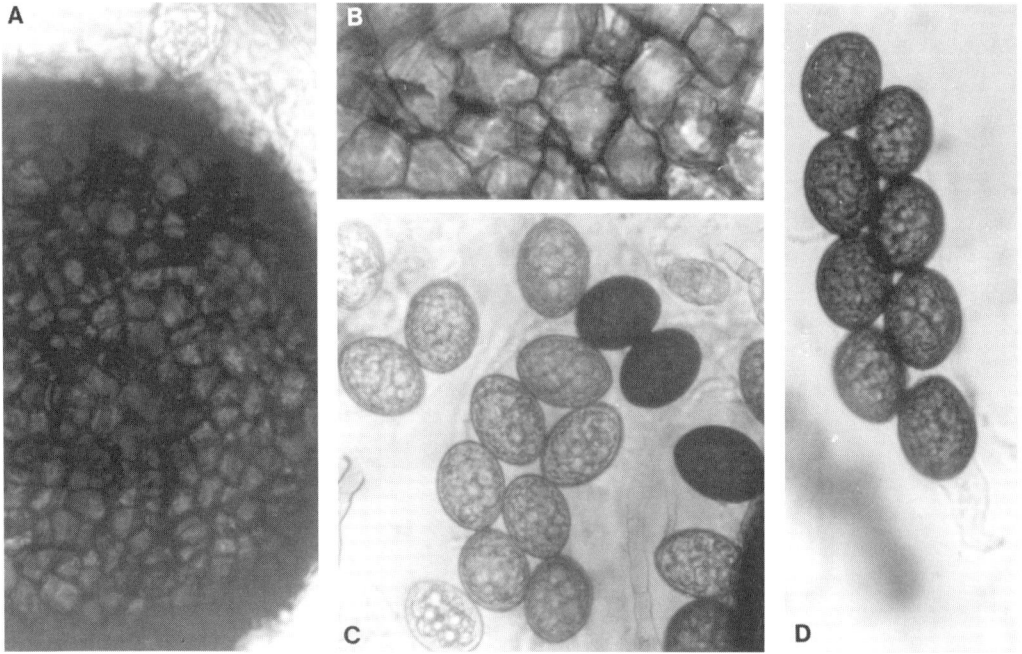

Anixiella reticulata. A: Part of clystothecium. B: Part of peridium. C: Ascospores. D: Ascospores in ascus.

Apiosordaria v. Arx & W. Gams

Nova Hedwigia 13:201, 1966.
Type species: *A. verruculosa* (Jensen) Arx & Gams

Apiosordaria verruculosa var. *maritima* (Apinis et Chesters) Arx et Gams

Anamorph: *Cladorrhinum* st.

References: Arx and Gams 1966; Mirza and Cain 1969; Stchigel et al. 2000; Udagawa and Furuya 1972; Watanabe 1971.

Morphology: Perithecia mostly embedded, black, flask-shaped, ovate, soft, apically papillate: peridium with soft skin, pseudoparenchymatous. Asci hyaline, clavate, 4-ascosporous. Ascospores 2-celled, composed of black, ovate, granulate, thick-walled cell, and hyaline, triangle cell. Paraphyses present.

Dimensions: Perithecia 210–340 × 220–290 µm: ostioles 30–35 µm in diameter. Asci 125–150 × 12.5–15 µm. Paraphyses 110–120 × 2.7–2.8 µm. Ascospores 31–35 × 13–14.8 µm: pigmented cells 20.5–22.5 × 13–14.8 µm: hyaline cells 10.5–12.5 × 7.2–7.8 µm.

Material: 69-300 (= ATCC 24150, Pineapple field soil, Okinawa, Japan).

Remarks: A key to the 16 soilborne species was provided by Stchigel et al. (2000).

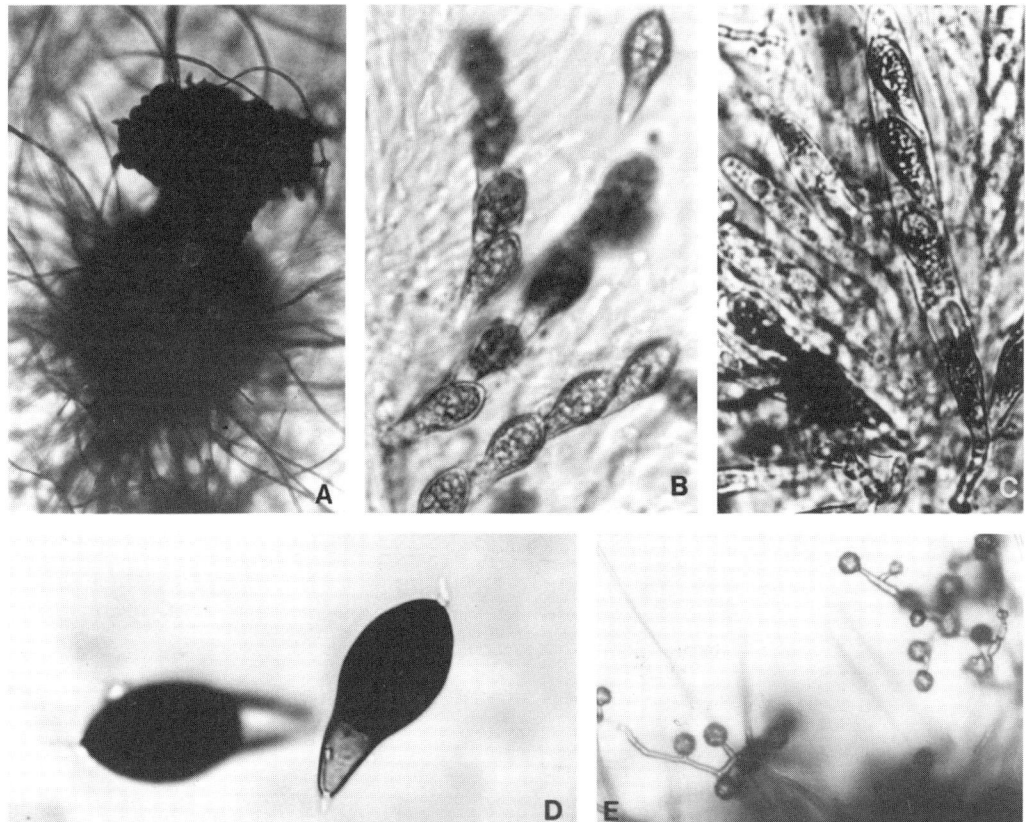

Apiosordaria verruculosa var. *maritima*. A: Perithecium. B,C: Asci and ascospores. D: Ascospores. E: *Cladorrhinum* st. (From Watanabe, T. 1971. *Trans. Mycol. Soc. Jpn.*, 12:35–47. With permission.)

Chaetomium Kunze

Mykol. Hefte 1:16, 1817.
Type species: *C. globosum* Kunze

Key to Species

1.	Terminal hairs	simple 2
		branched 10
2.	Terminal hairs	straight or undulate 3
		coiled 6
3.	Perithecia	ellipsoidal 4
		vase-shaped 5
4.	Ascospores	limoniform .... *C. globosum*
		ellipsoidal-fusiform .... *C. virescens*
5.	Anamorph	*Humicola*-like, ascospores 6–7 × 5 μm .... *C. homopilatum*
		absent, ascospores 7–10 × 5–6 μm .... *C. torulosum*
6.	Terminal hair	coiled only apically .... *C. aureum*
		coiled all over 7
7.	Asci	cylindrical, ascospores in one row .... *C. brasiliense*
		clavate, ascospores not in one row 8
8.	Ascospores	spindle-shaped, over 12 μm long .... *C. fusiforme*
		lemon-shaped, under 12 μm long 9
9.	Asci	over 70 μm long .... *C. cochliodes*
		under 50 μm long .... *C. spirale*
10.	Terminal hairs	all branched 11
		simple or branched 12
11.	Terminal hairs	branching repeatedly, curved .... *C. reflexum*
		dichotomously branched .... *C. erectum*
12.	Ascospores	mainly ovate .... *C. dolichotrichum*
		lemon-shaped .... *C. funicola*

Chaetomium aureum Chivers

Synonym: *C. cupreum* Ames

References: Ames 1949, 1963.

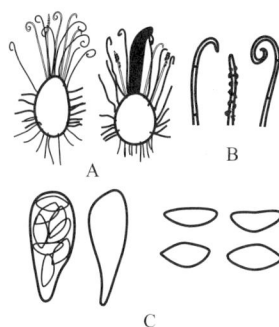

Morphology: Perithecia black, subglobose, ovate, covered with terminal hairs in the upper surface, rhizoidal basally: terminal hairs yellowish green, brown or reddish brown, rough superficially, septate, curled loosely at the apex or straight, homogeneous in width. Asci hyaline, clavate, with foot cell, evanescent, 8-ascosporous. Ascospores dark green, yellowish brown, spindle- or boat-shaped.

Dimensions: Perithecia 110–200 × 80–160 μm: terminal hairs over 130 μm long: 3–3.8 μm wide. Asci (23.7-) 37.5–40 × (-10) 11.2–12.5 μm. Ascospores 9.7–11.3 (-15) × 5–5.3 (-6.3) μm.

Material: 69-296, 69-305 (Pineapple field soil, Okinawa, Japan); 70-100 (Snap bean seed, Shinsyu-shinmachi, Nagano, Japan).

Remarks: Reddish brown terminal hairs and high frequency of boat-shaped ascospores may characterize *C. cupreum* (Arx et al., 1986).

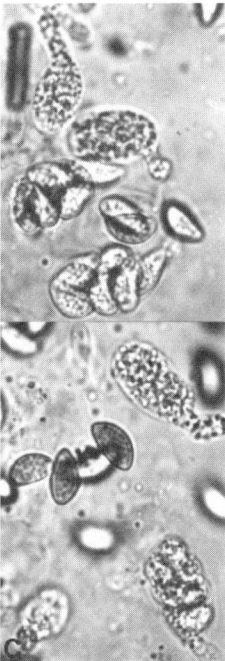

Chaetomium aureum. A: Part of perithecium. B: Part of terminal hairs. C: Asci and ascospores.

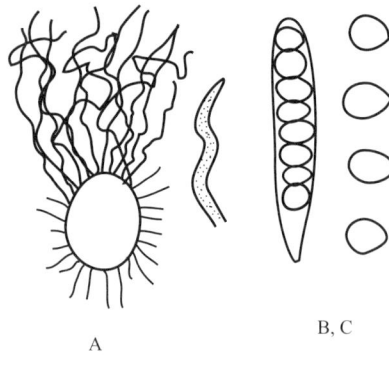

A   B, C

Chaetomium brasiliense Batista & Pontual

References: Ames 1963.

Morphology: Perithecia ovate, barrel-shaped, covered with terminal hairs in the upper surface: terminal hairs pale brown, curled spirally and curved, rough on the surface. Asci cylindrical, 8-ascosporous in one row. Ascospores dark green, ellipsoidal, apiculate at one end.

Dimensions: Perithecia ca. 100 × 80 μm: terminal hairs ca. 5.5 μm wide. Asci 47.5–48.8 × 6.2–7.5 μm. Ascospores 6.2–7.8 × 5.2–6.5 μm.

Material: 70-3, 70-23 (Snap bean seed, Shinsyu-shinmachi, Nagano, Japan).

Remarks: It was not confirmed whether asci are persistent or evanescent, but in literature, asci are rather persistent.

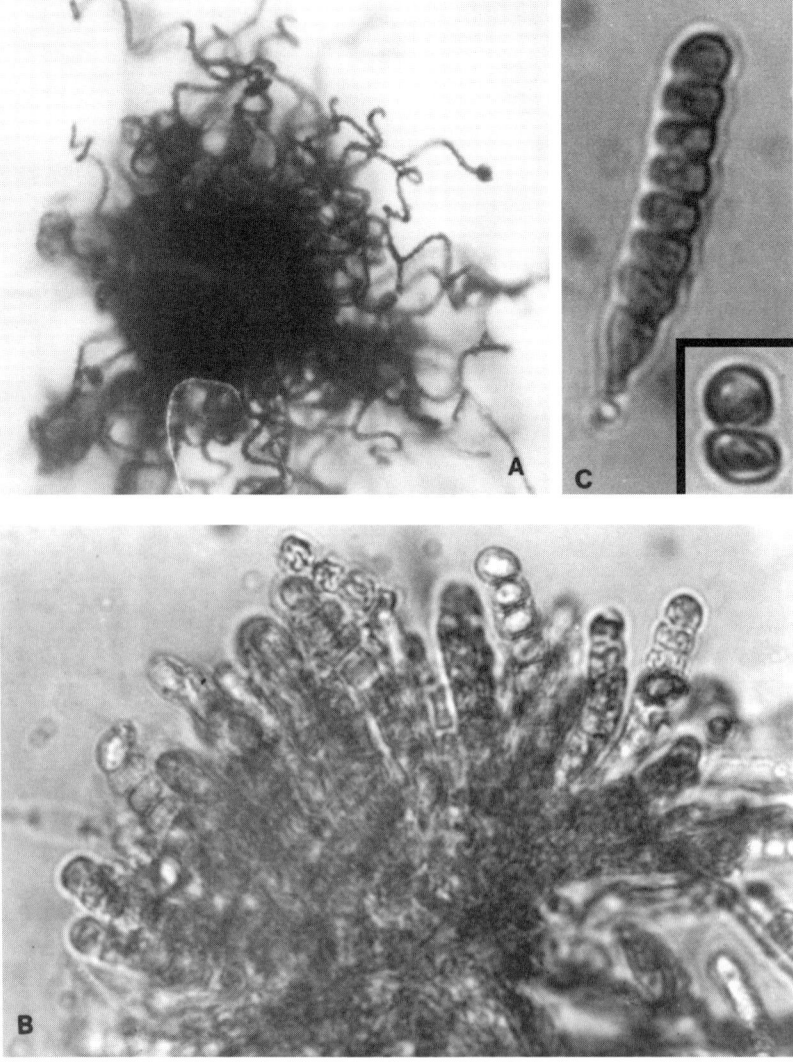

Chaetomium brasiliense.   A: Perithecium. B,C: Asci and ascospores.

Ascomycotina

Chaetomium cochliodes Palliser

References: Ames 1963; Watanabe et al. 1986a.

Morphology: Perithecia globose, subglobose, or barrel-shaped, covered with terminal hairs in the upper surface, rhizoidal basally: terminal hairs dark yellowish green, straight, wavy, curved, loosely curled spirally often 4 times or undulate. Asci hyaline, clavate, 8-ascosporous. Ascospores brown, lemon-shaped, apiculate at both ends.

Dimensions: Perithecia 180–280 × 190–225 (-280) µm. Asci 72.5–110 (-135) × 8–20 (-25) µm. Ascospores 7.5–10 × 6–8 µm.

Material: 70-24, 70-29 (Snap bean seed, Shinsyu-shinmachi, Nagano, Japan); 84-487 (Japanese red pine seed, Higashichikuma, Nagano, Japan).

Remarks: This fungus may be treated as a synonym of *C. globosum* (Arx et al., 1986).

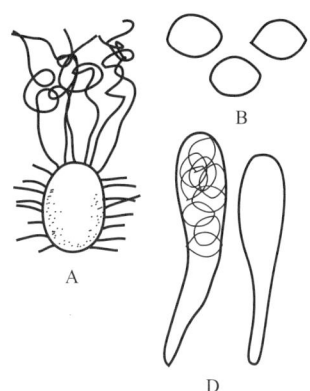

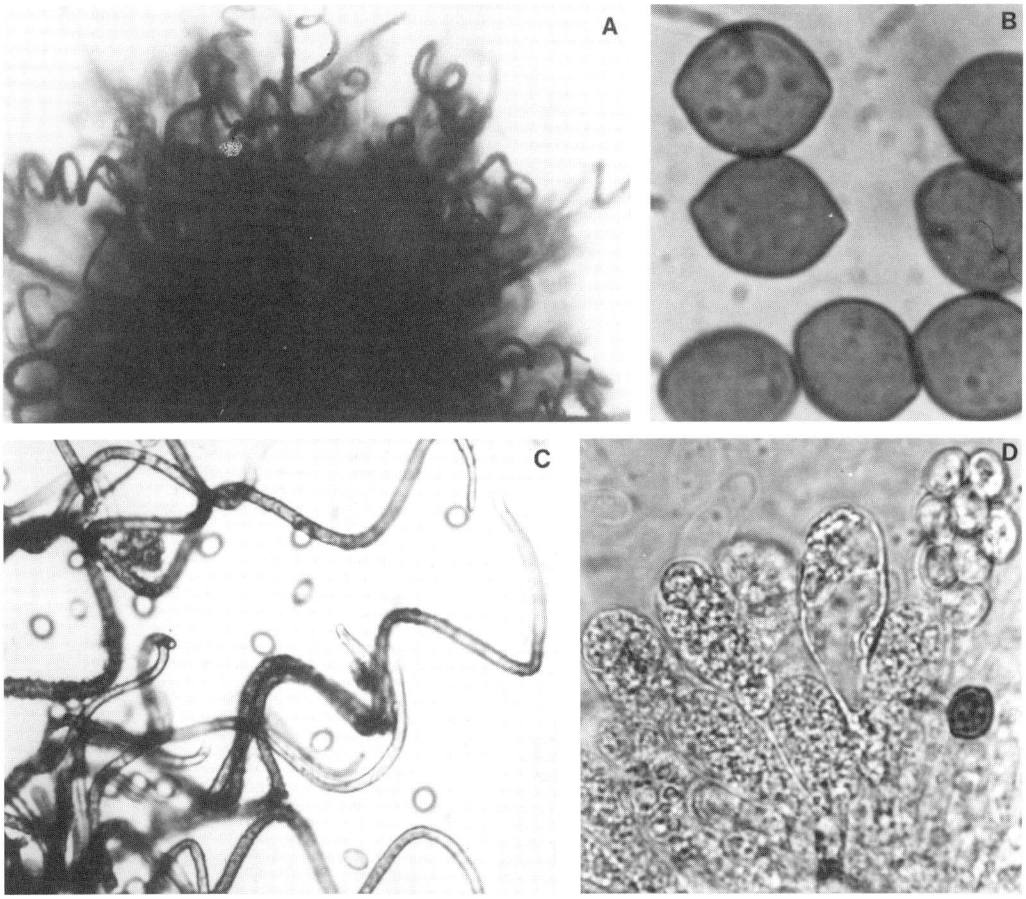

Chaetomium cochliodes. A: Terminal hairs on perithecium. B: Ascospores. C: Terminal hairs. D: Asci and ascospores.

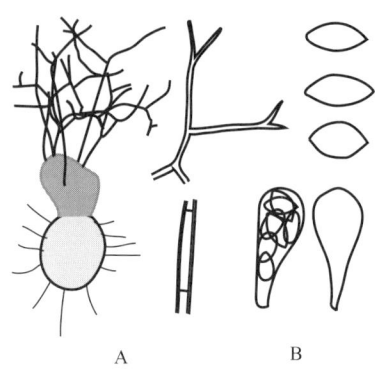

Chaetomium dolichotrichum Ames

References: Ames 1963; Greathouse and Ames 1945; Watanabe et al. 1986a.

Morphology: Perithecia black, ovate, subglobose or barrel-shaped, covered with terminal hairs on almost all the upper surface; terminal hairs dark brown, simple, straight, or curved, branched dichotomously. Asci hyaline, clavate, 8-ascosporous. Ascospores pale brown, ovate or ellipsoidal, apiculate at one or both ends.

Dimensions: Perithecia 70–110 (-125) × 65–100 (-110) μm: terminal hairs 2.5–5 μm wide. Asci (25-) 27.5–39.5 × 6.2–8.8 μm. Ascospores 4.5–6 (-7.5) × 2.5–3.8 (-4.5) μm.

Material: 70-104 (Snap bean seed, Shinsyu-shinmachi, Nagano, Japan), 83-26 (Japanese black pine seed, Kumano, Mie, Japan); 84-483, 84-494 (Japanese red pine seed, Higashichikuma, Nagano, Japan); 85-67, 85-81 (Flowering cherry seed, Hachioji, Tokyo, Japan); 86-35 (Japanese cedar seed, Kasama, Ibaraki, Japan); 86-36 (Japanese cedar seed, Nakashinkawa, Toyama, Japan).

Remarks: This fungus may be treated as a synonym of *C. funicola* Cooke (Arx et al., 1986).

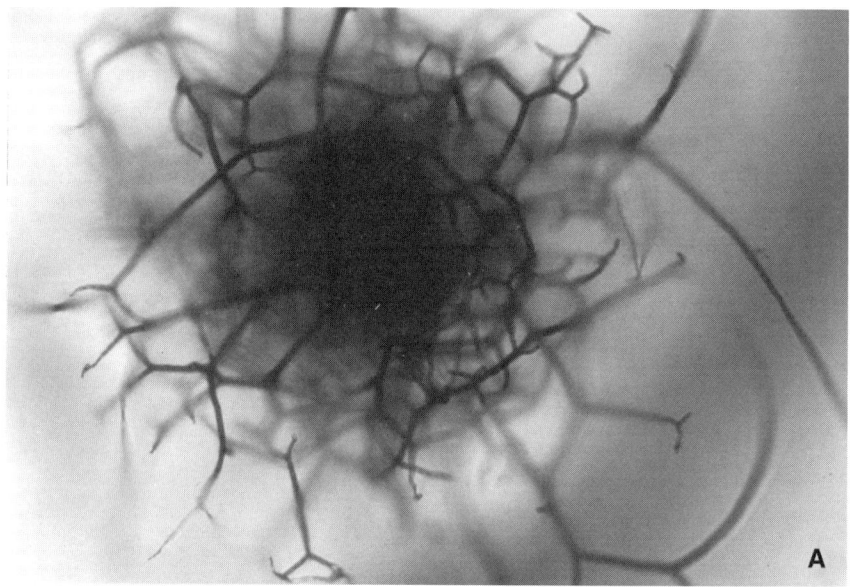

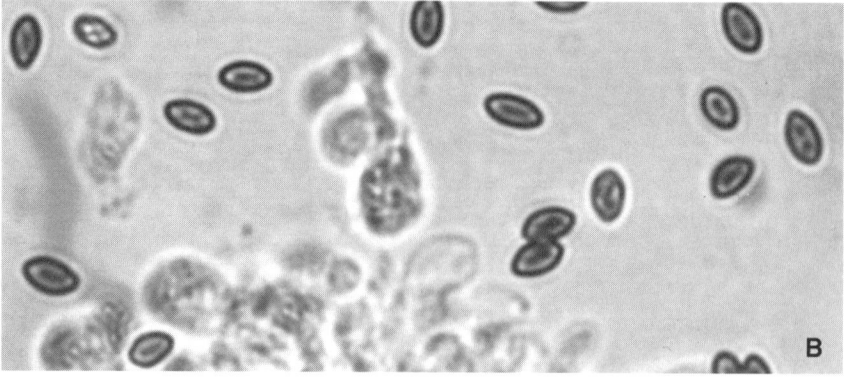

Chaetomium dolichotrichum.   A: Perithecium and terminal hairs. B: Asci and ascospores.

Chaetomium erectum Skolko & Groves

References: Ames 1963; Watanabe et al. 1986a.

Morphology: Perithecia globose, black, covered with terminal hairs in the upper surface: terminal hairs dark brown, curved, branched dichotomously and repeatedly. Asci hyaline, broadly ellipsoidal, tapering and sharpened basally, 8-ascosporous. Ascospores dark green, olive-colored, ellipsoidal, occasionally apiculate in both ends.

Dimensions: Perithecia 80–125 (-150) μm in diameter: terminal hairs 2.5–5 μm wide. Asci 13.7–15 × 8.7–10 μm. Ascospores 5–6.5 × 3.7–4.4 μm.

Material: 69-309 (Pineapple root, Okinawa, Japan); 84-489 (Japanese red pine seed, Higashichikuma, Nagano, Japan); 86-38 (Japanese black pine seed, Kyoto, Japan).

Remarks: This fungus differs from *C. reflexum* Skolko & Groves in having rather bluntly curved terminal hairs and broad ascospores.

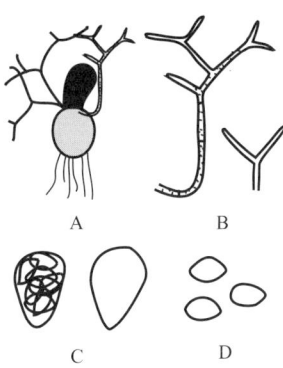

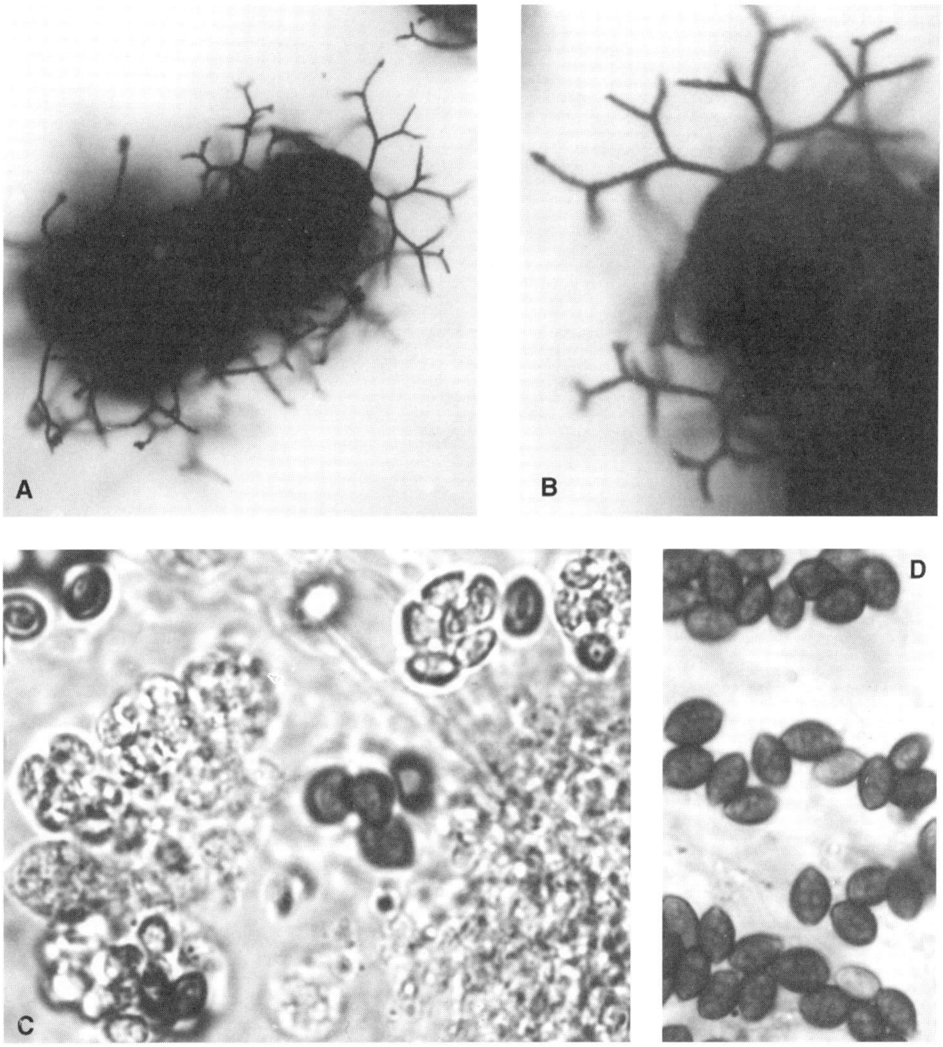

Chaetomium erectum. A,B: Perithecia and terminal hairs. C: Asci and ascospores. D: Ascospores.

Chaetomium funicola Cooke

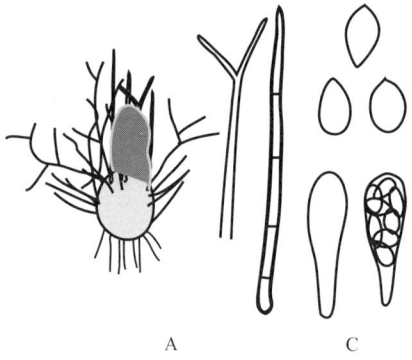

Synonym: *C. africanum* Ames

References: Ames 1963; Arx et al. 1986; Watanabe et al. 1986a.

Morphology: Perithecia brown, subglobose or ovate, covered with terminal hairs in the upper surface, rhizoidal basally. Terminal hairs brown, of two kinds, straight, seta-like and curved, branched dichotomously. Asci hyaline, clavate with foot cells, 8-ascosporous, evanescent. Ascospores grayish green, olive-colored, ellipsoidal or lemon-shaped, apiculate at one or both ends.

Dimensions: Perithecia 90–250 × 55–150 (-250) µm: terminal hairs 3.5–5 µm wide. Asci 20–37.5 × 6.2–10 µm. Ascospores 5–6.3 × 2.8–4.5 µm.

Material: 69-315 (Pineapple field soil, Okinawa, Japan); 70-93, 70-101 (Snap bean seed, Shinsyu-shinmachi, Nagano, Japan); 84-484 (Japanese red pine seed, Higashichikuma, Nagano, Japan); 84-490, 84-573 (Japanese red pine seed, Kamiina, Nagano, Japan).

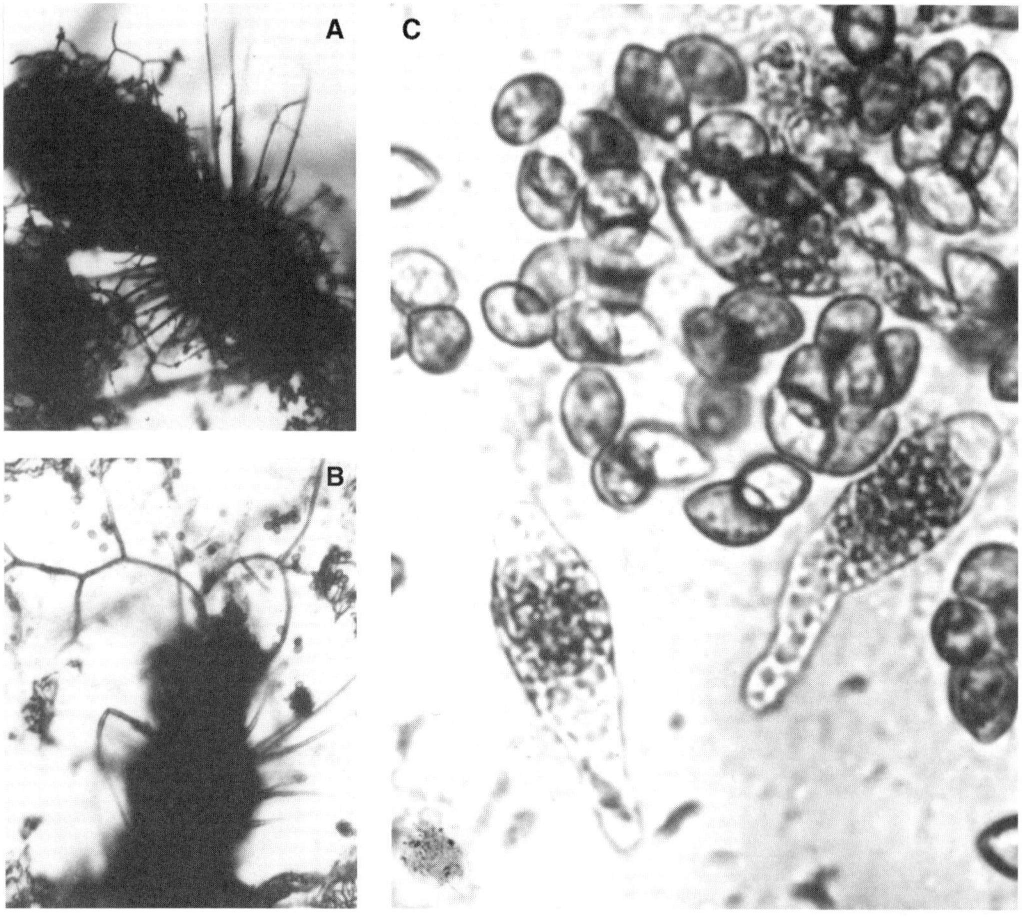

Chaetomium funicola. A,B: Perithecia and terminal hairs. C: Asci and ascospores.

Chaetomium fusiforme Chivers

References: Ames 1963; Arx et al. 1986.

Morphology: Perithecia brown or black, ovate or subglobose, almost covered with terminal hairs on the upper surface: terminal hairs brown, undulate or loosely curled with 1–3 spirals. Asci hyaline, clavate, 8-ascosporous, evanescent. Ascospores olive-colored, fusiform or ellipsoidal, occasionally with one long furrow.

Dimensions: Perithecia 170–240 × 120–220 µm: terminal hairs ca. 200 µm long, 2.5–3.8 µm wide. Asci 35–52.5 × 13.7–16.3 µm. Ascospores 12.5–14 × 7.5–8.8 µm.

Material: 69-306, 69-276 (Pineapple field soil, Okinawa, Japan).

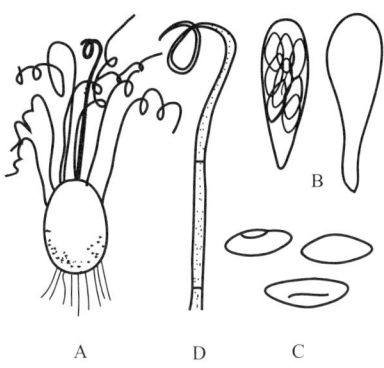

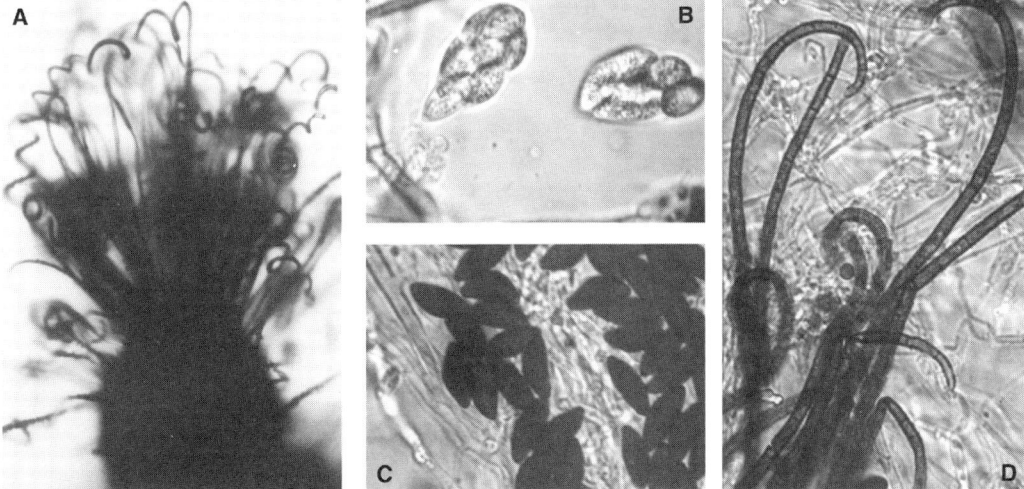

Chaetomium fusiforme. A: Perithecia and terminal hairs. B: Asci and ascospores. C: Ascospores. D: Part of the terminal hairs.

Chaetomium globosum Kunze : Fries

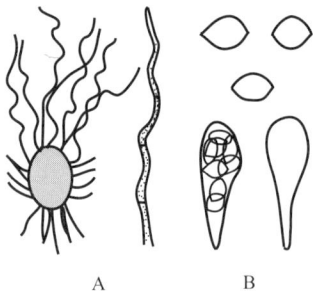

References: Ames 1963; Watanabe et al. 1986a.

Morphology: Perithecia brown, subglobose, ellipsoidal, almost covered with terminal hairs in the upper surface, rhizoidal basally: terminal hairs brown, rough on the surface, undulate, curved. Asci hyaline, clavate, 8-ascosporous, evanescent. Ascospores olive-colored, ellipsoidal or lemon-shaped, apiculate at both ends.

Dimensions: Perithecia (120-) 150–211 × (95-) 130–175 (-200) μm: terminal hairs 3.7–4.3 μm wide. Asci (40-) 45–75 × 12.5–13 (-19) μm. Ascospores (7.7-) 9–10.5 × 6.5–8.8 (-9.5) μm.

Material: 69-307, 69-314 (Pineapple field soil, Okinawa, Japan); 70-1180 (Pineapple field soil, Hachijo, Tokyo, Japan); 72-X77 (Sugarcane root, Taiwan, ROC); 83-25 (Japanese black pine seed, Kumano, Mie, Japan); 84-485, 84-493 (Japanese red pine seed, Higashichikuma, Nagano, Japan); 84-491 (Japanese red pine seed, Kamiina, Nagano, Japan); 86-34 (Japanese cypress seed, Higashichikuma, Nagano, Japan).

Remarks: This fungus differs from *C. cochliodes* Palliser in not forming coiled terminal hairs.

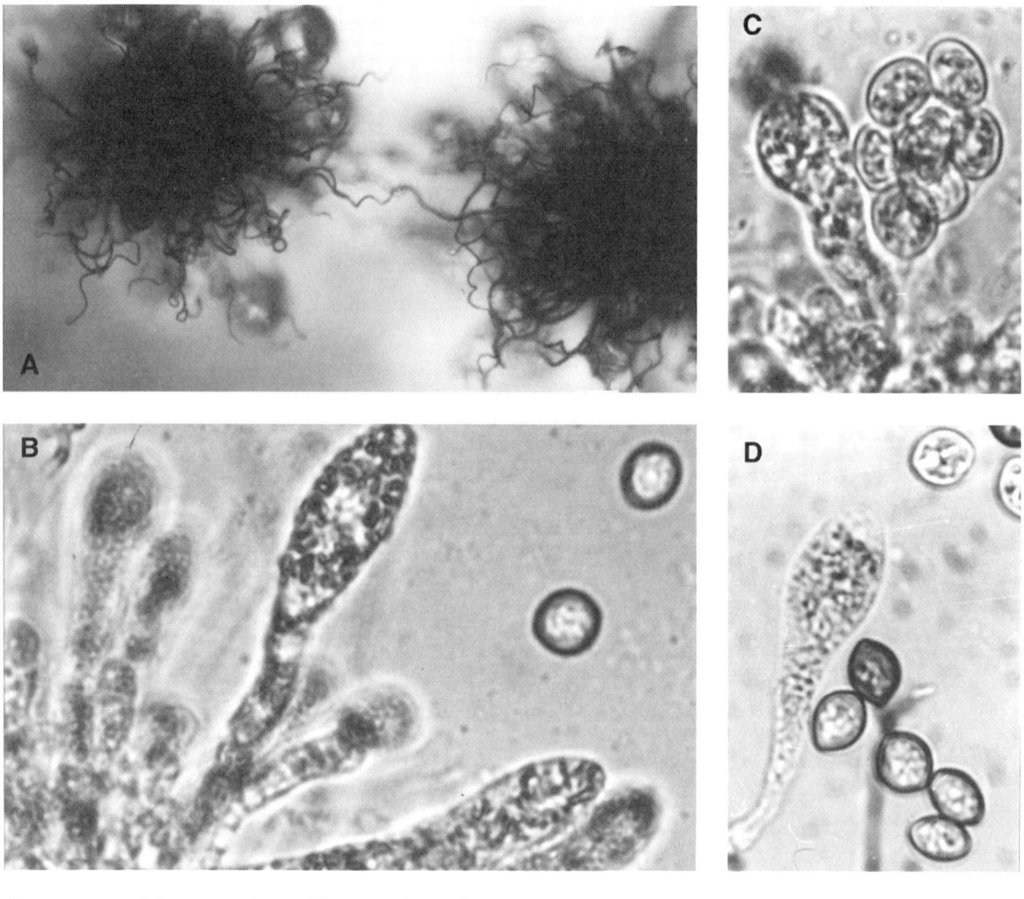

Chaetomium globosum. A: Perithecia and terminal hairs. B: Asci and ascospores.

Chaetomium homopilatum Omvik

References: Ames 1949; Arx et al. 1986; Skolko and Groves, 1953; Udagawa, 1960.

Morphology: Perithecia ellipsoidal, vase-shaped with well-developed neck, brown, covered with undulate terminal hairs. Terminal hairs pale yellowish brown to brown, whiplash-like, simple or rarely apically branched, roughened, crustaceous, septate. Asci hyaline, clavate, 8-ascosporous. Ascospores ellipsoidal, olive-colored, apiculate at both ends. Aleurioconidia dark brown, globose, smooth-walled.

Dimensions: Perithecia 100–225 × 80–90 µm. Terminal hairs 2.2–4 µm broad basally. Asci 48–54 × 16–20 µm. Ascospores 6–7 × 5–5.6 µm. Aleurioconidia mostly 7–9 µm in diameter.

Material: 98-21 (= MAFF 237997, Uncultivated soil, Tsukuba, Ibaraki, Japan).

Remarks: The ascospores are smaller in size than those of the original description (Ames, 1949).

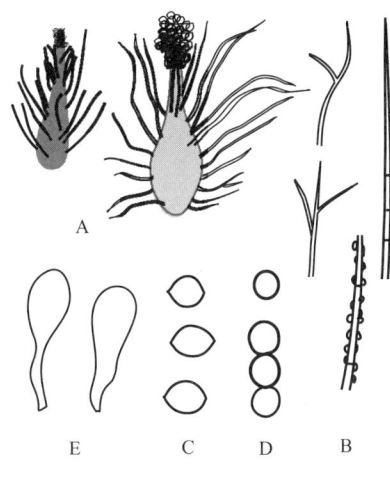

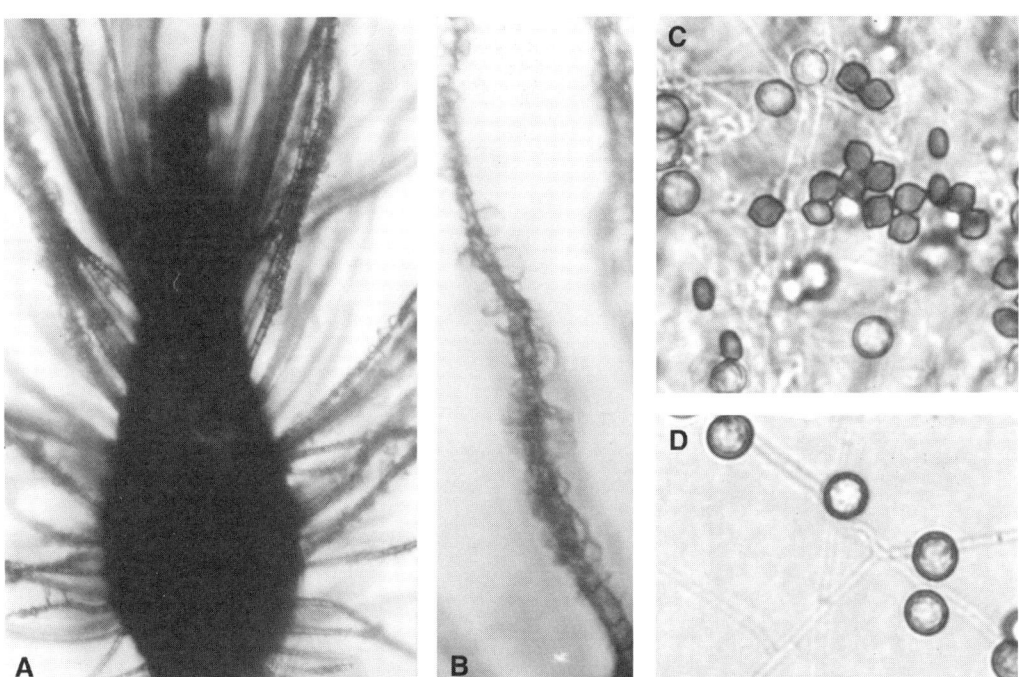

Chaetomium homopilatum. A: Perithecium. B: Crustaceous terminal hair. C: Ascospores and conidia. D: Conidia. E: Asci.

Chaetomium reflexum Skolko & Groves

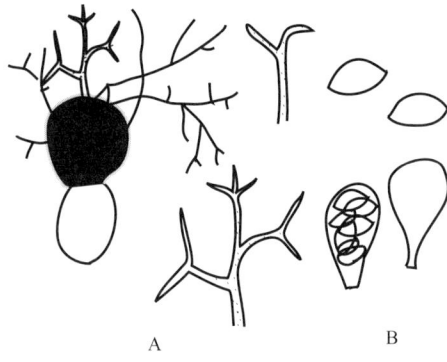

References: Ames 1963; Skolko and Groves 1948; Watanabe et al. 1986a.

Morphology: Perithecia black, globose, almost covered with terminal hairs in the upper surface: terminal hairs sharply curved dichotomously, branching repeatedly, rough on the surface. Asci hyaline, clavate, 8-ascosporous. Ascospores brown, olive-colored, ovate or ellipsoidal, apiculate at both ends.

Dimensions: Perithecia 40–100 µm in diameter: terminal hairs 2.7–4.8 µm wide. Asci ca. 20×10 µm. Ascospores (2.8-) 4.5–6.3×2.4–3.8 µm.

Material: 84-488 (Japanese red pine seed, Higashichikuma, Nagano, Japan).

Remarks: This fungus differs from *C. erectum* Skolko & Groves in having sharply curved terminal hairs and rather narrow ascospores.

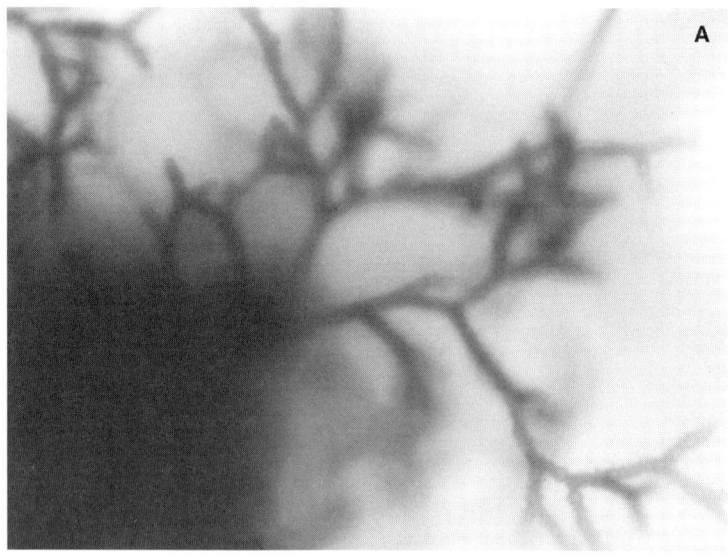

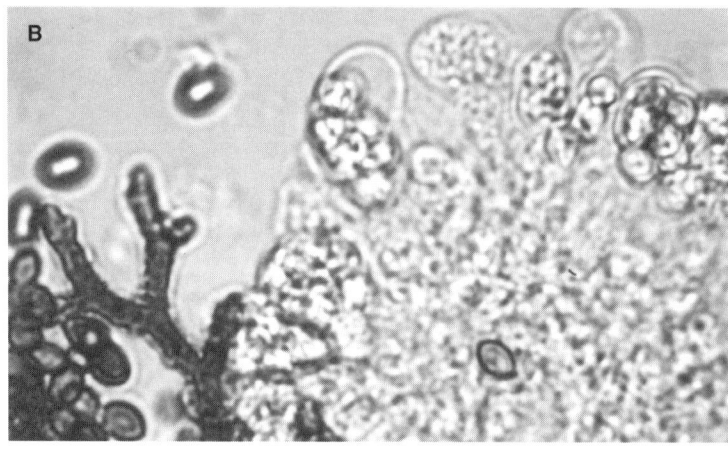

Chaetomium reflexum. A: Terminal hairs on perithecium. B: Terminal hair, asci, and ascospores.

Chaetomium spirale Zopf

References: Ames 1963.

Morphology: Perithecia dark brown, subglobose or barrel-shaped, covered with terminal hairs in the upper surface, rhizoidal basally: terminal hairs pale brown, rough on the surface, curled in 5–18 spirals or undulate. Asci hyaline, clavate, 8-ascosporus. Ascospores pale to grayish green, lemon- or spindle-shaped, apiculate at both ends.

Dimensions: Perithecia 200–350 × 150–290 µm: terminal hairs 3.7–6.3 µm wide. Asci 37.5–46.3 × 10–12.5 µm. Ascospores 8.5–10 (-12.5) × 6.2–8.8 µm.

Material: 69-277, 69-299 (Pineapple field soil, Okinawa, Japan); 70-45 (Snap bean seed, Shinsyu-shinmachi, Nagano, Japan); 73-176 (Strawberry root, Shizuoka, Japan).

Remarks: This fungus differs from *C. cochliodes* Palliser in having well-rolled spirals.

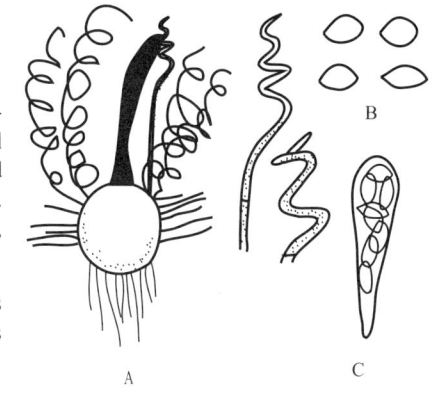

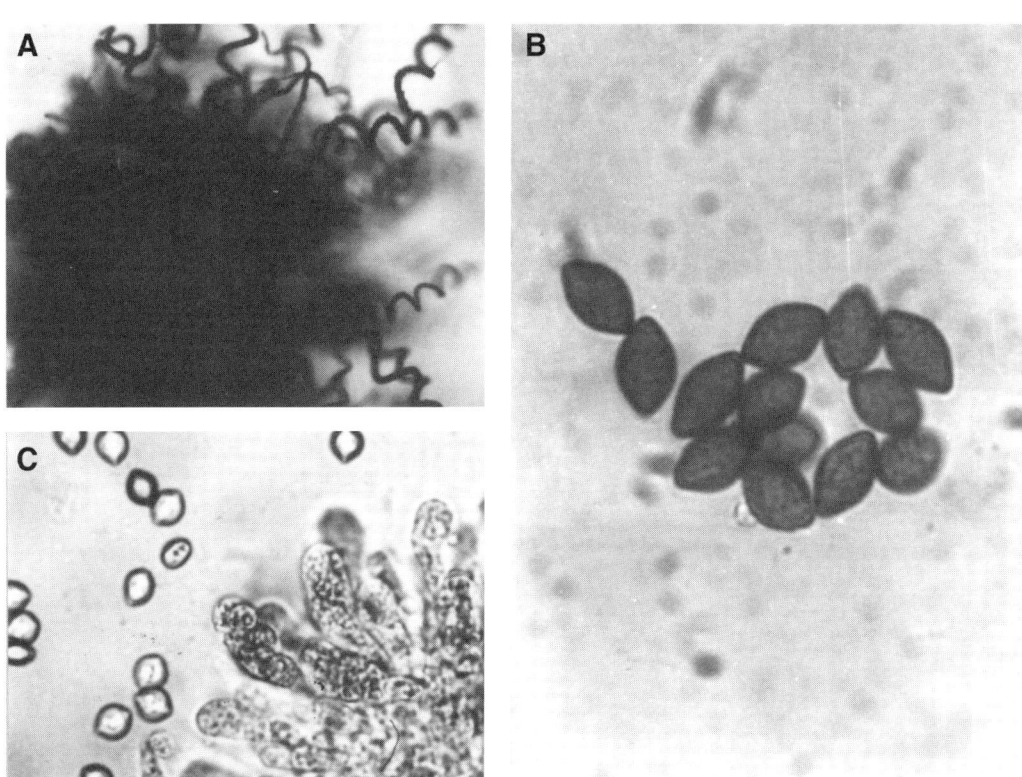

Chaetomium spirale. A: Perithecium and terminal hairs. B: Ascospores. C: Asci and ascospores.

Chaetomium torulosum Bainier

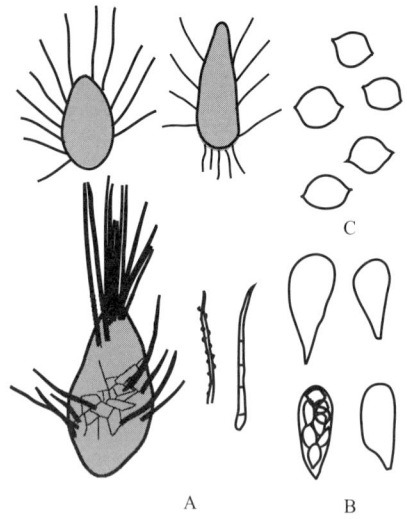

References: Ames 1949; Arx et al. 1986; Skolko and Groves, 1953; Udagawa, 1960.

Morphology: Perithecia black or dark brown, obclavate, obpyriform, flask-shaped, vase-shaped, or subglobose, ostiolate, covered with straight terminal hairs: terminal hairs crustaceous, rough, straight, undulate or slightly curved, very rarely apically branched. Lateral hairs sparse, simple, short. Asci hyaline, clavate or ovate, 8-ascosporous in two rows. Ascospores ellipsoidal or lemon-shaped, apiculate at both ends, pale grayish brown, usually guttulate.

Dimensions: Perithecia 70–175 × 50–90 µm: terminal hairs 70–100 × 4 µm. Asci 24–30 × 10 µm. Ascospores 7–10 × 3–6 µm.

Material: 99-460 (Forest soil, Agematsu, Shiojiri, Nagano, Japan).

Remarks: Among *C. torulosum* group forming the elongate flask-shaped perithecia bearing straight terminal hairs (less than 4 µm wide) and lemon-shaped (strongly umbonate at both ends) ascospores, including *C. seminudum* Ames and *C. homopilatum* Omvik, this fungus resembles *C. torulosum* most closely, but differs from it in forming smaller perithecia and ascospores with guttulation.

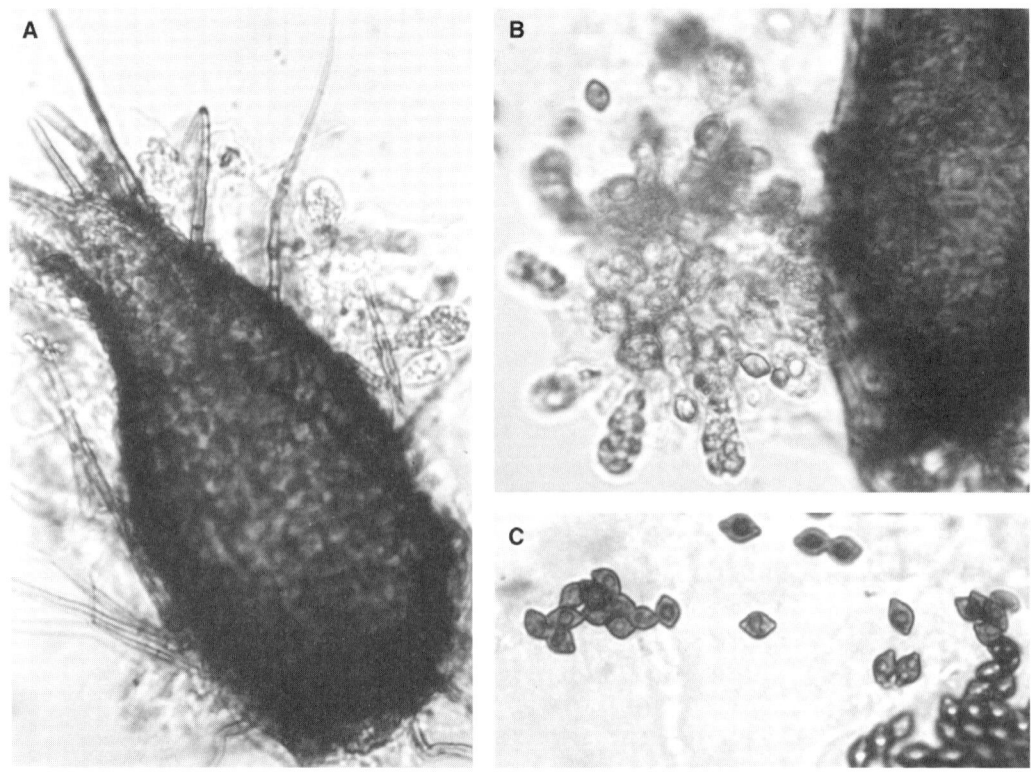

Chaetomium torulosum.   A: Perithecium. B: Asci and ascospores from crushed perithecium. C: Ascospores.

Chaetomium virescens (v. Arx) Udagawa

References: Udagawa 1980.

Morphology: Perithecia black, ellipsoidal, almost covered with terminal hairs all over the surface: terminal hairs dark brown, straight, rough superficially. Asci hyaline, clavate, pedicellate, 8-ascosporous, evanescent. Ascospores yellowish brown to dark green, ellipsoidal or boat-shaped, apiculate at both ends.

Dimensions: Perithecia 70–211 × 65–90 µm: terminal hairs over 140 µm long, 2–4 µm wide basally. Asci 24–34 × 10–12 µm. Ascospores 8.4–9 × 4–4.4 µm.

Material: 01-461 (Forest Soil, Chichijima, Ogasawara, Tokyo, Japan).

Remarks: This fungus resembles *C. aureum* morphologically, but production of dark grayish green pigment differentiates this species from *C. aureum* which produces reddish pigment.

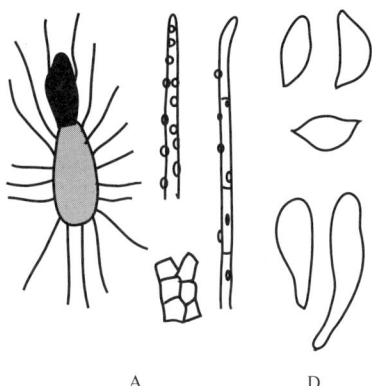

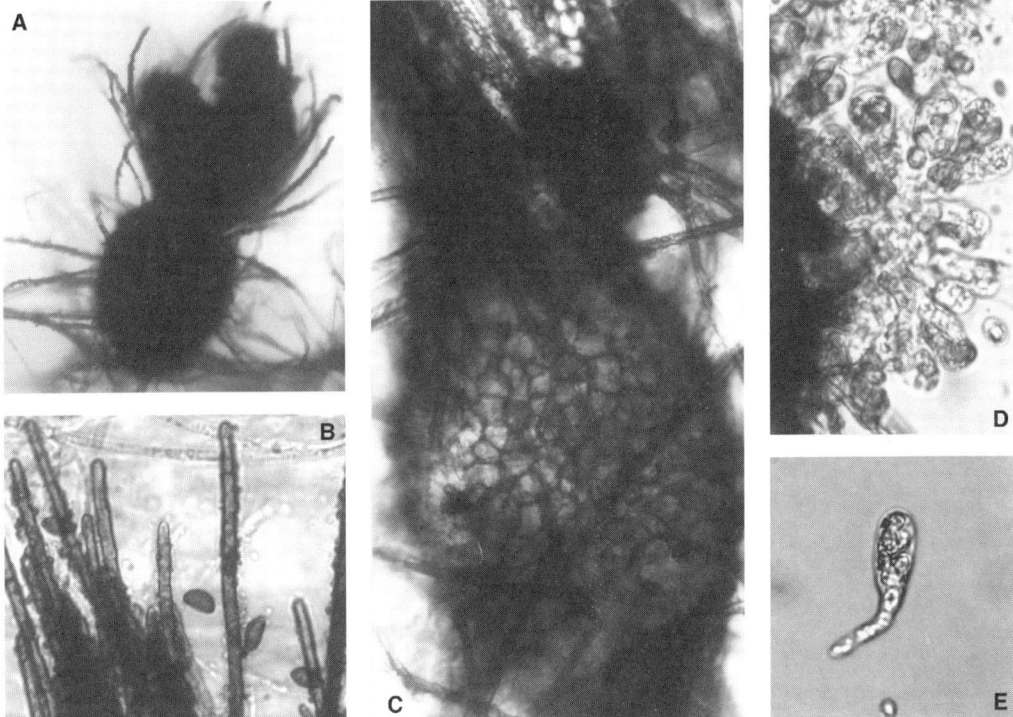

Chaetomium virescens. A,C: Perithecia. B: Terminal hairs and ascospores. D,E: Asci and ascospores.

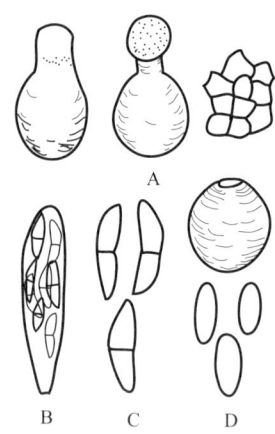

Didymella Sacc.

Syll. Fung. 1:545, 1882.
Type species: *D. exigua* (Niessl) Sacc.

Didymella effusa (Niessl) Sacc.

Anamorph: *Phoma* st.

References: Corbaz 1957; Sivanesan 1984; Watanabe et al. 1987a.

Morphology: Perithecia brown, globose or subglobose, necked conspicuously or ostiolate: peridium pseudoparenchymatous. Asci hyaline, cylindrical, bitunicate, 8-ascosporus. Ascospores hyaline, ellipsoidal, slightly curved, 2-celled, constricted at the septum.

Anamorph: pycnidial, conidia hyaline, cylindrical, 1-celled.

Material: 85-P51 (Paulownia root, Itapua, Paraguay).

Dimensions: Perithecia 75–125 µm in diameter: necks 40–55 × 40–50 µm. Asci 52–53 × 9–10 µm. Ascospores 12.5–15 × (2.7-) 4–5 µm. Conidia 5–8 × 2 µm.

Remarks: The subhyaline *Papulaspore*-like structures, 90–305 µm in diameter may be immature perithecia.

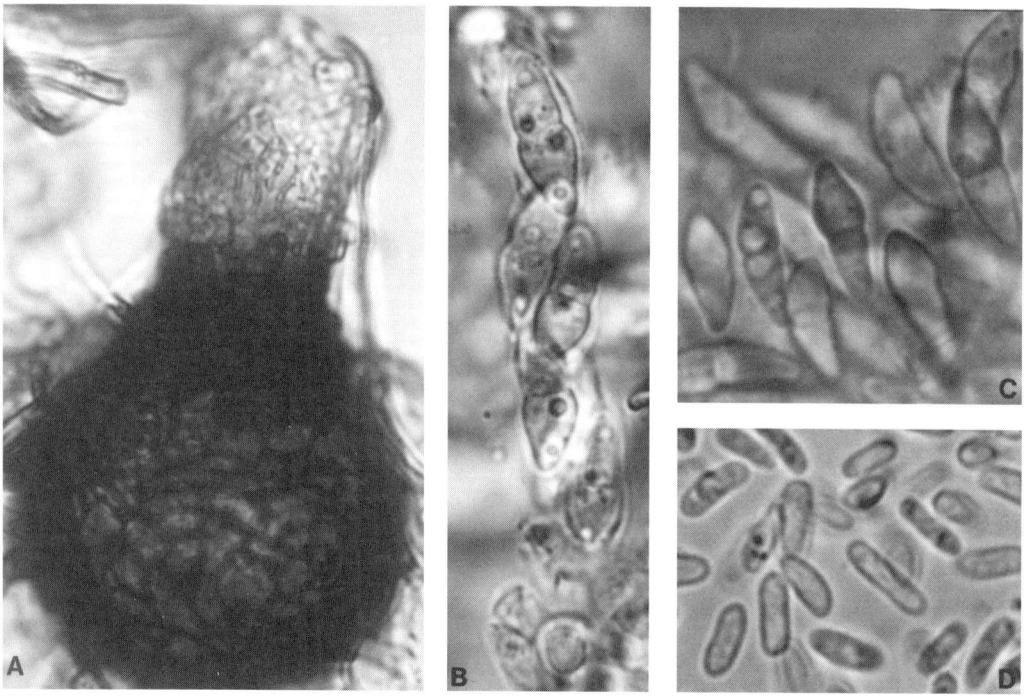

Didymella effusa.   A: Perithecium. B: Ascospores in ascus. C: Ascospores. D: Conidia in anamorph.

Eudarluca Speg.

Rev. Mus. La Plata 15:22, 1908.
Type species: *E. australis* Speg.

Eudarluca biconica Katumoto

References: Katumoto 1986; Watanabe 1989cd.

Morphology: Perithecia globose or subglobose, black. Asci clavate, bitunicate, 8-ascosporus in 2 rows. Paraphyses present. Ascospores hyaline, spindle-shaped, 2-celled, constricted at the septum, containing a few oil globules, bearing globose hyaline appendages in both ends.

Dimensions: Perithecia 115–290 μm in diameter. Asci 50–75 × 7–8 μm. Ascospores 15–20 × 3.5–4 μm: appendages 2–3 μm in diameter.

Material: 85-99, 85-100 (= IFO 32539, Flowering cherry seed, Hachioji, Tokyo, Japan).

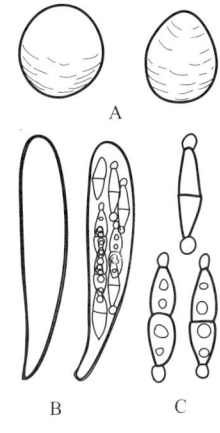

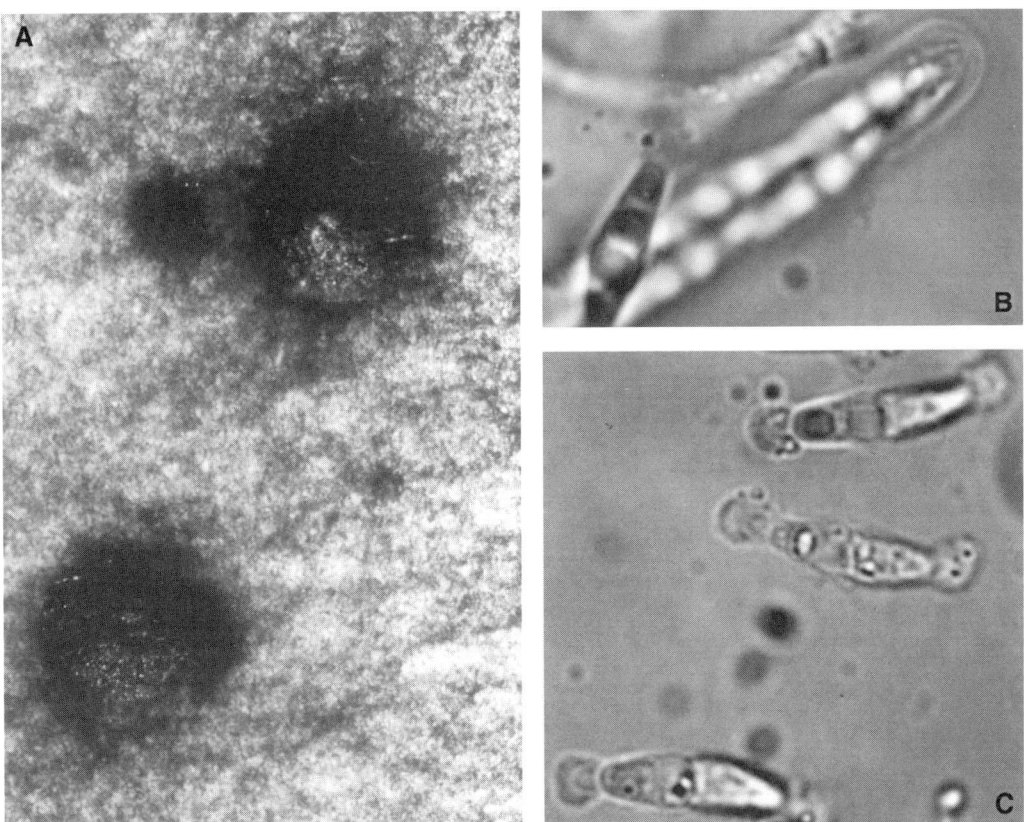

Eudarluca biconica. A: Perithecia. B: Ascospores in ascus. C: Ascospores. (From Watanabe, T. 1989d. *Trans. Mycol. Soc. Jpn.*, 30:395–400. With permission.)

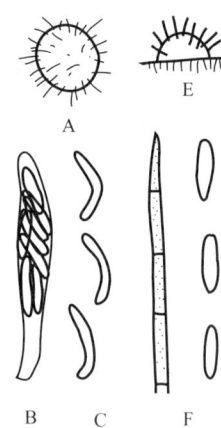

Glomerella Spauld. & Schrenk

Bull. U.S. Dep. Agric. 44:29, 1903.
Type species: *G. cingulata* (Stonem.) Spauld. & Schrenk

Glomerella glycines (Hori) Lehman et Wolf

Anamorph: *Colletotrichum destructivum* O'Gara

References: Kurata 1960; Tiffany and Gilman 1954.

Morphology: Perithecia dark brown, globose, covered all over with white hairs. Asci hyaline, cylindrical, curved.

Anamorph: sporodochial, pinkish, half-globose, mixed with dark brown setae. Conidia hyaline, cylindrical, 1-celled.

Dimensions: Perithecia 200–240 µm in diameter. Asci 80–90 × 9.5–10.5 µm. Ascospores 23.7–27.5 × 4.5–5 µm. Anamorph: sporodochia 250–360 µm in diameter: setae 130–170 × 5–5.3 µm. Conidia ca. 20 × 4.7–5.3 µm.

Material: 70-65 (Snap bean seed, Shinsyu-shinmachi, Nagano, Japan).

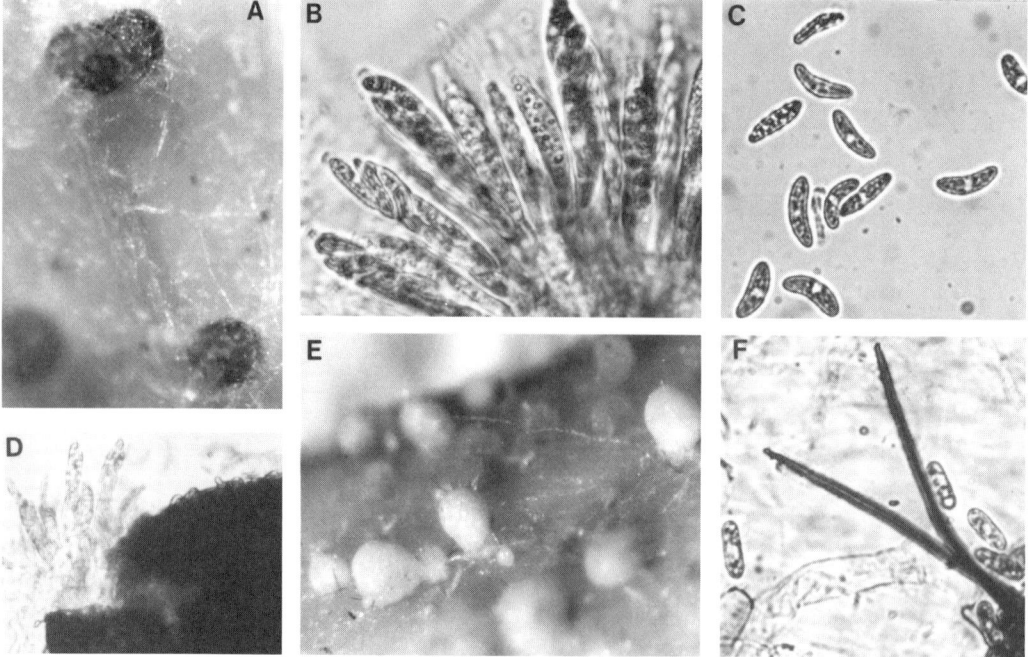

Glomerella glycines. A: Perithecia. B: Asci. C: Ascospores. D: Asci extruded from perithecium. E: Sporodochia. F: Setae and conidia.

Massarina Sacc.

Syll. Fung. 2:153, 1883.
Type species: *M. eburnea* (Tul.) Sacc.

Massarina sp. 1.

References: Bose 1961; Srinivasulu and Sathe 1972.

Morphology: Cleistothecia black, ellipsoidal or ovate: peridium brown, opaque, composed of soft skin. Asci cylindrical, bitunicate, 8-ascosporus. Paraphyses lacking. Ascospores yellowish, lunar-shaped or fusiform, often curved, 4-celled.

Dimensions: Cleistothecia 200–250 × 125–150 µm. Asci 55–75 × 6.3–7.5 µm. Ascospores 20–24 × 3.5–4 µm.

Material: 73-225 (Strawberry root, Shizuoka, Japan).

Remarks: This fungus is morphologically close to *M. dryopteri* Bose on the basis of conidial morphology and its dimensions. Another *Massarina* isolate (73-463) studied differs from this fungus in the number of ascospore septum and its dimensions.

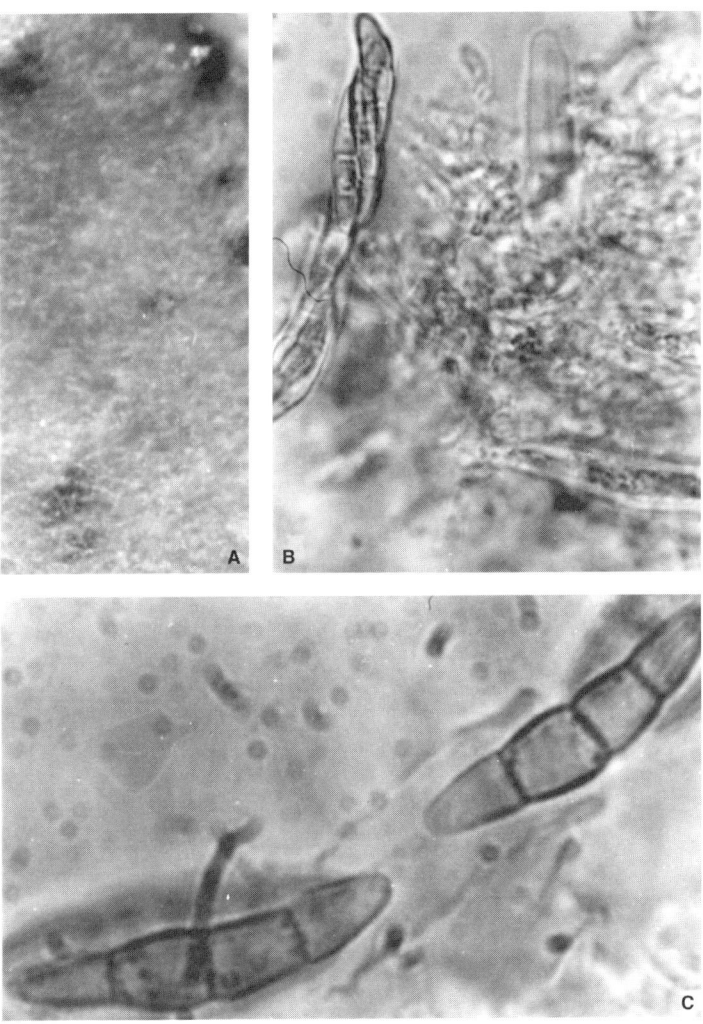

Massarina sp. 1.   A: Perithecia embedded in agar culture. B: Asci and ascospores. C: Ascospores.

Massarina sp. 2.

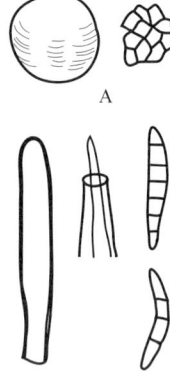

References: Bose 1961; Srinivasulu and Sathe 1972.

Morphology: Cleistothecia black, globose, subglobose or disc-shaped: peridium brown, opaque, with soft skin. Asci cylindrical, bitunicate, 8-ascosporus. Paraphyses lacking. Ascospores hyaline, fusiform, often curved, 4–7-celled.

Dimensions: Cleistothecia 225–300 µm in diameter. Asci 80–100 × 7.5–8.8 µm. Ascospores 27.5–32 × 3.8–5 µm.

Material: 73-463 (Strawberry field soil, Shizuoka, Japan).

Remarks: This fungus was identified as *Massarina* species on the basis of ascospore morphology and its dimensions, but differs from *Keissleriella gloeospora* (Berk. et Curt) Bose, closely related to this fungus which bears setae around ostioles.

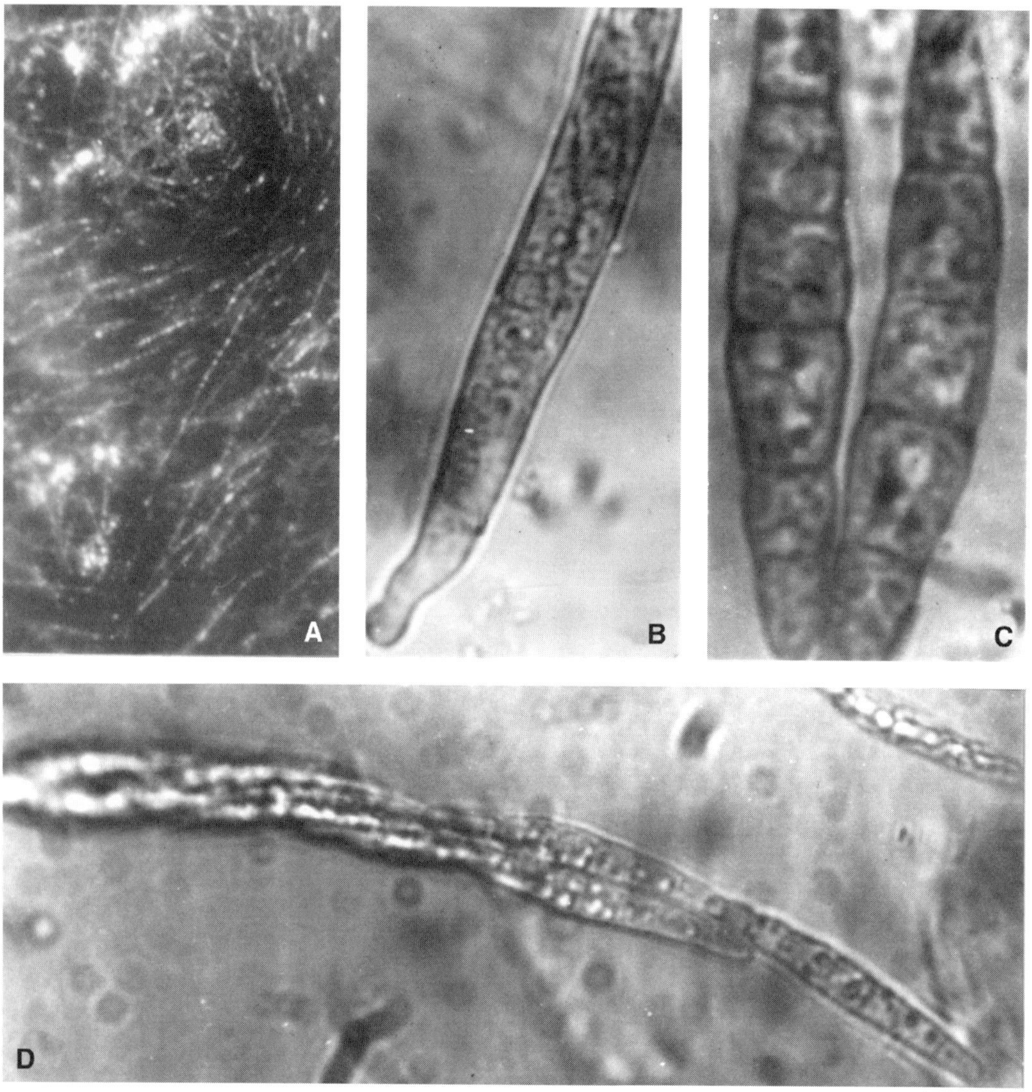

Massarina sp. 2. A: Cleistothecia embedded in agar culture. B: Part of ascus and ascospores. C: Ascospores. D: Ascospores in an ascus.

Microascus Zukal

Verh. Zool.-Bot. Ges. Wien 35:333, 1885.
Type species: *M. longirostris* Zukal

Microascus longirostris Zukal

Anamorph: *Scopulariopsis* st.

References: Arx 1975; Barron et al. 1961; Morton and Smith 1963; Udagawa 1963; Watanabe and Sato 1988.

Morphology: Peritheria black, solitary or aggregated, globose, subglobose or flask-shaped, often with long cylindrical neck apically. Asci hyaline, globose or ellipsoidal, branched, formed on catenulate ascigerous cells, unitunicate, 8-ascosporus. Paraphyses lacking. Ascospores hyaline or pale brown, boat-shaped, ellipsoidal, 1-celled.

Anamorph: conidiophores simple or branched with catenulate conidia apically on phialides with annelation. Conidia hyaline, ovate, 1-celled.

Dimensions: Peritheria (100-) 175–270 µm in diameter: necks 20–22.5 µm wide, up to 250 µm deep. Asci 8–9 × (4.5-) 6–7 µm. Ascospores 3.2–4 × 2.5–2.8 µm. Conidia 3–3.8 × 2.5 µm.

Material: 86-61 (Japanese cedar seed, Nakashinkawa, Toyama, Japan).

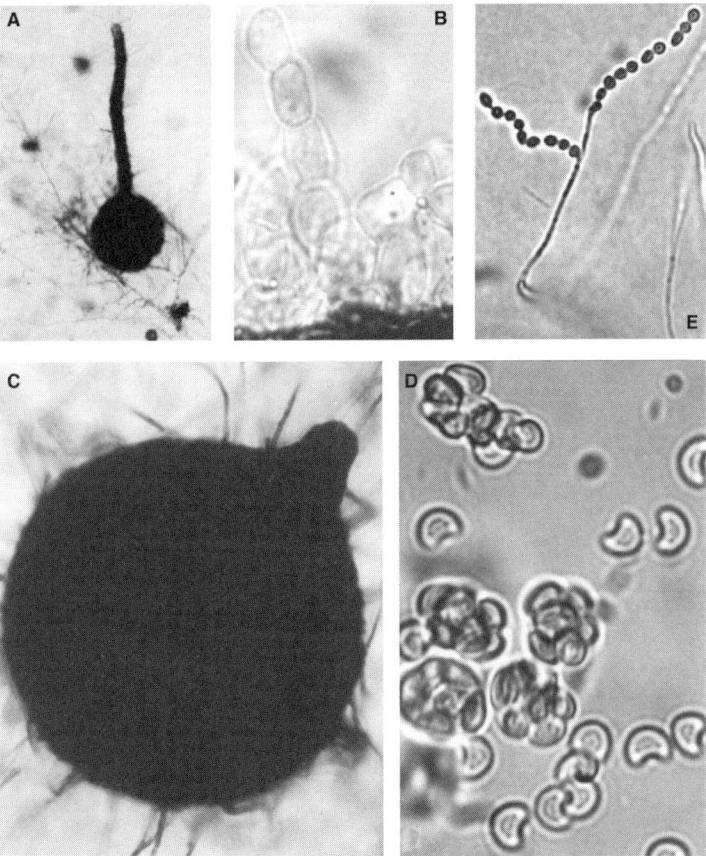

Microascus longirostris. A,D: Perithecia. B: Catenulate ascigerous hyphae. D: Asci and ascospores. E: Anamorph. (From Watanabe, T. and Sato, Y. 1988. *Trans. Mycol. Soc. Jpn.*, 29:143–150. With permission.)

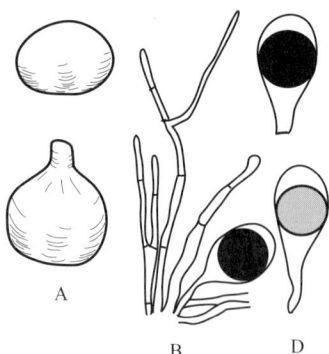

Monosporascus Pollack & Uecker

Mycologia 66:346, 1974.
Type species: *M. cannonballus* Pollack & Uecker

Monosporascus cannonballus Pollack & Uecker

References: Hawksworth and Ciccarone 1978; Pollack and Uecker 1974; Watanabe 1979.

Morphology: Perithecia black, globose or subglobose, smooth, non-ostiolate, ostiolate with irregular openings or necked conspicuously, often with darkened rings: peridium pale brown or brown, semitransparent or opaque. Asci hyaline, globose or flask-shaped, unitunicate, 1-ascosporus. Paraphyses hypha-like, hyaline, simple or branched. Ascospores black, globose, 1-celled.

Dimensions: Perithecia 222–568 µm in diameter: necks up to 148 µm tall; 74–148 µm wide basally. Asci 50–110 × 35–50 µm. Paraphyses 90–200 × 5–12.5 µm. Ascospores 32–47.5 µm in diameter.

Material: 78-2494 (Melon root, Tendo, Yamagata, Japan); 80-64, 80-71 (Melon root, Kagoshima, Japan).

Remarks: *Monosporascus monosporus* with small pores on the ascospore surface, and *M. eutypoides* with two ascospores per ascus are included in this genus (Hawksworth and Ciccarone, 1978).

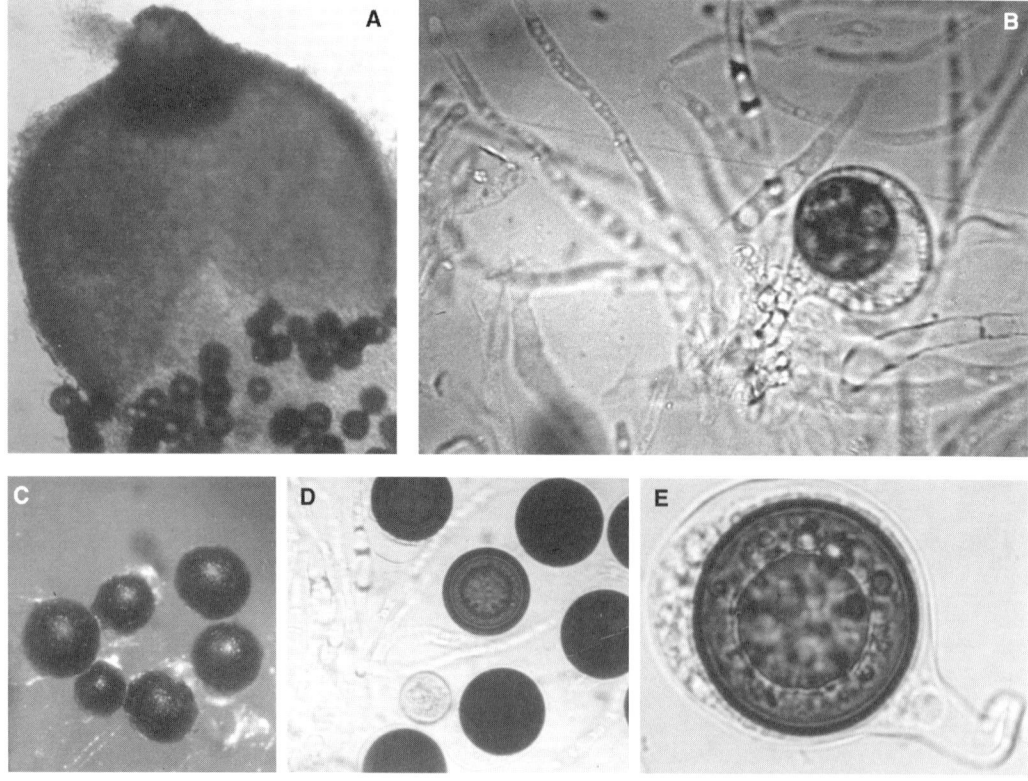

Monosporascus cannonballus. A: Crushed perithecium and ascospores. B: Immature ascus and ascospore, and paraphyses. C: Cultured perithecia. D–E: Asci and ascospores. (From Watanabe, T. 1979. *Trans. Mycol. Soc. Jpn.*, 20:312–316. With permission.)

Ascomycotina

Nectria Fr.

Summa Veg. Scand. 287, 1847.
Type species: *N. cinnabarina* (Tode : Fr.) Fr.

Key to Species

1. Ascospores   under 12 µm long .... *N. gliocladioides*
                over 18 µm long. 2

2. Anamorph   formed 3
               not formed .... *N. asakawaensis*

3. Anamorph   *Stachybotryna* with 1-celled conidial head .... *N. hachijyoensis*
               Penicillifer with catenulate 2-celled conidia .... *N. fragariae*

Nectria asakawaensis T. Watanabe

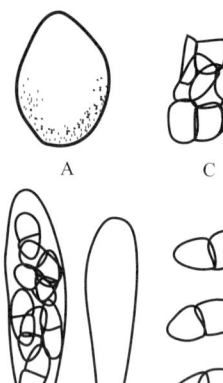

References: Perrin 1976; Rogerson and Samuels 1985; Rossman 1983; Samuels 1976; Watanabe 1990c.

Morphology: Perithecia mostly embedded, solitary, reddish brown, ovate, fleshy, ostiolate apically: peridium yellowish brown, with soft skin, pseudoparenchymatous. Asci hyaline, long-ellipsoidal or clavate, 8-ascosporus. Paraphyses lacking. Ascospores golden yellow, ellipsoidal, 2-celled, constricted at the septum. Anamorph not formed.

Dimensions: Perithecia 175–310 × 140–240 µm. Asci 75–87.5 × 15–20 µm. Ascospores 18.7–25 × 10–12.5 µm.

Material: 84-133 (From rhizomorph of *Armillaria mellea* associated with flowering cherry root, Hachioji, Tokyo, Japan).

Remarks: Colonies on PDA are dark brown, resupinate, almost nonaerial, apparently *Cylindrocarpon*-like, and the reverse dark brown.

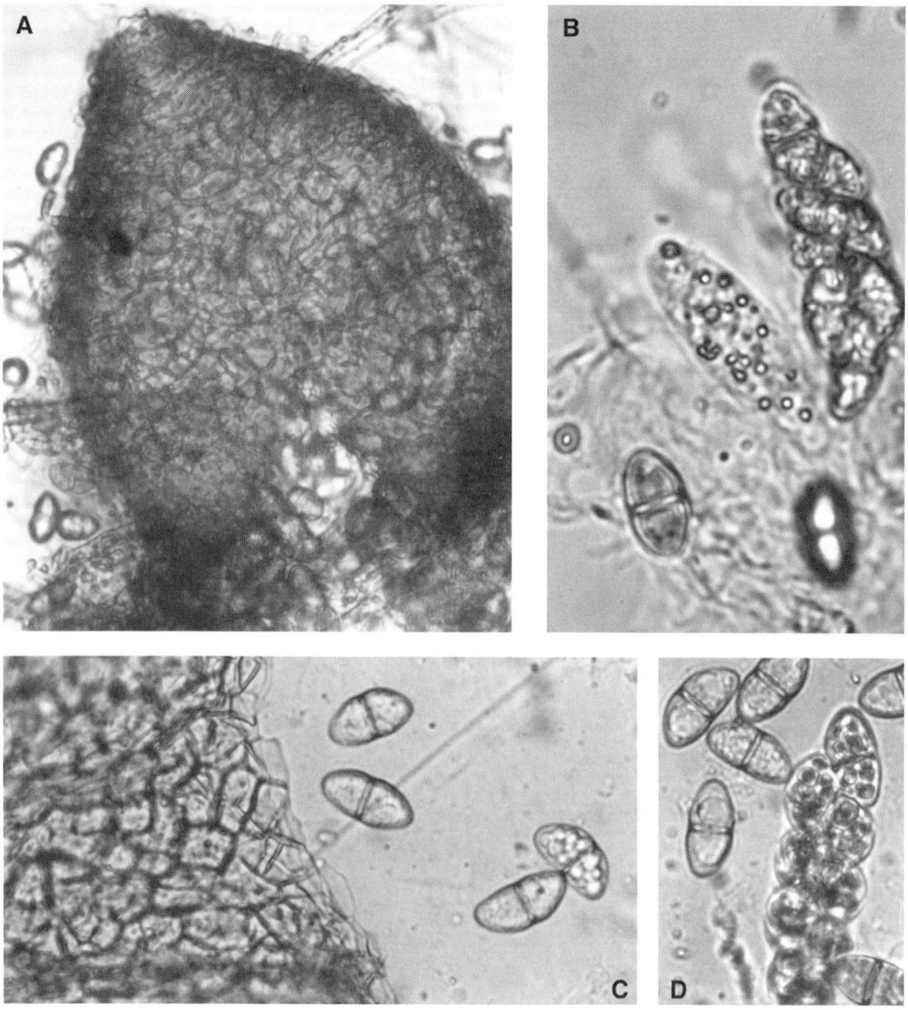

Nectria asakawaensis. A: Perithecium. B: Asci and ascospores. C: Peridium and ascospores. D: Asci and ascospores. (From Watanabe, T. 1990c. *Trans. Mycol. Soc. Jpn.*, 31:227–236. With permission.)

Nectria fragariae T. Watanabe

Anamorph: *Penicillifer fragariae* T. Watanabe

References: Booth 1959; Rogerson and Samuels 1985; Samuels 1988, 1989; Van Emden 1968; Watanabe 1990c.

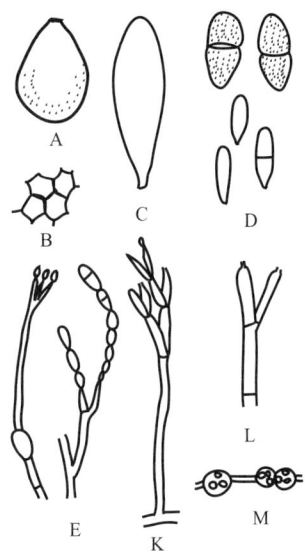

Morphology: Perithecia embedded, solitary, yellowish brown, ovate, ostiolate apically: peridium yellowish brown, with soft skin, pseudoparenchymatous. Asci hyaline, clavate, 8-ascosporus. Ascospores yellowish brown, ellipsoidal, 2-celled, constricted at the septum.

Anamorph: conidiophores erect, hyaline, simple or branched, with terminal phialides, often propagated (with internal nodes): phialides verticillate or alternate, bearing catenulate conidia or spore heads. Conidia hyaline, spindle-shaped, 1- or 2-celled, truncate at one end. Chlamydospores mostly intercalary, globose, brown, granulate, catenulate or in mass.

Dimensions: Perithecia 222–300 × 160–250 μm: ostioles ca. 35 μm. Asci 55–80 × 14–20 μm. Ascospores 21.3–25 × 10–10.5 μm. Anamorph: conidiophores 120–540 μm tall, 3.5–4.5 μm wide basally, 2.8–3.3 μm apically: primary branches 20–35 × 3–4.8 μm: phialides 20–25 × 3.5–5 μm. Conidia 13.8–17.5 × 4.8–5.3 μm. Chlamydospores 8.7–12.5 μm in diameter.

Material: 73-178 (Strawberry root, Shizuoka, Japan).

Remarks: Colonies on PDA are pale to grayish brown, resupinate, slightly aerial in the central part.

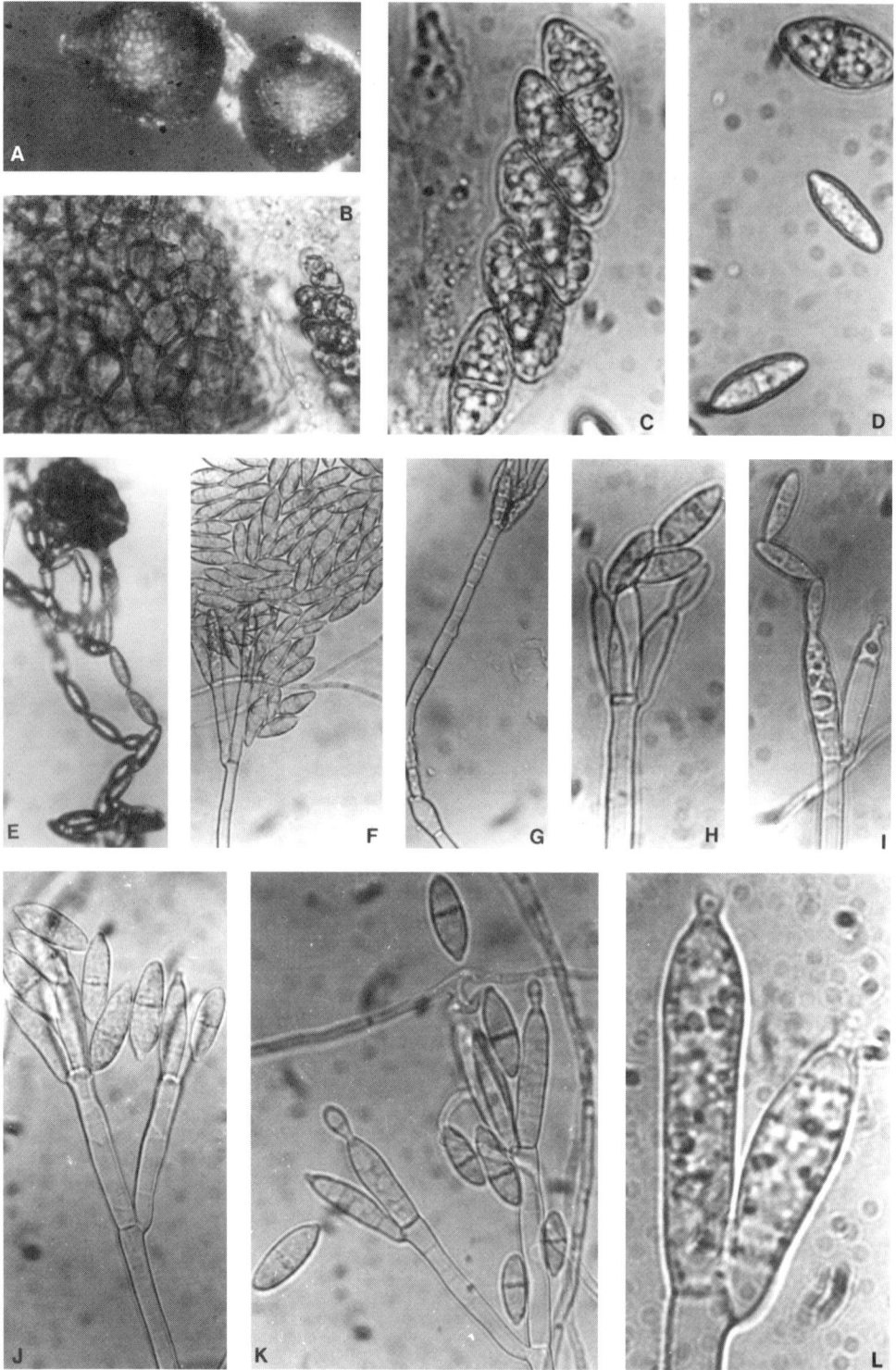

Nectria fragariae. A: Perithecia. B: Peridium, part of ascus and ascospores. C: Ascospores in ascus. D: Ascospore and conidia. E–H: Anamorph: catenulate conidia (E), part of conidiophores and conidia (F–K), and phialides (L), M: Chlamydospores. (From Watanabe, T. 1990c. *Trans. Mycol. Soc. Jpn.*, 31:227–236. With permission.)

Nectria gliocladioides Smalley et Hansen

Anamorph: *Gliocladium roseum* (Link) Bainier

References: Hanlin 1961; Smalley and Hansen 1957.

Morphology: Perithecia aggregate, reddish yellow, globose or ovate, ostiolate apically: peridium yellowish brown, with soft skin, pseudoparenchymatous. Asci hyaline, cylindrical, 8-ascosporus. Ascospores hyaline, ellipsoidal, 2-celled or rarely 3-celled, constricted at the septum.

Anamorph: conidiophores hyaline, erect, simple or branched with verticillate or penicillate phialides bearing single spore masses. Conidia hyaline, long-ellipsoidal, ovate, or subglobose. Chlamydospores brown, globose, thick-walled.

Dimensions: Perithecia (150-) 250–320 × 200–300 μm. Asci 40–58 × 6.2–7.5 μm. Ascospores 7.5–11.3 × 2.2–3.8 μm.

Material: 73-243 (Strawberry root, Shizuoka, Japan); 74-709 (Strawberry root, Mie, Japan).

Remarks: Anamorph state, *Gliocladium roseum* in Mitosporic fungi (Deuteromycotina) may be consulted.

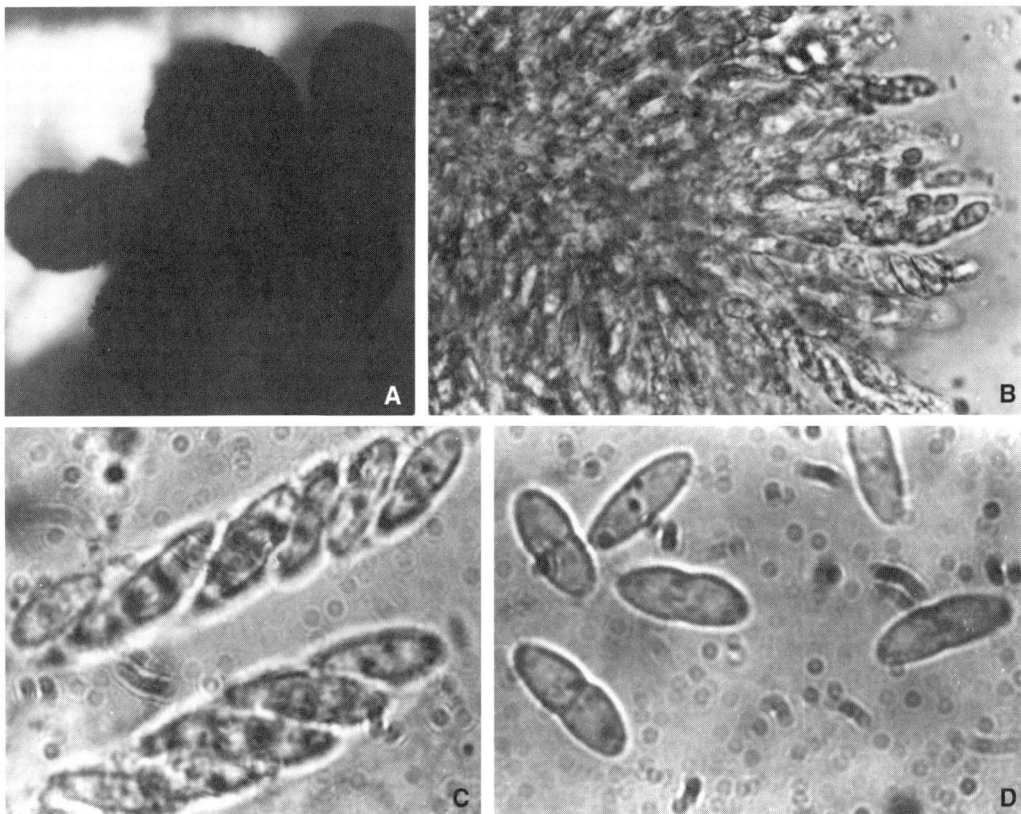

Nectria gliocladioides. A: Perithecia. B,C: Asci and ascospores. D: Ascospores. E: Anamorph; conidiophores and conidia. F: Chlamydospore.

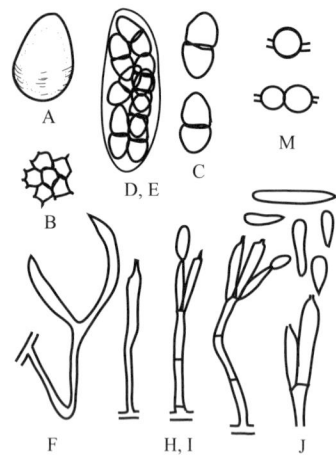

Nectria hachijoensis T. Watanabe

Anamorph: *Stachybotryna hachijoensis* T. Watanabe

References: Booth 1959; Tubaki and Yokoyama 1971; Watanabe 1990c.

Morphology: Perithecia distributed, solitary, reddish yellow, fleshy, ovate, ostiolate apically: peridium yellowish brown, with soft skin, pseudoparenchymatous. Asci hyaline long-ellipsoidal, 8-ascosporus. Paraphyses lacking. Ascospores yellow, ellipsoidal, 2-celled, constricted at the septum.

Anamorph: conidiophores hyaline, erect, simple or branched, with terminal phialides bearing single spore masses: phialides cylindrical, often inflated apically. Seta-like structures simple or branched, cylindrical, curved, tapering toward the end. Conidia hyaline, clavate, cylindrical, ovate or pear-shaped, 1-celled, often tapering toward the end. Chlamydospores globose, brown, single or twins.

Dimensions: Perithecia 170–200 µm in diameter. Asci 52.5–87.5 × 20–30 µm. Ascospores 20–22.5 × 10–11.3 µm. Anamorph: conidiophores 82.5–137.5 × 4.2–4.8 µm: phialides 17.5–26.3 × 4.5–5 µm. Spore masses 20–45 µm in diameter. Seta-like structures 87–175 µm long. Conidia 12.5–27.5 (-42.5) × 5–5.5 µm. Chlamydospores 8.7–10 µm in diameter.

Material: 70-1336 (= IFO 32545, Uncultivated soil, Hachijo, Tokyo, Japan).

Ascomycotina

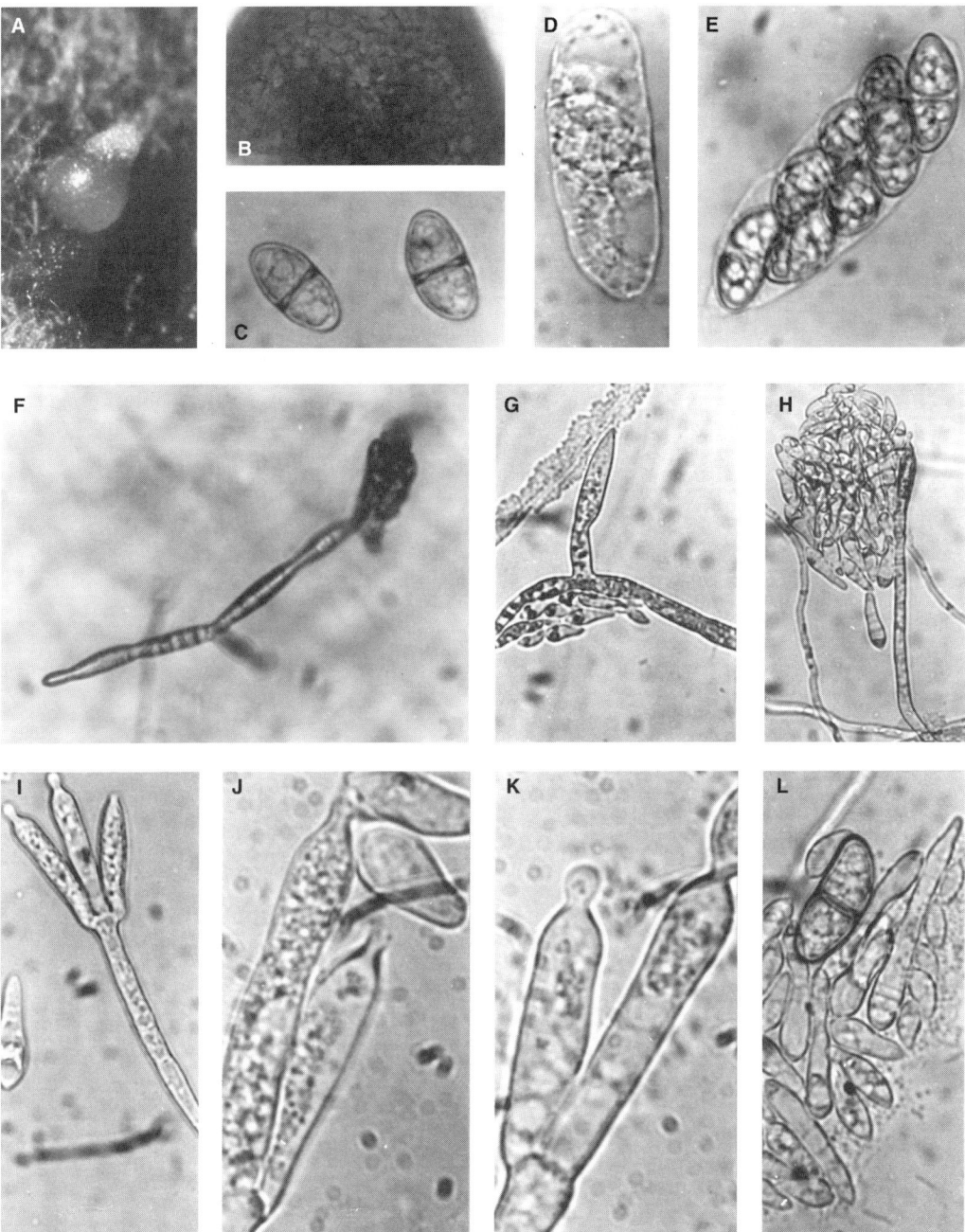

Nectria hachijoensis. A: Perithecium. B: Peridium. C: Ascospores. D,E: Asci and ascospores. F–K: Anamorph: Seta-like structures (F,G), conidiophores and conidia (H,I), phialides and conidia (J,K). L: Conidia and ascospore. M: Chlamydospore. (From Watanabe, T. 1990c. *Trans. Mycol. Soc. Jpn.*, 31:227–236. With permission.)

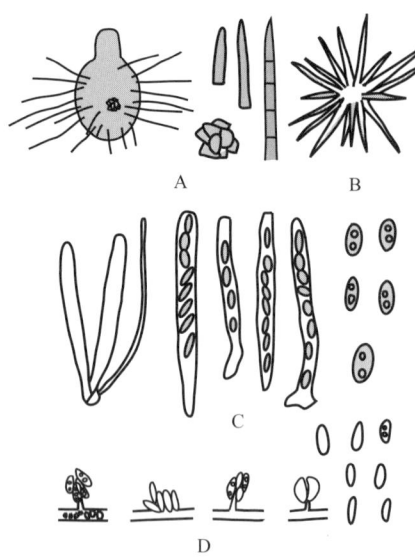

Phaeotrichosphaeria Sivanesan

Trans. Br. Mycol. Soc. 81:313, 1983.
Type species: *P. indica* Sivan. & N. D. Sharma

Phaeotrichosphaeria sp.

References: Sivanesan 1983.

Morphology: Perithecia black or dark brown, flask-shaped or subglobose, ostiolate, covered with stiff hairs or setae. Asci hyaline, cylindrical or clavate, 8-ascosporous in 1 row. Paraphyses hyaline, cylindrical, curved, thread-like. Ascospores ellipsoidal, brown, biguttulate.

Anamorph: *Aureobasidium*-like. Conidiophores lacking or very short. Conidia hyaline, cylindrical. Setal bodies with stiff setae: setae dark brown or black, septate.

Dimensions: Perithecia 275–300 × 175–200 μm: ostiolar regions 50 μm long, 60–80 μm wide. Asci 60–74 × 4–6 μm. Ascospores 4.4–7.6 × 3–3.4 μm. Paraphyses nearly 68–70 μm long, 1–2 μm wide. Conidia 5–6 × 1.2–2.4 μm. Setal bodies 90–180 μm including setae: setae 24–240 × 2.8–4 μm.

Material: 99-467 (Forest soil, Agematsu, Shiojiri, Nagano, Japan).

Remarks: This fungus characteristically forms setal perithecia with brown ascospores, and *Aureobasidium* anamorph.

Phaeotrichosphaeria sp.   A,B: Tomentose (A) and setose perithecia (B). C: Asci, paraphyses, and ascospores. D: Dark ascospores, and hyaline conidia in *Aureobasidium* anamorph.

Preussia Fuckel

Fungi Rhenan. Suppl., Fasc. 3:1750, 1866.
Type species: *P. funiculata* (Preuss) Fuckel

Preussia terricola Cain

References: Cain 1961a.

Morphology: Cleistothecia black, ovate or subglobose: peridium brown, opaque, with soft skin. Asci clavate, narrowing basally, often curved, bitunicate, 22–30 ascosporus cells in 3 rows in each ascus. Paraphyses thread-like, septate. Ascospores pale yellowish brown, cylindrical, 1–3-septate, conspicuously constricted at the septum, readily separable to each cell component: component cells cylindrical, homogeneous in size. Anamorph lacking.

Dimensions: Cleistothecia 98–247 μm in diameter. Asci 62.5–100 × 13.7–20 μm. Paraphyses 2–3 μm wide. Ascospores 24.4–32 × 5–5.5 μm: component cells 6.2–8 × 5–5.5 μm.

Material: 75-323 (Sorghum field soil, Ebina, Kanagawa, Japan).

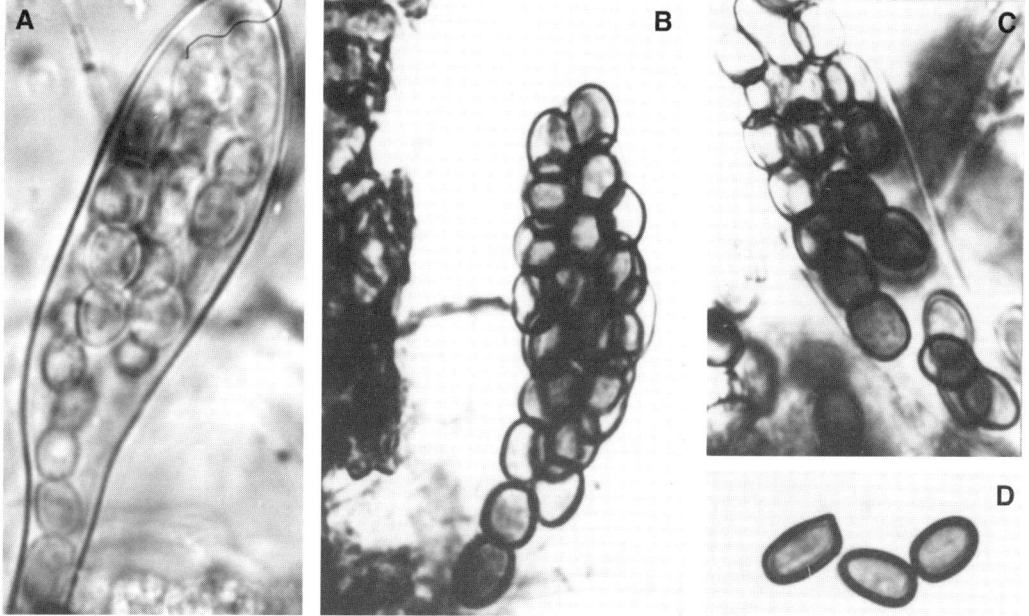

Preussia terricola. A–C: Asci and ascospores. D: Separated component cells of conidia. E: Cleistothecium.

Sordaria Ces. & de Not.

Comm. Soc. Critt. Ital. 1:225, 1863.
Type species: *S. fimicola* (Rob.) Ces. & de Not.

Key to Species

1.	Ascospores	nodulate .... *S. nodulifera*
		smooth 2
2.	Perithecia	over 200 µm in diameter, with asci over 125 µm long .... *S. fimicola*
		under 150 µm in diameter, with asci under 117.5 µm .... *S. tamaensis*

Sordaria fimicola (Rob.) Ces. et de Not.

References: Cain 1957; Cain and Groves 1948; Mehrotra 1969; Watanabe 1989d.

Morphology: Perithecia discrete or aggregated, dark brown or black, pear-shaped, flask-shaped, necked or papillate, covered with white hairs: peridium brown, pseudoparenchymatous. Asci cylindrical, 8-ascosporus, truncate apically, narrowed basally. Ascospores dark green or dark brown, ellipsoidal, usually with single germ pores, often covered with gelatinous sheath.

Dimensions: Perithecia 291–550 × 185–425 µm; necks 90–125 × 65–100 µm: ostioles 48.2–53.6 µm in diameter. Asci (75-) 127.5–202.5 × 13.7–20 (-22.5) µm. Ascospores 15–20 × 10–12.5 (-15) µm.

Material: 82-157 (Melon field soil, Tateyama, Chiba, Japan); 85-84 (= IFO 32550), 85-85 (Flowering cherry seed, Hachioji, Tokyo, Japan).

Remarks: Some of at least 25 species of the genus *Sordaria* are known as coprophilus fungi.

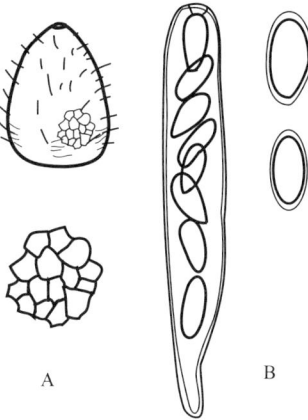

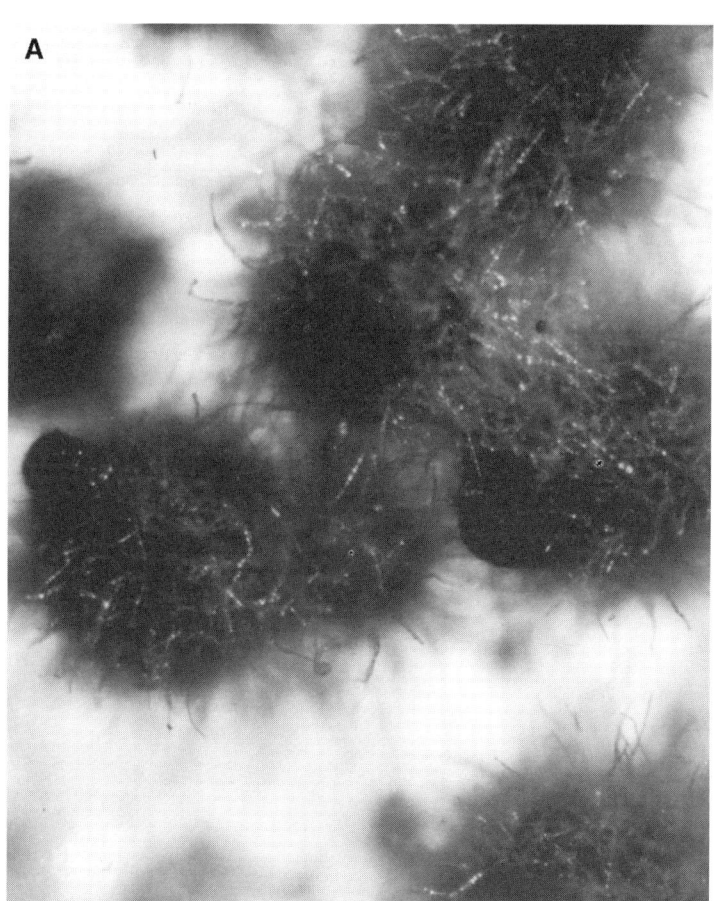

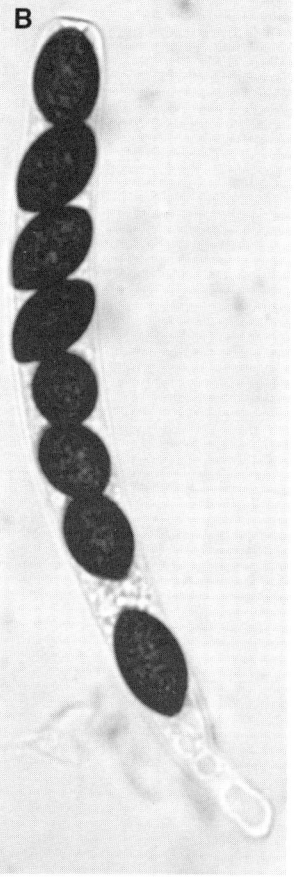

Sordaria fimicola. A: Perithecia. B: Ascospores in ascus. (From Watanabe, T. 1989d. *Trans. Mycol. Soc. Jpn.*, 30:395–400. With permission.)

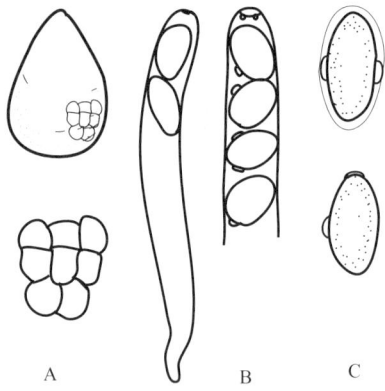

Sordaria nodulifera T. Watanabe

References: Barrasa et al. 1986; Boedijn 1962; Cailleux 1976; Das 1962; Guarro and Arx 1987; Olive and Fantini 1961; Watanabe 1989d.

Morphology: Perithecia dark brown or black, pear-shaped or ovate, ostiolate apically, covered with white hairs: peridium pale brown, pseudoparenchymatous. Asci cylindrical, 8-ascosporus, truncate apically, narrowed basally. Ascospores dark brown, ovate, with 2–4 nodules per ascospore, apiculate or germ pored at one end, often covered with gelatinous sheath.

Dimensions: Perithecia 350–605 × 200–421 µm: ostioles 40–66 µm wide. Asci 195–275 × 17.5–22.5 µm. Ascospores (17.5-) 20–24 × (13.7-) 15–16.3 (-21.3) µm: nodules 5–7.5 × 1.2–1.3 µm.

Material: 85-72 (= IFO 32551, Flowering cherry seed, Hachioji, Tokyo, Japan).

Remarks: This fungus is unique in having nodulate ascospores.

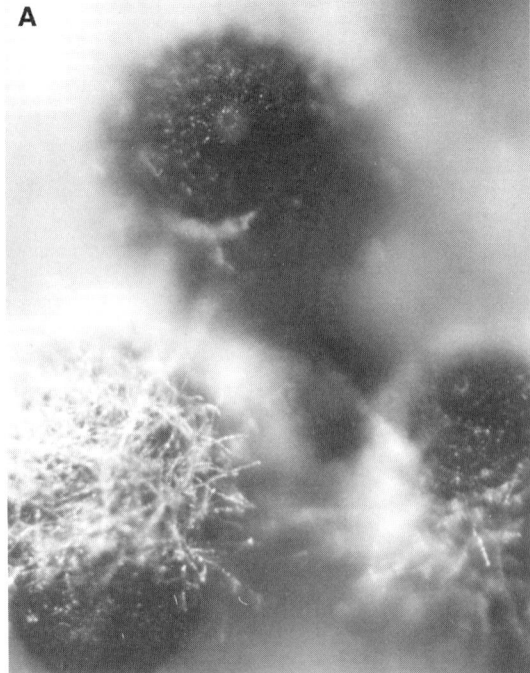
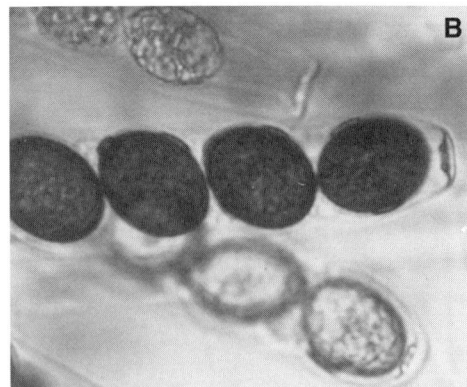
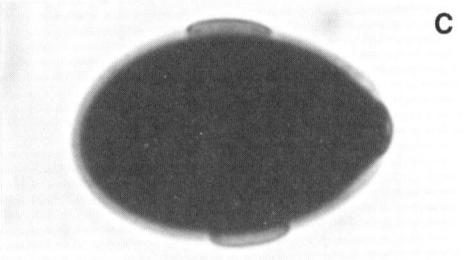

Sordaria nodulifera. A: Perithecia. B: Part of asci and ascospores. C: Ascospore. (From Watanabe, T. 1989d. *Trans. Mycol. Soc. Jpn.*, 30:395–400. With permission.)

Sordaria tamaensis T.Watanabe

References: Barrasa et al. 1986; Boedijn 1962; Cailleux 1976; Das 1962; Guarro and Arx 1987; Olive and Fantini 1961; Watanabe 1989d.

Morphology: Perithecia mostly embedded, brown, globose or subglobose, apically darkened, connected with rather thick hyphae: peridium pale brown, pseudoparenchymatous. Asci cylindrical, 8-ascosporus in 1 or 2 rows, truncate apically, narrowed basally. Ascospores dark brown, black, cylindrical, with germ pores at one or both ends, covered with gelatinous sheath.

Dimensions: Perithecia 50–150 μm in diameter. Asci 87.5–130 (-150) × 10–20 μm. Ascospores 15–20 × 7.2–12.5 μm.

Material: 85-89 (= IFO 32552, Flowering cherry seed, Hachioji, Tokyo, Japan).

Remarks: This fungus is characterized by small globose perithecia, and often twisted or curved cylindrical asci.

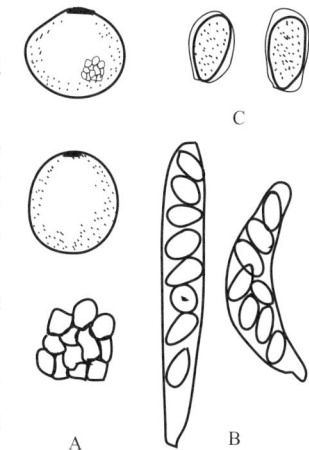

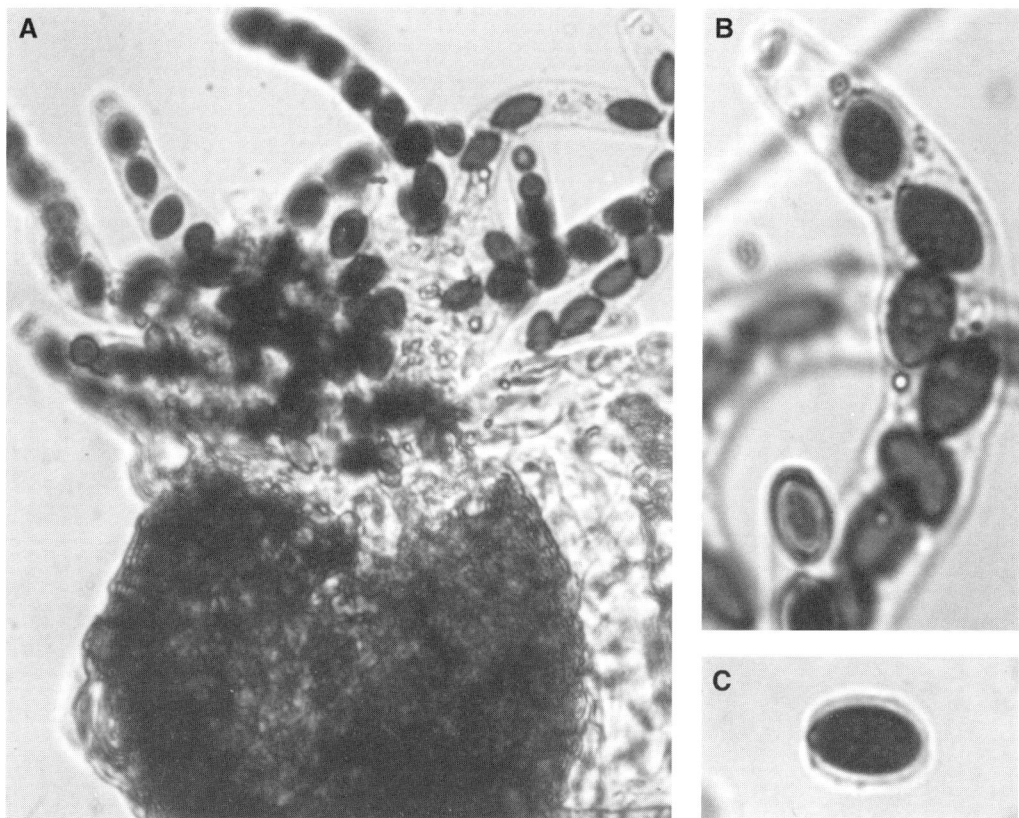

Sordaria tamaensis. A: Asci and ascospores extruded from crushed perithecium. B: Part of ascus and ascospores. C: Ascospore. (From Watanabe, T. 1989d. *Trans. Mycol. Soc. Jpn.*, 30:395–400. With permission.)

Thielavia Zopf

Verh. Bot. Ver. Brandenb. 18:101, 1876.
Type species: *T. basicola* Zopf

Thielavia terricola (Gilman et Abbott) Emmons

References: Booth 1961; Booth and Shipton 1966; Malloch and Cain 1973.

Morphology: Cleistothecia superficial, black, globose, covered with white hairs: peridium brown, double-membranous, pseudoparenchymatous. Asci ovate, clavate, narrowed basally with foot cells, 8-ascosporus, evanescent. Ascospores, olive-colored, dark green, ellipsoidal, with germ pores at one end, truncate at another end, blue lines recognizable. Anamorph lacking.

Dimensions: Cleistothecia (50-) 75–170 (-190) μm in diameter. Asci (20-) 27.5–35 (-40) × 13.7–17.5 (-19) μm. Ascospores 9–15.5 × (5-) 6.2–8 μm.

Material: 73-30 (Strawberry root, Shizuoka, Japan); 73-451 (Strawberry field soil, Shizuoka, Japan); 84-532 (Japanese red pine seed, Higashichikuma, Nagano, Japan); 86-64 (Japanese cedar seed, Kasama, Ibaraki, Japan).

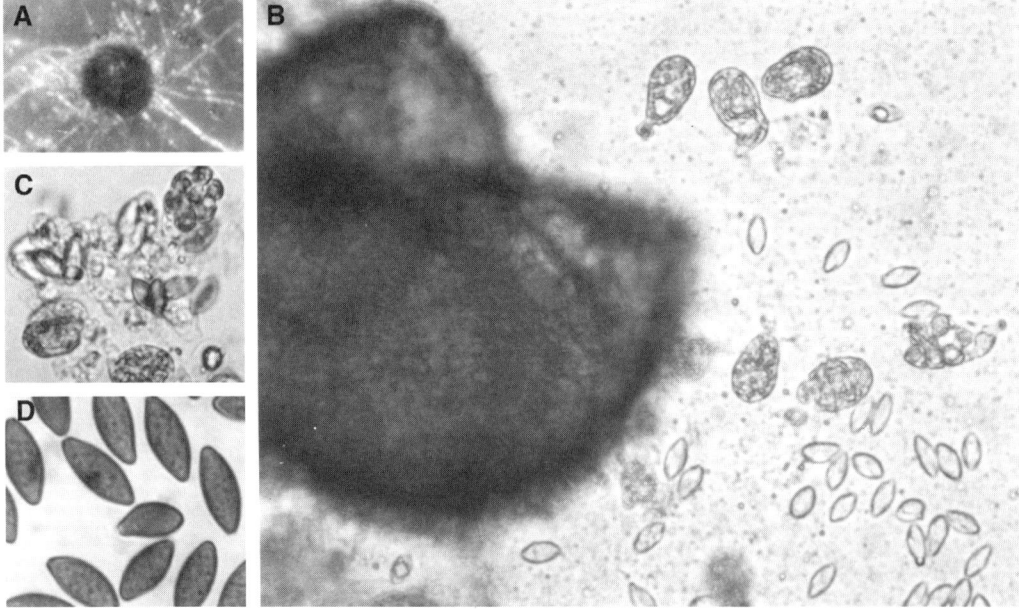

Thielavia terricola. A: Cleistothecium. B: Crushed cleistothecium, asci, and ascospores. C: Asci and ascospores. D: Ascospores.

Zopfiella Winter

Krypt. Fl. 1:56, 1887.
Type species: *Z. tabulata* (Zopf) Wint.

Key to Species

1. Ascospores   2–3-celled, curved.................................*Z. curvata*
                                    2-celled, uncurved..2

2. Ascospore pigmented cell   ellipsoidal.. *Z. latipes*
                                    triangle .... *Z. pilifera*

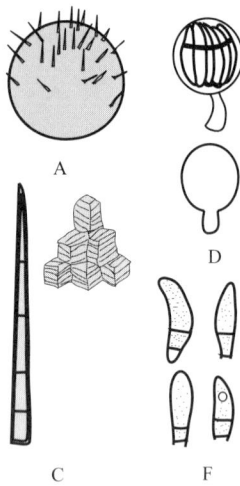

Zopfiella curvata (Fuckel) Winter

References: Malloch and Cain 1971; Udagawa and Horie 1974.

Morphology: Cleistothecia globose, black, covered with setae: peridium composed of 6-angled cephalathecoid tissues. Setae thick-walled, 4–6-septate, dark brown, tapered toward the end. Asci globose, evanescent, 8-ascosporus, often with pedicels. Ascospores curved, 3-celled, septate in the lower half, brown in the upper 2 cells, hyaline and often broken in the end cell.

Dimensions: Cleistothecia 210–342 µm in diameter. Setae 195–325 µm long, 4.5–6.3 µm wide basally. Asci 180–216 µm in diameter: pedicels 8.1–12.6 × 3.6–8.1 µm. Ascospores 13.5–19.8 × 5–5.8 µm.

Material: 92-0233 (Strawberry root, Shizuoka, Japan).

Remarks: The fungus fruited in old agar cultures together with needle-shaped crystals.

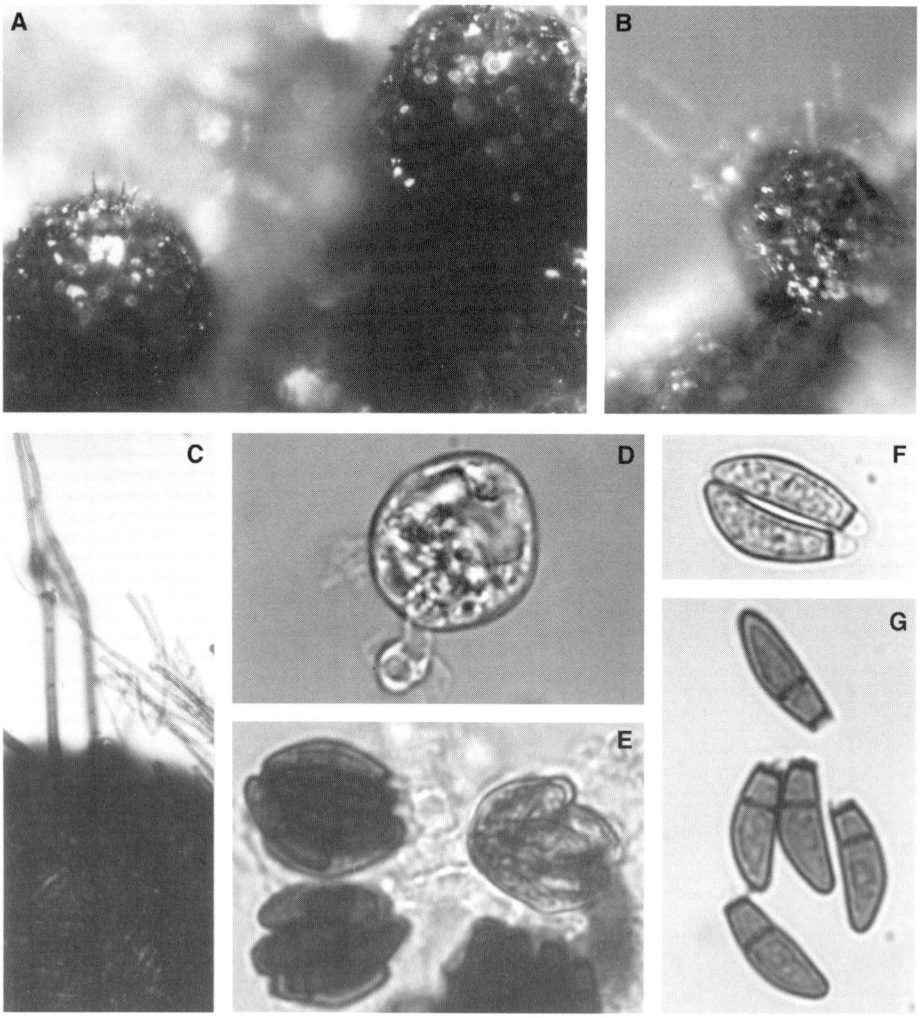

Zopfiella curvata. A,B: Cleistothecia on agar. C: Part of peridium and setae. D,E: Asci and ascospores (E). F: Ascospores.

Zopfiella latipes (Lunquist) Malloch et Cain

Anamorph: *Humicola* st.

References: Furuya and Udagawa 1973; Malloch and Cain 1971.

Morphology: Cleistothecia pale grayish green, globose, fleshy, covered with hyphae: peridium with soft skin, pseudoparenchymatous. Asci clavate, unitunicate, 8-ascosporus. Paraphyses lacking. Ascospores 2-celled, globulate, composed of ellipsoidal, apiculate, or germ-pored, dark green cell and triangle or cylindrical hyaline (rarely pigmented) cell which is often broken.

Anamorph: Conidiophores simple or lacking. Conidia hyaline, globose, ovate, or ellipsoidal, 1-celled.

Dimensions: Cleistothecia 160–200 µm in diameter. Asci 80–120 × 14–18 µm. Ascospores: pigmented cells 16–20 × 10–13 µm; hyaline cells 4–8 × 3.5–7 µm. Conidia: globose 3.5–5 µm in diameter; ovate or ellipsoidal 7.5–13.8 × 2–2.8 µm.

Material: 74-518, 74-539 (Strawberry root, Shizuoka, Japan).

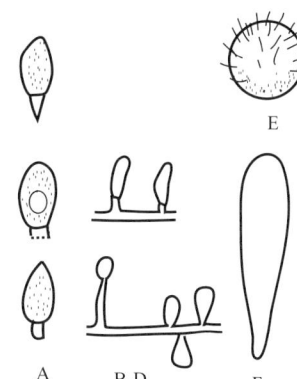

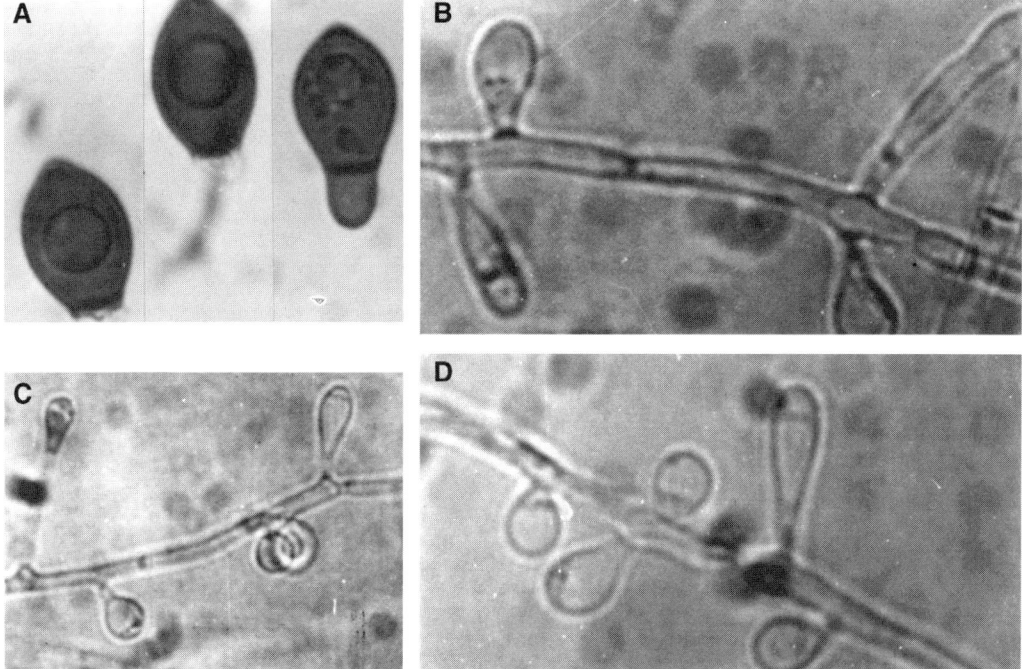

Zopfiella latipes. A: Ascospores. B–D: Anamorph. E: Cleistothecium. F: Ascus.

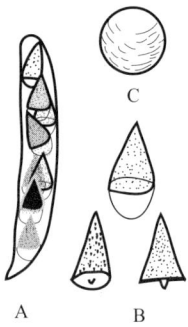

Zopfiella pilifera Udagawa et Furuya

References: Udagawa and Furuya 1972; Watanabe 1971.

Morphology: Cleistothecia black, globose, covered with hyphae: peridium pseudoparenchymatous. Asci clavate, cylindrical, 4-ascosporus. Ascospores 2-celled, composed of dark green triangle cell, and hyaline conical cell which is often lacking or transformed into an appendix. Anamorph lacking.

Dimensions: Cleistothecia 160–280 μm in diameter. Asci 100–120 × 15 μm. Ascospores: pigmented cells 17.5–24.8 × 10–10.5 μm; hyaline cells 7–7.8 μm long.

Material: 69-322 (Pineapple field soil, Okinawa, Japan).

Remarks: This fungus was originally recorded as *Podospora* sp. (Watanabe, 1971).

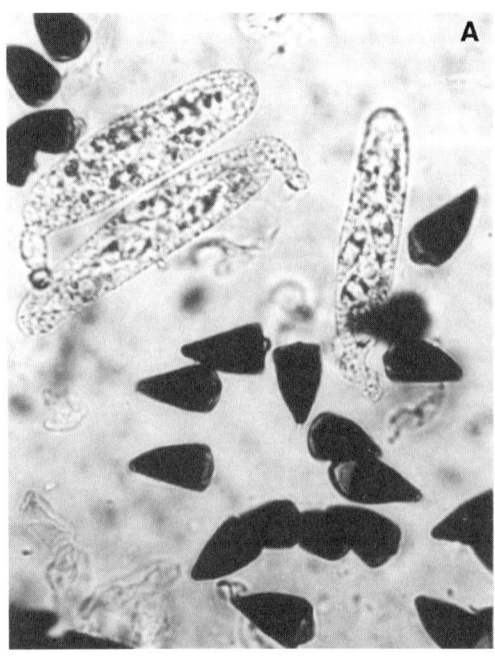

Zopfiella pilifera.   A,B: Asci and ascospores. C: Cleistothecium. (From Watanabe, T. 1971. *Trans. Mycol. Soc. Jpn.*, 12:35–47. With permission.)

MORPHOLOGY OF SOIL FUNGI

Basidiomycotina

Basidiomycetous fungi

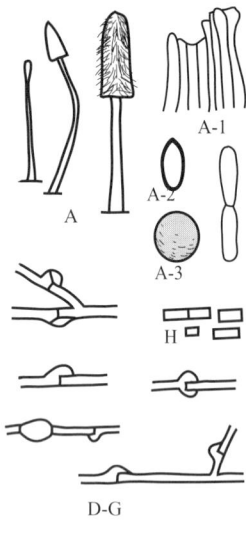

References: Warcup 1959; Warcup and Talbot 1962, 1963; Watanabe 1986.

Morphology: Basidiomycetous fungi are generally not fruitful in agar cultures, except *Coprinus* spp. which often fruited *in vitro* (Figures A,B). Many Basidiomycetous fungi have hyphae with clamp-connections (Figures D–I), but others do not. Rhizomorphs are often formed *in vitro* by *Armillaria mellea* (Vahl:Fr.) Kumm. (Figure C). However, some fungi form arthroconidia (Figure H) or chlamydospores. Because of nonfruiting *in vitro*, identification is very difficult, although their colonies are clearly differentiated to one another.

Dimensions: *Coprinus* sp. (Isolate 85-17): hyphae 2.5–7 µm broad: pileus 8 mm long: stipe 35–60 mm tall, ca. 1 mm wide basally: basidiospores 5.7–9.5 × 5–6.3 µm: sclerotia 105–155 µm in diameter.

Material: 73-50, 73-83, 73-145 (Strawberry root, Shizuoka, Japan); 82-735 (Bell pepper field soil, Wakayama, Japan); 84-481 (Japanese red pine seed, Saihaku, Tottori, Japan); 84-595 (Japanese red pine seed, Kamiina, Nagano, Japan); 85-P101 (Paulownia root, Itapua, Paraguay); 85-17 (Flowering cherry seed, Hachioji, Tokyo, Japan); 85-153 (Flowering cherry root, Hachioji, Tokyo, Japan).

Remarks: To identify unknown Basidiomycetous fungi, it is essential to compare them with already known isolates culturally or cultures derived from natural fruiting structures. Trials for inducing fruiting in agar cultures with soils under various atmospheres are recommended (Warcup, 1959; Warcup and Talbot, 1962; 1963).

Basidiomycotina

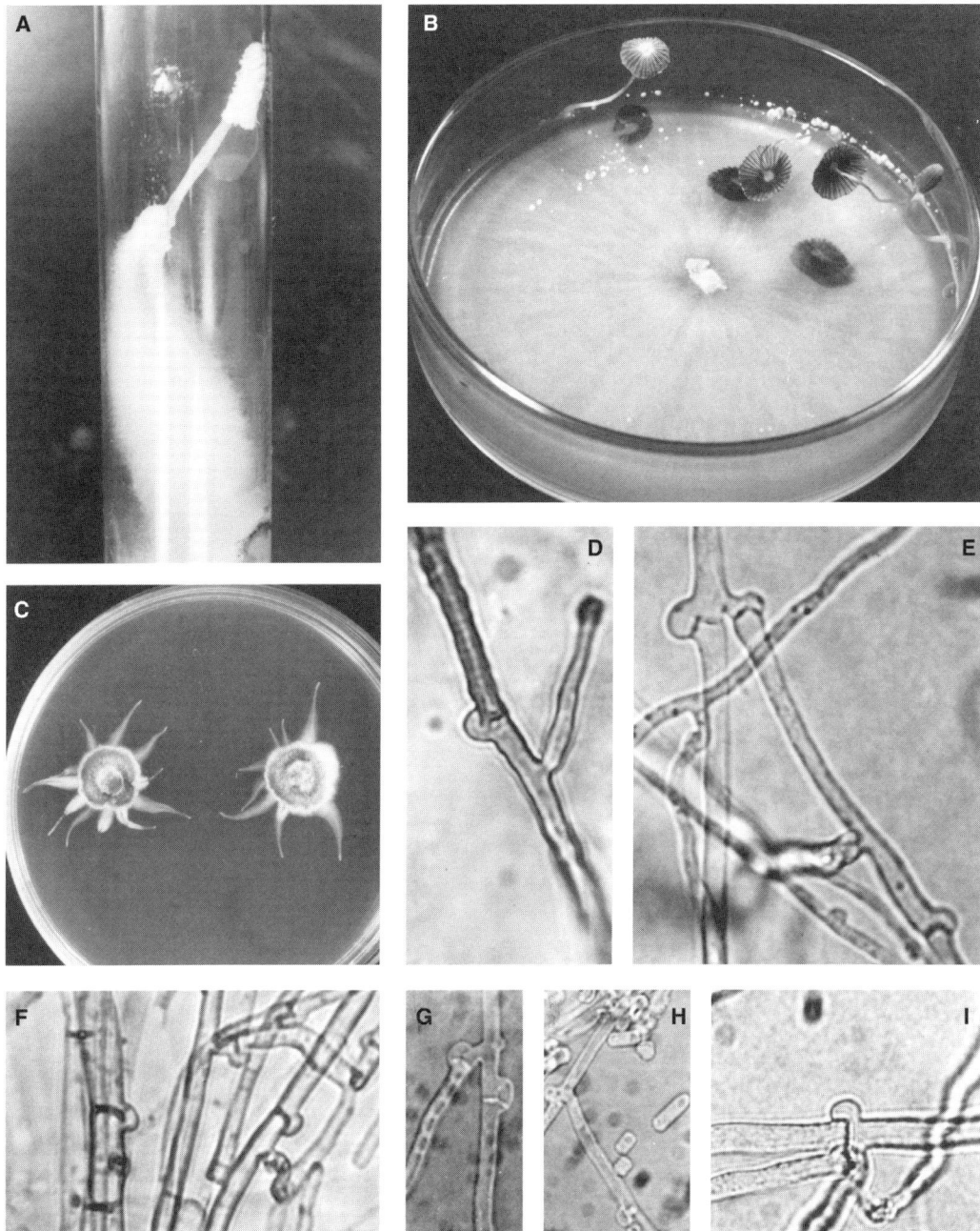

Basidiomycetous fungi. A,B: *Coprinus* spp. sporulated on agar (A: Isolate 85-17, B: 82-735). C: Rhizomorph formation of *Armillaria mellea* (Isolate 85-153) on agar culture. D–G,I: Hyphae with clamp connections on agar (D: Isolate 84-481; E: 84-595, F: P101, G: 73-50, I: 84-145). H: Hyphae and arthrospores (Isolate 73-83).

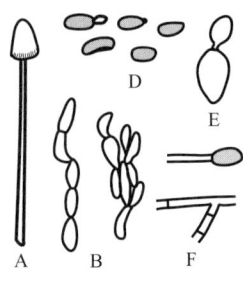

Coprinus Pers.

Tent. Disp. 62, 1797.

Type species: *C. comatus* (Mill) Pers.

References: Imazeki and Hongo 1987.

Morphology: Caps bell-shaped, pale yellowish brown centrally, grayish marginally, composed of pale brown spherical cells, radially lined. Stipes white, gradually tapering toward apexes. Basidia clavate. Basidiospores brown, cylindrical, slightly curved occasionally. Hyphae hyaline to subhyaline, not clumped, constricted at the branches, with close septa.

Dimensions: Cap 10 mm tall, 8 mm broad: component cells 24–26 × 12–14 μm. Stipe 60 mm tall, 2 mm wide basally, 1.2 mm wide apically. Basidiospores 6–7 × 3–4 μm. Hyphae 2–5 μm broad.

Material: 99-1 (= MAFF 347998, uncultivated soil, Tsukuba, Ibaraki, Japan).

Remarks: This fungus is characterized by dark basidiospores, bell-shaped caps, gills deliquesced into a black inky liquid on maturing from below upward. Basidia not well discernible.

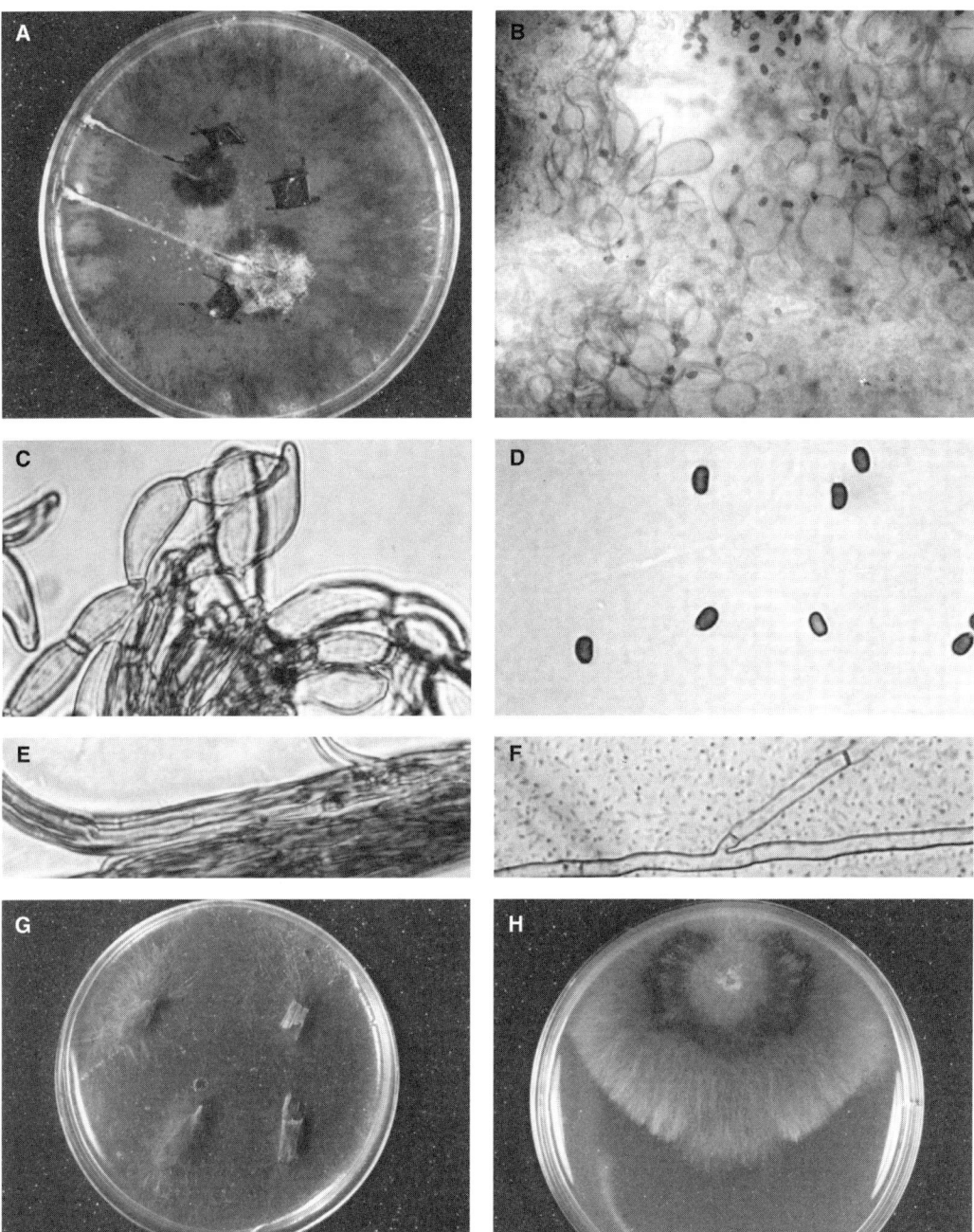

Coprinus sp. A: Colony on PDA with the residues of two fruit bodies. B,C: Part of cap component cells and basidiospores. D: Basidiospores. E: Part of stipe tissues. F: Unclumped hyphae. F: The fungus on the sectioned pieces of chopsticks recovered from soil. G,H: Eight-day-old colony on PDA.

MORPHOLOGY OF SOIL FUNGI

Deuteromycotina (Motosporic Fungi)

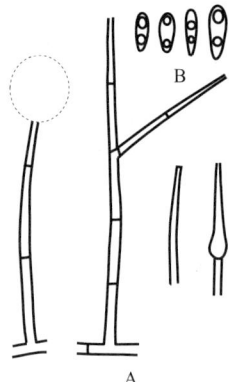

Acremonium Link

Syst. Mycol. 1:44, 1821.
Type species: *A. alternatum* Link

Acremonium macroclavatum T. Watanabe

References: Gams 1971; Gams et al. 1984, Watanabe et al., 2001b.

Morphology: Conidiophores (phialides) erect, simple, septate, hyaline, gradually tapering toward apexes with terminal slimy conidial masses. Conidia phialosporous, clavate, cylindrical or ellipsoidal, hyaline, 1-celled, usually biguttulate. Hyphae often crustaceous.

Dimensions: Conidiophores 20 to over 200 μm long, 2–4 μm wide basally. Conidial masses 12–40 μm in diameter. Conidia (5-) 8–14 × 2–3.6 (-4) μm.

Material: 00-50 (= MAFF 238162, Forest soil, Mt. Chibusa, the Bonin Islands, Tokyo, Japan).

Remarks: This fungus resembles *A. obclavatum* W. Gams (Gams et al., 1984) in forming exclusively solitary phialides and catenulate clavate conidia in imbricate chains, but differs from the latter in forming conidial heads and larger guttulate conidia.

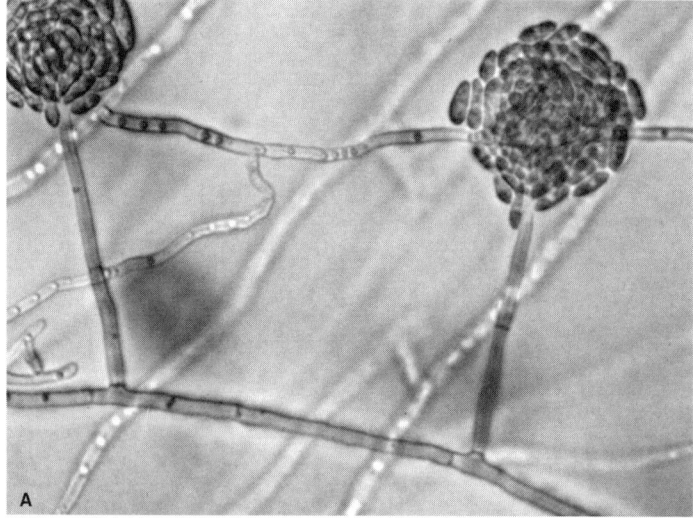

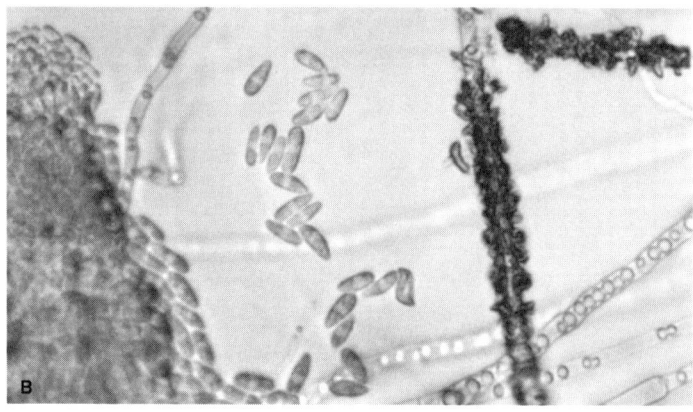

Acremonium macroclavatum. A: Conidiophores and spore masses. B: Conidia and crustaceous hyphae. (From Watanabe, T. et al. 2001b. *Mycoscience.* In press. With permission.)

Acremonium sp.

Synonym: *Cephalosporium* sp.

References: Gams 1971; Obayashi and Watanabe 1987.

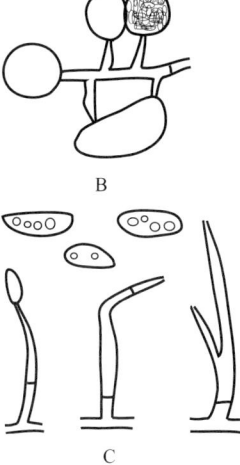

Morphology: Conidiophores (phialides) hyaline, erect, simple or branched, gradually tapering from base toward apex, bearing spherical spore masses terminally. Conidia phialosporous, hyaline, 1-celled, cylindrical or long-ellipsoidal, pointed at one or both ends, with 3–7 oil globules.

Dimensions: Conidiophores (phialides) 17.5–37.5 (-50.4) × 3.2–4 µm. Spore masses 27.5–45 µm in diameter. Conidia 6–8.5 (-9.3) × 2.1–2.8 (-4) µm.

Material: 87-3 (Radish roots, Miura, Kanagawa, Japan).

Remarks: This fungus causes round, brown lesions on radish roots. For a few plant pathogens such as *C. gramineum* Nisikado et Ikata and *C. gregatum* Allington et Chamberlain, the genus name *Cephalosporium* is more commonly used.

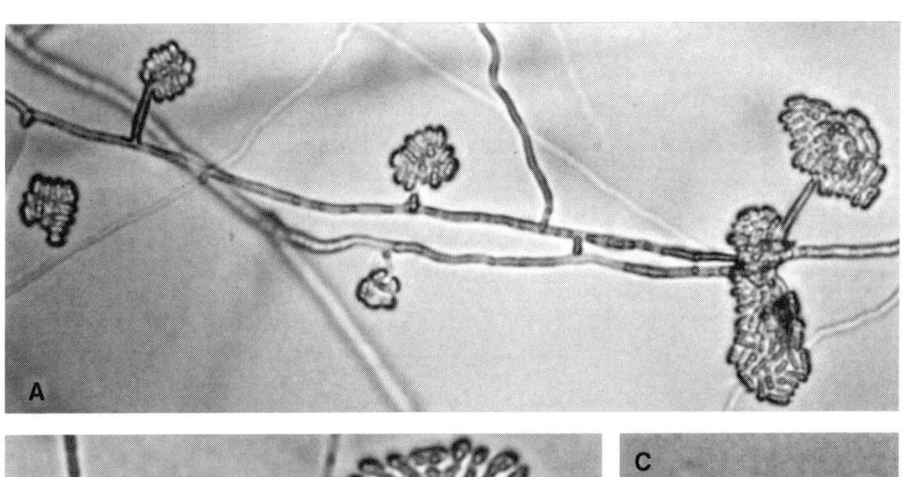

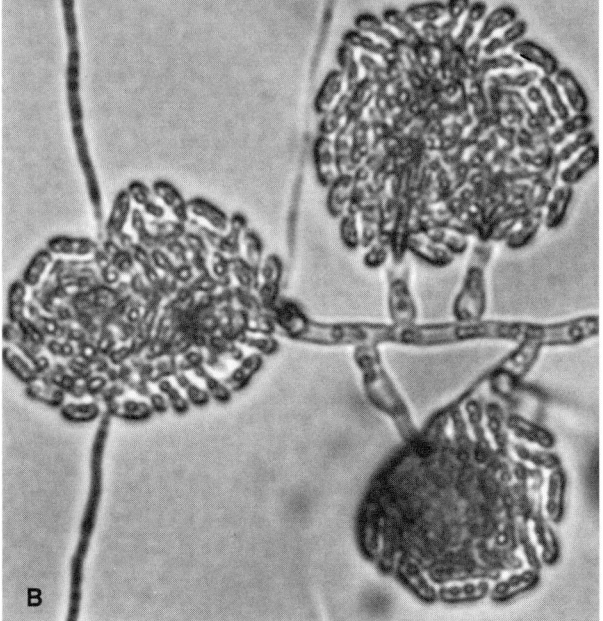

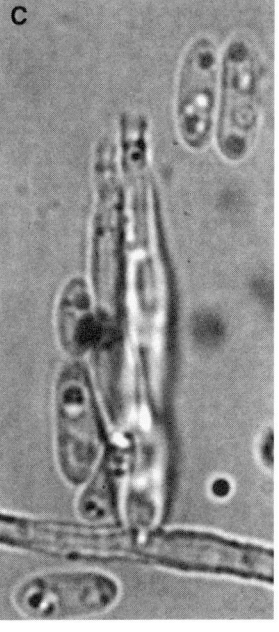

Acremonium sp.   A,B: Conidiophores and spore masses. C: Conidiophores and conidia.

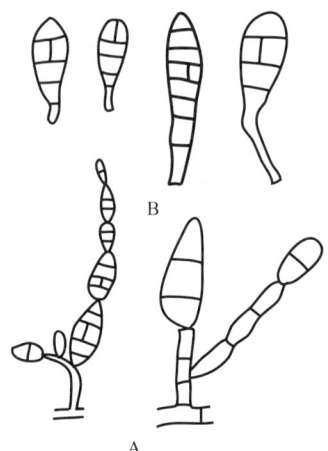

Alternaria Nees : Fr.

Syst. Mycol. 1:46, 1821.
Type species: *A. alternata* (Fr.) Keissler

Alternaria alternata (Fr.) Keissler

References: Ellis 1971, 1976; Simmons 1967; Watanabe 1975d; Watanabe et al. 1986a.

Morphology: Conidiophores pale brown, simple or branched, bearing catenulate conidia at the apex and apical fertile parts. Conidia catenulate, mostly up to 9 in a chain, often branched. Conidia porosporous, acropetally developed, (dark) brown, cylindrical or spindle-shaped, often with cylindrical beaks, muriform composed of 3–4 (-8) transverse walls and 1–2 longitudinal walls.

Dimensions: Conidiophores 17–40 (-139) × 3–3.9 µm. Conidia 18–45 (-70.5) × 6.5–15.5 (-17) µm: beaks 2.5–35 × 7–7.5 µm.

Material: 70-58 (Snap bean seed, Nagano, Japan); 72-X143 (Sugarcane roots, Taiwan, ROC); 83-22 (Japanese red pine seed, Hiroshima, Japan); 85-90 (Flowering cherry seed, Hachioji, Tokyo, Japan); 86-28 (Japanese cedar seed, Kasama, Ibaraki, Japan).

Remarks: Morphologies and dimensions of the fungus are variable in isolates studied.

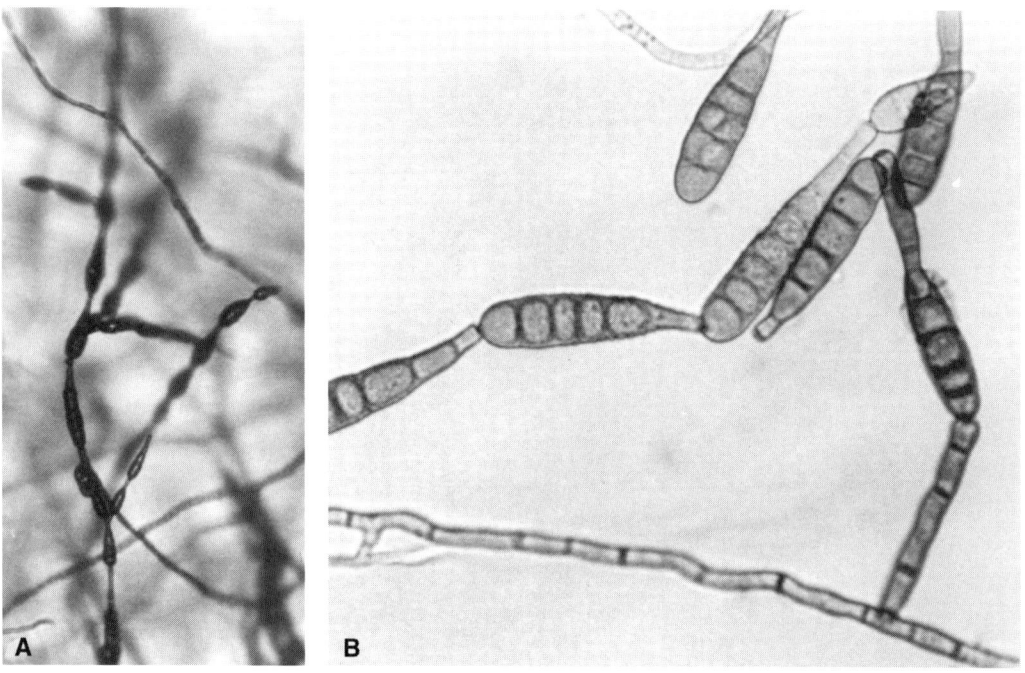

Alternaria alternata. A,B: Conidiophores and conidia. (From Watanabe, T. 1975d. *Trans. Mycol. Soc. Jpn.*, 16:264–267. With permission.)

Apiocarpella H. Sydow & Sydow

Ann. Mycol. 17:43, 1919.
Type species: *A. macrospora* (Speg.) H. & P. Syd.

Apiocarpella sp.

References: Sprague 1948.

Morphology: Pycnidia black, subglobose, covered with white hairs, peridium hard. Conidiophores hyaline, simple but united basally, bearing conidia apically. Conidia hyaline or subhyaline, ovate or spindle-shaped, rounded apically and pointed basally, usually 2- (rarely 3-) to 4-celled, unequal in apical and basal cell size.

Dimensions: Pycnidia 470–670 × 430–536 µm. Conidia (8.7-) 11–16.3 (-19.8) × (3-) 4.7–6.3 µm.

Material: 85-24, 85-93 (Flowering cherry seed, Hachioji, Tokyo, Japan).

Remarks: The genus name *Apiocarpella* Melnik is adopted by Farr et al. (1989).

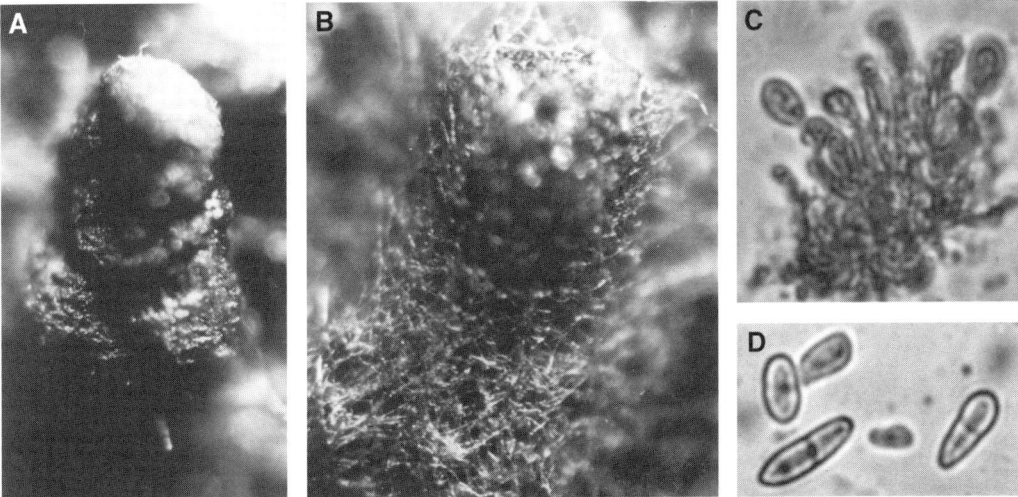

Apiocarpella sp. A,B: Pycnidia. C: Conidiophores and conidia. D: Conidia.

A, B

Arthrinium Kunze : Fr.

Syst. Mycol. 3:376, 1831.
Type species: *A. caricicola* Kunze : Fr.

Arthrinium st.

Teleomorph: *Apiospora montagnei* Sacc.

References: Cooke 1954; Ellis 1965; Watanabe et al. 1986a.

Morphology: Conidiophores erect, simple, bearing 1 to several conidia apically or near apex. Conidia blastosporus, brown, globose, ellipsoidal or lenticular with hyaline or subhyaline rim or germ slit.

Dimensions: Conidia 5–7.5 (-8.8) μm in maximum diameter, 3.7–4.5 (-6.3) μm thick.

Material: 70-95 (Snap bean seed, Nagano, Japan); 83-27, 84-469 (Japanese red pine seed, Hiba, Hiroshima, Japan); 85-94, 85-111 (Flowering cherry seed, Hachioji, Tokyo, Japan); 86-33 (Japanese cypress seed, Higashichikuma, Nagano, Japan).

Remarks: Conidia of this fungus are smaller, under 8 μm in diameter, compared with those of some 20 species of *Arthrinium* so far described.

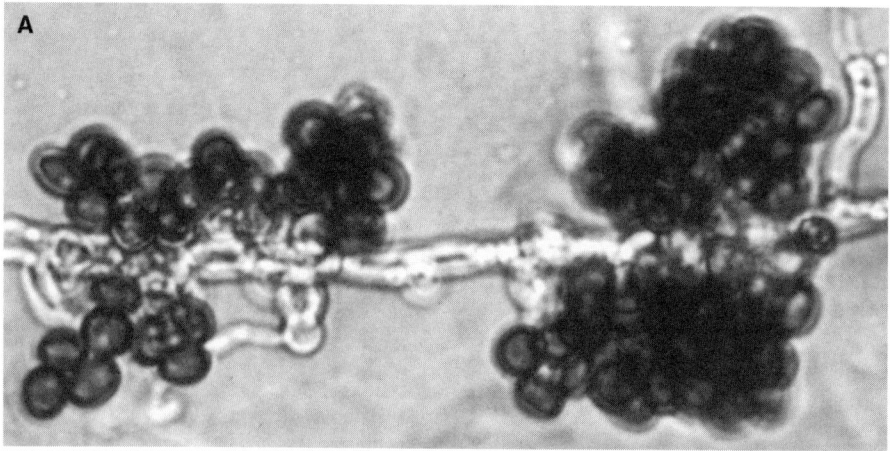

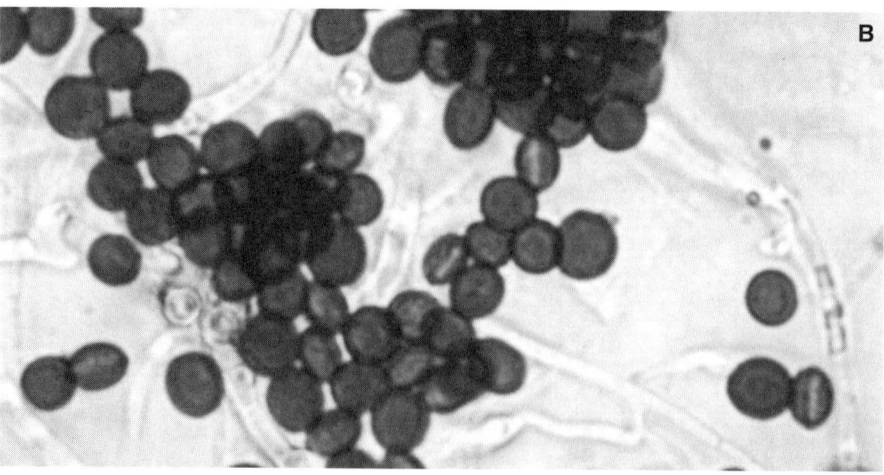

Arthrinium state of *Apiospora montagnei*.   A: Conidia and conidiogenous cells. B: Conidia.

Arthrobotrys Corda

Prachtflora p. 43, 1839.
Type species: *A. superba* Corda

Arthrobotrys oligospora Fresenius

References: Cooke and Godfrey 1964; Drechsler 1937.

Morphology: Conidiophores hyaline, erect, simple or rarely branched, bearing 2–6 conidia sympodially on sterigmata in the apical parts. Conidia sympodulosporous, hyaline, ovate, 2-celled, composed of large apical cells and small basal cells, apiculate and truncate at base.

Dimensions: Conidiophores 200–575 × 7–8 (-12.5) μm. Conidia 17.5–22.5 × 10.2–13.8 μm.

Material: 85-5, 85-113 (Flowering cherry seed, Hachioji, Tokyo, Japan).

Remarks: This fungus is one of nematode trappers.

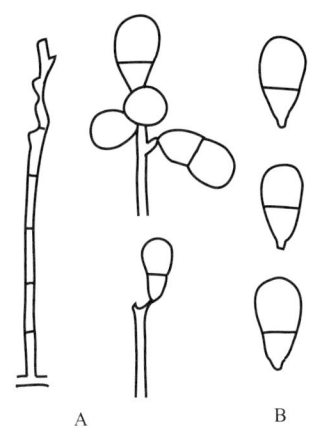

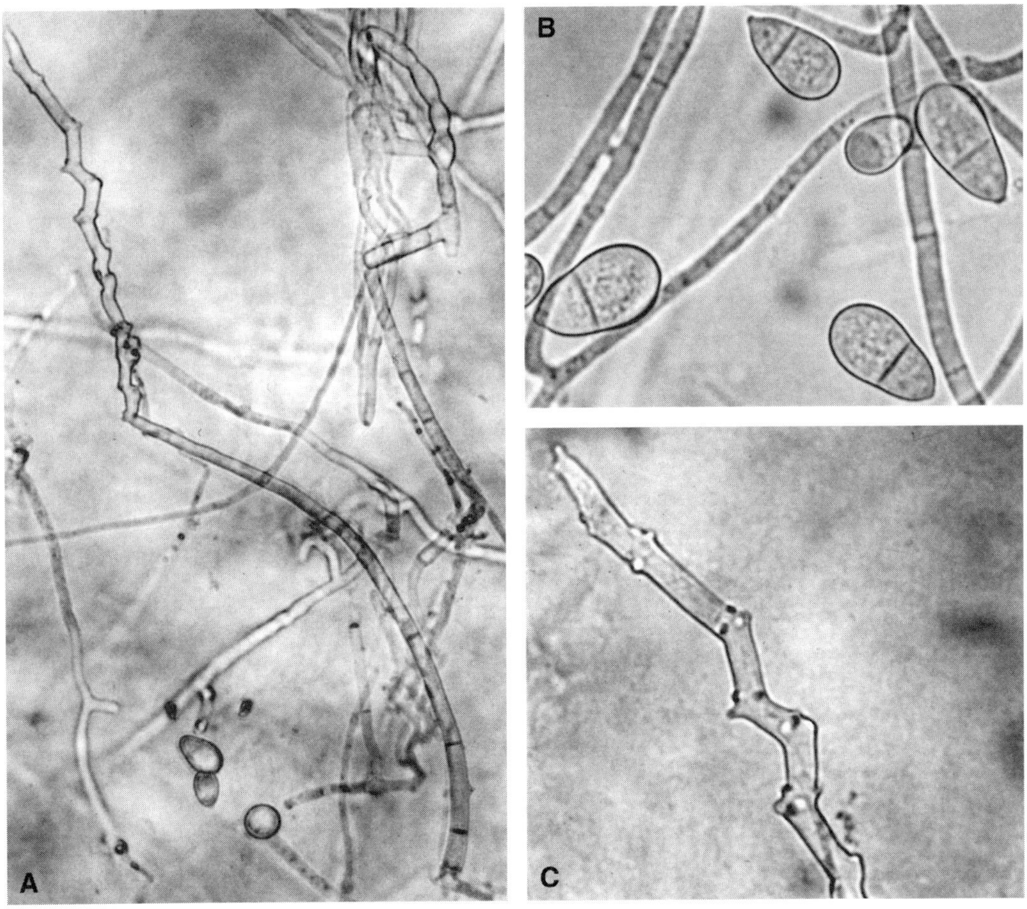

Arthrobotrys oligospora.   A,C: Conidiophores. B: Conidia.

Aspergillus Mich. : Fr.

Syst. Mycol. 3:385, 1832.
Type species: *A. flavus* Link : Fr.

Key to Species

1. Spore masses   green...2
                     not green..5

2. Spore masses   cylindrical, dark green..3
                     radiate, yellowish green...............................*A. parasiticus*

3. Conidia   under 3 µm in diameter..4
                     4.2–6.3 µm in diameter..................................*A. brevipes*

4. Spore masses   175–244 µm long..*A. fumigatus*
                     under 160 µm long .... *Aspergillus* sp. Sect. *Clavati*

5. Spore masses   black..*A. niger*
                     brown.................................. *Aspergillus* sp. Sect. *Wentii*

Aspergillus brevipes Smith

References: Raper and Fennell 1965.

Morphology: Conidiophores hyaline, simple, occasionally thick-walled, inflated globosely or ellipsoidally at the apex (called vesicles), bearing spore heads composed of catenulate conidia borne on uniseriate phialides on vesicles: conidial heads dark green, loosely columnar. Conidia phialosporous, pale brown to yellowish brown, globose, delicately rough at the surface.

Dimensions: Conidiophores 35–125 × 5–6.8 µm: vesicles 12.5–16.3 µm; phialides 7.5–10 × 3.7–4.5 µm. Conidial heads 140–325 × 15–20 µm. Conidia 4.2–6.3 µm in diameter.

Material: 73-462 (Strawberry field soil, Shizuoka, Japan).

Remarks: Colonies on agar cultures are grayish green.

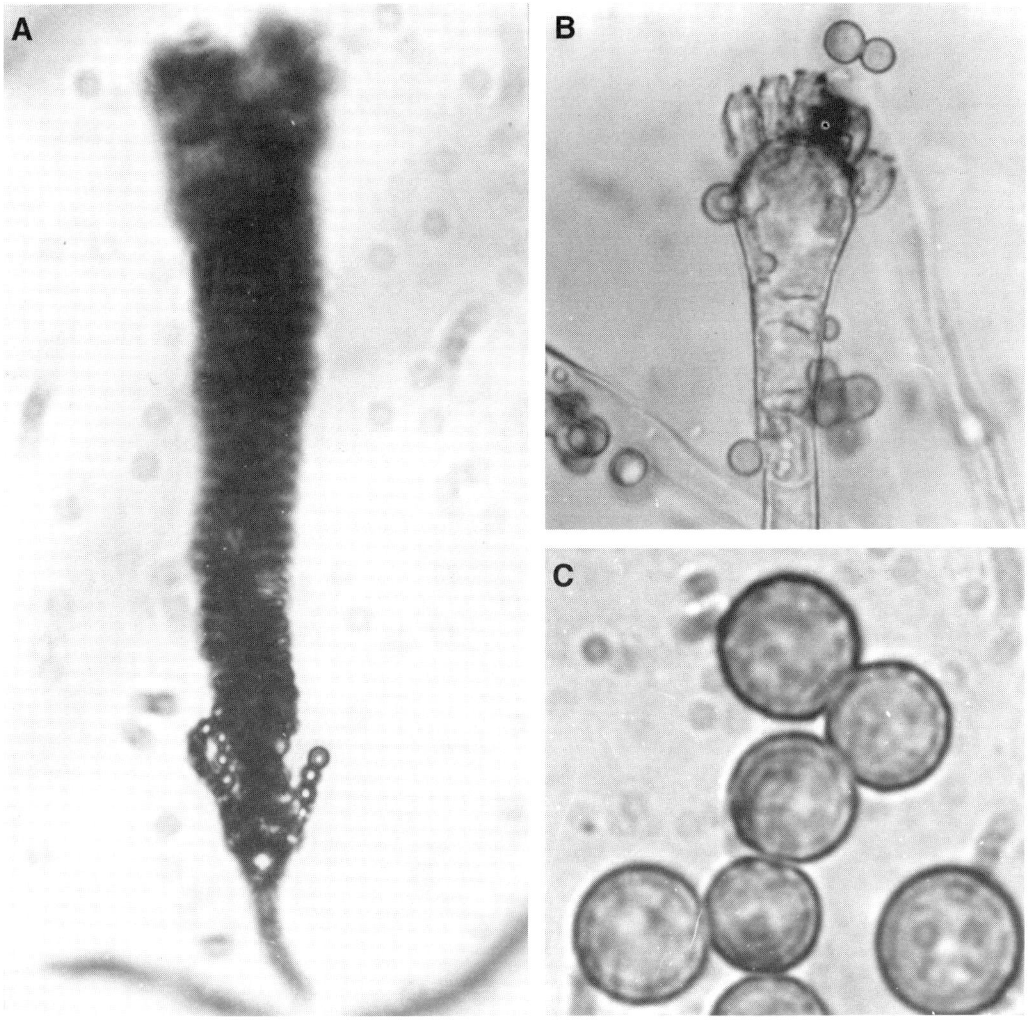

Aspergillus brevipes. A: Conidiophore and spore mass. B: Vesicle, phialides, and conidia. C: Conidia.

Aspergillus fumigatus Fresenius

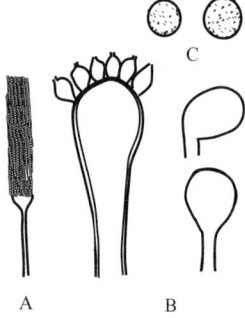

References: Raper and Fennell 1965; Watanabe 1975d.

Morphology: Conidiophores hyaline, simple, thin in cell wall, inflated clavately at the apex forming often nodded vesicles, bearing conidial heads composed of catenulate conidia borne on uniseriate phialides on pale brown vesicles: conidial heads dark bluish green, columnar. Conidia phialosporous, pale green, globose, slightly echinulate.

Dimensions: Conidiophores 55–125 μm tall: vesicles (12.5-) 13.3–14.6 (-16.3) μm in diameter: phialides 3.6–4.9 × 2.5 μm. Conidial heads 175.1–243.3 × 15–55 μm. Conidia 2.4–2.7 μm in diameter.

Material: 72-X29 (Sugarcane roots, Taiwan, ROC).

Remarks: Colonies on agar cultures are dark green.

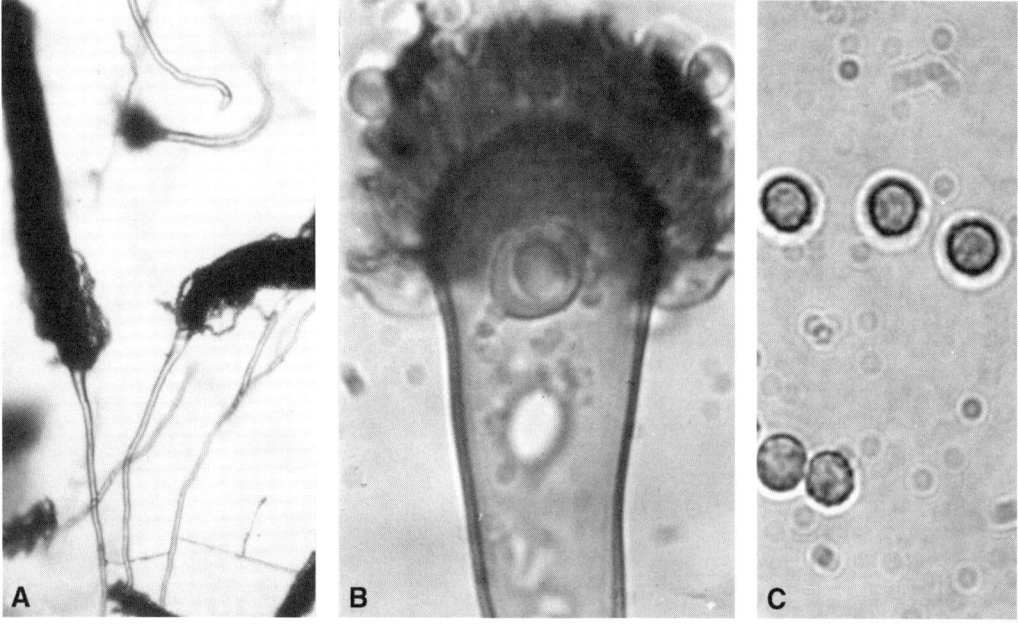

Aspergillus fumigatus. A: Conidiophores and spore mass. B: Vesicle and phialides. C: Conidia. (From Watanabe, T. 1975d. *Trans. Mycol. Soc. Jpn.*, 16:264–267. With permission.)

Aspergillus niger van Tiegh.

References: Raper and Fennell 1965; Watanabe et al. 1986a.

Morphology: Conidiophores hyaline or pale brown, erect, simple, thick-walled, with foot cells basally, inflated at the apex forming globose vesicles, bearing conidial heads split into over 4 loose conidial columns with over 4 fragments apically, composed of catenulate conidia (over 15 conidia/chain) borne on uniseriate or biseriate phialides on pale brown, globose vesicles and phialides acutely tapered at apex. Conidia phialosporous, brown, black in mass, globose, minutely echinulate.

Dimensions: Conidiophores over 740 μm × 10–14 μm: vesicles 55–75 μm long: phialides 5–13.8 μm long. Conidia (2.7-) 3.7–4.5 μm in diameter.

Material: 69-303 (Pineapple field soil, Okinawa, Japan); 70-1493 (Pineapple field soil, Hachijo, Tokyo, Japan); 84-464 (Japanese red pine seed, Higashichikuma, Nagano, Japan).

Remarks: Colonies on agar cultures are typically black homogeneously.

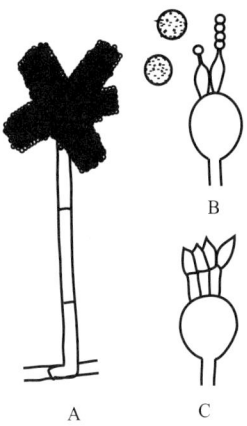

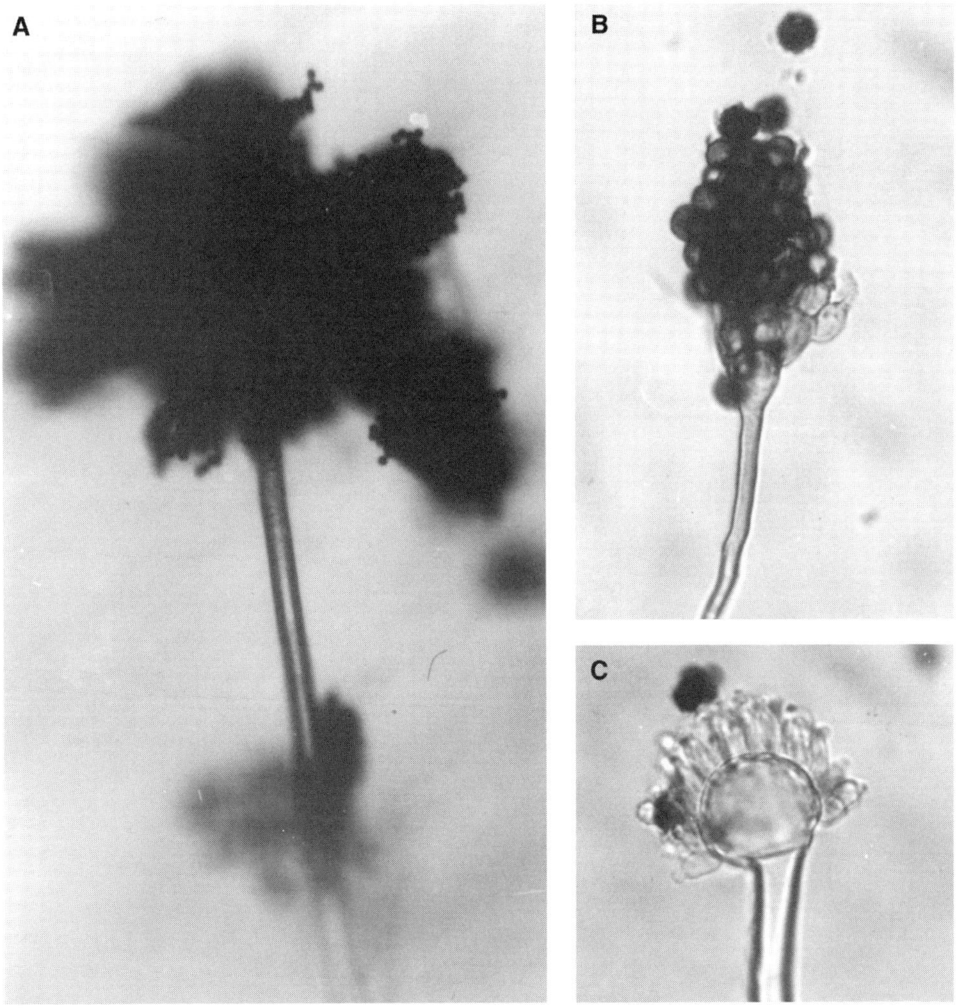

Aspergillus niger.   A: Conidiophore and spore mass. B,C: Vesicles, phialides, and conidia (B).

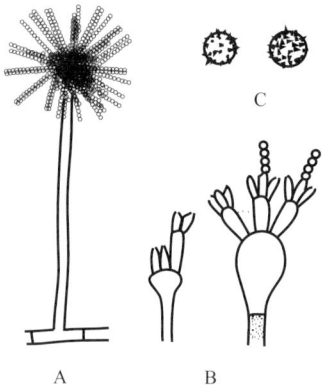

Aspergillus parasiticus Speare

References: Raper and Fennell 1965; Watanabe 1971.

Morphology: Conidiophores erect, simple, rough in the surface, with foot cells basally, inflated at the apex forming globose vesicles, bearing radiate conidial heads composed of catenulate conidia borne on uniseriate or rarely biseriate phialides: conidial heads yellowish green, radiate, columnar. Conidia phialosporous, pale green, globose, echinulate.

Dimensions: Conidiophores over 400 μm tall × 10.5–13.5 μm: vesicles 24–40 μm in diameter: phialides 6.2–16.3 × 3.5–4.3 μm. Conidial heads 160–250 μm in diameter. Conidia 3.7–5.5 μm in diameter.

Material: 69-329 (Pineapple field soil, Okinawa, Japan).

Remarks: Colonies on Czapek agar are dark yellowish green and velvety.

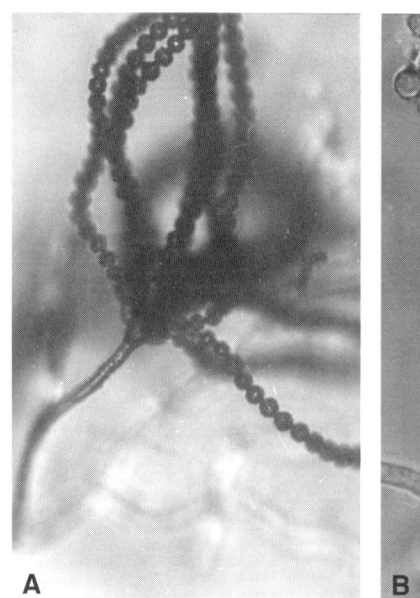
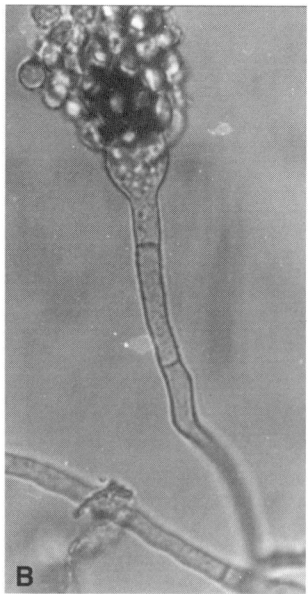
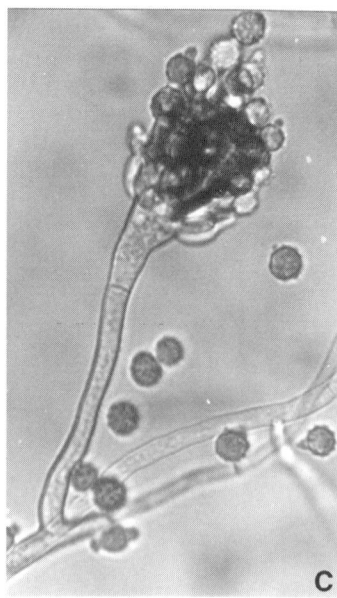

Aspergillus parasiticus. A: Conidiophore and catenulate conidia. B,C: Vesicles, phialides, and conidia. (From Watanabe, T. 1971. *Trans. Mycol. Soc. Jpn.*, 12:35–47. With permission.)

Aspergillus sp., Sect. *Clavati*

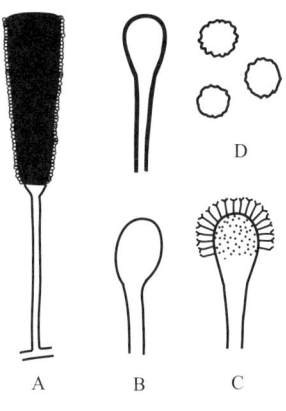

References: Raper and Fennell 1965; Samson and Pitt 1985, 1990; Watanabe and Sato 1988.

Morphology: Conidiophores, pale brown, erect, simple or rarely branched, smooth in the surface, bearing dark green cylindrical or clavate conidial heads composed of densely catenulate conidia borne on uniseriate phialides developed on subglobose or clavate vesicles. Conidia phialosporous, pale green, globose, rough in the surface.

Dimensions: Conidiophores 125–230 µm tall: vesicles 20–30 µm in diameter: phialides 6–6.5 × 2 µm. Conidial heads 125–160 × 42–70 µm. Conidia 2.5–2.8 µm in diameter.

Material: 86-31 (Japanese cedar seed, Nakashinkawa, Toyama, Japan).

Remarks: Cultures on agar cultures are yellowish green.

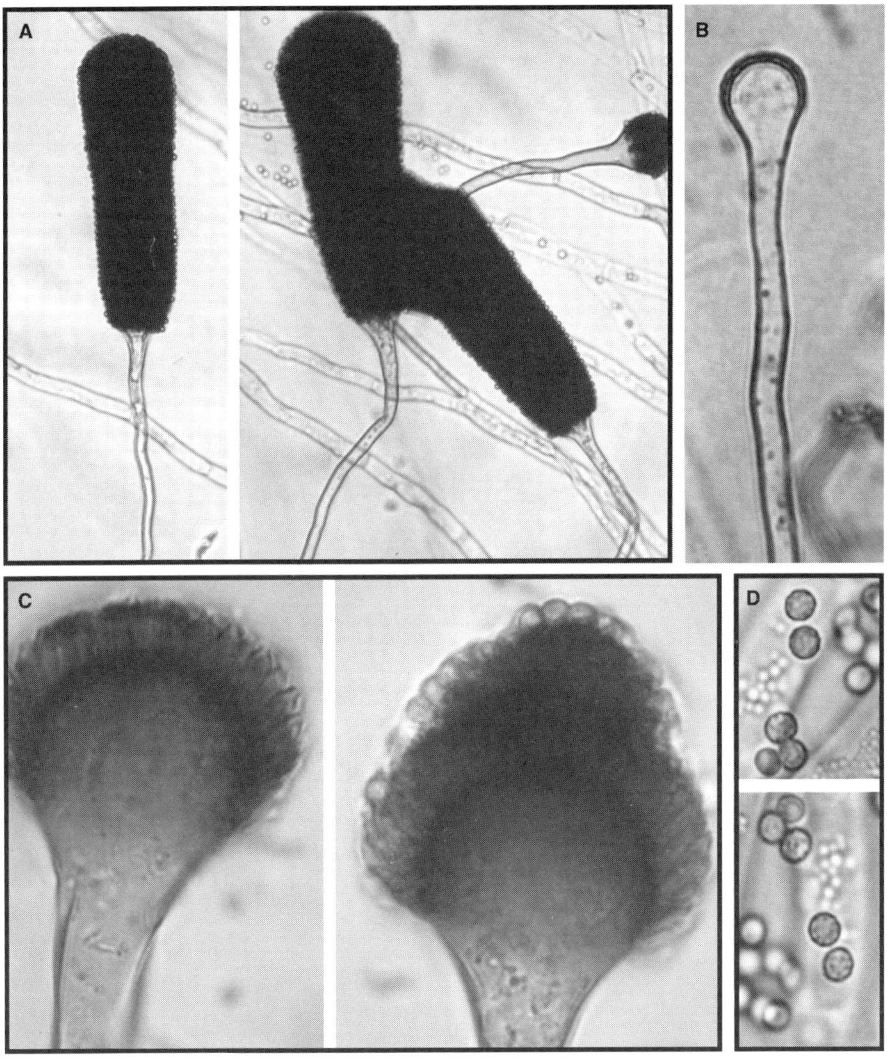

Aspergillus sp., Sect. *Clavati*. A: Conidiophore and spore mass. B: Conidiophore with terminal vesicle. C: Phialides. D: Conidia. (From Watanabe, T. and Sato, T. 1988. *Trans. Mycol. Soc. Jpn.*, 29:143–150. With permission.)

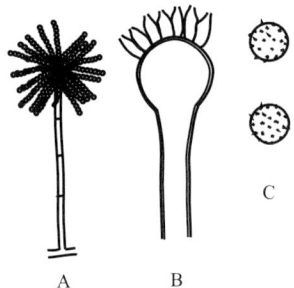

Aspergillus sp., Sect. *Wentii*

References: Domsch et al. 1980; Samson and Pitt 1985, 1990; Watanabe et al. 1986a.

Morphology: Conidiophores pale brown, erect, simple, smooth in the surface, bearing grayish brown spherical spore masses composed of catenulate conidia borne on uniseriate phialides developed on pale brown globose vesicles. Conidia phialosporous, yellowish brown, globose or ellipsoidal, rough in the surface.

Dimensions: Conidiophores ca. 365 × 5 µm: vesicles 12.5–20 µm in diameter: phialides 6–6.5 × 2 µm. Conidia 3.6–5.1 (-6.3) µm in diameter.

Material: 84-463, 84-467 (Japanese red pine seed, Higashichikuma, Nagano, Japan).

Remarks: Cultures on agar cultures are pale yellowish to yellowish green.

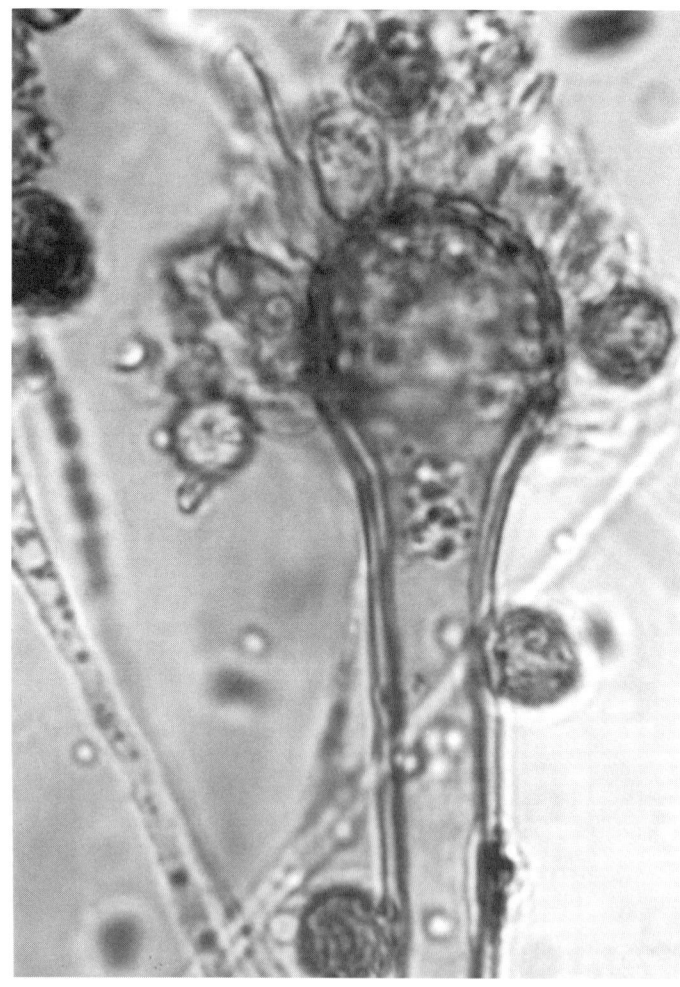

Aspergillus sp., Sect. *Wentii*. A–C: Conidiophore, vesicle, and conidia.

Aureobasidium Viala & Boyer

C. R. Hebd. Séanc. Acad. Sci., Paris 112:1149, 1891.
Type species: *A. pullulans* (de Bary) Arnaud

Aureobasidium pullulans (de Bary) Arnaud

References: Cooke 1959, 1962; Ellis 1971; Hermanides-Nijhof 1977; Watanabe et al. 1986a.

Morphology: Conidiophores lacking. Conidia blastosporous, lateral directly on hyphae, forming spore masses on hyphae, hyaline or brown, long-ellipsoidal or subglobose, 1- or 2-celled, often apiculate at one or both ends.

Dimensions: Conidia: cylindrical 6.2–13 × 2.5–7.5 μm: subglobose 2.5–3.8 μm in diameter.

Material: 83-50 (Japanese red pine seed, Hiba, Hiroshima, Japan); 83-67 (Japanese cedar seed, Yoshino, Nara, Japan); 83-70 (Japanese cypress seed, Kiso, Nagano, Japan), 84-473 (Japanese red pine seed, Kamiina, Nagano, Japan); 84-474 (Japanese red pine seed, Saihaku, Tottori, Japan); 85-6, 85-7 (Flowering cherry seed, Hachioji, Tokyo, Japan).

Remarks: Conidia are apparently arthrosporous. Cultures on PDA are resupinate, spreading, often black yeast-like.

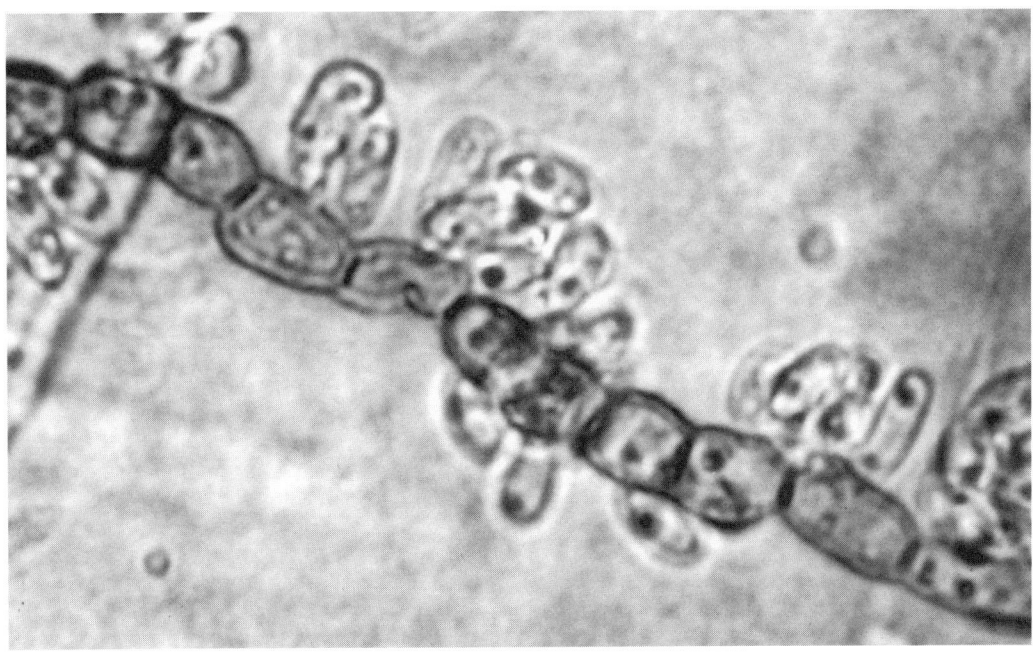

Aureobasidium pullulans. Hypha and conidia.

Aureobasidium sp.

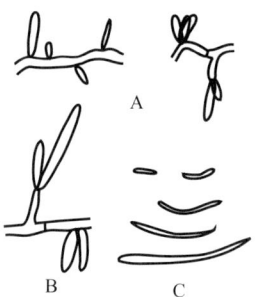

References: Cooke 1959, 1962; Ellis 1971; Hermanides-Nijhof 1977; Watanabe et al. 1986a.

Morphology: Conidiophores lacking. Conidia blastosporous, lateral directly on hyphae, forming spore masses on hyphae, hyaline, cylindrical, 1-celled.

Dimensions: Conidia 32–54 × 2–5 µm.

Material: 99-5 (Cucumber seed, Tateyama, Chiba, Japan).

Remarks: Cultures on PDA are resupinate, spreading, often black yeast-like.

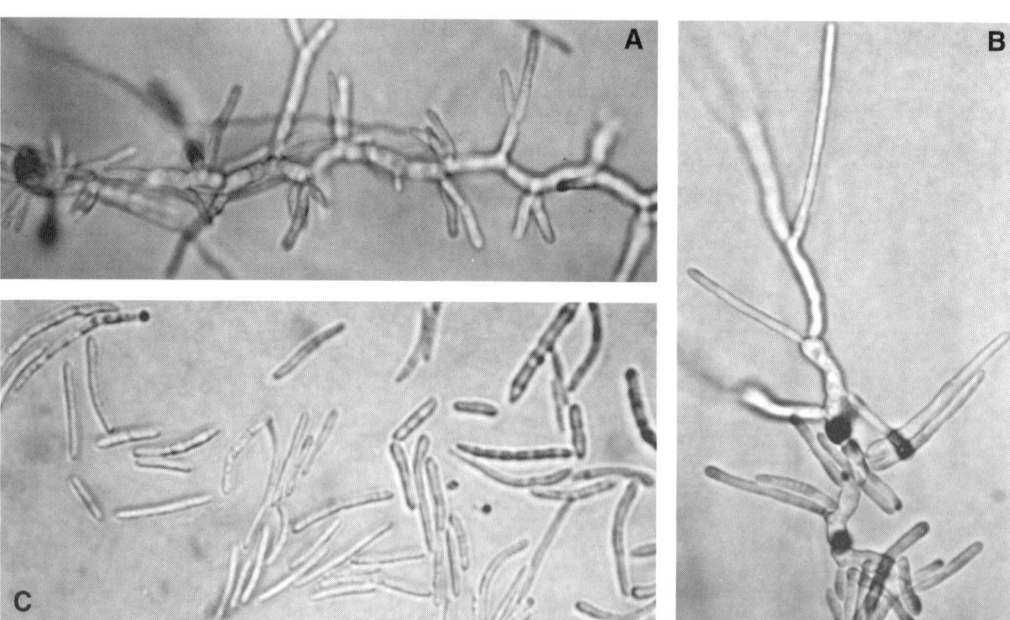

Aureobasidium sp.   A,B: Hypha and conidia. C: Detached conidia.

Basipetospora Cole & Kendrick

Can. J. Bot. 46:991, 1968.

Type species: *B. rubra* Cole & Kendrick

Basipetospora rubra Cole & Kendrick

Teleomorph: *Monascus ruber* van Tieghem = *Backusia terricola* Thirum., Whiteh. & Mathur

References: Cole and Kendrick 1968; Thirumalacher et al. 1964.

Morphology: Conidiophores and hyphae not well differentiated. Conidia aleuriosporous or meristem arthrosporous, formed by inflation of apical hyphal cells, followed by septation and further growth, up to 13 conidia in a chain developed basipetally, formed on agar surface or aerially, hyaline or pale brown, 1-celled, globose or subglobose, smooth, readily detached singly or in chains of 2–3 spores together, rarely truncate at one end.

Dimensions: Hyphae 1.8–6 μm wide. Conidia 7.5–12.5 μm in diameter.

Material: 86-65 (Japanese black pine seed, Tango, Kyoto, Japan).

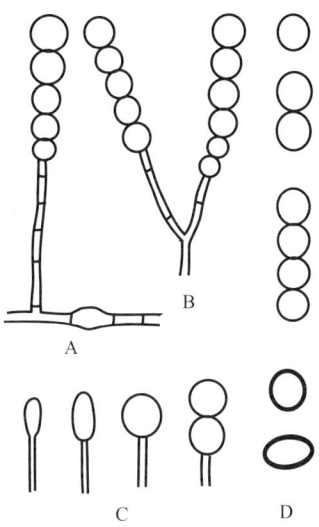

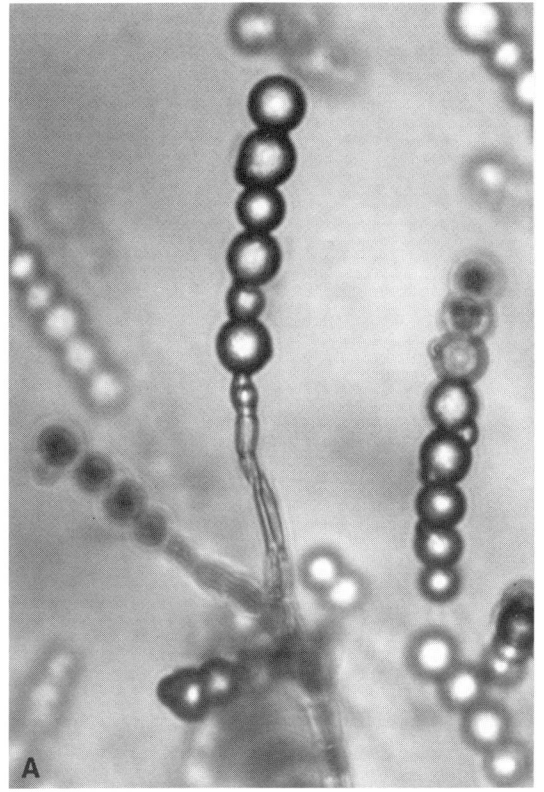
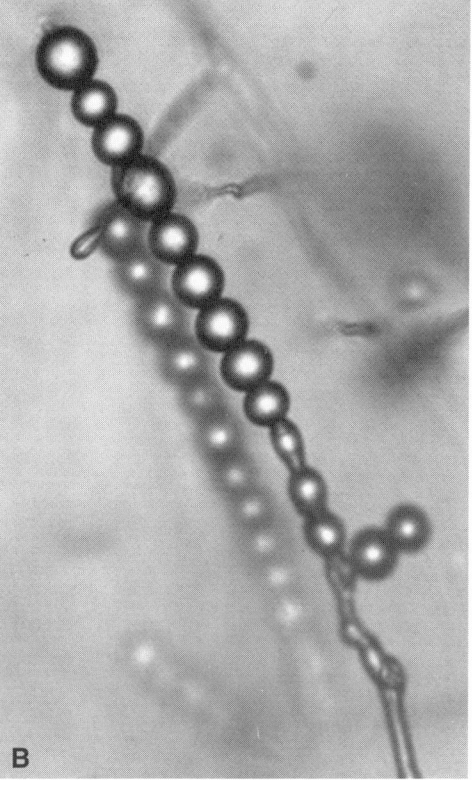

Basipetospora rubra.   A,B: Catenulate conidia. C: Sporulation process. D: Thick-walled conidia.

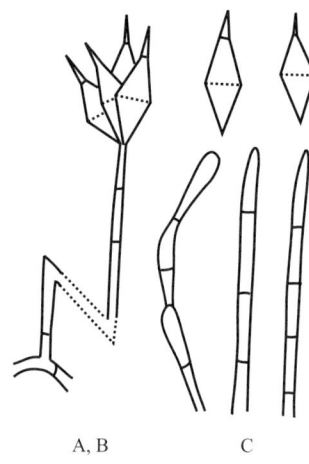

Beltrania Penzig

Nuovo G. Bot. Ital. 14:72, 1882.
Type species: *B. rhombica* Penzig

Beltrania rhombica Penzig

References: Ellis 1971; Pirozynski 1963; Watanabe 1971.

Morphology: Conidiophores pale brown, simple, occasionally proliferated forming nodes, inflated slightly at the apex, bearing 3–4 conidia sympodially. Setae erect, simple, dark brown, elongated over conidiophores. Conidia sympodulosporous, pale brown, 1-celled, biconical, with single apical protuberances.

Dimensions: Conidiophores over 250 × 3.7–5.5 μm. Conidia 22.5–28 × 8.7–10 μm: appendages 3.2–6 μm long. Setae over 350 × 2.7–3.3 μm.

A, B   C

Material: 69-312 (Pineapple field soil, Okinawa, Japan).

Remarks: The fungus sporulated well on corn meal agar.

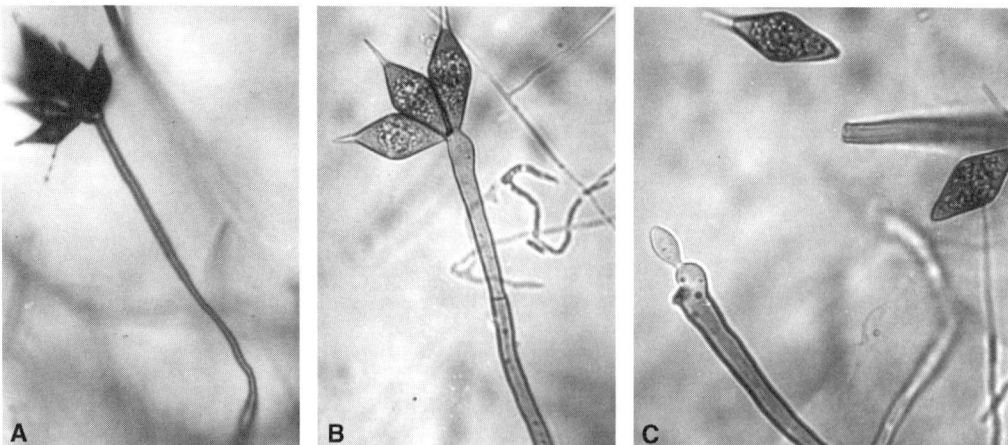

Beltrania rhombica. A,B: Conidiophores and conidia. C: Apex of conidiophore and conidia. (From Watanabe, T. 1971. *Trans. Mycol. Soc. Jpn.*, 12:35–47. With permission.)

Biporalis Shoemaker

Can. J. Bot. 37:879, 1959.
Type species: *B. maydis* (Nisikado) Shoemaker

Key to Species

1. Conidia over 28 μm long... 2

   15–26.3 μm long .... *B. australiensis*

2. Conidia over 70 μm long, with basal hilum.......................... *B. holmi*

   under 60 μm long, without basal hilum .... *B. sacchari*

Biporalis australiensis (M. B. Ellis) Tsuda & Ueda

Synonym: *Drechslera australiensis* M. B. Ellis

Teleomorph: *Cochliobolus australiensis* (Tsuda & Ueyama) Alcorn

References: Ellis 1971; Shoemaker 1959; Tsuda and Ueyama 1981.

Morphology: Conidiophores brown, erect, simple or branched, bearing conidia apically and laterally on apical fertile parts. Conidia porosporous, solitary, pale brown, ellipsoidal, usually 4-celled, occasionally inflated apically, without hilum basally.

Dimensions: Conidiophores 60–150 × 3.7–5 µm. Conidia 15–26.3 × 7–8.8 µm.

Material: 85-P17 (Paulownia root, Itapua, Paraguay).

Remarks: Conidia of this fungus are characteristically small in size.

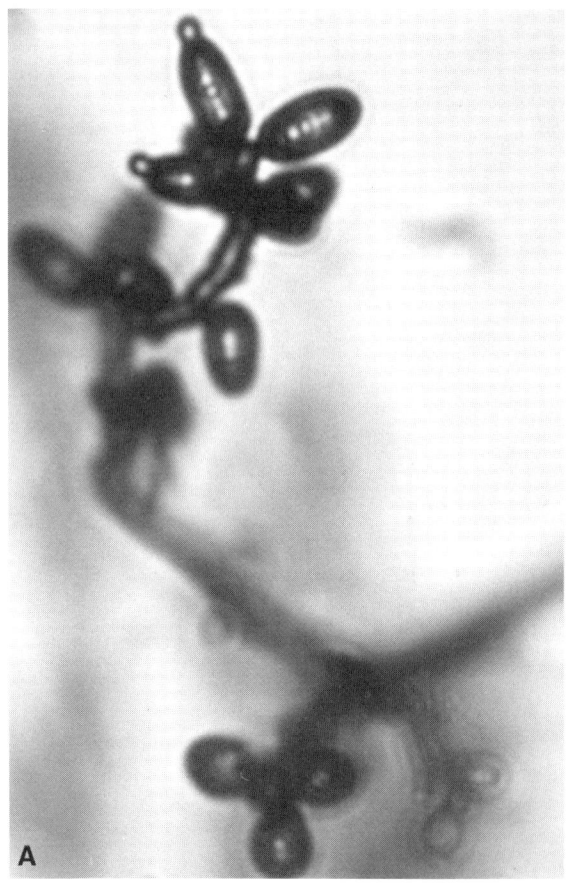
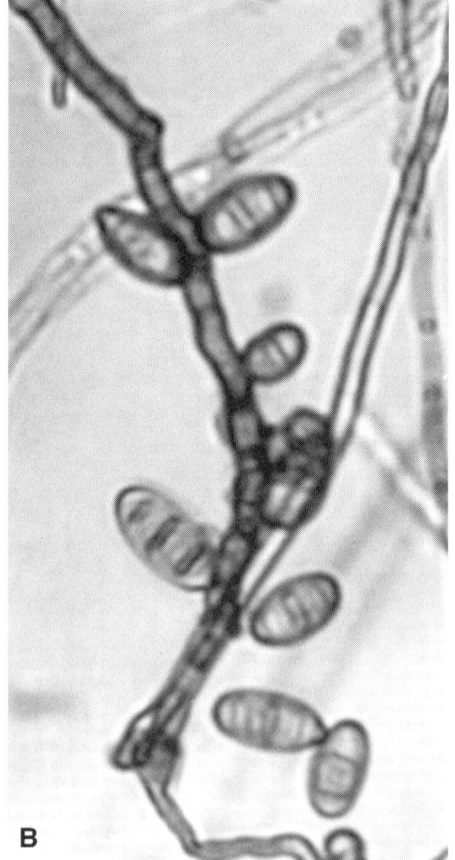

Biporalis australiensis. A,B: Conidiophores and conidia.

Biporalis holmii (Luttrell) Shoemaker

Synonym: *Helminthosporium holmii* Luttrell

References: Luttrell 1963; Shoemaker 1959; Watanabe 1975d.

Morphology: Conidiophores brown, erect, simple, bearing conidia at the apical fertile area, conspicuously porous after detachment of conidia. Conidia porosporous, dark brown, thick-walled, subellipsoidal, inflated basally, mainly 8-celled, with distinctly clear septa near both ends, and hilum basally.

Dimensions: Conidiophores 300–500 × 5–5.2 µm. Conidia 70–102.2 × 17.5–27.5 µm.

Material: 72-X 14 (Sugarcane root, Taiwan, ROC).

Remarks: Although this fungus was described as *Helminthosporium holmii* Luttrell (Watanabe, 1975), it may be better classified as *B. holmii* following Shoemaker (1959), because of formation of fusoid conidia and germination pattern by 1 germ tube from each end.

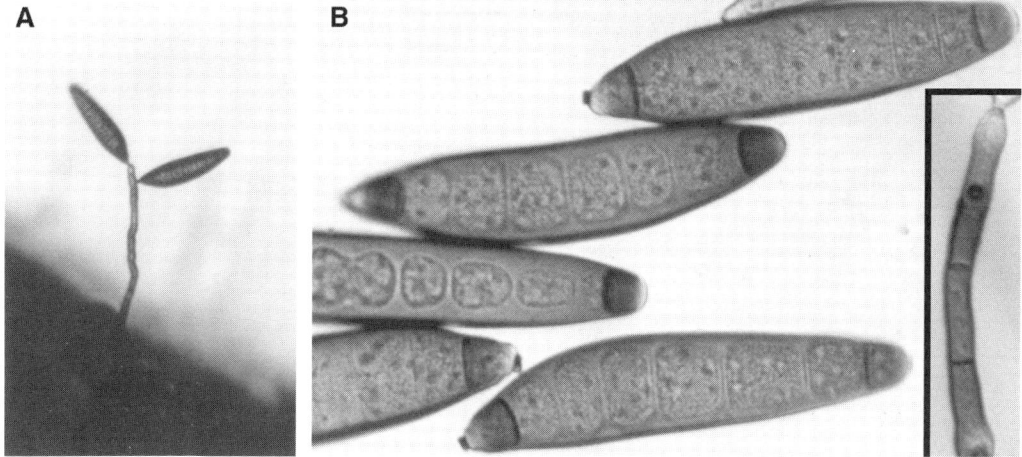

Biporalis holmii. A,B: Conidiophores and conidia. C: Part of germ tube and end cell of conidium. (From Watanabe, T. 1975d. *Trans. Mycol. Soc. Jpn.*, 16:264–267. With permission.)

Biporalis sacchari (Butler) Shoemaker

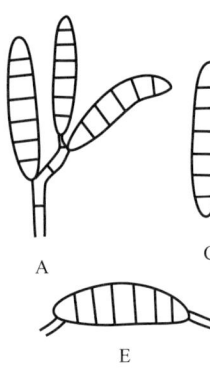

Synonym: *Helminthosporium sacchari* Butler

References: Martin et al. (Eds.) 1961; Sprague 1950; Watanabe 1975d.

Morphology: Conidiophores pale brown, erect, simple, often twisted, bearing conidia in the apical fertile parts. Conidia porosporous, pale brown, cylindrical, straight or curved partially, mainly 6- to 7-celled, without hilum basally.

Dimensions: Conidiophores 75.3–100 × 5 µm. Conidia 28.7–50 × 8.7–11.3 µm.

Material: 72-X 13 (Sugarcane root, Taiwan, ROC).

Remarks: This fungus was previously described as *Helminthosporium sacchari* (Watanabe 1975). It usually germinated from each end cell, but rarely from one of the central cells.

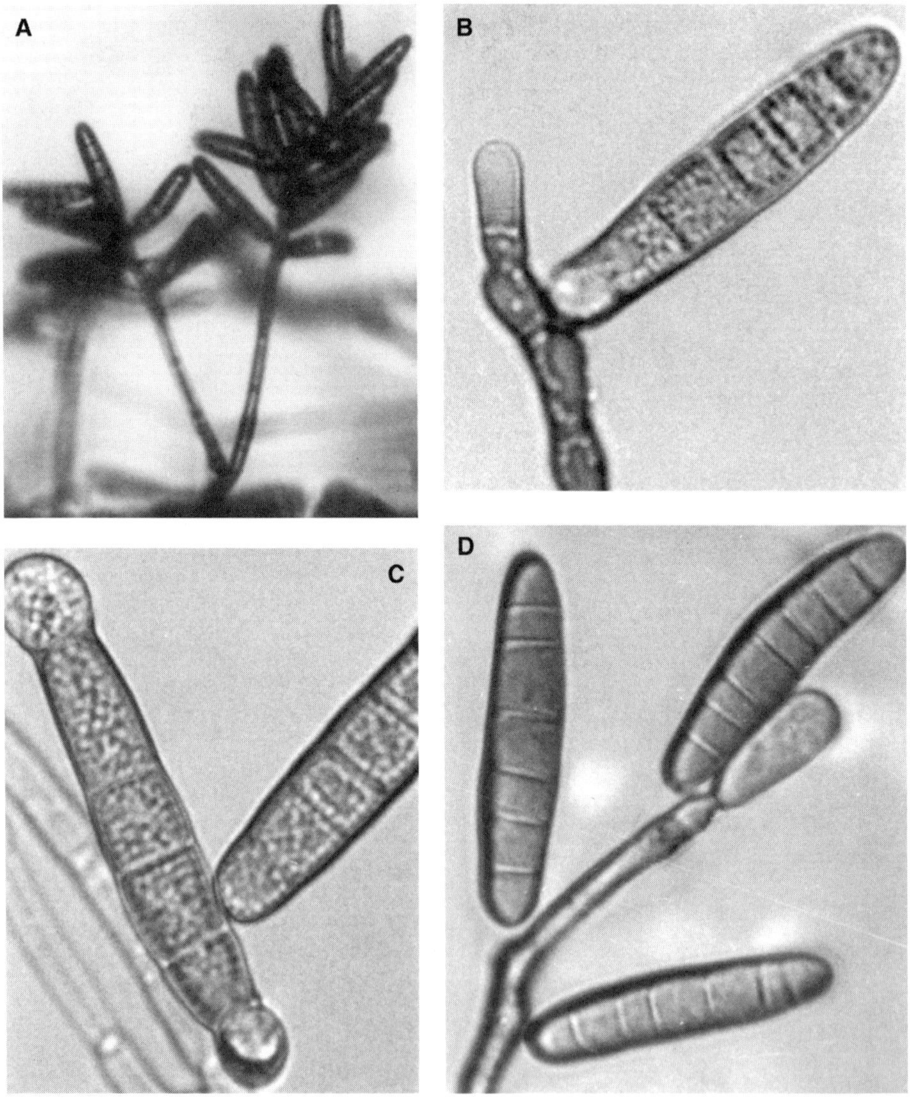

Biporalis sacchari. A–D: Conidiophores and conidia. E: Germ tubes from both end cells of germinated conidium. (From Watanabe, T. 1975d. *Trans. Mycol. Soc. Jpn.*, 16:264–267. With permission.)

Bispora Corda

Icon. Fung. 1:9, 1837.
Type species: *B. antennata* (Pers. : Fr.) Mason.

Bispora betulina (Corda) Hughes

References: Ellis 1971; Hughes 1958; Tubaki and Ito 1975.

Morphology: Conidiophores pale brown, erect, short, simple, bearing catenulate conidia at the apex. Conidia blastosporous, apically developed, brown, ellipsoidal, usually 2-celled, with thick, dark brown septum.

Dimensions: Conidiophores 6.5–69 × 2.5–4.5 µm. Conidia 12.5–17 × 4.5–5 µm.

Material: 75-155 (Gentian seed, Nagano, Japan).

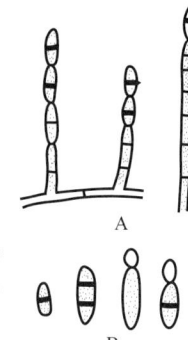

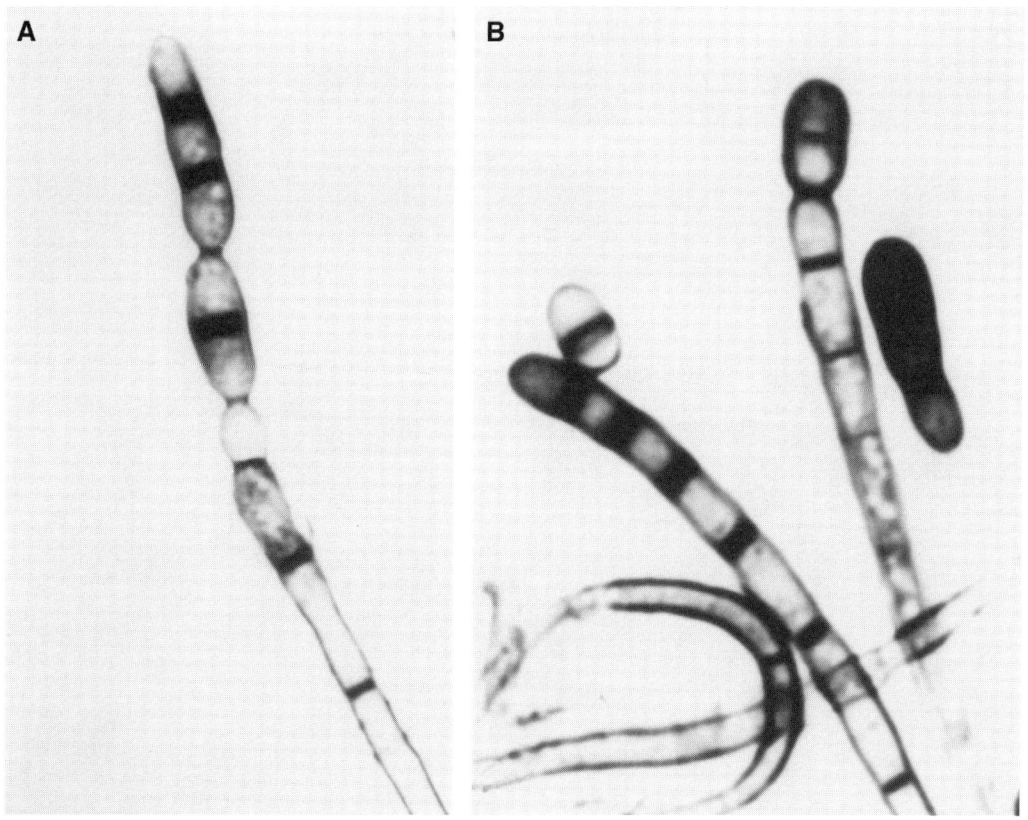

Bispora betulina.   A,B: Conidiophores and conidia.

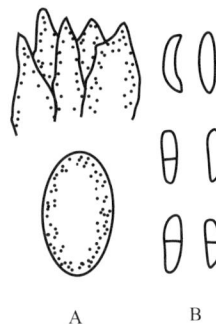

Botryodiplodia Sacc.

Syll. Fung. 3:377, 1884.
Type species: *B. juglandicola* (Schw.) Sacc.

Botryodiplodia sp.

References: Sutton 1980; Watanabe et al. 1986a; Zambettakis 1954.

Morphology: Pycnidia brown, subglobose or flask-shaped, aggregated, often spore masses oozed out from apical ostioles. Conidiophores hyaline, simple, aggregated. Conidia solitary, hyaline, ellipsoidal or cylindrical, usually 2-celled, constricted slightly at the septum.

Material: Pycnidia 210 × 150 µm. Conidia 7.5–17.5 × 2.2–4.5 µm.

Remarks: Among more than 25 species of *Botryodiplodia* described (Arx, 1981), *B. theobromae* Pat. (synonym: *Lasiodiplodia theobromae* [Pat.] Griffon & Maubl.) may be very common in warmer areas with its teleomorph, *Botryosphaeria rhodina* (Berk. & Curt.) v. Arx.

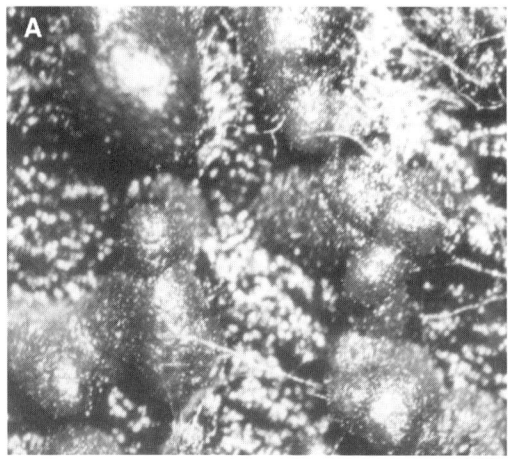
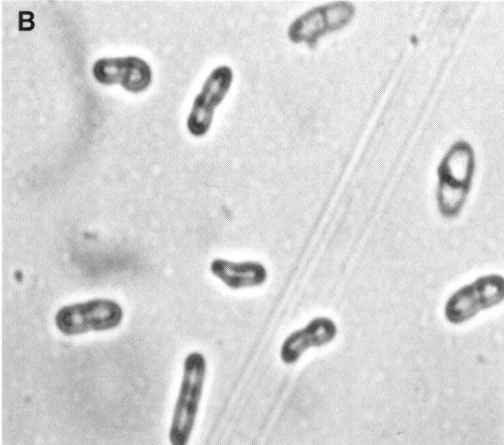

Botryodiplodia sp.   A: Pycnidia. B: Conidia.

Deuteromycotina (Motosporic Fungi)

Botryotrichum Sacc. & March

Bull. Soc. R. Bot. Belg. 24:66, 1885.
Type species: *B. piluliferum* Sacc. & March

Botryotrichum piluliferum Sacc. & March

References: Downing 1953; Hammill 1970.

Morphology: Conidiophores and setae grown together. Conidiophores hyaline, simple or branched, bearing catenulate conidia or spore masses apically in simple conidiophores, and aleurioconidia in branched conidiophores. Conidia of two kinds: phialosporus, hyaline, cylindrical; and aleuriosporous, brown, globose. Setae 1- to 2-septate basally, brown, with hyaline sharpened apexes.

Dimensions: Conidiophores mostly 20 μm tall. Setae 280–300 × 4.7–4.8 μm. Conidia phialosporous 2.5–3 × 1–1.3 μm; aleuriosporous 10–15 μm in diameter.

Material: 77-72 (Strawberry field soil, Tokyo, Japan); 78-2017, 78-2521 (Strawberry field soil, Tokyo, Japan).

Remarks: Colonies on agar cultures are grayish white with yellowish brown tint.

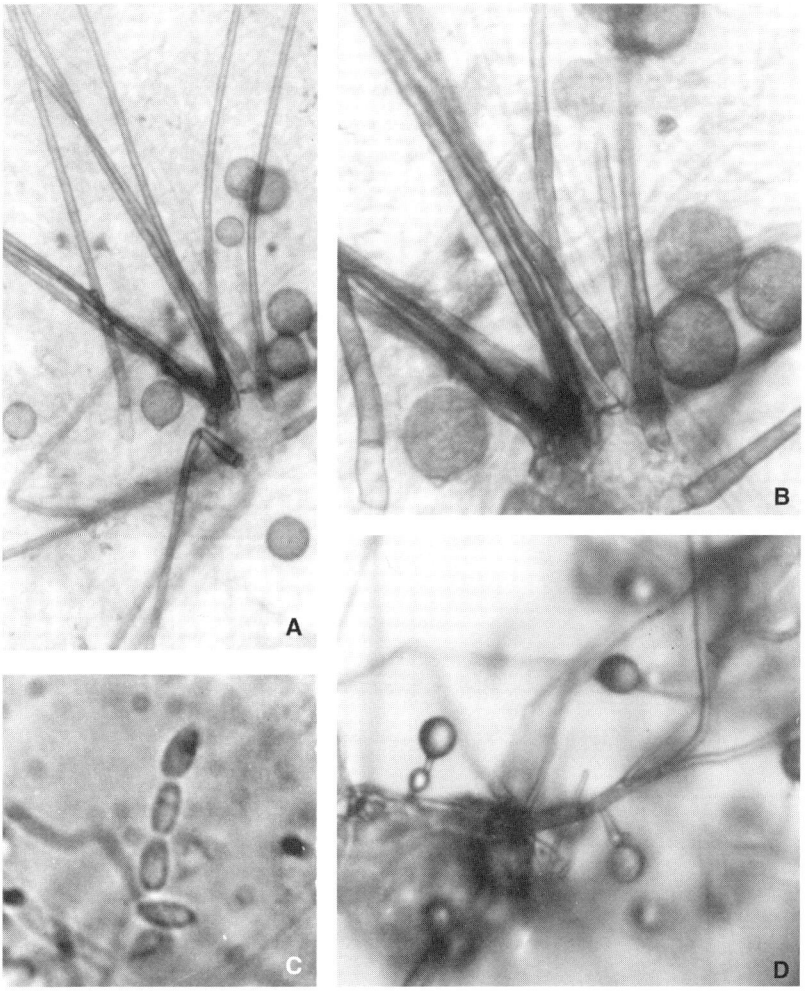

Botryotrichum piluliferum.   A,B: Setae and aleurioconidia. C: Phialoconidia. D: Formatin of aleurioconidia.

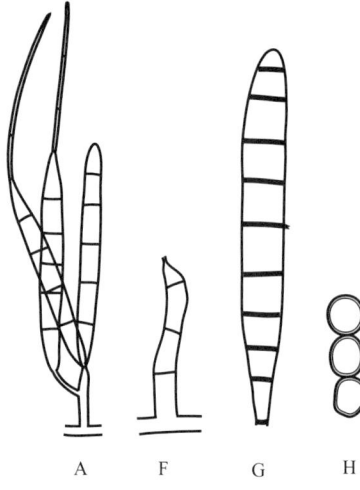

A   F   G   H

Camposporium Harkn.

Bull. Calif. Acad. Sci. 1:37–38, 1884.
Type species: *C. antennatum* Harkn.

Camposporium laundonii M.B. Ellis

References: Ellis 1976; Hughes 1954a; Watanabe 1993a.

Morphology: Conidiophores pale brown, erect, occasionally curved, twisted irregularly 0- to 3-septate, bearing conidia on cylindrical denticles sympodially developed on apical fertile parts. Conidia aleuriosporous, solitary, brown or dark brown with greenish tint, hyaline or subhyaline at both end cells, straight, or slightly curved, smooth, clavate or long-ellipsoidal, tapering toward the ends, 6- to 11-septate (mainly 9-septate) often with an apical appendage, truncate basally or with basal scars. Appendages hyaline, simple, smooth, 1- to 5-septate, tapering toward the apex. Chlamydospores brown to dark brown, globose, granular, with oil globules, catenulate or in aggregates.

Dimensions: Conidiophores 30–50 × 4.5–8.1 µm: denticles up to 7.5 µm long, 2–3 µm wide. Conidia (72-) 75–97.2 (-135) × 10–12.6 (-16.2) µm: basal scars 3.1–3.8 µm broad: apical appendages up to 167.5 × 2.5–3.3 µm. Chlamydospores 7.2–17.5 µm in diameter.

Material: 92-51 (= IFO 32523, root of *Aralia elata*, Matsumoto, Nagano, Japan).

Remarks: A key of 13 *Camposporium* species was prepared by Watanabe (1993a).

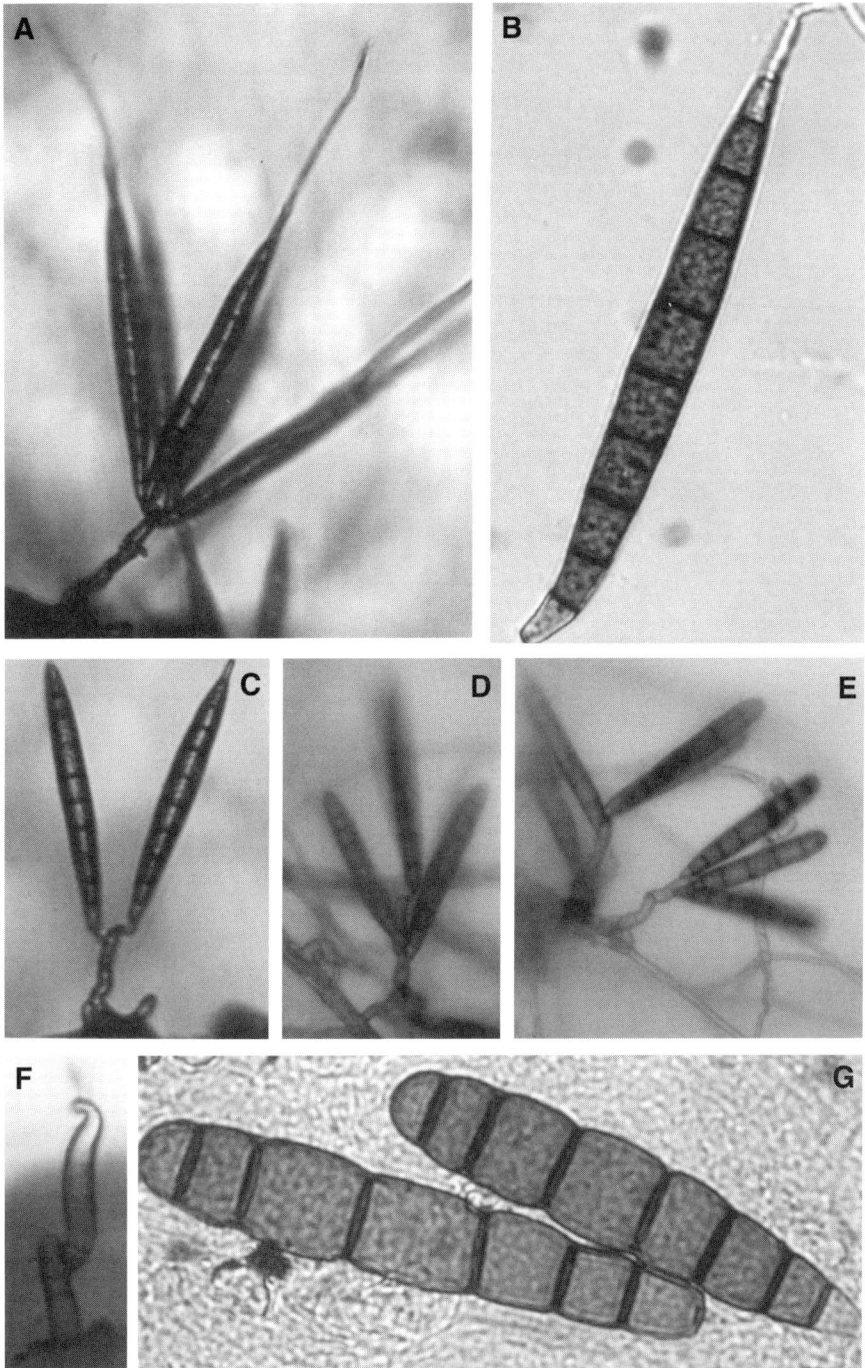

Camposporium laundonii. A,C: Sporulation on root segment of *Aralia elata* (substrate). B: Conidium on substrate. D,E: Sporulation on agar. F: Conidiophore on substrate. G: Conidia formed on agar culture. H: Catenulate chlamydospores. (From Watanabe, T. 1983b. *Trans. Mycol. Soc. Jpn.*, 24:25–33. With permission.)

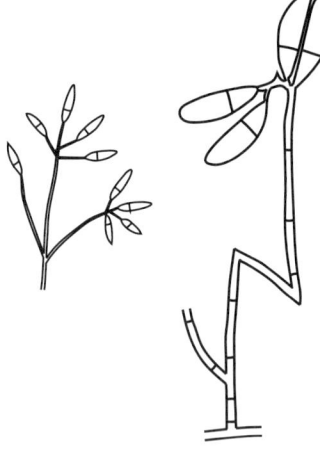

Candelabrella Rifai & R. C. Cooke

Trans. Br. Mycol. Soc. 49:160, 1966.
Type species: *C. javanica* Rifai & R. C. Cooke

Candelabrella musiformis (Drechsler) Rifai et R. C. Cooke

Synonym: *Arthrobotrys musiformis* Drechsler

References: Rifai and Cooke 1966.

Morphology: Conidiophores hyaline, simple or branched, gradually tapering from base toward apex, bearing conidia at branchlets sympodially developed on the apical fertile part. Conidia sympodulosporous, hyaline, long-ellipsoidal, unequally 2-celled with small basal cell. Chlamydospores formed.

Dimensions: Conidiophores 180–495 × 6.2–6.3 µm: branchlets ca. 12.5 × 2.5 µm. Conidia 25–44.5 × 6.5–10 µm.

Material: 69-292 (Pineapple field soil, Okinawa, Japan).

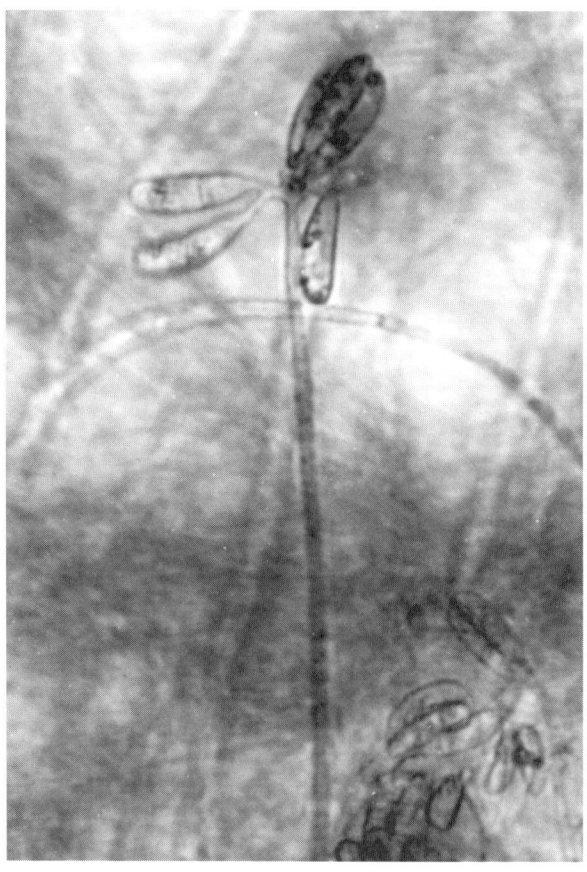

Candelabrella musiformis. Conidiophore and conidia.

Candelabrella sp.

References: Rifai and Cooke 1966.

Morphology: Conidiophores hyaline, erect, tapering from base toward apex, bearing 2–4 conidia at sterigmata developed sympodially on the apical fertile part. Conidia sympodulosporous, hyaline, cylindrical, 2-celled with narrowed basal end. Chlamydospores dark brown, globose.

Dimensions: Conidiophores 45–135 × 2.5 µm. Conidia 22.5–42.5 × 3.5–6.3 µm. Chlamydospores 12.5–13.8 µm in diameter.

Material: 74-525 (Strawberry root, Shizuoka, Japan).

Remarks: Conidiophores of this fungus are shorter, and conidia are narrower than those of *C. musiformis*.

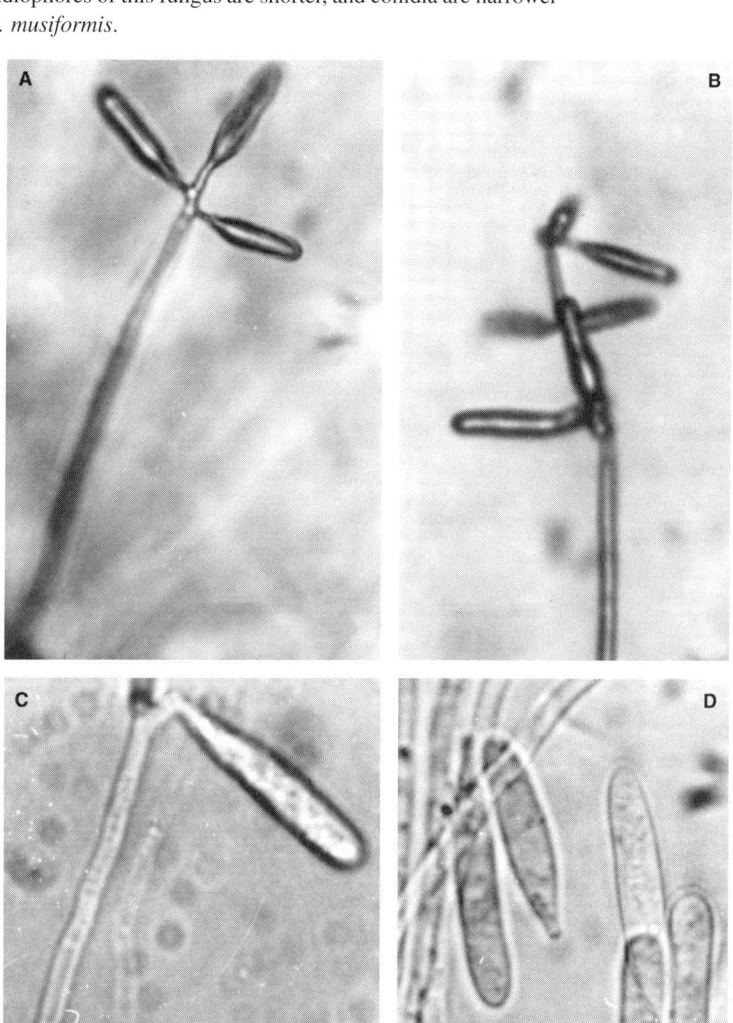

Candelabrella sp.   A–C: Conidiophores. D: Conidia. E: Chlamydospore.

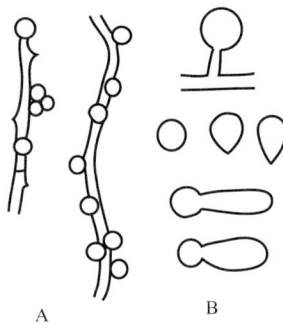

Candida Berkhout

De shimmelgeslachaten *Monilia, Oidium, Oospora* en *Tolura*, Thesis Utrecht (1923).
Type species: *C. albicans* (Robin) Berkhout

Candida sp.

References: Lodder (Ed.) 1970; Watanabe et al. 1986a.

Morphology: Conidiophores not developed. Conidia blastosporous, apical or lateral, directly on hyphae or sterigmata on hyphae, hyaline, globose, apiculate at one end, proliferated by budding.

Dimensions: Hyphae 2.9–4.8 µm wide. Conidia 3.8–4.9 µm in diameter.

Material: 72-X133 (Sugarcane root, Taiwan, ROC); 83-87 (Japanese red pine seed, Saihaku, Tottori, Japan); 84-569 (Japanese black pine seed, Kumano, Wakayama, Japan); 85-71 (Flowering cherry seed, Hachioji, Tokyo, Japan).

Remarks: Colonies on PDA are yeast-like with crystalline substances deposited. The isolates from forest seeds are roughly separated into pinkish and white types. Their conidial dimensions are various, but may be mostly 3.7–10 × 1–2.4 µm.

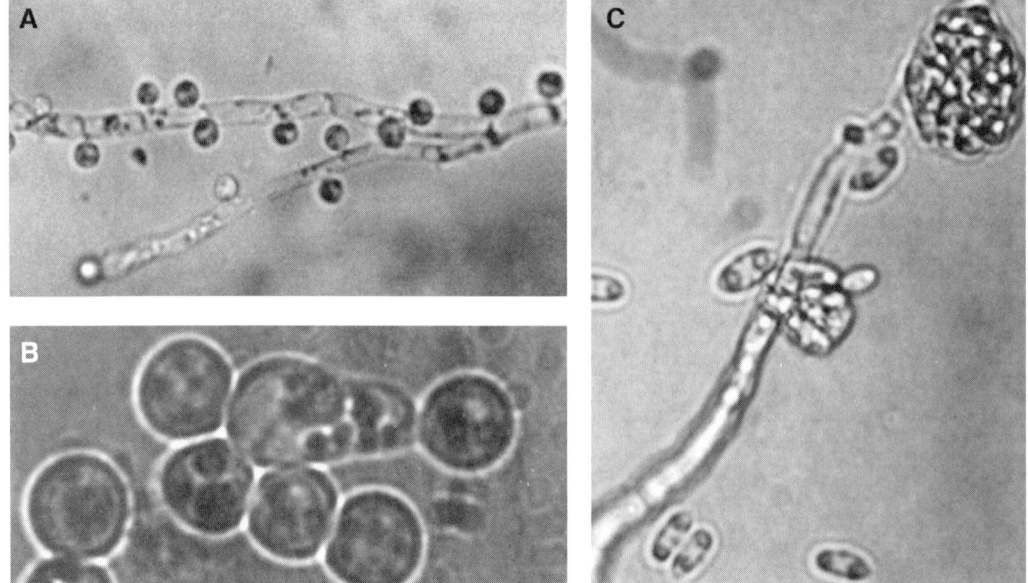

Candida sp.　A–C: Hyphae and conidia. B: Conidia and budding.

Cephaliophora Thaxter

Bot. Gaz. 35:153, 1903.
Type species: *C. tropica* Thaxter

Cephaliophora irregularis Thaxter

References: Matsushima 1975; Thaxter 1903; Tubaki 1956.

Morphology: Conidiophores short, simple, bearing nearly 10 conidia at subglobose or clavate inflated apical cells. Conidia botryoblastosporous, hyaline, or pale brown, pear-, turnip-, fan-, or Y-shaped, usually 2-celled, rarely 3-celled with extremely large apical, and small septate basal cell.

Dimensions: Conidiophores 16.2–70 × 4.5–20 μm. Spore massses 55–70 μm in diameter. Conidia 21.2–23.8 μm in diameter.

Material: 75-182 (*Lagenaria* root, Miyazaki, Japan).

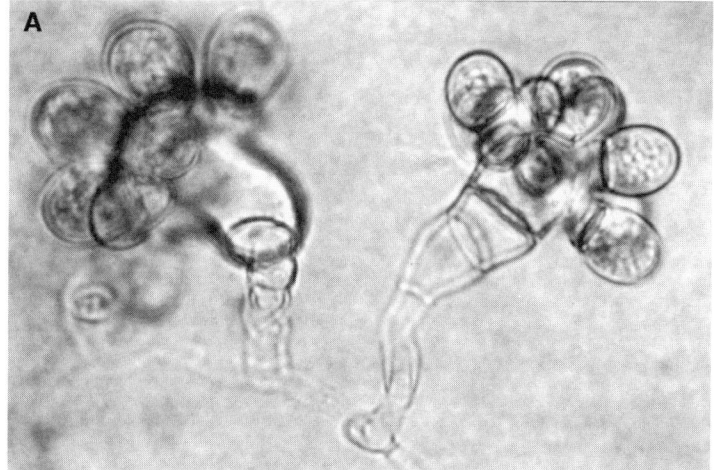

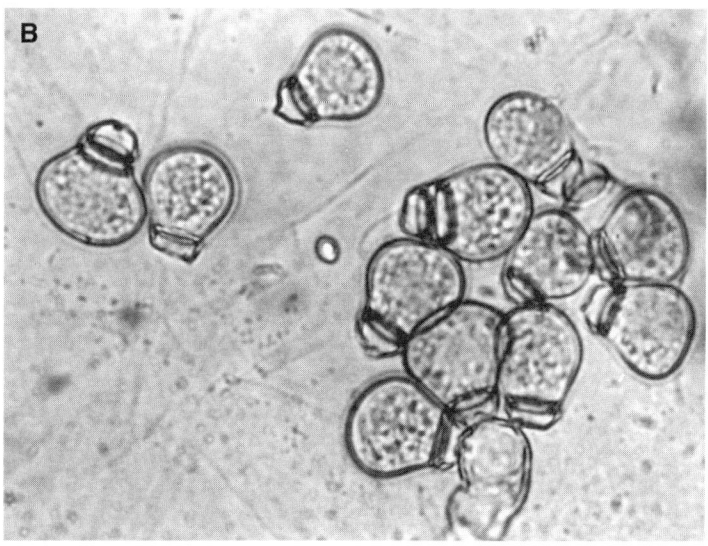

Cephaliophora irregularis. A: Conidiophores and conidia. B: Conidia.

Cephaliophora tropica Thaxter

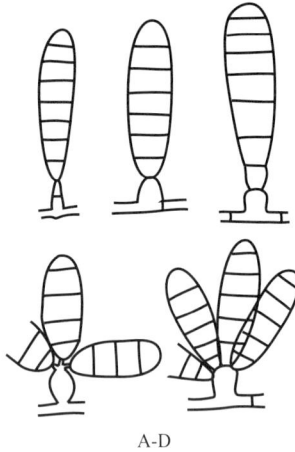

References: Ellis 1971; Matsushima 1975; Takada and Udagawa 1956; Thaxter 1903; Wolf 1949.

Morphology: Conidiophores erect, very short, simple, globose or cylindrical bearing 1–5 conidia at the apical fertile part. Conidia botryoblasporous, hyaline, pale brown, cylindrical, ellipsoidal, hyaline excluding pale brown septation area, usually 5- to 7-septate.

Dimensions: Conidiophores 10–12.5 × 7.5–12.5 µm. Conidia 55–70 × 17.5–24 µm.

Material: 73-30 (Strawberry field soil, Shizuoka, Japan); 82-125 (Melon root, Futtsu, Chiba, Japan).

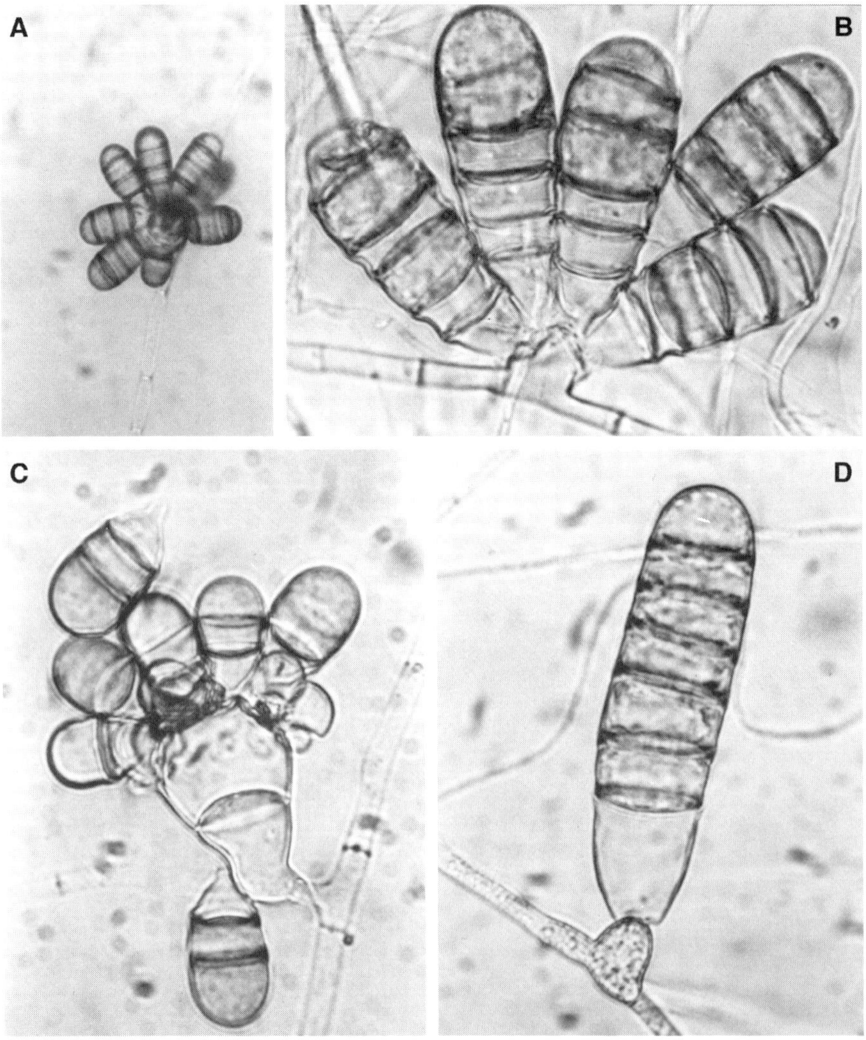

A-D

Cephaliophora tropica. A–D: Conidiophores and conidia.

Chaetomella Fuckel

Symb. Mycol. P. 401, 1870.
Type species: *C. oblonga* Fuckel

Chaetomella sp.

References: Ramchandra Rao & Baheker 1964; Stolk 1963; Sutton & Sarbhoy 1976.

Morphology: Pycnidia black, superficial, ellipsoidal, setose: setae dark, bristle, gradually tapering toward coiled apex. Conidiophores branched, septate. Conidia hyaline, 1-celled, cylindrical or fusiform, occasionally slightly curved.

Dimensions: Pycnidia mostly 200–250 × 170 μm: setae mostly 70 μm long. Conidiophores 20–25 × 1–1.5 μm. Conidia 6–8.5 × 2–2.5 μm.

Material: 75-P-110 (Paulownia root, Pirapo, Paraguay).

Remarks: Although this fungus is close to *Amerosporium* Speg. on the basis of setose pycnidia and aseptate conidia, its conidiophores were branched and no raphe was observed on its pycnidia. The conidiophores of *Amerosporium* are simple, and no raphe is present.

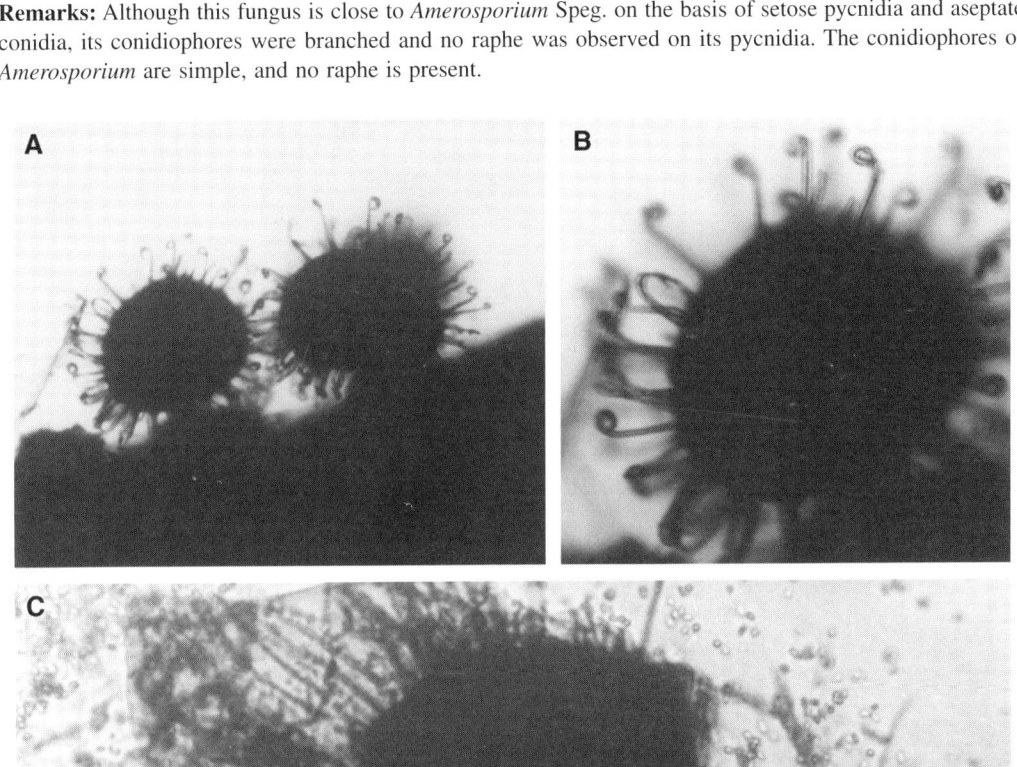

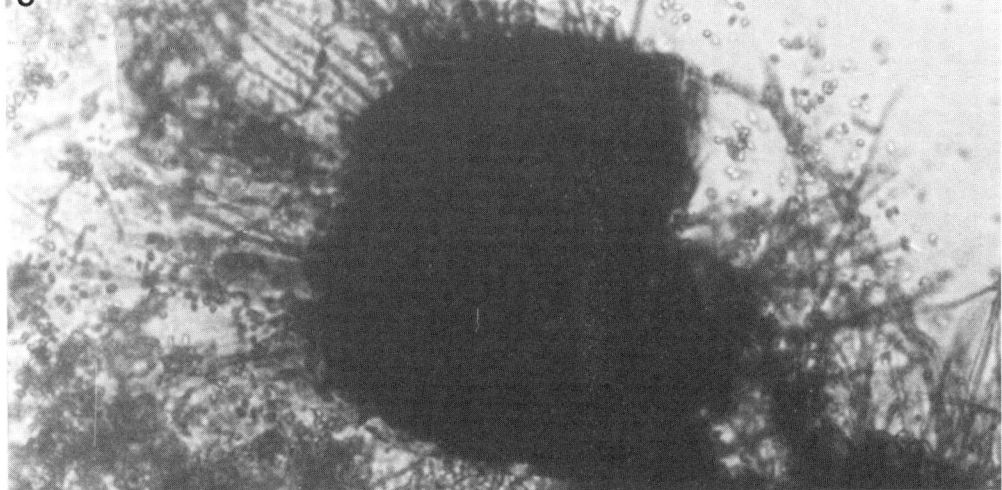

Chaetomella sp.   A,B: Setose pycnidia. C: Crushed pycnidium and extruded conidia. D: Conidiophore.

Chalara (Corda) Rabenh.

Krypt. Fl. 1:38, 1844.
Type species: *C. fusidioides* (Corda) Rabenh.

Chalara thielavioides Peyron.

Synonym: *Chalaropsis thielavioides* Peyron.

References: Domsch et al. 1980; Ellis 1971; Nag Raj and Kendrick 1975; Sugiyama 1968.

Morphology: Conidiophores simple, pale brown, forming conidia from apical cylindrical phialides. Conidia phialosporous, cylindrical, hyaline, 1-celled. Chlamydospores solitary, globose, brown.

Dimensions: Conidiophores 58–62.5 µm long. Conidia 9.5–10 × 3 µm. Chlamydospores 7.5–11 µm in diameter.

Material: 70-1096 (Pineapple field soil, Hachijo, Tokyo, Japan).

Remarks: This fungus has *Rhizoctonia*-like hyphae with constriction near branching area.

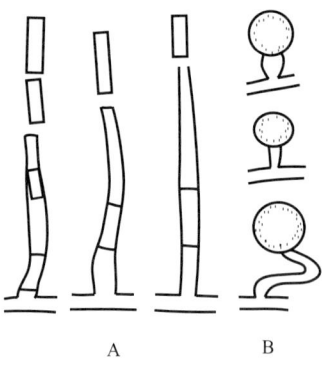

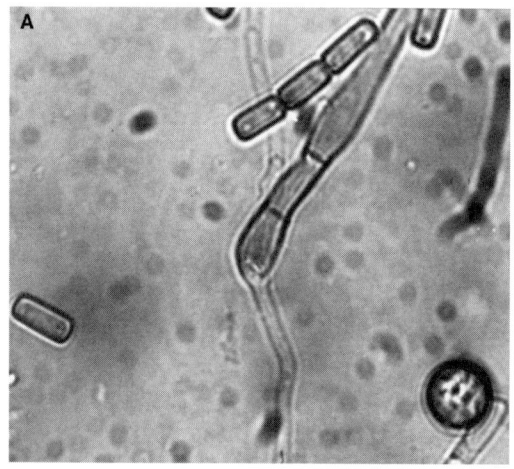

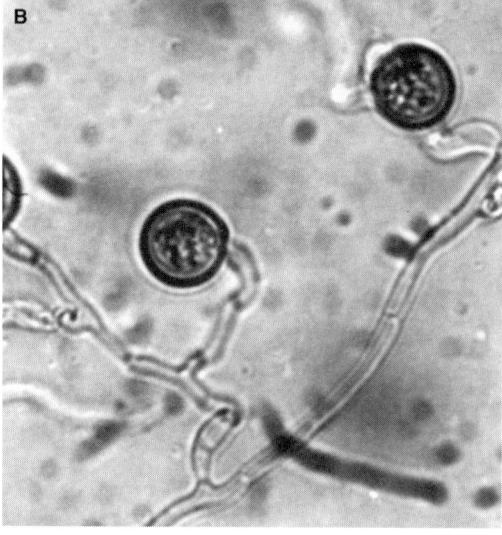

Chalara thielavioides. A: Conidiophore, conidia, and chlamydospore. B: Chlamydospores.

Chloridium Link

Syst. Mycol. 1:46, 1821, Figure 76a.
Type species: *C. viride* (Link) Hughes

Chloridium virescens (Pers. Ex Pers.) W. Gams & Hol.-Jech. var. *chlamydosporum* (van Beyma) W. Gams & Hol.-Jech.

References: Gams & Holubová-Jechová 1976; Tubaki 1963.

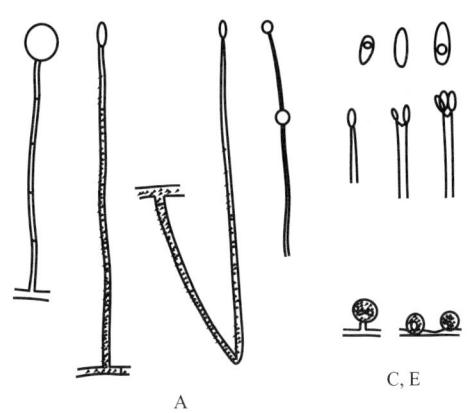

Morphology: Conidiophores erect, simple, septate, brown, gradually tapering toward hyaline apexes with terminal phialides, and slimy conidial heads, proliferate from the phialides. Conidia phialosporous, cylindrical or ellipsoidal, hyaline, 1-celled, often developed together in twos. Chlamydospores lateral, intercalary, or borne on short chlamydosporephores.

Dimensions: Conidiophores 60–290 × 2–3 µm. Conidial heads 10–12 µm in diameter. Conidia 3.6–5 × 1.8–2.2 µm. Chlamydospores (3-) 4–6 µm in diameter.

Material: 99-470 (Forest soil, Agematsu, Nagano, Japan); 99-478 (Forest soil, Agematsu, Nagano, Japan).

Remarks: This fungus resembles the genus *Gonytrichum*, forming spore masses at the apex of conidiophores, but they may be differentiated whether conidiophores are simple or branched.

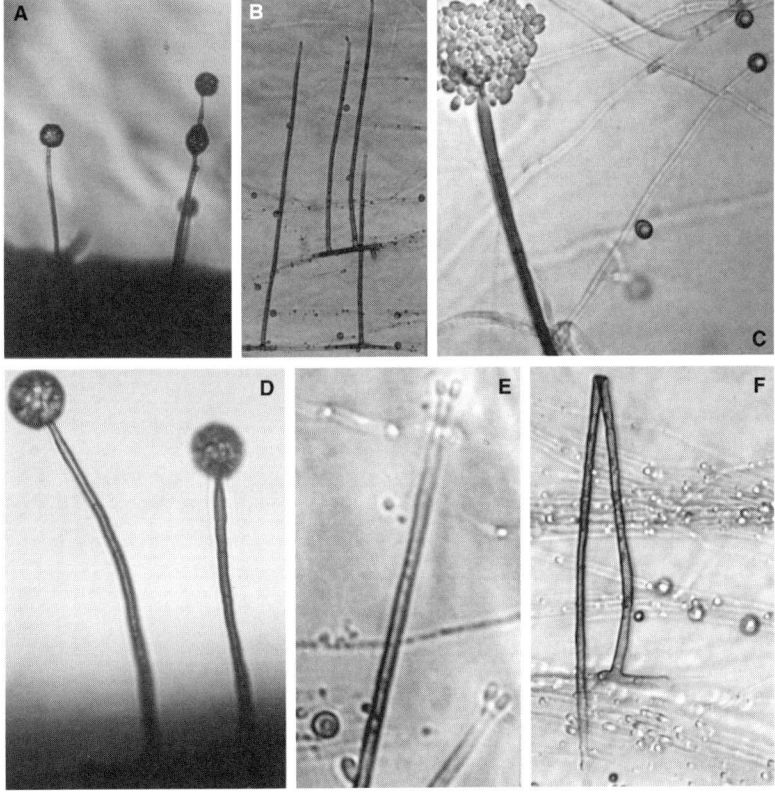

Chloridium virescens var. *chlamydosporum.* A,D: Conidiophores and spore masses in simple (A,D) or proliferated conidiophore (A). B: Conidiophores on dehiscence of spore masses. C: Apexes of conidiophores, conidia, and chlamydospores. E: Conidiophore apexes bearing two conidia at each apex, and chlamydospores. F: Conidiophore after dehiscence of spore mass.

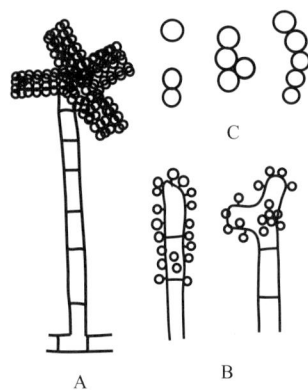

Chromelosporium Corda

in Sturm, Deutschl. Fl., Pilze 3, 3:81, 1833.
Type species: *C. ochraceum* Corda

Chromelosporium fulvum (Link) McGinty, Hennebert & Korf

Teleomorph: *Peziza ostracoderma* Korf

References: Domsch et al. 1980a,b; Fergus 1960; Hennebert 1973.

Morphology: Conidiophores hyaline, erect, broad, septate, branched once or twice at the apex, bearing spore masses, at sterigmata on the fertile portions of branches. Conidia blastosporous, pale yellowish brown, globose, smooth, 1-celled.

Dimensions: Conidiophores 125–1350 × 7.5–15 µm. Conidia 5–7.8 µm in diameter.

Material: 76-572 (Gentian root, Shimoina, Nagano, Japan).

Remarks: *Ostracoderma* may be a synonym of this fungus, which forms closed fruiting structures (Hennebert, 1973).

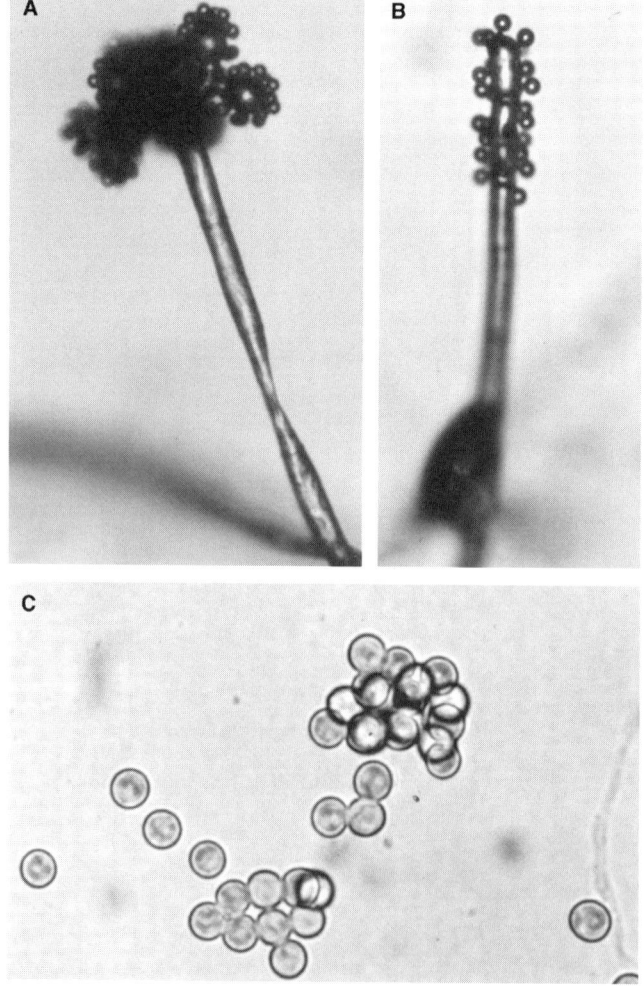

Chromelosporium fulvum. A,B: Conidiophores and conidia. C: Conidia.

Chrysosporium Corda

in Sturm, Deutschl. Fl., Pilze 3, 3:13:85, 1833.
Type species: *C. merdarium* (Link ex Grev.) Carmichael

Chrysosporium keratinophilum (Frey) Carmichael

References: Carmichael 1962; Watanabe 1975.

Morphology: Conidiophores lacking, or very short, simple or branched, hyaline, bearing conidia at apical parts, or rarely directly on hyphae. Conidia aleuriosporous, solitary (rarely, 2–3 conidia in a chain), globose, ellipsoidal, 1-celled, cylindrical, apiculate or truncate at one end, pale yellow, readily detachable.

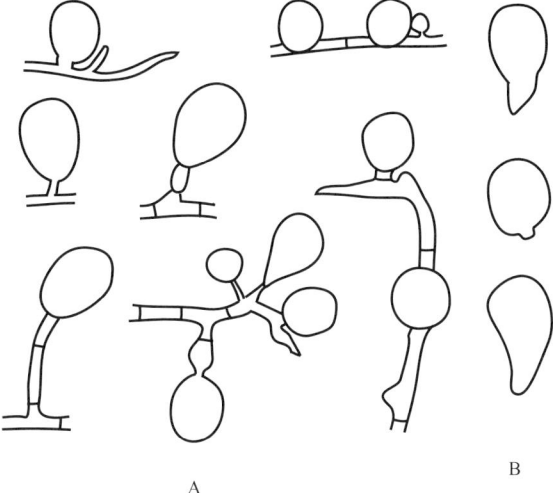

Dimensions: Conidia (8.7-) 12.3–16.3 × 8.5–12.7 µm.

Material: 72-X37-63 (Sugarcane root, Taiwan, ROC); 73-255 (Strawberry field soil, Shizuoka, Japan); 74-704 (Strawberry root, Mie, Japan).

Remarks: Cultures on PDA are pale yellowish brown, resupinate, flat.

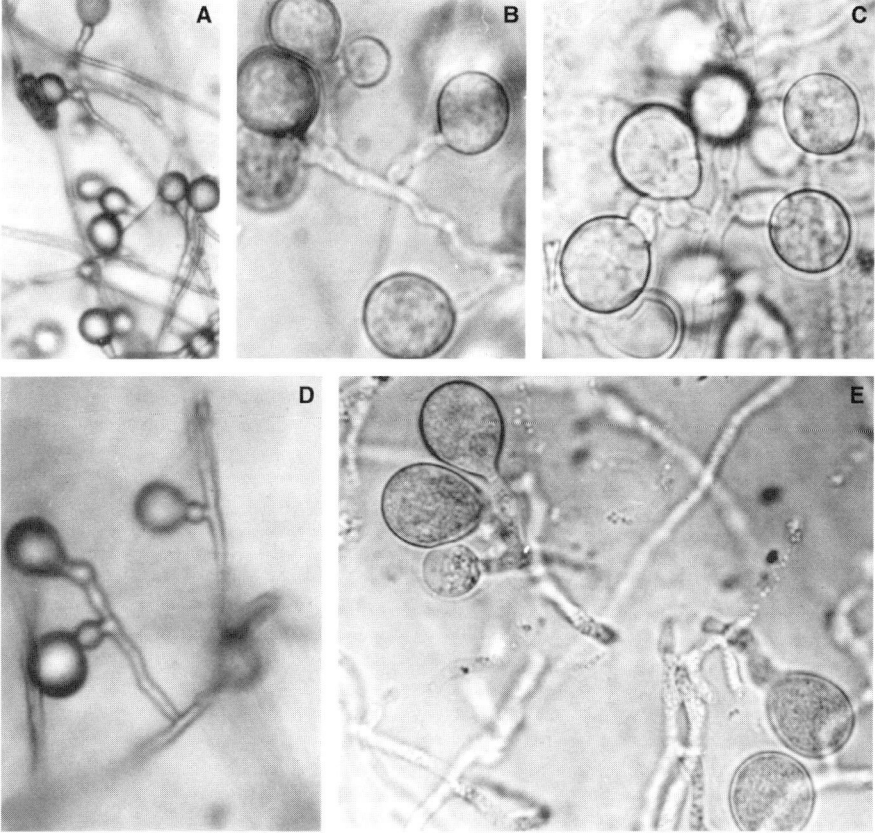

Chrysosporium keratinophilum. A,B: Conidiophores and conidia. (From Watanabe, T. 1975c. *Trans. Mycol. Soc. Jpn.*, 16:149–182. With permission.)

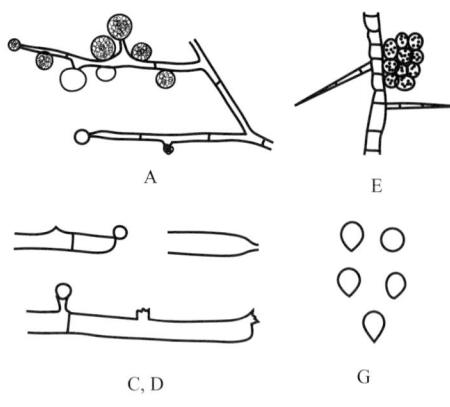

Cladorrhinum Sacc. & March

Bull. Soc. Bot. Belg. 24:64, 1885.
Type species: *C. foecundissimum* Sacc. & March

Cladorrhinum bulbillosum Gams & Mouch

References: Arx and Gams 1966; Gams and Domsch 1969; Domsch et al. 1980; Mouchacca and Gams 1993; Watanabe 1975d.

Morphology: Conidiophores hyaline or pale brown, erect, branched, bearing spore masses composed of over 30 conidia on phialides developed on branches. Conidia phialosporous, hyaline, globose, apiculate at one end. Sclerotia pale brown, globose, often with setae nearby.

Dimensions: Conidiophores up to 174 μm long, 3.2–4.4 μm wide. Spore masses 7.5–12.5 μm in diameter. Conidia 2–2.8 μm in diameter. Sclerotia 120–300 μm in diameter.

Material: 72-X10 (= CBS 267.76, Strawberry root, Taiwan, ROC); 81-466 (Grassland soil, Hokkaido, Japan); 85-20, 85-21 (Flowering cherry seed, Hachioji, Tokyo, Japan).

Remarks: Colonies on PDA are black initially, changing to grayish black and floury with age. The genus name is also spelled as *Cladorhinum* (Hawksworth et al., 1983).

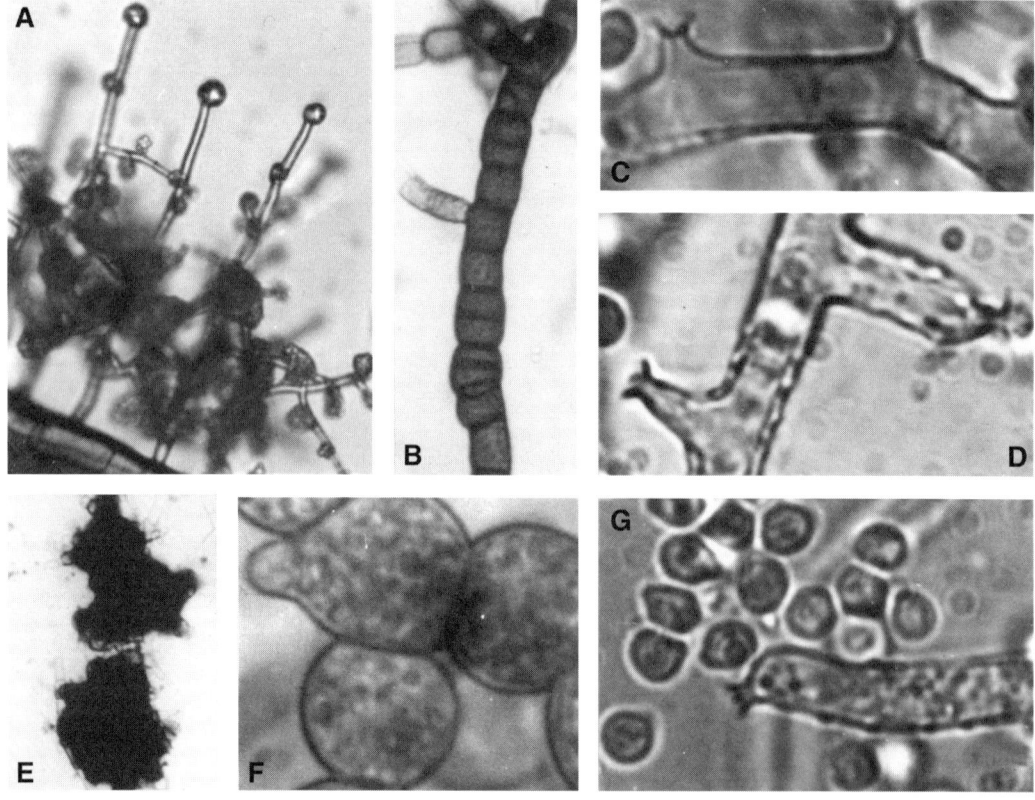

Cladorrhinum bulbillosum. A: Conidiophores and spore masses. B: Hypha. C,D: Phialides. E: Sclerotia. F: Component cells of sclerotium. G: Phialides and conidia. (From Watanabe, T. 1975d. *Trans. Mycol. Soc. Jpn.*, 16:264–267. With permission.)

Cladorrhinum samala (Subram. & Lodha) W. Gams & Mouchacca

References: Arx and Gams 1966; Domsch et al. 1980; Gams and Domsch 1969; Mouchacca and Gams 1993; Watanabe 1975d.

Morphology: Conidiophores hyaline or pale brown, erect, branched, bearing spore masses composed of 30 conidia on phialides developed on branches. Conidia phialosporous, hyaline, globose, apiculate at one end. Chlamydospores pale brown, globose.

Dimensions: Conidiophores 46–87 × 4 µm. Spore masses 10–20 µm in diameter. Conidia 2.5–2.8 µm. Chlamydospores 7.2–11.8 µm in diameter.

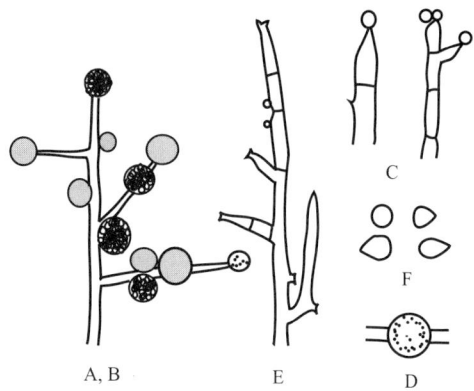

Material: 70-1175 (Pineapple field soil, Hachijo, Tokyo, Japan); 72-X155-1010 (= CBS 266.76, Sugarcane root, Taiwan, ROC); 73-60 (Strawberry root, Shizuoka, Japan).

Remarks: This fungus differs from *C. bulbillosum* in forming chlamydospores. Intrahyphal hyphae are often observed in this fungus.

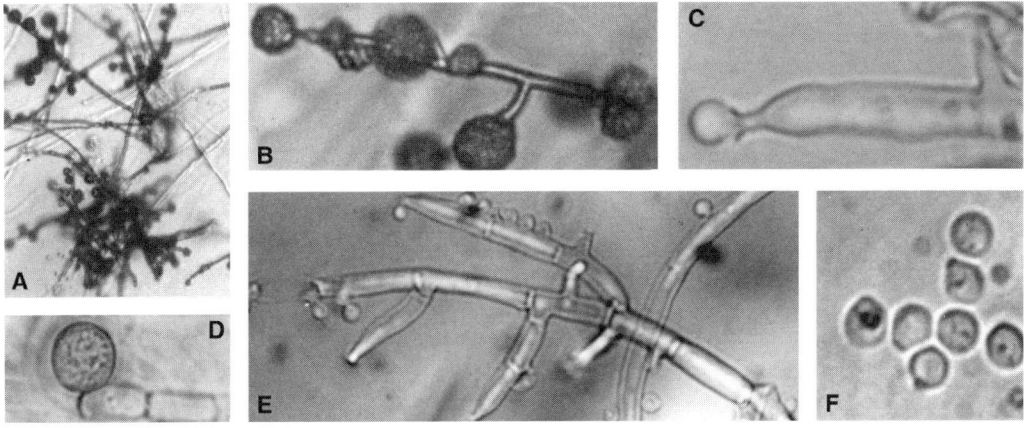

Cladorrhinum samala. A,B: Conidiophores and spore masses. C: Phialides and conidia. D: Chlamydospore. E: Conidiophores. F: Conidia. (From Watanabe, T. 1975d. *Trans. Mycol. Soc. Jpn.*, 16:264–267. With permission.)

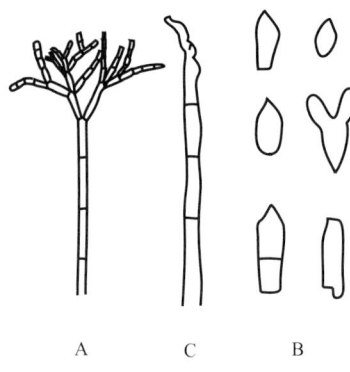

Cladosporioum Link

Syst. Mycol. 3:368, 1832.
Type species: *C. herbarum* Link ex Fr.

Cladosporium cladosporioides (Fresn.) de Vries

References: De Vries 1952; Ellis 1971.

Morphology: Conidiophores pale brown, erect, branched 2–3 times, at the apical parts, bearing catenulate conidia in each branch. Conidia blastosporous, often not well differentiated from branches, hyaline or pale brown, ovate, ellipsoidal, cylindrical, subglobose, irregular in shape, apiculate at one end, often truncate at another end.

Dimensions: Conidiophores 100–224 × 2.5–3.8 μm: branches 10–22.5 × 2.5–3 μm. Conidia ovate 2.5–4.9 × 2–3 μm: cylindrical (4.5-) 7.5–12.7 (-22.5) × (2.7-) 3.6–4.2 μm.

Material: 70-1035 (Pineapple field soil, Hachijo, Tokyo, Japan); 72-X-149 (Sugarcane root, Taiwan, ROC); 83-46 (Japanese black pine seed, Kumano, Mie, Japan); 85-120 (Flowering cherry seed, Hachioji, Tokyo, Japan).

Remarks: This fungus is characterized by erect conidiophores without nodes and aseptate or inconspicuously septate nonspherical conidia, under 40 μm long.

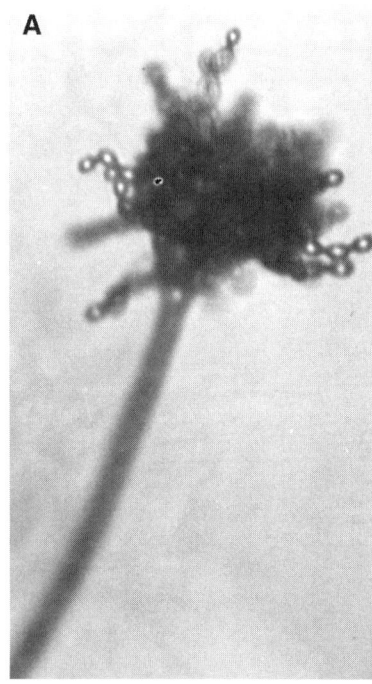
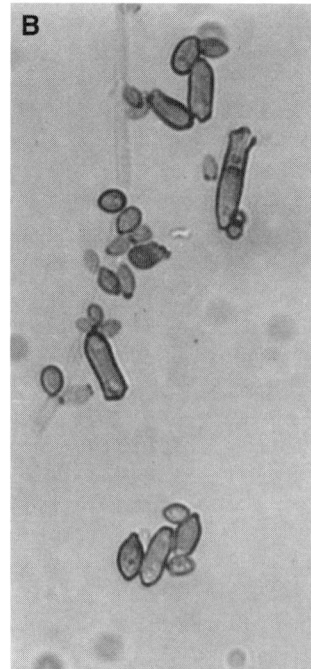
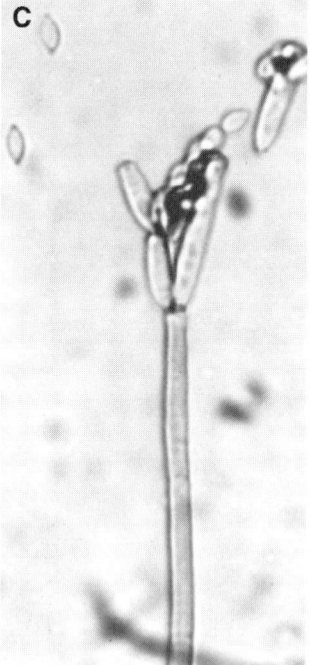

Cladosporium cladosporioides.   A,C: Conidiophores and conidia. B: Conidia.

Codinaea Maire

Publ. Inst. Bot. Barcelona 3:15, 1937.
Type species: *C. aristata* Maire

Codinaea parva Hughes & Kendrick

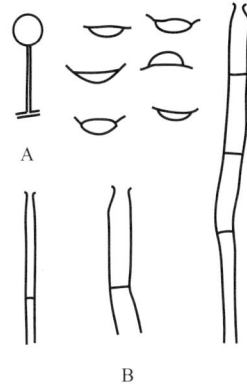

References: Hewings and Crane 1981; Hughes and Kendrick 1968.

Morphology: Conidiophores (phialides) pale brown, erect, simple or rarely branched once, bearing spore masses at the straight or slightly curved apex. Conidia phialosporous, hyaline, lunar-shaped, with a setula at each end.

Dimensions: Conidiophores 50–187.5 × 3.7–5 μm. Conidia 8.5–13 × (2.5-) 3–3.5 (-4.5) μm: setulae 2.5–5 (-6.3) μm.

Material: 69-443 (Pineapple field soil, Ishigaki, Okinawa, Japan).

Remarks: No setulae were observed in the anamorph of *Chaetosphaeria talbotii* Hughes.

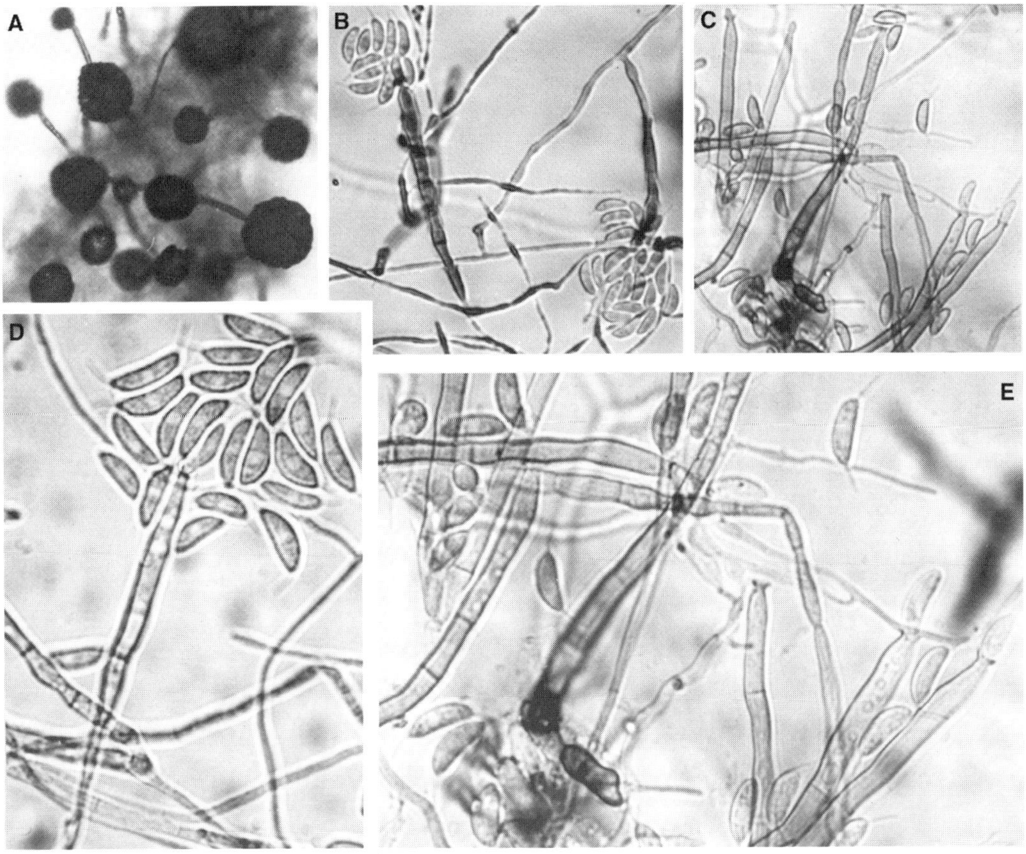

Codinaea parva. A: Conidiophores and spore masses. B–E: Conidiophores and conidia.

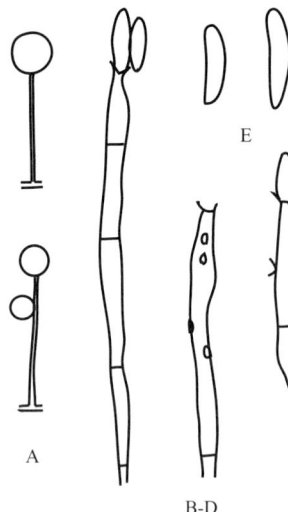

Codinaea state of *Chaetosphaeria talbotii*

Teleomorph: *Chaetosphaeria talbotii* Hughes, Kendrick et Shoemaker

References: Hewings and Crane 1981; Hughes and Kendrick 1968.

Morphology: Conidiophores (phialides) pale brown, erect, bearing spore masses (over 30 conidia per spore mass) at the apex and its vicinity. Conidia phialosporous, hyaline, 1-celled, cylindrical, boat-shaped or slightly curved.

Dimensions: Conidiophores 68.1–389.2 × (3-) 3.8–5 µm: openings ca. 2 µm wide. Spore masses 10–17.5 µm in diameter. Conidia (5-) 9.5–13.4 (-16.3) × 2.5–3.4 (-4.3) µm.

Material: 69-501 (Pineapple root, Ishigaki, Okinawa, Japan); 72-X123 (= ATCC 32909 Sugarcane root, Taiwan, ROC); 73-270 (Strawberry root, Shizuoka, Japan).

Remarks: The conidia of this fungus are slightly broader than those of the original description.

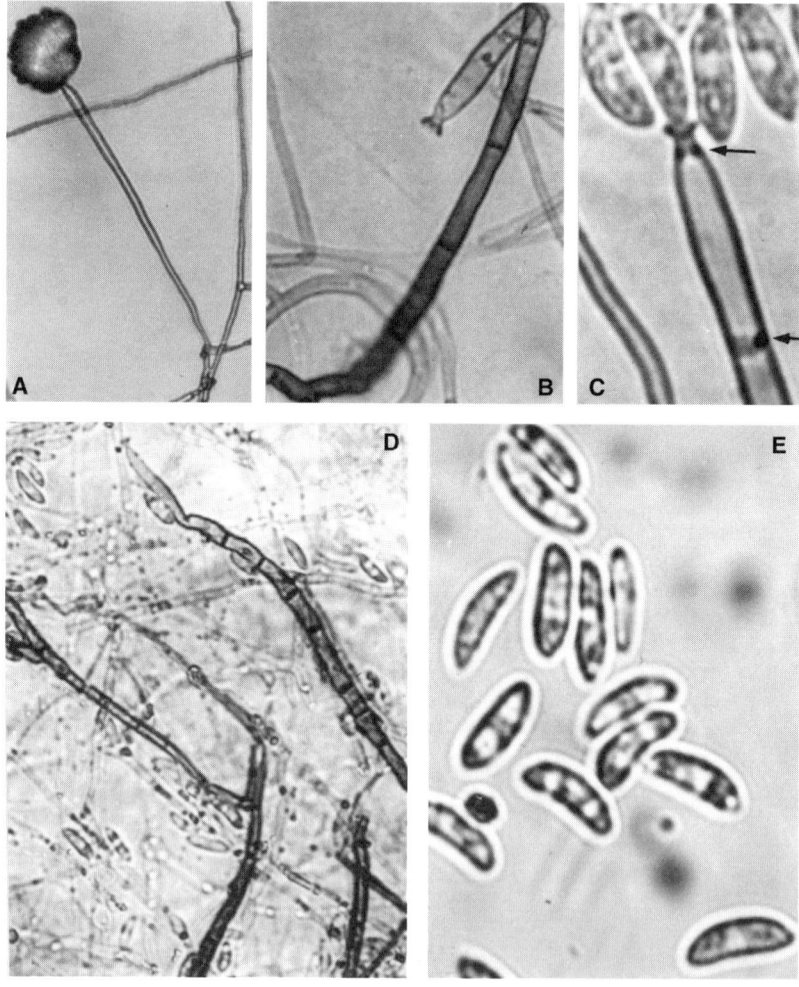

Codinea state of *Chaetosphaeria talbotii*. A: Conidiophore and spore mass. B: Conidiophore. C: Phialide and conidia. D: Conidiophores and conidia. E: Conidia.

Colletotrichum Corda

in Sturm, Deutschl. Fl., Pilze 3:41, 1831.
Type species: *C. dematium* (Fr.) Grove

Key to Species

1. Ascocarps   formed in single culture.................... *Colletotrichum destructivum*
   (ref. *Glomerella glycines*, p. 154)
                        not formed 2

2. Conidia   lunar-shaped... 3
                        cylindrical... 4

3. Sclerotia   formed .... *C. truncatum*
                        not formed .... *C. falcatum*

4. Conidia   under 4 μm wide, setose sclerotia formed .... *C. coccodes*
                        over 4 μm wide, sclerotia not formed................ *C. lindemuthianum*

Colletotrichum coccodes (Wallr.) Hughes

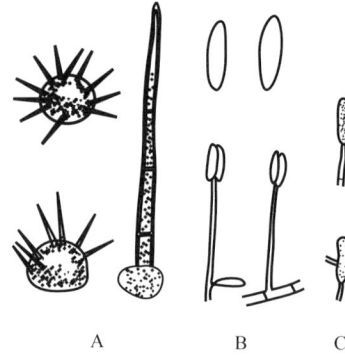

Type species: *C. atramentarium* (Berk. & Br.) Taubenh.

References: Chesters and Hornby 1965.

Morphology: Sporodochia in cultures hemispherical, composed of conidiophores, setae, and spore masses. Conidiophores free from sporodochia hyaline, simple, erect, bearing usually 2 conidia at the apex of phialides. Conidia phialosporous, hyaline, cylindrical. Setae brown, thick-walled, sharp at the apex. Appressoria brown, cylindrical.

Dimensions: Conidiophores 36–55 µm tall. Conidia 12–18.3 × 3.5–4 µm. Sclerotia 110–160 µm in diameter: setae 75–125 × 5 µm.

Material: 70-5 (Snap bean seed, Shinsyu-shinmachi, Nagano, Japan).

Remarks: Setose sclerotia characterize this fungus. Setae are formed on sclerotia on natural media including snap bean straws, but not on agar media.

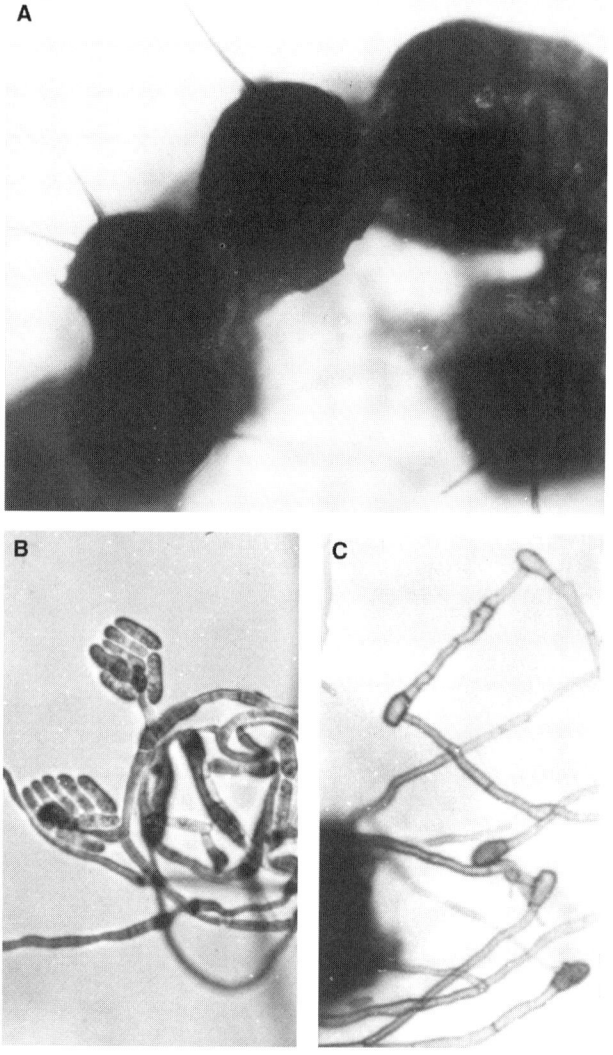

Colletotrichum coccodes. A: Sclerotia with setae. B: Conidiophores, conidia and hyphae. C: Appressoria.

Colletotrichum falcatum Went

Teleomorph: *Glomerella tucumanensis* (Speg.) Arx & E. Muller

References: Morris 1956; Watanabe 1975b.

Morphology: Sporodochia hemispherical, pale yellow, with about 6 setae in cultures. Conidiophores hyaline or pale brown, simple, erect, irregular, aggregated in mass. Conidia phialosporous, hyaline, 1-celled, lunar-shaped. Setae dark green, sharp at the apex. Appressoria thick-walled, dark brown, subglobose.

Dimensions: Conidia 15–33.5 × 4.6–5.1 µm. Appressoria 5.5–10 µm in diameter.

Material: 72-X33, 72-X40 (Sugarcane root, Taiwan, ROC).

Remarks: *Physalospora tucumanensis* Spegazzini is often used as the teleomorph name. Conidia of this fungus appear porosporus or blastosporus.

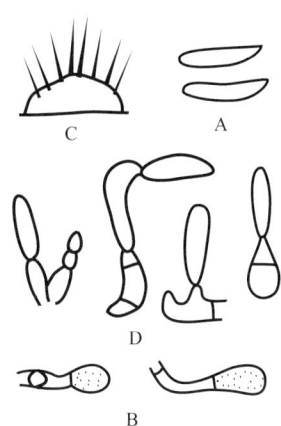

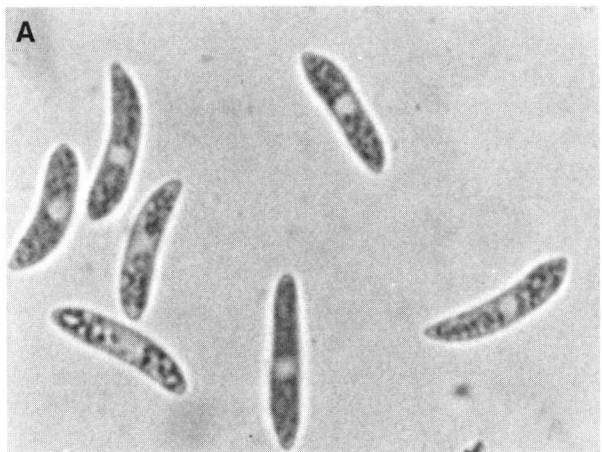
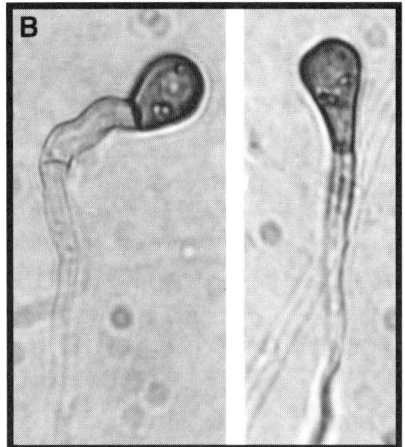

Colletotrichum falcatum. A: Conidia. B: Appressoria. C: Sporodochium. D: Conidiophores and conidia. (From Watanabe, T. 1975b. *Trans. Mycol. Soc. Jpn.*, 16:28–35. With permission.)

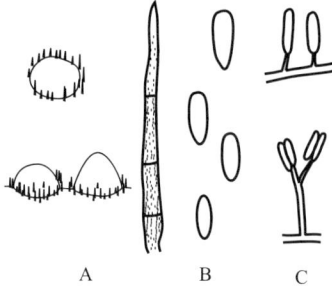

Colletotrichum lindemuthianum (Sacchardo et Magnus) Briosi et Cavara

Teleomorph: *Glomerella lindemuthiana* Shear

References: Arx 1957; Sutton 1980; Watanabe 1972b.

Morphology: Acervuli only formed on natural media containing bean straws and agar. Sporodochia in cultures, pale brown, hemispherical, composed of conidiophores and conidia, and indistinct setae. Conidiophores free from sporodochia hyaline, simple or branched, erect, bearing 2–3 conidia at phialides. Conidia phialosporous, hyaline, 1-celled, cylindrical. Setae brown, thick-walled, sharp at the apex, 2- to 3-septate.

Dimensions: Conidia 13–17.5 × 4.7–5.3 µm. Sporodochia 200–330 µm in diameter. Setae 65–107.5 × 3.7–6 µm.

Material: 70-112 (Snap bean seed, Shinsyu-shinmachi, Nagano, Japan).

Remarks: Setae on the sporodochia are not conspicuous.

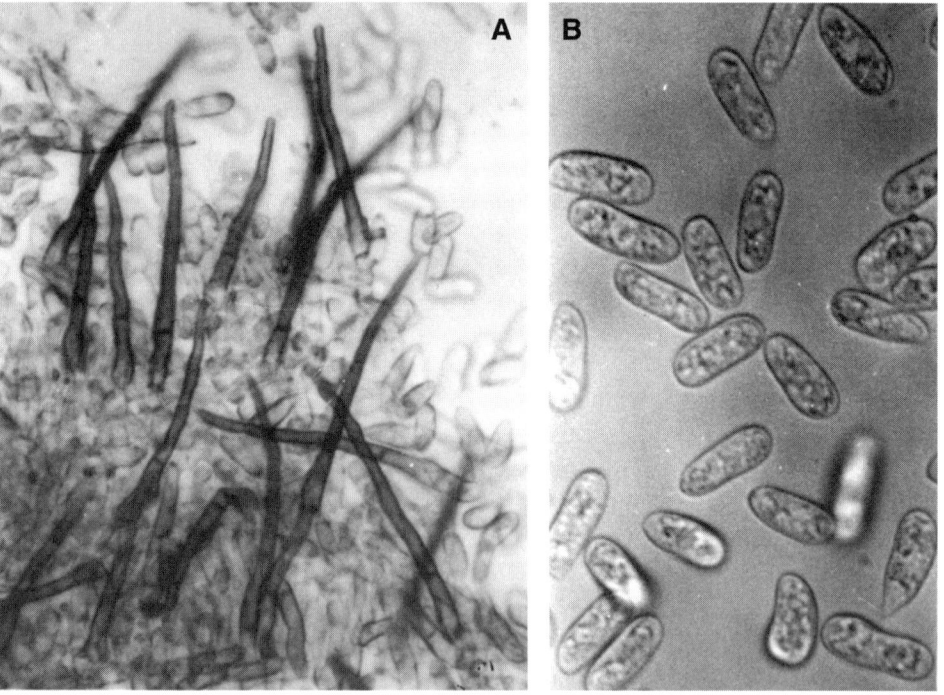

Colletotrichum lindemuthianum. A: Conidia and sporodochium with setae. B: Conidia. C: Conidiophores and conidia. (From Watanabe, T. 1972a. *Ann. Phytopathol. Soc. Jpn.*, 38:106–110. With permission.)

Colletotrichum truncatum (Schweinitz) Andrus et Moore

References: Arx 1957; Sutton 1980; Watanabe 1972b.

Morphology: Acervuli only formed on natural media containing bean straws and agar. Sporodochia in cultures pinkish, hemispherical, composed of conidiophores, conidia, and setae. Conidiophores free from sporodochia hyaline, simple or branched, erect, bearing 2–3 conidia at the phialides. Conidia phialosporous, hyaline, 1-celled, lunar-shaped. Setae brown, thick-walled, 2-septate. Sclerotia small, black.

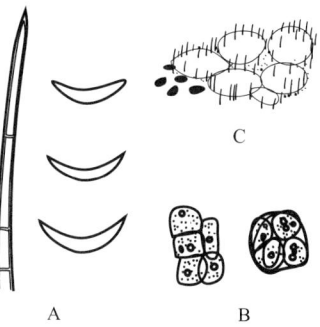

Dimensions: Sporodochia 80–110 μm in diameter. Conidia 20–22.5 × 4.7–5 μm. Setae 122.5–150 × 4.5–5 μm. Sclerotia 32.5–50 μm in diameter.

Material: 70-97 (Snap bean seed, Shinsyu-shinmachi, Nagano, Japan).

Remarks: Numerous setae were formed on sporodochia together with black sclerotia. This fungus is often cited as forma speciales or a variety of *C. dematium* (Pers.: Fr.) Grove (Holliday, 1989).

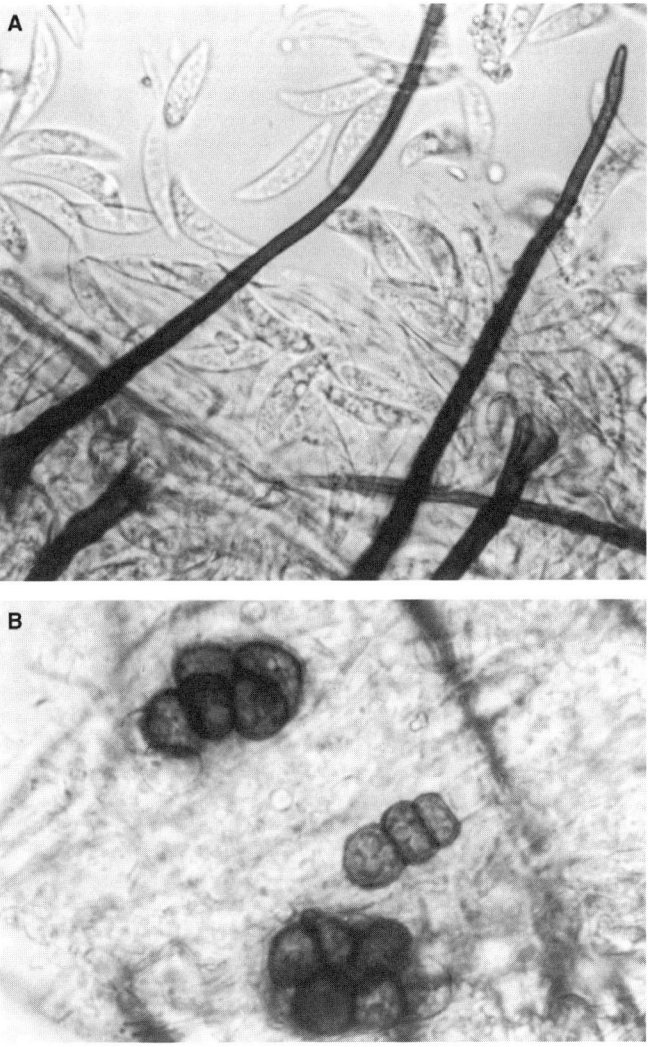

Colletotrichum truncatum. A. Conidia and setae. B: Sclerotia. C: Sporodochia with setae and sclerotia. (From Watanabe, T. 1972a. *Ann. Phytopathol. Soc. Jpn.*, 38:106–110. With permission.)

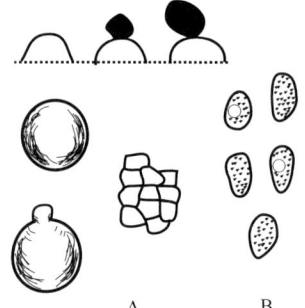

Coniothyrium Corda

Icon. Fung. 4:38, 1840.

Type species: *C. palmarum* Corda

Coniothyrium fuckelii Sacc.

Teleomorph: *Diapleella coniothyrium* (Fuckel) Barr

References: Bestagno et al. 1958; Domsch et al. 1980a,b; Gams and Domsch 1969.

Morphology: Pycnidia superficial or half-embedded, brown to black, globose occasionally necked or ostiolate conspicuously, often black spore masses oozed out from apical ostiolar regions: peridium dark brown, pseudoparenchymatous. Conidia pale brown or brown, ovate, 1-celled, often globulate.

Dimensions: Pycnidia 320–460 μm in diameter. Conidia 3–4.8 (5.4) × (1.8-) 2.5–3.3 (-3.6) μm.

Material: 74-671 (Strawberry root, Shizuoka, Japan); 83-10 (Japanese cypress, Isshi, Mie, Japan).

Remarks: *Leptosphaeria coniothyrium* (Fuckel) Sacc., the teleomorph binomial, is also commonly used. Black spore ooze readily extruded on pycnidia in cultures.

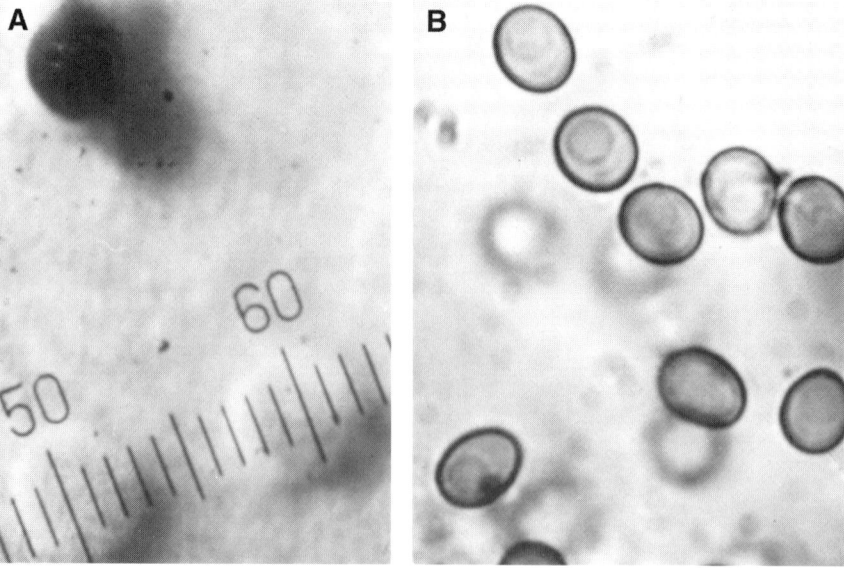

Coniothyrium fuckelii. A: Pycnidia and spore mass extruded. B: Conidia.

Corynespora Güssow

Z. Pflanzenkr. 16:13, 1906.
Type species: *C. cassiicola* (Berk. & Curt.) Wei

Corynespora cassiicola (Berk. and Curt.) Wei

References: Ellis 1963, 1971; Sato and Kitazawa 1980; Wei 1950.

Morphology: Conidiophores hyaline or pale brown, simple, erect, 3- to 7-celled. Conidia porosporous, terminal, solitary or 2–3 conidia in a chain acropetally developed, brown, cylindrical, long-ellipsoidal, often curved, occasionally branched, usually 9- to 18-celled, with conspicuous hilum basally.

Dimensions: Conidiophores 35 over 250 × 7.5–10 μm. Conidia 165–180 × 17.5–22.5 μm.

Material: 83-191 (Soybean root, Shiojiri, Nagano, Japan).

Remarks: Colonies on PDA are yellowish gray green, zonate. This fungus is a root and stem rot pathogen of soybean.

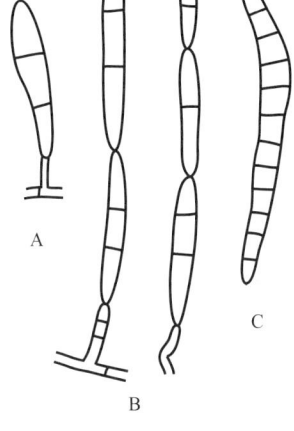

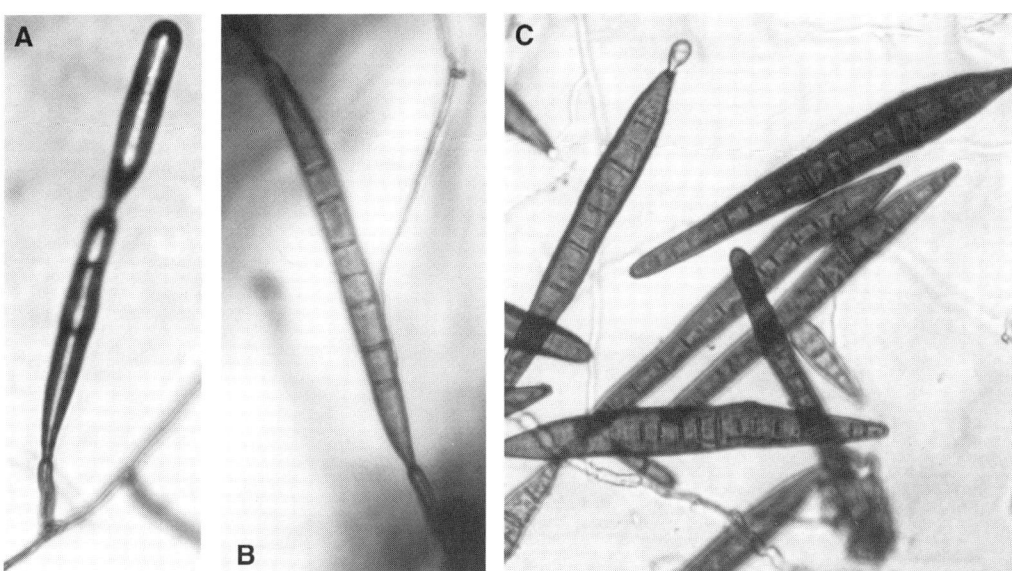

Corynespora cassiicola. A,B: Conidiophore and conidia. C: Conidia.

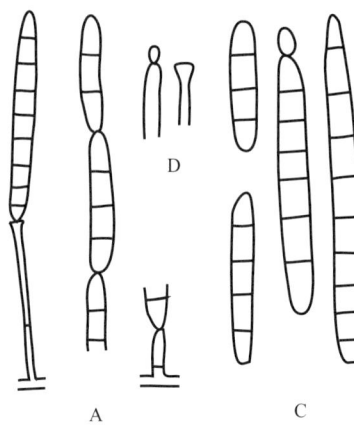

Corynespora citricola Ellis

References: Ellis 1963; Wei 1950.

Morphology: Conidiophores pale brown, simple, erect, 1- to 2-celled, occasionally slightly inflated at the apex, bearing conidia apically. Conidia porosporus, developed in a chain of 2–3 conidia apically, (pale) brown, cylindrical or long-ellipsoidal, usually 4- to 17-celled, truncate basally.

Dimensions: Conidiophores 75–210 × 2.5–4 µm. Conidia (23.7-) 51.2–210 (-227.5) × 4.7–7.5 (-10) µm.

Material: 70-1821 (Pineapple root, Hachijo, Tokyo, Japan).

Remarks: Conidia of this fungus are narrower than those of *C. cassiicola*.

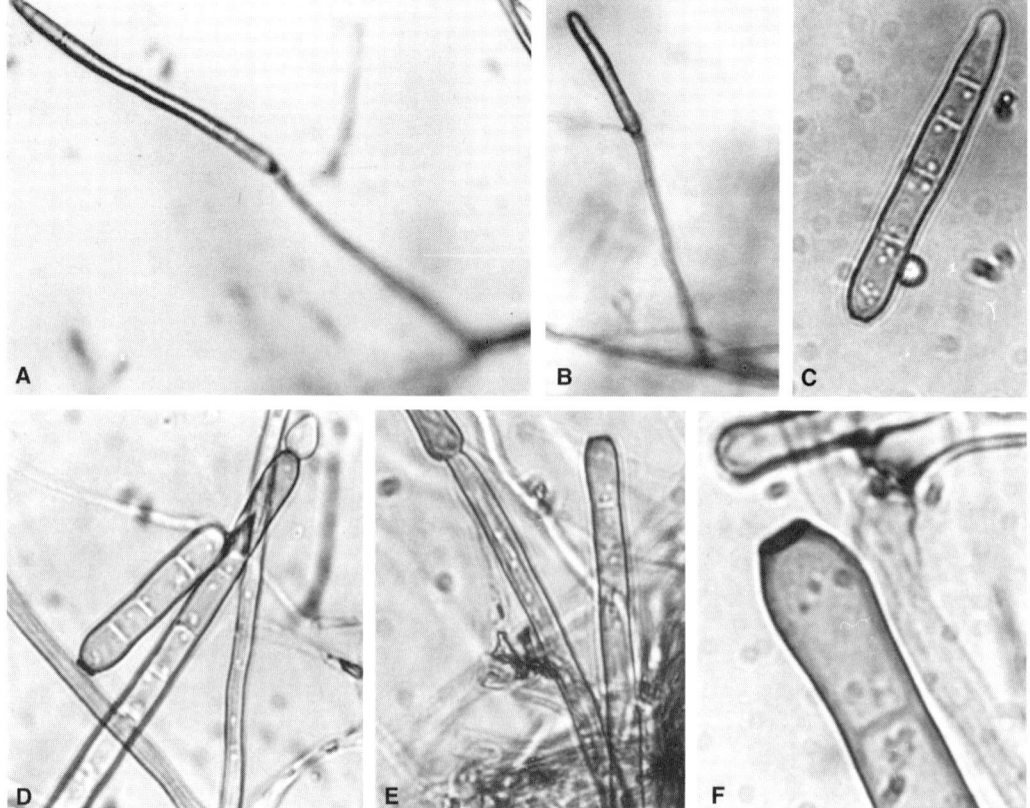

Corynespora citricola. A,B: Conidiophores and conidia. C: Conidia. D: Apical part of conidiophore and conidia. E,F: Apex of conidiophore.

Curvularia Boedijn

Bull. Jard. Bot. Buitenzorg 3, 13(1):127, 1933.
Type species: C. lunata (Wakker) Boedijn

Key to Species

1. Conidia	warted	*C. tuberculata*
	smooth	2
2. Conidium hilum	conspicuous	3
	lacking or indistinct	4
3. Conidia	mainly 4-celled, chlamydospores formed	*C. clavata*
	4- to 5-celled, stroma formed	*C. protubereta*
4. Conidia	mainly 4-celled	5
	5-celled	7
5. Conidia	with large penultimate cell, subellipsoidal	*C. lunata*
	with a septum centrally located	6
6. Conidia	ellipsoidal, over 8.5 µm wide	*C. brachyspora*
	cylindrical or ellipsoidal under 8 µm wide	*C. pallescens*
7. Conidia	often curved in one side	*C. prasadii*
	uncurved, or curved in both sides	8
8. Conidia	almost uncurved, large and dark in central cells	*C. affinis*
	curved totally	*C. senegalensis*

Curvularia affinis Boedijn

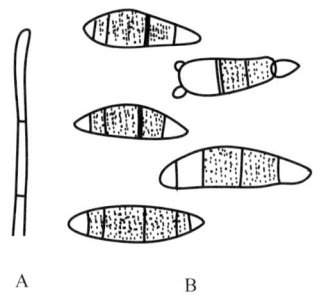

References: Ellis 1966, 1971.

Morphology: Conidiophores pale brown, erect, simple or branched, bearing conidia apically and laterally on apical fertile parts. Conidia porosporous, long ellipsoidal, almost uncurved, mainly 5-celled, large and more pigmented in 1 central cell, without hilum basally. Stroma in agar cultures simple.

Dimensions: Conidiophores 60–75 × 3–4.3 µm. Conidia 32.5–35 × 7.5–11 µm.

Material: 73-417 (Strawberry root, Shizuoka, Japan).

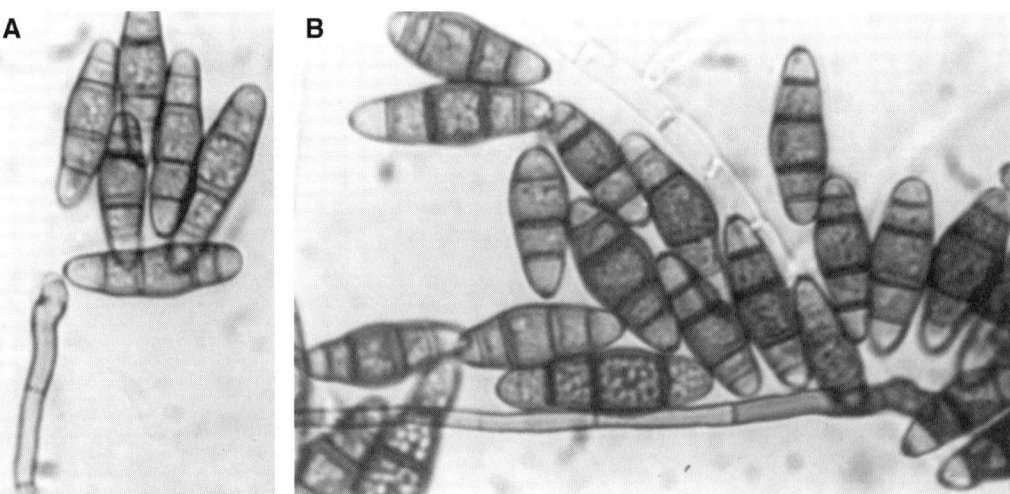

Curvularia affinis. A,B: Conidiophores and conidia.

Curvularia brachyspora Boedijn

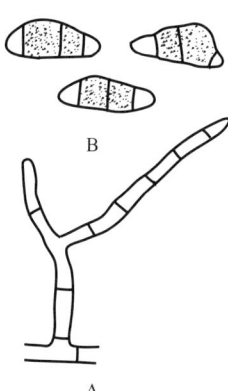

References: Ellis 1966, 1971.

Morphology: Conidiophores pale brown, erect, simple, bearing conidia apically and laterally. Conidia porosporous, ellipsoidal, mainly 4-celled, with a septum centrally located, more pigmented and large in 2 central cells, without hilum basally.

Dimensions: Conidiophores 80–187.5 × 4.5–5 µm. Conidia 21.2–23.8 × 8.7–11.3 µm.

Material: 85-P16 (Paulownia root, Itapua, Paraguay).

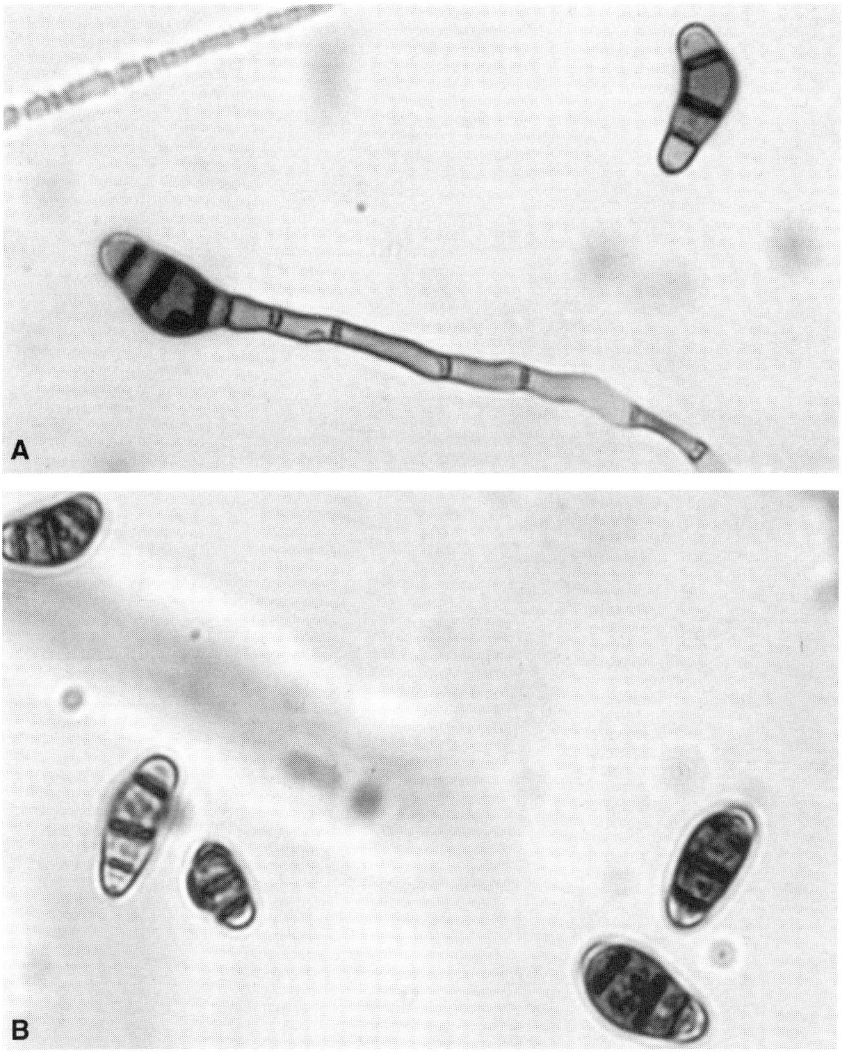

Curvularia brachyspora. A,B: Conidiophore and conidia.

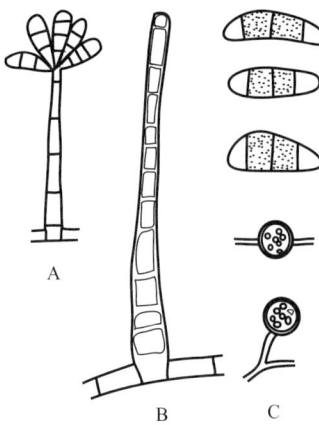

Curvularia clavata Jain

References: Ellis 1966, 1971; Jain 1962.

Morphology: Conidiophores dark brown, simple, erect, thick-walled, bearing conidia apically and laterally. Conidia porosporous, subellipsoidal, cylindrical, mainly 4-celled, darker brown and larger in central 2 cells, with hilum basally. Chlamydospores globose, pale to dark brown, often granulate, intercalary.

Dimensions: Conidiophores 90–142.5 × 5–10 μm. Conidia 21.3–28.8 × 7.3–8.8 μm. Chlamydospores 10–11.3 μm in diameter.

Material: 73-459 (Strawberry root, Tottori, Japan).

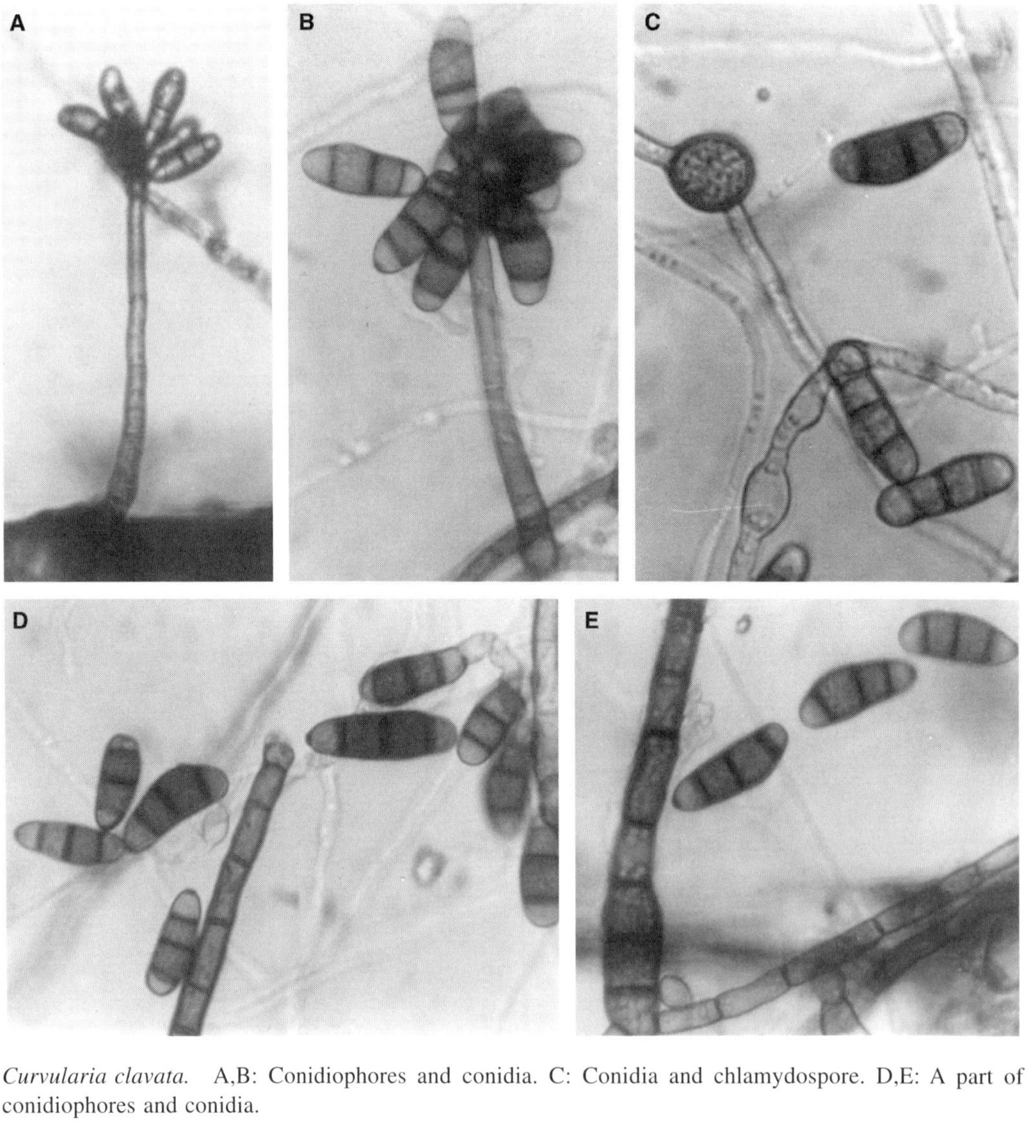

Curvularia clavata. A,B: Conidiophores and conidia. C: Conidia and chlamydospore. D,E: A part of conidiophores and conidia.

Curvularia lunata (Wakker) Boedijn

Teleomorph: *Cochliobolus lunatus* Nelson et Haasis

References: Ellis 1966, 1971; Nelson 1964.

Morphology: Conidiophores erect, brown, simple or branched, straight or curved, bearing conidia apically and laterally, conspicuous pores left after detachment of conidia. Conidia porosporous, subellipsoidal, mostly 4-celled, darker brown in 2-central cells, especially curved, larger in the penultimate cells, with indistinct hilum basally.

Dimensions: Conidiophores 160–300 × 3.7–5 µm. Conidia (17.5-) 18.7–23 (-30) × 8.7–12.5 (-14) µm.

Material: 70-1051 (Pineapple root, Hachijo, Tokyo, Japan); 74-690 (Strawberry root, Mie, Japan).

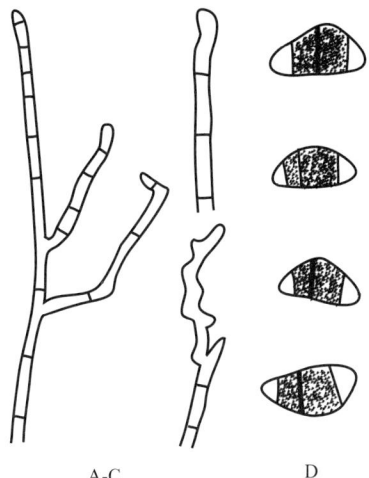

A–C   D

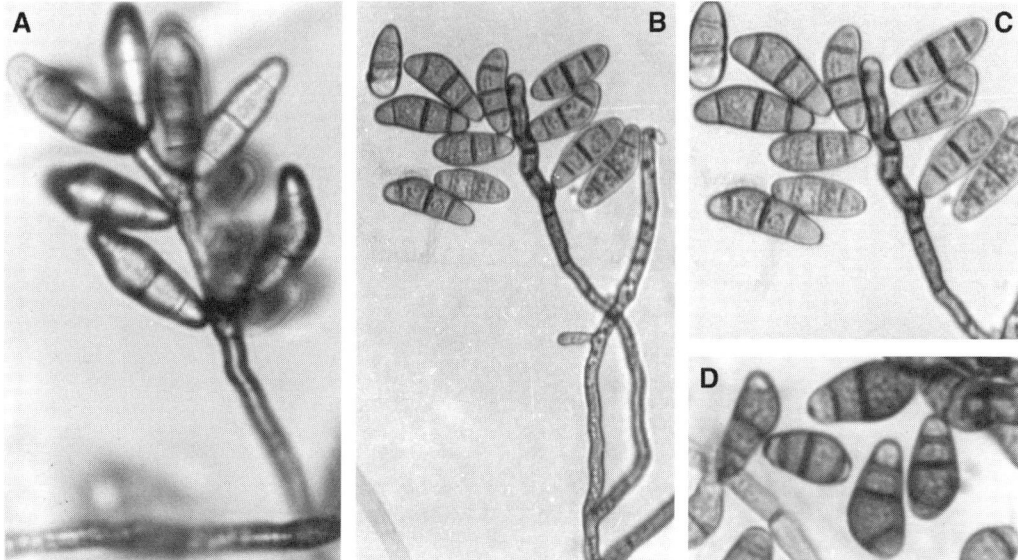

Curvularia lunata.   A–C: Conidiophores and conidia. D: Conidia.

Curvularia pallescens Boedijn

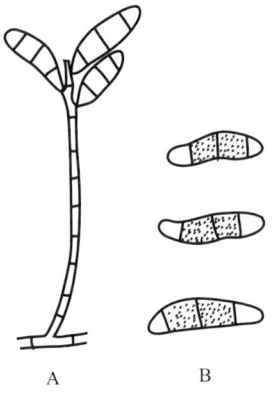

Teleomorph: *Cochliobolus pallescens* (Tsuda & Ueyama) Sivan.

References: Ellis 1966, 1971; Sivanesan 1987.

Morphology: Conidiophores pale brown, erect, simple or branched, bearing conidia apically and laterally. Conidia porosporous, pale brown, cylindrical or ellipsoidal, mainly 4-celled, with a septum located centrally, without hilum basally.

Dimensions: Conidiophores (11.2-) 46.1–243 × 3.5–4 µm. Conidia (17.5-) 22.3–26.8 × 6–8 (-10) µm.

Material: 72-X13-174 (= ATCC 32910, Sugarcane root, Taiwan, ROC); 85-P53 (Paulownia root, Itapua, Paraguay).

Remarks: Conidia of this fungus are pale brown, rather slim.

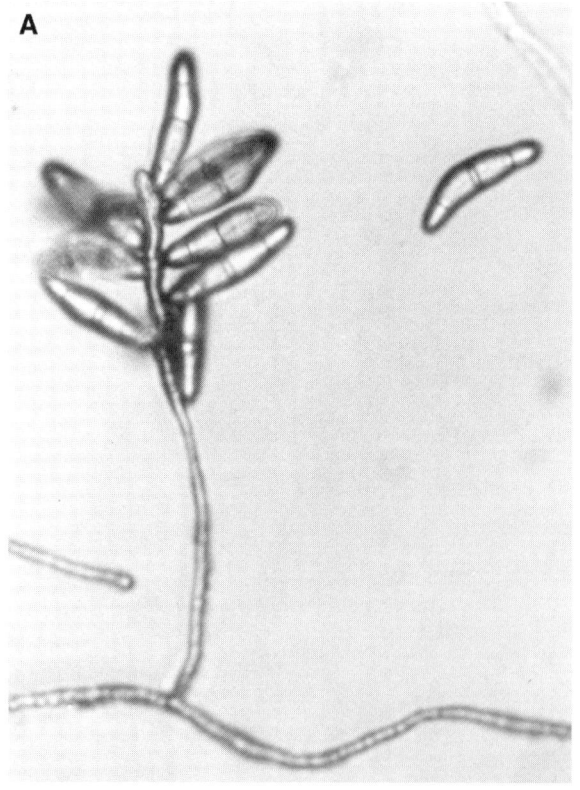
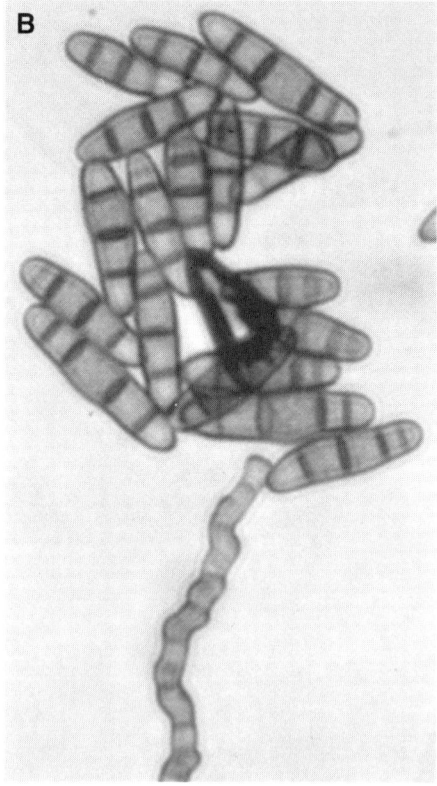

Curvularia pallescens. A,B: Conidiophores and conidia.

Curvularia prasadii R. L. & B. L. Mathur

References: Ellis 1966, 1971.

Morphology: Conidiophores brown, erect, simple or branched, densely septate at apical fertile portions, bearing conidia apically and laterally. Conidia porosporous, long-ellipsoidal, often curved in one side, brown, mainly 4- to 5-celled, with indistinct hilum basally.

Dimensions: Conidiophores 70–195 × 5–6.5 µm. Conidia 20–32.5 × 7.5–12.5 µm.

Material: 85-22 (Flowering cherry seed, Hachioji, Tokyo, Japan).

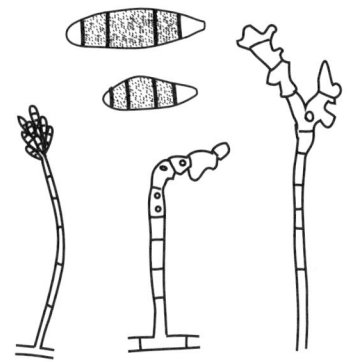

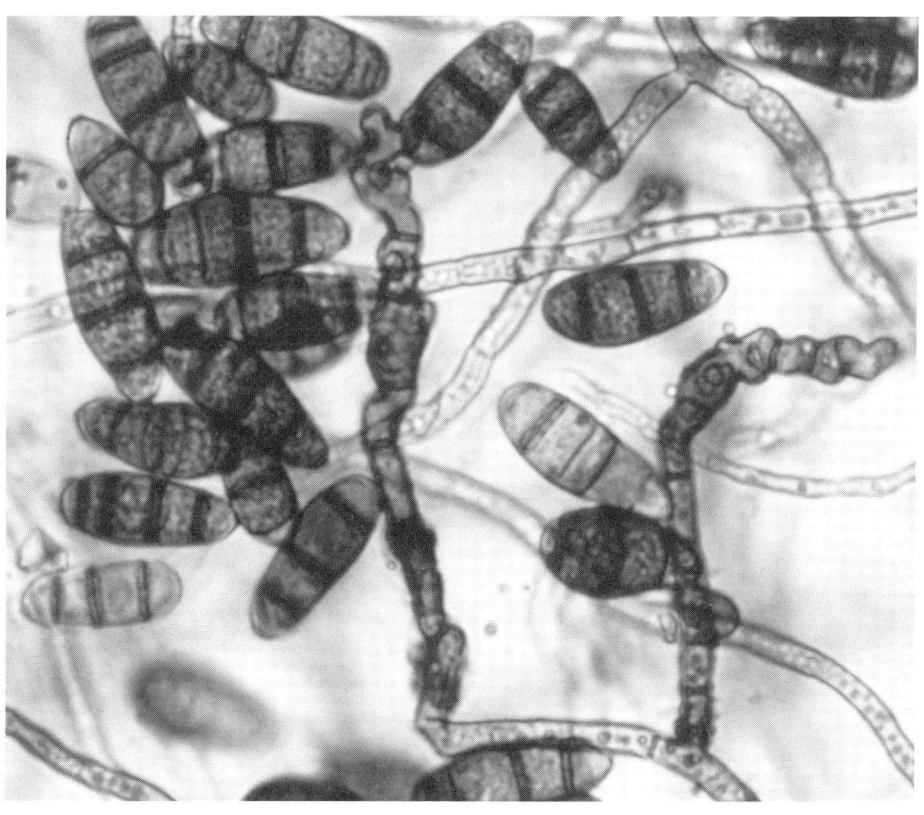

Curvularia prasadii. Conidiophores and conidia.

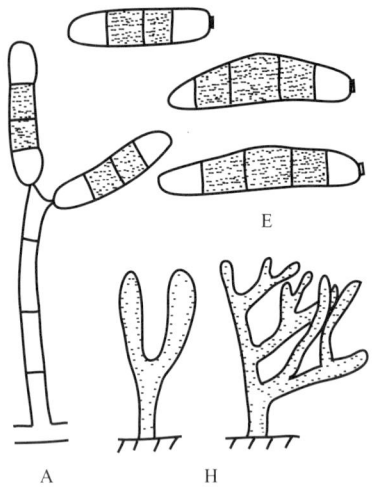

Curvularia protuberata Nelson & Hodges

References: Ellis 1966, 1971; Nelson and Hodges 1965.

Morphology: Conidiophores pale brown, simple, erect, bearing conidia apically and laterally. Conidia porosporous, 4- to 5-celled, with septum centrally located, darker brown in 2–3 central cells, hyaline or pale brown in both end cells, with hilum basally. Chlamydospores globose, brown, often catenulate. Stroma often dendroid, erect, in old agar cultures.

Dimensions: Conidiophores 38–105 × 2.5–5 µm. Conidia 17–35 × (5.5-) 6–9 (-11.5) µm. Chlamydospores 7.5–8.8 µm in diameter. Stroma up to 5 mm high.

Material: 76-55 (Melon seed, Shizuoka, Japan); 77-25 (= IFO 32536, Strawberry root, Nishigahara, Tokyo, Japan).

Remarks: Stroma formation may be significant for identification for a certain isolate with mainly 4-celled conidia.

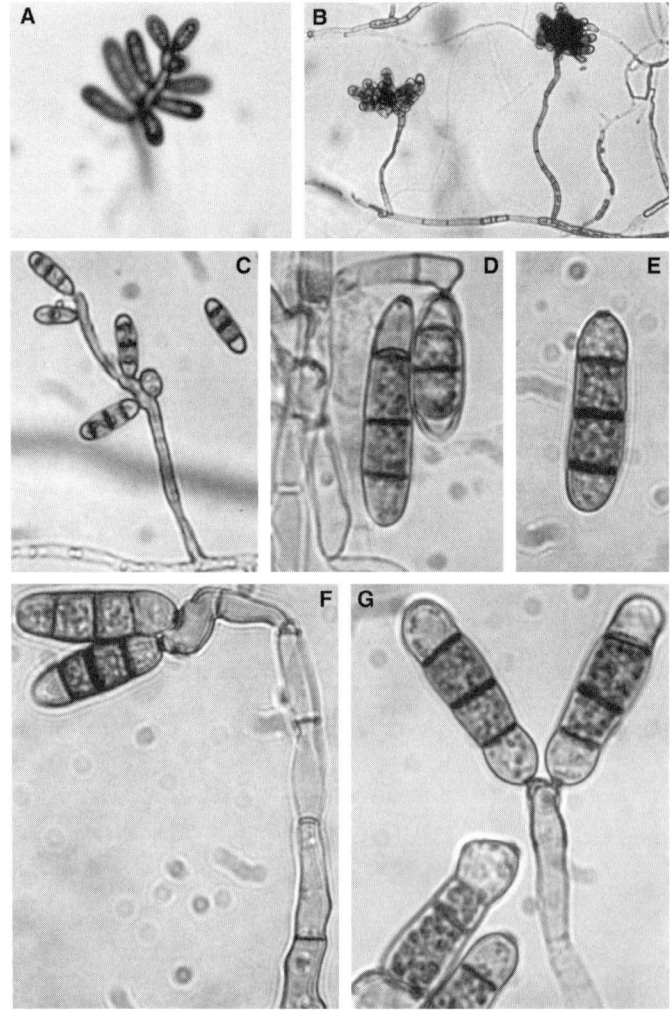

Curvularia protuberata. A–D,F,G: Conidiophores and conidia. E: Conidia. H: Stroma.

Curvularia senegalensis (Speg.) C. V. Subramanian

References: Ellis 1966, 1976; Watanabe et al. 1986a.

Morphology: Conidiophores erect, pale brown, simple or branched, bearing conidia apically and laterally. Conidia porosporous, cylindrical, pale brown, mostly 5-celled, darker brown and larger in central cells, without hilum basally.

Dimensions: Conidiophores 19.4–116.7 (-175) × (2.5-) 3.7–3.8 (-5) µm. Conidia (20-) 23–27.5 (-32.5) × 6.8–8.7 (-11.3) µm.

Material: 72-X12-115 (= ATCC 32911, Sugarcane root, Taiwan, ROC); 84-502 (Japanese black pine seed, Kumano, Mie, Japan); 85-P9 (Paulownia root, Itapua, Paraguay).

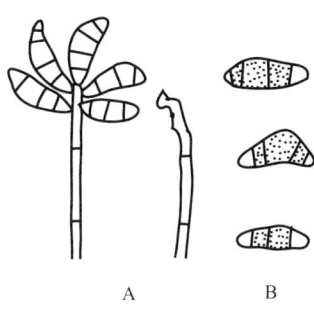

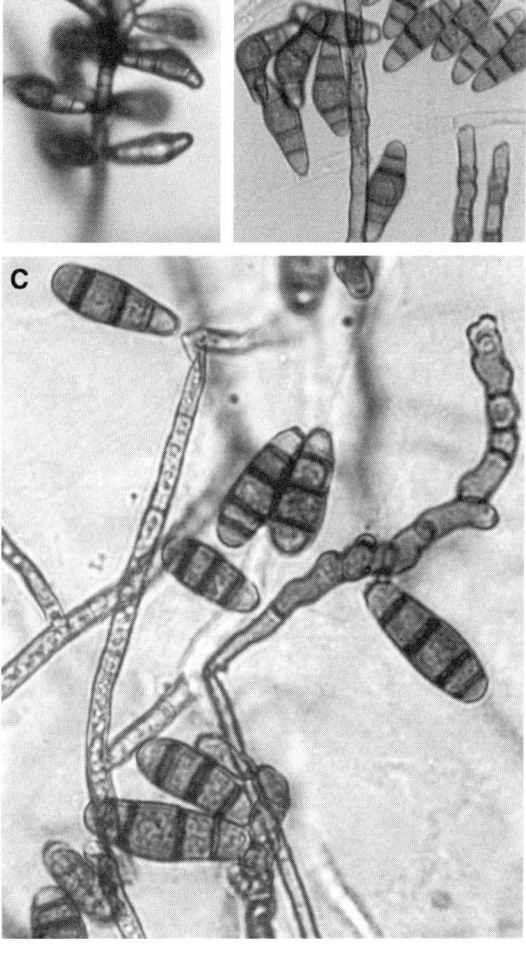

Curvularia senegalensis. A–C: Conidiophores and conidia.

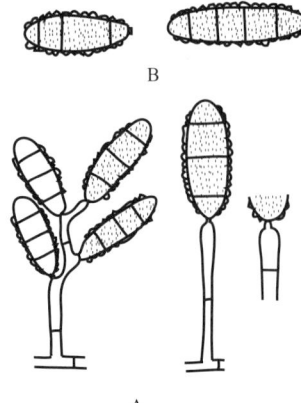

Curvularia tuberculata Jain

Teleomorph: *Cochliobolus tuberculatus* Sivan.

References: Ellis 1971; Jain 1962.

Morphology: Conidiophores erect, (pale) brown, simple, conspicuously curved, often zig zag-shaped, rarely branched, bearing conidia apically and laterally. Conidia porosporous, cylindrical or long-ellipsoidal, dark brown, warted all over, mainly 4-celled, more pigmented and larger in central 2 cells, with or without hilum.

Dimensions: Conidiophores 75–145 × 2.8–5 μm. Conidia 37.5–47.5 × 10–20 μm.

Material: 79-137 (Taro root, Naha, Okinawa, Japan).

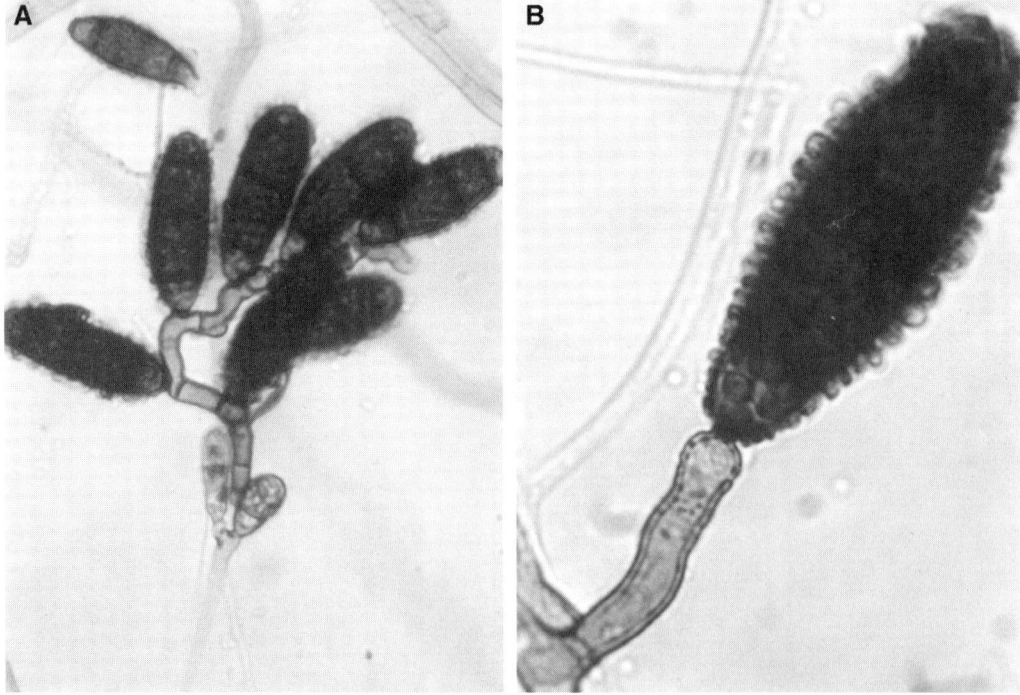

Curvularia tuberculata. A,B: Conidiophores and conidia.

Cylindrocarpon Wollenw.

Phytopathology 3:225, 1913.
Type species: *C. cylindroides* Wollenw.

Key to Species

1. Macroconidia   mostly 4-celled... 2
        not so 3

2. Microconidia   subglobose or ovate .... *C. obtusisporum*
        cylindrical or ellipsoidal .... *C. destructans*

3. Macroconidia   up to 6-celled, mostly curved .... *C. olidum*
        straight or slightly curved 4

4. Conidia   cylindrical, mostly 6-celled, slightly curved................*C. janthothele*
        clavate, 5- to 6-celled................................. *C. boninense*

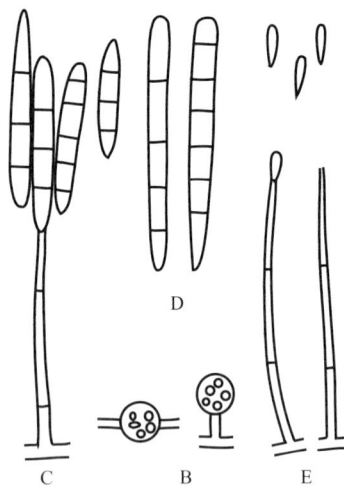

Cylindrocarpon boninense T. Watanabe

References: Arx 1982; Booth 1966; Watanabe et al. 2001b.

Morphology: Conidiophores (phialides) simple, hyaline, erect, septate, bearing terminal spore masses composed of macro- and microconidia. Conidia phialosporous, hyaline, of two kinds: macroconidia clavate, cylindrical or ellipsoidal, mostly 4-5-septate, and microconidia ellipsoidal, apiculate at one end, rather straight in one side, curved in another side. Chlamydospores globose, single, pale brown, granulate.

Dimensions: Conidiophores 60–200 μm long or more, 3.2–5 μm wide basally, 1.2–1.6 μm wide apically. Macroconidia 20–98 × 5–7 μm. Microconidia 7–30 × 2.6–4 μm. Chlamydospores 9–18 μm in diameter.

Material: 00-62 (= MAFF 238163, Sandy soil, Miyukigahama, Hahajima, the Bonin Islands, Tokyo, Japan).

Remarks: Colonies on PDA are nonaerial, yellowish brown.

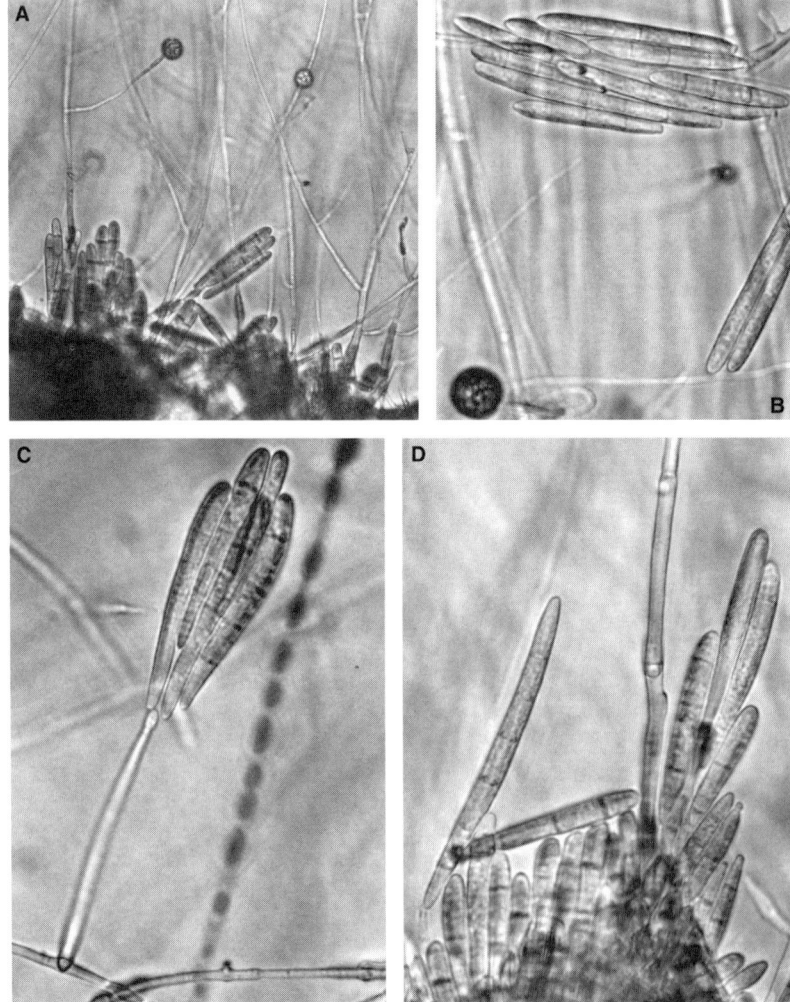

Cylindrocarpon boninense. A,B: Conidia and chlamydospores. C: Conidiophore and macroconidia. D: Macroconidia. E: Conidiophore and microconidia.

Cylindrocarpon destructans (Zins.) Scholten

Teleomorph: *Nectria radicicola* Gerlach & Nilsson

References: Booth 1966; Gerlach 1958.

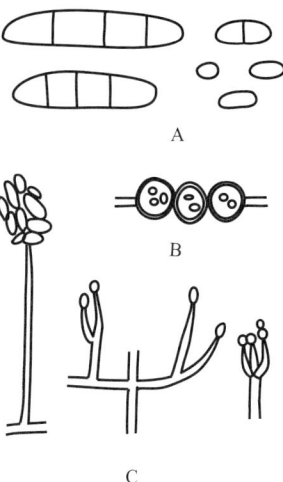

Morphology: Conidiophores erect, simple or branched, bearing spore masses at apical phialides. Conidia terminal, of two kinds: macroconidia, cylindrical, mainly 4-celled, and microconidia, cylindrical, 1-celled. Chlamydospores yellowish brown, ovate to ellipsoidal, a few in a chain.

Dimensions: Conidiophores ca. 65 µm tall: phialides 7.5–12.5 × 2.2–3.8 µm. Conidia: macroconidia (22.5-) 25–55 × 4.8–10 µm and microconidia 2.7–10.5 (-12.5) × 2.2–5.0 µm. Chlamydospores 8.7–12.5 (-15) µm in diameter.

Material: 73-114 (Strawberry root, Shizuoka, Japan); 77-18, 77-183 (Strawberry root, Nishigahara, Tokyo, Japan); 85-P54 (Paulownia root, Itapua, Paraguay).

Remarks: The isolates with broad macroconidia are often given the variety name *crassum*.

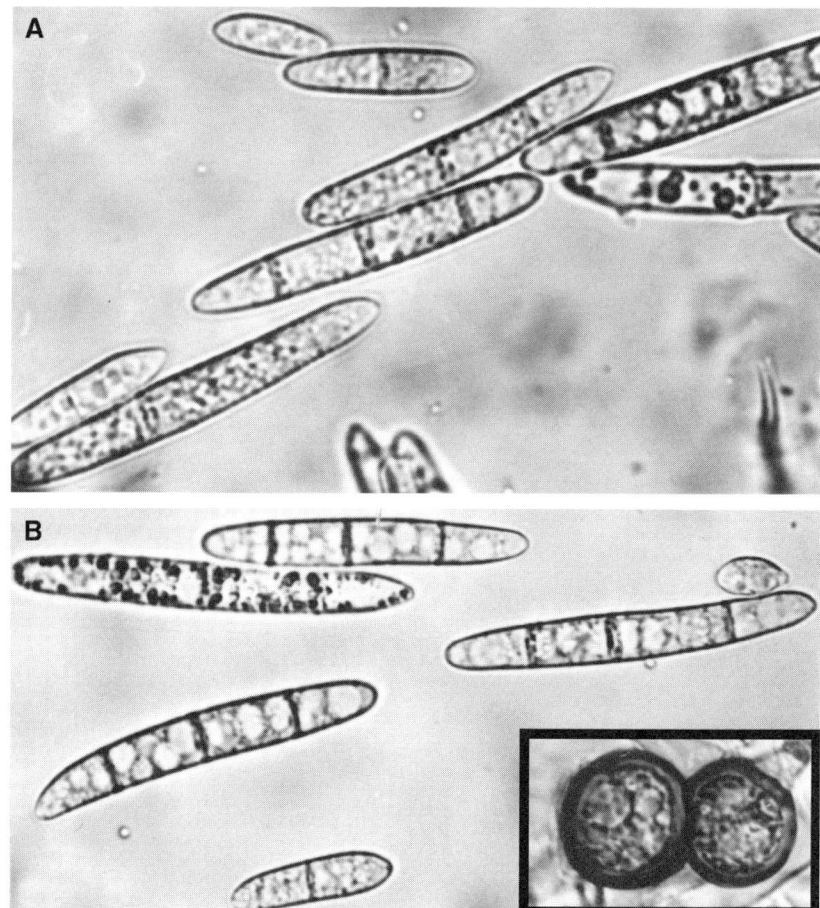

Cylindrocarpon destructans. A,B: Macroconidia, microconidia, and chlamydospores (B). C: Conidiophores and microconidia.

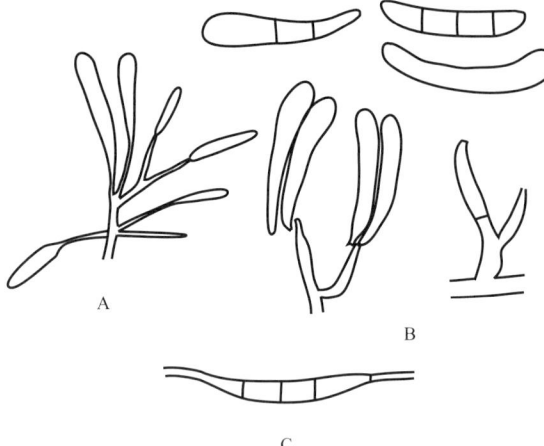

Cylindrocarpon janthothele Wollenw.

References: Arx 1982; Booth 1966.

Morphology: Conidiophores branched, brown, erect, bearing conidial masses at the terminal phialides. Conidia phialosporous, cylindrical, often curved, mostly 6-celled. No microconidia and chlamydospores observed.

Dimensions: Conidiophores 44–50 μm tall. Conidia 48–60 × 6–8 μm.

Material: 98-14 (= MAFF 238002, Uncultivated soil, Tsukuba, Japan).

Remarks: Colonies on PDA are centrally aerial, yellowish brown, purplish-tinted, zonate, marginally resupinate; with age, dark purplish brown.

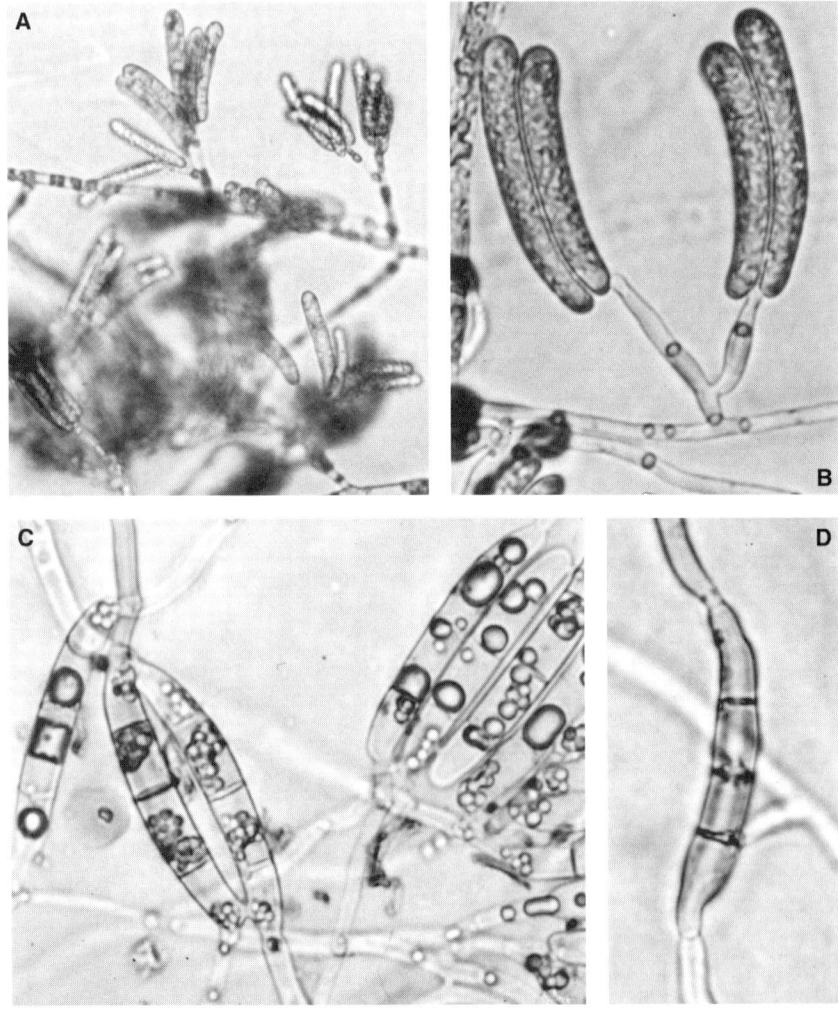

Cylindrocarpon janthothele. A: Habit. B: Conidiophore and conidia. C,D: Germination of conidia with germ tubes from both ends.

Cylindrocarpon obtusisporum (Cooke & Harkness) Wollenw.

References: Booth 1966.

Morphology: Conidiophores: for macroconidia erect, simple or branched, bearing spore masses at apical phialides generally; for microconidia more complicated, often penicillate. Conidia phialosporous, terminal, of two kinds: macroconidia, mainly 4-celled, cylindrical, slightly curved, and microconidia, subglobose or ovate, 1-celled, apiculate at one end. Chlamydospores yellowish brown, globose, occasionally catenulate, a few in a chain, or aggregated.

Dimensions: Conidiophores 60–85 × 2.7–3 µm: phialides 7.5–17.5 × 2.5–3.8 µm. Macroconidia 30.0–46.3 × (6.5-) 7.5–10 µm. Microconidia 3.2–12.5 × 2.5–3.8 (-7.5) µm. Chlamydospores 6.2–13.8 µm in diameter.

Material: 70-1287 (Pineapple field soil, Hachijo, Japan); 73-144, 73-267 (Strawberry root, Shizuoka, Japan); 74-823 (Strawberry root, Saitama, Japan).

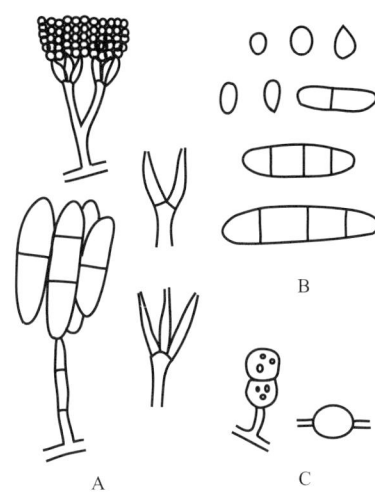

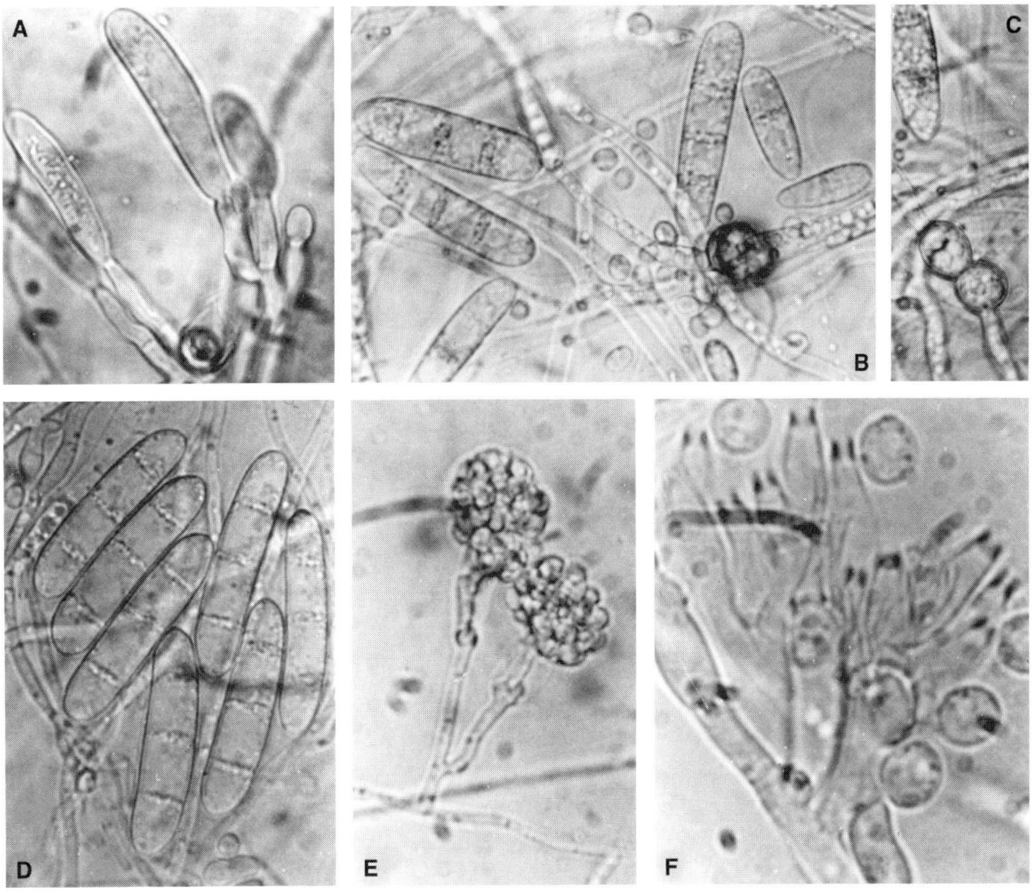

Cylindrocarpon obtusisporum. A: Part of conidiophores and macroconidia. B: Macroconidia and microconidia. C: Part of macroconidia and chlamydospores. D: Macroconidia. E,F: Conidiophores with terminal phialides and microconidia.

Cylindrocarpon olidum (Wollenw.) Wollenw.

References: Booth 1966; Cormack 1937; Gerlach 1959; Watanabe 1992f.

Morphology: Conidiophores erect, simple or branched, bearing conidia at phialides at apexes of branches. Conidia phialosporous, terminal, of two kinds: macroconidia 3- to 6-celled, mainly 4-celled, cylindrical, curved; and microconidia ovate, cylindrical or clavate, 1-celled, slightly apiculate at one end. Chlamydospores yellowish brown, globose.

Dimensions: Conidiophores 22.5–62.5 µm tall: phialides 15 µm tall. Conidia: macroconidia 40.0–67.5 (-70) × (5-) 6.5–8.8 µm, and microconidia 4.5–10 × 2.7–5 µm. Chlamydospores 7.5–21 µm in diameter.

Material: 70-1061 (Pineapple field soil, Hachijo, Tokyo, Japan); 77-96, 77-127, 77-149 (Strawberry root, Nishigahara, Tokyo, Japan).

Remarks: Smell of soil in cultures is very characteristic. This fungus may be given the variety name *olidum* on the basis of formation of microconidia.

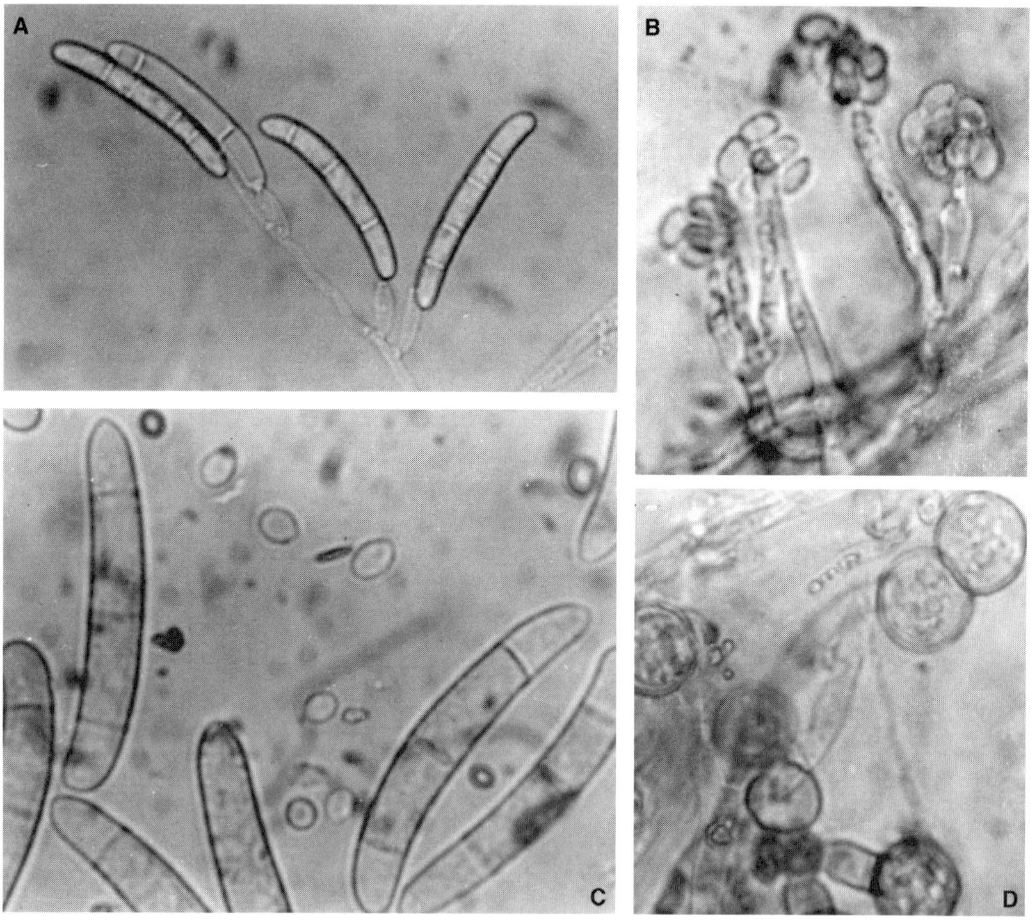

Cylindrocarpon olidum. A: Conidiophores and macroconidia. B: Conidiophores and microconidia. C: Macroconidia and microconidia. D: Chlamydospores. (From Watanabe, T. 1992f. *Trans. Mycol. Soc. Jpn.*, 33:231–236. With permission.)

Cylindrocladium Morgan

Bot. Gaz. 17:191, 1892.
Type species: *C. scoparium* Morgan

Key to Species

1.	Conidia	2-celled	2
		4-celled	6
2.	Stipes and terminal vesicles	present	3
		absent	*C. tenue*
3.	Terminal vesicles	globose, sclerotia formed	*C. floridanum*
		not so	4
4.	Vesicles	long-ellipsoidal, sclerotium-like structures formed	*C. scoparium*
		not so, chlamydospores formed	5
5.	Chlamydospores	over 3 spores in a chain	*C. camelliae*
		solitary or in twins	*C. parvum*
6.	Terminal vesicles	subglobose, constricted	*Cylindrocladium* sp.
		clavate or ellipsoidal	*C. colhounii*

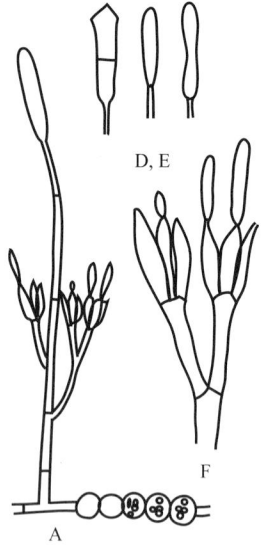

Cylindrocladium camelliae Venkataramani & Venkata Ram

References: Peerally 1991; Terashita 1969; Watanabe 1993c.

Morphology: Conidiophores erect, branched once or twice verticillately in the central parts, elongating primary and secondary branches, bearing spore masses at the phialides on the branches, with stipes and terminal vesicles: vesicles clavate, lanceolate, or cylindrical. Conidia phialosporous, cylindrical, 2-celled. Chlamydospores over 3 spores in a chain.

Dimensions: Conidiophores (including stipes and terminal vesicles) 52.5–295 (-390) × 2.5–5 µm: primary branches 11.2–23 × 2–4 µm; secondary branches 10–17 × 2.5–3.8 µm: phialides 11–25 × 1.7–3.6 µm: stipes (from the highest branch area to the vesicles) 48–150 (-310) × 2.5–5 µm: vesicles 7.5–62.5 × 2.5–5.8 µm. Conidia 6–17 × 1.5–3 µm. Chlamydospores 11.2–20 µm in diameter.

Material: 92-202 (= IFO 32528, Root of *Phellodendron amurense*, Meguro, Tokyo, Japan).

Deuteromycotina (Motosporic Fungi)

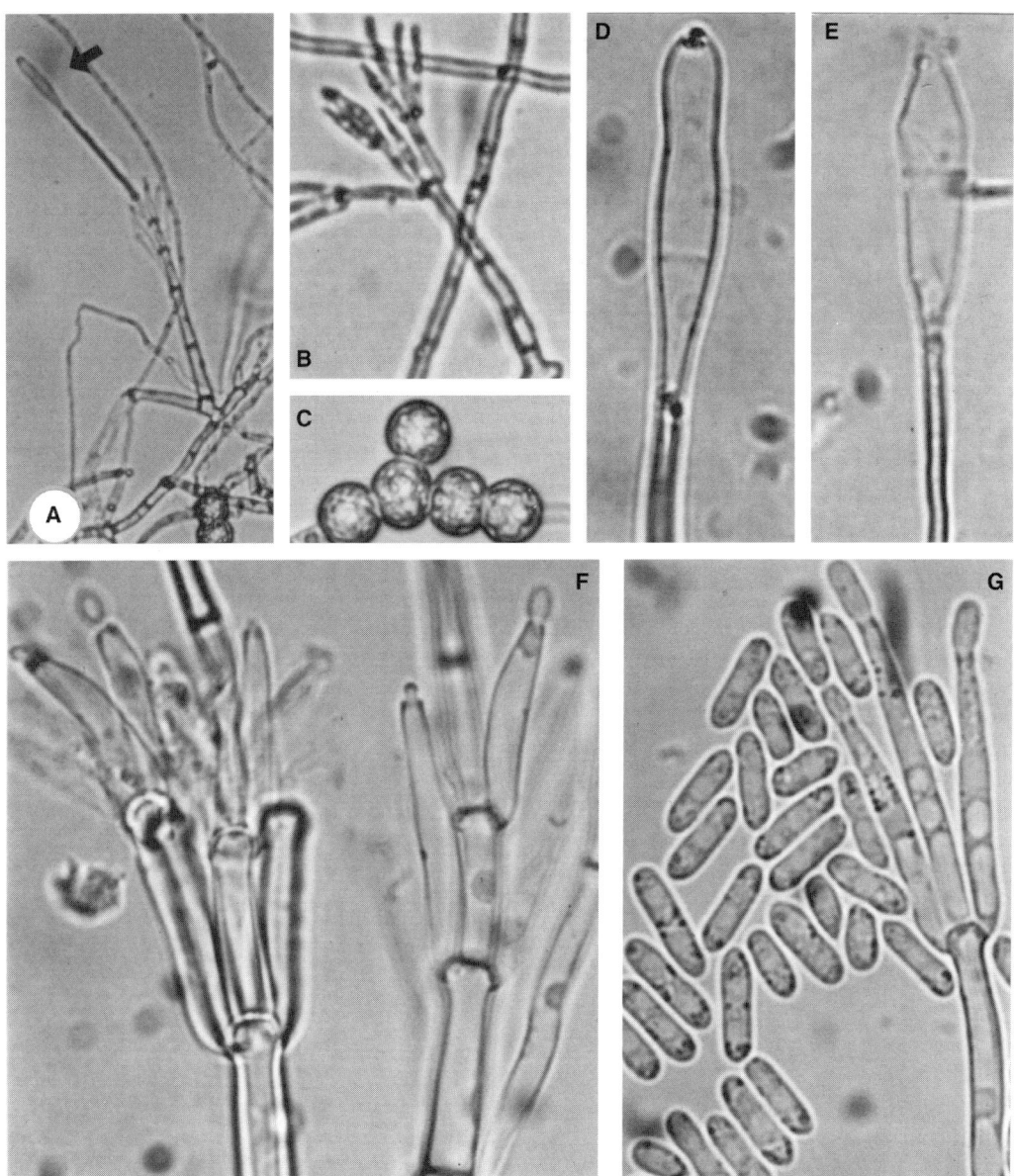

Cylindrocladium camelliae. A: Conidiophore with stipe and vesicle (arrow) and chlamydospores. B,F: Part of the conidiophores and branches. C: Chlamydospores. D:E: Vesicles. G: Part of the conidiophore and conidia. (From Watanabe, T. 1994a. *Mycologia*, 86:151–156. With permission.)

Cylindrocladium colhounii Peerally

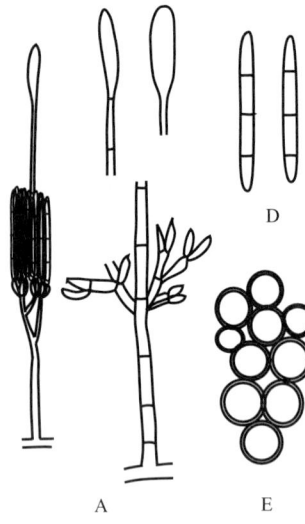

Teleomorph: *Calonectria colhounii* Peerally

References: Peerally 1991; Watanabe 1993c.

Morphology: Conidiophores erect, branched once or twice, elongating primary and secondary branches, bearing spore masses at the phialides on the branches, with stipes and terminal vesicles: vesicles clavate or ellipsoidal. Conidia phialosporous, hyaline, cylindrical, 4-celled. Chlamydospores globose, reddish brown, catenulate or aggregated.

Dimensions: Conidiophores (including stipes and terminal vesicles) 215–475 × 6–10 μm: primary branches 10–42.5 × 4.7–6.3 μm: secondary branches 8.7–17.5 × 3.5–5.4 μm: tertiary branches 7.5–15 × 2.5–5 μm: phialides 7.5–17.5 × 2.5–6.2 μm: stipes (from the highest branch area to the vesicles) 140–290 μm long: vesicles 12.5–65 × 2–4.5 μm. Conidia 45–77.5 × 4.7–6 μm. Chlamydospores 15–20 μm in diameter.

Material: 92-260 (= IFO 32530, Root of *Phellodendron amurense*, Meguro, Tokyo, Japan).

Remarks: Teleomorph of this fungus was not observed.

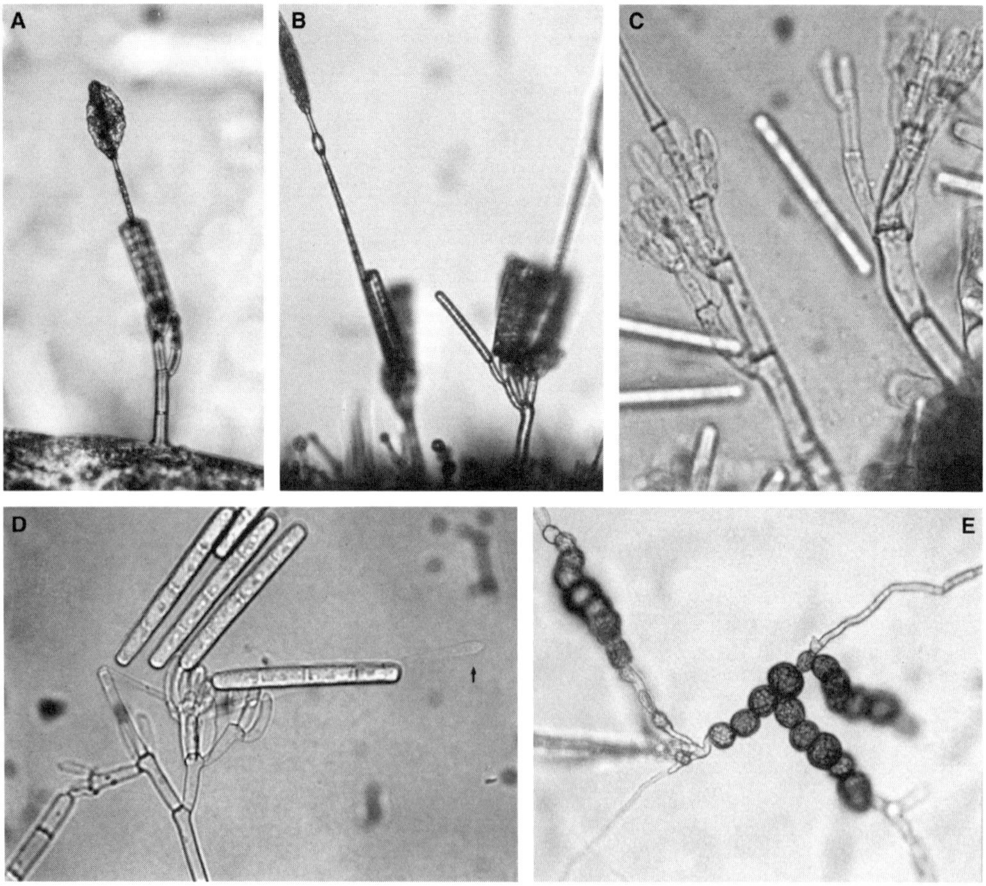

Cylindrocladium colhounii. A,B: Conidiophores, spore masses, stipes, and vesicles. C,D: Part of the conidiophores, vesicle (arrow), and conidia. E: Catenulate chlamydospores. (From Watanabe, T. 1994a. *Mycologia*, 86:151–156. With permission.)

Cylindrocladium floridanum Sobers et Seymour

Teleomorph: *Calonectria kyotoensis* Terashita.

References: Peerally 1974; Sobers 1972; Sobers and Seymour 1967; Terashita 1968.

Morphology: Conidiophores erect, branched once or twice in the central parts, bearing spore masses at phialides on the branches, with stipes and terminal vesicles: vesicles globose or subglobose. Conidia phialosporous, 2-celled, cylindrical. Sclerotia dark brown, globose, granulate.

Dimensions: Conidiophores ca. 170 × 5 µm: primary branches 12.5–20 × 3.7–5 µm: phialides 11.2–17.5 × 3–4.3 µm: vesicles 5–8 × 3.5–6.5 µm. Conidia 47.5–55 × 4.5–5 µm. Sclerotia 72.5–110 × 55–85 µm.

Material: 73-231 (= IFO 32532, Strawberry roots, Shizuoka, Japan); 76-77 (= IFO 32531, Soybean root, Nagano, Japan).

Remarks: This fungus resembles *C. scoparium* Morgan morphologically with ovate or ellipsoidal vesicles. On the basis of priority, it is claimed to be the anamorph of *Calonectria kyotoensis* by Sobers (1972) and Peerally (1974).

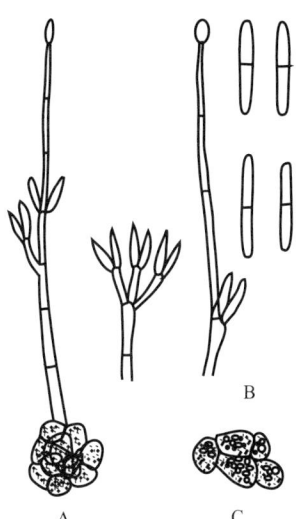

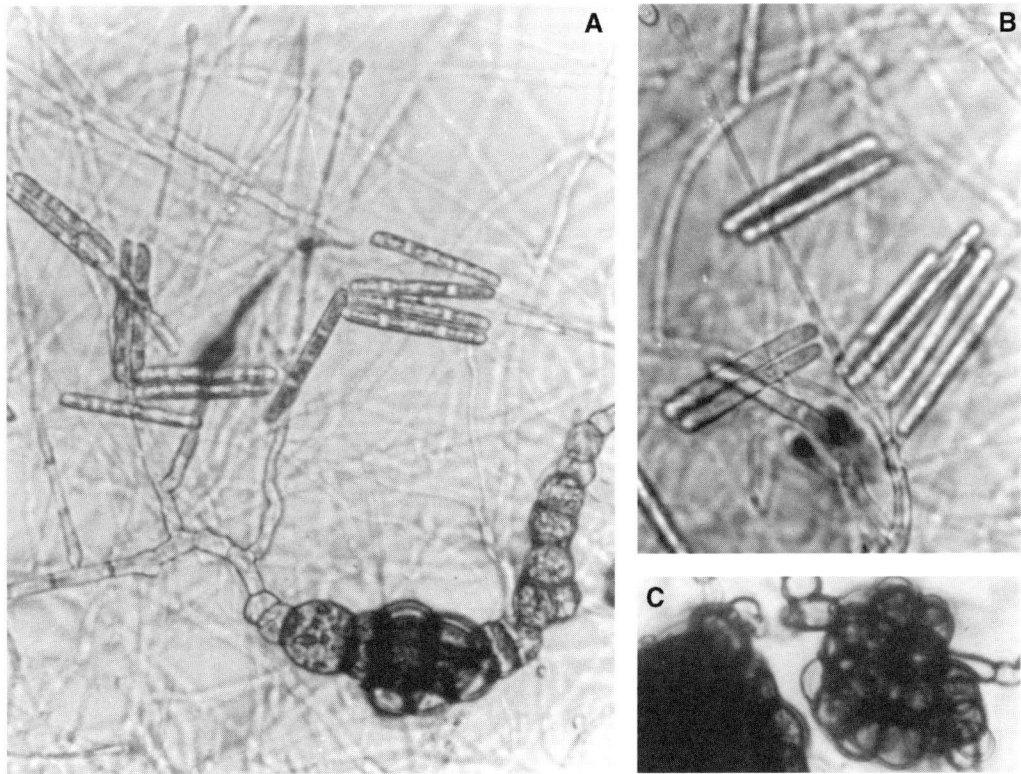

Cylindrocladium floridanum. A,B: Conidiophores, stipes, vesicles, conidia, and sclerotium (A). B: Conidia and stipe. C: Sclerotia.

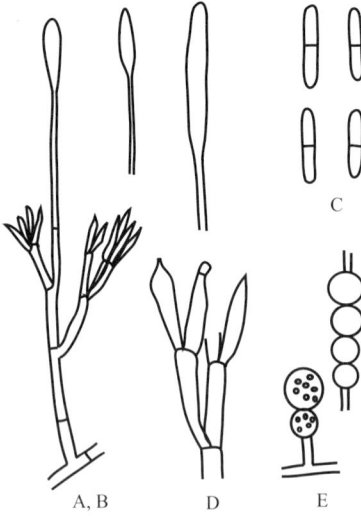

Cylindrocladium parvum Anderson

References: Anderson 1919; Boedijn and Reitsma 1950; Domsch et al. 1980a,b; Watanabe 1992f.

Morphology: Conidiophores erect, branched once or twice mainly verticillately, elongating primary and secondary branches, bearing spore masses at the phialides on the branches, with stipes and terminal vesicles: vesicles cylindrical. Conidia phialosporous, cylindrical, 1- to 2-celled. Chlamydospores yellowish brown, globose, often catenulate.

Dimensions: Conidiophores 120–175 × 5 µm: primary branches 17.5–19 × 2.5 µm: secondary branches 6–12.3 µm long: phialides 6.5–12 × 2.5 µm: vesicles 35–62.5 × 4–5.3 µm. Conidia 13–17 × 1.9–3 µm. Chlamydospores 7.5–12.5 µm in diameter.

Material: 87-150 (= IFO 32534, Flowering cherry orchard soil, Hachioji, Tokyo, Japan).

Remarks: This fungus resembles *C. clavatum* Hodges & May and *C. pteridis* Wolf on the basis of cylindrical or clavate vesicles and 2-celled conidia, but dimensions of the respective organs are different.

Deuteromycotina (Motosporic Fungi)

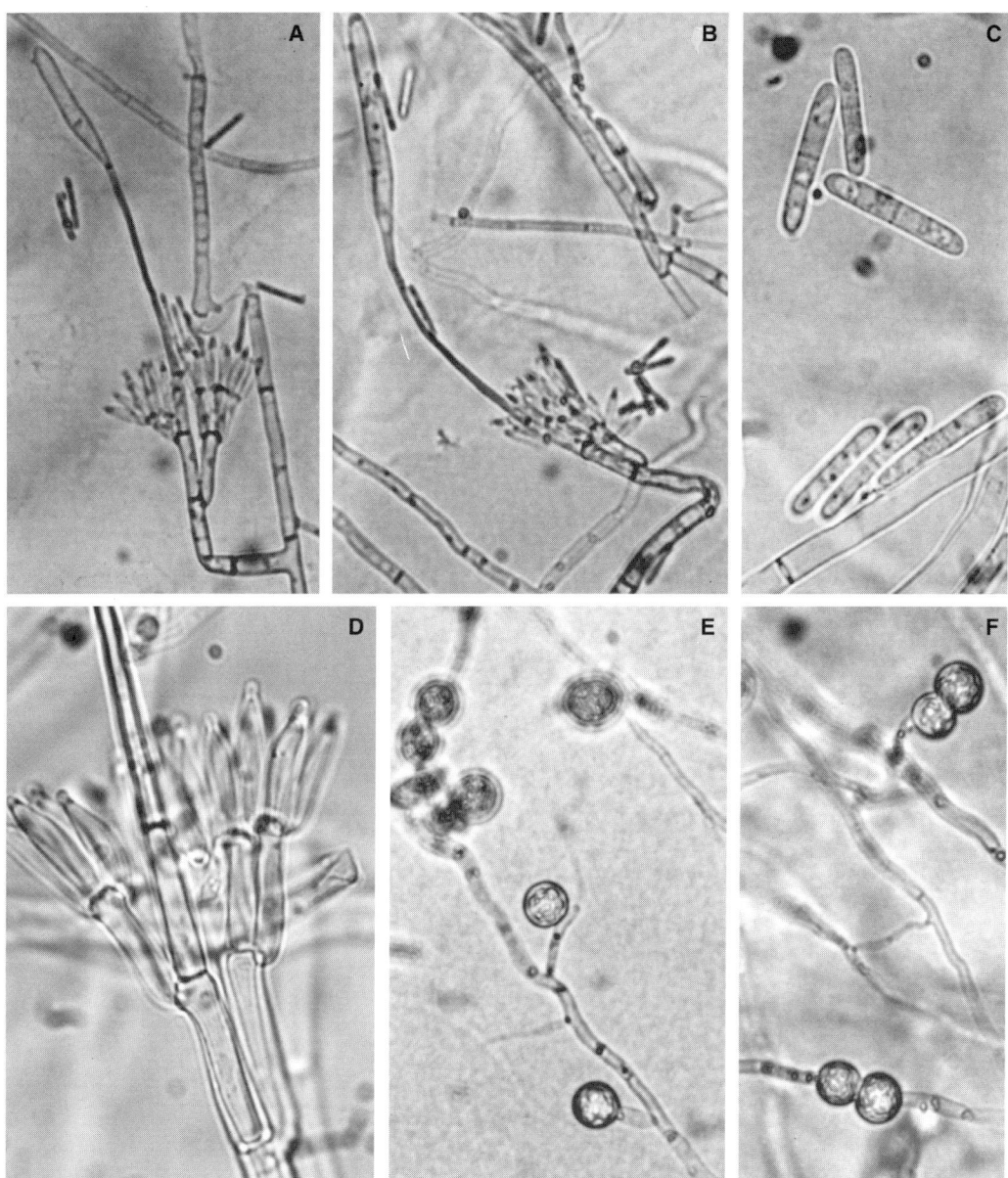

Cylindrocladium parvum. A,B: Conidiophores, stipes, vesicles, and conidia. C: Conidia. D: Branches and phialides. E,F: Chlamydospores. (From Watanabe, T. 1992f. *Trans. Mycol. Soc. Jpn.*, 33:231–236. With permission.)

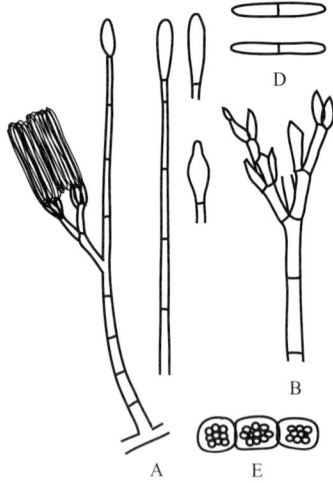

Cylindrocladium scoparium Morgan

References: Boedijn and Reitsma 1950; Boesewinkel 1974; Rossman 1983; Sobers and Seymour 1967; Sobers 1972; Terashita 1968.

Morphology: Conidiophores hyaline, erect, gradually tapering from base toward the apex, branched 1–4 times in the central parts, with stipes and terminal vesicles, bearing spore masses at the phialides on the branches: vesicles ellipsoidal or spindle-shaped. Conidia phialosporous, hyaline, cylindrical, 1- to 2-celled. Chlamydospores dark brown, globose, catenulate or aggregated.

Dimensions: Conidiophores 187.5–262.5 × 6.2–7.5 µm: primary branches 12.5–22.5 × 4.5–5 µm: secondary branches 11.2–17.5 × 4.5–6.3 µm: phialides 7.5–15 × 4–5 µm: vesicles 17.5–23.8 × 5–7.5 µm. Conidia 45–56.3 × 4.5–5 µm. Chlamydospores 15–25 µm in diameter.

Material: 92-118 (= IFO 32535, Eucalyptus field soil, Mikkabi, Shizuoka, Japan).

Remarks: This fungus is also claimed to be the anamorph of *Calonectria kyotoensis* Terashita (Rossman, 1983).

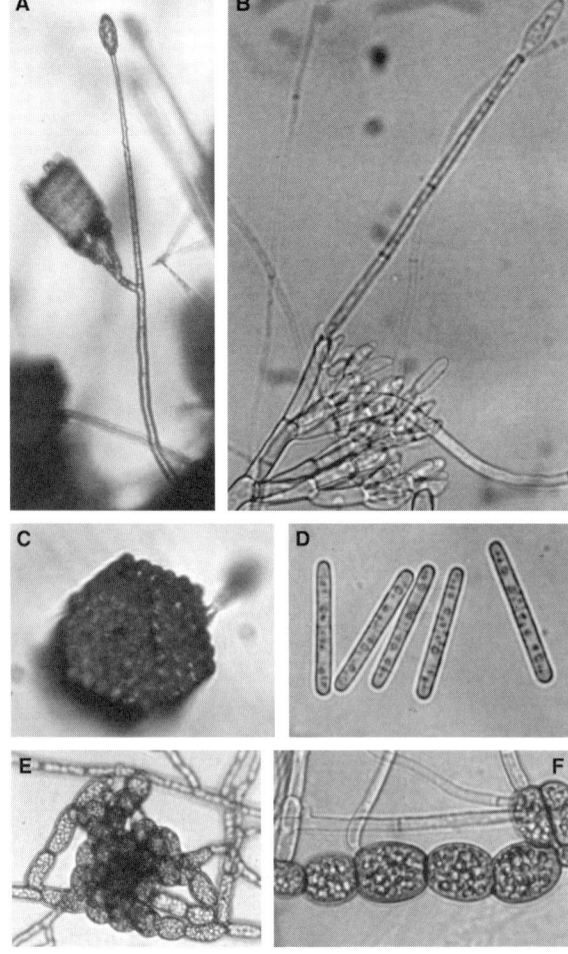

Cylindrocladium scoparium. A: Conidiophore with spore mass, stipe, and vesicle. B: Conidiophore, branches, and phialides. C: Overview of spore mass. D: Conidia. E,F: Chlamydospores and sclerotium-like structure.

Cylindrocladium tenue (Bugnicourt) T. Watanabe

References: Booth and Murray 1960; Peerally 1991; Watanabe 1993c.

Morphology: Conidiophores hyaline, erect, branched 1–3 times, mainly verticillate or penicillate at the apical parts, bearing spore masses at phialides on the respective branches, without stipes and terminal vesicles characteristic of most of *Cylindrocladium* species. Conidia phialosporous, hyaline, cylindrical, 2-celled. Chlamydospores globose, dark brown, catenulate.

Dimensions: Conidiophores 60–205 × 3.7–7.5 μm: primary branches 10–25 × 2.5–5 μm: secondary branches 10–27.5 × 2–4.8 μm: tertiary branches 10–14 × 2–3 μm: phialides 9–27 × 1.9–3.6 μm. Spore masses 30–90 μm in diameter. Conidia 15–25.2 × 1.5–2.6 μm. Chlamydospores 12.5–22.5 μm in diameter.

Material: 92-246 (= IFO 32533, Root of *Phellodendron amurense*, Meguro, Tokyo, Japan).

Remarks: This fungus looks like *Gliocladium* and smells of soil in agar cultures.

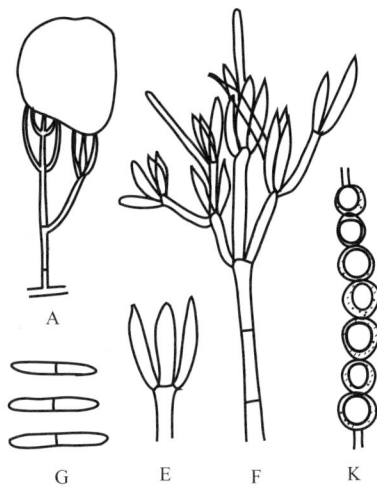

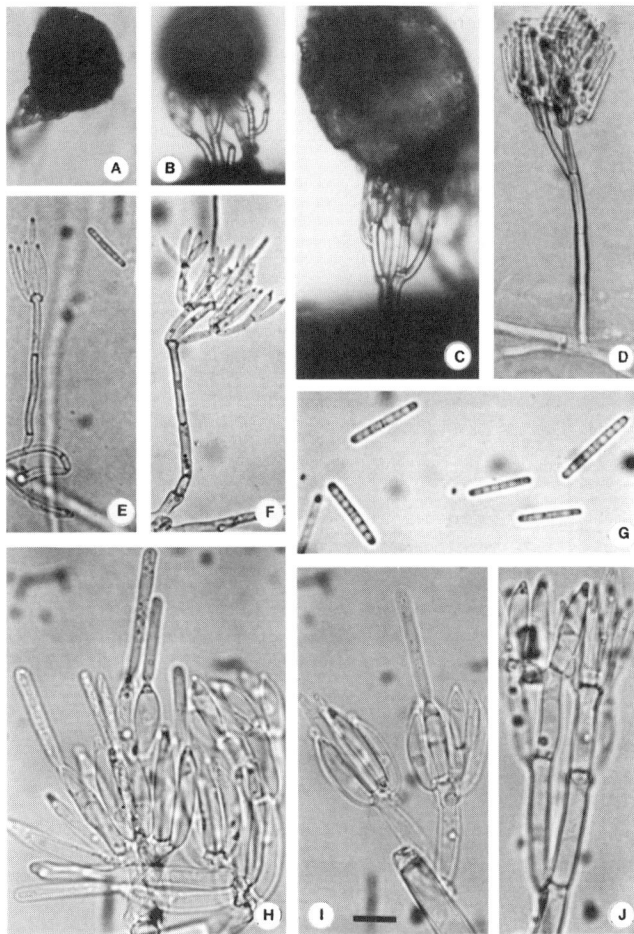

Cylindrocladium tenue.   A–D: Conidiophores and spore masses. E,F: Conidiophores and conidia. G: Conidia. H–J: Branches, phialides, and conidia. K: Catenulate chlamydospores. Note conidia borne on phialides (H,I). (From Watanabe, T. 1994a. *Mycologia*, 86:151–156. With permission.)

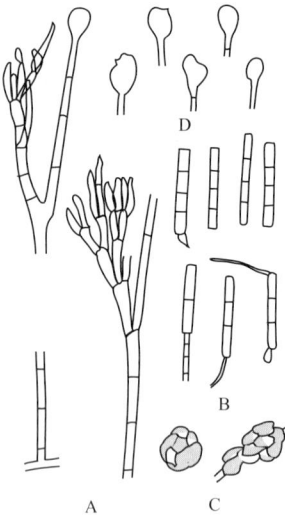

Cylindrocladium sp.

References: Crous and Wingfield 1994; Peerally 1991; Watanabe 2001c.

Morphology: Conidiophores hyaline, erect, branched 1–3 times, mainly penicillate at the fertile portions, bearing spore masses at phialides on the respective branches, with stipes and terminal vesicles characteristic of most of *Cylindrocladium* species. Vesicles globose to subglobose, often constricted or furrowed. Conidia phialosporous, hyaline, cylindrical, 1 to 4-celled (mainly 4-celled). Chlamydospores globose, reddish brown, thick-walled, catenulate. Sclerotia dark reddish brown, spherical or irregular, composed of catenulate and aggregated chlamydospores.

Dimensions: Conidiophores mostly 140–330 μm tall, nearly 6–8 μm wide basally, 3–4 μm wide apically: primary branches 15 μm long: secondary branches 20 μm long: tertiary branches 12.5 μm long: phialides 7–19 × 2.4–4 μm: vesicles 14–20 × 6–14 μm. Conidia 32–48 × 3–4 μm. Chlamydospores 12.5–26 μm in diameter. Microsclerotia 65–115 μm or more in diameter.

Material: 00-52 (= MAFF 238171, Forest soil, Chibusa-yama, Hahajima, the Bonin Islands, Tokyo, Japan).

Remarks: This fungus resembles *C. citri* (Fawcett & Klotz) Boedijn & Reitsma morphologically except for formation of chlamydospores and microsclerotia, although these structures were described and illustrated by Crous and Wingfield (1994).

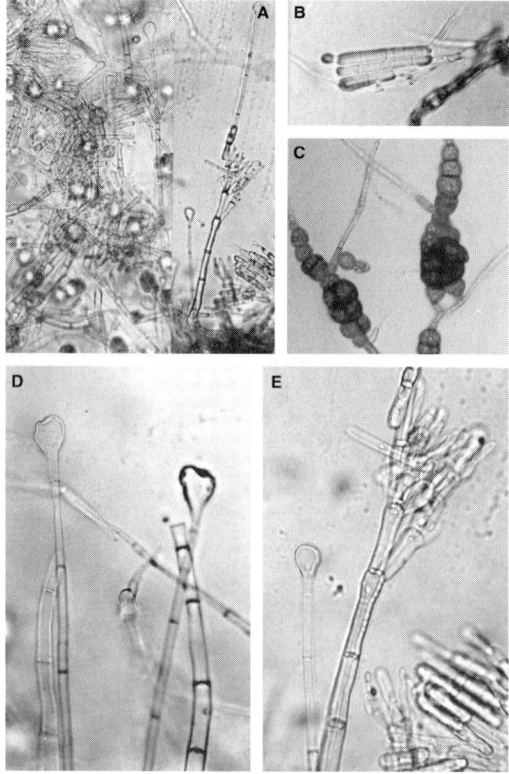

Cylindrocladium sp. A: Conidiophores with penicillate fertile portions, stipes and vesicles, and conidia. B: Conidia germinated or not germinated. C: Catenulate chlamydospores forming microsclerotia. D,E: Part of stipes and vesicles (E,F), penicillate fertile portion, and detached conidia (E). (From Watanabe, T. et al. 2001c. *Mycoscience*. In press. With permission.)

Cytospora Ehrenb. : Fr.

Syst. Mycol. 2:540, 1823.
Type species: *C. betulina* Ehrenb. : Fr.

Cytospora sacchari Butl.

References: Butler 1906; Matsumoto 1952; Sutton 1980; Watanabe 1975b.

Morphology: Pycnidia flask-shaped, covered with yellowish white hairs, with long cylindrical black necks; peridium brown, hard, undifferentiated. Conidiophores simple or branched, bearing conidia. Conidia solitary or catenulate, hyaline, 1-celled, long-ellipsoidal.

Dimensions: Pycnidia 500–900 × 460–840 µm: necks 200–700 × 140–240 µm. Conidiophores 7.5–18.8 × 1.7–2.5 µm. Conidia 2.7–5 × 0.7–0.9 µm.

Material: 72-X153-814 (= ATCC 32322, Sugarcane root, Taiwan, ROC).

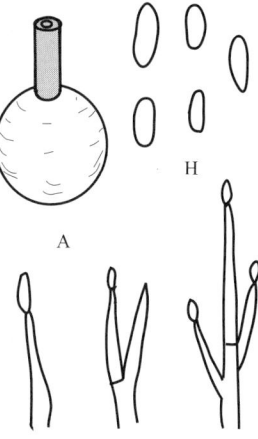

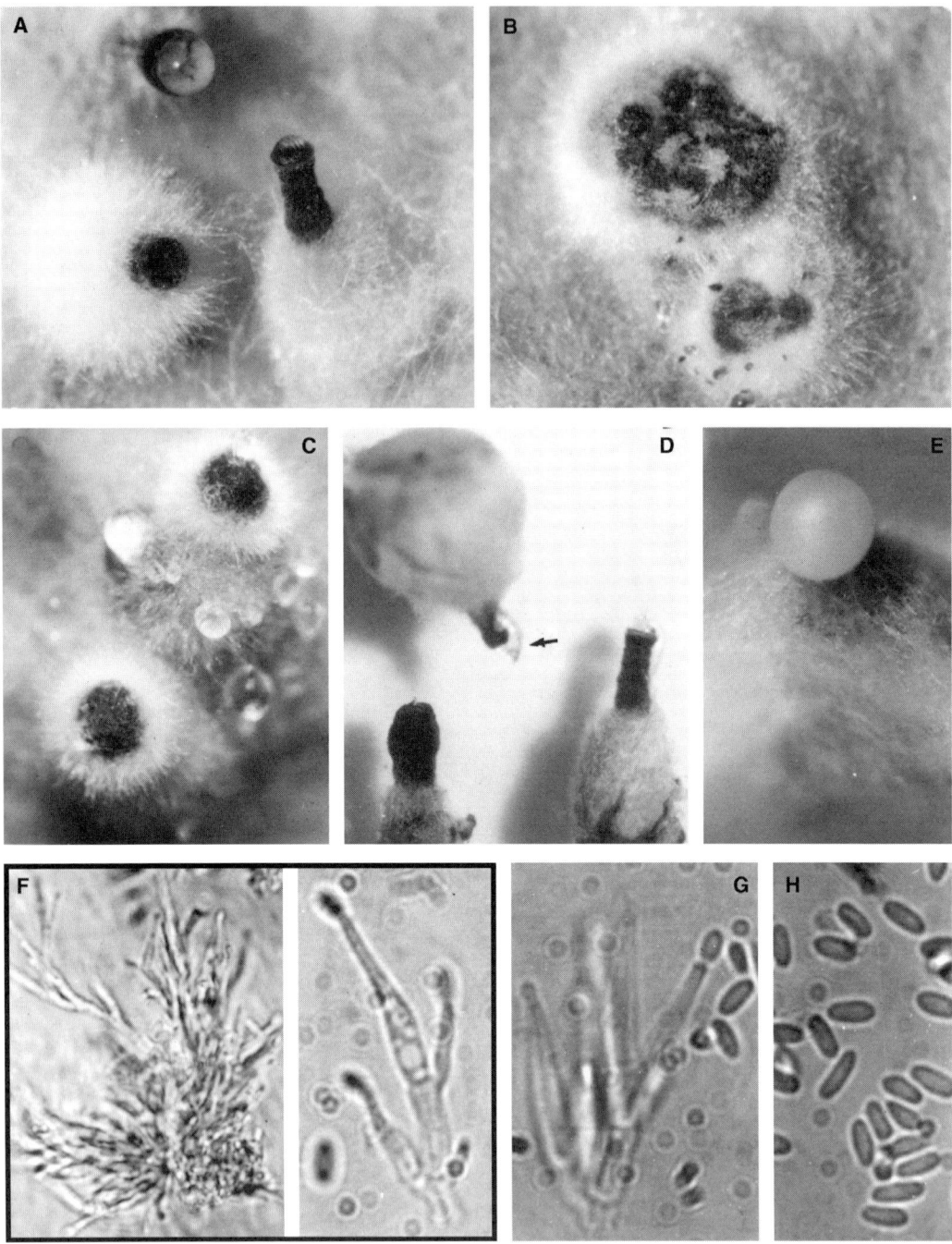

Cytospora sacchari. A–D: Pycnidia and spore mass on apical neck (arrow in D). E: Spore mass on apical pycnidium. F,G: Conidiophores and conidia. H: Conidia. (From Watanabe, T. 1975b. *Trans. Mycol. Soc. Jpn.*, 16:28–35. With permission.)

Dactylaria Sacc.

Michelia 2:20, 1880.
Type species: *D. purpurella* Sacc.

Dactylaria candidula (Hohn.) Bhatt & Kendrick

References: Bhatt and Kendrick 1968; Hoog 1985; Timms and Hepden 1965.

Morphology: Conidiophores hyaline, simple, erect, bearing 2–8 conidia sympodially at sterigmata on apical fertile parts. Conidia sympodulosporous, hyaline, cylindrical, 2-celled, rarely 3- to 4-celled, with septum at the median.

Dimensions: Conidiophores up to 28 μm tall. Conidia 22.5–28.8 × 0.7–1.3 μm.

Material: 73-139, 73-268 (Strawberry root, Shizuoka, Japan).

Remarks: The conidia of this fungus are slightly narrower than those in the original description.

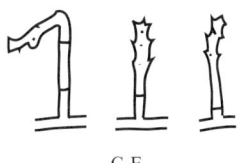

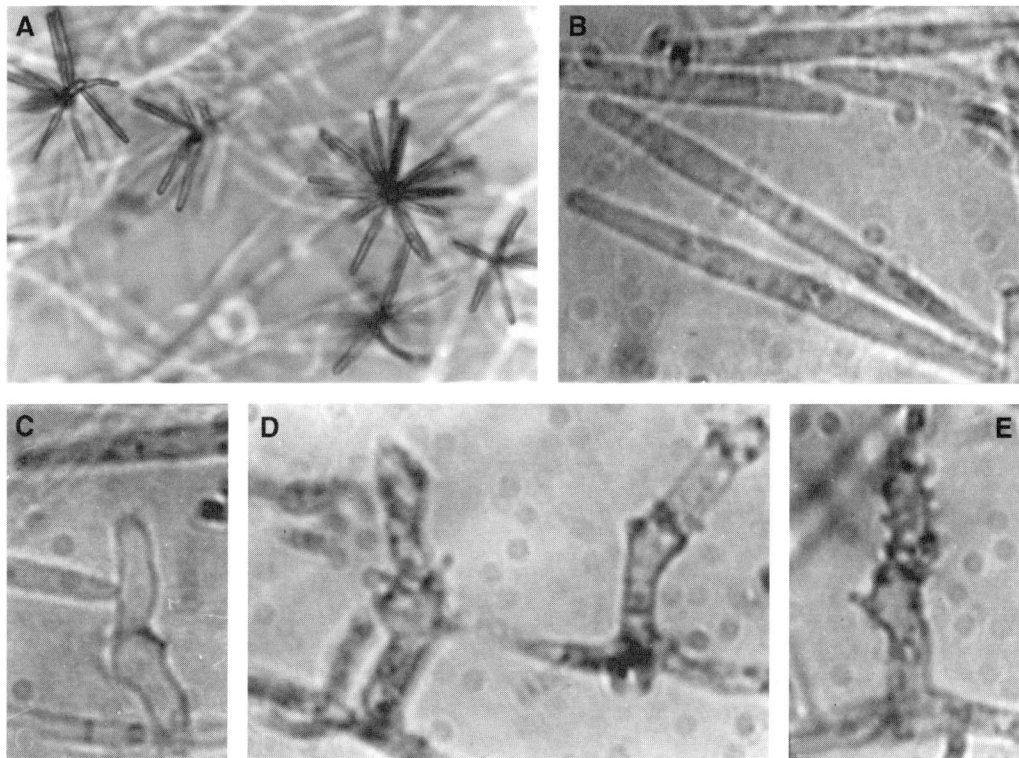

Dactylaria candidula. A: Condiophores and conidia. B: Conidia. C–E: Conidiophores.

Dactylaria naviculiformis Matsushima

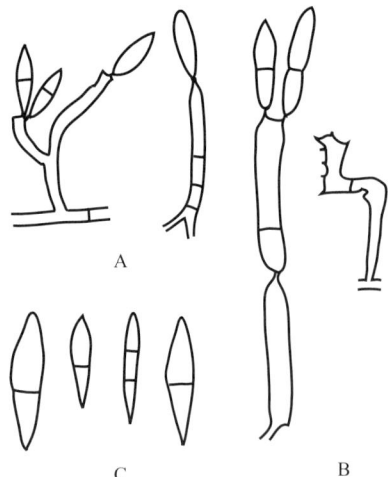

References: Hoog 1985; Matsushima 1975.

Morphology: Conidiophores simple, erect, hyaline or lightly pigmented, often proliferated, bearing conidia apically and laterally on short sterigmata on apical fertile parts. Conidia sympodulosporous, fusiform, often 1- to 3-septate, but usually 1-septate.

Dimensions: Conidiophores 20–30 μm tall, 3.2 μm wide. Conidia 11–28 × 3–4 μm.

Material: 98-63 (= MAFF 238004, Uncultivated soil, Tsukuba, Japan).

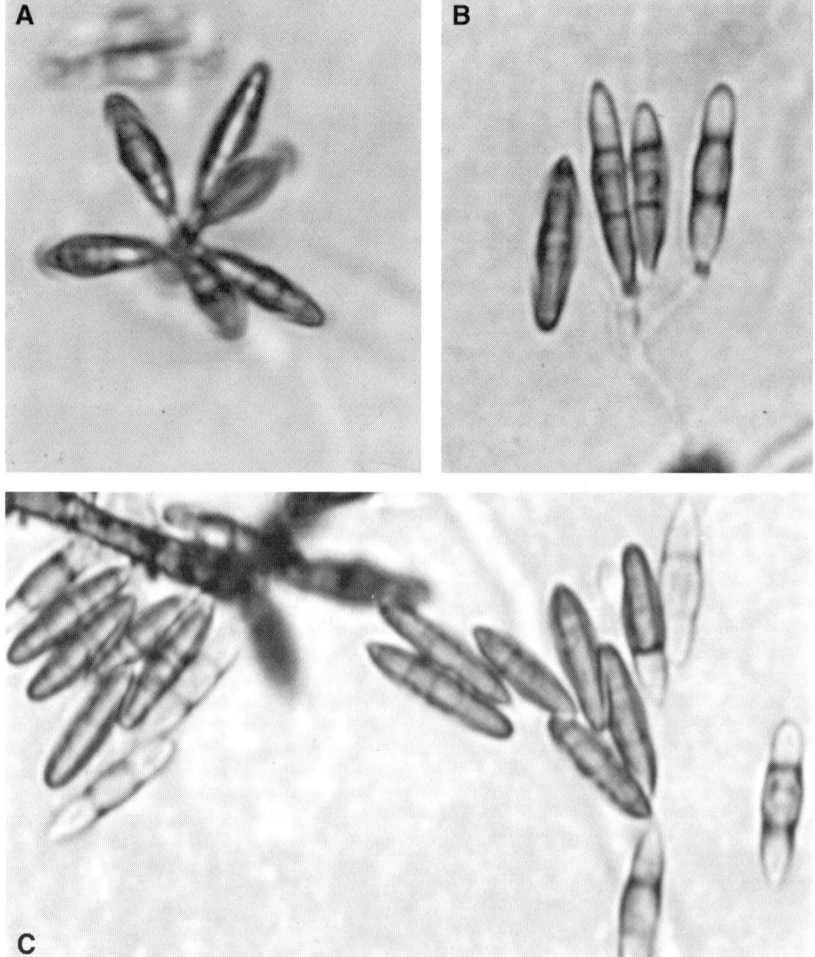

Dactylaria naviculiformis. A,B: Conidiophore and conidia. C: Detached conidia.

Dactylella Grove

J. Bot. Lond. 22:195, 1884.
Type species: *D. minuta* Grove

Dactylella chichisimensis T. Watanabe

References: Cooke and Dickinson 1965; Dowsett et al. 1984; Subramanian 1963; Watanabe et al. 2001c.

Morphology: Conidiophores erect, simple, hyaline, septate, gradually tapering toward hyaline apexes with single terminal conidia. Conidia aleurosporous, clavate, cylindrical or fusoid, hyaline, thin-walled, 3- to 9-septate. Microsclerotia embedded, pale yellow, nearly spherical. Chlamydospores in chains, granulate, pale yellow.

Dimensions: Conidiophores 24–86 × 2.4–3 μm. Conidia 26–56 × 5–6.4 μm. Microsclerotia mostly 125–200 μm. Chlamydospores 9–11 μm in diameter.

Material: 00-315 (= MAFF238165, Forest soil, Mikazukiyama, Chichijima, the Bonin Islands, Tokyo, Japan).

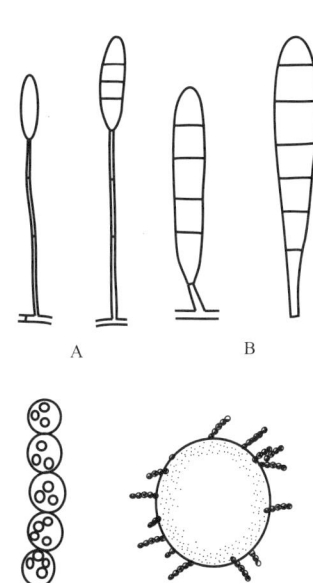

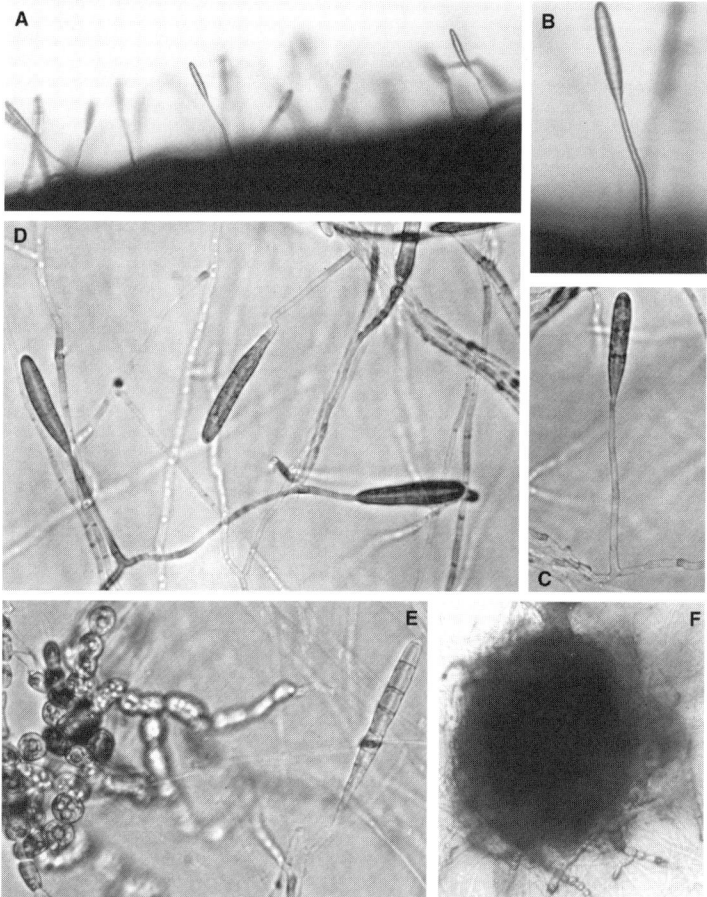

Dactylella chichisimensis. A: Habit. B–D: Conidiophores and aleurioconidia. E: Chlamydospore and aleurioconidia. F: Sclerotium. (From Watanabe, T. et al. 2001c. *Mycoscience*. In press. With permission.)

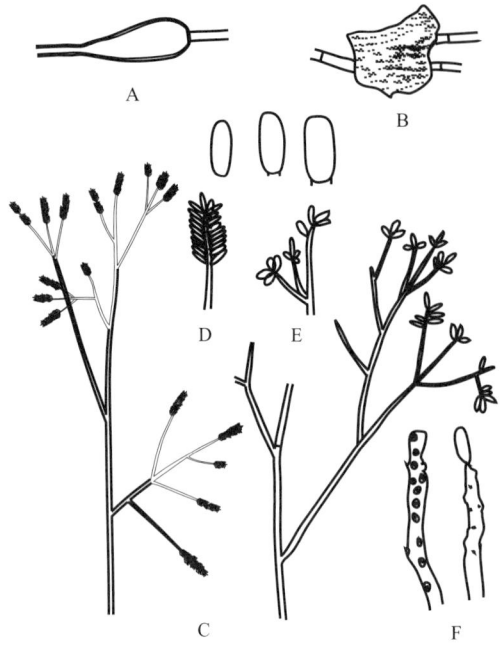

Dematophora Hartig

Unters. Forstbot. Inst. München 3:95, 1883.
Type species: *D. necatrix* Hartig
Teleomorph: *Rosellinia necatrix* (Hartig) Berl. : Prill.

Dematophora necatrix Hartig

Teleomorph: *Rosellinia necatrix* (Hartig) Berl. : Prill.

References: Ellis 1971; Francis 1985; Watanabe 1992b.

Morphology: Mycelia in cultures white initially, gradually becoming partially brown with age. Hyphae with pear-shaped swellings at the septum. Sporulated at the pigmented parts of the old cultures. Conidiophores brown, erect, branched once or twice, bearing conidia in 2 rows in the apical fertile parts. Conidia sympodulosporous, hyaline, long-ellipsoidal, 1-celled, readily detached from conidiophores.

Dimensions: Conidiophores ca. 500 µm high: fertile portions 30–65 × 7–8 µm. Conidia 3.5–5.5 × 1.6–2.3 µm. Sclerotia ca. 50 µm in diameter. Pear-shaped hyphal cells 7.5–10 µm wide.

Material: 84-373 (= IFO 32537); 84-374 (Populus root, Sakaide, Ehime, Japan).

Remarks: Known as the white root rot pathogen of various plants. Sporulation *in vitro* may be due to variation. No synnemata which occur in nature were observed in cultures.

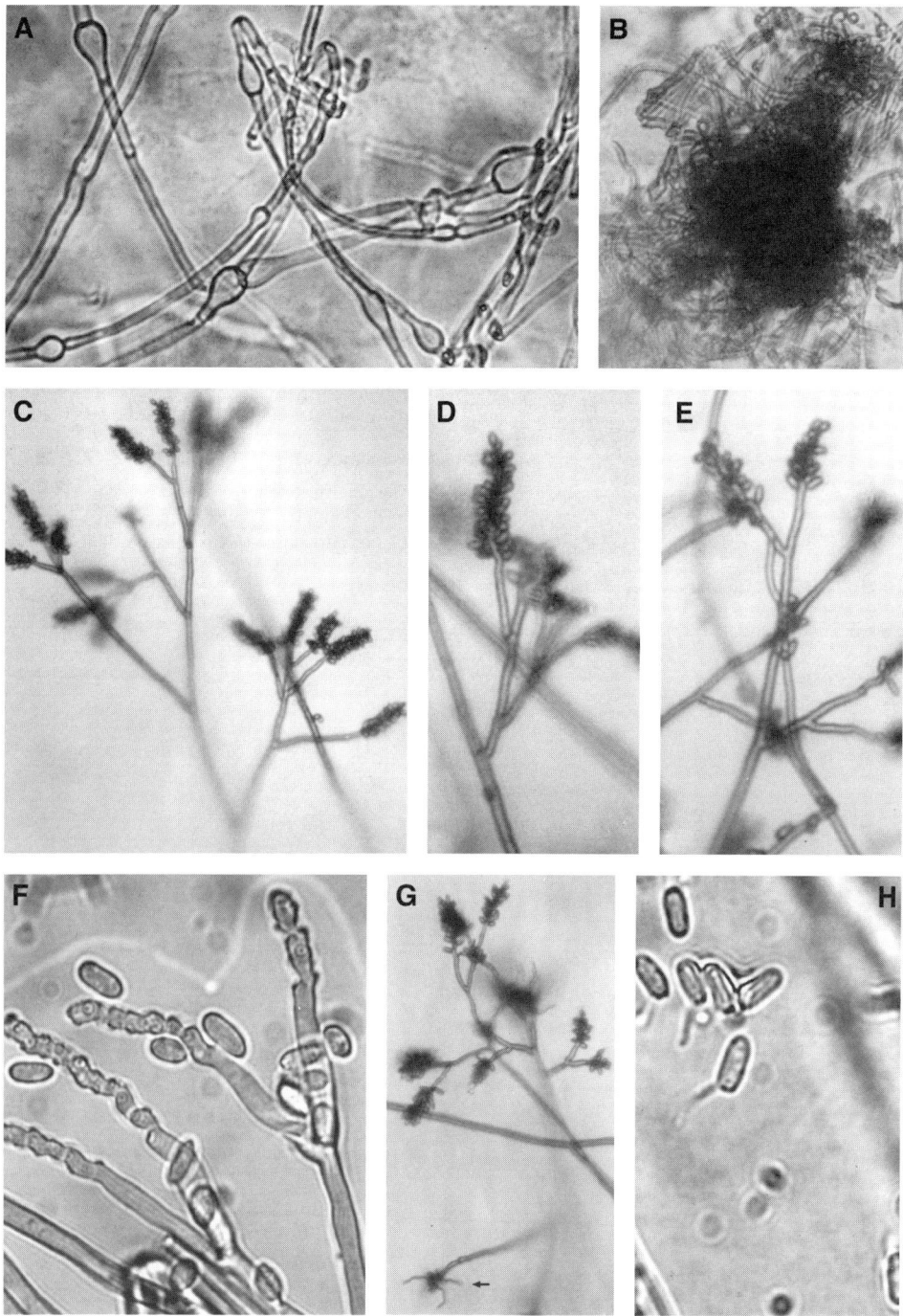

Dematophora necatrix. A: Hyphae with pear-shaped cells. B: Sclerotia. C–E: Conidiophores and conidia. F: Fertile portions of conidiophores and conidia. G: Germinated conidia (arrow) on conidiophores. H: Germinated conidia. (From Watanabe, T. 1992b. *Ann. Phytopathol. Soc. Jpn.*, 58:65–71. With permission.)

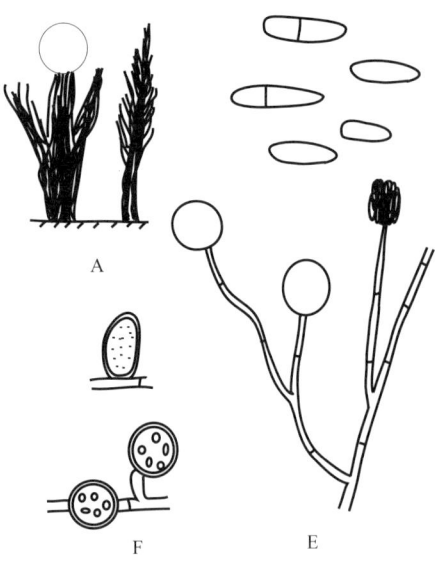

Didymostilbe P. Henn.

Hedwigia 41:148, 902.
Type species: *D. coffeae* P. Henn.

Didymostilbe sp.

References: Crane and Hewings 1982.

Morphology: Synnemata composed of united conidiophores, erect, pink or reddish brown, branched apically, apparently feather-like. Conidiophores free and elongated from synnemata, branched dichotomously, bearing spore masses at the apexes. Conidia phialosporous, hyaline, often greenish-tinted, 2-celled, cylindrical, often apiculate at one end. Chlamydospores golden brown, globose, granulate.

Dimensions: Synnemata over 3 mm tall; ca. 0.5 mm wide. Conidiophore branches 85–115 × 4–4.5 μm. Spore masses 20–25.5 μm in diameter. Conidia 8.8–22.5 (-30) × 4.5–7.5 (-10) μm. Chlamydospores 9–11.3 μm in diameter.

Material: 73-470 (Strawberry root, Shizuoka, Japan).

Remarks: This fungus is morphologically close to *D. eichleriana* Bres. and Sacc.

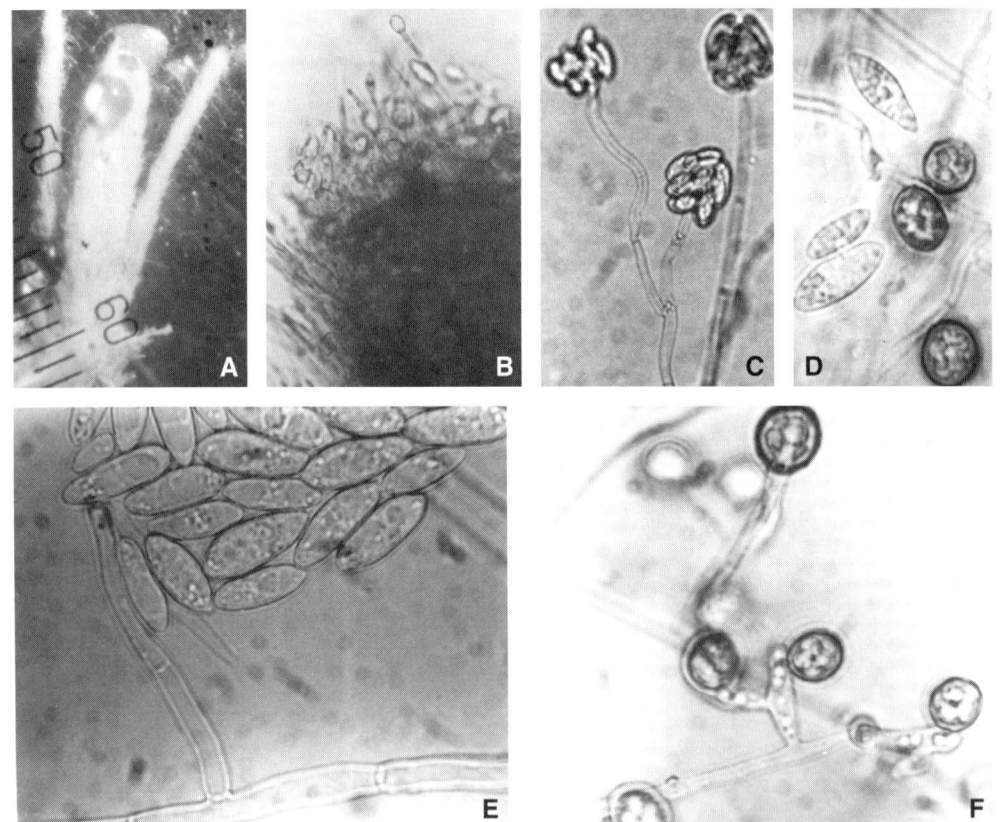

Didymostilbe sp.   A: Synnema. B: Sporulation on synnema. C: Independent conidiophores irrespective of synnemata, and spore masses. D: Immature conidia and chlamydospores. E: Conidiophore and conidia. F: Chlamydospores.

Diplodia Fr.

Summa Veg. Scand. 416, 1849.
Type species: *D. mutila* Fr.

Diplodia frumenti Ellis & Everh.

Teleomorph: *Botryosphaeria festucae* (Lib.) Arx & E. Muller

References: Eddins 1930; Sutton 1980; Watanabe et al. 1986a.

Morphology: Pycnidia globose, brown, usually with cylindrical necks, covered with aerial hyphae, peridium grayish brown. Conidiophores hyaline, simple, gradually tapering toward apexes. Conidia brown, 2-celled, with thick transverse septum and indistinct longitudinal patterns, often mixed with hyaline, immature conidia.

Dimensions: Pycnidia up to 3 mm in diameter. Conidiophores ca. 22 × 4 μm. Conidia 18–24.5 × 10–13 μm.

Material: 69-297, 69-438 (Pineapple root, Okinawa, Japan); 83-75 (Japanese black pine seed, Kumano, Mie, Japan); 84-496, 84-497, 84-500 (Japanese red pine seed, Saihaku, Tottori, Japan); 85-P9 (Paulownia root, Itapua, Paraguay).

Remarks: Aseptate, hyaline, immature conidia are often mixed up with mature 2-celled dark brown conidia.

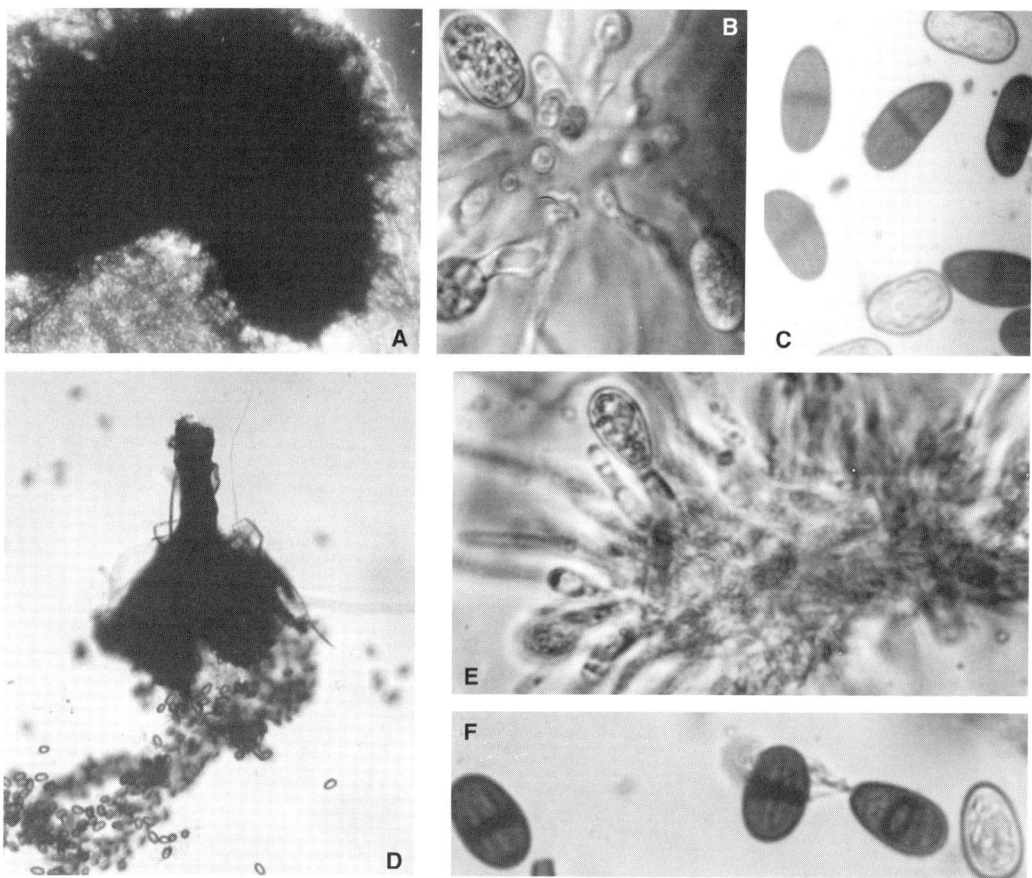

Diplodia frumenti. A,D: Crushed pycnidia. B,E: Conidiophores and conidia. C,F: Conidia (A–C: Isolate 69-438; D–F: Isolate P9).

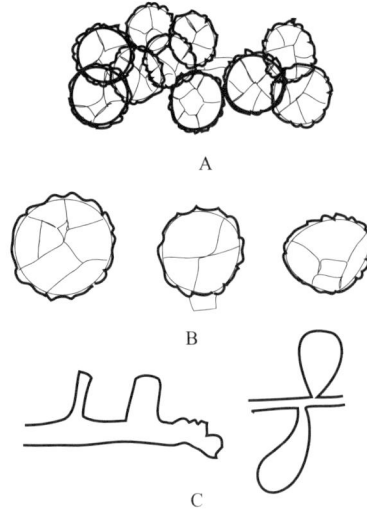

Epicoccum Link : Fr.

Syst. Mycol. 3:466, 1832.
Type species: *E. nigrum* Link = *E. purpurascens* Ehrenb. : Schlecht.

Epicoccum purpurascens Ehrenb. : Schlecht.

Synonym: *E. nigrum* Link

References: Domsch et al. 1980a,b; Schol-Schwarz 1959; Watanabe et al. 1986a.

Morphology: Conidiophores yellowish brown, short, not well differentiated from ordinary hyphae. Conidia solitary or aggregated forming sporodochia, brown, globose, rough on the surface, occasionally with cylindrical hilum, muriform composed of transverse and longitudinal septa.

Dimensions: Conidiophores up to 20 µm long. Conidia 15–20 µm in diameter.

Material: 84-526 (Japanese red pine seed, Hiba, Hiroshima, Japan).

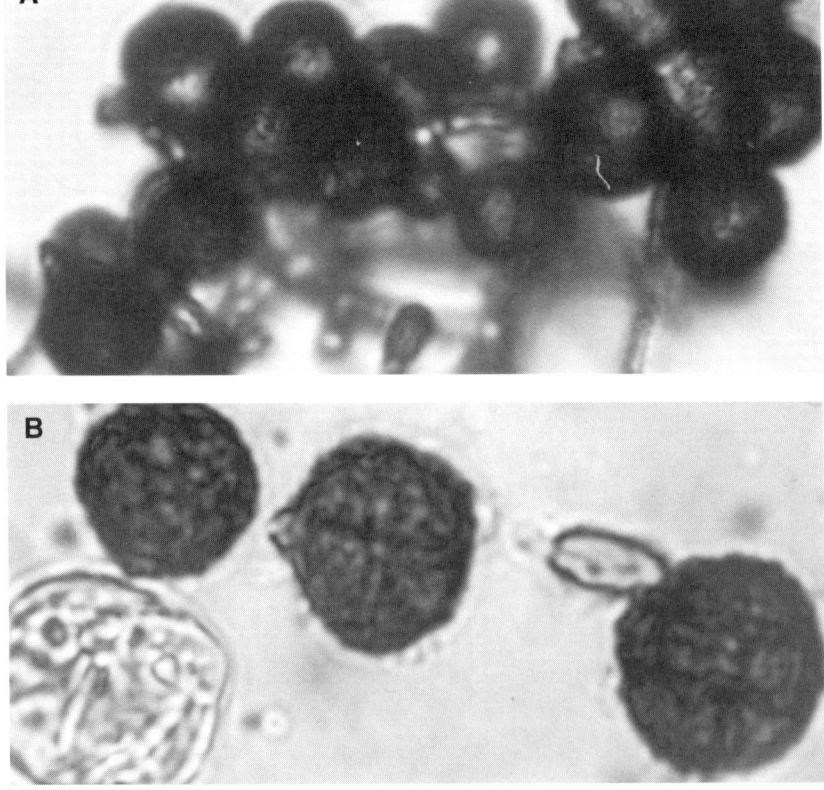

Epicoccum purpurascens.   A: Spore mass. B: Conidia. C: Fertile portions.

Fumago Pers.

Myc. Eur. 1:9, 1822.
Lectotype: *F. vagans* Pers.

Fumago sp.

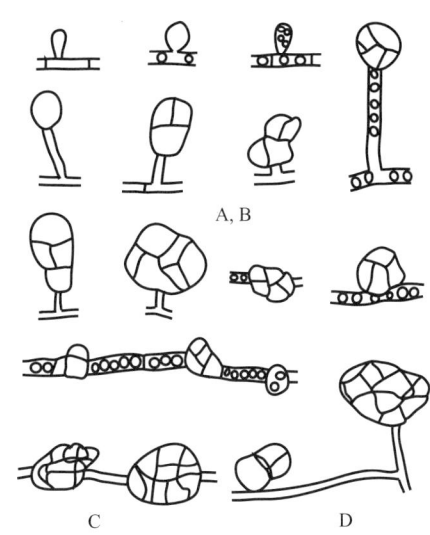

References: Friend 1965; Hughes 1976, 1983; Sahni 1964.

Morphology: Papulospore-like spores borne intercalarly on ordinary hyphae or terminally on short conidiophores, reddish brown or dark brown, subglobose or irregular, muriform when immature, composed of several component cells. Hyphae often rich in oil globules.

Dimensions: Muriform spores mostly 14–28 µm in diameter. Sporephore 4–32 × 2–3.6 µm. Hyphae 3–4.4 µm wide.

Material: 00-76 (Uncultivated soil, Chichijima, the Bonin Islands, Tokyo, Japan).

Remarks: Colonies on PDA are dark gray, velvety, homogeneous. The genus *Fumago* is selected to accommodate a member of soil fungi only with hyphae and dark brown muriform spores simply or in chains, although its validity is often controversial. This fungus may resemble *Sarcinella* Sacc. with hyphopodiate hyphae. However, *Fumago* and *Sarcinella* are commonly known as members of sooty molds with a few other hyphopodiate fungi. *Fumago* is also listed as a member of soil fungi (Ranzoni, 1968).

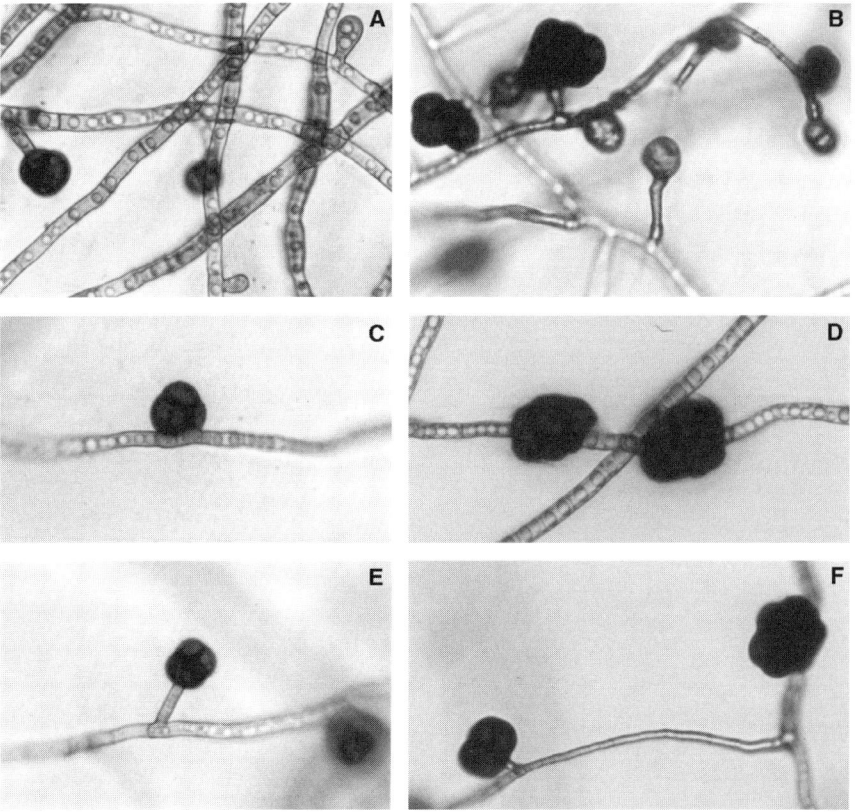

Fumago sp. A,B: Muriform spores and hyphae. C,D: Intercalary muriform spores. E,F: Both apical and intercalary muriform spores.

Fusarium Link : Fr.

Syst. Mycol. 1:16, 1821.
Type species: *F. roseum* Link : Gray = *F. sambucinum* Fuckel

Key to Species

1.	Microconidia	not formed .... *F. ciliatum*
		formed 2
2.	Microconidia	catenulate, chlamydospores lacking .... *F. moniliforme*
		not catenulate, chlamydospores formed 3
3.	Chlamydospores	borne in a chain of over 3 spores .... *F. roseum*
		solitary or twins 4
4.	Conidiophores	longer than macroconidium length by a few times..................... 5
		shorter than macroconidium width .... *F. oxysporum*
5.	Conidiophores	mostly branched, taller than 100 μm..................... *F. ventricosum*
		not so, mostly simple, shorter .... *F. solani*

Fusarium ciliatum Link

References: Booth 1971; Gerlach and Nirenberg 1982; Nelson et al. 1983.

Morphology: Conidiophores simple (monophialidic), resupinate, rarely branched apically, with massive and terminal macroconidia, forming sporodochia. Macroconidia hyaline, extremely slender, lunate, mostly 3- to 6-septate. No microconidia and chlamydospores observed.

Dimensions: Conidiophores 10–20 × 3.2–5 µm. Conidia 40–56 × 2.2–3.2 µm.

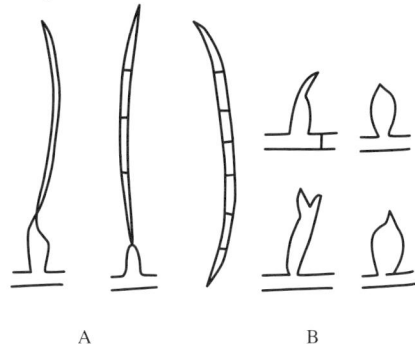

A   B

Material: 00-54 (Forest soil, Mt. Chibusa, Hahajima, the Bonin Islands, Tokyo, Japan).

Remarks: Colonies on PDA are homogeneous, yellowish brown centrally, white marginally, resupinate.

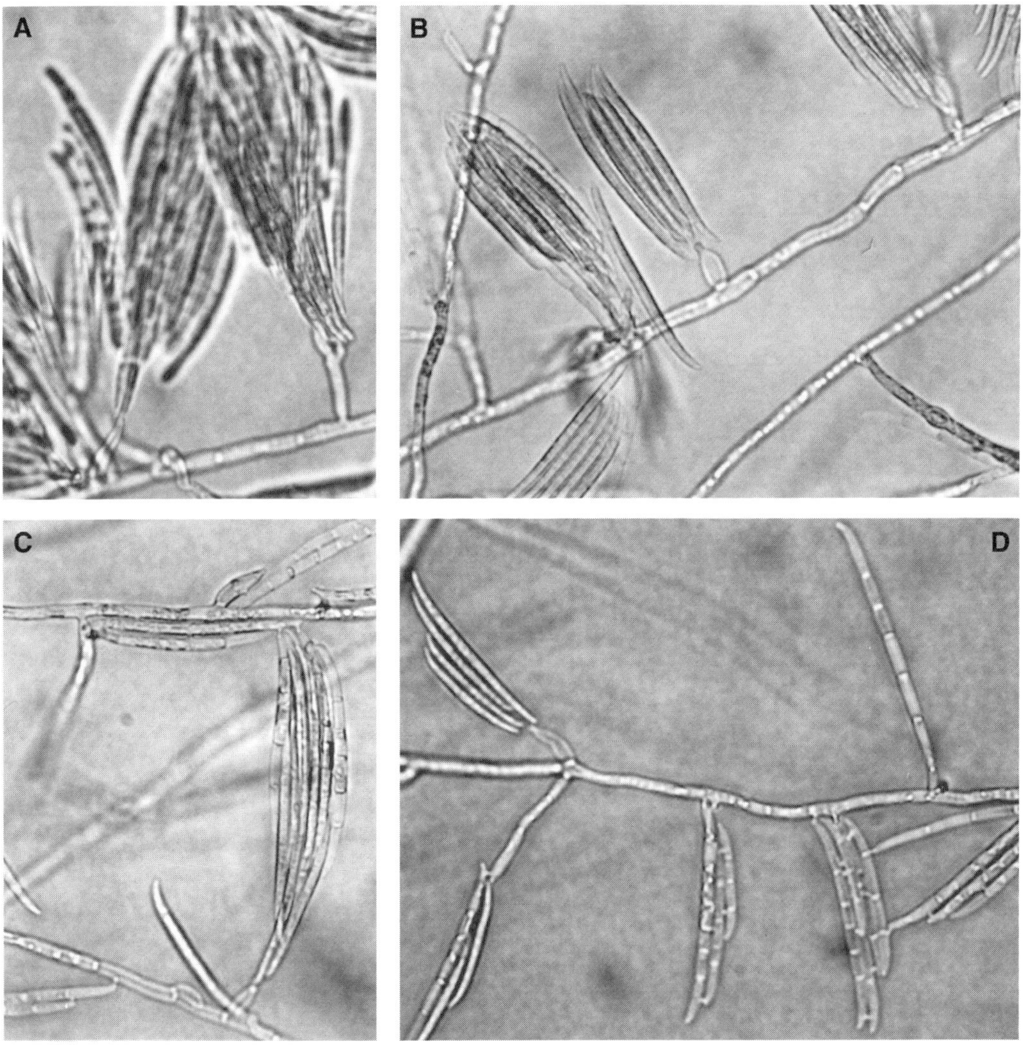

Fusarium ciliatum.   A–D: Conidiophores and conidia.

Fusarium moniliforme (Sheldon) emend. Snyder & Hansen

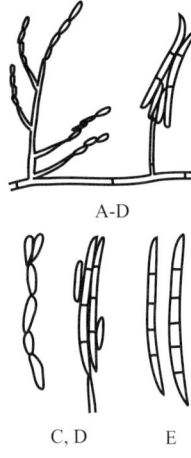

A-D

C, D   E

Teleomorph: *Gibberella fujikuroi* (Sawada) Ito apud Ito et Kimura

References: Booth 1971; Matuo et al. 1980; Nelson et al. 1981, 1983; Snyder and Hansen 1945; Watanabe 1975d; Watanabe et al. 1986a.

Morphology: Conidiophores hyaline, simple or branched, bearing conidia in chains and/or spore masses at the apexes of branches. Conidia phialosporous, hyaline, of two kinds: macroconidia borne in spore masses, boat-shaped, with slightly curved apical cells, hooked foot cells, and 2 central cylindrical cells, mainly 4- to 5-celled, and microconidia hyaline, ellipsoidal or ovate, apiculate at one end. Chlamydospores not formed.

Dimensions: Conidia: macroconidia 26.4–38.9 × 2.4–3.7 µm: microconidia 7.2–12 × 2.4–3.2 µm.

Material: 72-X83 (Sugarcane root, Taiwan, ROC); 84-503 (Japanese red pine seed, Saihaku, Tottori, Japan); 86-39 (Japanese cedar seed, Kasama, Ibaraki, Japan).

Remarks: Known as the pathogen of Bakanae disease of rice.

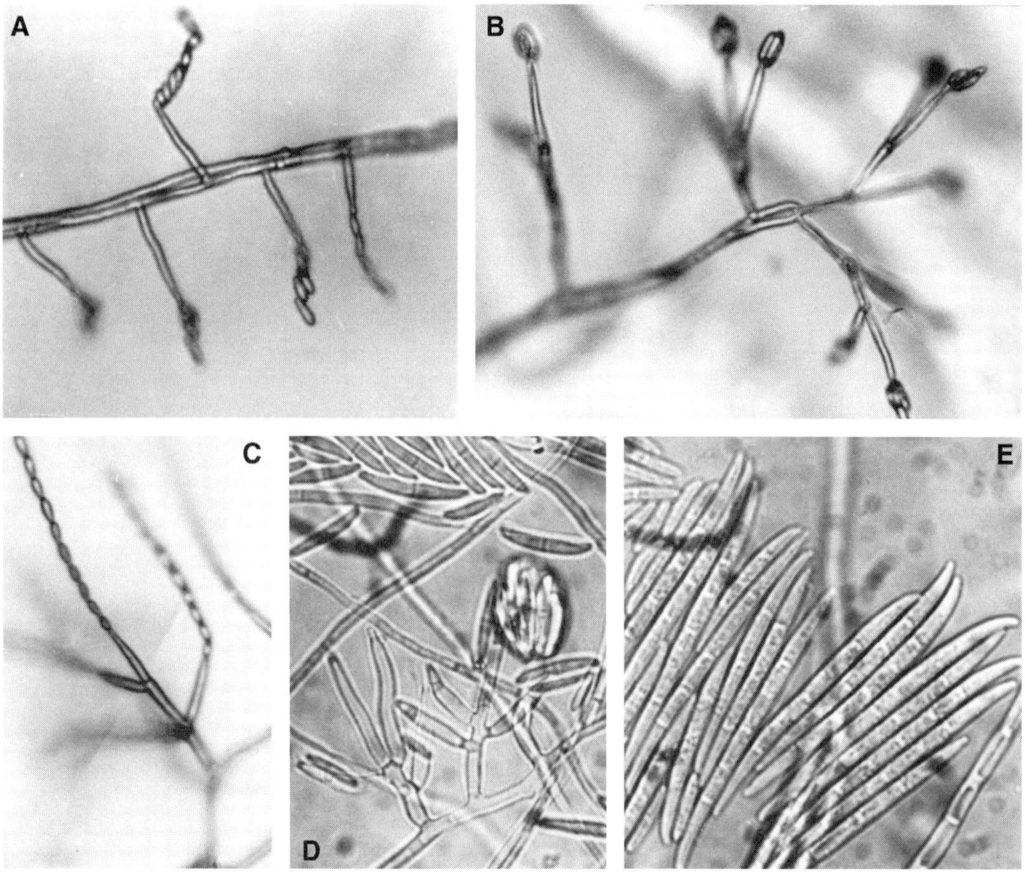

Fusarium moniliforme. A–C: Conidiophores and microconidia. D: Conidiophores and macroconidia. E: Macroconidia. (From Watanabe, T. 1975d. *Trans. Mycol. Soc. Jpn.*, 16:264–267. With permission.)

Fusarium oxysporum (Schl.) emend. Snyder & Hansen

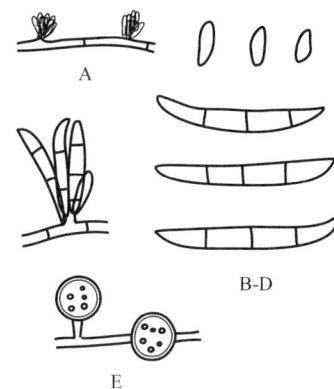

References: Booth 1971; Matuo et al. 1980; Nelson et al. 1981, 1983; Snyder and Hansen 1940; Watanabe 1975d.

Morphology: Conidiophores hyaline, simple, short or not well differentiated from hyphae, bearing spore masses at the apexes. Conidia phialosporous, hyaline, of two kinds: macroconidia boat-shaped, with slightly tapering apical cells and hooked basal cells, 4-celled; and microconidia ellipsoidal, 1-celled. Chlamydospores brown, globose, usually solitary.

Dimensions: Conidia: macroconidia (17.5-) 29.1–45 × 2.9–4.7 µm: microconidia 6–15.8 × 1.9–3.7 (-5) µm. Chlamydospores (5.3-) 10.2–15 µm in diameter.

Material: 70-1389, 70-1424 (Pineapple field soil, Hachijo, Tokyo, Japan); 72-X5 (Sugarcane root, Taiwan, ROC); 83-66 (Japanese cedar seed, Yoshino, Nara, Japan); 85-28, 85-77 (Flowering cherry seed, Hachioji, Tokyo, Japan).

Remarks: More than 54 formae speciales have been recorded (Holliday 1989).

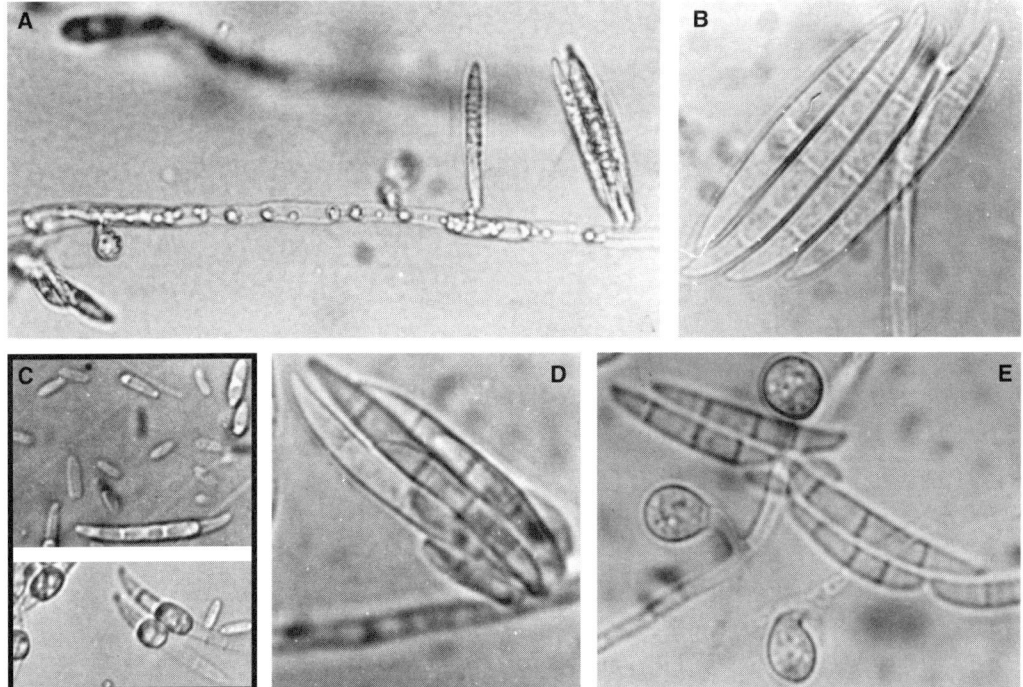

Fusarium oxysporum. A: Hypha, macroconidia, microconidia, and chlamydospores. B: Macroconidia. C: Macroconidia, microconidia, and chlamydospores. D: Macroconidia. E: Macroconidia and chlamydospores. (From Watanabe, T. 1975d. *Trans. Mycol. Soc. Jpn.*, 16:264–267. With permission.)

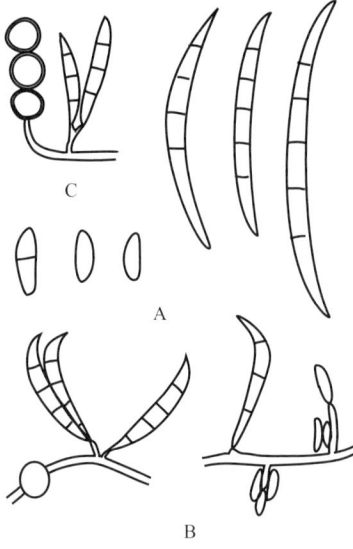

Fusarium roseum (Lk.) emend. Snyder & Hansen

References: Booth 1971; Matuo et al. 1980; Nelson et al. 1981, 1983; Snyder and Hansen 1945; Watanabe 1975d.

Morphology: Conidiophores hyaline, simple, short or indistinct from ordinary hyphae, bearing spore masses at the apexes. Conidia phialosporous, hyaline, of two kinds: macroconidia lunar- or boat-shaped with curved apical cells, 2 central, curved cells and hooked foot cells, 4- to 6-celled; and microconidia cylindrical, 1- to 2-celled. Chlamydospores brown, globose, over 3 spores in a chain.

Dimensions: Conidia: macroconidia 24.5–45 (-105) × 4–5 (-7.5) µm, and microconidia 5–17.1 × 1.7–6.1 µm. Chlamydospores (6.2-) 10.2–15 µm in diameter.

Material: 70-1350 (Pineapple field soil, Hachijo, Tokyo, Japan); 72-X5 (Sugarcane root, Taiwan, ROC); 85-30, 85-31 (Flowering cherry seed, Hachioji, Tokyo, Japan).

Remarks: At least two types, i.e., yellow and pinkish colony types, are commonly observed. Dimensions of conidia are various in isolates studied.

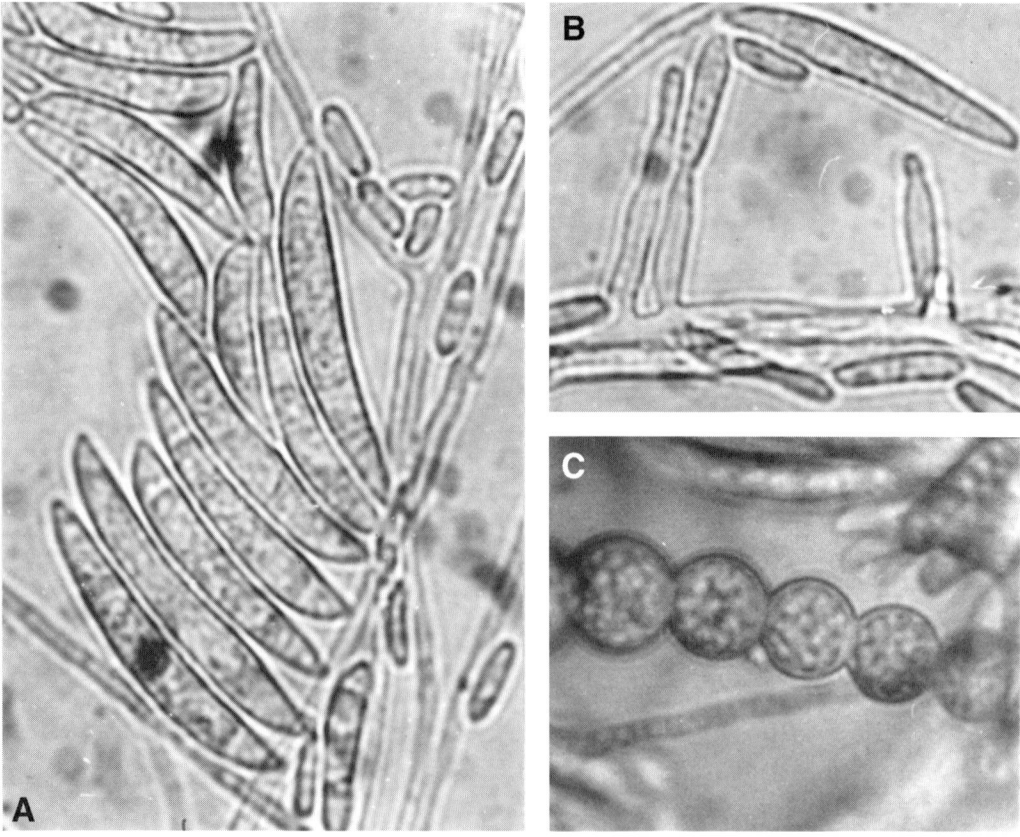

Fusarium roseum. A: Macroconidia and microconidia. B: Conidiophores and conidia. C: Catenulate chlamydospores. (From Watanabe, T. 1975d. *Trans. Mycol. Soc. Jpn.*, 16:264–267. With permission.)

Fusarium solani (Mart.) App. et Wr. emend. Snyder & Hansen

Teleomorph: *Nectria haematococca* Berk. & Broome

References: Booth 1971; Matuo et al. 1980; Nelson et al. 1981, 1983; Snyder and Hansen 1941; Watanabe 1975d.

Morphology: Conidiophores hyaline, simple, bearing spore masses at the apexes, as tall as the length of macroconidia by a few times. Conidia phialosporous, hyaline, of two kinds: macroconidia with slightly curved apical cells, 2 cylindrical central cells, often slightly curved in one side, and hooked foot cells, usually 3- to 5-celled; and microconidia cylindrical, 1- to 2-celled. Chlamydospores brown, globose, and usually solitary.

Dimensions: Conidiophores 50–165 × 2.4–4.3 µm. Spore masses 10–25 µm in diameter. Conidia: macroconidia (26.2-) 31.5–59.4 × 4.6–6.2 (-6.8) µm: microconidia 7.2–15 (-17.5) × 2.4–3.9 (-6.3) µm. Chlamydospores 6–7.3 µm in diameter.

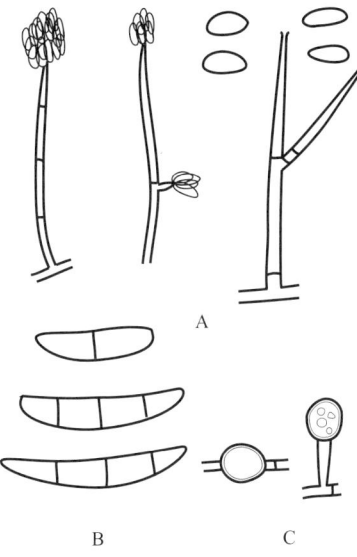

Material: 70-1239, 70-1559 (Pineapple field soil, Hachijo, Tokyo, Japan); 74-X34-911 (Sugarcane root, Taiwan, ROC); 85-33, 85-38 (Flowering cherry seed, Hachioji, Tokyo, Japan); 85-13R5 (Paulownia root, Itapua, Paraguay).

Remarks: There are at least 18 formae speciales in *F. solani* (Holliday, 1989). Heterothallic isolates of this species are common, whereas homothalic isolates have been very rarely reported.

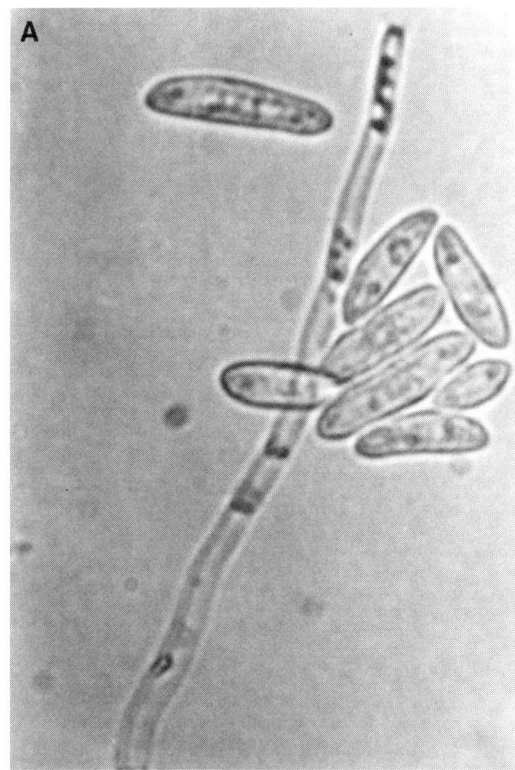
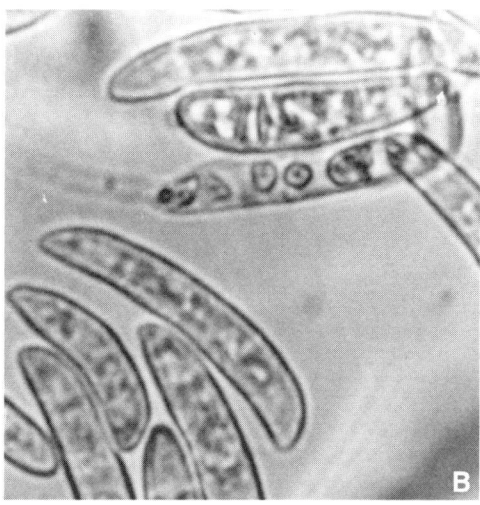
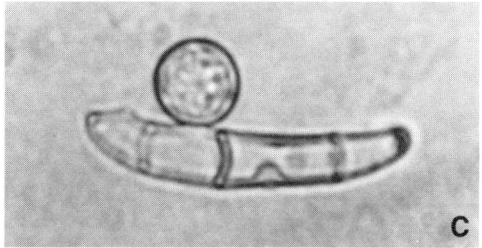

Fusarium solani. A: Conidiophore and microconidia. B: Macroconidia. C: Chlamydospore formed on macroconidium. (From Watanabe, T. 1975d. *Trans. Mycol. Soc. Jpn.*, 16:264–267. With permission.)

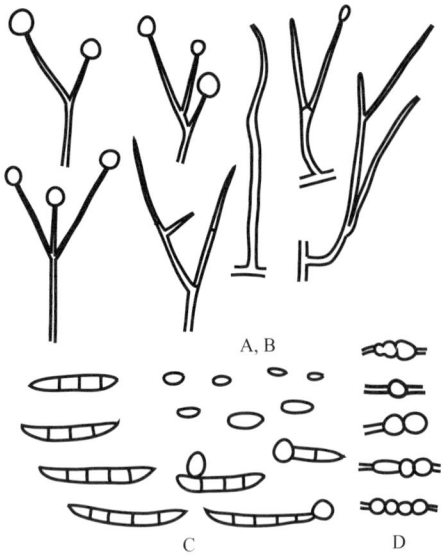

Fusarium ventricosum Appel & Wollenw.

Teleomorph: *Nectria ventricosa* C. Booth

References: Booth 1971; Gerlach and Nierenberg 1982; Nelson et al. 1983.

Morphology: Conidiophores hyaline, erect, long, mostly branched verticillately, alternately or variously branched, rarely simple, bearing spore masses apically at each branch. Conidia of two kinds: macroconidia hyaline, lunate-shaped, long ellipsoidal, 4-5-celled; and microconidia hyaline, 1-celled. Chlamydospores yellowish brown, single or 2–4 in a chain.

Dimensions: Conidiophores mostly 125–150 μm long. Branches 32.5–90 μm long. Spore masses 10–25 μm in diameter. Macroconidia 23.7–47.5 × 3.7–6.3 μm. Microconidia 3.7–11.3 × 1.5–5.0 μm. Chlamydospores 6.2–8.8 μm in diameter.

Material: 83-33 (Flowering cherry seed, Asakawa, Tokyo, Japan); 92-240 (*Phellodendron* roots, Tokyo, Japan).

Remarks: Colonies on PDA are nonaerial, pale yellowish brown or pinkish brown, and zonate. This fungus is often treated as a synonym of *F. solani*, but on the basis of elongated and branched conidiophores, it is treated as an independent species in this work.

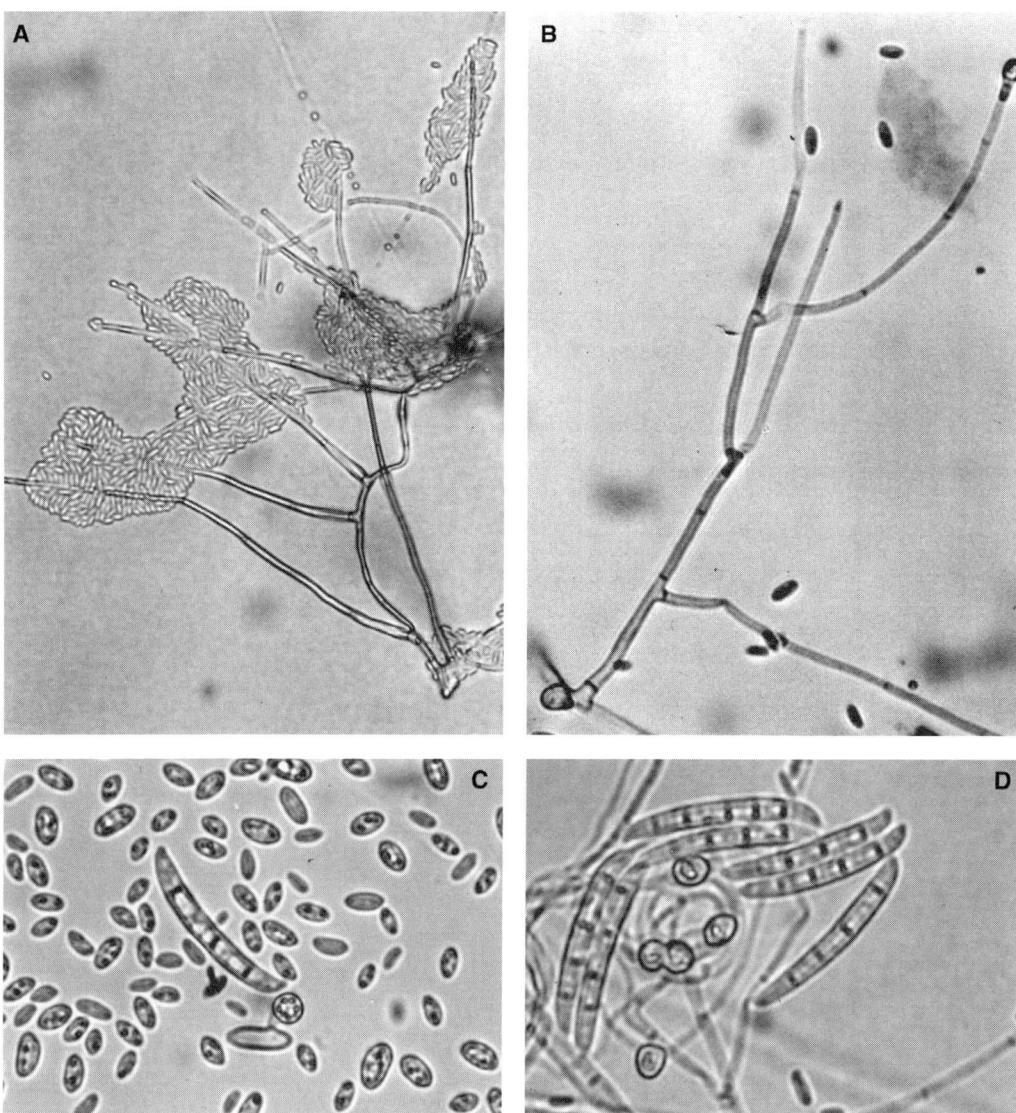

Fusarium ventricosum. A: Conidiophore and microconidia. B: Conidiophore with elongated branches and microconidia. C: Macroconidium and microconidia. Note a microconidium bearing chlamydospore. D: Macroconidia and chlamydospores.

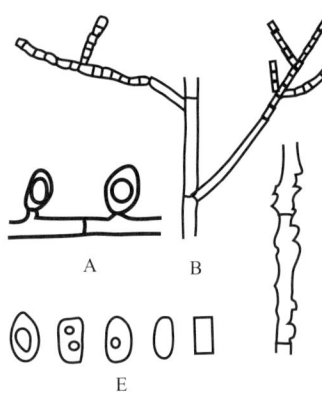

Geotrichum Link : Pers.

Mycol. Eur. 1:26, 1822.
Type species: *G. candidum* Link : Pers.

Geotrichum candidum Link

References: Carmichael 1957; Watanabe 1975d.

Morphology: Conidiophores lacking. Conidia arthrosporous, terminal or intercalary, aerial on the agar surface from creeping hyphae, hyaline, cylindrical, barrel-shaped, subglobose, 1-celled, often guttulate. Chlamydospores subglobose, solitary, borne at the sterigmata on the hyphae.

Dimensions: Conidia (3.7-) 4.8–12.5 (-13.8) × (1.7-) 2.4–5 µm. Chlamydospores 4.3–6.1 µm in diameter.

Material: 72-X8-76 (Sugarcane root, Taiwan, ROC), 81-416 (Forest soil, Rausu, Hokkaido, Japan).

Remarks: Several isolates examined are almost similar to one another, forming white, nonaerial colonies. Some cultures are fragrant.

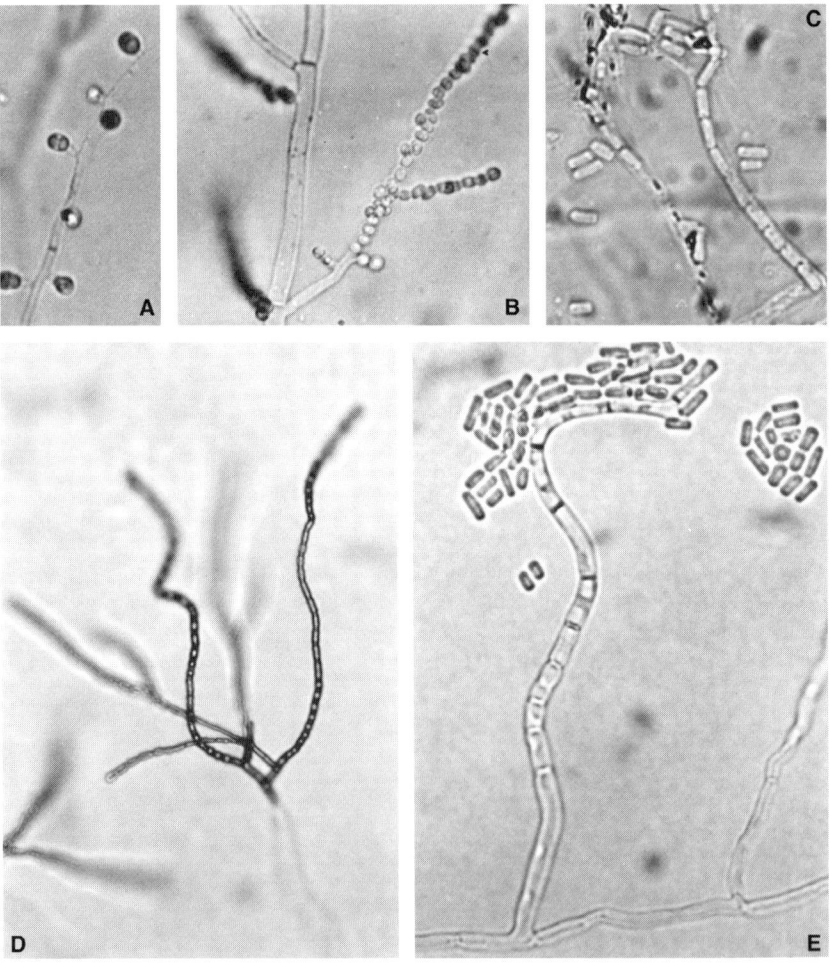

Geotrichum candidum. A: Hypha and chlamydospores. B–E: Conidia and sporulation (B,D,E). (From Watanabe, T. 1975d. *Trans. Mycol. Soc. Jpn.*, 16:264–267. With permission.)

Gliocladium Corda

Icones Fung. 4:30, 1840.
Type species: *G. penicillioides* Corda

Key to Species

1. Conidiophores   penicillate 2
                   penicillate and verticillate 3

2. Phialides   amplified, colonies dark green .... *G. virens*
                   gradually tapering from base toward apex, colonies brown .... *G. viride*

3. Conidiophores   90–185 µm tall, colonies grayish green .... *G. catenulatum*
                   21–37 µm tall, colonies pink .... *G. roseum*

Gliocladium catenulatum Gilman & Abbott

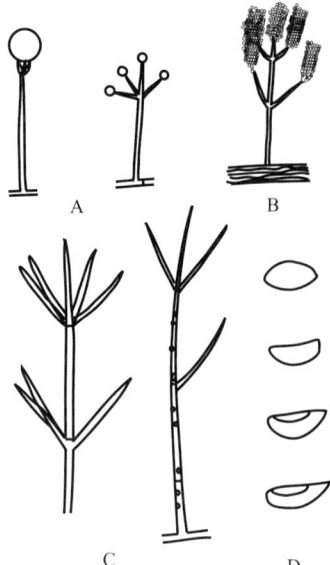

References: Domsch et al. 1980; Raper and Thom 1949; Watanabe 1975d.

Morphology: Conidiophores hyaline, erect, granular and rough on the surface, branched, bearing spore masses on phialides at the apex: phialides verticillate or penicillate, gradually tapering toward apex. Conidia hyaline, pinkish in mass, spindle-shaped, ellipsoidal or boat-shaped, 1-celled. Chlamydospores brown, globose, often a few in a chain.

Dimensions: Conidiophores 90–185 × 3.5–4 µm: phialides 17.5–25 × 2.4–2.7 µm. Spore masses 25–50 µm in diameter. Conidia 3.5–5 (-6.1) × 2.7–4 µm.

Material: 72-X92-606 (= ATCC 32913, Sugarcane root, Taiwan, ROC).

Remarks: The fungus was previously described as *G. fimbriatum* Gilman & Abbot (Watanabe, 1975d). Colonies on agar cultures are pale yellow or grayish green with pinkish tint.

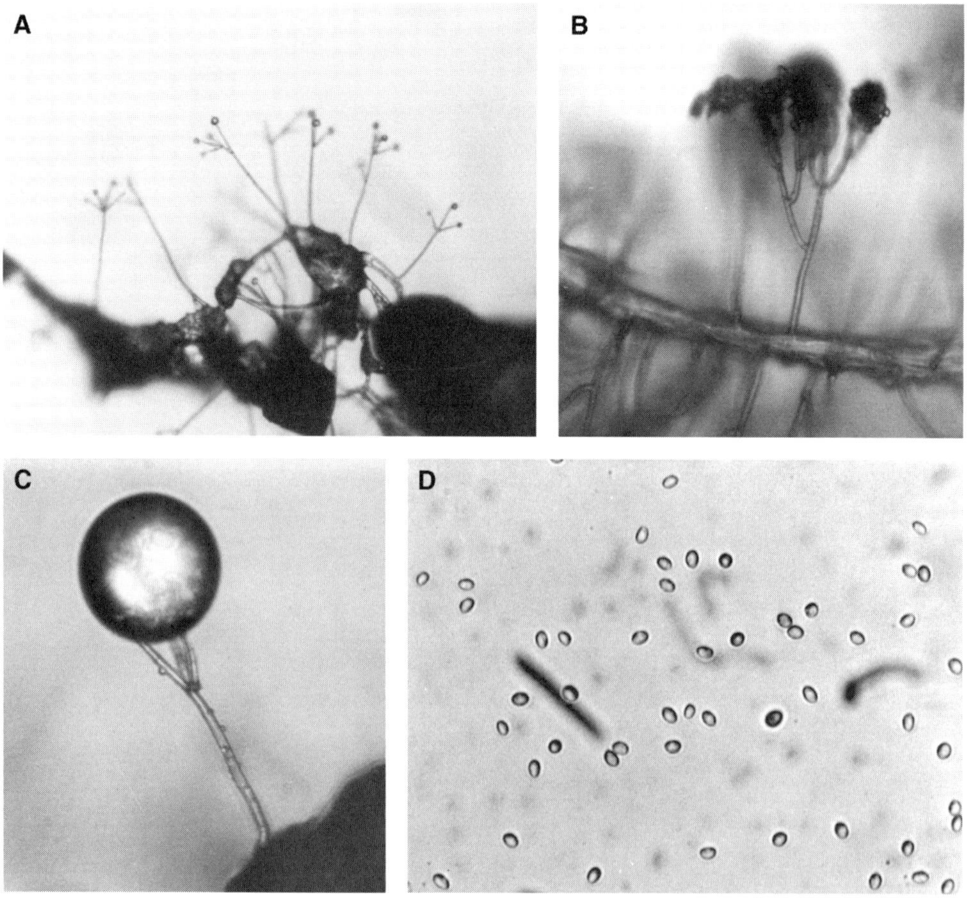

Gliocladium catenulatum. A: Conidiophores with verticillate branches and spore masses. B: Conidiophore with penicilliate branches, and catenulate conidia. C: Spore mass formed on verticillate phialides. D: Conidia. (From Watanabe, T. 1975d. *Trans. Mycol. Soc. Jpn.*, 16:264–267. With permission.)

Gliocladium roseum (Link) Bainier

Teleomorph: *Nectria gliocladioides* Smalley et Hansen

References: Domsch et al. 1980a,b; Raper and Thom 1949; Watanabe 1975; Watanabe et al. 1986a.

Morphology: Conidiophores hyaline, erect, simple or branched, bearing spore masses on phialides on the apex of branches: phialides single, verticillate or penicillate. Conidia phialosporous, hyaline, pinkish in mass, long-ellipsoidal to ovate, 1-celled, slightly apiculate at one end. Chlamydospores brown, globose to ellipsoidal, thick-walled.

Dimensions: Conidiophores 21.8 to over 36.5 × 2.1–2.7 μm: primary branches 12.1–19.5 × 2.4–2.5 μm: phialides 13.3–24.3 × 2.4–3.2 μm. Conidia (4.3-) 5.1–6.1 (-9) × (1.7-) 2.6–3.9 μm. Chlamydospores 5.3–7.3 μm in diameter.

Material: 72-X9 (Sugarcane root, Taiwan, ROC), 84-544, 84-545 (Japanese red pine seed, Saihaku, Tottori, Japan).

Remarks: Colonies on agar cultures are yellowish pink with concentric zonation.

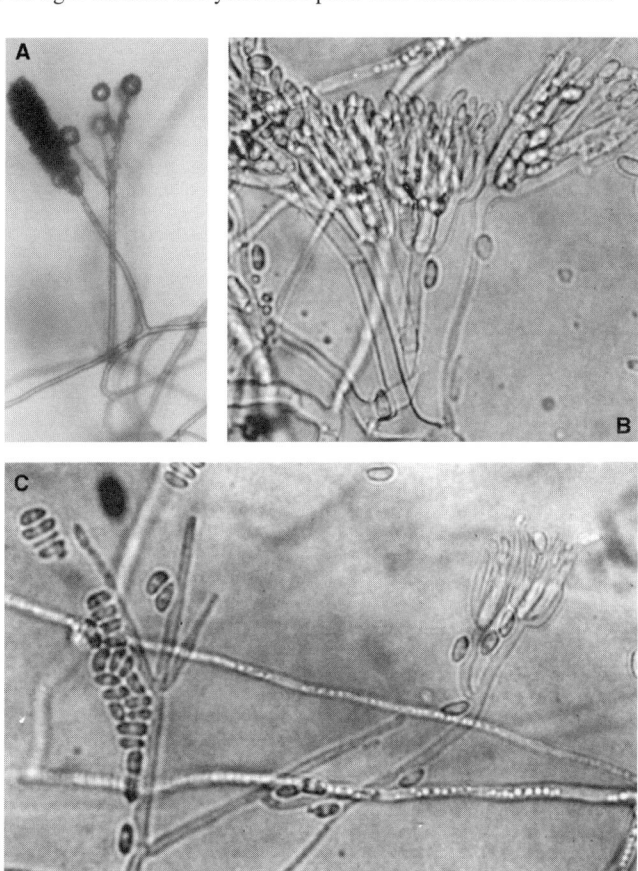

Gliocladium roseum. A: Conidiophores and spore masses. B: Conidiophores and conidia. C: Part of the conidiophores and conidia. (From Watanabe, T. 1975c. *Trans. Mycol. Soc. Jpn.*, 16:149–182. With permission.)

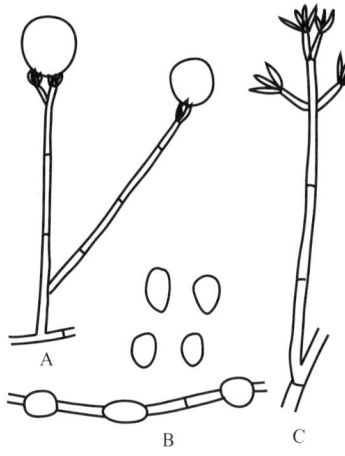

Gliocladium virens Miller, Giddens & Foster

References: Domsch et al. 1980; Watanabe 1975d.

Morphology: Conidiophores hyaline, erect, simple or branched oppositely or verticillately, especially in metula, septate, bearing spore masses at phialides on the apical branches. Conidia pale green, dark green in mass, broadly ellipsoidal or subglobose, 1-celled, apiculate at one end. Chlamydospores globose or subglobose.

Dimensions: Conidiophores 62.5–158 μm tall: primary branches (connected to phialides) 6–14.6 μm long: phialides 7.2–12.5 × 2.2–4.3 μm. Spore masses 25–65 μm in diameter. Conidia 2.7–5 × 2.2–3.8 μm. Chlamydospores 7.7–11.2 μm in diameter.

Material: 72-X23-107 (Sugarcane root, Taiwan, ROC).

Remarks: The cultures on agar are dark green in color.

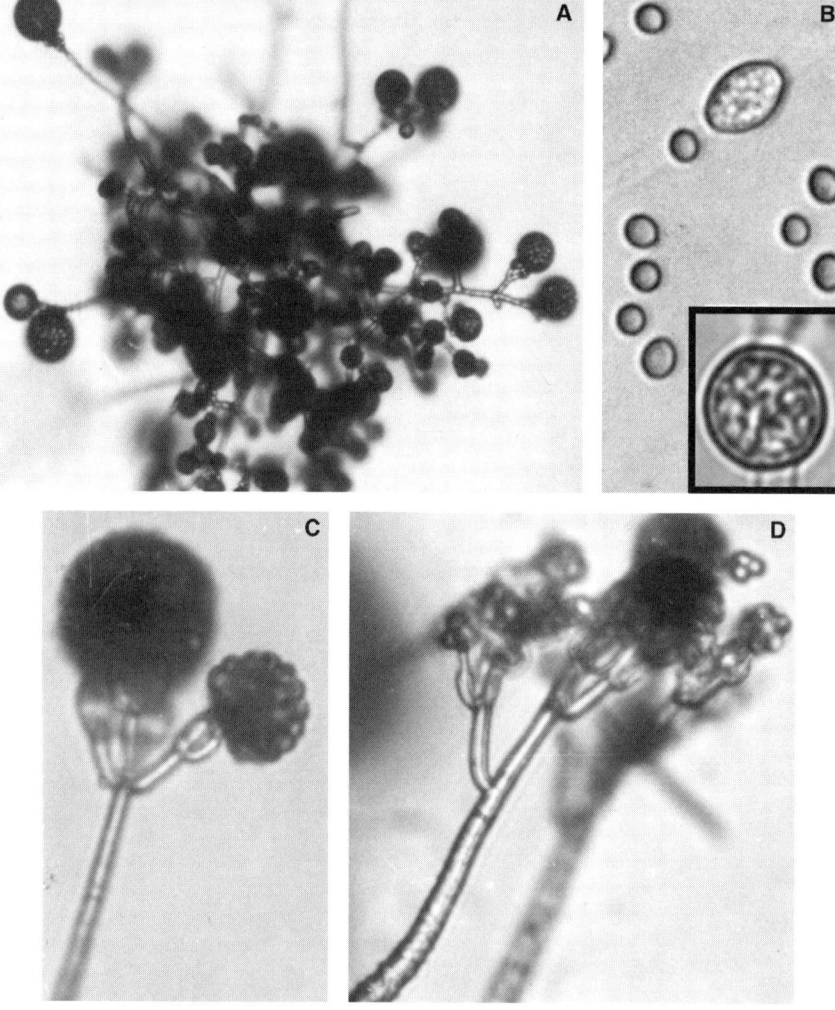

Gliocladium virens. A: Conidiophores and spore masses. B: Conidia and chlamydospores. C,D: Conidiophores with phialides and conidia. (From Watanabe, T. 1975d. *Trans. Mycol. Soc. Jpn.*, 16:264–267. With permission.)

Gliocladium viride Matr.

Synonym: *Gliocladium deliquescens* (Sopp) Gillman et Abbott

References: Raper and Thom 1949; Watanabe 1975d.

Morphology: Conidiophores hyaline, erect, simple or branched 1–3 times oppositely or verticillately especially in metula, bearing spore masses at the phialides on the apical branches. Conidia hyaline or pale green, ellipsoidal, 1-celled. Chlamydospores lacking.

Dimensions: Conidiophores 59.5–201.5 × 3.6–3.7 µm: primary branches (connected to phialides) 14.5–18.2 × 2.4–2.5 µm: phialides 14–23.1 × 2.4–2.7 µm. Spore masses 35–65 µm in diameter. Conidia 5.1–7.3 × 2.4–4.4 µm.

Material: 72-X31-177 (= ATCC 32912, Sugarcane root, Taiwan, ROC).

Remarks: Cultures on agar are pale yellowish brown with concentric zonation. The conidia are longer than those reported. It is also morphologically close to *G. penicilloides*, which forms conidia in chains and spore masses.

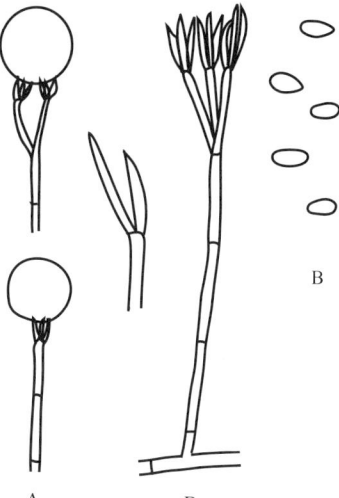

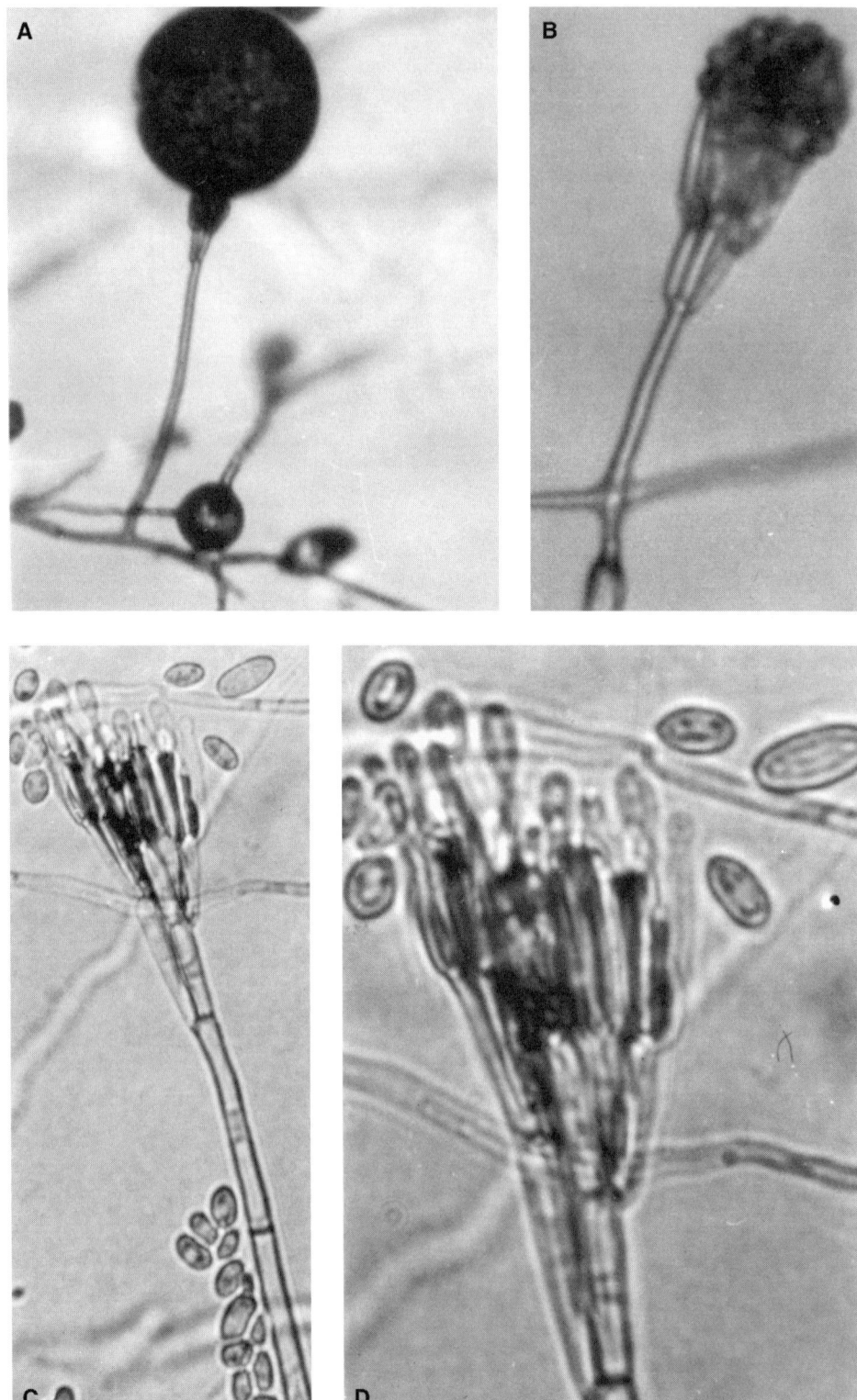

Gliocladium viride. A,B: Conidiophores and spore masses. C,D: Conidiophores with fertile portions and conidia. (From Watanabe, T. 1975d. *Trans. Mycol. Soc. Jpn.*, 16:264–267. With permission.)

Gloeocercospora Bain & Edgerton : Deighton

Trans. Br. Mycol. Soc. 57:358, 1971.
Type species: *G. sorghi* Bain & Edgerton

Gloeocercospora sorghi Bain & Edgerton

References: Bain and Edgerton 1943; Watanabe and Hashimoto 1978.

Morphology: Sporodochia pale brown, very sticky, apparently millet jelly-like. Conidiophores hyaline, simple or branched, united for sporodochium formation. Conidia hyaline, mainly filiform, gradually tapering from base toward apex, often curved, but clavate or long spindle-shaped when immature, 2- to 25-celled. Sclerotia black, globose, smooth, hard.

Dimensions: Sporodochia 98.8–296.4 µm in diameter. Conidia 15–24.5 × 2.5-4.5 µm. Sclerotia 74–230 µm in diameter.

Material: 74-291, 74-292 (Sorghum seed, Ebina, Kanagawa, Japan).

Remarks: This fungus, known as the pathogen of zonate leaf spot on sorghum is seed- and soilborne. Sclerotia germinate in forming sporodochia directly (Watanabe and Hashimoto, 1978).

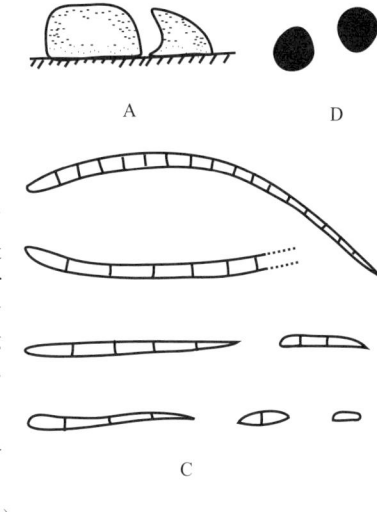

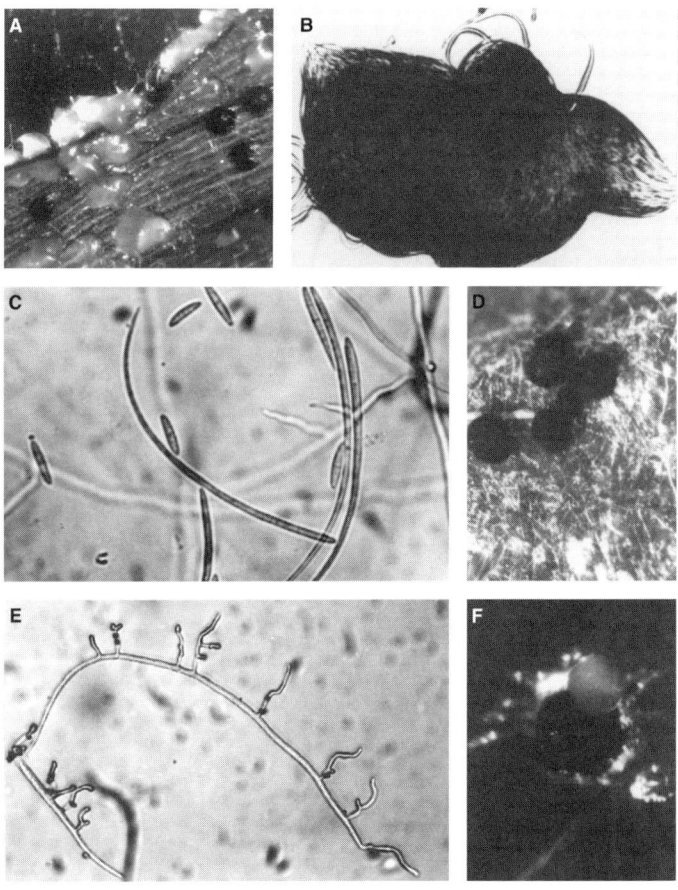

Gloeocercospora sorghi. A: Sporodochia and sclerotia formed on natural medium containing sorghum straw. B: Sporodochium. C: Conidia. D: Sclerotia. E: Germinated conidia. F: Sporodochium on sclerotium. (From Watanabe, T. and Hashimoto, K. 1978. *Ann. Phytopathol. Soc. Jpn.*, 44:633–640. With permission.)

Gonatobotrys Corda

Prachtflora 9, 1839.
Type species: *G. simplex* Corda

Gonatobotrys sp.

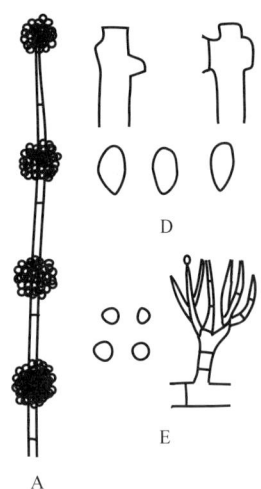

References: Barron 1968; Walker and Minter 1981.

Morphology: Two types of sporulation occurred: botryoblastosporous type, conidiophores dark brown, erect, thick-walled, simple, bearing spore masses in apical fertile parts (nodes) of more than 4 positions, partially residual after detachment of conidia; and phialosporous type, conidiophores short, bearing conidia in verticillate phialides. Conidia: botryoblastosporous, larger, slightly pigmented, ovate; and phialosporous, hyaline, globose.

Dimensions: Conidiophores over 1 mm tall: node 187.6–214.4 × 6.2–8.8 µm. Conidia: botryoblastosporous 8.7–12.5 (-15) × 6.5–10 µm, and phialosporous 2.5 µm in diameter.

Material: 85-18, 85-19 (Flowering cherry seed, Hachioji, Tokyo, Japan).

Remarks: This fungus is morphologically close to *G. simplex*, except for phialospore formation.

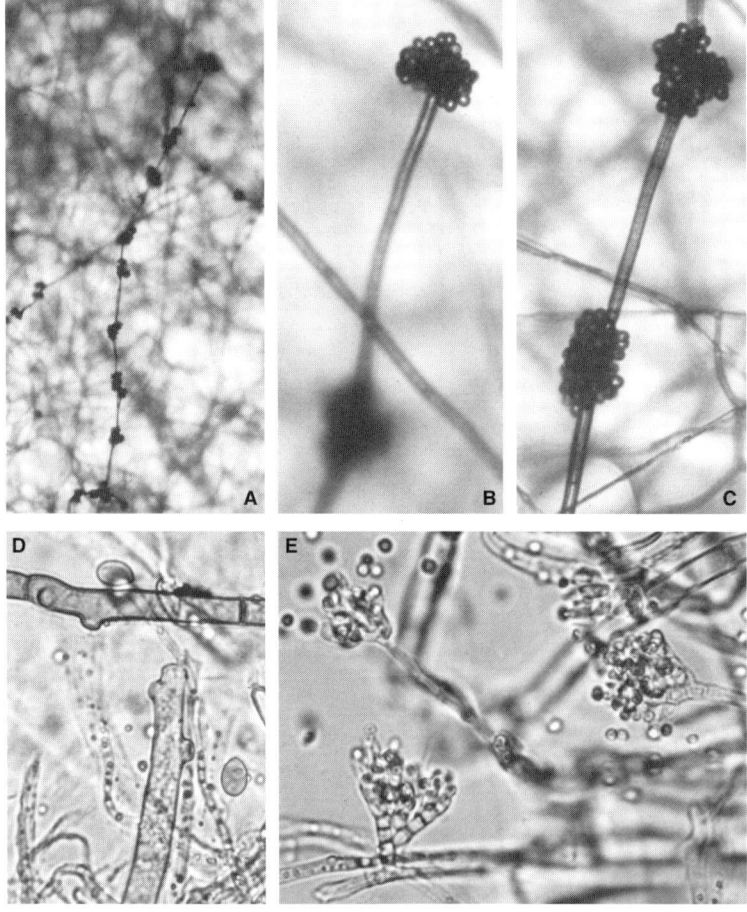

Gonatobotrys sp. A–C: Conidiophores and spore masses. D: A portion of the apical conidiophore and macroconidia. E: Conidiophores and microconidia.

Gonytrichum Nees : Leman

Dict. Sci. Nat., Paris 19:209, 1821.
Type species: *C. caesium* Nees : Leman.

Gonytrichum chlamydosporium Barron & Bhatt

References: Barron and Bhatt 1967; Gams and Holubová-Jechová 1976; Hughes 1951b; Watanabe 1975.

Morphology: Conidiophores pale brown, erect, mostly simple, often proliferated, gradually tapering from base toward apex, branched in central 1–2 positions alternately or verticillately, bearing spore masses at the phialides, occasionally inflated characteristically at basal branches. Conidia phialosporous, pale brown, ellipsoidal, 1-celled. Chlamydospores rarely formed, brown, subglobose.

Dimensions: Conidiophores 32.5–107.5 (-150) × 2.5–3 (-10) μm: phialides 12.5–15 μm long. Spore masses 12.5–15 (-20) μm in diameter. Conidia 3.2–5.0 × 2.2–2.8 μm. Chlamydospores 6.2–7.5 × (3-) 3.7–4 μm.

Material: 72-X61-353 (= ATCC 32914, Sugarcane root, Taiwan, ROC), 98-64 (= MAFF 238006, Uncultivated soil, Tsukuba, Ibaraki, Japan).

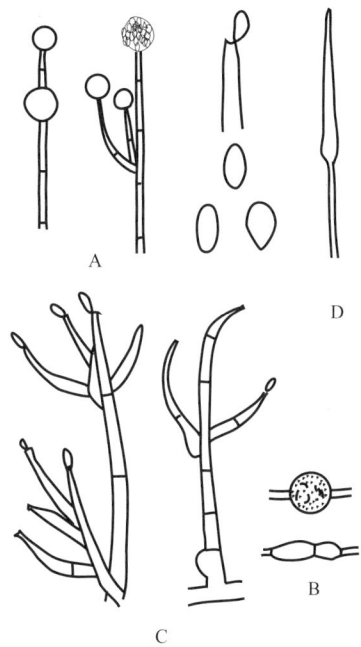

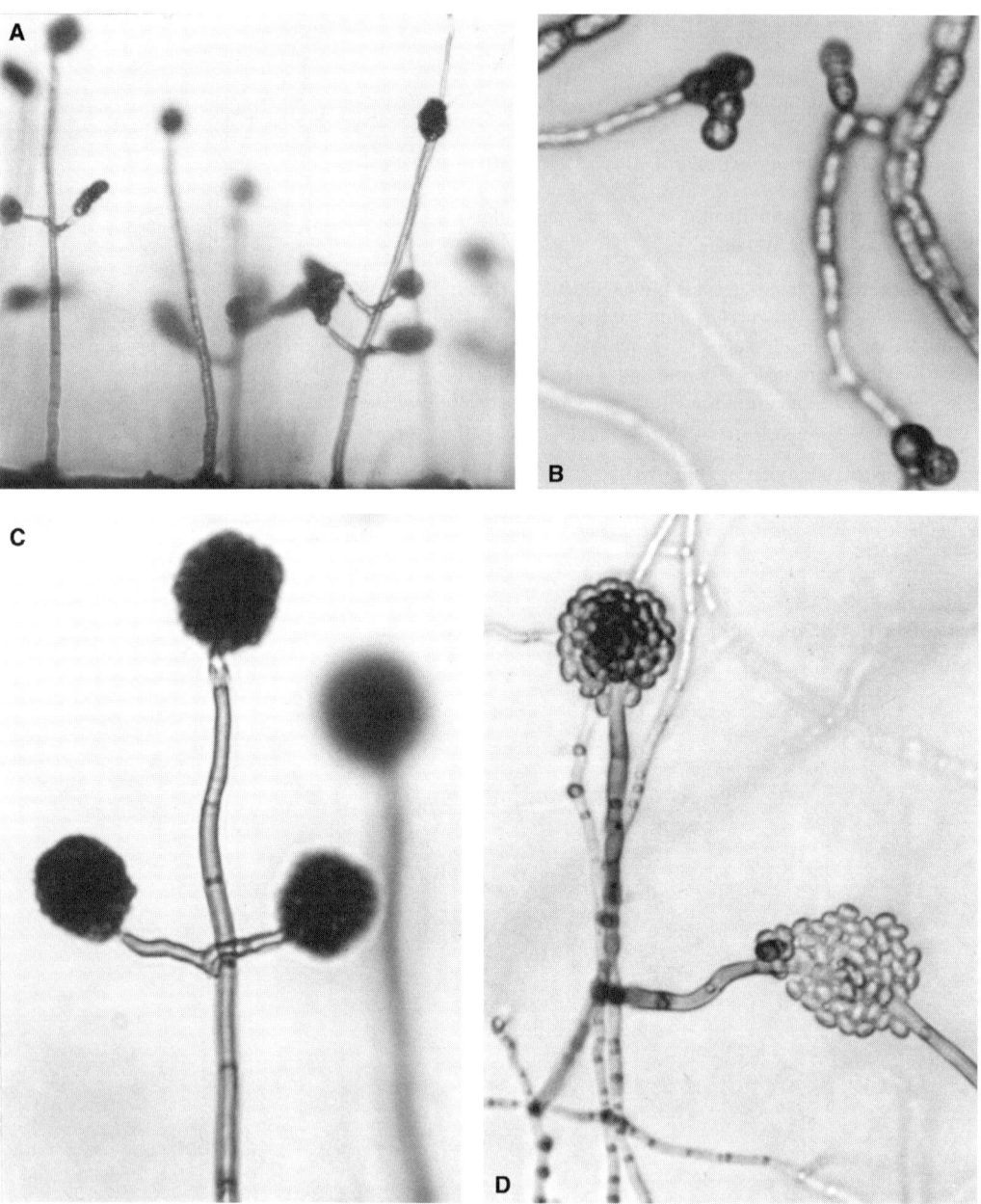

Gonytrichum chlamydosporium. A–C: Conidiophores and spore masses. B: Chlamydospores. D: Conidiophore and conidia. (From Watanabe, T. 1975c. *Trans. Mycol. Soc. Jpn.*, 16:149–182. With permission.)

Gonytrichum macrocladum (Sacc.) Hughes

References: Barron and Bhatt 1967; Hughes 1951b.

Morphology: Conidiophores brown and rhizoidal basally, subhyaline or hyaline apically, erect, gradually tapering from base toward apex, several-septate, branched in lower positions of more than 3 locations developed from collar-like hyphae more or less verticillately, often proliferated, bearing spore masses at the hyaline phialides, characteristically surrounding conidiophores. Conidia phialosporous, subhyaline to pale brown, ovate to ellipsoidal, 1-celled, with single oil droplets.

Dimensions: Conidiophores 200–220 μm long, 6–10 μm basally, 3 μm apically: phialides 13–28 μm long, 3–4 μm basally. Spore masses 12–16 μm in diameter. Conidia 4–5 × 2.2–3 μm.

Material: 00-329 (Coprinus-associate, Shintone, Chiba, Japan).

Remarks: This fungus, known as a representative member of soil fungi was isolated as a *Coprinus*-associate in this study. Sterile hyphal elongations were observed from the upper positions in fresh cultures (A).

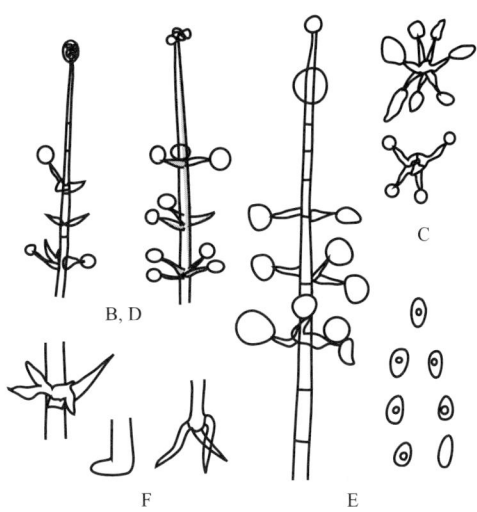

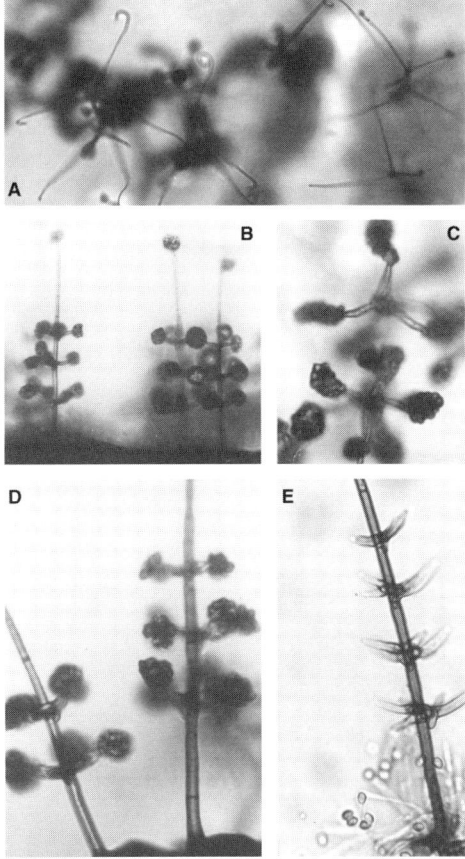

Gonytrichum macrocladum.   A: Apical sterile branches formed on natural substrate. B,D: Conidiophores and spore masses. C: Overview of phialides and conidia. E: Conidiophore with phialides and conidia. F: Rhizoids.

Hainesia Ellis & Sacc.

Syll. Fung. 3:699, 1884.
Type species: *H. lythri* (Desm.) Höhnel

Hainesia lythri (Desm.) Höhnel

Teleomorph: *Pezizella lythri* (Desmazieres) Shear et Dodge

References: Arx 1981; Barron 1968; Strong and Strong 1921.

Morphology: Conidiocarps of two kinds: pycnidial, globose, fleshy when immature, with black, hard peridium; and sporodochium-like, composed of branched conidiophores and conidia, cup-shaped with spore masses apically, brown when mature. Conidiophores pale brown, branched, united. Conidia hyaline, 1-celled, boat-shaped.

Dimensions: Pycnidia closed: up to ca. 400 μm in diameter: opened 49 to over 55 μm tall, ca. 15 μm wide at base, 65–74.1 μm wide at apex. Spore masses 65–123.5 μm in diameter. Conidiophores ca. 40 μm tall. Conidia 6.2–7.8 × 1.5–2.3 μm.

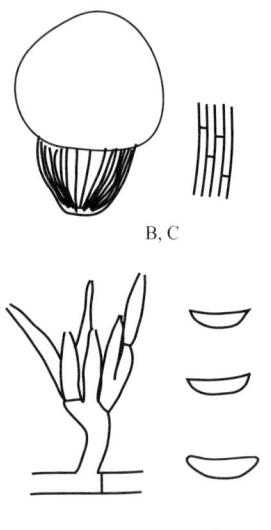

Material: 73-439 (Strawberry root, Mie, Japan).

Remarks: *Pezizella oenotherae* or *Discohainesia oenotherae* (Cooke & Ellis) Nannf. may be also used as the teleomorph binomials.

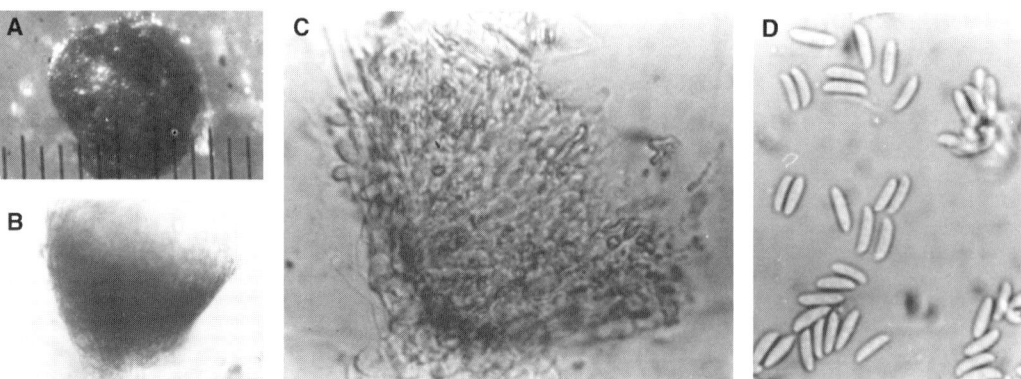

Hainesia lythri. A: Closed pycnidium. B,C: Opened pycnidia. D: Conidia. E: Conidiophores.

Hansfordia Hughes

Mycol. Pap. 43:15–24, 1951.
Type species: *H. ovalispora* Hughes

Hansfordia biophila (Cif.) M. B. Ellis

References: Arx 1981; Ellis 1976; Hughes 1951.

Morphology: Conidiophores brown, erect, rough on the surface, branched apically, bearing 2 to several conidia terminally and laterally at the apical parts of the branches, denticulate after detachment of conidia. Conidia sympodulosporous, (pale) brown, ovate, cylindrical, 1-celled. Sclerotium-like structures brown, composed of aggregates of barrel-shaped, chlamydospore-like cells.

Dimensions: Conidiophores 52.5–250 × 1.2–3 µm. Conidia (3.9-) 5–8 × 1.9–3 µm.

Material: 69-424 (Pineapple leaf near soil, Nakijin, Okinawa, Japan).

Remarks: This fungus was isolated from leaf partially embedded in soil, and treated as a member of soil fungi. Cultures on PDA are pale brown with blackish fragments distributed all over. *Dicyma* Boulanger may be a synonym of this genus (Arx, 1981).

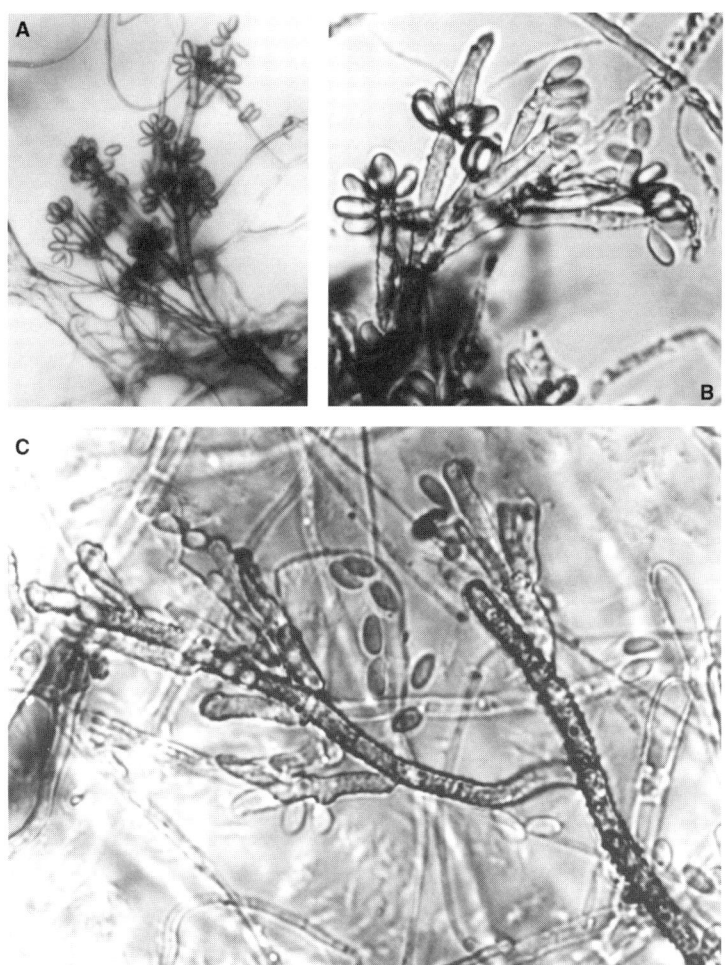

Hansfordia biophila. A–C: Conidiophores and conidia.

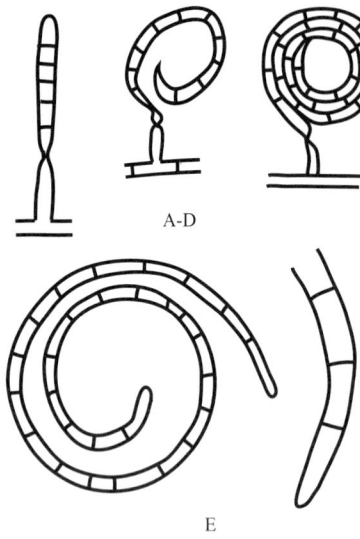

A–D

E

Helicomyces Link

Linn. Spec. Pl. 1:131, 1824.
Type species: *H. roseus* Link

Helicomyces roseus Link

References: Goos 1985; Moore 1955, 1957.

Morphology: Conidiophores not differentiated from ordinary hyphae (micronematous), short, simple, hyaline to subhyaline, bearing single conidia laterally on hyphae. Conidia sympodulosporous, hyaline to subhyaline, white in mass, tightly coiled in 1 plane mostly 3 times, conidial filaments elongated when wet (hygroscopic), numerously septate, often with over 28 septations.

Dimensions: Conidia 30–35 μm in diameter: conidial filament 3–3.6 μm wide.

Material: 93-408 (*Phellodendron amurense* root, Meguro, Tokyo, Japan); 98-50 (= MAFF 238007, Uncultivated soil, Tsukuba, Ibaraki, Japan).

Remarks: There are at least 44 genera with helicoid conidia, including *Helicogoosia* Hol.-Jech. (Holubová-Jechová, 1991).

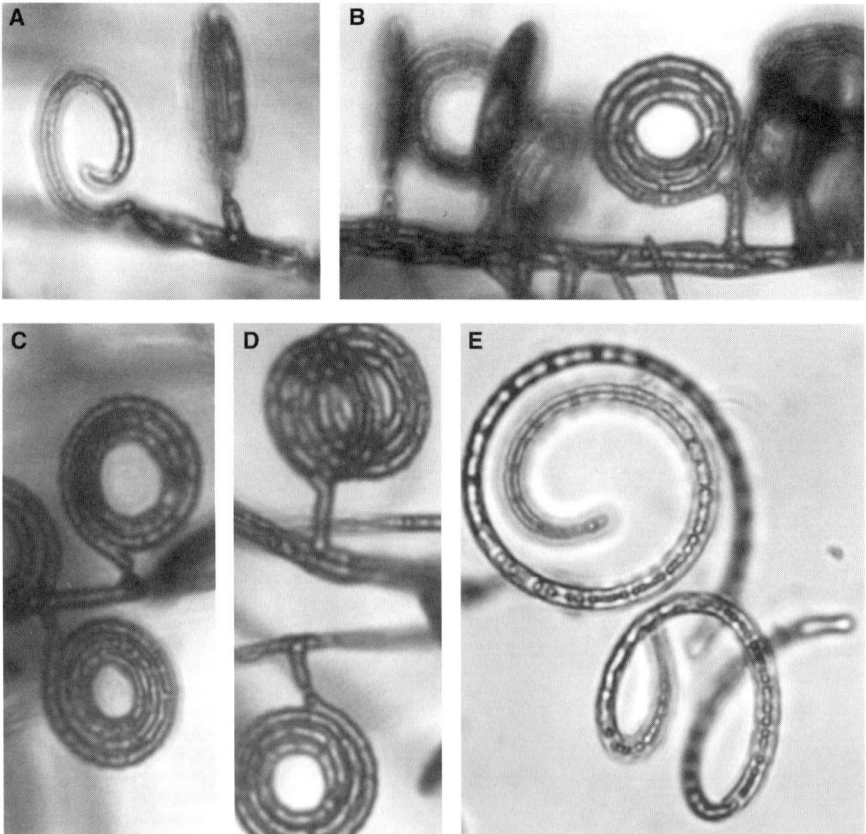

Helicomyces roseus. A–D: Conidia and hyphae with short conidiophores. E: Elongated conidium after mounting in water.

Helminthosporium Link : Fr.

Syst. Mycol. 3:359, 1832.
Type species: *H. velutinum* Link : Fr.

Helminthosporium solani Durieu et Montagne

Synonym: *Spondylocladium atrovirenz* Harz

References: Barron 1968; Ellis 1971.

Morphology: Conidiophores pale brown, erect, simple, slightly curved, bearing conidia apically and laterally on the fertile apical parts in alternate, opposite, or verticillate fashion. Conidia porosporous, brown, clavate, long-ellipsoidal, tapering from base toward apex, partially curved, usually 3- to 7-celled, without hilum basally.

Dimensions: Conidiophores 175–350 × 5–6.3 μm. Conidia 27.5–60 × 7.5–8.8 μm.

Material: 75-212 (Potato tuber, Hokkaido, Japan); 75-213 (Potato tuber, Sanpougahara, Shizuoka, Japan); 75-214 (Potato tuber, Yasaidanchi, Chiba, Japan).

Remarks: Known as the pathogen of silver scurf of potato.

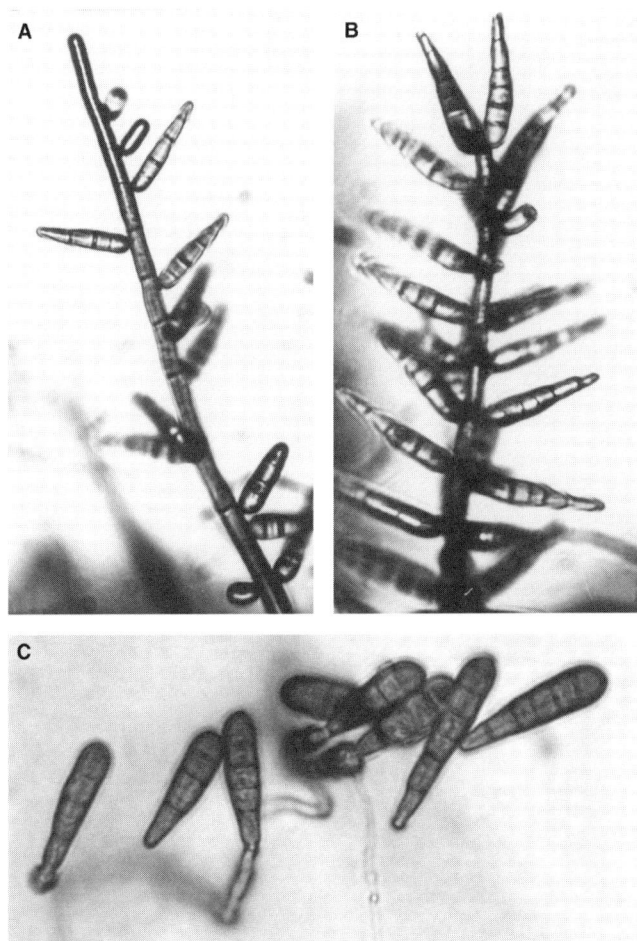

Helminthosporium solani.   A,B: Conidiophores and conidia. C: Conidia.

Humicola Traaen

Nyt Mag. Naturvid. 32:20, 1914.
Type species: *H. fuscoatra* Traaen

Key to Species

1.	Aleurioconidia	under 6 µm in diameter.......................... *H. dimorphospora*
		over 6 µm in diameter.. 2
2.	Aleurioconidia	mostly over 3 in a chain, with lunate phialoconidia......... *H. tainanensis*
		usually solitary, with ovate phialoconidia............................ 3
3.	Alurioconidia	under 10 µm in diameter.. 4
		over 11 µm in diameter................................. *H. grisea*
4.	Phialoconidia	catenulate... *H. fuscoatra*
		aggregated in spore masses............................ *Humicola* sp.

Humicola dimorphospora Roxon & Jong

References: Roxon and Jong 1974; Watanabe et al. 1986a.

Morphology: Conidia aleuriosporous, borne directly on hyphae or sterigmata, extended from hyphae, solitary, or aggregated, dark brown, ovate, globose or subglobose, thick-walled, 1-celled, apiculate or truncate at one end.

Dimensions: Conidia 4.8–5.4 × 3.6–4.5 µm.

Material: 84-582 (Japanese black pine seed, Okawa, Kagawa, Japan).

Remarks: The fungus grows slowly (less than 2 cm in the colony diameter at 25°C for 20 days), and the colonies are black in color. No sympodulosporous hyaline conidia, characteristic for *Sporothrix* and this species, were observed.

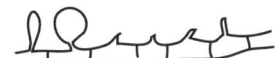

A

B

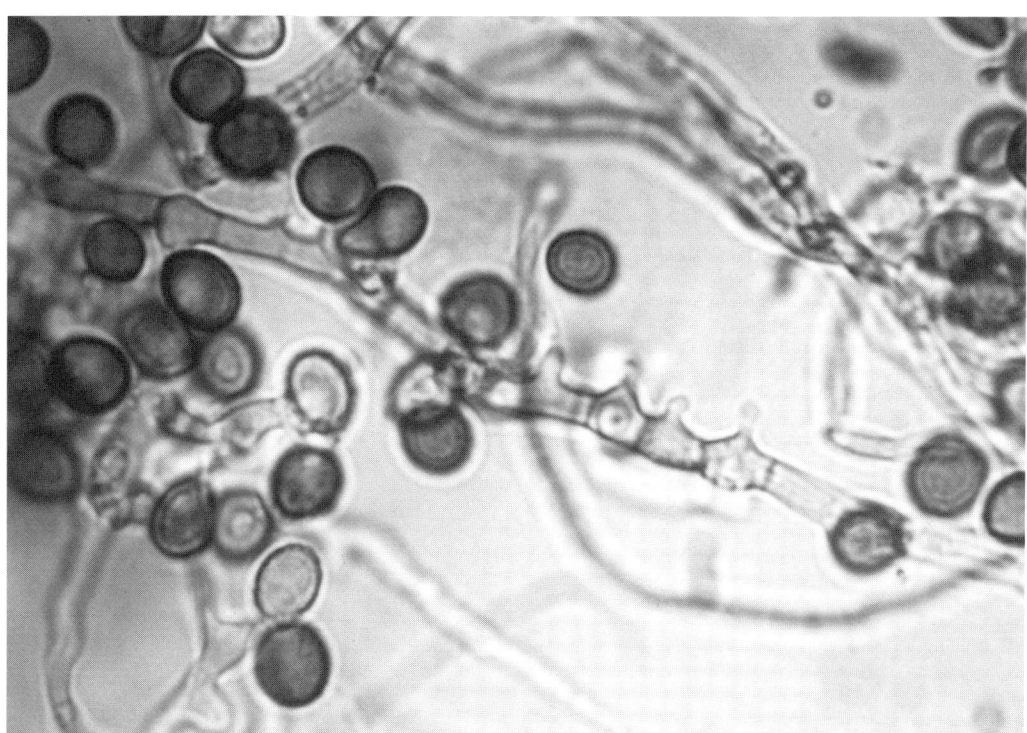

Humicola dimorphospora. Hyphae and aleurioconidia.

Humicola fuscoatra Traaen

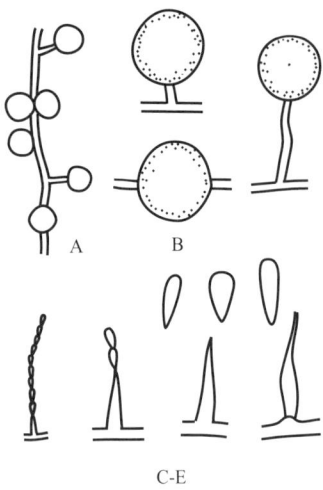

References: Watanabe 1975d; White and Downing 1953.

Morphology: Conidia of two kinds: aleuriosporous, and phialosporous. Aleurioconidia borne directly, or on sterigmata extended from creeping hyphae; pale brown, globose or subglobose. Phialoconidia borne on simple, erect conidiophores (phialides), catenulate, hyaline, long-ovate, apiculate at one end.

Dimensions: Conidiophores 13.3–20.7 × 2–2.5 μm. Conidia aleuriosporous 7.2–10 (-14) μm in diameter, and phialosporous 1.7–3.7 × 0.4–1 μm.

Material: 69-298 (Pineapple field soil, Okinawa, Japan); 70-1099 (Pineapple field soil, Hachijo, Tokyo, Japan); 72-X68-1 (= ATCC 32915); 72-X72-1, 72-341 (= ATCC 32916) (Sugarcane root, Taiwan, ROC); 73-110 (Strawberry root, Shizuoka, Japan); 85-66 (Flowering cherry seed, Hachioji, Tokyo, Japan), 85-P26 (Paulownia root, Itapua, Paraguay).

Remarks: This fungus differs from *H. grisea* Traaen in forming the pale brown, rather small aleurioconidia, usually under 10 μm in diameter.

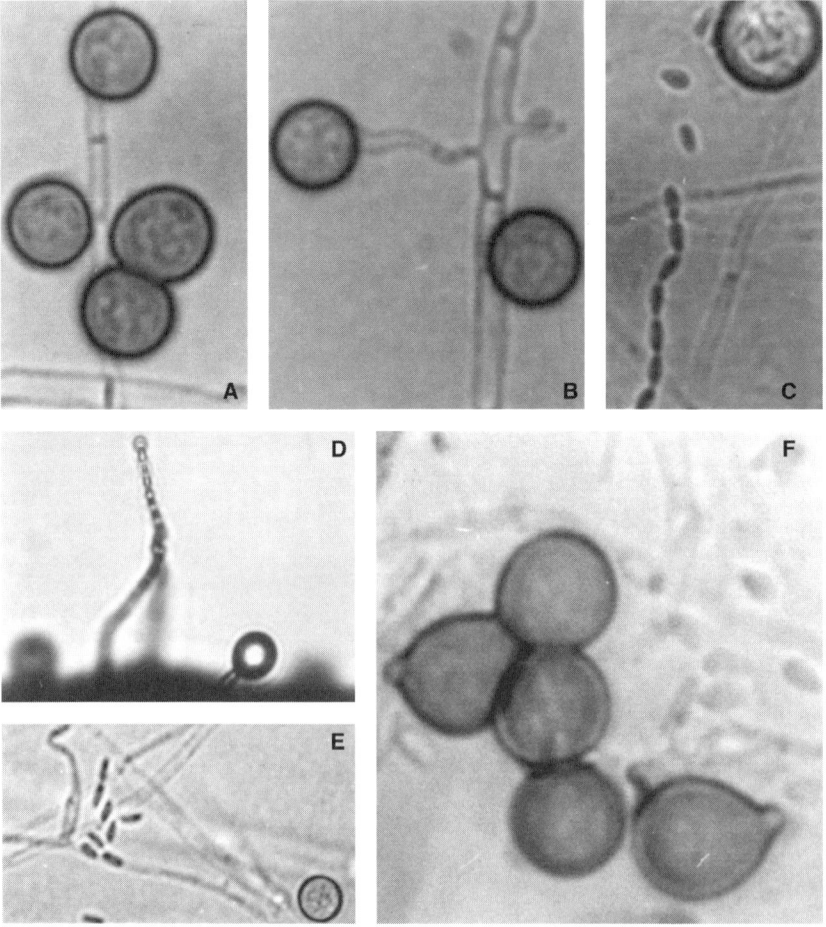

Humicola fuscoatra. A,B: Hyphae and aleurioconidia. C–E: Sporulation of phialoconidia and aleurioconidia. F: Aleurioconidia and phialoconidia. (From Watanabe, T. 1975d. *Trans. Mycol. Soc. Jpn.*, 16:264–267. With permission.)

Humicola grisea Traaen

References: White and Downing 1953.

Morphology: Conidia of two kinds: aleuriosporous and phialosporous. Aleurioconidia borne directly, or on sterigmata extended from creeping hyphae; dark brown, globose or subglobose. Phialoconidia borne on simple, erect conidiophores (phialides), catenulate, hyaline, broadly ovate, apiculate at one end.

Dimensions: Conidia: aleuriosporous (11.5-) 12–16.3 µm in diameter, and phialosporous 1.7–3 × 1.2–2 µm.

Material: 69-354 (Pineapple field soil, Okinawa, Japan); 70-1099 (Pineapple field soil, Hachijo, Japan); 73-163 (Strawberry root, Shizuoka, Japan).

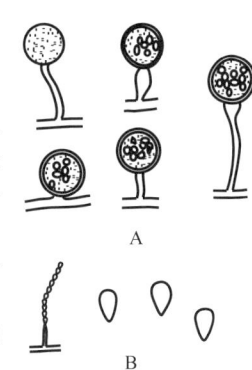

Remarks: This fungus is morphologically close to *H. fuscoatra*, but different in forming more pigmented darker brown, and larger aleurioconidia than the latter. Phialoconidia are rarely formed, and often neglected as a morphological criterion, but they are broadly ovate.

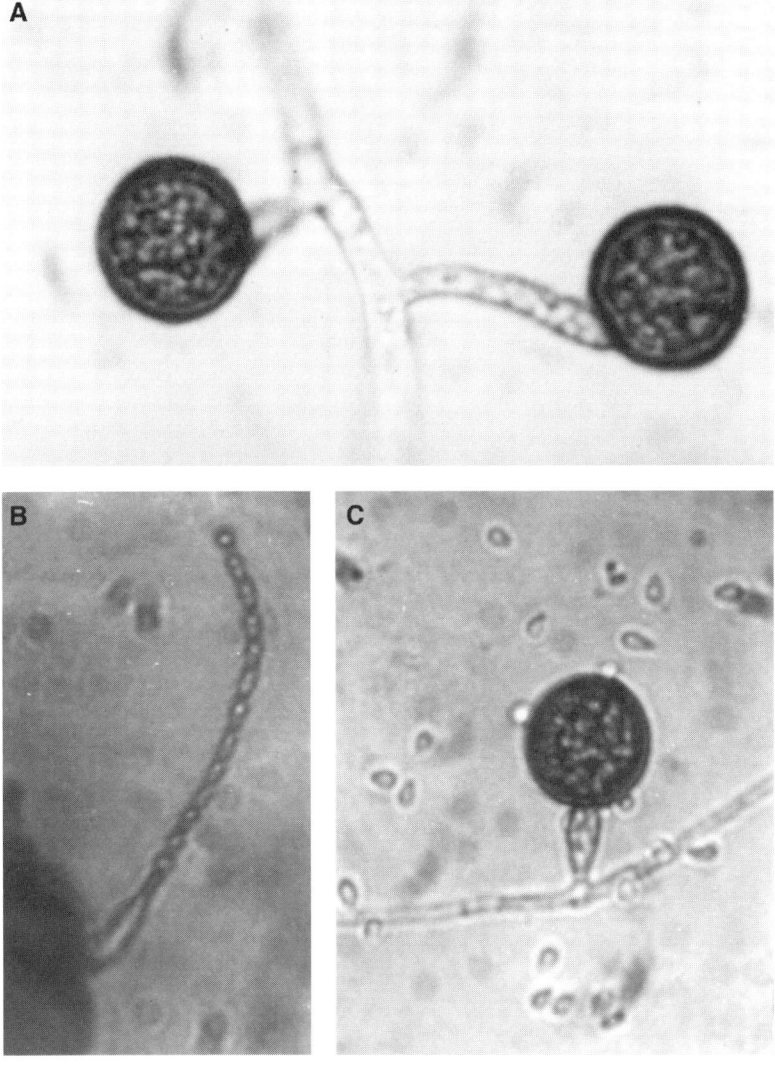

Humicola grisea. A,C: Conidiophores and aleurioconidia. B: Phialoconidia.

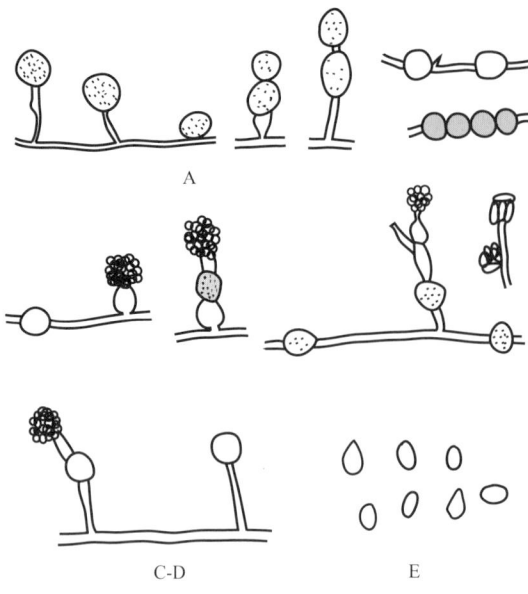

Humicola sp.

References: Barron 1968; Domsch et al. 1980a,b.

Morphology: Conidia of two kinds; aleurioconidia single or in twos, or rarely catenulate, in a short chain, globose or spindle-shaped, verrucose, pale brown, terminal or intercalary; phialoconidia aggregated in a spore mass, borne directly on hyphae or on aleurioconidia, hyaline, ovate, ellipsoidal, cylindrical or barrel-shaped, occasionally apiculate at one end.

Dimensions: Aleurioconidia globose 5–6.5 (-7.5) μm in diameter; spindle-shaped, 6.5–8.8 × 4.5–6.8 μm. Phialoconidia 2–2.3 (-3.5) × (1.2-) 2.3–2.5 μm. Phialide, mostly 3.8–12.5 μm long.

Materials: 76-55 (Gentian root, Chino, Nagano, Japan).

Remarks: This fungus is characterized in forming small aleurioconidia and phialoconidia in spore masses.

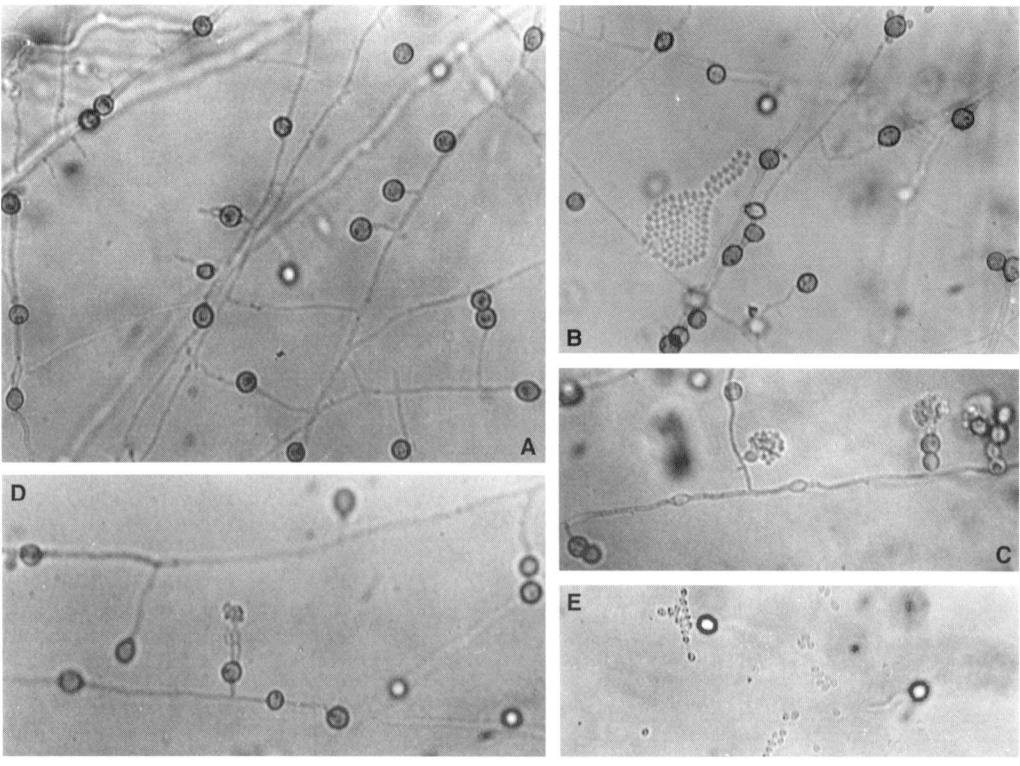

Humicola sp. A: Aleurioconidia and hyphae. B: Aleurioconidia and a mass of phialoconidia formed on hypha. C,D: Phialides and phialoconidia formed on aleurioconidia, and aleurioconidia. E: Phialoconidia.

Humicola tainanensis T. Watanabe

Synonym: *Microdochium tainanense* (T. Watanabe, de Hoog and Hermanides-Nijhof)

References: de Hoog and Hermanides-Nijhof 1977; Domsch et al. 1980a,b; Watanabe 1975d.

Morphology: Conidiophores hyaline, erect, simple or branched, short, stout, bearing 2 types of conidia: aleurioconidia and phialoconidia. Aleurioconidia pale brown, subglobose, ellipsoidal or irregular, granulate, solitary or catenulate on creeping hyphae. Phialoconidia hyaline, lunar-shaped, 1- or 2-celled.

Dimensions: Conidiophores 2.5–5 (-11.3) × 1.2–2.5 (-4.5) μm. Conidia: aleuriosporous 8.2–14.6 μm in diameter, and phialosporous 6.2–12.5 × 1.2–3 μm.

Material: 72-X58 (= ATCC 32917 = CBS 269.76, Sugarcane root, Taiwan, ROC).

Remarks: Colonies on agar are dark gray to black.

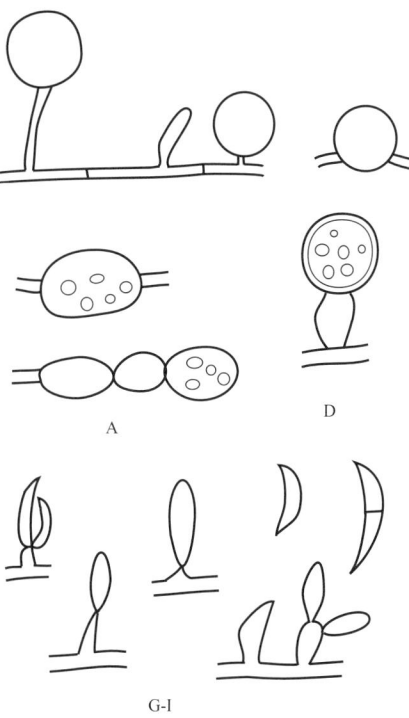

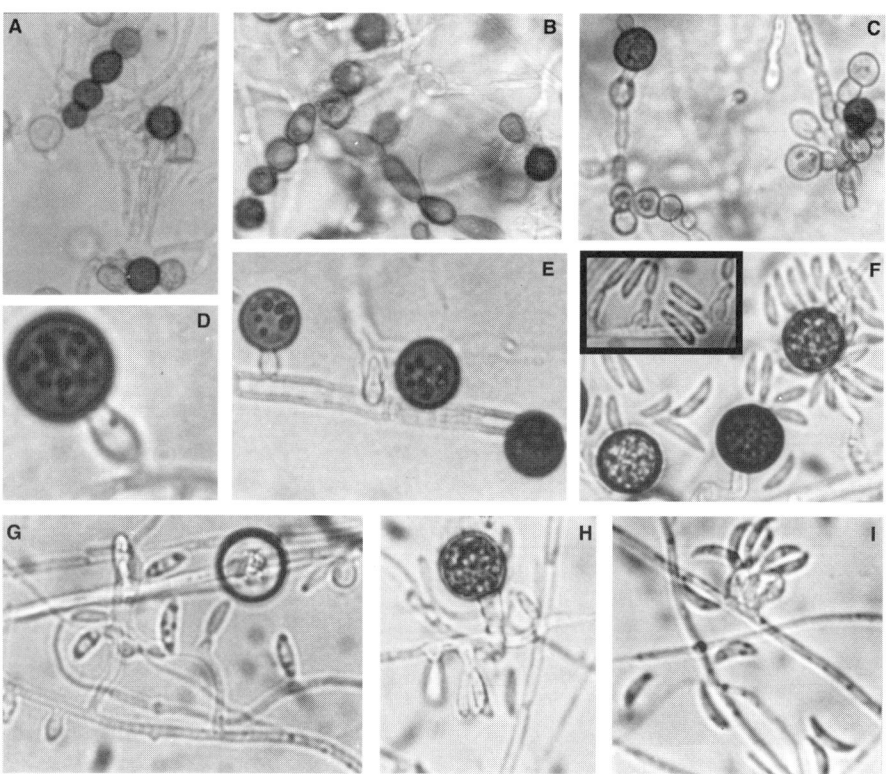

Humicola tainanensis. A–C: Catenulate aleurioconidia. D,E: Solitary aleurioconidia. F: Conidiophores, aleurioconidia, and phialoconidia. G–I: Conidiophores and phialoconidia. (From Watanabe, T. 1975d. *Trans. Mycol. Soc. Jpn.*, 16:264–267. With permission.)

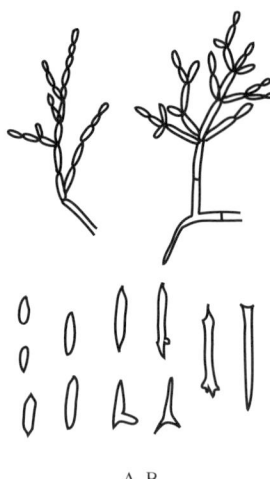

A, B

Hyalodendron Diddens

Zentbl. Bakt. Parasitkde., Abt. 2, 90:315, 1934.
Type species: *H. lignicola* Diddens

Hyalodendron sp.

References: Barron 1968; Hoog 1979.

Morphology: Conidiophores hyaline, erect, branched in the central parts, and dendroid-like, bearing catenulate conidia at the apex of the branches. Conidia blastosporous, indistinct between conidia and branches, hyaline, cylindrical, spindle-shaped, ovate or various in shape.

Dimensions: Conidiophores ca. 20–35 × 2–2.5 μm. Conidia 6.2–25 × 1–2.5 μm.

Material: 73-141 (Strawberry root, Shizuoka, Japan).

Remarks: This fungus is morphologically close to the genus *Cladosporium*, but it lacks pigmentation.

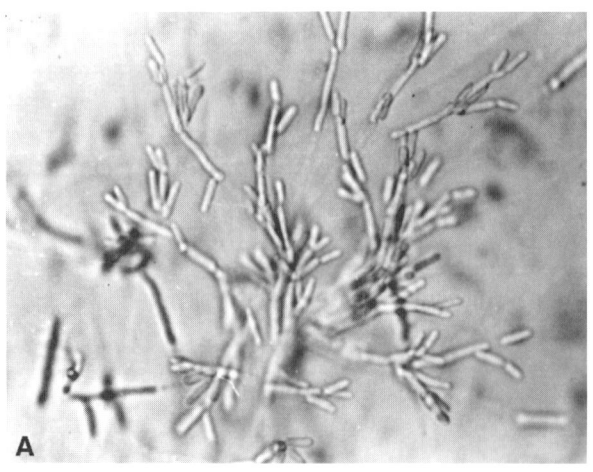
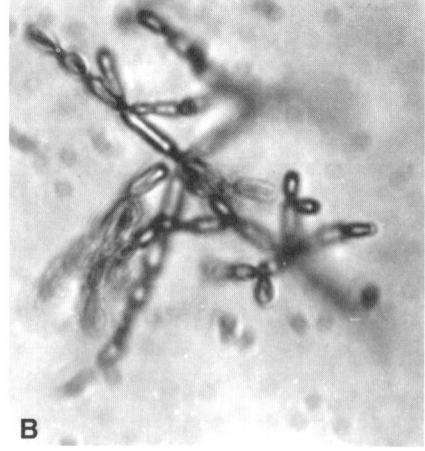

Hyalodendron sp.   A,B: Conidiophores and conidia.

Hyphodiscosia Lodha & Chandra Reddy

Trans. Br. Mycol. Soc. 62:418, 1974.
Type species: *H. jaipurensis* Lodha & Chandra Reddy

Hyphodiscosia radicicola T. Watanabe

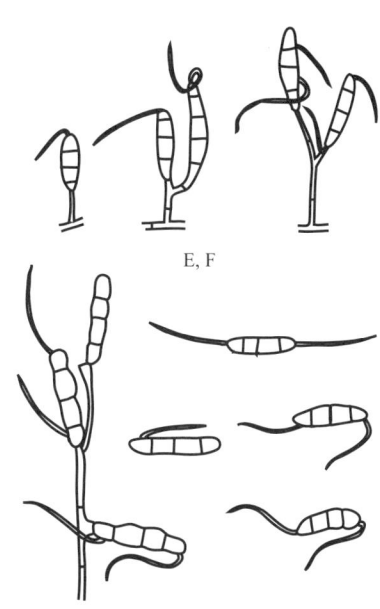

References: Bollard 1950; Lodha and Chandra Reddy 1974; Watanabe 1992a.

Morphology: Conidiophores not well differentiated from hyphae (micronematous or semimicronematous), hyaline or slightly pigmented, branched, bearing conidia at the apex of short branchlets, often aggregated, forming sporodochia. Conidia aleuriosporous, hyaline or pale yellowish green, cylindrical, often curved, occasionally truncate basally, usually 4-celled, slightly constricted at the septum, with 1–2 filiform, simple, often curved, cellular appendages usually from subapical and/or basal lateral cell.

Dimensions: Conidiophores up to 81 μm tall. Conidia 11.5–20 × 2.5–3.8 μm: cellular appendages 4.5–18 × 0.2–0.3 μm.

Material: 77-94 (= IFO 32540, Strawberry root, Nishigahara, Tokyo, Japan).

Remarks: Cultures on PDA are grayish green, resupinate, reverse dark green to black.

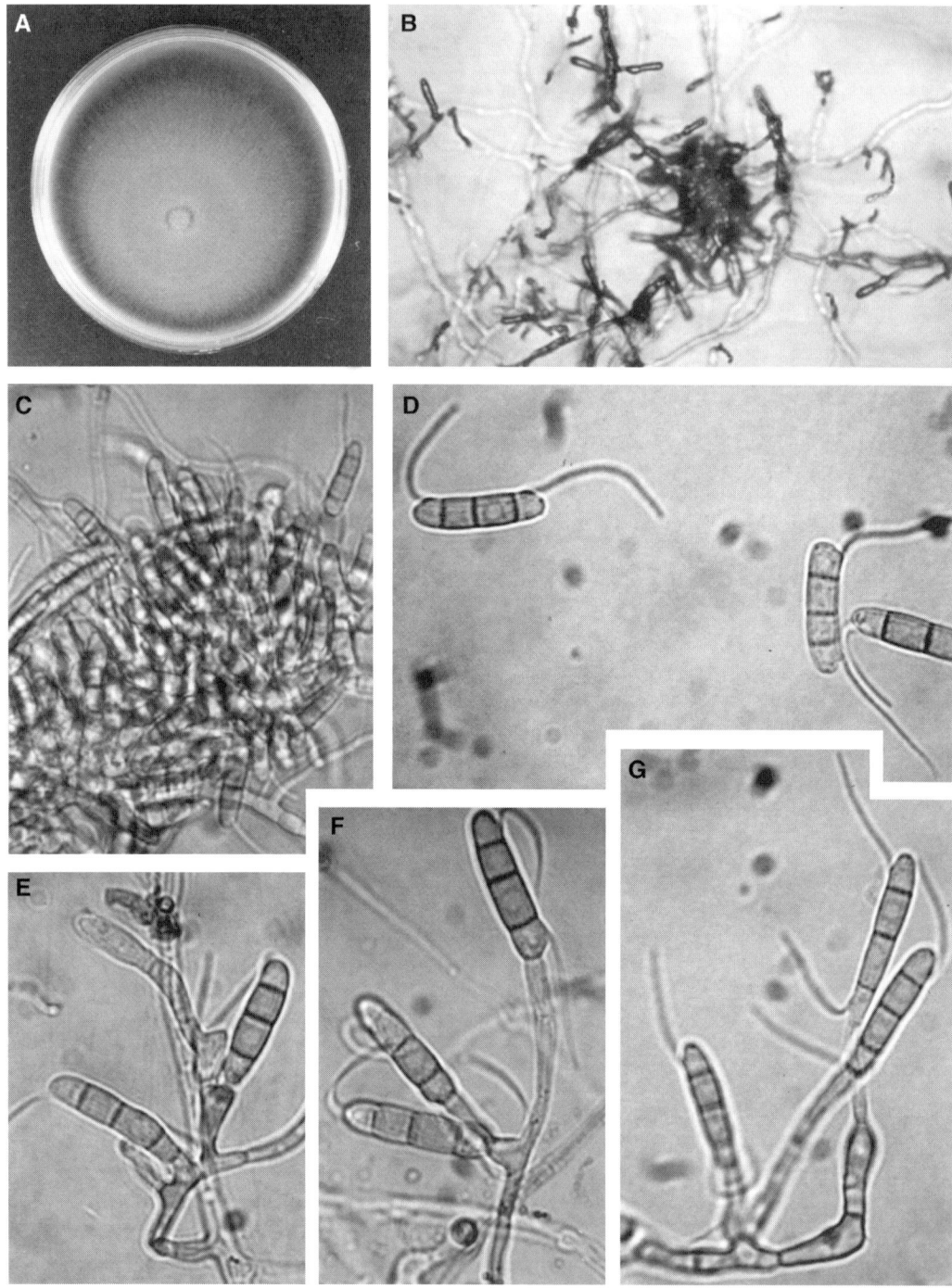

Hyphodiscosia radicicola.   A: PDA colony. B,C: Sporodochia and conidia. D: Conidia. E–G: Conidiophores and conidia. (From Watanabe, T. 1992a. *Mycologia*, 84:113–116. With permission.)

Macrophomina Petrak

Annl. Mycol. 21:314, 1923.

Type species: *M. philippinensis* Petrak = *Tiarosporella phaseoli* (Maubl.) van der Aa = *Macrophoma phaseolina* Tassi

Macrophomina phaseolina (Tassi) Goidanich

Synonyms: *M. phaseoli* (Maubl.) Ashby, *Rhizoctonia bataicola* (Taubenhaus) Butler, *Sclerotium bataticola* Taubenhaus

References: Ashby 1927; Sutton 1980; Watanabe 1972.

Morphology: Pycnidia black, globose, ostiolate apically. Conidiophores hyaline, simple, cylindrical, narrowing apically. Conidia hyaline, cylindrical, 1-celled. Microsclerotia black, homogeneous in size.

Dimensions: Pycnidia 130–230 µm in diameter: ostioles 15–35 µm in diameter. Conidiophores 13–23 × 3–6 µm. Conidia 14–35 × 6–11.5 µm. Sclerotia 60–120 µm in diameter. Hyphae 2.5–7.5 µm wide.

Material: 70-113, 70-114, 70-134 (Snap bean seed, Shinsyu-shinmachi, Nagano, Japan); 70-1028, 70-1858 (Pineapple field soil, Hachijo, Tokyo, Japan); 73-15 (Strawberry field soil, Shizuoka, Japan); 85-PR15R (Paulownia root, Itapua, Paraguay).

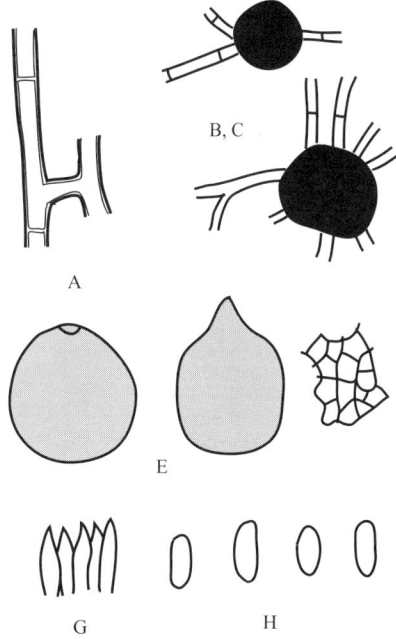

Remarks: Known as the pathogen of ashy stem blight or charcoal rot of snap bean and various plants. Agar cultures are black and homogeneous because of microsclerotia formed all over. Pycnidia are rarely formed on natural media including bean straw agar (Watanabe, 1972).

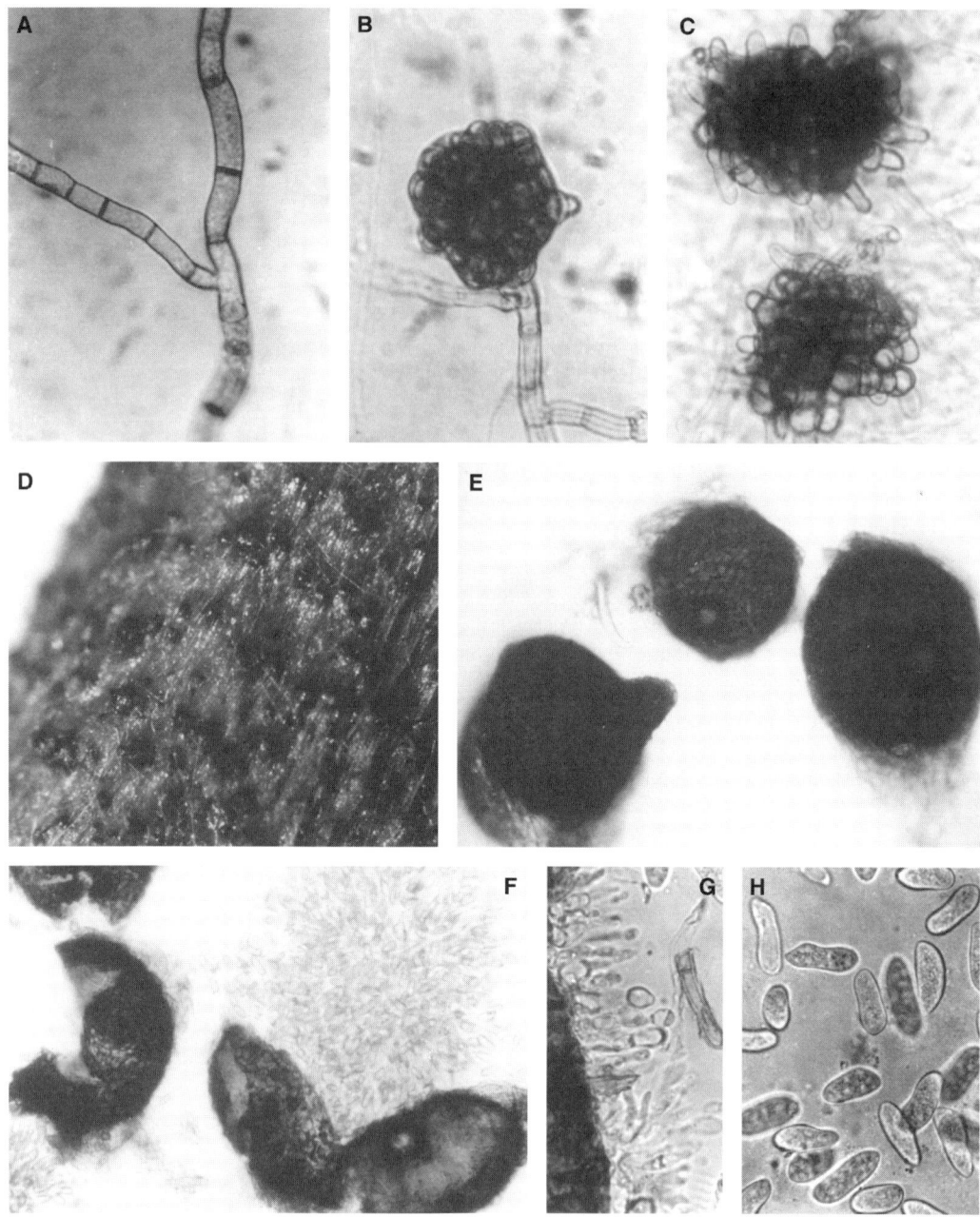

Macrophomoina phaseolina. A: Hypha. B,C: Sclerotia on agar culture. D: Sclerotia on bean straw in natural media. E: Pycnidia. F: Crushed pycnidia and conidia. G: Conidiophores and conidia. H: Conidia. (From Watanabe, T. 1972a. *Ann. Phytopathol. Soc. Jpn.*, 38:106–110. With permission.)

Mammaria Ces.

Flora 12:207, 1854.
Type species: *M. echinobotryoides* Ces.

Mammaria sp.

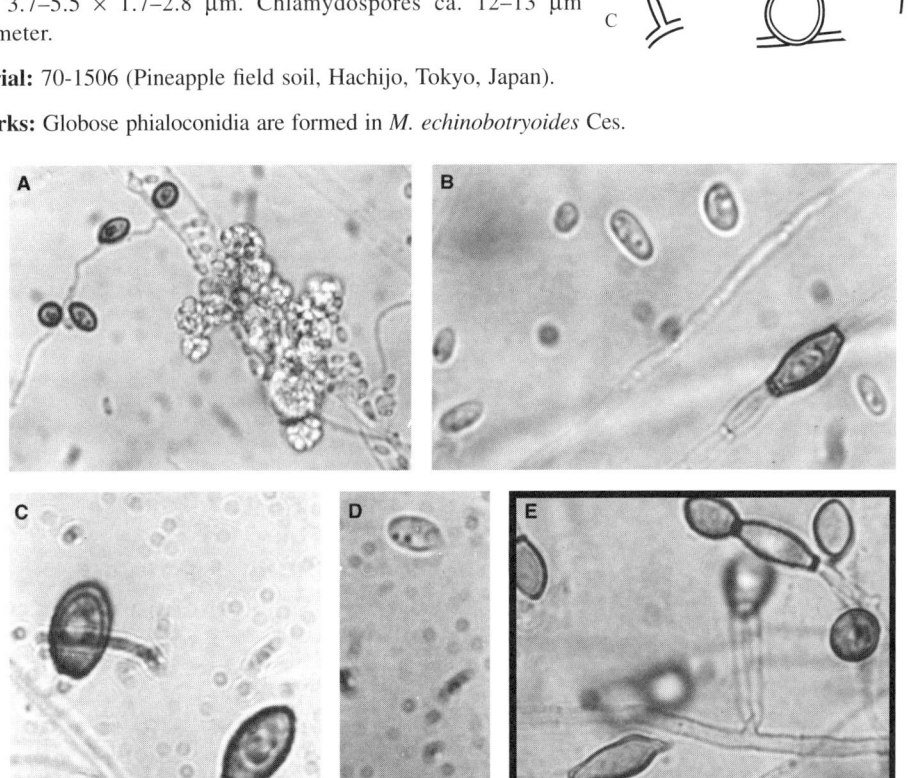

References: Hennebert 1968; Sugiyama 1969.

Morphology: Conidia of two kinds: aleuriosporous and phialosporous. Aleurioconidia common, brown, spindle-shaped, ellipsoidal or subglobose, occasionally formed in twos, with hyaline central slit in spindle-shaped spores, occasionally double-walled, borne directly on hyphae or sterigmata extended from hyphae. Phialoconidia hyaline, cylindrical, 1-celled, often aggregated, forming spore masses. Chlamydospores rarely formed, solitary, thick-walled, globose.

Dimensions: Conidia: aleuriosporous, spindle-shaped 6.2–10.5 × 3–3.8 µm, globose 3.7–5 (-13) µm in diameter, and phialospores (2.5-) 3.7–5.5 × 1.7–2.8 µm. Chlamydospores ca. 12–13 µm in diameter.

Material: 70-1506 (Pineapple field soil, Hachijo, Tokyo, Japan).

Remarks: Globose phialoconidia are formed in *M. echinobotryoides* Ces.

Mammaria sp.   A: Spore masses of phialoconidia and aleurioconidia. B: Phialoconidia and aleurioconidium. C: Aleurioconidia. D,E: Phialoconidia and conidiophores.

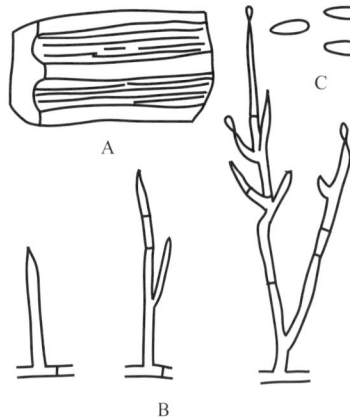

Metarhizium Sorokin

Les Maladies des Plantes, etc. 2:268, 1883.
Type species: *M. anisopliae* (Metschn.) Sorokin

Metarhizium anisopliae (Metschn.) Sorokin

References: Pope 1944; Tulloch 1976; Watanabe 1975d.

Morphology: Conidiophores often united forming sporodochia, simple or branched, bearing catenulate conidia at the phialides in the branches: phialides apically pointed. Conidia phialosporous, hyaline, cylindrical, 1-celled.

Dimensions: Conidiophores over 40–80 μm tall: phialides 9.7–35.3 × 2.1–2.7 μm. Conidia 6.3–9.8 × 2.1–2.5 μm.

Material: 72-X74 (Sugarcane root, Taiwan, ROC).

Remarks: Colonies on PDA are yellowish green.

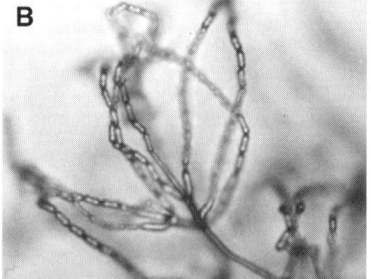
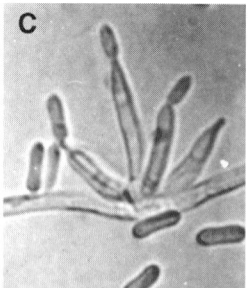

Metarhizium anisopliae. A: Sporodochia. B,C: Conidiophores, phialides, and conidia. (From Watanabe, T. 1975d. *Trans. Mycol. Soc. Jpn.*, 16:264–267. With permission.)

Microsphaeropsis Höhn.

Hedwigia 59:267, 1917.
Type species: *M. olivacea* (Bonord.) Höhn.
Synonym: *Coniothyrium olivaceum* Bonord.

Microsphaeropsis sp.

References: Sutton 1980.

Morphology: Pycnidia globose, white when immature, pigmented gradually with age, covered with pigmented hairs, brown when mature, with black spore masses extruded from indistinct apical ostioles: peridium yellowish brown, pseudoparenchymatous. Conidia globose, smooth, hyaline or slightly pigmented initially, brown, 6- to 7-angled, subglobose, rough at the margin, 1-celled, thick-walled when mature.

Dimensions: Pycnidia ca. 250 µm. Conidia 7.5–8 µm in diameter.

Material: 69-313 (Pineapple field soil, Ishigaki, Okinawa, Japan).

Remarks: Cultures on PDA are pale brown with black spots distributed, reverse pale reddish brown. Sporogenous cells not observed. However, its angular globose conidia are characteristic morphologically. This fungus is different from any of the 14 genera of *Sphaeropsidales* with globose conidia recorded by Sutton (1980). The genus *Microsphaeropsis* is introduced conveniently by Sutton (1977), but it is close to the genus *Coniothyrium* morphologically.

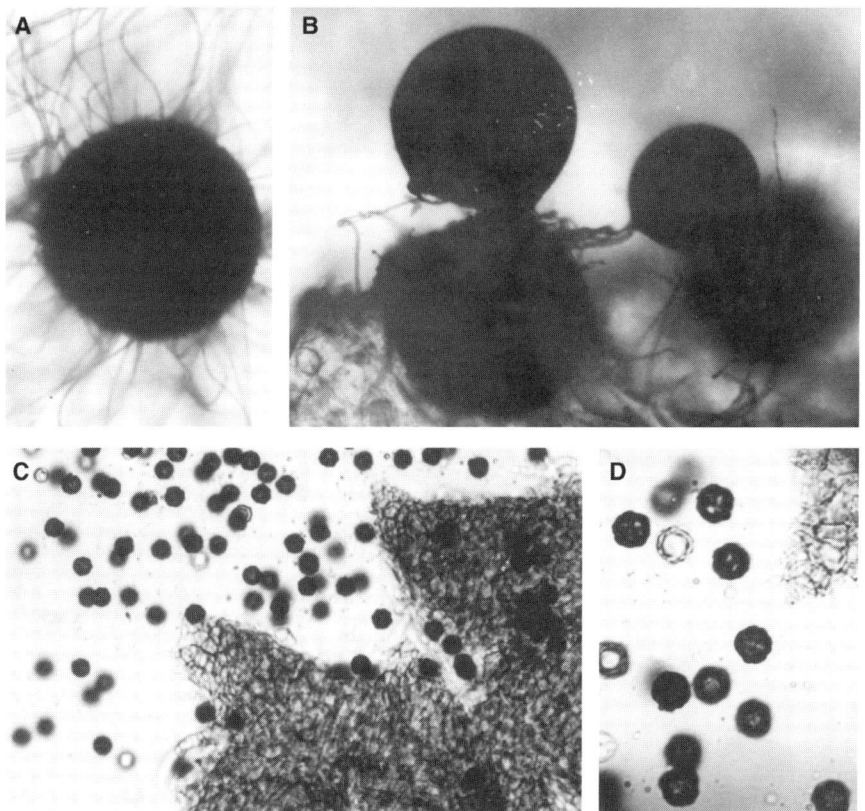

Microsphaeropsis sp.   A: Pycnidium. B: Spore masses extruded from pycnidia. C,D: Peridium and conidia.

Monacrosporium Oudem.

Ned. Kruidk. Arch. 2, 4:250, 1885.
Type species: *M. elegans* Oudem.

Key to Species

1. Conidia   mostly borne singly at the apical conidiophore 2

                     not so .... *M. sclerohyphum*

2. Conidia   with conspicuously large central cells
                     (larger than half of the total conidia)................... *M. bembicodes*

                     not so .... *M. ellipsosporum*

Monacrosporium bembicodes (Drechsler) Subram.

Synonym: *Dactylella bembicodes* Drechsler

References: Cooke and Dickinson 1965; Cooke and Godfrey 1964; Drechsler 1937; Subramanian 1963.

Morphology: Conidiophores hyaline, simple, erect, gradually tapering from base toward apex, bearing single conidia apically. Conidia aleuriosporous hyaline, 4-celled, spindle-shaped, with a central barrel-shaped, larger cell. Chlamydospores yellowish brown, globose, often catenulate.

Dimensions: Conidiophores 130–410 × 4–7.5 µm. Conidia 33.7–50 × 15–22 µm. Chlamydospores 11.2–20 (-25) µm in diameter.

Material: 69-337 (Pineapple field soil, Okinawa, Japan); 70-1331 (Mountainous forest soil, Hachijo, Tokyo, Japan); 73-292, 73-298 (Strawberry root, Shizuoka, Japan).

Remarks: This fungus is very common in strawberry roots and the rhizosphere soils.

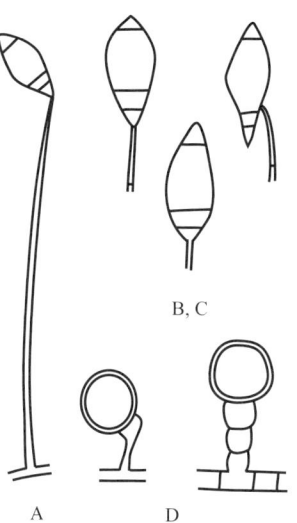

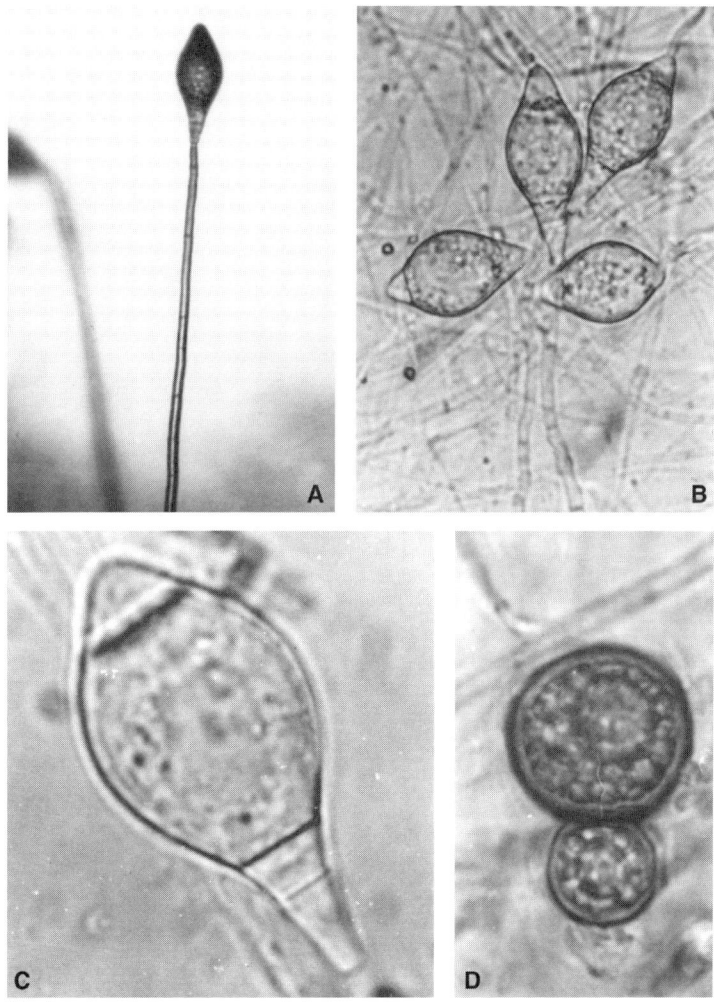

Monacrosporium bembicodes.   A: Conidiophores and conidia. B,C: Conidia. D: Chlamydospores.

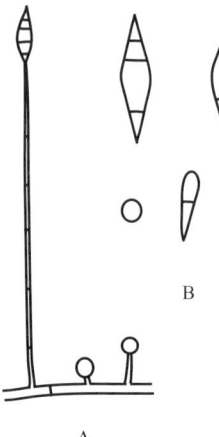

Monacrosporium ellipsosporum (Grove) Cooke & Dickinson

References: Cooke and Dickinson 1965; Drechsler 1937.

Morphology: Conidiophores erect, simple, hyaline, tapering from base toward apex, bearing conidia apically. Conidia aleuriosporous, 2- to 5-celled, terminal, hyaline, spindle-shaped with a larger central cell. Chlamydospore-like structures called predaceous organs or adhesive knobs globose, borne on short stipes directly connected to the hyphae.

Dimensions: Conidiophores ca. 200 μm tall. Conidia 30–50 × 10–22.5 μm. Predaceous organs ca. 5 μm in diameter.

Material: 83-186 (Cultivated soil, Iwaki, Fukushima, Japan).

Remarks: This fungus is also close to *M. phymatopagum* (Drechsler) Subram. on the basis of conidial morphology, but the latter fungus forms ellipsoidal predaceous organs without stipes.

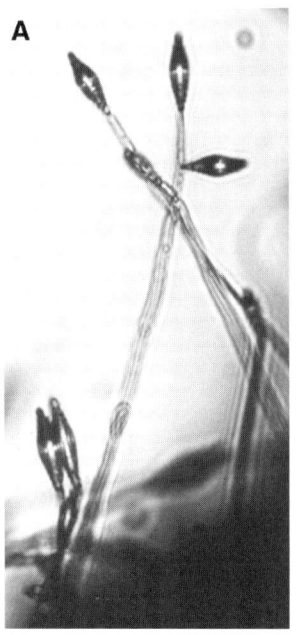
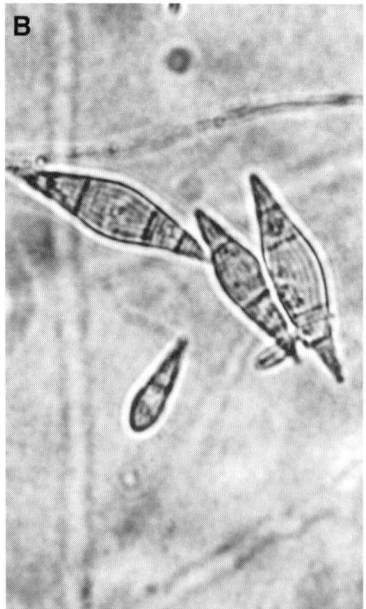
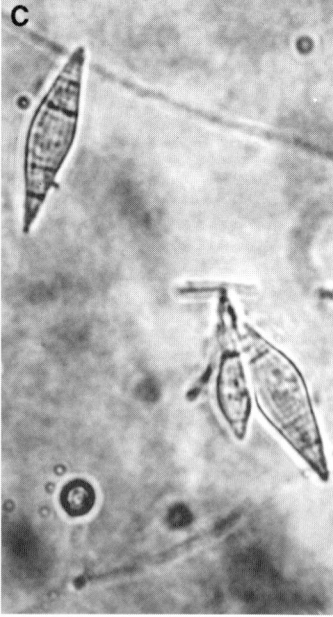

Monacroaporium ellipsosporum. A: Conidiophores and conidia. B: Conidia and predaceous organ.

Monacrosporium sclerohyphum (Drechsler) Xing-Z Liu & K.-Q. Zhang

References: Cooke and Godfrey 1964; Drechsler 1950; Liu and Zhang 1994.

Morphology: Conidiophores hyaline, erect, simple and branched, bearing conidia singly, or two to several conidia, on pedicels in pairs or verticillately in apical fertile portions. Conidia hyaline, fusoid, mostly 4-septate with single large central cells. Predaceous organs globose, occasionally borne on conidiophores and conidia. Chamydospores globose catenulate.

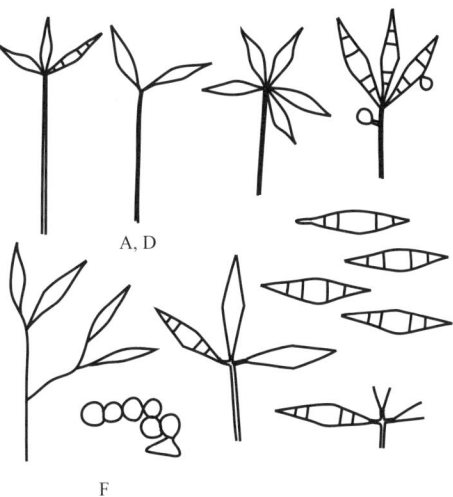

Dimensions: Conidiophores mostly 200–300 µm tall, 2 µm broad apically, 4 µm wide basally: pedicel 4–7 µm long. Conidia 21–44 × 6–10 µm. Predaceous organs 6–8 × 5 µm. Chlamydospores mostly 7–8 µm in diameter.

Material: 01-483 (Beach sand, Chichijima, the Bonin Islands, Tokyo, Japan).

Remarks: Two to several conidia characteristically formed on conidiophores terminally and in a group together with catenulate chlamydospores. Colonies on PDA are pinkish and zonate, similar to the colonies of *Pestalotia* or *Mortierella*.

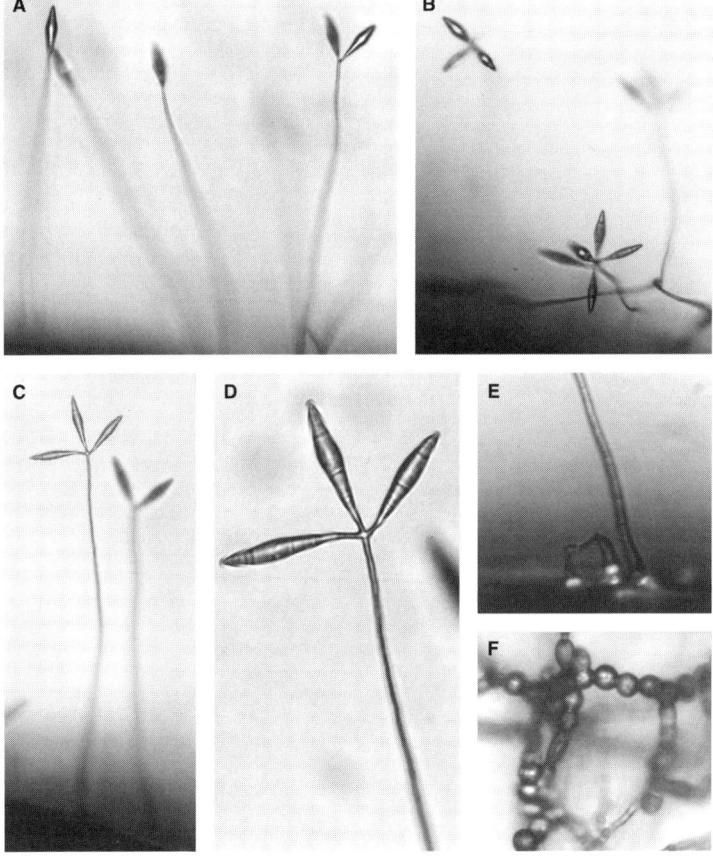

Monacrosporium sclerohyphum. A–C: Conidiophore and conidia. D: Enlargement of C. E: Conidiophore at the base. F: Catenulate chlamydospores.

Monilia Pers. : Fr.

Syst. Mycol. 3:409, 1832.
Type species: *M. fructigena* Pers. : Fr.

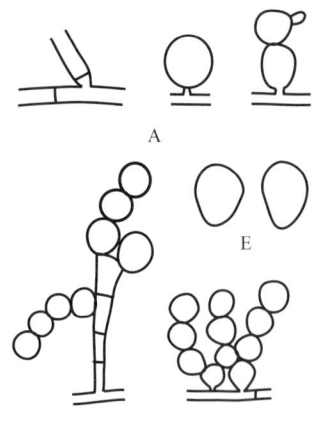

Monilia pruinosa Cooke & Masse

References: Domsch and Gams 1972; Gilman 1957; Watanabe 1975d.

Morphology: Hyphae hyaline, nonaerial, creeping, with nearly right-angled branches, constricted at basal side branches. Conidia aleuriosporous or blastosporous, borne directly on hyphae or at sterigmata extended from hyphae, hyaline, globose or subglobose, 1-celled, mainly catenulate, branched, often detachable.

Dimensions: Aggregates of monilioid cells 48.6–146 µm in diameter; component cells 12.1–19.2 µm.

Material: 70-1476 (Pineapple field soil, Hachijo, Tokyo, Japan); 72-X38 (Sugarcane root, Taiwan, ROC).

Remarks: This fungus is morphologically close to *Rhizoctonia* on the basis of hyphal morphology and catenulate conidium-like structures, except for formation of white colonies. The latter structures are just like monilioid cells in *Rhizoctonia*. Further work is necessary for determining its taxonomical position.

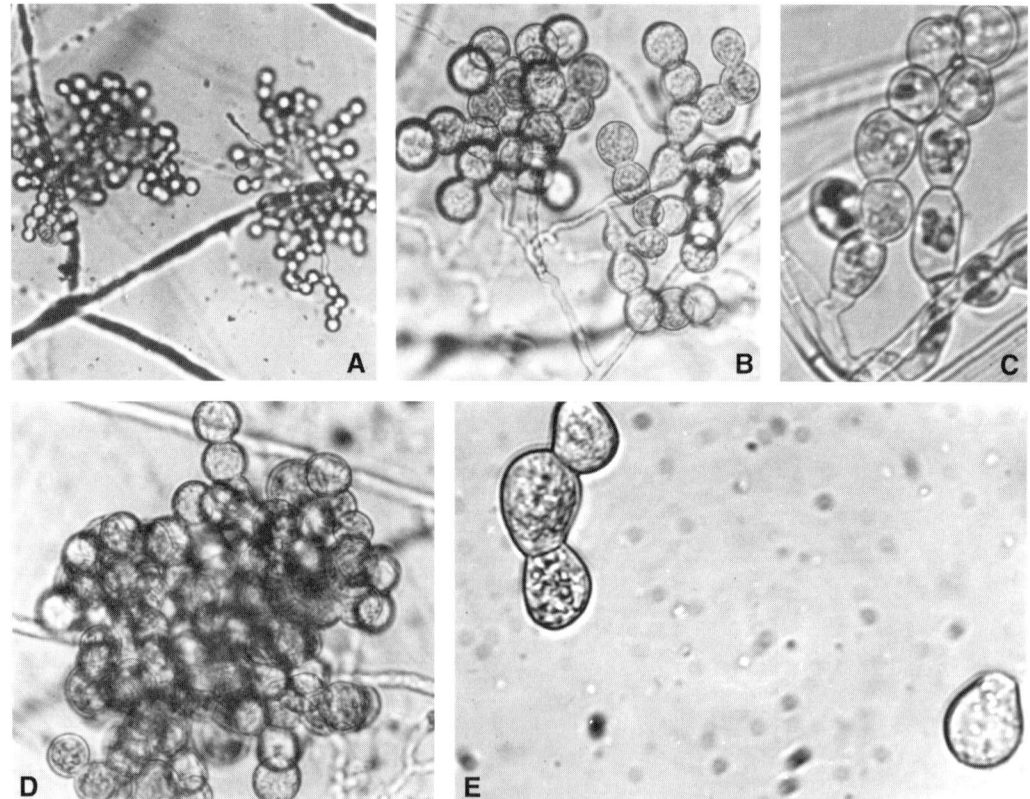

Monilia pruinosa. A–D: Hyphae and monilioid cells. E: Detached monilioid cells. (From Watanabe, T. 1975d. *Trans. Mycol. Soc. Jpn.*, 16:264–267. With permission.)

Monilia sp.

References: Domsch and Gams 1972; Gilman 1957.

Morphology: Conidiophores lacking or indistinct, not well differentiated from hyphae. Hyphae hyaline, nonaerial, creeping, bearing catenulate conidia developed by budding directly from hyphae: conidial chains simple or branched. Conidia hyaline, blastosporous, 1-celled, ellipsoidal.

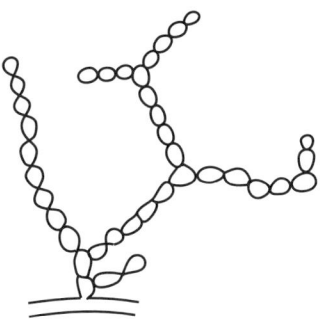

Dimensions: Conidia 8.7–12.5 × 5.5–8 µm.

Material: 77-302 (Uncultivated field soil, Nishigahara, Tokyo, Japan).

Remarks: This fungus is abundant in autoclaved field soil.

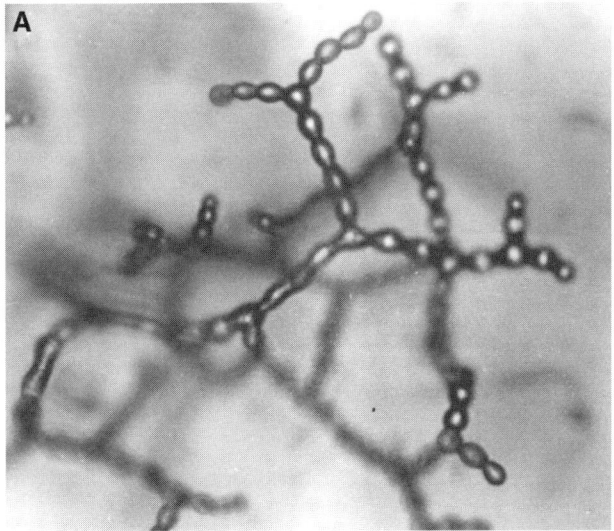
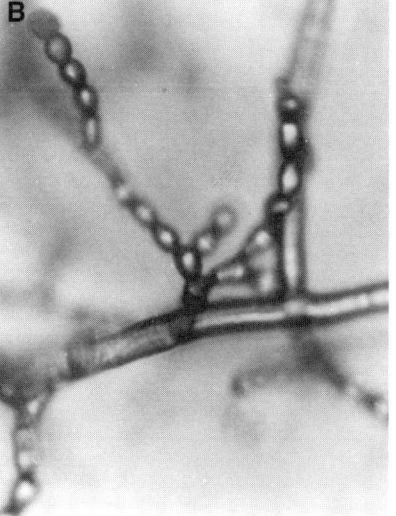

Monilia sp.   A,B: Hyphae and catenulate conidia.

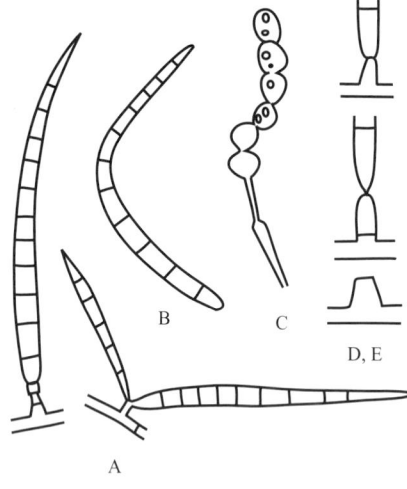

Mycocentrospora Deighton

Taxon 21:716, 1972.
Type species: *M. acerina* (Hartig) Deighton

Mycocentrospora acerina (Hartig) Deighton

Synonyms: *Centrospora acerina* (Hartig) Newhall; *Ansatospora acerina* (Ostw.) Newhall

References: Neergaard and Newhall 1951; Newhall 1946; Petersen 1962.

Morphology: Conidiophores very short or indistinct, simple or branched, not well differentiated from hyphae, bearing single conidia apically. Conidia aleuriosporous, hyaline or pale brown, cylindrical, curved, occasionally branched, attenuated toward apex, separated with tiny connecting cells from conidiophores, over 20-celled, broadest at the 2 to 4 septa. Chlamydospore-like structures globose, brown, catenulate.

Dimensions: Conidiophores 2.5–11.3 × 2–4.5 µm. Conidia 90–220 × 4.2–5 µm. Chlamydospores ca. 10 µm in diameter.

Material: 73-168, 73-244 (Strawberry root, Shizuoka, Japan); 74-849 (Strawberry root, Otone, Saitama, Japan).

Remarks: In certain isolates, a single filiform appendix is extended from the basal cells. Known as the pathogen of carrot and celery and it is also one of aquatic fungi.

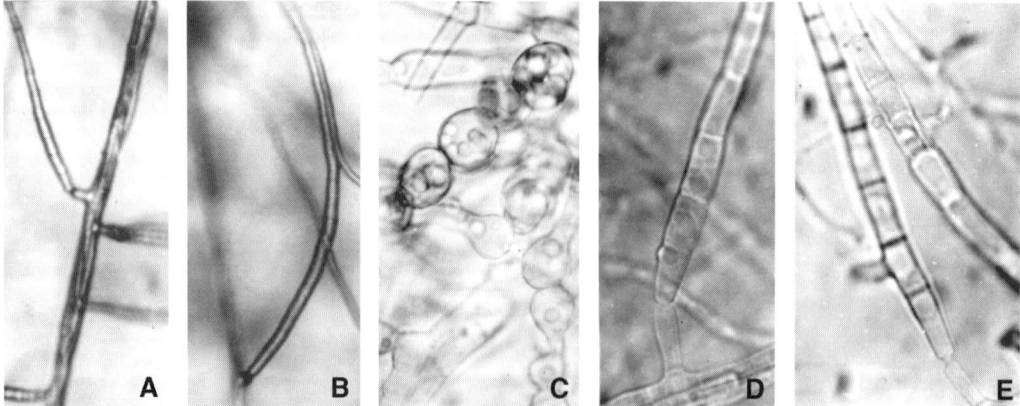

Mycocentrospora acerina.   A,B: Conidiophores and conidia. C: Chlamydospores. D,E: Conidiophores and part of conidia.

Mycoleptodiscus Ostazeski

Mycologia 59:970, 1967.

Type species: *M. terrestris* (Gerd.) Ostazeski

Synonym: *Leptodiscus terrestris* Gerdemann

Mycoleptodiscus terrestris (Gerd.) Ostazeski

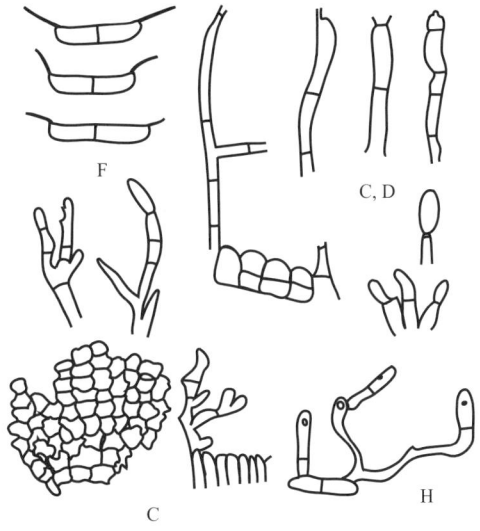

References: Alcorn 1994; Gerdemann 1953; Ostazeski 1967; Sutton and Alcorn 1990; Watanabe et al. 1996.

Morphology: Sporodochia superficial, yellowish cream-colored, composed of cells radially arranged, and simple or branched conidiophores. Conidiogenous cells phialidic or blastic, mostly branched, subhyaline to pale brown, septate, bearing conidia sympodially. Conidia hyaline, mostly boat-shaped, or cylindrical with rounded tips, one side more sharpened than the other, aseptate or 1-septate, straight or very rarely curved, bearing two filiform appendages at both ends laterally on one side, but often lacking appendages. Sclerotia black, spherical, subspherical, fusiform, irregular in shape, often elongated, or aggregated, embedded, tissues composed of hyphae with subspherical thick-walled cells. Appressoria subhyaline or pale brown, clavate or cylindrical, straight or bent, aseptate or 1-septate, single or in a short chain, commonly with a single pore.

Dimensions: Sporodochia up to 1100 µm. Conidiophores 32.5–62.5 µm tall. Conidia 25–32.5 (-33.8) × (4.8-) 5–6.3 µm: appendages 3.7–15 × 0.1–0.2 (-0.3) µm. Sclerotia up to 2 mm in diameter: tissue-component thick-walled cells 6.5–20 µm in diameter (each). Appressoria 16–30.2 × 5.4–8.7 µm.

Material: 96-SP 22 (= ATCC 200587, Black pepper roots, Sierra Prieta, Dominican Republic).

Remarks: The fungus is characterized by superficial sporodochial conidiomata, constituted mostly of thick-walled radiate stroma with dark brown conidiogenous cells, one-cell-layer thick, and a prominent circular aperture in the upper wall; hyaline, cylindrical to falcate, 0- to 2-septate conidia bearing apical and sometimes basal cellular unbranched appendages. Germ tubes from the conidia often formed characteristically clavate or cylindrical appressoria at their apexes within 5 days after treatment.

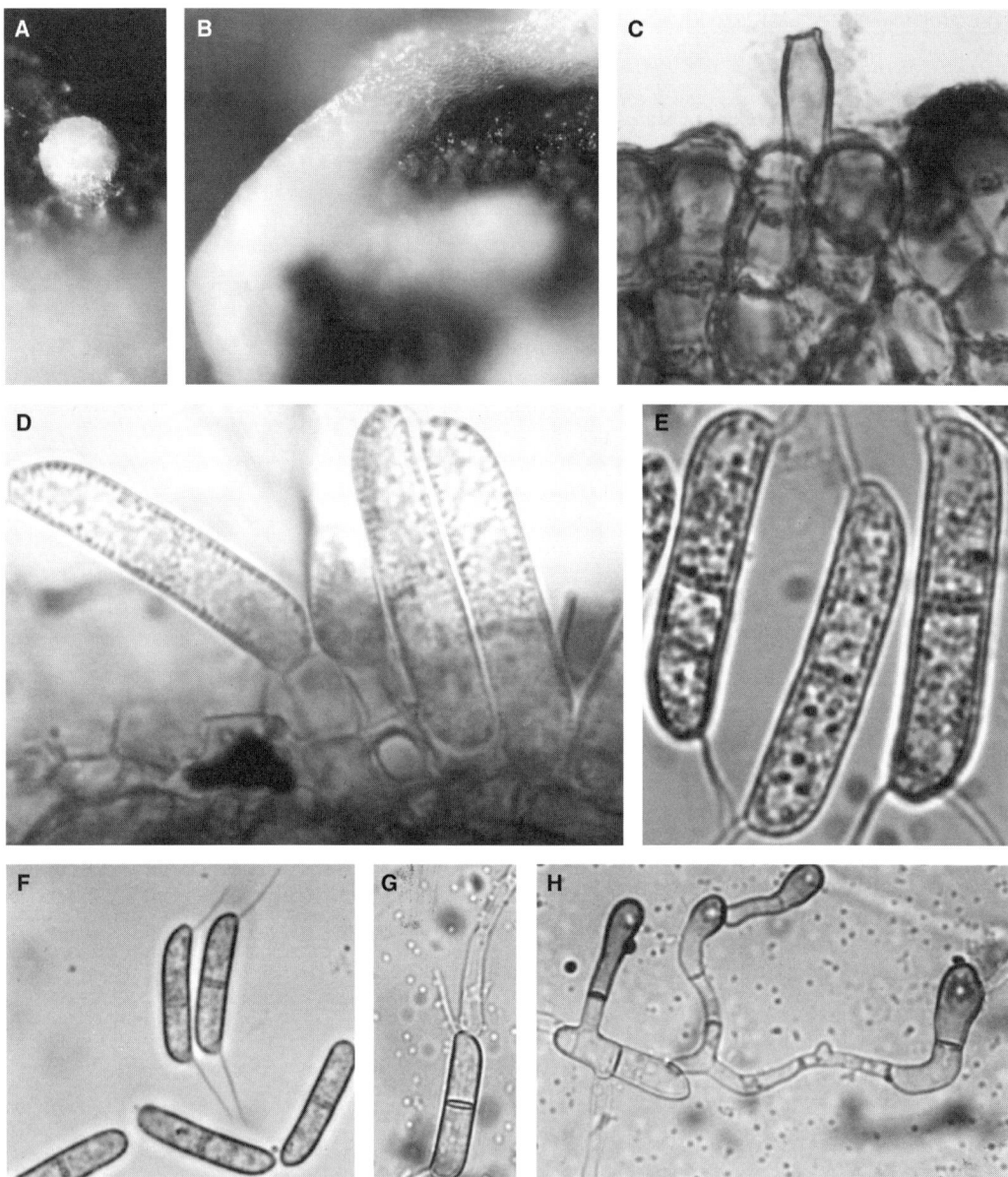

Mycoleptodiscus terrestris.   A,B: Young (A) and mature (B) sporodochia. C: Stroma with extruded phialide. D: Young conidia attached to the sporogenous cells by short stalks. E,F: Conidia. Note 1-septate (E,F) and aseptate conidia (F). G,H: Germination of conidia on wet slides 5 days after treatment at 25°C. Note formation of appressoria on elongated germ tubes (H).

Myrioconium Syd.

Ann. Mycol. 10:449, 1912.
Type species: *M. scirpicola* (Ferd. & Winge) Ferd. & Winge

Myrioconium sp.

References: Berthet 1964.

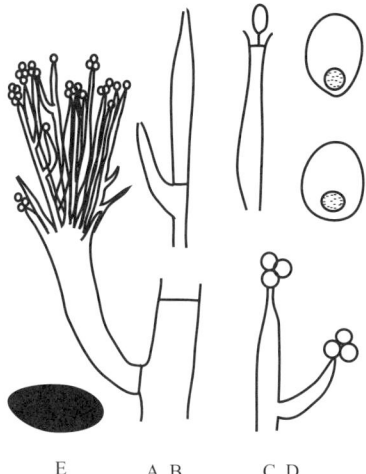

Morphology: Conidiophores united, branched irregularly, bearing spore masses at the apical phialides. Conidia pale brown, globose, apiculate at one end, with single oil globules. Sclerotia or stroma black, disc-shaped or subspherical.

Dimensions: Spore masses ca. 7–8 µm in diameter. Conidia 2–2.3 µm in diameter. Phialides ca. 14–15 × 2.5–3 µm. Sclerotia 296–371 µm in diameter.

E A, B C, D

Material: 77-97 (Strawberry root, Nishigahara, Tokyo, Japan).

Remarks: *Sclerotinia folicola* Cash & Dividson forms the anamorph like this fungus. Intrahyphal hyphae were often observed in this fungus.

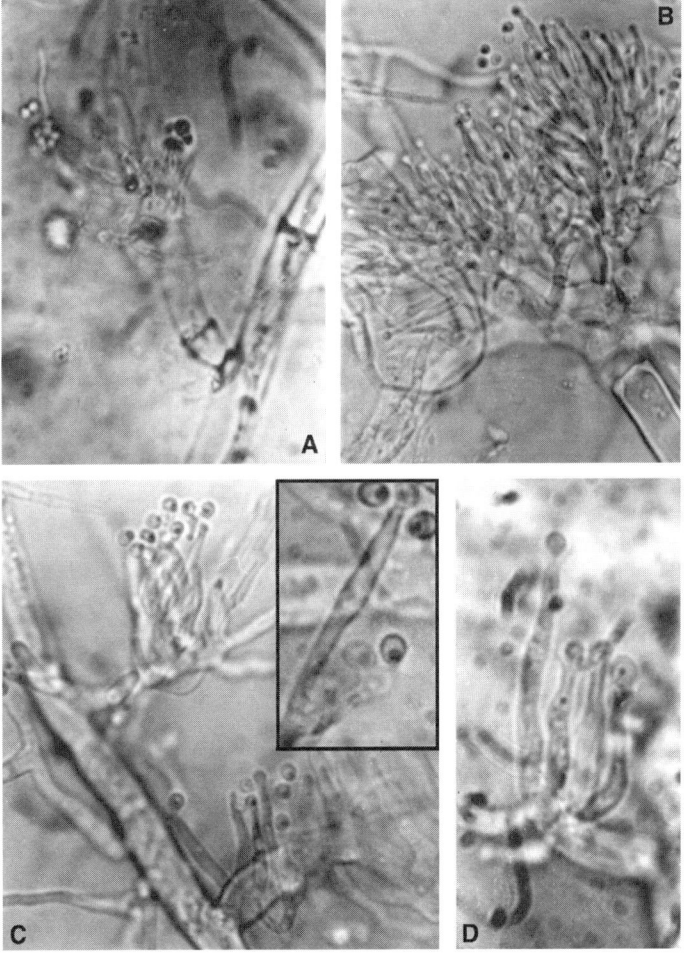

Myrioconium sp.   A–C: Conidiophores and conidia. D: Conidia and phialides. E: Sclerotium.

Myrothecium Tode : Fr.

Syst. Mycol. 3:216, 1829.
Type species: *M. roridum* Tode : Fr.

Key to Species

1. Setae or seta-like conidiophores formed.................................*M. dimorphum*
   not formed...2

2. Conidia striate......................................*M. cinctum*
   not so.......................................*M. verrucaria*

Myrothecium cinctum (Corda) Sacc.

Synonym: *Myrothecium striatisporum* Preston

References: Ellis 1971; Preston 1948; Watanabe 1975d.

Morphology: Conidiophores aggregated, forming dark green hemispherical sporodochia, branched, bearing spore masses at the apical phialides. Conidia pale green or pale yellowish green, spindle-shaped, with 8–12 stripes, 1-celled, truncate at one end, apiculate at another end.

Dimensions: Sporodochia 0.5–1.5 mm in diameter. Conidia 6.8–8.3 × 2.4–5.4 µm.

Material: 72-X136-815 (= ATCC 32918, Sugarcane root, Taiwan, ROC); 77-225 (Strawberry root, Nishigahara, Tokyo, Japan).

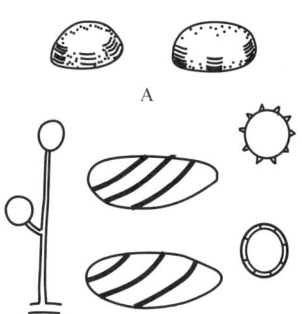

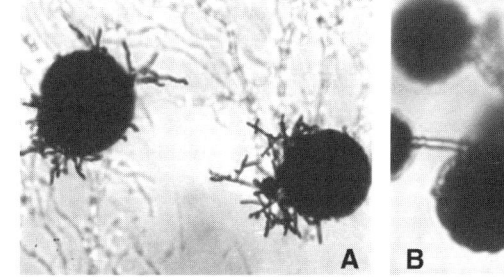
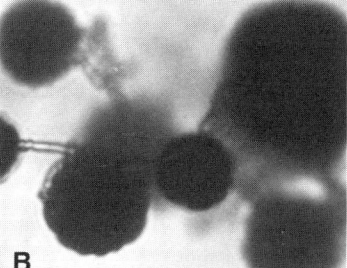
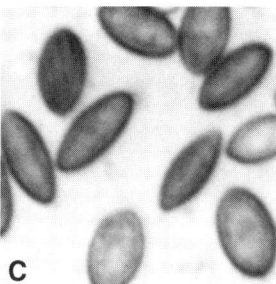

Myrothecium cinctum. A: Sporodochia. B: Conidiophores and conidia. (From Watanabe, T. 1975d. *Trans. Mycol. Soc. Jpn.*, 16:264–267. With permission.)

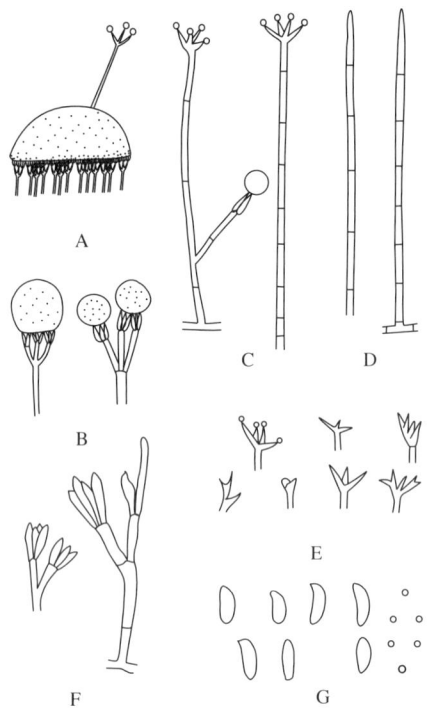

Myrothecium dimorphum T. Watanabe

Morphology: Sporodochia mixed with setae and setae-type conidiophores. Sporodochia: conidiophores erect, mostly aggregated, with biverticillate phialides with spore masses: phialides cylindrical or gradually tapered toward apexes. Setae-type conidiophores erect, simple, tall, dichotomously branched once or twice at the apexes with tiny simple conidia, usually 6–7 septate. Seta straight, simple, septate, brown, gradually tapering toward apex. Conidia phialosporous, of two kinds: hyaline or pale green ellipsoidal, curved, often appeared to be comma-shaped, 1-celled; and globose, 1-celled.

Dimensions: Sporodochial conidiophores mostly 48 μm tall, 3–4 μm wide basally: metulae 10–11 μm long: phialides 6–12 μm long. Setae-type conidiophores 200–275 μm tall, 30 μm wide basally, 1.6–2 μm apically: phialides 8 μm long. Conidia: cylindrical or comma-shaped mostly 6–12 × 2.4 μm; globose nearly 2 μm in diameter.

Material: 01-250 (Sand, Ohgiura, Chichijima, the Bonin Islands, Tokyo, Japan).

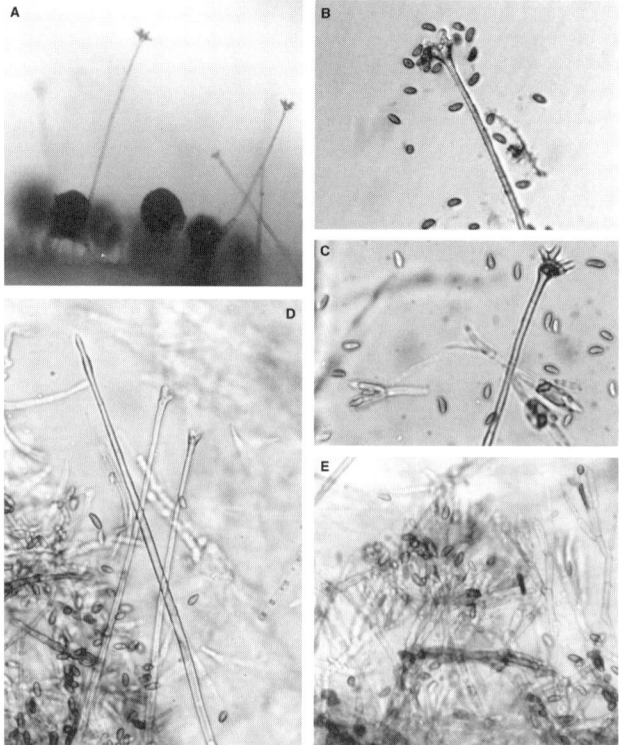

Myrothecium dimorphum. A: Sporodochia with erect seta-like conidiophores. B,C: Apical fertile portions of seta-like conidiophores bearing microconidia, macroconidia, and phialides (C). D: Seta-like conidiophore, setae, and conidia. E: Phialides and macroconidia detached. (From Watanabe, T. 1975d. *Trans. Mycol. Soc. Jpn.*, 16:264–267. With permission.)

Myrothecium verrucaria (Alb. & Schwein) Ditmar : Fr.

References: Ellis 1971; Preston 1943; Tulloch 1972; Watanabe et al. 1986a.

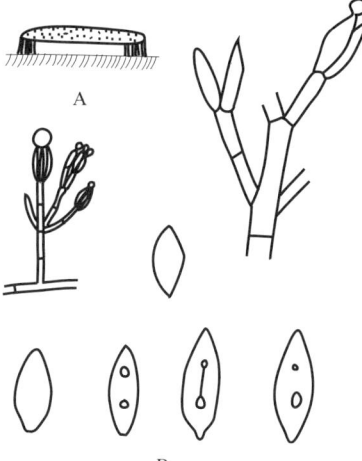

Morphology: Conidiophores usually aggregated, forming sporodochia, branched, bearing spore masses at the apical phialides. Sporodochia cylindrical, black, conspicuous with white aerial hyphae marginally. Conidia hyaline, pale brown, spindle-shaped or ellipsoidal, 1-celled, truncate at one end, apiculate at another end, occasionally guttulate.

Dimensions: Conidia 5.4–8.8 × (1.8-) 2.5–3.3 µm.

Material: 70-15 (Snap bean seed, Shinsyu-shinmachi, Nagano, Japan); 74-526 (Strawberry root, Shizuoka, Japan); 83-44, 83-45 (Japanese black pine seed, Kumano, Mie, Japan).

Remarks: Setae were not formed together with sporodochia.

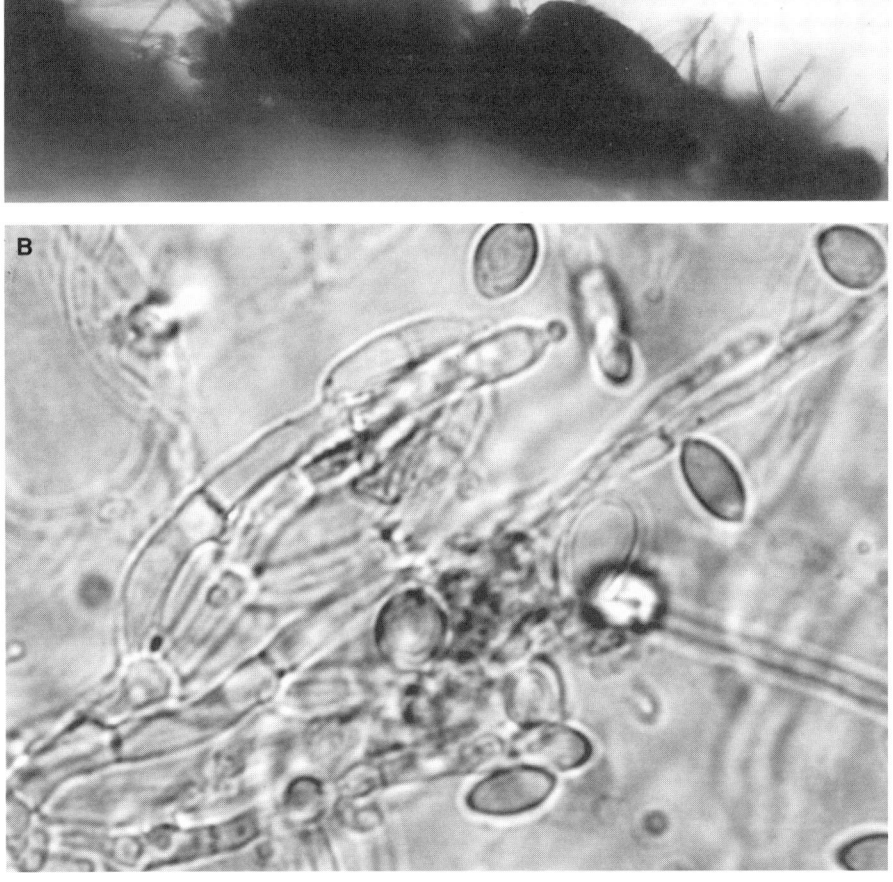

Myrothecium verrucaria. A: Sporodochia. B: Conidiophores and conidia.

Naranus T. Watanabe

Mycol. Res. Trans. 99:806, 1995.

Naranus cryptomeriae T. Watanabe

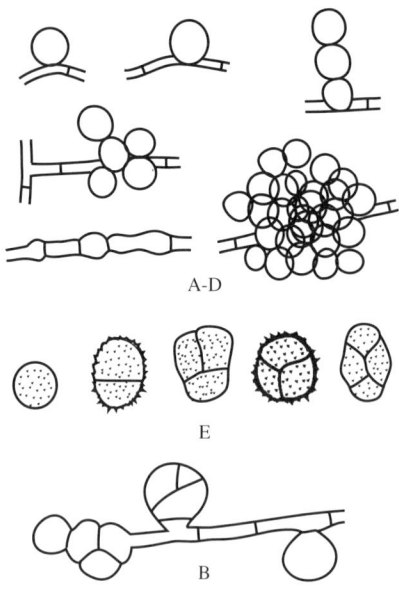

References: Watanabe 1995.

Morphology: Mycelium composed of branched, brown, septate hyphae. Conidiophores lacking. Conidiogenous cells blastic, integrated in hyphae, intercalary or terminal. Conidia blastosporous, globose or subglobose, brown or brownish, smooth, verrucose or echinulate, constricted at septum, muriform up to 4-celled, catenulate or aggregated into spherical conidial masses, readily detached separately without remains of the conidiogenous cells.

Dimensions: Hyphae, up to 5 μm wide. Conidia 6.5–13 (-15.3) μm in diameter: spherical conidial masses, nearly 70–80 μm in diameter.

Material: TW 83-83 (= MAFF 425384, Seeds of *Cryptomeria japonica*, Yoshino, Nara, Japan).

Remarks: This fungus is characterized by spherical or subspherical conidial aggregates composed of globose or subglobose brown muriform blastoconidia directly developed intercalarly on aerial hyphae.

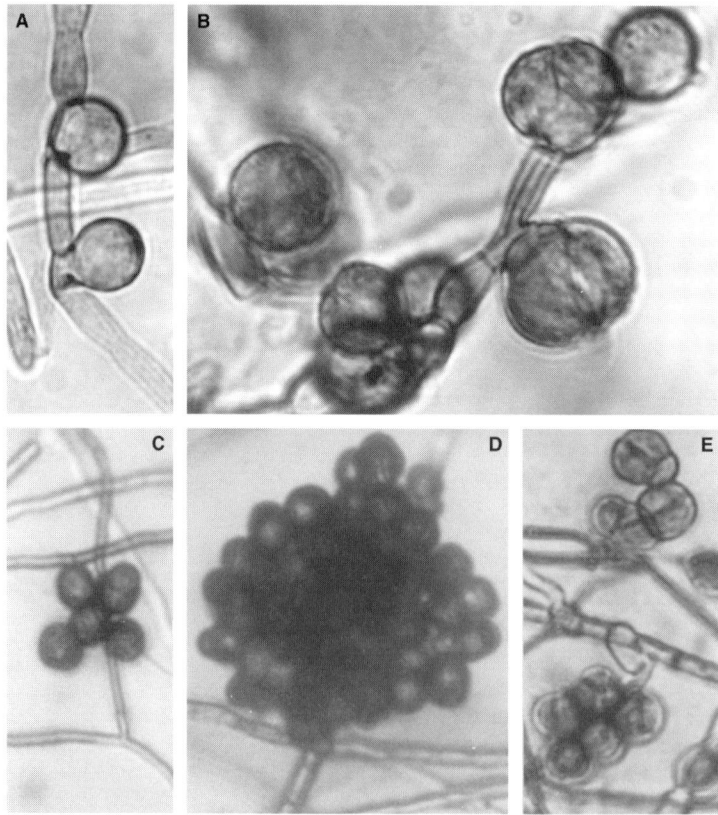

Naranus cryptomeriae. A,B: Initiation of hyphal swellings. C,D: Catenulate and aggregated conidia. E: Detached conidia. (From Watanabe, T. 1995. *Mycol. Res.*, 99:806–808. With permission.)

Neta Shearer & Crane

Mycologia 63:239, 1971.
Type species: *N. patuxentica* Shearer & Crane.

Neta quadriguttata (Mats.) de Hoog

References: Matsushima 1975; Hoog 1985.

Morphology: Two types of sporulation occurred; sympodulosporous, and aleuriosporous. In the former, conidiophores hyaline, simple, mostly short, often inflated apically, pedicelate with 1 to several conidia. Conidia hyaline, 1- to 3-septate, usually curved, allantoid, cylindrical. Aleurioconidia in the latter, borne directly on hyphae or on short aleuriophores, spherical, muriform, but often just black without visible septa.

Dimensions: Conidiophores 6–16 × 3–4 μm: pedicel 1.2–1.6 μm long. Conidia 16–22 × 4–5.2 μm.

Material: 01-481 (Kitakou, Hahajima, the Bonin Islands, Tokyo, Japan).

Remarks: In two types of sporulation, the sympodulosporous type is just like *Dactylaria*, and the aleuriosporous type is just like *Monodictys*.

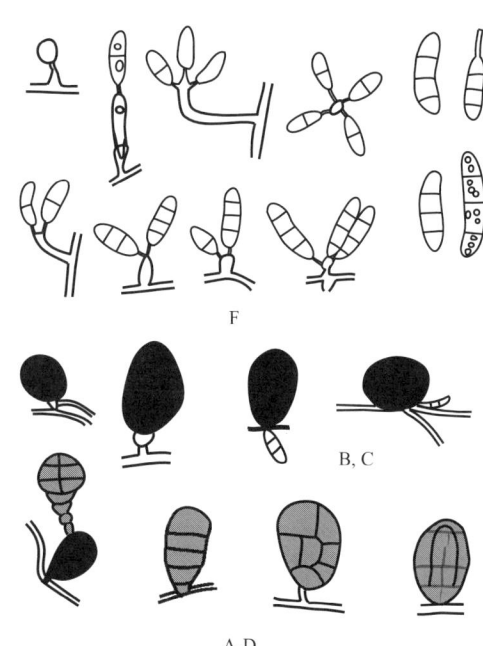

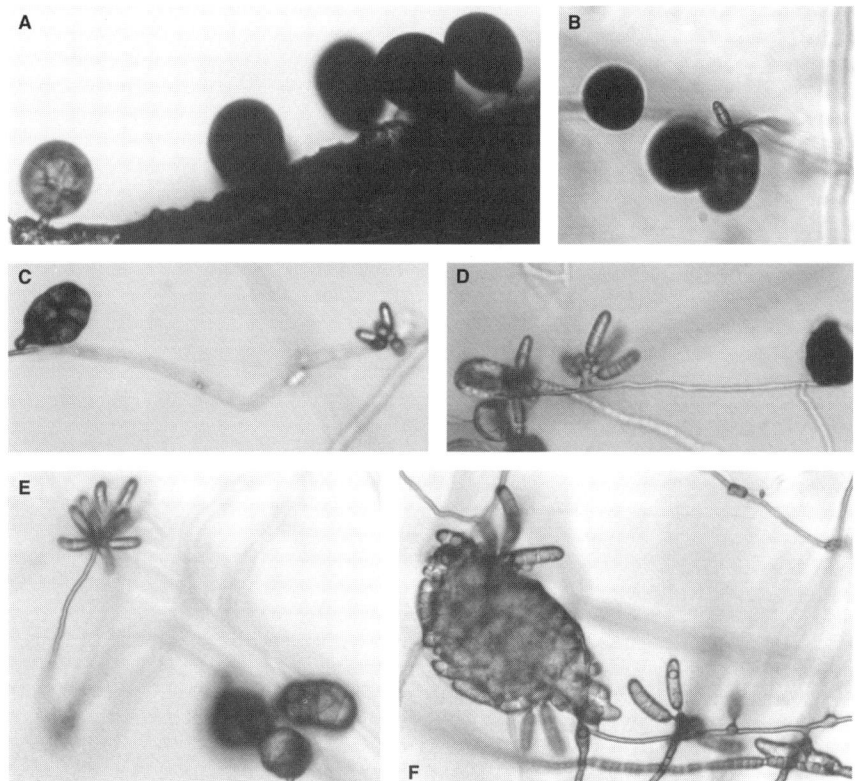

Neta quadriguttata. A: Habit showing aleurioconidia. B: Aleurioconidia and one allantoid hyaline conidium. C–E: Aleurioconidia and allantoid conidia formed on the identical hyphae. F: Mass of allantoid conidia.

Nigrospora Zimmerm.

Zentbl. Bakt. Parasitkde. Abt. 2, 8:220, 1902.
Type species: *N. panici* Zimmerm.

Nigrospora oryzae (Berkeley et Broome) Petch

Teleomorph: *Khuskia oryzae* Hudson

References: Ellis 1971; Hudson 1963; Watanabe and Tsudome 1970; Watanabe et al. 1986a.

Morphology: Conidiophores simple, hyaline, globose, bearing single conidia apically. Conidia aleuriosporous, black, subglobose or disc-shaped, occasionally apiculate in the upper part.

Dimensions: Conidia 11.2–15 (-16.3) × 7.5–10 (-11.3) µm.

Material: 69-321, 69-324 (Pineapple field soil, Okinawa, Japan); 83-63 (Japanese black pine seed, Hiba, Hiroshima, Japan); 85-45, 85-46 (Flowering cherry seed, Hachioji, Tokyo, Japan); 86-42, 86-43 (Japanese cypress seed, Kasama, Ibaraki, Japan).

Remarks: Five species have been differentiated on the basis of conidial dimensions in the genus *Nigrospora,* with a certain degree of duplication to one another (Domsch et al., 1980a).

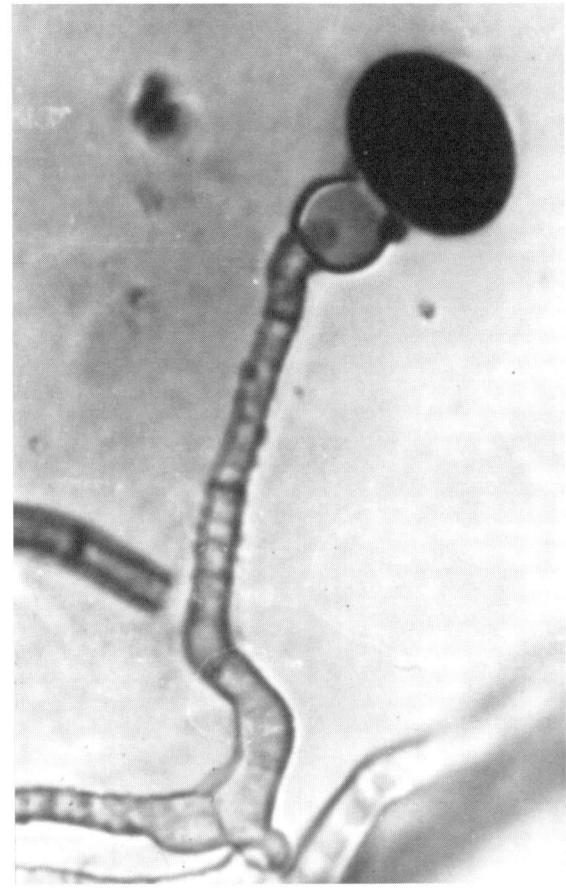

Nigrospora oryzae. Conidiophore and conidium. (From Watanabe, T. and Tsudome, K. 1970. *Trans. Mycol. Soc. Jpn.*, 11:64–71. With permission.)

Nigrospora sacchari (Speg.) Mason

References: Ellis 1971; Watanabe et al. 1986a.

Morphology: Conidiophores simple, hyaline, globose, bearing single conidia apically. Conidia aleuriosporous, black, subglobose or disc-shaped, occasionally apiculate in the upper part.

Dimensions: Conidia 15–22.5 (-25) × 12.5–17.5 µm.

Material: 70-1841 (Pineapple field soil, Hachijo, Tokyo, Japan); 84-509 (Japanese black pine seed, Okawa, Kagawa, Japan).

Remarks: In the isolate 84-509, conidia are occasionally over 20 µm in diameter, but in the isolate 70-1841, usually under 17 µm. The latter fungus is close to *N. sphaerica* (Sacc.) Mason.

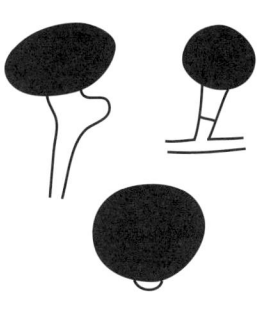

A, B

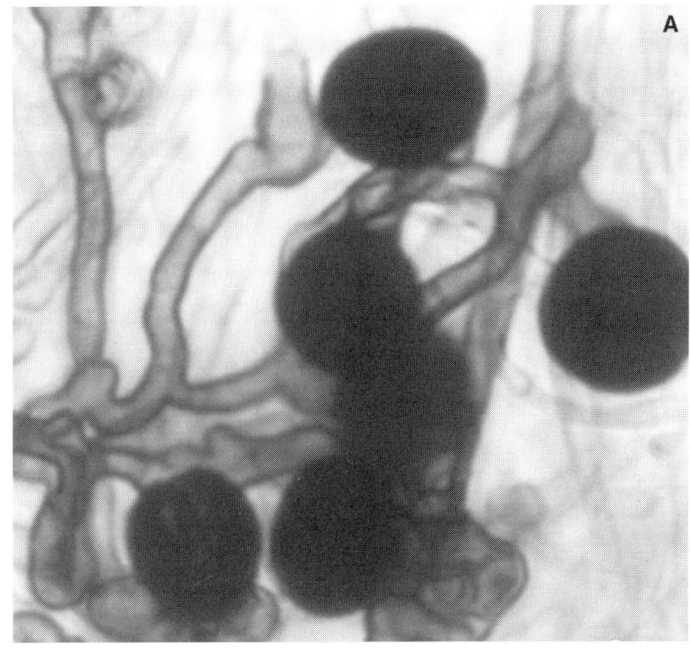

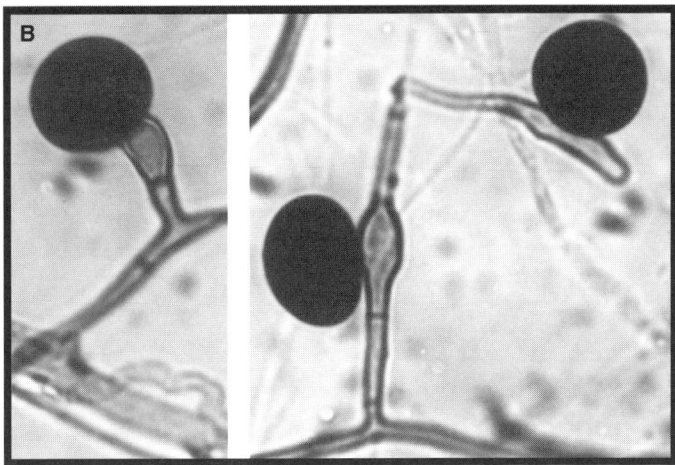

Nigrospora sacchari. A,B: Conidiophores and conidia (A: Isolate 84-509; B: 70-1841).

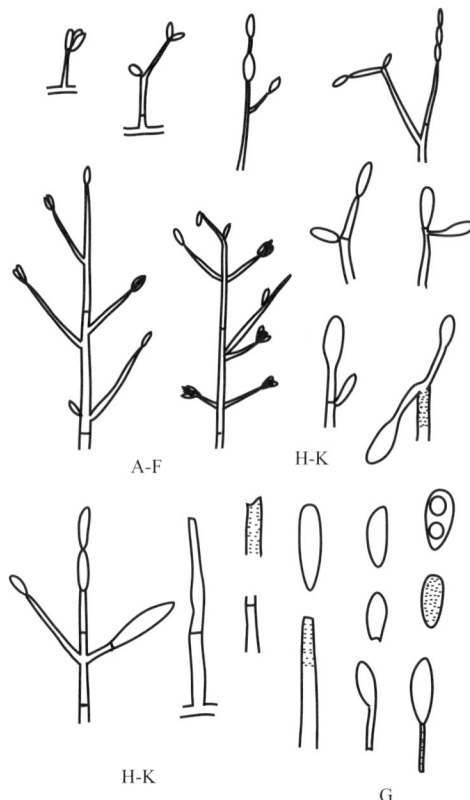

Nodulisporium Preuss

Herb. Viv. Mycol. no. 1272, 1849.
Type species: *N. ochraceum* Preuss.

Nodulisporium melonis T. Watanabe & M. Sato

References: Barron 1968; Ellis 1971; Hoog and Hermanides-Nijhof 1977; Smith 1962; Watanabe and Sato 1995.

Morphology: Conidiophores erect, hyaline, verticillate, or irregularly branched apically, bearing a few conidia at sterigmata on apical fertile portions, denticulate after detachment of conidia, often proliferated from conidia on conidiophores. Conidia sympodulosporous, hyaline, 1-celled, ellipsoidal or irregular in shape, with scars or part of conidiophores basally.

Dimensions: Conidiophores 103–140 μm tall, 1.2–2.3 μm wide basally, 0.7–1.8 μm wide apically. Conidia 2.5–13.8 × 1.2–3.0 μm: appendages 3–10 × 0.4–1.0 μm.

Material: 75-355 (Melon root, Shizuoka, Japan).

Remarks: This fungus is morphologically close to *Hansfordia* or *Calcalisporium*, the anamorph of *Xylaria* or *Hypoxylon*.

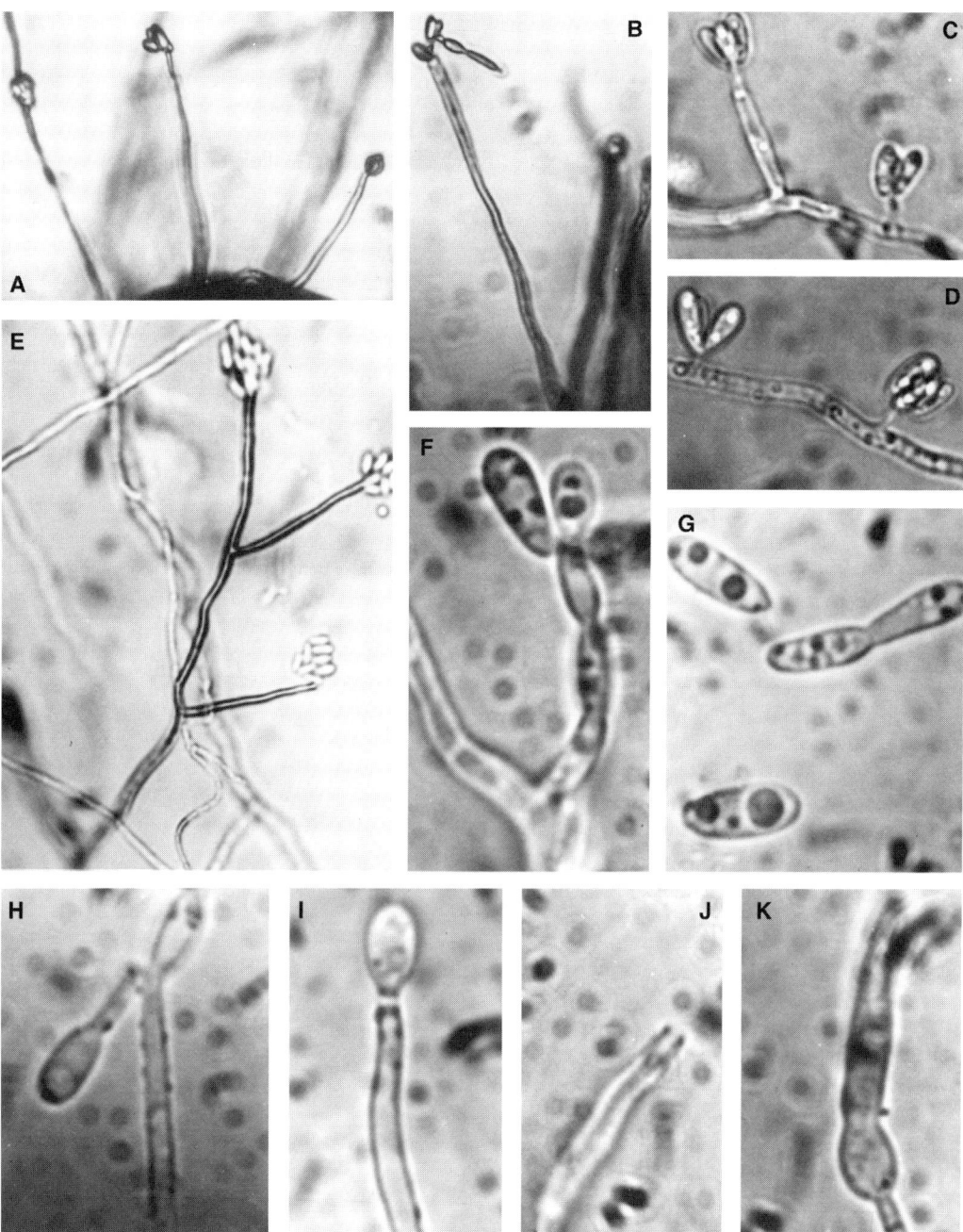

Nodulisporium melonis. A–F: Conidiophores and conidia. G: Conidia. H–K: Apical fertile portions of conidiophores. (From Watanabe, T. and Sato, M. 1995. *Ann. Phytopathol. Soc. Jpn.*, 61:330–333. With permission.)

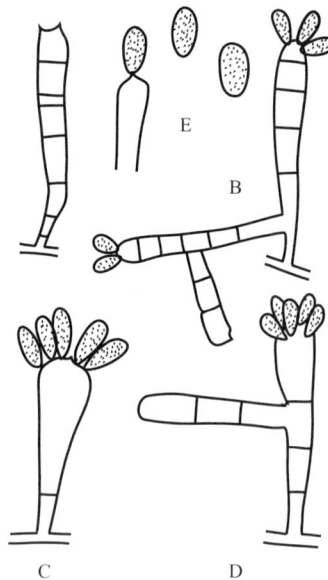

Oedocephalum Preuss

Linnaea 24:131, 1851.
Type species: *O. elegans* Preuss

Oedocephalum nayoroense T. Watanabe

References: Baksi 1950, 1951; Dodge 1937; Hughes 1953a; Stalpers 1974; Thaxter 1891; Watanabe 1991.

Morphology: Conidiophores erect, simple or rarely branched, clavate, bearing 1–16 conidia at sterigmata borne on apical fertile portions, on which appear basidia and basidiospores. Conidia yellowish brown or brown, ellipsoidal, rough or minutely echinulate on the surface, 1-celled. No clamp connection present on hyphae.

Dimensions: Conidiophores 92.5–195 × 12.5–20 µm. Conidia 11.5–15 × 6–7.5 µm.

Material: 81-498 (= IFO 32546, Potato field soil, Nayoro, Hokkaido, Japan).

Remarks: This fungus is also close to the genus *Spiniger* morphologically.

Deuteromycotina (Motosporic Fungi)

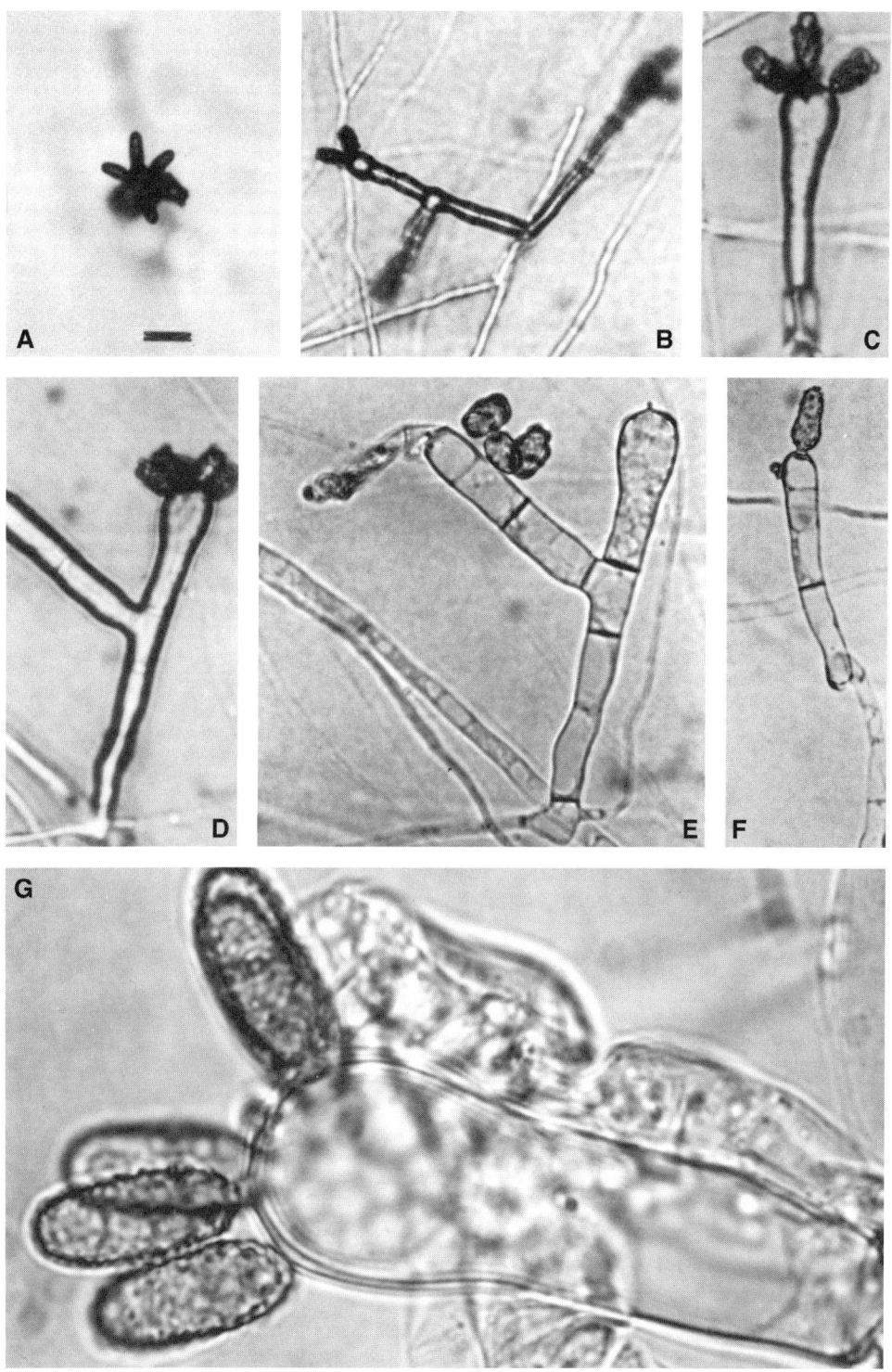

Oedocephalum nayoroense. A–G: Conidiophores and conidia. (From Watanabe, T. 1991. *Mycologia*, 83:524–529. With permission.)

Oidiodendron Robak

Nyt. Mag. Naturvid. 71:243, 1932.
Type species: *O. fuscum* Robak = *O. tenuissimum* (Peck) Hughes

Key to Species

1. Conidia thick-walled, lenticular. .... *O. cerealis*
   not thick-walled, not lenticular 2

2. Conidia cylindrical. .... *O. citrinum*
   globose, various in shape. .... *O. flavum*

Oidiodendron cerealis (Thum) Barron

References: Barron 1962; Ellis 1971; Watanabe et al. 1986a.

Morphology: Conidiophores brown, erect, branched oppositely or alternately in the median, bearing catenulate conidia basipetally developed at the apex, apparently dendroid-like, denticulate after detachment of conidia. Conidia sympodulosporous, brown, thick-walled, globose, lenticular or disc-shaped surrounded by ring, 1-celled.

Dimensions: Conidiophores ca. 24 µm tall. Conidia 2.1–3.8 µm in diameter; 1.8–2 µm thick.

Material: 84-506, 84-507, 84-572 (Japanese black pine seed, Okawa, Kagawa, Japan); 84-606 (Japanese red pine seed, Kamiina, Nagano, Japan).

A, B

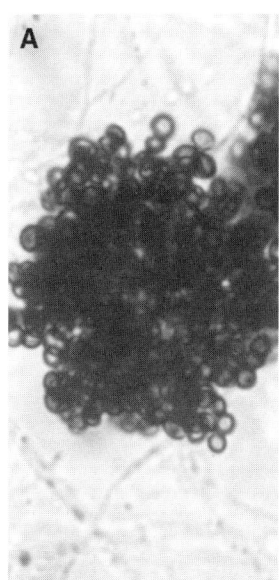 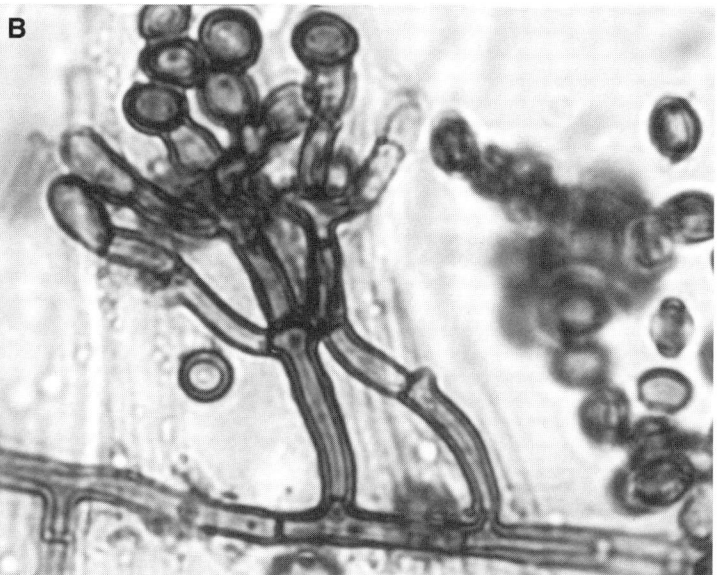

Oidiodendron cerealis. A,B: Conidiophores and conidia.

Oidiodendron citrinum Barron

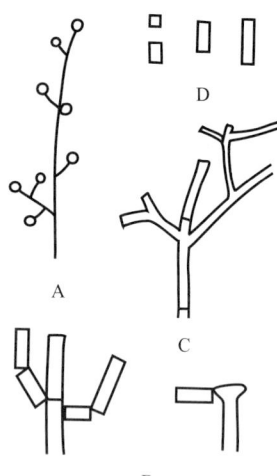

References: Barron 1962; Watanabe et al. 1987a.

Morphology: Conidiophores erect, branched sympodially in the median, bearing catenulate conidia apically, with crevice among catenulate conidia, apparently dendroid-like. Conidia arthrosporous, hyaline, cylindrical, 1-celled.

Dimensions: Conidiophores ca. 160 μm tall. Spore masses ca. 10 μm in diameter. Conidia (1.7-) 3.7–5 (-7) × 1.3–2 μm.

Material: 85-P115 (Paulownia root, Misiones, Argentina).

Remarks: Colonies on PDA are resupinate, whitish yellow.

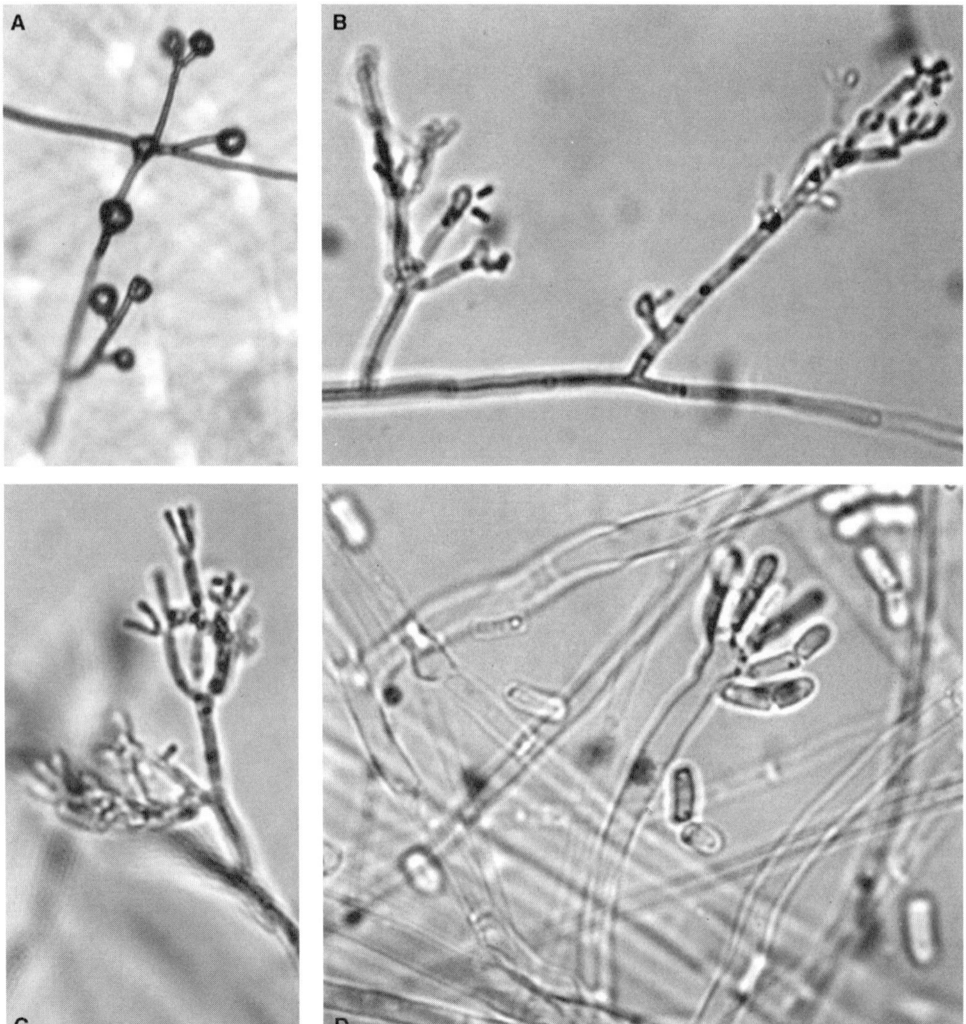

Oidiodendron citrinum. A–D: Conidiophores and conidia.

Oidiodendron flavum Szilvinyi

References: Barron 1962; Watanabe et al. 1986a.

Morphology: Conidiophores brown, erect, branched alternately, oppositely or rarely verticillately in the median, bearing conidia apically, apparently dendroid-like, denticulate after detachment of conidia. Conidia arthrosporous, hyaline, globose or irregular, occasionally with a fragment of conidiophores, 1-celled, readily detached.

Dimensions: Conidiophores 62.5–87.5 µm tall, 20–37.5 µm tall from base to branching site. Conidia 1.8–3.8 (-6.3) µm in diameter.

Material: 84-475, 84-571 (Japanese black pine seed, Higashichikuma, Nagano, Japan); 84-570, 84-579, 84-610 (Japanese red pine seed, Kamiina, Nagano, Japan); 85-117 (Flowering cherry seed, Hachioji, Tokyo, Japan).

Remarks: Conidia of Isolate 85-117 are mostly 6–6.3 µm in diameter. Colonies on PDA are limited in growth, yellowish brown, nonaerial, and partially ropy.

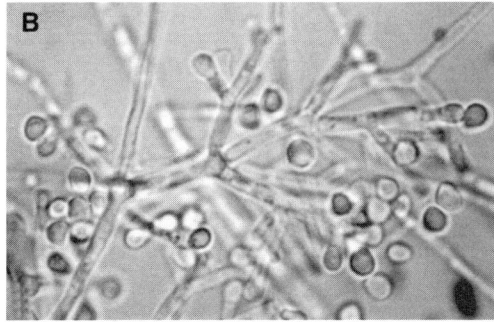

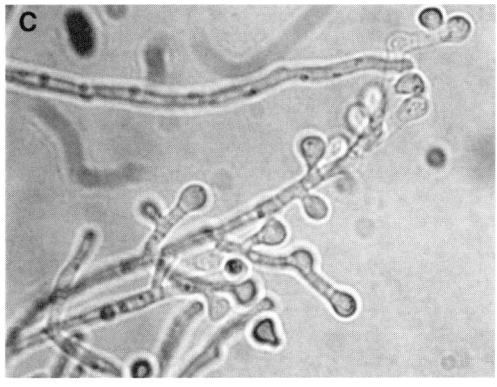

Oidiodendron flavum. A–C: Conidiophores and conidia.

Paecilomyces Bain.

Bull. Soc. Mycol. Fr. 23:26, 1907.
Type species: *P. variotii* Bain.

Key to Species

1. Phialides   verticillate. . . . . *P. puntonii*
                    nonverticillate. 2

2. Conidia   mainly spindle-shaped, over 5 μm long 3
                     not so 4

3. Conidia   over 2.5 μm wide. . . . . *P. roseolus*
                     under 2.3 μm wide. . . . . *P. javanicus*

4. Conidia   globose or subglobose 5
                     not so 6

5. Conidia   of two kinds: subglobose and ovate, over 3.2 μm long . . . . *P. variabilis*
                     subglobose, under 3 μm long . . . . *P. victoriae*

6. Phialides   short and thick, with broader median . . . . *P. inflatus*
                     slender 7

7. Colonies   yellowish brown. . . . . *P. farinosus*
                     pinkish . . . . *P. perscinus*

Paecilomyces farinosus (Holm : Gray) A. H. S. Brown & G. Smith

Teleomorph: *Cordyceps memorabilis* Ces.

References: Brown and Smith 1957; Domsch et al. 1980a,b.

Morphology: Conidiophores (phialides) simple or rarely branched, erect, hyaline, 1-septate basally, tapering from base toward apex, bearing over 10 catenulate conidia apically. Conidia phialosporous, hyaline, ovate, 1-celled, slightly apiculate at one end.

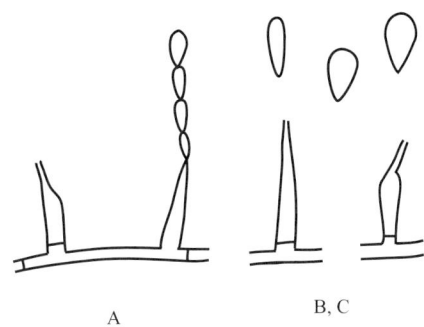

Dimensions: Conidiophores 12.5–20 × 2–2.5 µm. Conidia 2–3.8 × 1–1.5 µm.

Material: 74-669 (Strawberry root, Shizuoka, Japan).

Remarks: Colonies on agar are fluffy, yellowish brown.

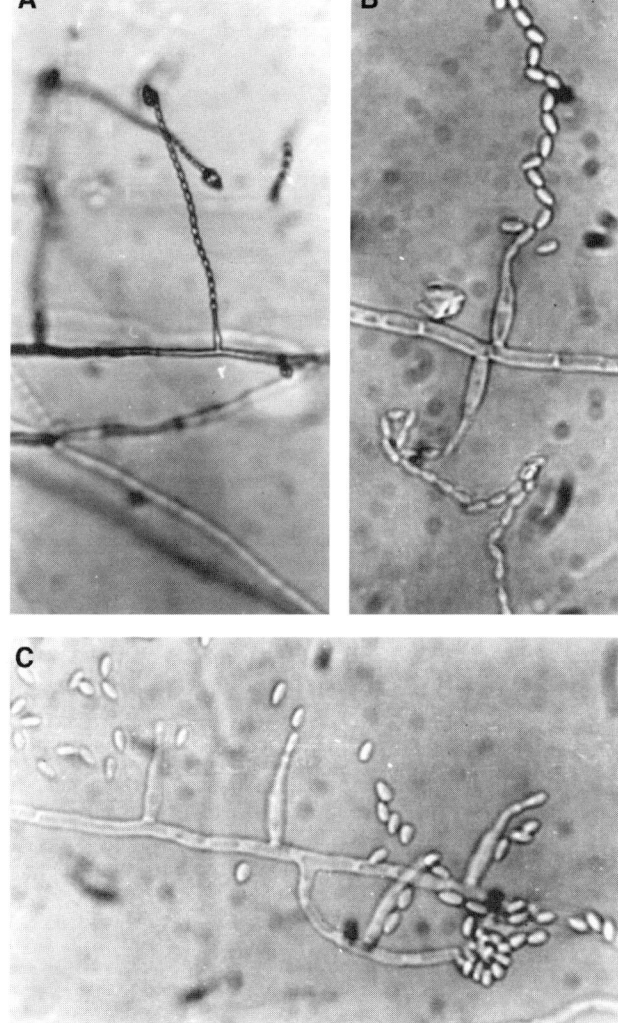

Paecilomyces farinosus. A–C: Conidiophores and conidia.

Paecilomyces inflatus (Burnside) Carmichael

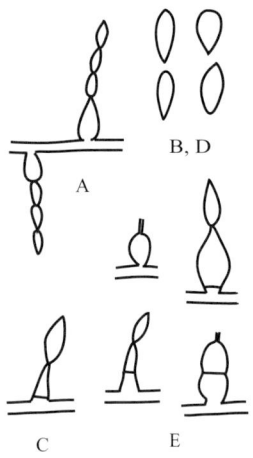

References: Onions and Barron 1967; Watanabe 1975d.

Morphology: Conidiophores (phialides) hyaline, erect, simple, short, thick in the median, rarely 1-septate, constricted basally, tapering toward apex, bearing catenulate conidia apically. Conidia phialosporous, terminal, hyaline, ovate, spindle-shaped, minutely echinulate, 1-celled, apiculate at one end.

Dimensions: Conidiophores 6–13.4 × 2.9–3.9 μm. Conidia 4.8–5.6 × 2.4–3.7 μm.

Material: 72-X30-142 (= ATCC 32919, Sugarcane root, Taiwan, ROC).

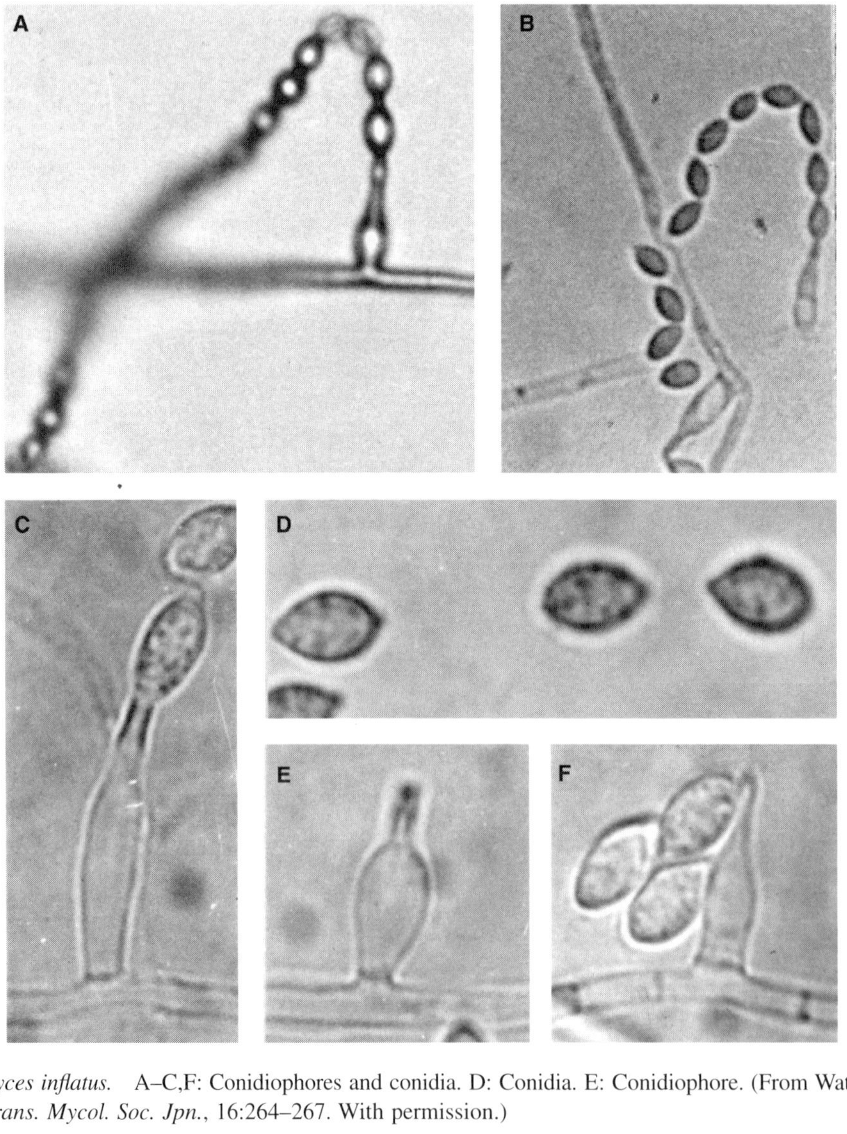

Paecilomyces inflatus. A–C,F: Conidiophores and conidia. D: Conidia. E: Conidiophore. (From Watanabe, T. 1975d. *Trans. Mycol. Soc. Jpn.*, 16:264–267. With permission.)

Paecilomyces javanicus (Friedrichs et Bally) Brown & Smith

References: Brown and Smith 1957.

Morphology: Conidiophores hyaline erect, simple or branched, bearing catenulate conidia at opposite or verticillate apical phialides: cylindrical, gradually tapering from the median toward the end. Conidia phialosporous, hyaline, spindle-shaped, 1-celled, apiculate at both ends.

Dimensions: Conidiophores 12.5–50 × 2.7–3.8 μm: phialides 14–25 × 2.5–3.3 μm. Conidia 5–6.5 × 2–2.3 μm.

Material: 73-447 (Strawberry field soil, Shizuoka, Japan).

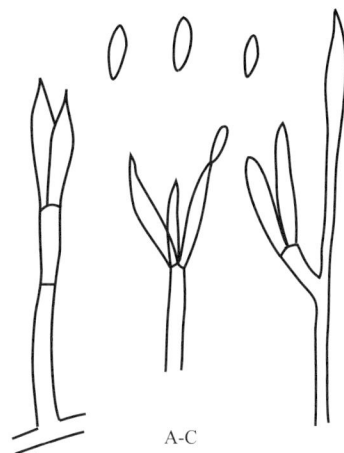

A-C

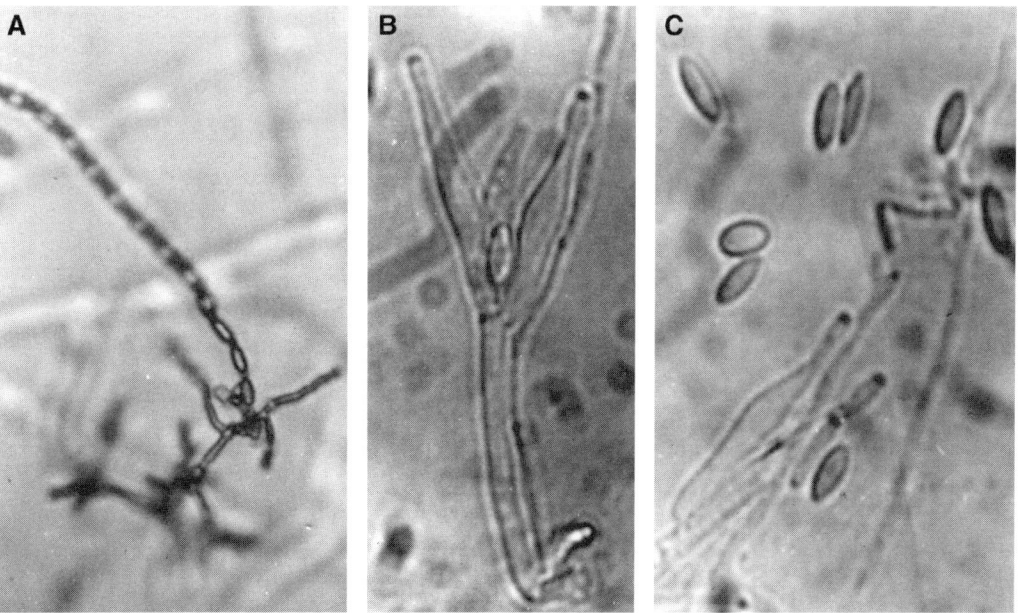

Paecilomyces javanicus.   A,B: Conidiophores and conidia. C: Phialides and conidia.

Paecilomyces persicinus Nicot

References: Onions and Barron 1967; Watanabe 1975d.

Morphology: Conidiophores (phialides) hyaline, erect, simple or branched bearing catenulate conidia or spore masses apically. Conidia phialosporous, hyaline, ovate, 1-celled, minutely echinulate, truncate or apiculate at one end. Occasionally, specific hyphal cells observed that are chlamydospore-like, constricted in the middle, and inflated at both ends.

Dimensions: Conidiophores 11.2–40 × 2–2.5 µm. Conidia 2.5–3 × 1.2–1.8 µm.

Material: 72-X101-489 (= ATCC 32920, Sugarcane root, Taiwan, ROC).

Remarks: Colonies in agar cultures are pinkish.

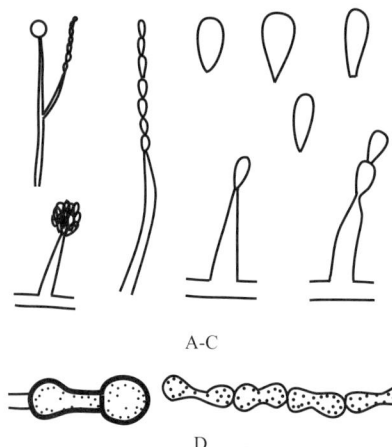

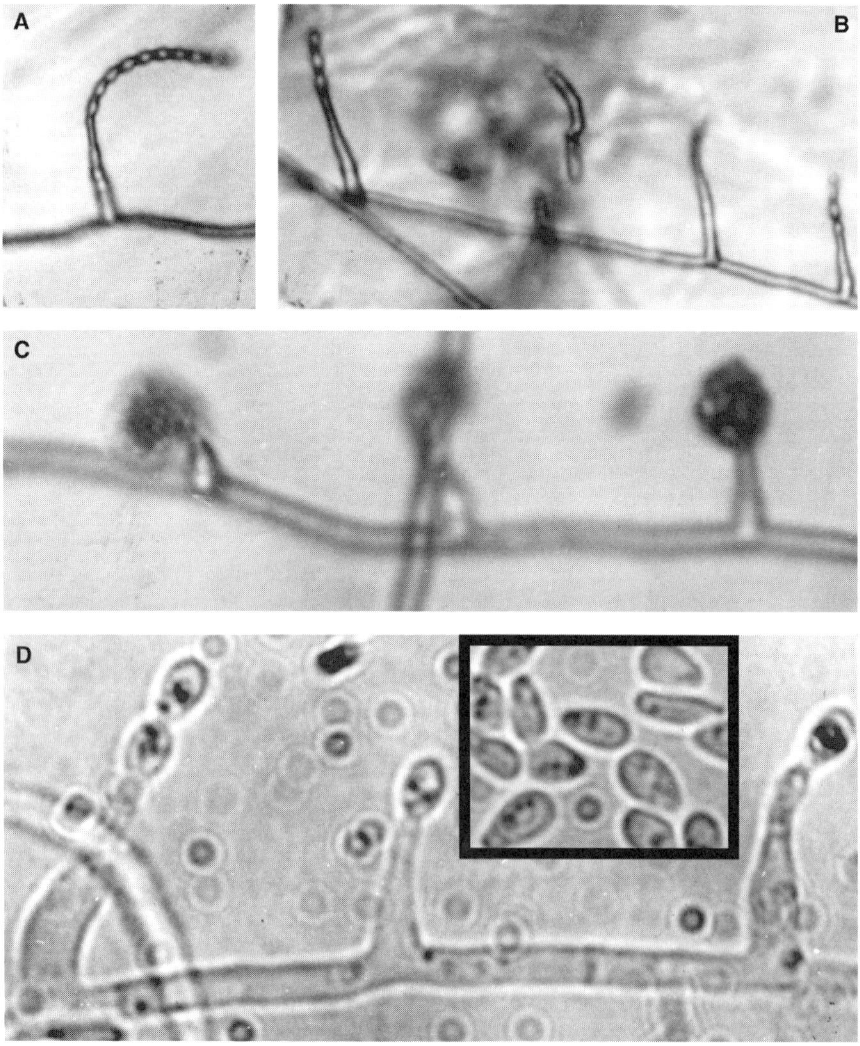

Paecilomyces persicinus. A–C: Conidiophores and conidia. D: Chlamydospore-like structures. (From Watanabe, T. 1975d. *Trans. Mycol. Soc. Jpn.*, 16:264–267. With permission.)

Paecilomyces puntoni (Vuill.) Nannizzi

References: Raper and Thom 1949; Samson 1974.

Morphology: Conidiophores hyaline, erect, branched with verticillate phialides in the apical 2 or 3 fertile portions, bearing over 20 conidia in each phialide. Conidia phialosporous, hyaline, spindle-shaped, ovate to ellipsoidal. Hyphae and conidiophores often aggregated, forming synnemata.

Dimensions: Conidiophores 100–310 × 4.5–7.5 µm: phialides 12.5–26.3 × 2.7–3 µm. Conidia 4.5–6.3 × 2.5–2.8 µm.

Material: 70-1233 (Pineapple root, Hachijo, Tokyo, Japan).

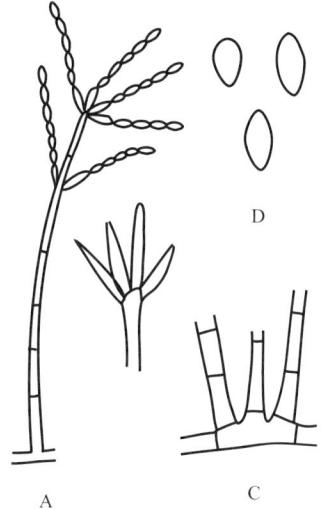

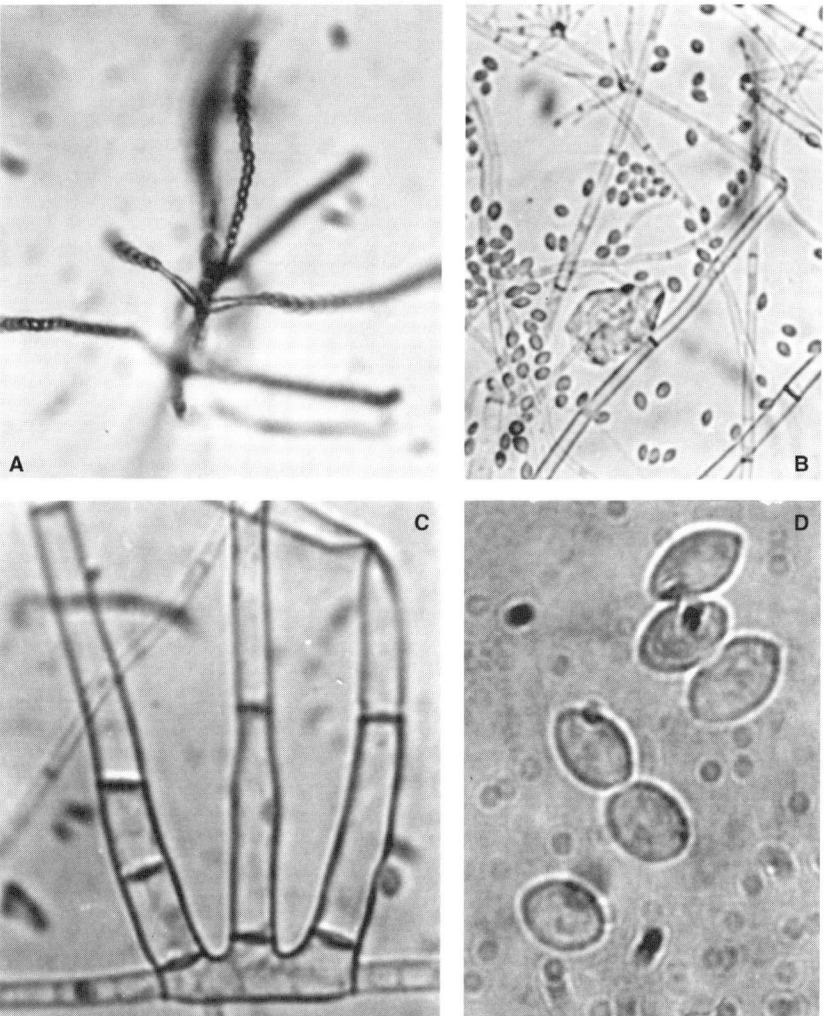

Paecilomyces puntoni. A,B: Conidiophores and conidia. C: Basal parts of conidiophores. D: Conidia.

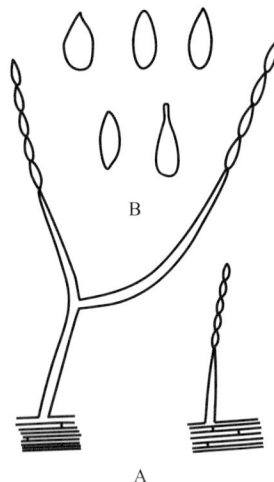

Paecilomyces roseolus Smith

Synonym: *Acremonium roseolum* (Smith) Gams

References: Onions and Barron 1967; Smith 1962; Watanabe et al. 1986a.

Morphology: Conidiophores hyaline, erect, simple or branched in the median, bearing catenulate conidia apically on branches (phialides). Conidia phialosporous, hyaline, spindle-shaped, ovate, pear-shaped, lemon-shaped, 1-celled, pointed at an apex. Hyphal strands composed of united hyphae formed frequently.

Dimensions: Conidiophores 12.5–55.8 × 2.5 µm. Conidia 5.2–7 × 2.5–2.8 µm.

Material: 84-510 (Japanese red pine seed, Hiba, Hiroshima, Japan); 86-44 (Japanese cedar seed, Nakashinkawa, Toyama, Japan).

Remarks: Colonies on PDA are pinkish, nonaerial.

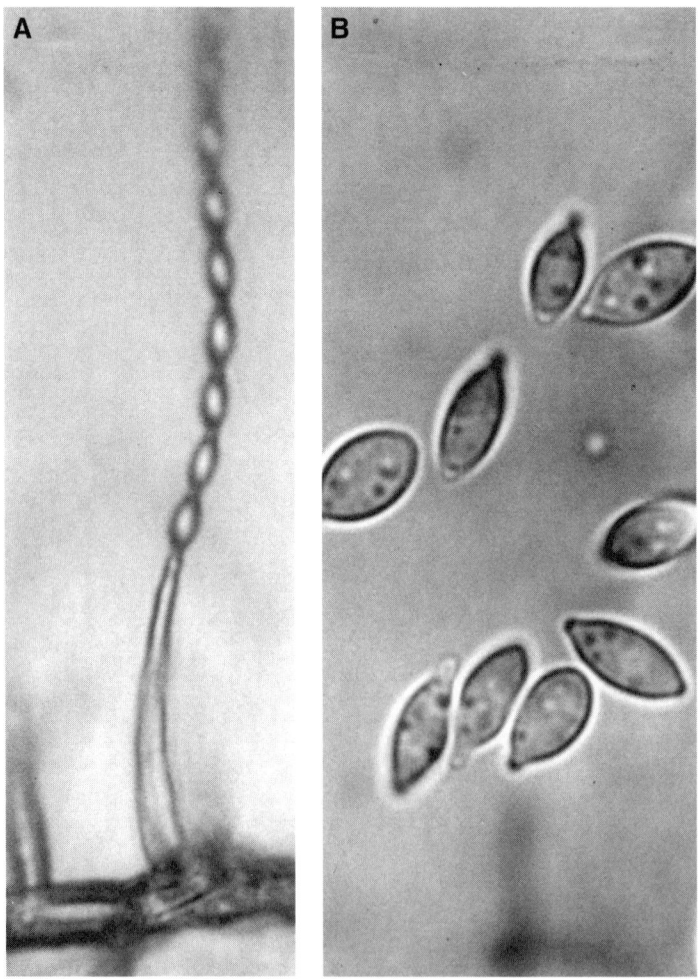

Paecilomyces roseolus.   A: Conidiophore and catenulate conidia. B: Conidia.

Paecilomyces variabilis Barron

References: Barron 1961; Onions and Barron 1967; Watanabe 1975d.

Morphology: Conidiophores (phialides) hyaline, erect, gradually tapering from base toward apex, 0- to 3-septate, bearing catenulate conidia apically, often developed from hyphal strands. Conidia phialosporous, globose, subglobose, ovate, hyaline, 1-celled, especially minutely echinulate in globose conidia.

Dimensions: Conidiophores 23–43.8 × 2.1–2.7 µm. Conidia: ovate 3.2–4.9 × 2.4–3 µm, and globose 3.6–4.7 µm in diameter.

Material: 72-X95-54 (= ATCC 32921, Sugarcane root, Taiwan, ROC).

Remarks: This fungus characteristically forms the two kinds of conidia and mycelial strands.

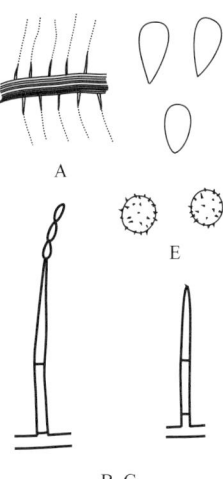

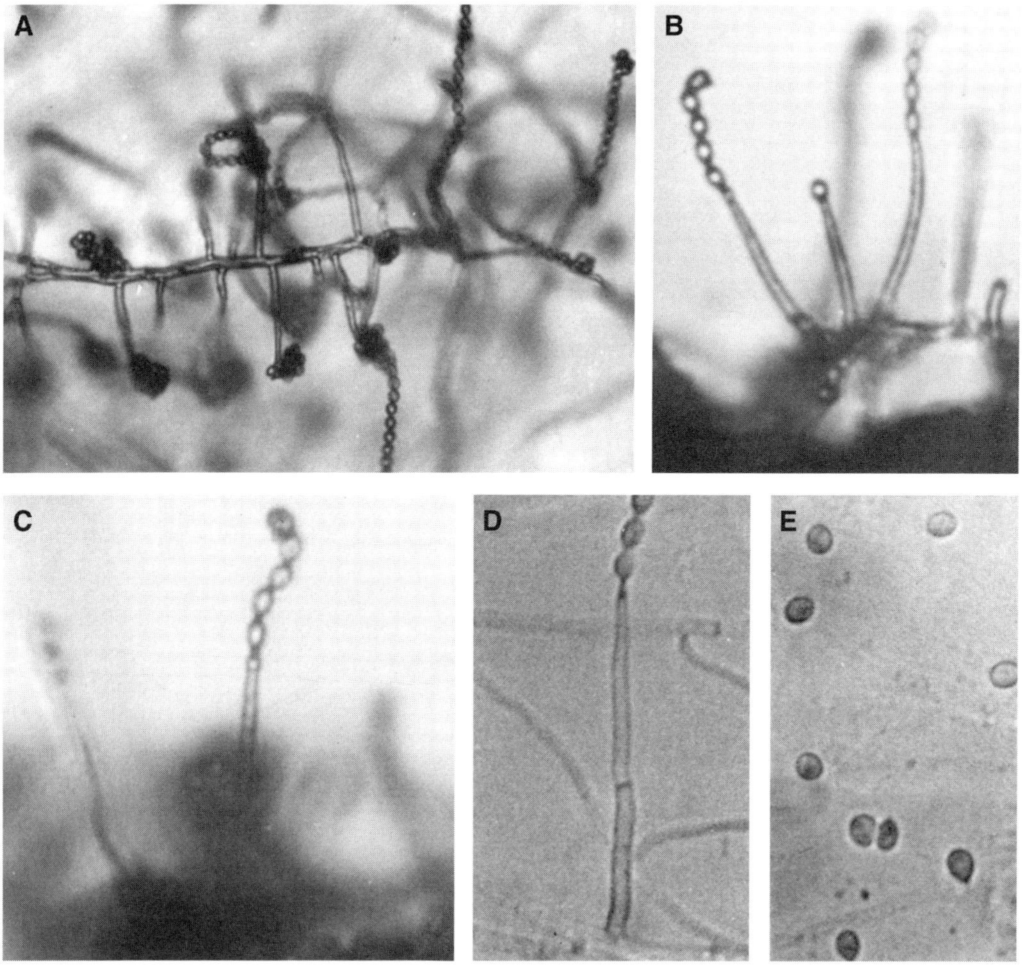

Paecilomyces variabilis. A–D: Conidiophores and conidia. E: Conidia. (From Watanabe, T. 1975d. *Trans. Mycol. Soc. Jpn.*, 16:264–267. With permission.)

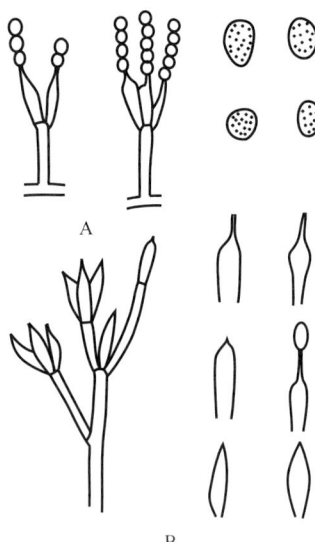

Paecilomyces victoriae (Szilvinyi) A. H. S. Brown & G. Smith

References: Brown and Smith 1957.

Morphology: Conidiophores hyaline, erect, branched apically, bearing catenulate conidia on terminal phialides: phialides opposite, or occasionally verticillate, ampulliform with cylindrical base and acutely pointed in the median. Conidia phialosporous, terminal, subglobose or broadly ellipsoidal, slightly rough or minutely echinulate on the surface.

Dimensions: Conidiophores 22.5–64 μm: primary branches 17–18 μm long: phialides 6.3–22 × 2.7–2.8 μm. Conidia 2.5–3 μm in diameter.

Material: 69-280, 69-327 (Pineapple field soil, Okinawa, Japan).

Remarks: Colonies on Czapek agar are white all over, or slightly yellowish green tinted, but this fungus is different from the original description in forming the rough or minutely echinulate conidia.

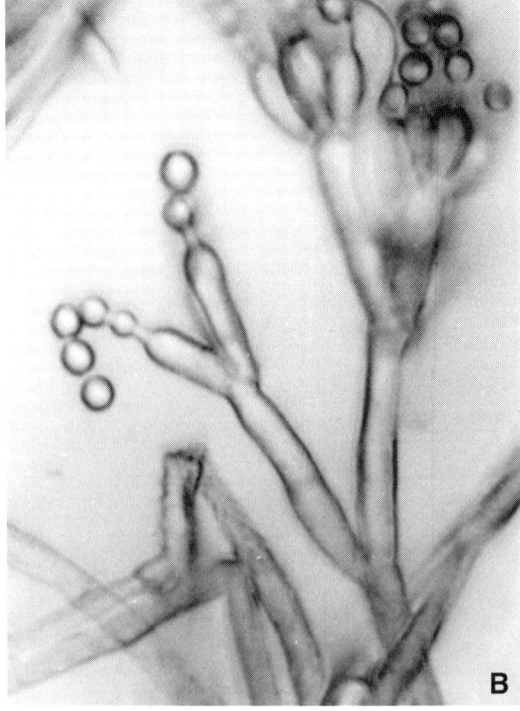

Paecilomyces victoriae. A,B: Conidiophores and conidia.

Papulaspora Preuss

Linnaea 24:113, 1851.
Type species: *P. sepedonioides* Preuss

Key to Species

1. Papulospores with setae . . . .*P. nishigaharanus*
   Papulospores without setae 2

2. Papulospores dark brown, 68–165 µm in diameter. . . . . *P. pannosa*
   Papulospores hyaline, pale or yellowish brown. 3

3. Papulospores under 80 µm. 4
   Papulospores over 100 µm in diameter, smooth 5

4. Papulospores under 30.5 µm in diameter . . . . *Papulaspora* sp. 1
   Papulospores 41–80 µm in diameter. . . . . *P. irregularis*

5. Arthrospores formed . . . . *Papulaspora* sp. 2
   Arthrospores not formed . . . . *P. pallidula*

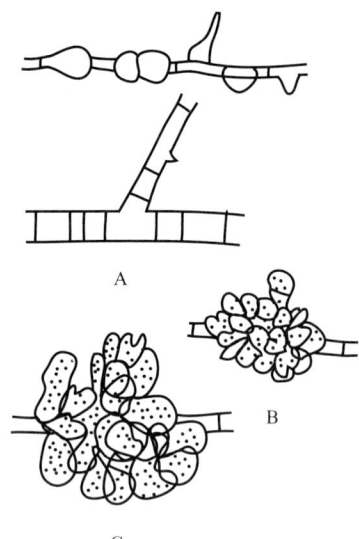

Papulaspora irregularis Hotson

References: Hotson 1917; Watanabe 1975d.

Morphology: Papulospores hyaline, slightly yellowish, globose or subglobose, apparently soft sclerotium-like, composed of a few original cells and surrounding numerous cells, with rough margin. Hyphae hyaline, or yellowish tinted, uneven or characteristically inflated, often densely septate.

Dimensions: Papulospores 41–60 (-80) μm in diameter: component cells up to 15 μm in diameter. Hyphae 5.5–6.8 μm broad: septum length 11.6–16.6 μm long.

Material: 72-X121 (= ATCC 34345, Sugarcane root, Taiwan, ROC).

Remarks: Papulospores and microsclerotia are occasionally quite difficult to differentiate morphologically.

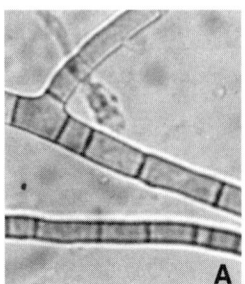
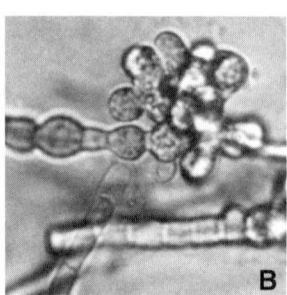
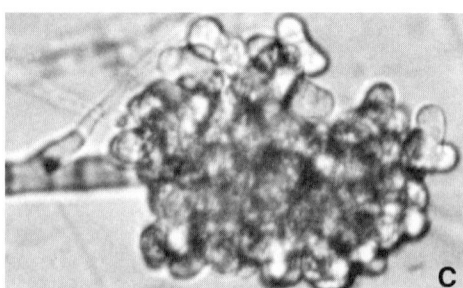

Papulaspora irregularis. A: Hyphae. B,C: Papulospores and component cells. (From Watanabe, T. 1975d. *Trans. Mycol. Soc. Jpn.*, 16:264–267. With permission.)

Papulaspora nishigaharanus T. Watanabe

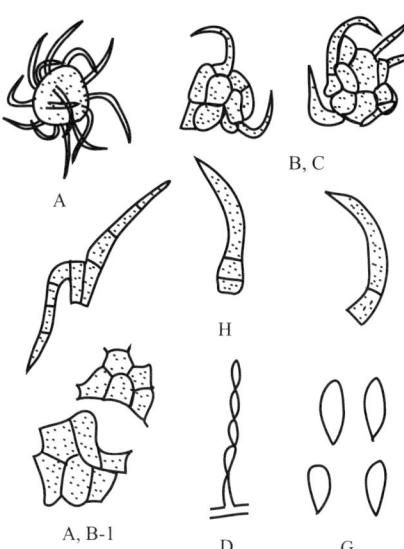

References: Hennebert 1963; Hotson 1912, 1917, 1942; Watanabe 1991; Weresub and LeClair 1971.

Morphology: Spores dimorphic, i.e., papulosporous, and phialosporous. Papulospores globose, brown, soft sclerotium-like, terminal or intercalary, covered with setose cells: setae usually 1-septate, subhyaline to pale brown, thick-walled, curved, rough, tapering toward apex. Conidiophores for phialoconidia hyaline, erect, simple, rarely branched, bearing nearly 10 catenulate conidia apically. Conidia phialosporous, hyaline, 1-celled, obovoid, apiculate at one end.

Dimensions: Papulospores (excluding setae) 15–45 µm in diameter: setae 15–40 × 1.5–2.8 µm. Conidiophores ca. 5 µm tall, ca. 2 µm wide at the base. Conidia 2.1–3.2 × 1–1.7 µm.

Material: 73-1071 (= IFO 32547, Cultivated soil, Nishigahara, Tokyo, Japan).

Remarks: This fungus is unique in the combination of setose papulospores and phialoconidia. Colonies on PDA are whitish gray or pale yellowish brown, slightly aerial.

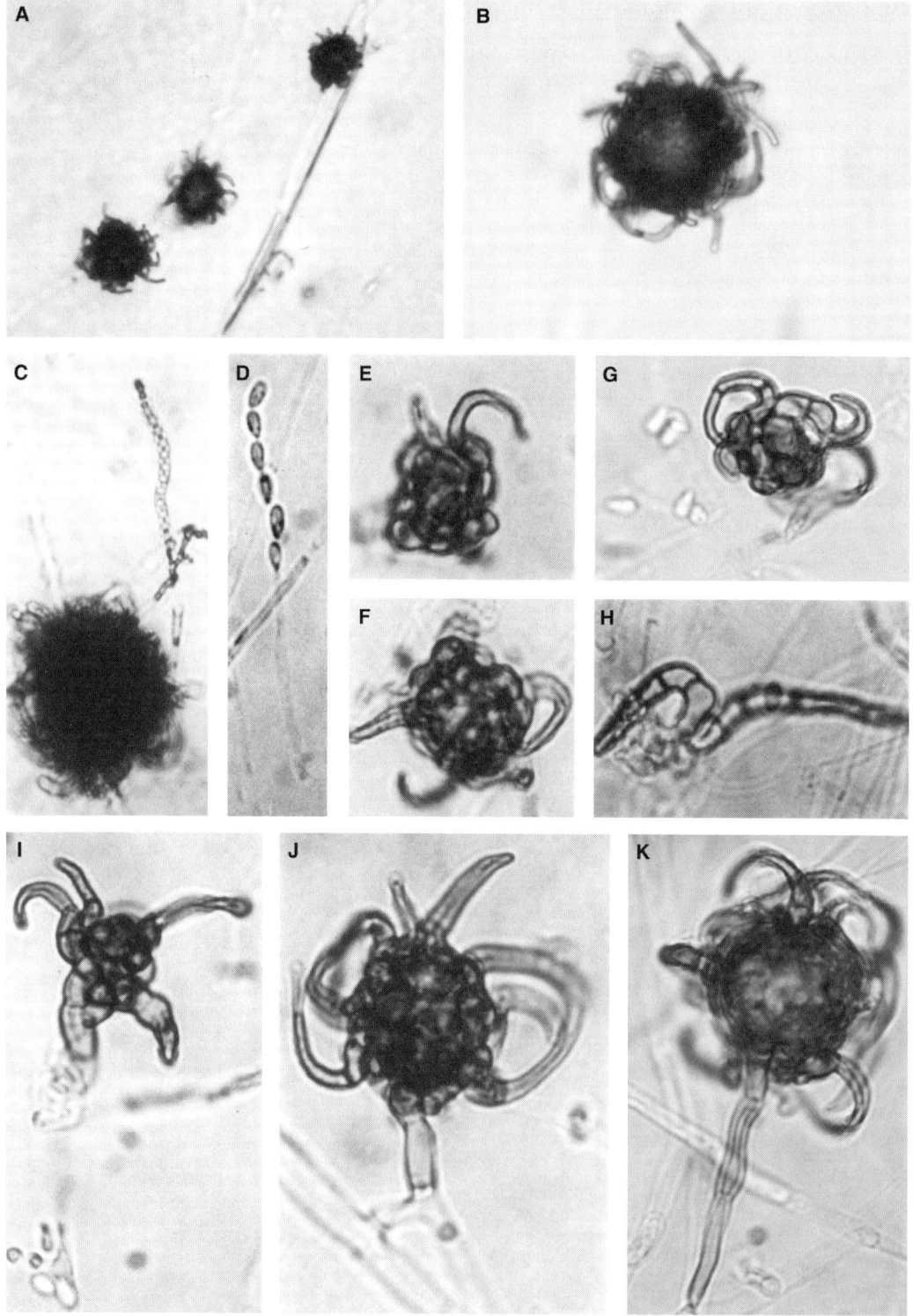

Papulaspora nishigaharanus. A,B: Papulospores and component cells (B-1). C: Papulospore and catenulate conidia. D: Conidiophores and conidia. E–K: Process of papulospore formation. (From Watanabe, T. 1991. *Mycologia*, 83:524–529. With permission.)

Papulaspora pallidula Hotson

References: Hotson 1917; Watanabe 1975d.

Morphology: Papulospores yellowish brown, globose or subglobose, apparently soft sclerotium-like, composed of discernible numerous component cells, smooth marginally. No conidia formed.

Dimensions: Papulospores (65-) 75–200 μm in diameter.

Material: 72-X44 (= ATCC 34346, Sugarcane root, Taiwan, ROC); 73-469 (Strawberry field soil, Tottori, Japan).

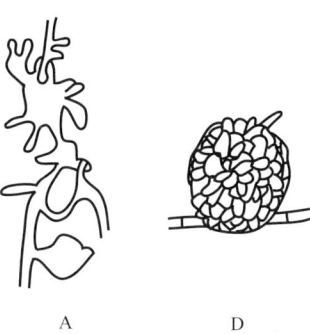

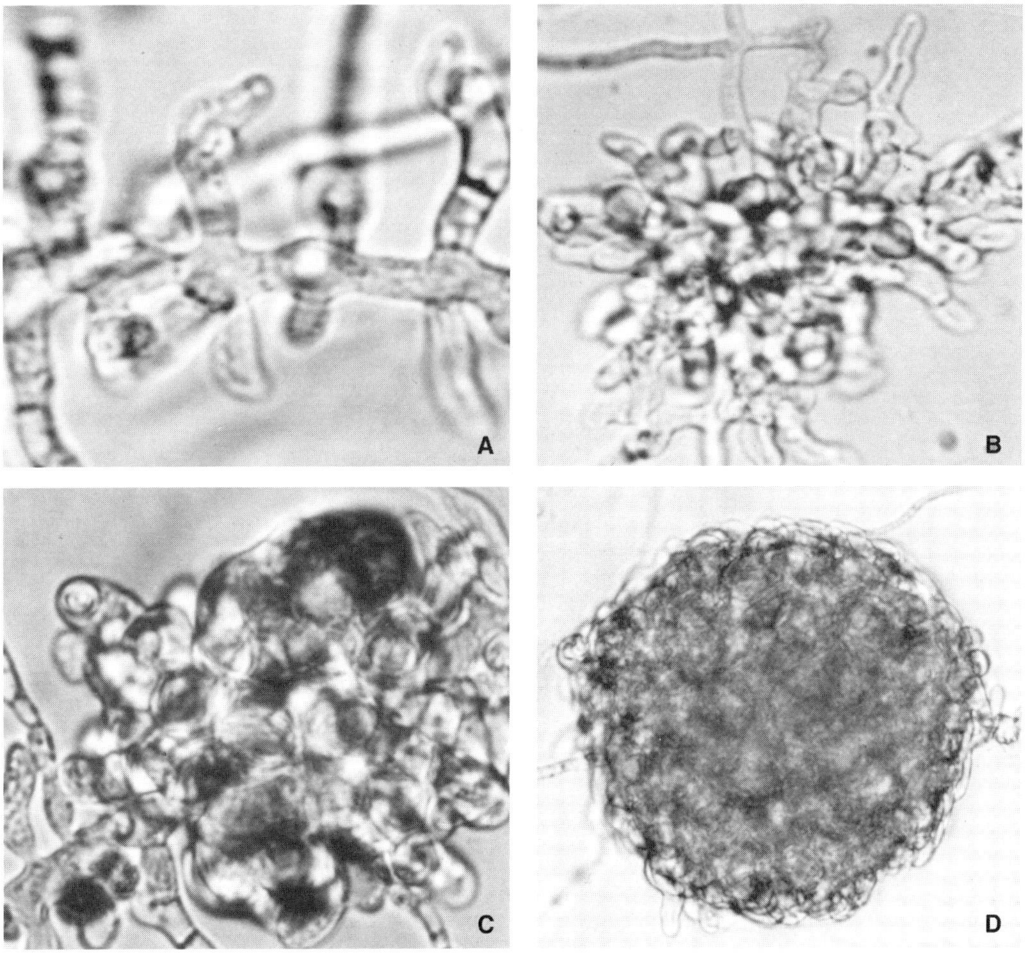

Papulaspora pallidula. A–C: Process of papulospore formation. D: Papulospore. (From Watanabe, T. 1975d. *Trans. Mycol. Soc. Jpn.*, 16:264–267. With permission.)

Papulaspora pannosa Hotson

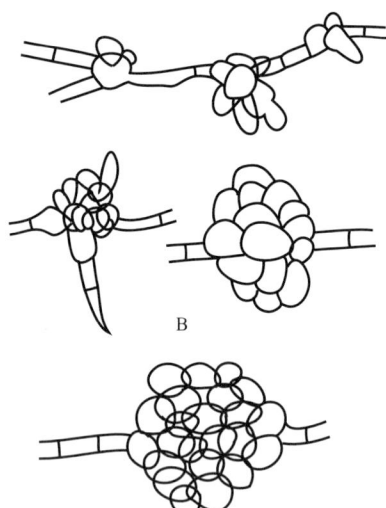

References: Hotson 1917; Watanabe 1975d.

Morphology: Papulospores formed on and in agar media or aerially, showing apparently fluffy appearance, smoky brown or dark brown, globose or subglobose, apparently soft sclerotium-like, composed of numerous, unrecognizable component cells, rough marginally. Hyphae hyaline, or yellowish-green tinted, gradually becoming brown, often spherical in shape, granulate inside.

Dimensions: Papulospores 68–165 µm in diameter. Hyphae ca. 10 µm wide: septum length 12.5–25 µm long.

Material: 72-X6-88 (= ATCC 34347, Sugarcane root, Taiwan, ROC).

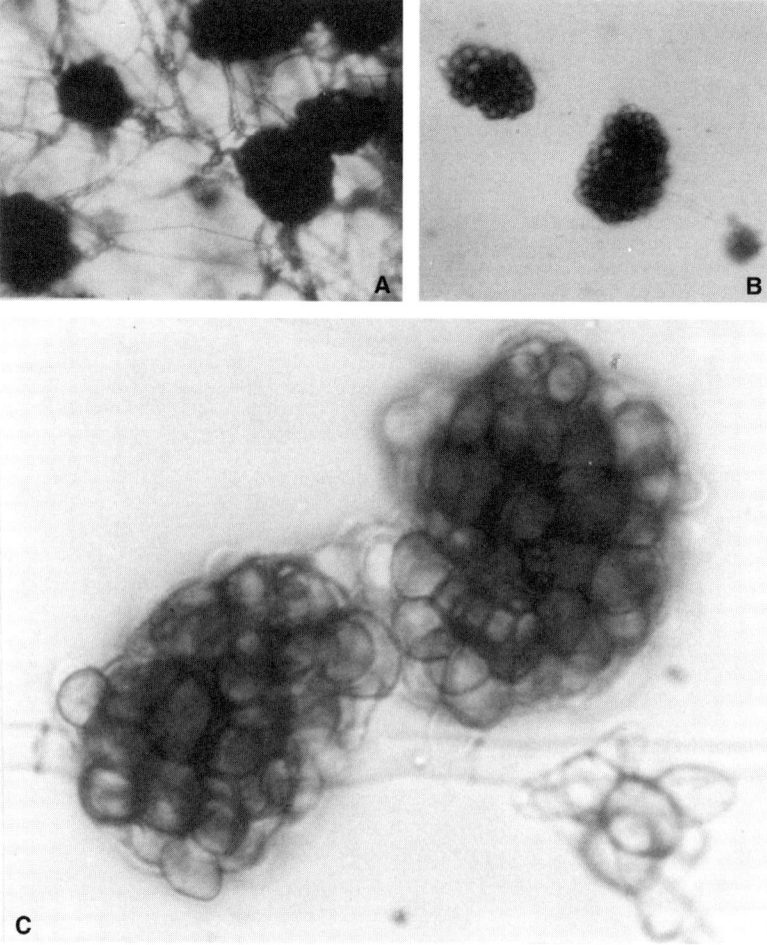

Papulaspora pannosa. A–C: Papulospores formed on aerial hyphae (A), and on agar medium (B,C). (From Watanabe, T. 1975d. *Trans. Mycol. Soc. Jpn.*, 16:264–267. With permission.)

Papulaspora sp. 1.

References: Hotson 1912, 1917, 1942; Weresub and LeClair 1971.

Morphology: Spores dimorphic, i.e., papulosporous, and phialosporous. Papulospores borne intercalarly on ordinary hyphae, or terminally on short conidiophores undifferentiated from hyphae, pale brown or brown, subglobose composed of globose component cells. Conidiophores phialidic, hyaline, erect, gradually tapering toward apex or acutely narrowed in the middle, 1-septate. Conidia phialosporus, hyaline, ovate, apiculate at one end.

Dimensions: Papulospores 15–25 (-30.5) μm in diameter. Phialides (7.5-) 10–17.5 (-22.5) × 2–2.5 μm. Conidia 2.2–2.5 × 0.7–1 μm.

Material: 73-1071 (Uncultivated soil, Nishigahara, Tokyo, Japan), 74-535; -615; -675 (Strawberry root, Shizuoka, Japan).

Remarks: Colonies on PDA are yellowish brown, with irregular concentric zonation. Papulospores of this fungus were always formed on agar cultures, but conidia not always formed.

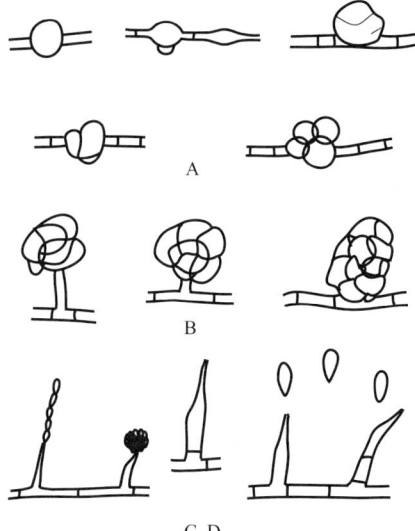

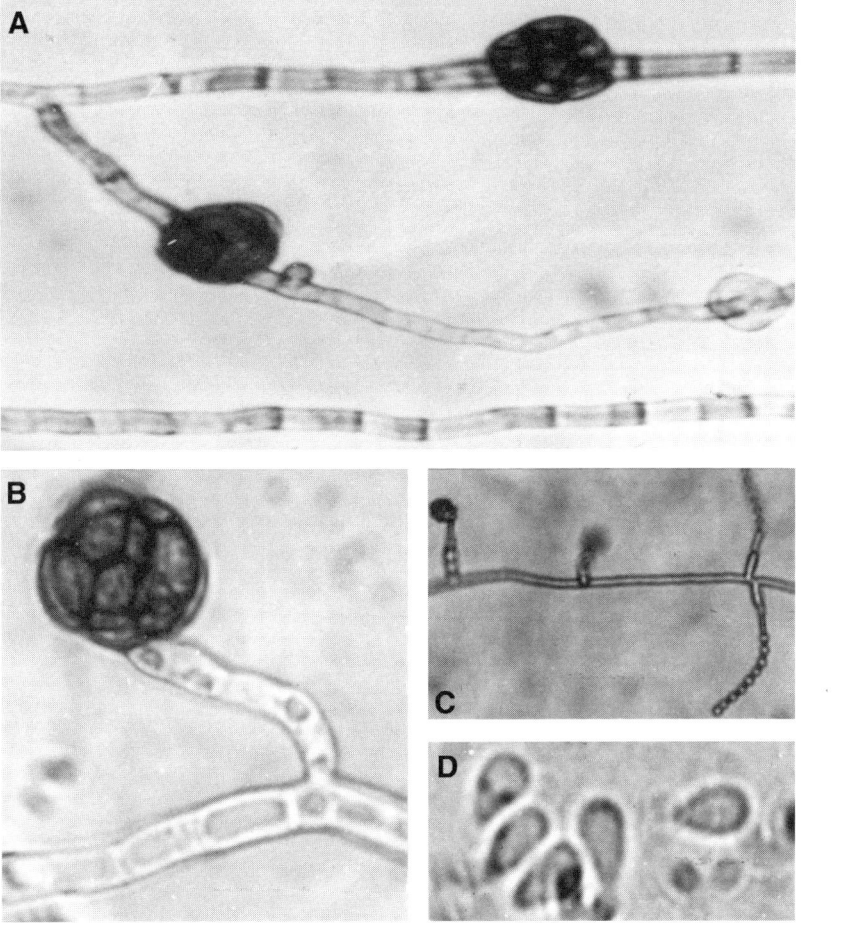

Papulaspora sp. 1.   A,B: Papulospores and hyphae. C: Phialides and conidia. D: Conidia.

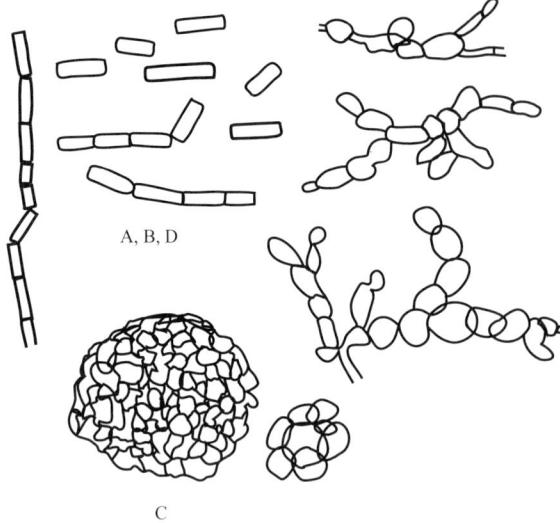

Papulaspora sp. 2.

References: Hotson 1912, 1917, 1942; Weresub and LeClair 1971.

Morphology: Spores dimorphic, i.e., papulosporous, and arthrosporous. Papulospores borne directly on ordinary hyphae, or terminally on short conidiophores undifferentiated from hyphae, embedded, yellowish brown or brown, fleshy, subglobose composed of globose component cells, smooth marginally. Arthrospores hyaline, erect, borne directly on hyphae, aerial. Conidia arthrosporous, hyaline, cylindrical.

Dimensions: Papulospores 100–225 µm in diameter: component cells 11.2–17.5 µm in diameter. Conidia 11.2–27.5 × 5–7 µm.

Material: 73-467 (Uncultivated soil, Nishigahara, Tokyo, Japan); 83-415 (Strawberry root, Shizuoka, Japan).

Remarks: Colonies on PDA are dark yellowish green.

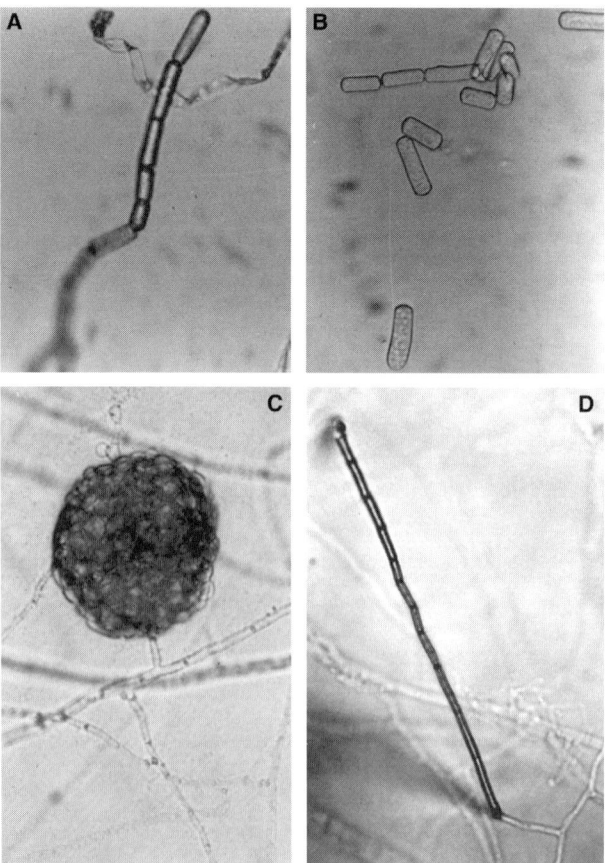

Papulaspora sp. 2. A,B,D: Arthrospores in chains (A,B) and detached (B). C: Papulaspore.

Penicillium Link

Mag. Ges. Natur. Freunde, Berlin, 3:16, 1809.
Type species: *P. expansum* Link

Key to Species

1. Colonies on Czapex agar	reddish tinted 2	
	not so 3	
2. Conidia	ellipsoidal, 2.8–3.5 µm in diameter .... *P. corylophilum*	
	globose, 2.5–2.7 µm in diameter..................... *P. restriculosum*	
3. Conidia	ellipsoidal, apiculate at one end .... *P. janthinellum*	
	globose... 4	
4. Conidiophores	over 100 µm long, with ampulliform phialides .... *P. nigricans*	
	under 100 µm, with pen-shaped phialides .... *P. lanosum*	

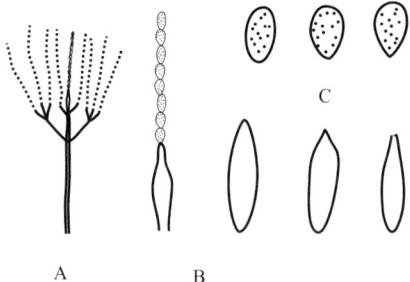

Penicillium corylophilum Dierckx

References: Raper and Thom 1949; Samson and Pitt 1985, 1990; Tzean et al. 1994; Watanabe 1975d.

Morphology: Conidiophores hyaline, erect, branched penicillately at the apexes with 2–3 metula, verticillate phialides on each metula, and rather aggregated, compact, conidial heads composed of catenulate conidia on each phialide: phialides tapering gradually or cylindrical with pointed tips. Conidia phialosporous, hyaline or yellowish brown in mass, ellipsoidal or ovate, 1-celled, rough or minutely echinulate on the surface.

Dimensions: Conidiophores 120–220 μm long: primary branches 10–12.5 × 2.5 μm: phialides 10.5–12.5 × 2.5 μm. Conidia 2.7–3.5 × 2.2–2.3 μm.

Material: 72-X45 (Sugarcane root, Taiwan, ROC).

Remarks: Cultures on Czapek agar are lanose, white with pinkish tint, reverse the same with rather intensified yellowish tint. Growth rate: under 1 cm in diameter in 10 days after incubation at 25°C.

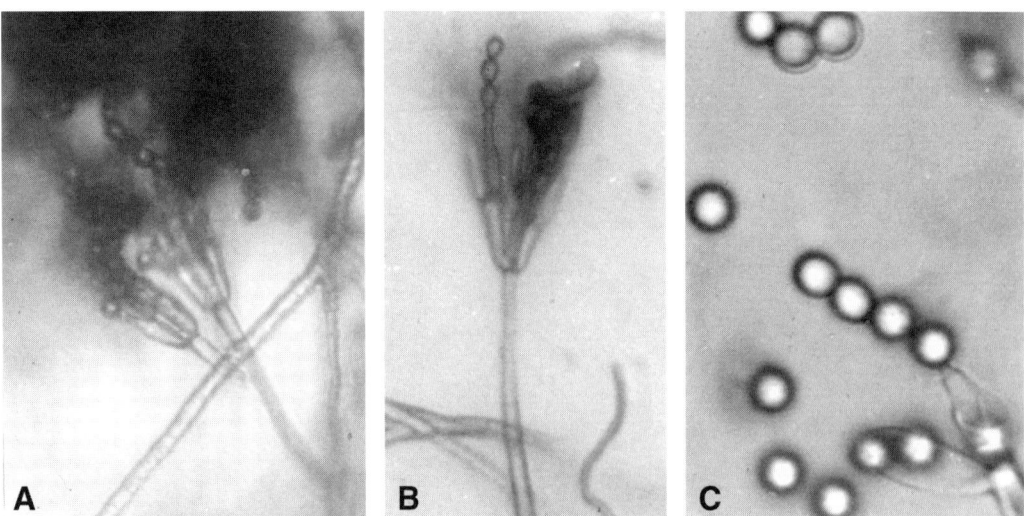

Penicillium corylophilum. A–C: Conidiophores, phialides, and conidia. (From Watanabe, T. 1975d. *Trans. Mycol. Soc. Jpn.*, 16:264–267. With permission.)

Penicillium janthinellum Biourge

References: Raper and Thom 1949; Samson and Pitt 1985, 1990; Tzean et al. 1994; Watanabe 1975d.

Morphology: Conidiophores hyaline, erect, branched penicillately at the apexes with verticillate metula, terminal phialides and catenulate conidia on each phialide, forming rather divergent conidial heads: phialides pen-pointed with abruptly tapered tips. Conidia phialosporous, pale green, dark in mass, ellipsoidal or subglobose, 1-celled, smooth, apiculate at one end.

Dimensions: Conidiophores 100 to over 250 × 2.5 µm: primary branches 12.5 × 2.5 µm: phialides 7.5–12.5 × 2–2.5 µm. Conidia 2.5–3.2 × 1.7–2.5 µm.

Material: 72-X135 (Sugarcane root, Taiwan, ROC).

Remarks: Colonies on Czapek agar are velvety, pale grayish green on the surface: reverse pale yellowish brown. Growth rate: 2–3 cm in diameter in 10 days after incubation at 25°C.

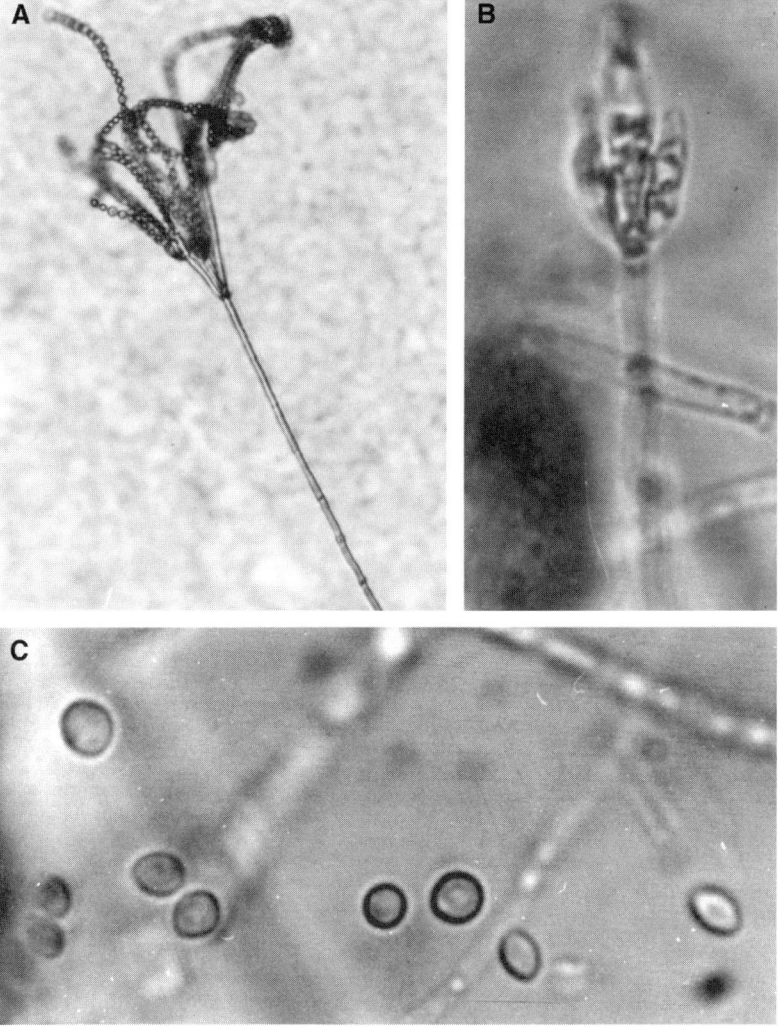

Penicillium janthinellum. A: Conidiophores and conidia. B: Phialides. C: Conidia. (From Watanabe, T. 1975d. *Trans. Mycol. Soc. Jpn.*, 16:264–267. With permission.)

Penicillium lanosum Westling

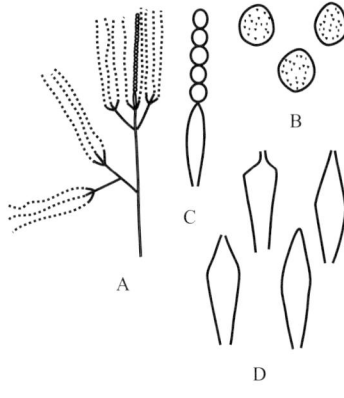

References: Raper and Thom 1949; Samson and Pitt 1985, 1990; Watanabe 1975d.

Morphology: Conidiophores hyaline, erect, developed from aerial hyphae, branched penicillately at the apexes with primary and secondary metula, verticillate phialides and catenulate conidia in each phialide, forming rather open-spaced yellowish green conidial heads: phialides lanceolate or abruptly sharpened. Conidia phialosporous, pale green, dark in mass, globose to subglobose, 1-celled, minutely echinulate on the surface.

Dimensions: Conidiophores 12.5–62.5 × 2.5–2.8 µm: primary branches 7.5–9.5 × 2.7–2.8 µm: phialides 10–13.8 × 2.5–3 µm. Conidia 2.7–4 µm in diameter.

Material: 72-X76 (Sugarcane root, Taiwan, ROC).

Remarks: Cultures on Czapek agar are fluffy, bright yellowish green with bluish green tint, funiculose with bundles of hyphae, reverse yellowish pink with reddish purple tint. Rather good in growth.

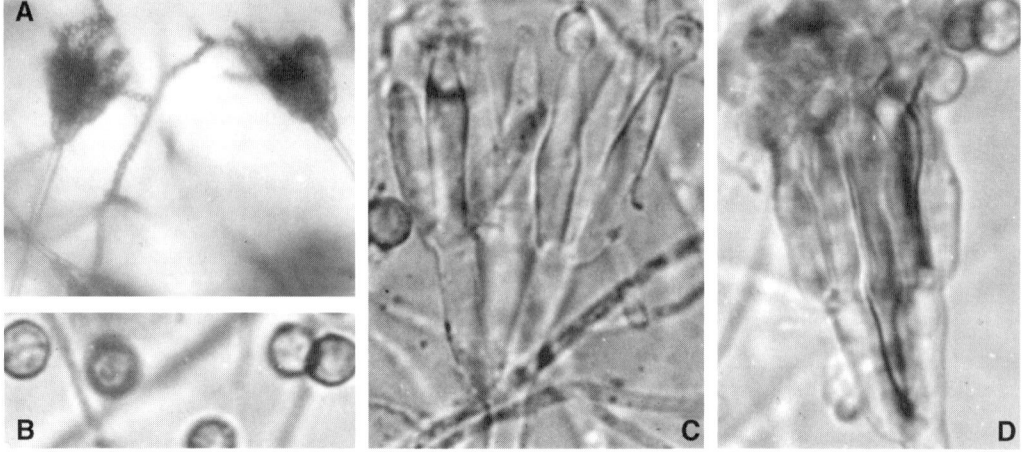

Penicillium lanosum. A: Conidiophores and conidia. B: Conidia. C,D: Phialides and conidia. (From Watanabe, T. 1975d. *Trans. Mycol. Soc. Jpn.*, 16:264–267. With permission.)

Penicillium nigricans (Bainier) Thom

References: Raper and Thom 1949; Samson and Pitt 1985, 1990; Watanabe 1975d.

Morphology: Conidiophores hyaline, erect, slightly rough, developed from aerial hyphae, branched penicillately at the apexes, with primary or secondary metula, verticillate phialides and catenulate conidia at each phialide, forming cylindrical columnar grayish green conidial heads: phialides pen-pointed with abruptly sharpened tips. Conidia phialosporous, pale green, globose or subglobose, 1-celled, verrucose or minutely echinulate on the surface.

Dimensions: Conidiophores 120–270 μm tall: primary branches 10–12.5 μm long: phialides 8.7–13.8 × 2.5–2.8 μm. Conidia 2.7–3.2 (-4) μm in diameter.

Material: 72-X88 (Sugarcane root, Taiwan, ROC).

Remarks: Colonies on Czapek agar are velvety, good in growth, bright grayish green with whitish tint, reverse pale yellowish brown.

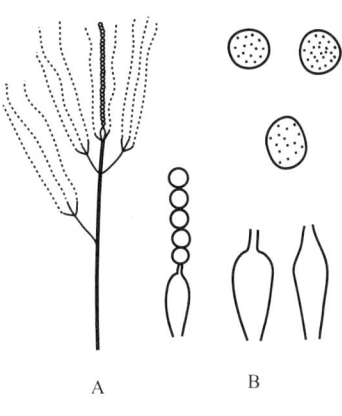

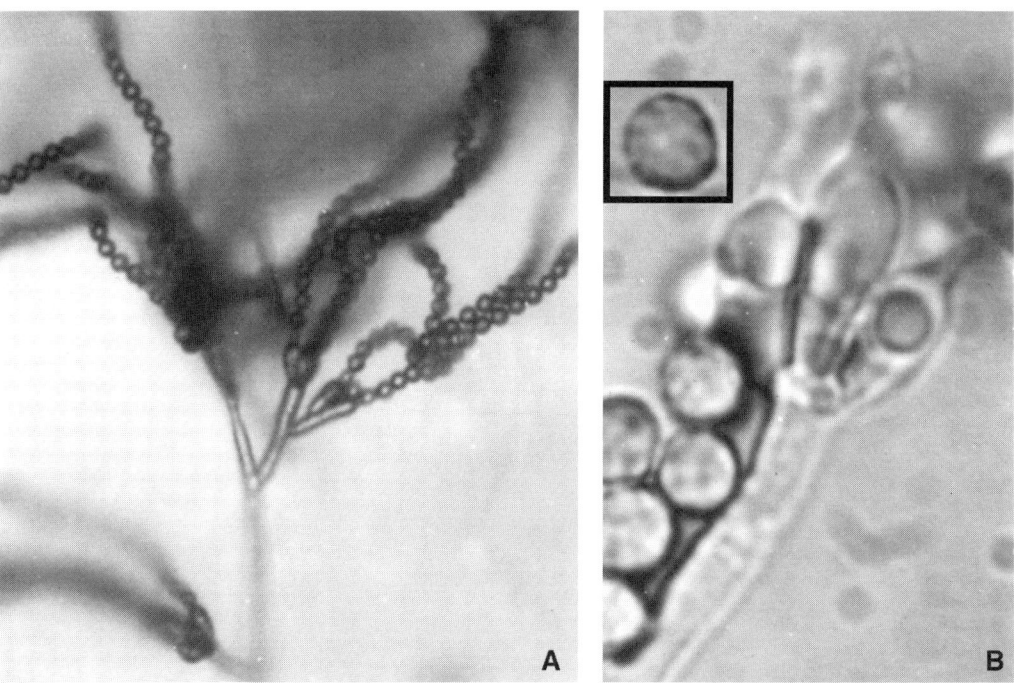

Penicillium nigricans.   A: Conidiophores and conidia. B: Phialides and conidia. (From Watanabe, T. 1975d. *Trans. Mycol. Soc. Jpn.*, 16:264–267. With permission.)

Penicillium resticulosum Birkinshaw, Raistrick, & Smith

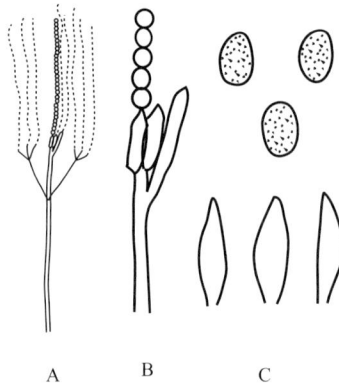

References: Raper and Thom 1949; Samson and Pitt 1985, 1990; Watanabe 1975d.

Morphology: Conidiophores hyaline, erect, branched penicillately at the apexes with 2–3 metula, 3–4 verticillate phialides and catenulate conidia in each phialide, forming rather compact cylindrical grayish green conidial heads: phialides lanceolate. Conidia pale green, dark brown in mass, subglobose, minutely echinulate on the surface.

Dimensions: Conidiophores 60–130 µm tall: primary branches 10–12.8 × 2.5–2.8 µm: phialides 10–12.5 × 2.5 µm. Conidia 2.1–2.8 µm in diameter.

Material: 72-X75-646 (Sugarcane root, Taiwan, ROC).

Remarks: Cultures on Czapek agar are fluffy, good in growth, funiculose with bundles of aerial hyphae, reddish brown with greenish tint, reverse clear yellowish red.

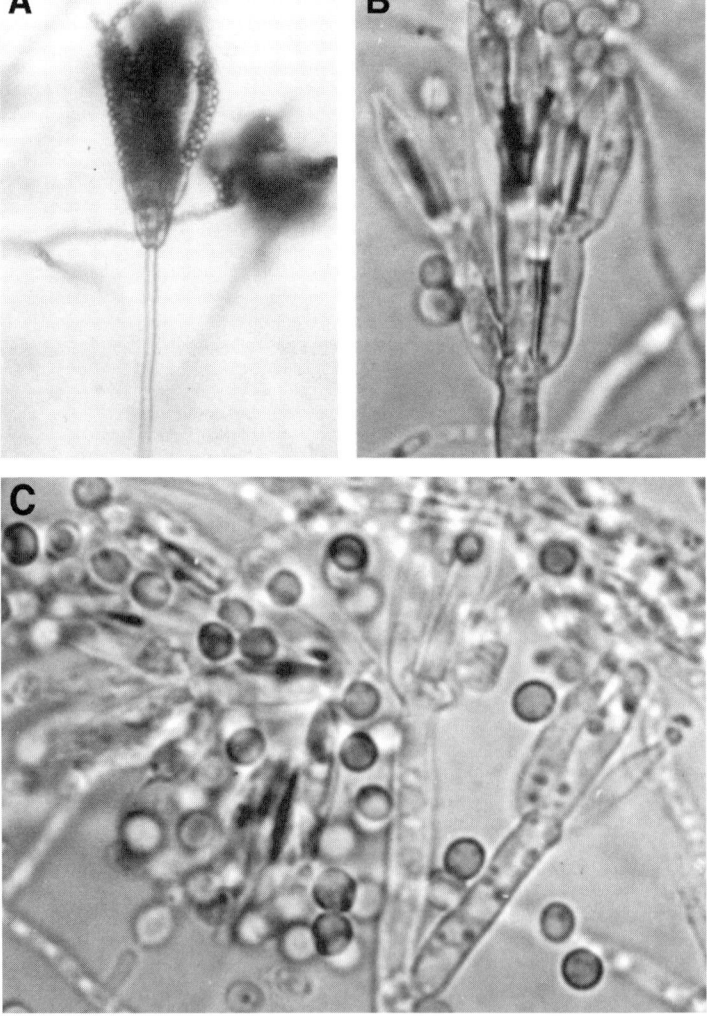

Penicillium resticulosum. A: Conidiophores and conidia. B,C: Phialides and conidia. (From Watanabe, T. 1975d. *Trans. Mycol. Soc. Jpn.*, 16:264–267. With permission.)

Periconia Tode

Syn. Fung. Carol. Sup. 125, 1822.
Type species: *P. lichenoides* Tode : Mérat

Key to Species

1. Conidia   conspicuously echinulate, with over 2 µm long
   protuberances................................. *P. macrospinosum*
   indistinctly echinulate 2

2. Conidia   13–17 µm in diameter................................ *P. byssoides*
   5–11 µm in diameter........................... *P. saraswatipurensis*

Periconia byssoides Pers. : Mérat

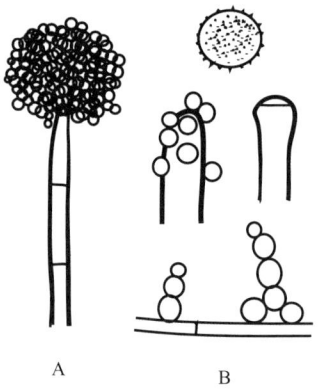

References: Mason and Ellis 1953; Ellis 1971; Watanabe and Sato 1988.

Morphology: Conidiophores dark brown, erect, simple, thick-walled, bearing globose spore heads at the apexes, composed of catenulate conidia developed acropetally on terminal fertile portions, or sporulated directly on hyphae. Conidia blastosporous, brown, globose, minutely echinulate.

Dimensions: Conidiophores 375–1100 × 10–16.3 µm. Spore masses ca. 175 µm in diameter. Conidia 13–17 µm in diameter.

Material: 86-40 (Japanese cedar seed, Kasama, Ibaraki, Japan).

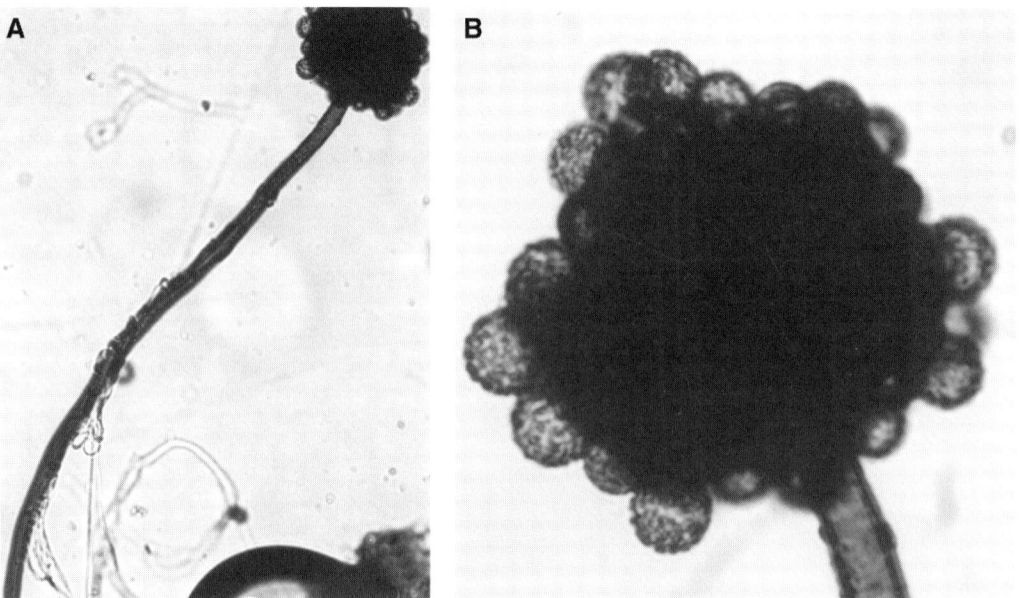

Periconia byssoides. A,B: Conidiophores and conidia. (From Watanabe, T. and Sato, Y. 1988. *Trans. Mycol. Soc. Jpn.*, 29:143–150. With permission.)

Periconia macrospinosa Lefebvre et Johnson

References: Ellis 1971; Lefebvre et al. 1949; Mason and Ellis 1953; Watanabe 1975d.

Morphology: Conidiophores brown to reddish brown, erect, simple, bearing aggregates of conidia on terminal fertile portions: conidial chains often branched. Conidia blastosporous, dark brown, globose, echinulate with well-developed protuberances. Chlamydospores often formed.

Dimensions: Conidiophores (60-) 109.3–291.6 × (5.5-) 6–9.3 μm. Conidia (including protuberances) 17–29.2 μm in diameter: protuberances 2.4–4.9 μm long.

Material: 72-X39-320 (= ATCC 32922, Sugarcane root, Taiwan, ROC); 77-125 (Strawberry root, Nishigahara, Tokyo, Japan).

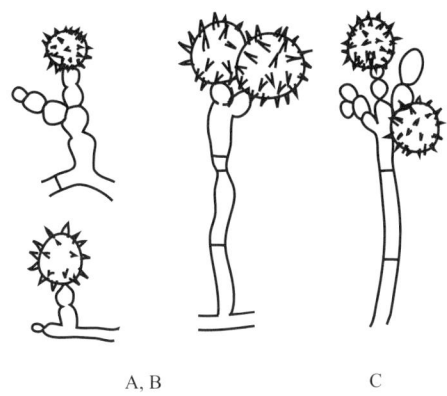

A, B   C

Periconia macrospinosa. A,B: Conidiophores and conidia. C,D: Apical fertile part of conidiophore and conidia. (From Watanabe, T. 1975d. *Trans. Mycol. Soc. Jpn.*, 16:264–267. With permission.)

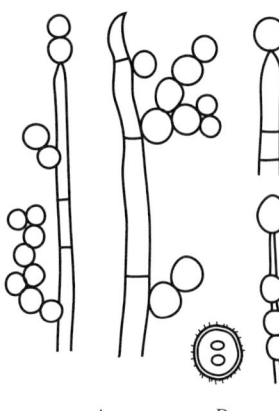

Periconia saraswatipurensis Bilgrami

References: Ellis 1971, 1976; Watanabe et al. 1987a.

Morphology: Conidiophores brown, erect, simple, bearing aggregates of conidia on apical and subapical fertile portions: conidial chains often branched. Conidia blastosporous, reddish brown, globose, indistinctly echinulate with numerous minute protuberances.

Dimensions: Conidiophores 150–175 × 3.7–3.8 μm. Conidia 5–11 μm in diameter.

Material: 85-P24 (Paulownia root, Itapua, Paraguay).

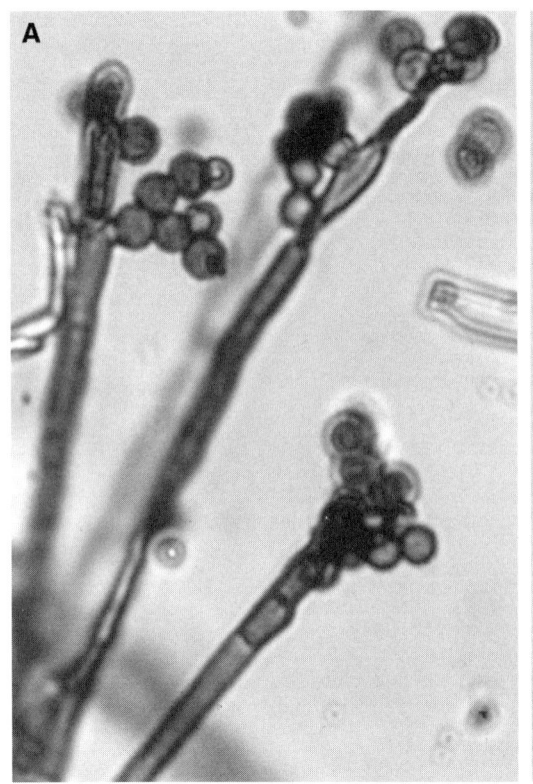
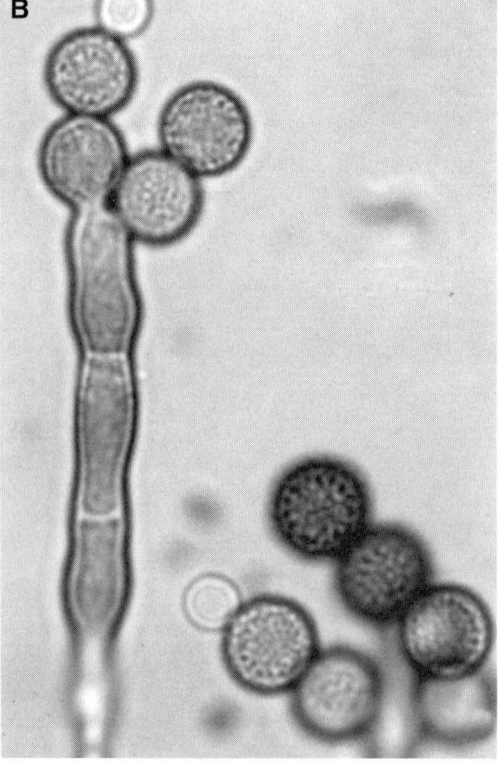

Periconia saraswatipurensis.   A,B: Conidiophores and conidia.

Pestalotia de Not.

Mem. R. Acad. Sci. Torino 2, 3:80, 1839.
Type species: *P. pezizoides* de Not.

Pestalotia spp.

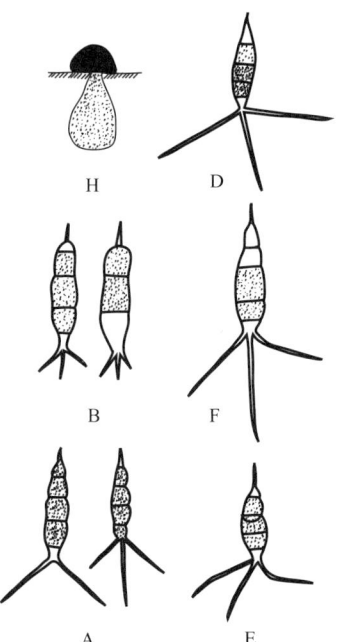

Synonym: *Pestalotiopsis* spp.

References: Dube and Bilgrami 1966; Guba 1932, 1961; Steyaert 1949, 1955, 1956; Sutton 1969; Watanabe et al. 1986a.

Morphology: Spodochia (spore masses) on agar cultures hemispherical, mucilagenous, black, occasionally leaked around sporodochia. Conidiophores short, simple. Conidia spindle-shaped or ellipsoidal, 4- to 5-celled, with 2–3 central, pigmented cells (especially darker in 2 cells), with 2–4 appendages (setulae) in apical cells and 1 short appendage (pedicel) in basal cells.

Dimensions: Conidia (excluding apical and basal appendages) 19.7–50 × 5.0–10 µm: pigmented cells (7.5-) 10–20 µm long: apical appendages (4.7-) 10–30 µm long: basal appendages 1.2–12.5 µm long.

Material: 69-319 (Pineapple field soil, Okinawa, Japan); 70-1752 (Pineapple field soil, Hachijo, Tokyo, Japan); 83-1, 83-2, 83-5 (Japanese red pine seed, Hiba, Hiroshima, Japan); 83-3 (Japanese black pine seed, Kumano, Mie, Japan); 83-4 (Japanese cedar seed, Yoshino, Nara, Japan); 85-47, 85-48 (Flowering cherry seed, Hachioji, Tokyo, Japan).

Remarks: On the basis of the number of component cells of conidia, the genus *Pestalotia* may be further split into *Pestalotia* (6 cells), *Pestalotiopsis* Steyaert. (5 cells), and *Truncatella* Steyaert (4 cells).

In agar cultures, no acervuli are generally formed and thus, it is rather difficult to identify unknowns by comparing the morphology of known species. Among *Pestalotia* isolates with almost similar conidial dimensions obtained from Japanese pine seeds, conidia are formed with central larger and more darkly pigmented cells in the isolate 83-2 (A), with short setulae (B) in the 83-1, with thick and black central septum (C) in the 85-5, and with constriction at the septum in the 83-3 (D). Isolate 69-319 formed conidia with predominantly 2 setulae (G).

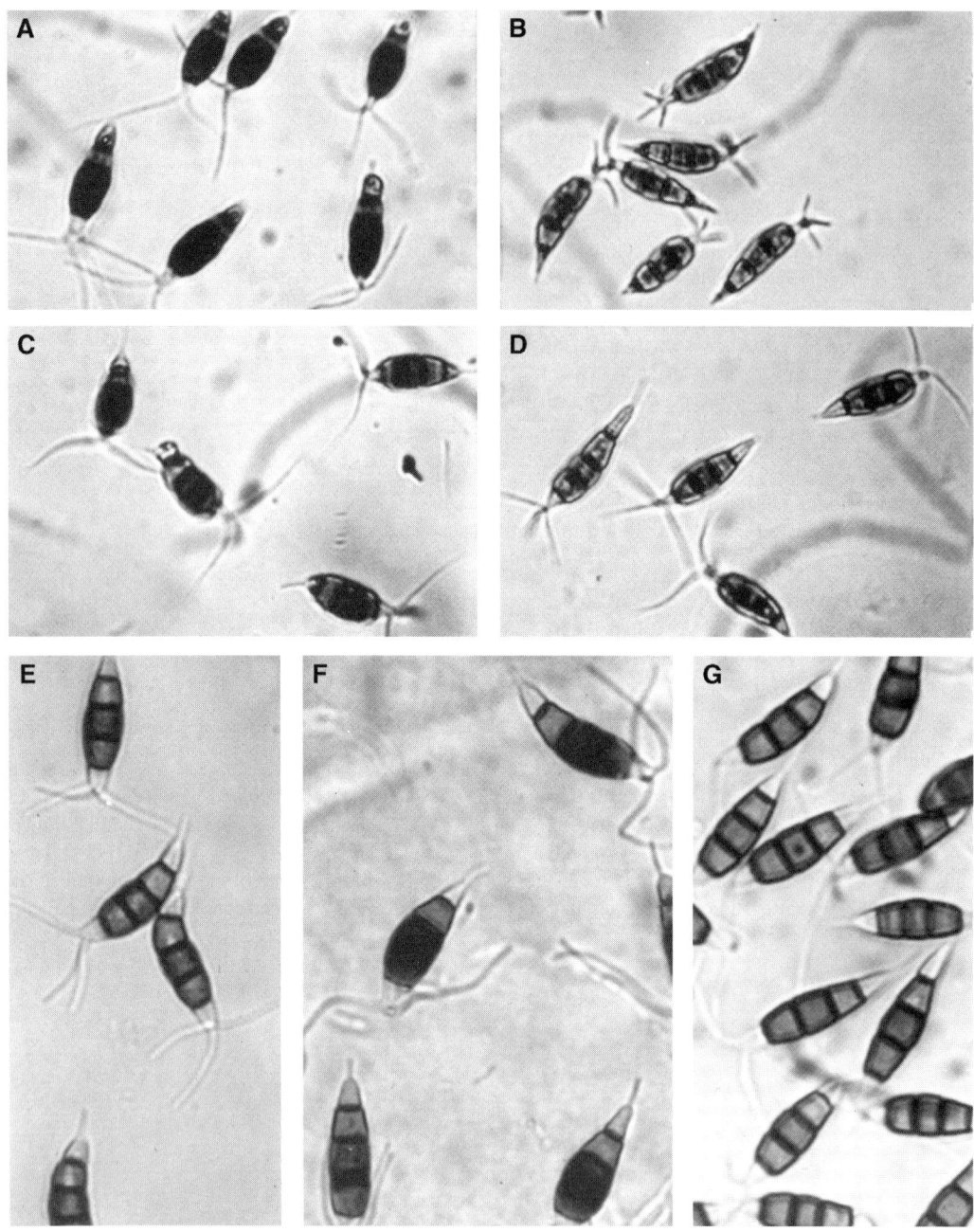

Pestalotia spp. Conidia (A: 83-2, B: 83-1, C: 83-5, D: 83-3, E: 85-88, F: 85-49, G: 69-319).

Pestalotia sp.

Synonym: *Pestalotiopsis* sp.

References: Dube and Bilgrami 1966; Guba 1932, 1961; Steyaert 1949, 1955, 1956; Sutton 1969.

Morphology: Sporodochia (spore masses) on agar cultures hemispherical, mucilagenous, black. Conidiophores indistinct. Conidia ellipsoidal, 4- to 5-celled, with 3 central, pigmented cells, with 3–6 (mainly 4) appendages (setulae) in apical cells and 1 short appendage (pedicel) in basal cells that are often lacking. Chlamydospores globose, thick-walled, borne in one or two cells of conidia with age.

Dimensions: Conidia (excluding apical and basal appendages) 14–18 × 5–6 μm: pigmented cells 10–14 μm long: apical appendages 26–44 × 0.8–1 μm. Chlamydospores 7–9 μm in diameter.

Material: 01-464 (Forest soil, Ogasawara, Tokyo, Japan).

Remarks: Chlamydospore formation is noteworthy.

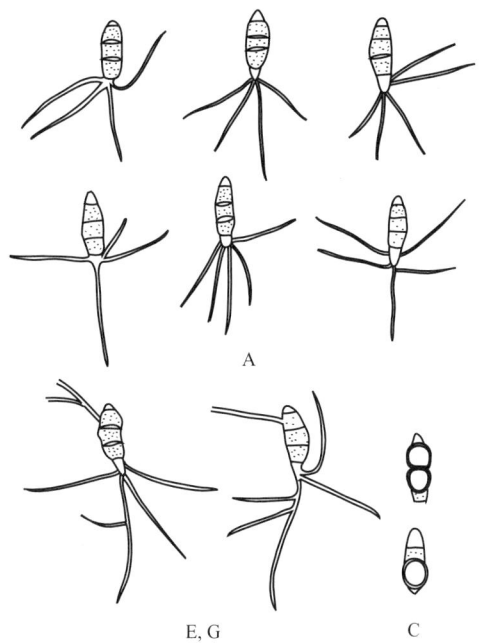

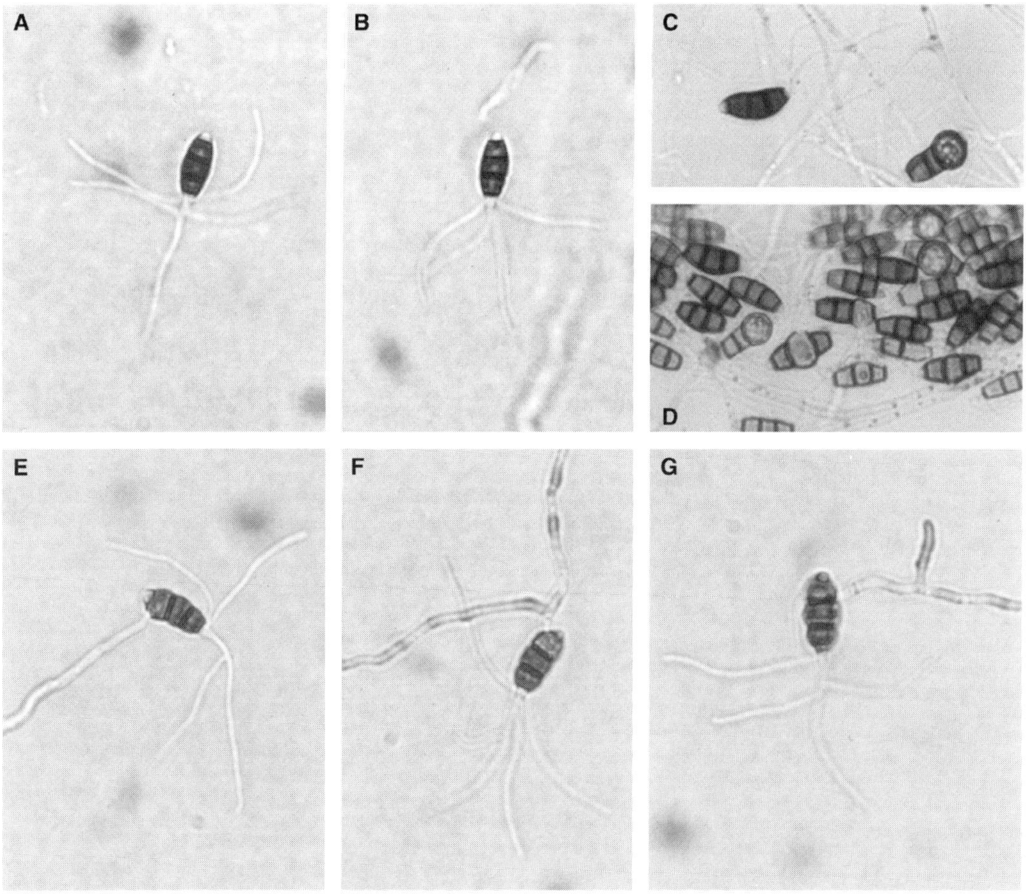

Pestalotia sp.   A,B: Conidia. C,D: Chlamydospores and conidia. E–G: Germination of conidia.

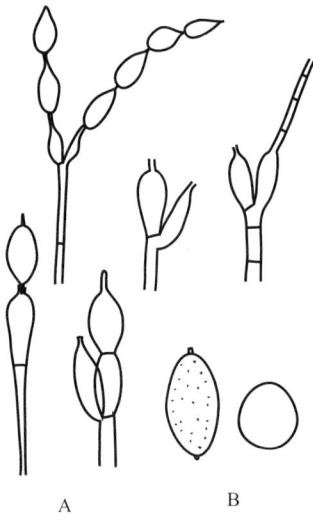

Phialomyces Misra & Talbot

Can. J. Bot. 42:1287, 1964.
Type species: *P. macrosporus* Misra & Talbot

Phialomyces macrosporus Misra & Talbot

References: Matsushima 1975; Misra and Talbot 1964.

Morphology: Conidiophores hyaline, erect, simple, bearing over 20 catenulate conidia basipetally developed on terminal phialides: phialides simple or branched oppositely or verticillately, often proliferated, flask-shaped. Conidia phialosporous, dark brown, ellipsoidal, often with short appendixes at both ends, minutely echinulate all over.

Dimensions: Conidiophores 90 to over 275 µm tall: ca. 5.5 µm wide at base: phialides 25–30 µm long. Conidia 27.5–32.5 × 20–23.8 µm.

Material: 86-93 (Pinus forest soil, Tsukuba, Ibaraki, Japan).

Remarks: Only one species is known in this genus (Hawksworth et al., 1995).

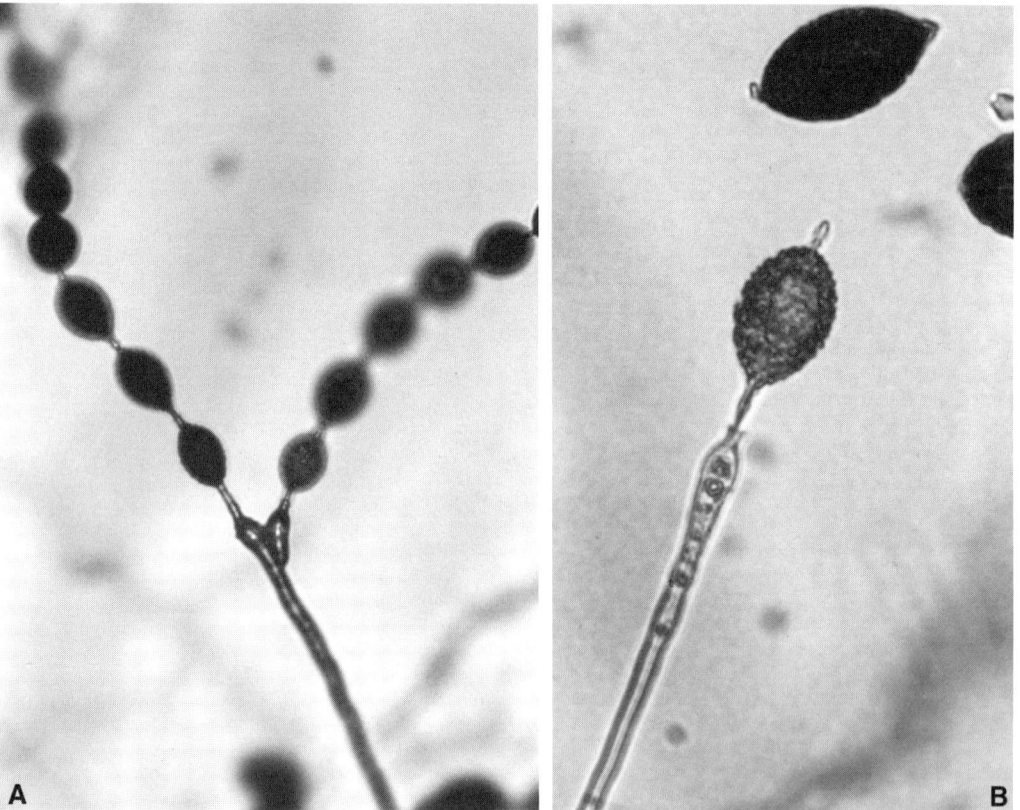

Phialomyces macrosporus.   A,B: Conidiophores, phialides, and conidia.

Phialophora Medlar

Mycologia 7:200, 1915.
Type species: *P. verrucosa* Medlar

Key to Species

1. Conidia	lunar-shaped	*P. radicicola*
	not so	2
2. Conidia	globose or subglobose	3
	not so	7
3. Conidia	of two kinds: globose and ovate	*P. richardsiae*
	globose	4
4. Sclerotia	formed	Phialophora sp. 2
	not formed	5
5. Phialides	often constricted in the median	*P. cyclaminis*
	not so	6
6. Phialides	short, thick, indistinctly necked, conidia 1.7–4.5 μm in diameter	*P. atrovirens*
	ampulliform or various in shape, conidia under 2 μm in diameter	Phialophora sp. 1
7. Conidia	6.3–12.6 μm long	*P. malorum*
	under 6 μm long	8
8. Phialides	aggregated, with indistinct collarette, conidia mainly cylindrical	*P. cinerescens*
	not aggregated, with conspicuous collarette, conidia mainly ovate	*P. fastigiata*

Phialophora atrovirens (Beyma) Schol-Schwarz

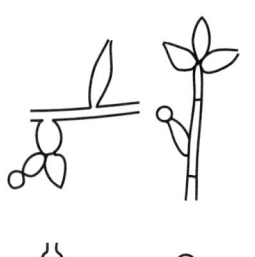

References: Schol-Schwarz 1970; Watanabe et al. 1986a.

Morphology: Conidiophores (phialides) pale brown, erect, simple or branched, aggregated, totally short and thick, bearing spore masses apically: phialides with indistinct collarette. Conidia hyaline, globose, guttulate with one large oil globule. Chlamydospores brown, globose, catenulate.

Dimensions: Phialides 4.5–10.8 × 2–3.5 µm. Conidia 1.7–4.5 µm in diameter. Chlamydospores 6.2–8.8 µm in diameter.

Material: 84-524 (Japanese black pine seed, Okawa, Kagawa, Japan).

Remarks: Short and thick phialides are characteristic of this fungus.

A, B

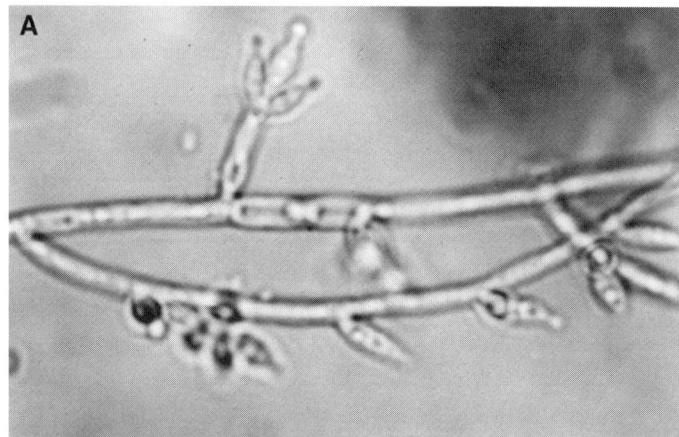

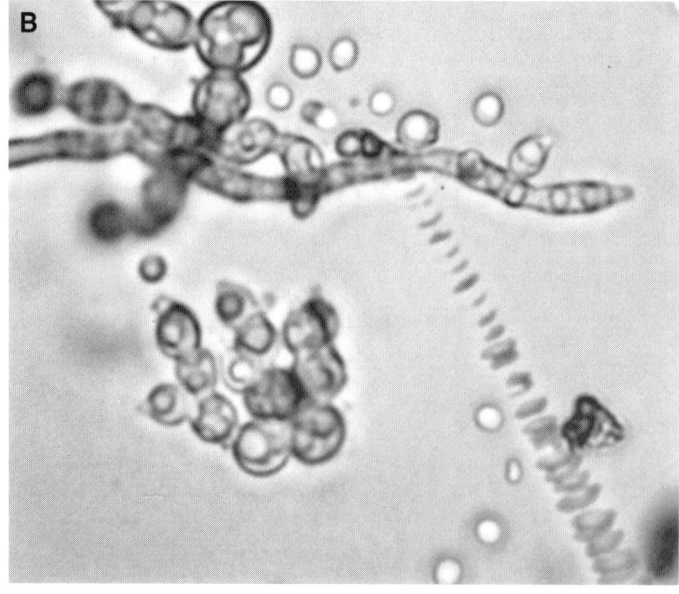

Phialophora atrovirens. A,B: Conidiophores, phialides, and conidia.

Phialophora cinerescens (Wollenw.) van Beyma

References: Schol-Schwarz 1970; Watanabe et al. 1986a.

Morphology: Conidiophores (phialides) brown, erect, simple or branched, aggregated in clusters, bearing spore masses apically: phialides with indistinct collarette. Conidia hyaline, cylindrical or ovate, 1-celled, slightly apiculate at one end.

Dimensions: Conidiophores 11.5–40 × 2.2–2.5 µm: phialides 5.5–12.5 µm long. Conidia 3.7–5 × 1.8–2.6 µm.

Material: 86-47 (Japanese cedar seed, Kasama, Ibaraki, Japan); 86-49 (Japanese black pine seed, Kyoto, Japan).

Remarks: Colonies on PDA are dark gray, more or less restricted in growth.

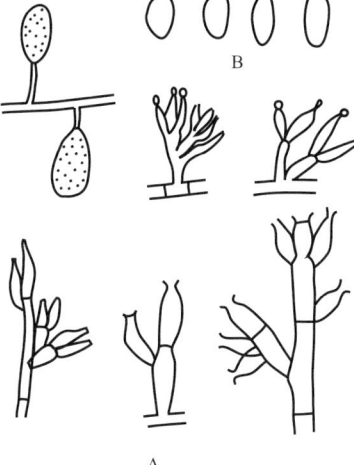

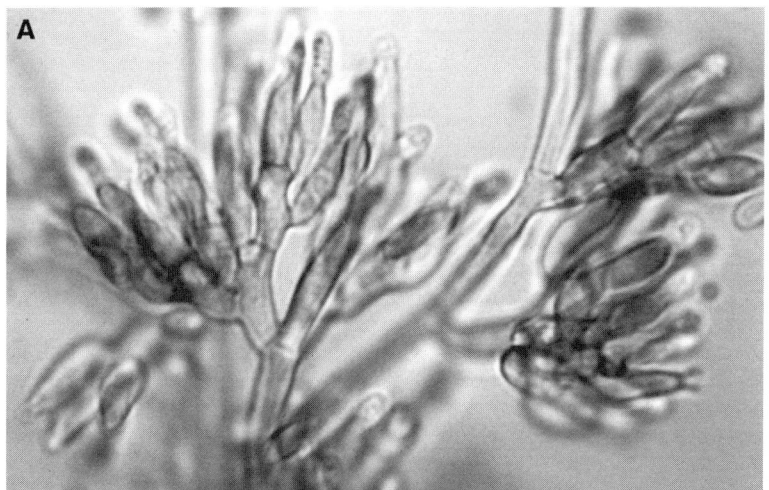

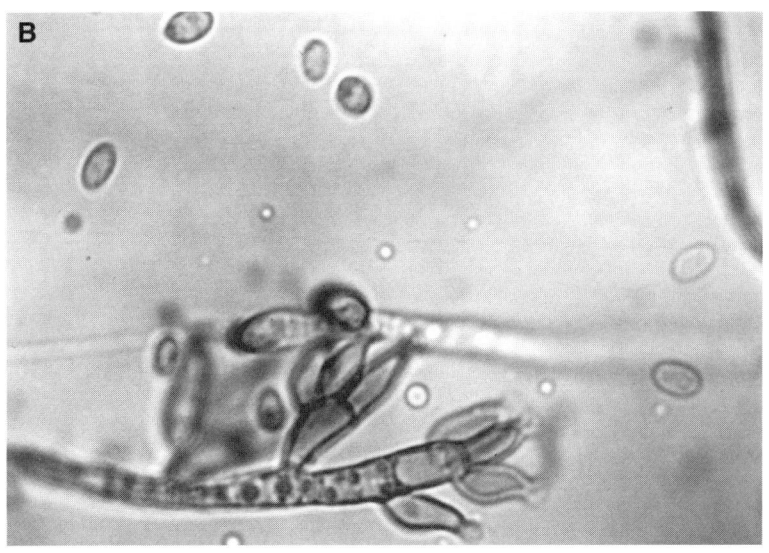

Phialophora cinerescens. A,B: Conidiophores, phialides, and conidia.

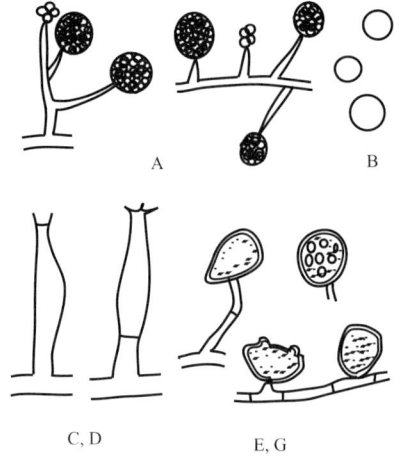

Phialophora cyclaminis van Beyma

References: Schol-Schwarz 1970; Watanabe 1975d; Watanabe et al. 1986a.

Morphology: Conidiophores (phialides) brown, erect, simple or branched, septate or aseptate, bearing spore masses apically, occasionally proliferated from spore masses: phialides often constricted in the median, with conspicuous collarette. Conidia hyaline, globose. Chlamydospores brown, subglobose or irregular in shape, solitary or occasionally twins, usually granular.

Dimensions: Conidiophores (10-) 17.5–26.3 (-36) × (2-) 2.5–2.8 µm: collarettes ca. 2 µm wide, 1 µm deep. Spore masses 9–20 µm in diameter. Conidia (1.2-) 1.5–2.5 µm in diameter. Chlamydospores 9.7–13.4 × 7.2–11 µm.

Material: 72-X142-988 (= ATCC 32923, Sugarcane root, Taiwan, ROC); 74-682 (Strawberry root, Shizuoka, Japan); 85-P32 (Paulownia root, Itapua, Paraguay).

Remarks: No chlamydospores were recorded in the original description.

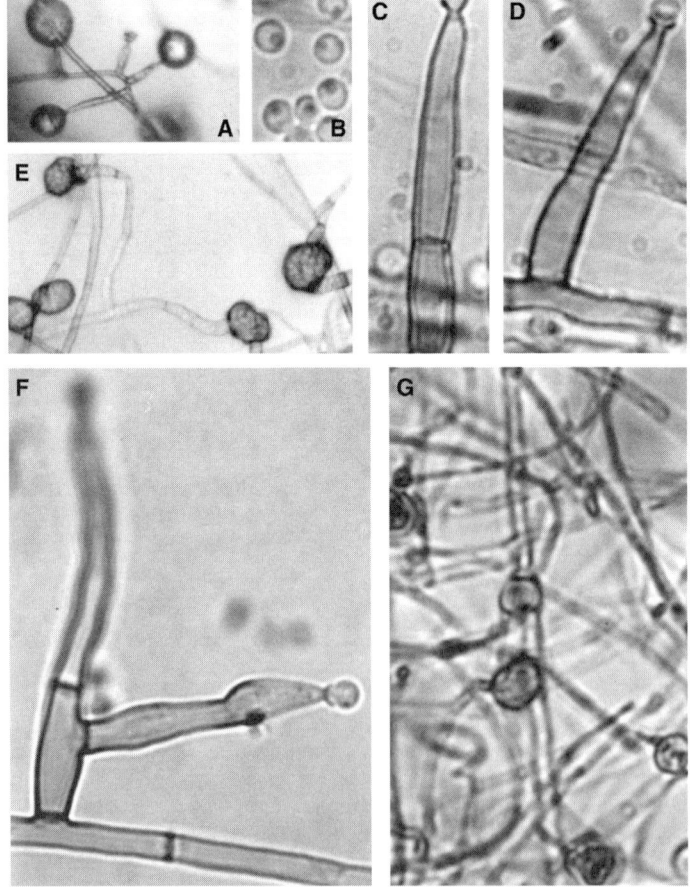

Phialophora cyclaminis. A: Conidiophores (phialides) and spore masses. B: Conidia. C,D,F: Phialides. E,G: Chlamydospores (A–E: Isolate 72-X142; F,G: 85-P32). (From Watanabe, T. 1975d. *Trans. Mycol. Soc. Jpn.*, 16:264–267. With permission.)

Phialophora fastigiata (Lagerberg & Melin) Conant

References: Cole and Kendrick 1973; Schol-Schwarz 1970; Watanabe et al. 1986a.

Morphology: Conidiophores (phialides) pale brown, erect, simple or branched, bearing spore masses apically: phialides with conspicuous collarette. Conidia hyaline, ovoid, 1-celled.

Dimensions: Conidiophores 11.7–30 × 2.2–2.7 µm: phialides 7.5–11.7 × 2.2–2.7 µm. Conidia 2.5–5.4 (-6) × 1.7–3.8 µm.

Material: 84-574, 84-575 (Japanese red pine seed, Saihaku, Tottori, Japan); 84-604, 84-606 (Japanese red pine seed, Kamiina, Nagano, Japan).

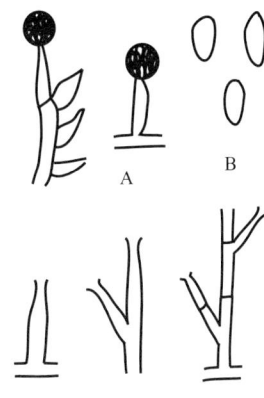

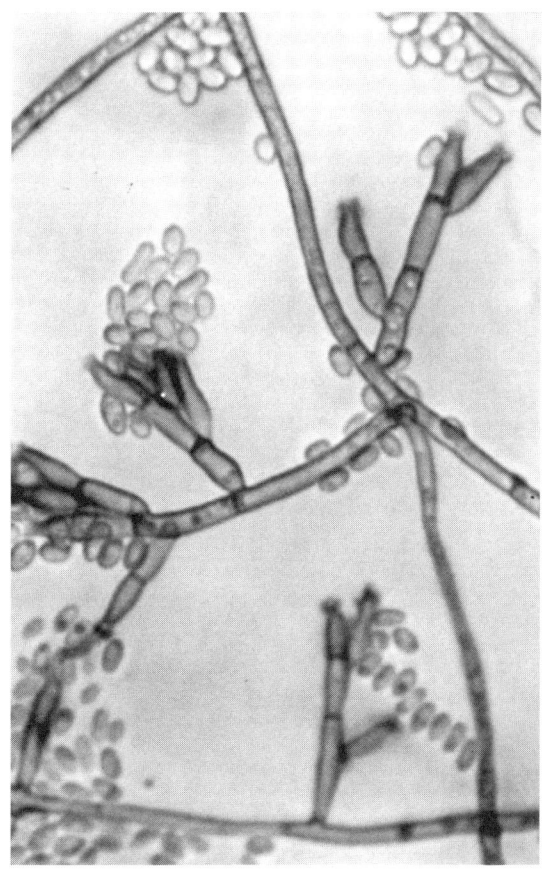

Phialophora fastigiata. Conidiophores, phialides, and conidia.

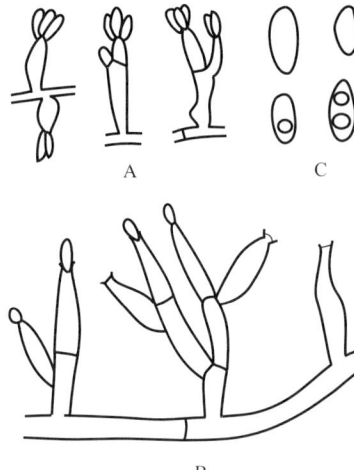

Phialophora malorum (Kidd & Beaumont) McColloch

References: Schol-Schwarz 1970; Watanabe et al. 1986a.

Morphology: Conidiophores (phialides) hyaline or slightly pigmented, simple or branched, bearing spore masses apically: phialides with indistinct collarette. Conidia hyaline, ovate, broadly ellipsoidal, 1-celled, guttulate with 1–2 oil globules.

Dimensions: Phialides ca. 12 µm. Conidia 6.3–12.6 × 2.1–2.2 µm.

Material: 83-37 (Japanese red pine seed, Hiba, Hiroshima, Japan).

Remarks: Colonies on PDA are grayish green, nonaerial.

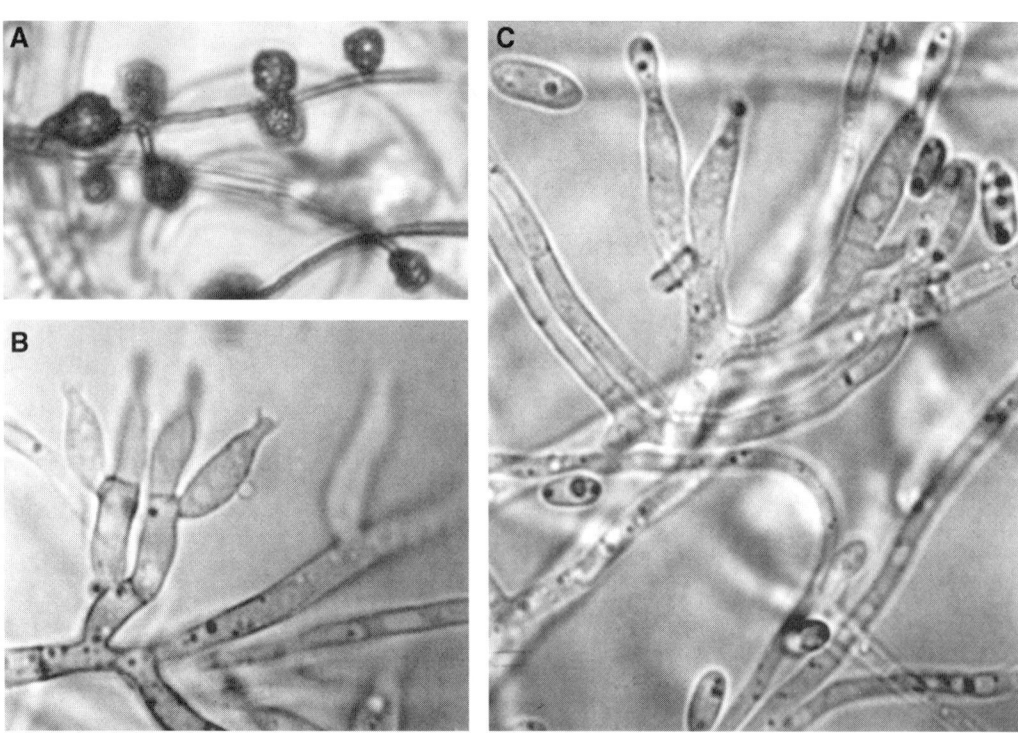

Phialophora malorum. A: Conidiophores and spore masses. B: Conidiophores. C: Phialides and conidia.

Phialophora radicicola Cain

References: Cain 1952; McKeen 1952a; Scott 1970; Watanabe 1975d.

Morphology: Conidiophores (phialides) brown, branched, bearing spore masses apically: phialides with indistinct collarette. Conidia hyaline, lunar-shaped, boat-shaped, 1-celled. Chlamydospores brown, globose, catenulate. Hyphae often undulate.

Dimensions: Conidiophores ca. 94 μm tall: phialides 7.5–11.3 μm long. Spore masses 6–7 μm in diameter. Conidia 4.8–8.3 × 1.4–2.5 μm. Chlamydospores 14–15 μm in diameter.

Material: 72-X146-1048 (= ATCC 32924, Sugarcane root, Taiwan, Japan).

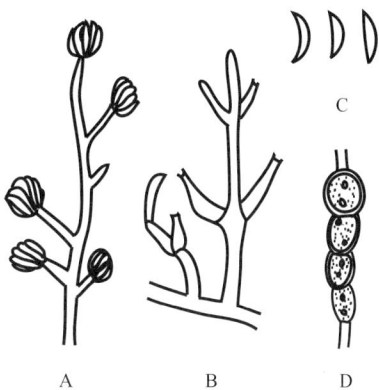

Remarks: Colonies on agar cultures are dark gray to black. This fungus is morphologically close to *Gaeumannomyces graminis* (Sacc.) v. Arx & Olivier. Comparative studies of both fungi were conducted by Decon (1973, 1974).

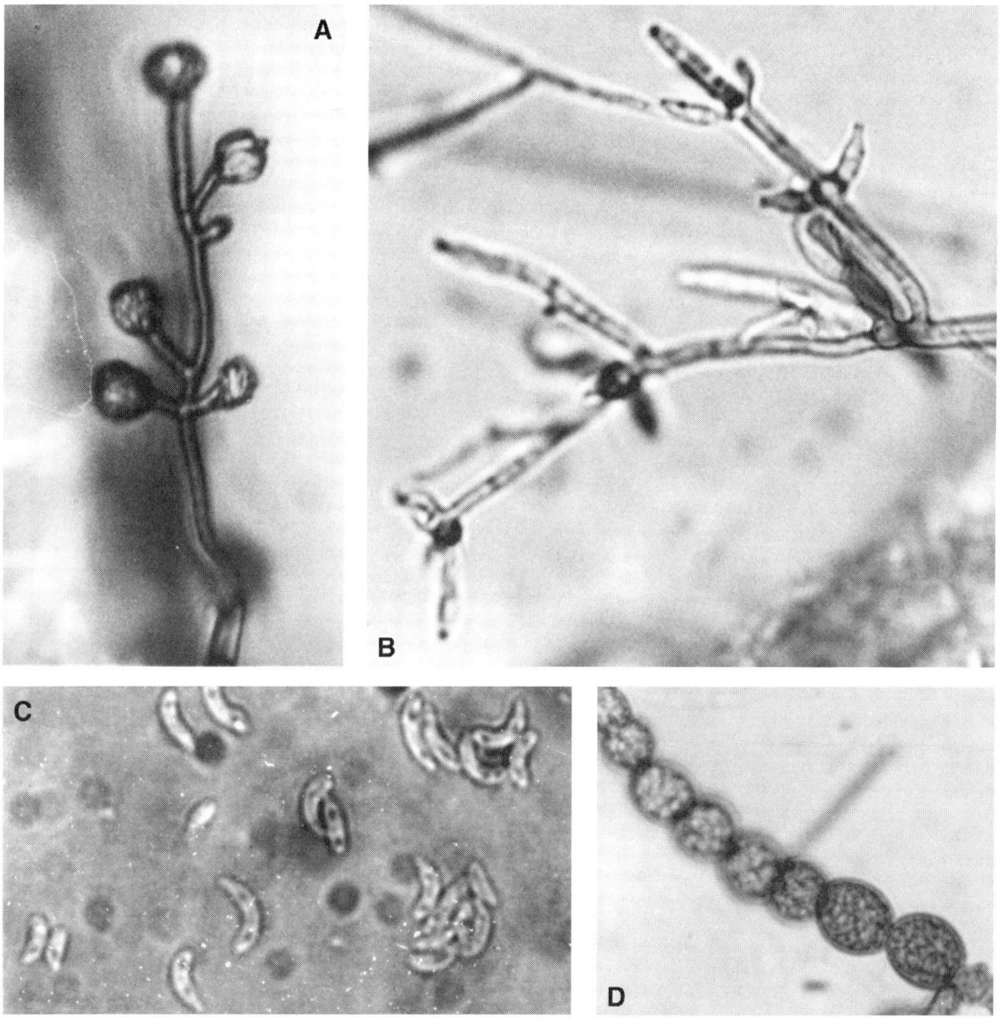

Phialophora radicicola. A: Conidiophores and conidia. B: Conidiophores. C: Conidia. D: Catenulate chlamydospores. (From Watanabe, T. 1975d. *Trans. Mycol. Soc. Jpn.*, 16:264–267. With permission.)

Phialophora richardsiae (Melin et Nannfeldt) Conant

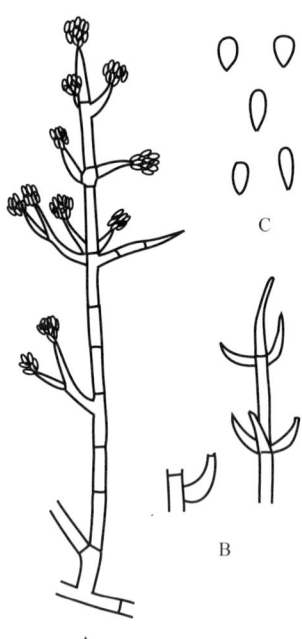

References: Cole and Kendrick 1973; Schol-Schwarz 1970; Watanabe 1975d.

Morphology: Conidiophores (phialides) pale brown, erect, bearing spore masses on verticillate phialides in a few positions at the apexes: phialides with indistinct collarette. Conidia hyaline, 1-celled, of two kinds: globose and ovate to ellipsoidal.

Dimensions: Conidiophores ca. 70 μm tall: phialides 12.5–17.5 μm long. Spore masses 11.2–27.5 μm in diameter. Conidia: ovate 3.7–4.3 × 2–2.8 μm, and globose 3.7–5.5 in diameter.

Material: 72-X148 (= ATCC 32925, Sugarcane root, Taiwan, ROC).

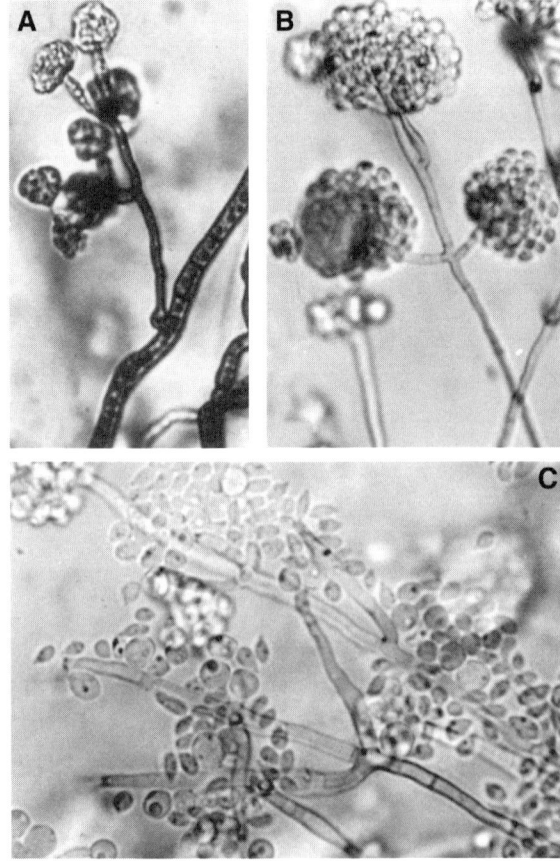

Phialophora richardsiae. A,B: Conidiophores, phialides, and spore masses. C: Ovate and globose conidia. (From Watanabe, T. 1975d. *Trans. Mycol. Soc. Jpn.*, 16:264–267. With permission.)

Phialophora sp. 1.

References: Cole and Kendrick 1973; Schol-Schwarz 1970.

Morphology: Conidiophores (phialides) pale brown, erect, simple or branched, developed often on aerial hyahae, bearing catenulate conidia (often over 165 µm long) or spore masses: phialides ampulliform (flask-shaped) or cylindrical with conspicuous collarette. Conidia hyaline, subglobose, truncate at one end.

Dimensions: Conidiophores 7–11.3 × 3–4.8 µm: collarettes 1.2–2.3 µm wide. Conidia 1.2–2.0 µm in diameter.

Material: 73-410 (Strawberry root, Tsu, Mie, Japan).

Remarks: This fungus is morphologically close to *P. verrucosa* Medlar, in forming flask-shaped phialides with conspicuous collarettes, but their conidia are globose with truncate base, and ellipsoidal, respectively.

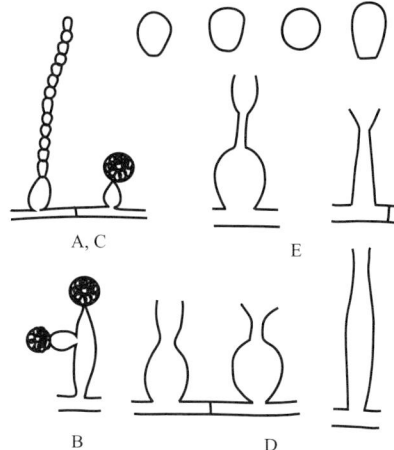

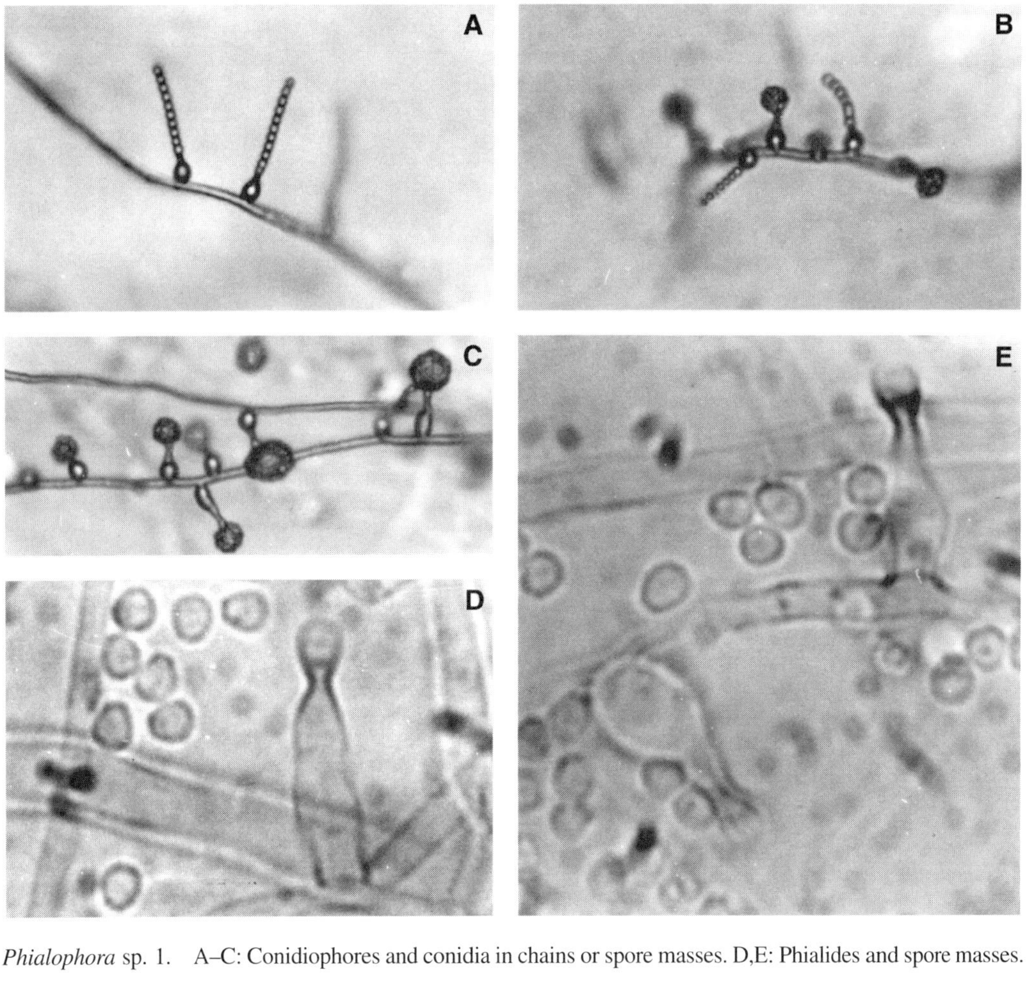

Phialophora sp. 1.   A–C: Conidiophores and conidia in chains or spore masses. D,E: Phialides and spore masses.

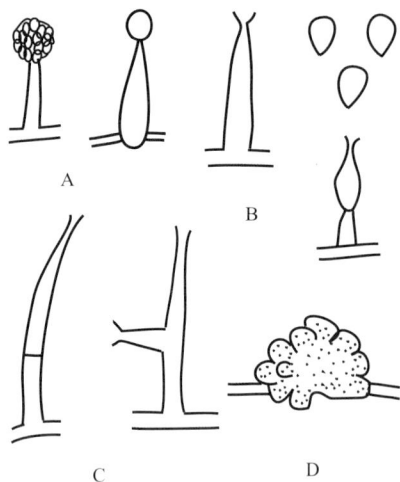

Phialophora sp. 2.

References: Cole and Kendrick 1973; Schol-Schwarz 1970.

Morphology: Conidiophores (phialides) pale brown, erect, simple or branched, bearing spore masses apically: phialides tapering from base toward apex with well-developed collarette. Conidia hyaline, subglobose, apiculate at one end. Sclerotia brown.

Dimensions: Phialides 12.5–30 × 2–3.8 µm: collarettes ca. 2 µm wide, 0.7 µm deep. Conidia 1.7–2.5 µm in diameter. Sclerotia 30–75 µm in diameter.

Material: 74-750 (Strawberry root, Mie, Japan).

Remarks: This fungus is morphologically close to *P. cyclaminis* except for sclerotium formation.

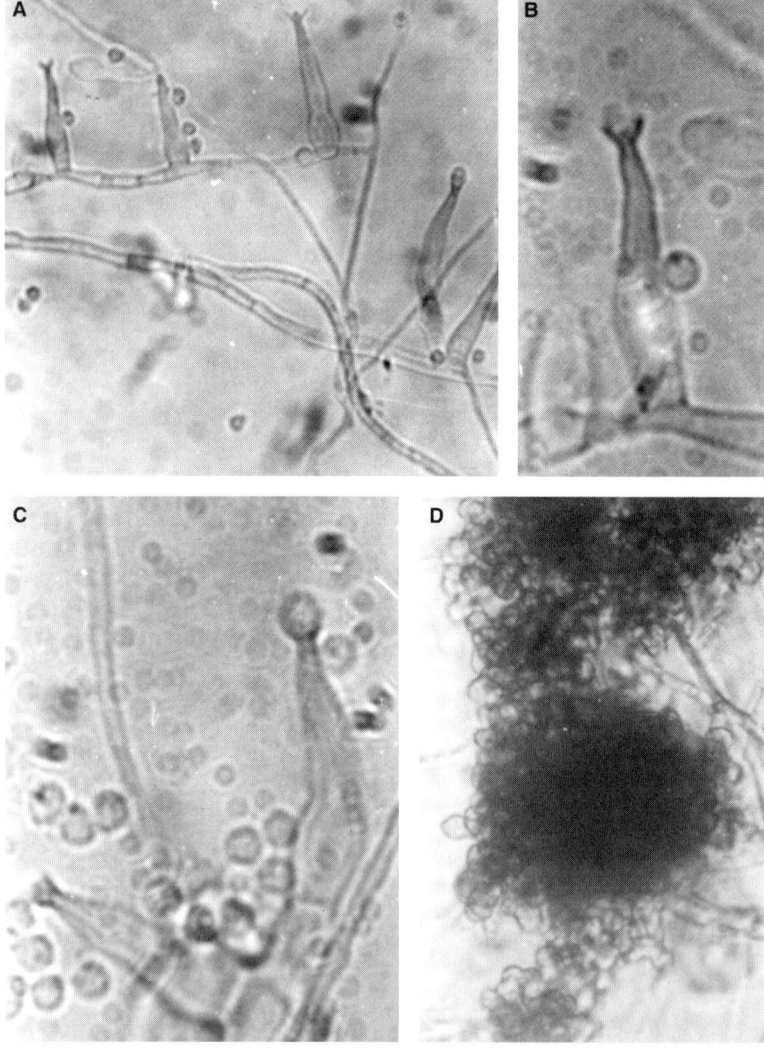

Phialophora sp. 2.   A–C: Conidiophores (phialides) and conidia. D: Sclerotia.

Phoma Sacc.

Michelia 2:4, 1880.
Type species: *P. herbarum* Westend

Key to Species

1. Chlamydospores muriform. .... *P. glomerata*
   not so 2

2. Chlamydospores catenulate .... *P. medicaginis* var. *pinodella*
   not so .... *Phoma* sp.

Remarks: Among over 2000 species were described in *Phoma*, nearly 40 are valid (Hawksworth et al., 1995).

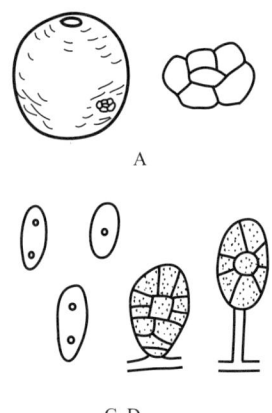

Phoma glomerata (Corda) Wollenw. & Hochapf.

References: Dorenbosch 1970; Sutton 1980; Watanabe et al. 1986a; White and Morgan-Jones 1987.

Morphology: Pycnidia globose or subglobose, dark brown, conspicuously ostiolate: peridium dark brown, pseudoparenchymatous. Conidia hyaline, ellipsoidal, 1-celled. Chlamydospores dark brown, subglobose, muriform with transverse and longitudinal septa rather irregularly.

Dimensions: Pycnidia 50–150 µm: ostioles 7–8 µm wide. Conidia 4.5–5.3 × 1.7–2.5 µm. Chlamydospores 15–32.5 (-42.5) × (10-) 13.8–16.8 (-20) µm.

Materials: 77-160 (Strawberry root, Shizuoka, Japan); 83-17 (Japanese red pine seed, Hiba, Hiroshima, Japan).

Remarks: Conidiophores were indistinct. This fungus is well characterized by pycnidia, and chlamydospores that resemble conidia of genus *Alternaria*.

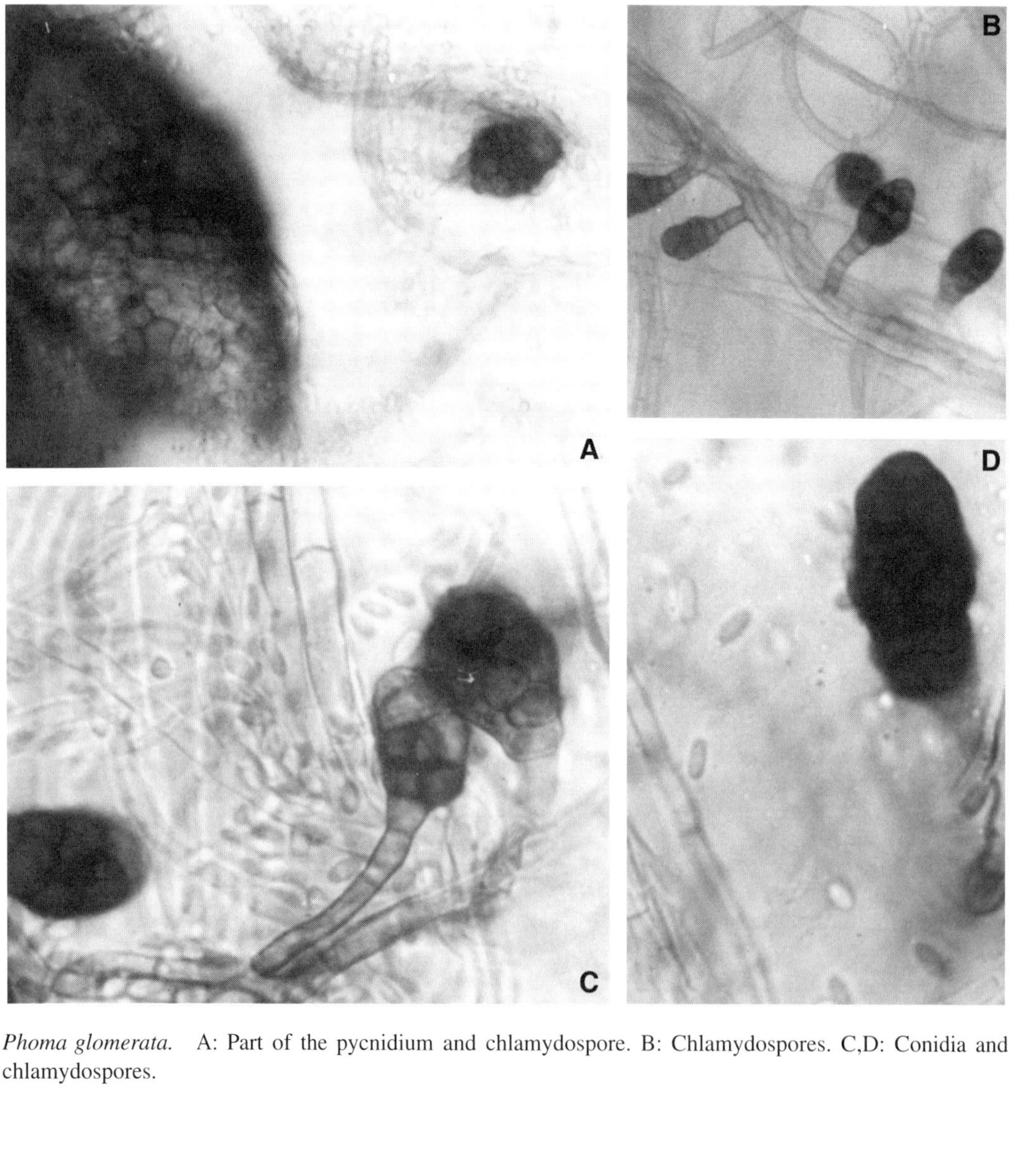

Phoma glomerata. A: Part of the pycnidium and chlamydospore. B: Chlamydospores. C,D: Conidia and chlamydospores.

Phoma medicaginis Malbr. & Roum. var. *pinodella* (L. K. Jones) Boerema

References: Dorenbosch 1970; Watanabe et al. 1987a.

Morphology: Pycnidia brown, globose or subglobose with well-developed cylindrical neck: peridium brown, pseudoparenchymatous. Conidia hyaline, cylindrical, 1-celled, with 2 oil globules. Chlamydospores brown, globose, catenulate.

Dimensions: Pycnidia 50–90 µm in diameter: necks 50–80 × 20–22.5 µm. Conidia 3.5–5.5 × 1.2–2.5 µm. Chlamydospores 7.5–11.3 µm in diameter.

Material: 85-P102 (Paulownia root, Misiones, Argentina).

Remarks: Conidiophores were indistinct. This fungus is characterized by pycnidia and catenulate chlamydospores.

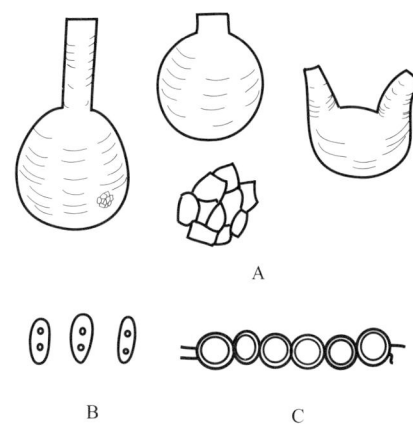

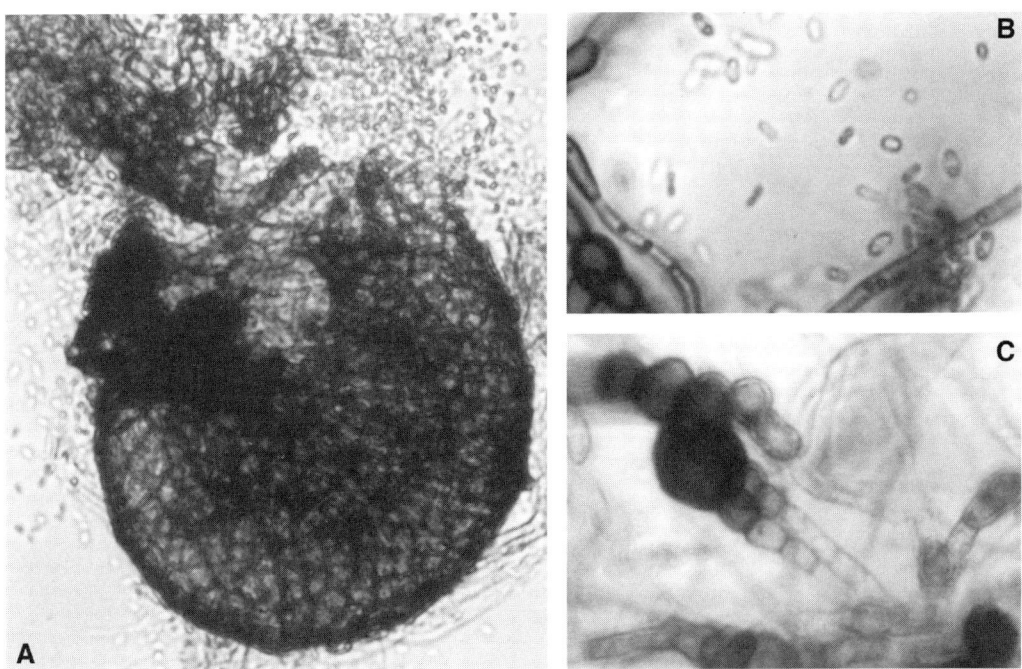

Phoma medicaginis.   A: Pycnidium and conidia. B: Conidia. C: Chlamydospores.

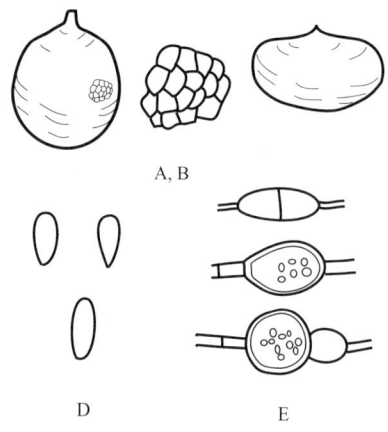

Phoma sp.

References: Dennis 1946; Dorenbosch 1970; Sutton 1964, 1980; Watanabe 1975b.

Morphology: Pycnidia globose, subglobose, or disc-shaped, conspicuously ostiolate: peridium dark brown, pseudoparenchymatous. Conidia hyaline, ellipsoidal, 1-celled. Chlamydospores solitary, dark brown, granulate, thick-walled.

Dimensions: Pycnidia 107–146 × 97.3–107 µm: ostioles 4.8–5 µm in diameter. Conidia 4.8–6.1 × 2.1–2.7 µm. Chlamydospores 10.2–18.3 µm in diameter.

Materials: 72-X24 (Sugarcane root, Taiwan, ROC).

Remarks: Conidiophores were indistinct.

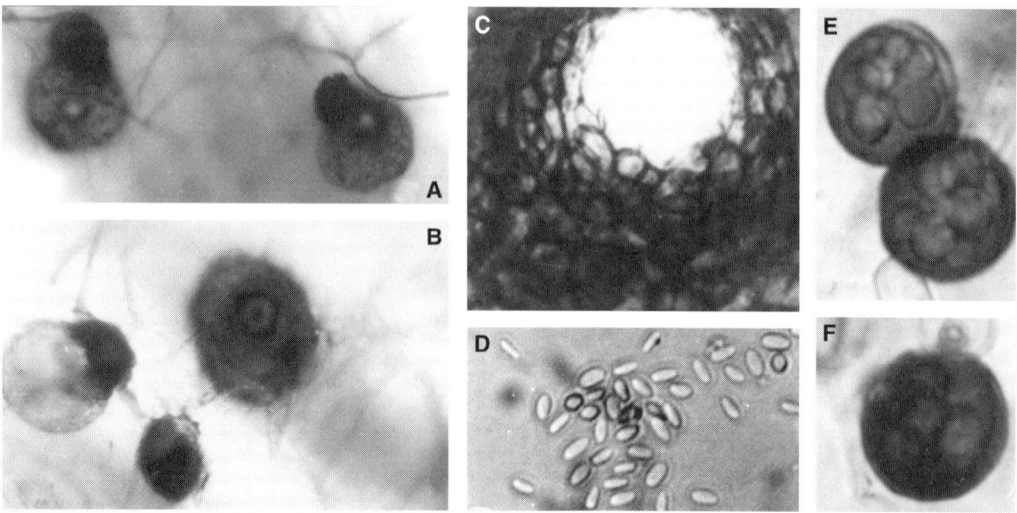

Phoma sp. A,B: Pycnidia. C: Pycnidial ostiole. D: Conidia. E,F: Chlamydospores. (From Watanabe, T. 1975b. *Trans. Mycol. Soc. Jpn.*, 16:28–35. With permission.)

Phomopsis (Sacc.) Bubák

Ann. Mycol. 3:166, 1905.
Type species: *P. oblonga* (Desm.) Hohnel

Phomopsis spp.

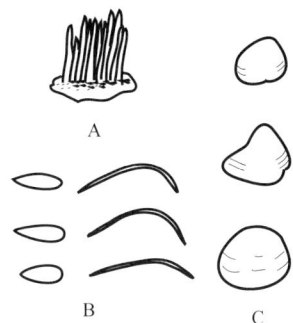

References: Hahn 1930; Hobbs et al. 1985; Sutton 1980; Watanabe et al. 1986a.

Morphology: Pycnidia half-embedded in agar media, solitary or aggregated, globose, irregular with indistinct ostioles: peridium dark brown, pseudoparenchymatous. Conidia hyaline, 1-celled, dimorphic: the alpha type, often borne in pale pink spore masses, spindle-shaped or ellipsoidal; and the beta type, borne in white spore masses, filiform, curved.

Dimensions: Pycnidia ca. 450×800 μm. Conidiophores ca. 25×1.2 μm. Conidia: alpha type $(5.7-) 8.4-12.5 \times 1.8-2.5$ μm, and beta type $16.5-28 \times 0.5-1.3 (-1.6)$ μm.

Materials: 83-9 (Japanese black pine seed, Kumano, Mie, Japan); 83-33 (Japanese cedar seed, Yoshino, Nara, Japan); 84-517, 84-518 (Japanese red pine seed, Saihaku, Tottori, Japan).

Remarks: Nearly 100 species are distributed widely (Hawksworth et al., 1996). The two conidial types are variously produced in frequency by the isolates studied.

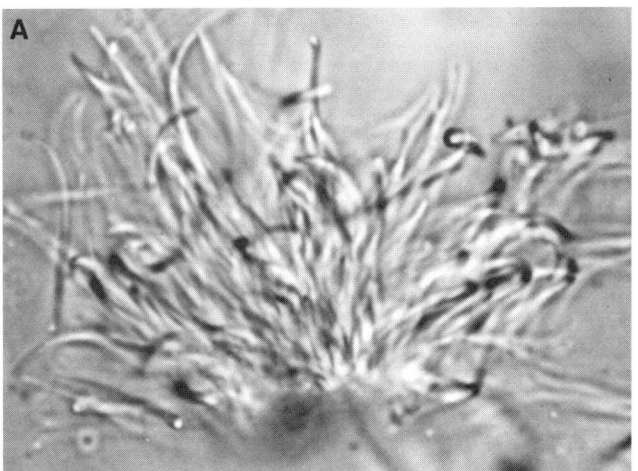
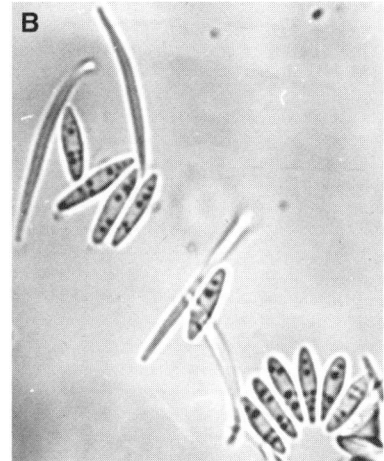

Phomopsis sp.　A: Conidiophores. B: Two kinds of conidia. C: Pycnidia.

Pithomyces Berk. & Br.

J. Linn. Soc. London 14:100, 1873.
Type species: *P. flavus* Berk. & Br.

Pithomyces chartarum (Berk. & Curt) M. B. Ellis

References: Ellis 1960, 1971; Watanabe et al. 1986a, 1987a.

Morphology: Conidiophores hyaline, simple, short, bearing single conidia apically. Conidia aleuriosporous, brown, ovate or ellipsoidal, muriform, usually composed of 3 transverse septa and 1–2 longitudinal septa, smooth marginally, constricted at or near cross septa.

Dimensions: Conidia (20-) 21.2–25 × 12.5–15 (-17.5) µm.

Materials: 84-585 (Japanese black pine seed, Okawa, Kagawa, Japan); 85-P23 (Paulownia root, Itapua, Paraguay).

Remarks: Colonies on PDA are dark green to black, resupinate with zonation.

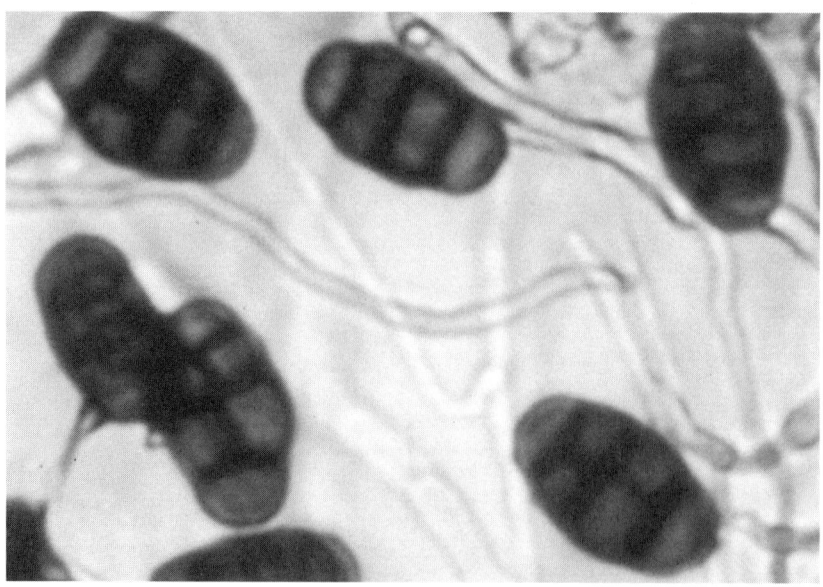

Pithomyces chartarum. Conidia.

Pithomyces maydicus (Sacc.) M. B. Ellis

References: Ellis 1960, 1971; Watanabe et al. 1986a.

Morphology: Conidiophores hyaline, simple, short, bearing single conidia apically. Conidia aleuriosporous, brown, broadly ellipsoidal, muriform composed of usually 2 transverse septa and 1 longitudinal septum, smooth marginally.

Dimensions: Conidiophores ca. 9 μm long. Conidia 20–23.4 × 12.5–12.6 μm.

Materials: 84-516 (Japanese black pine seed, Okawa, Kagawa, Japan).

Remarks: The number of transverse septum of *P. maydicus* and *P. chartarum* is 2 and 3, respectively.

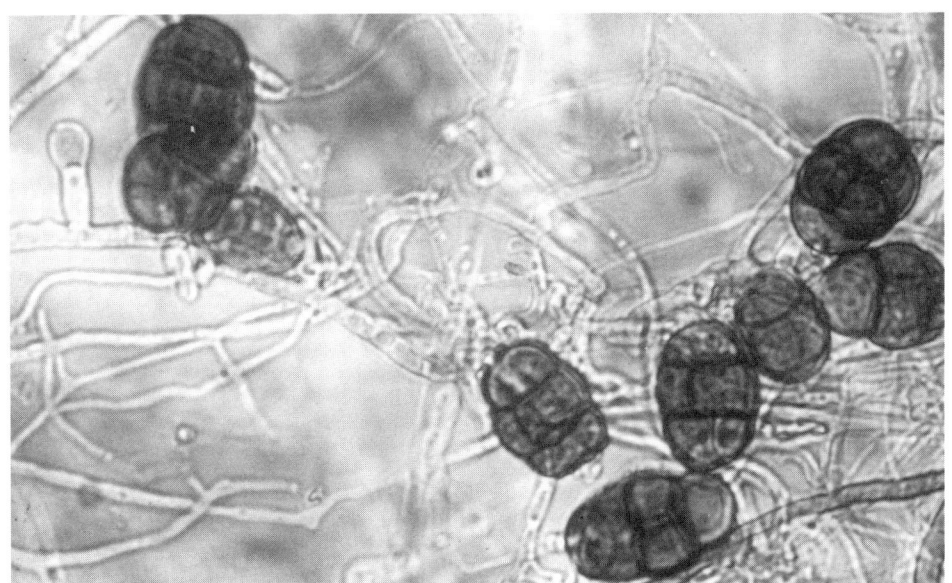

Pithomyces maydicus. Conidia and conidiophore.

Pyrenochaeta De Not.

Micromycetes Ital. 5:15, 1845.
Type species: *P. nobilis* De Not.

Key to Species

1. Conidia   globose .... *P. globosa*

                    ellipsoidal 2

2. Pycnidia   well ostiolate, conidiophores ampulliform with round basis,
                        conidia cylindrical .... *P. gentianicola*

                    indistinctly ostiolate, conidiophores tapering from base
                        toward apex, conidia broadly ellipsoidal .... *P. terrestris*

Pyrenochaeta gentianicola T. Watanabe

References: Barnett and Hunter 1987; Watanabe and Imamura 1977, 1995.

Morphology: Pycnidia globose or subglobose, well-necked with setae around: peridium brown, pseudoparenchymatous: setae brown, thick-walled, tapering from base toward apex, septate. Conidiophores hyaline, simple, ampulliform with abruptly sharpened or narrowed tips from the median, occasionally septate. Conidia hyaline, long-ellipsoidal, 1-celled with 2 oil globules.

Dimensions: Pycnidia (82-) 107–297 (-333.5) μm in diameter: necks 49–210 × 39.5–75 μm: setae (12-) 35–200 μm long: 3–7.5 μm wide basally: 2–3 μm wide apically. Conidia 2.8–5.2 × 0.5–1.5 (-2.5) μm. Conidiophores 4.7–9.5 × 2.8–4.8 μm, 0.9 μm wide apically. Chlamydospores 15–26.3 μm in diameter. Sclerotia 27.5–95 μm in diameter.

Materials: 76-501 (Gentian root, Chino, Nagano, Japan).

Remarks: Colonies on PDA are dark yellowish green. This fungus causes brown root rot on cultivated gentian in Japan.

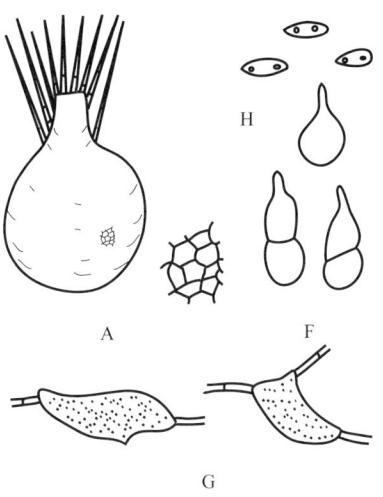

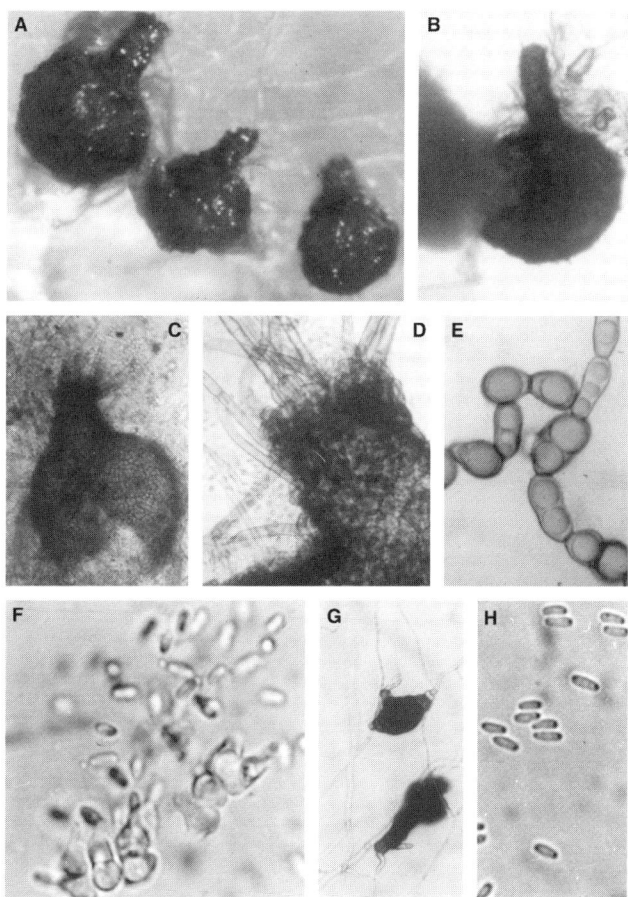

Pyrenochaeta gentianicola. A,B: Pycnidia. C,D: Pycnidia and setae. E: Hyphae with rich oil globules. F: Conidiophores and conidia. G: Sclerotia. H: Conidia.

Pyrenochaeta globosa T. Watanabe

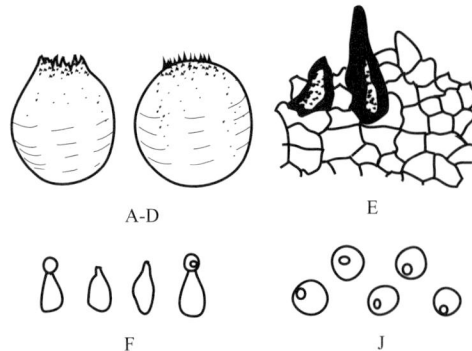

A-D E

F J

References: Sutton 1980; Watanabe 1992c.

Morphology: Pycnidia brown, spherical, semiellipsoidal, flask-shaped, with 10–20 setae around apical ostioles, surrounded with thick hyphae basally: peridium pale brown, double membranous, pseudoparenchymatous. Setae dark brown, thick-walled, short. Conidiophores simple, short and thick, nonseptate. Conidia phialosporous, hyaline, globose, slightly angular.

Dimensions: Pycnidia 50–87.5 × 30–72.5 μm: setae up to 26 μm long, 4.5–6 μm wide. Conidiophores 4–7.5 × 2–3 μm. Conidia 2–2.4 μm in diameter.

Materials: 84-523 (= IFO 32549, Japanese black pine seed, Okawa, Kagawa, Japan).

Remarks: Setae around ostioles of pycnidia and globose conidia are very characteristic for this species, although the conidiophores are simple. Branched conidiophores may be more significant than setae around ostioles of pycnidia in literature.

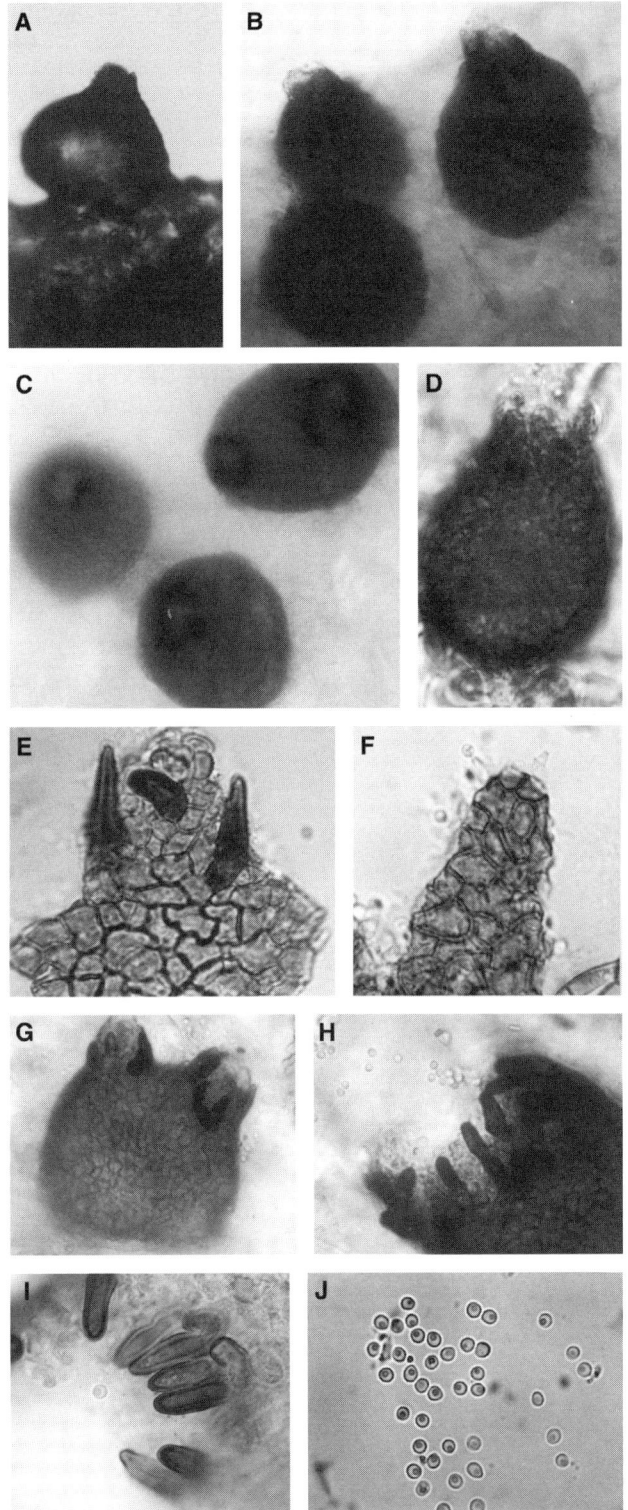

Pyrenochaeta globosa. A–F: Pycnidia. E: Part of peridium and setae. F: Conidiophores (phialides) on peridium and conidia. G–I: Part of pycnidia and setae. J: Conidia. (From Watanabe, T. 1992c. *Trans. Mycol. Soc. Jpn.*, 33:21–24. With permission.)

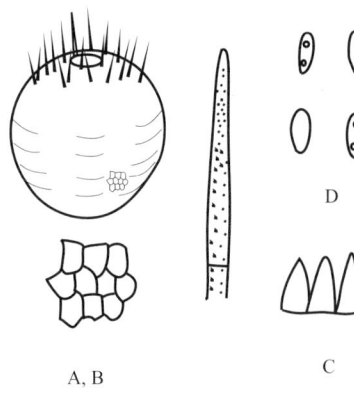

A, B C

Pyrenochaeta terrestris (Hansen) Gorenz, Walker & Larson

Synonym: *Phoma terrestris* E. M. Hansen

References: Gorenz et al. 1948; Sutton 1980; Watanabe 1975b, 1995.

Morphology: Pycnidia globose or subglobose, ostiolate or necked apically with setae around, covered with hyphae all over: peridium dark brown, pseudoparenchymatous: setae brown, thick-walled, tapering from base toward apex, 1- to 6-septate. Conidiophores simple, hyaline, short, tapering toward apex. Conidia hyaline, ellipsoidal or cylindrical, 1-celled.

Dimensions: Pycnidia 175–249 μm in diameter: ostioles ca. 20 μm in diameter: setae ca. 50 μm long. Conidia 4.1–5.2 × 1.4–2.5 μm.

Materials: 72-X141-897 (= ATCC 32327, Sugarcane root, Taiwan, ROC); 77-167 (Strawberry root, Shizuoka, Japan).

Remarks: Cultures on PDA are grayish, often pinkish tinted.

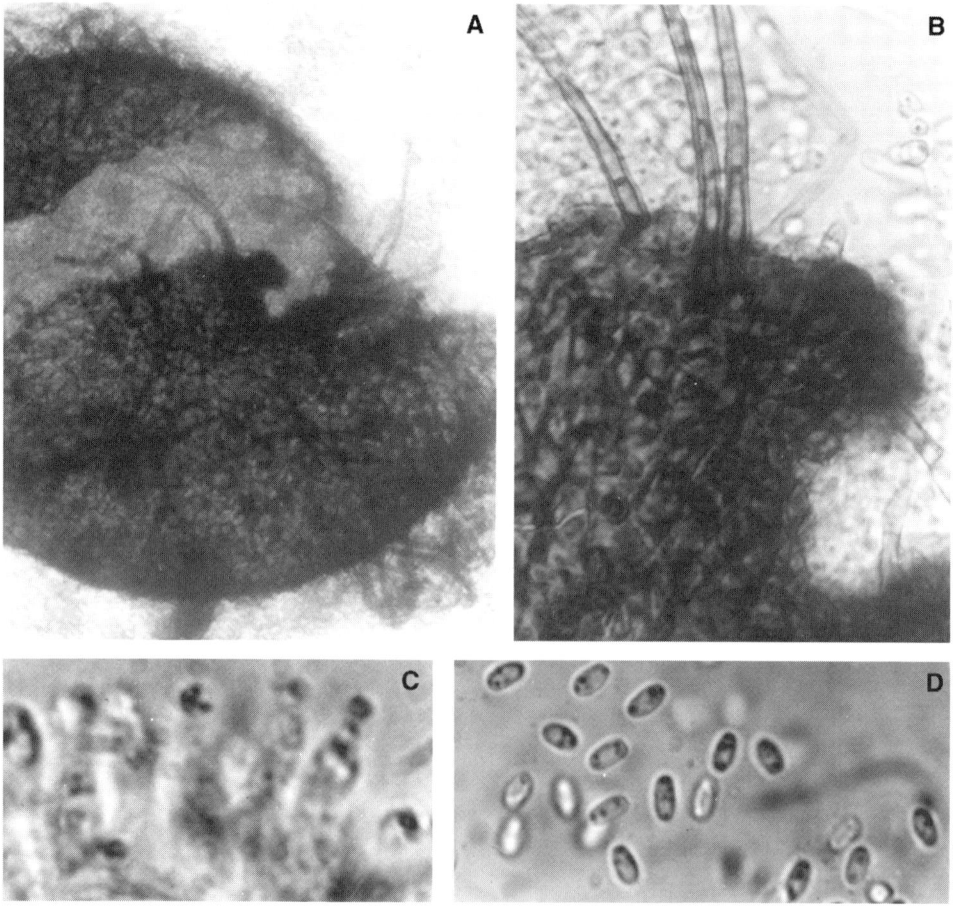

Pyrenochaeta terrestris. A: Pycnidium crushed. B: Setae around ostiolar region. C: Conidiophores. D: Conidia. (From Watanabe, T. 1975b. *Trans. Mycol. Soc. Jpn.*, 16:28–35. With permission.)

Ramichloridium Stahel : de Hoog

Stud. Mycol. 15:59, 1977.
Type species: *R. apiculatum* (Miller et al.) de Hoog

Ramichloridium anceps (Sacc. & Ellis) de Hoog

Synonym: *Rhinocladiella anceps* (Sacc. & Ellis) Hughes

References: Hoog and Hermanides-Nijhof 1977; Schol-Schwarz 1968; Watanabe et al. 1986a, 1987a.

Morphology: Conidiophores pale brown, erect, simple or branched, bearing conidia in whorls in numerous levels of apical and central parts, gradually tapering toward apex, denticulate after detachment of conidia. Conidia sympodulosporous, terminal and lateral, hyaline or pale brown, lacrymoid, ovate, ellipsoidal, apiculate at one end, 1-celled.

Dimensions: Conidiophores 160–235 × 1.6–3.3 μm. Conidia 3.6–6.5 × 1.6–4.5 μm.

Material: 84-525 (Japanese black pine seed, Okawa, Kagawa, Japan); 85-13R-2 (Paulownia root, Itapua, Paraguay).

Remarks: Conidiophores of *R. anceps* and *R. sublatum* de Hoog are over 160 μm long, and under 100 μm long, respectively.

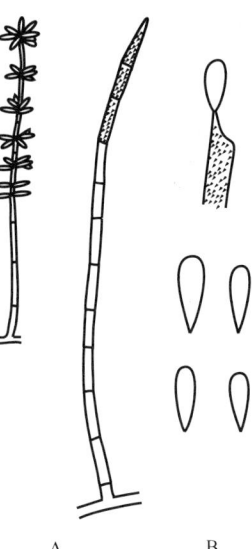

A   B

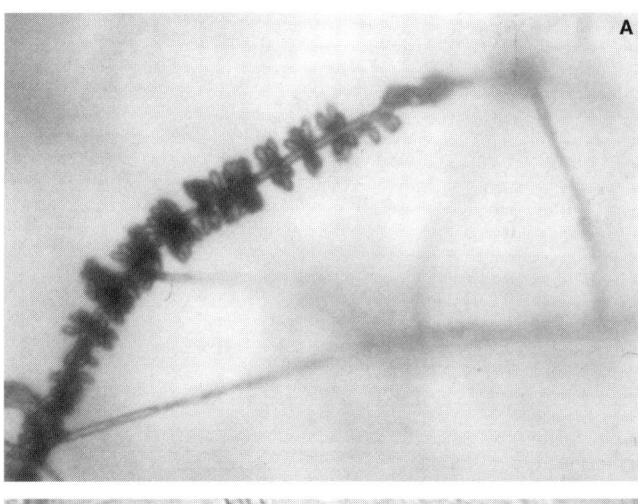

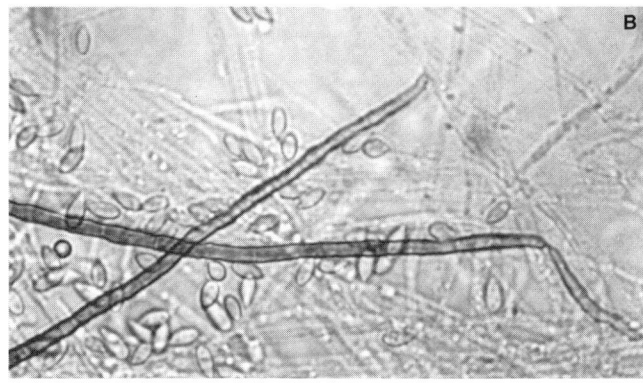

Ramichloridium anceps. A,B: Conidiophores and conidia. (From Watanabe, T. et al. 1987a. *Trans. Mycol. Soc. Jpn.*, 28:453–469. With permission.)

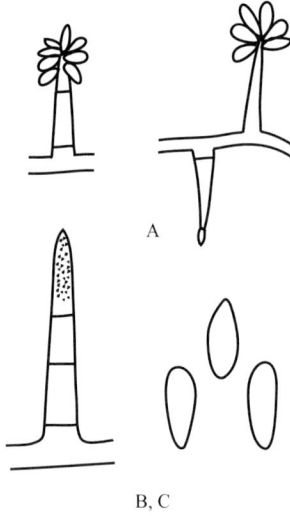

Ramichloridium subulatum de Hoog

References: Hoog and Hermanides-Nijhof 1977; Hoog et al. 1983; Watanabe and Sato 1988.

Morphology: Conidiophores pale brown, simple or rarely branched, gradually tapering toward apex, bearing conidia in whorls in more than 3 levels of the apical and central fertile portions, denticulate after detachment of conidia. Conidia sympodulosporous, terminal or lateral, hyaline or subhyaline, long-ellipsoidal or ovate, apiculate or truncate at one end, 1-celled.

Dimensions: Conidiophores 22.5–55 (-100) × (2.2-) 3.5–5.5 µm. Conidia (2.5-) 5–6.3 (-10) × (1.5-) 2.2–2.8 (-3.3) µm.

Material: 74-550 (Strawberry root, Shizuoka, Japan); 77-235 (Strawberry root, Nishigahara, Tokyo, Japan); 83-71 (Japanese cedar seed, Yoshino, Nara, Japan).

Remarks: A key of the *Ramichloridium* species was prepared by Hoog et al. (1983).

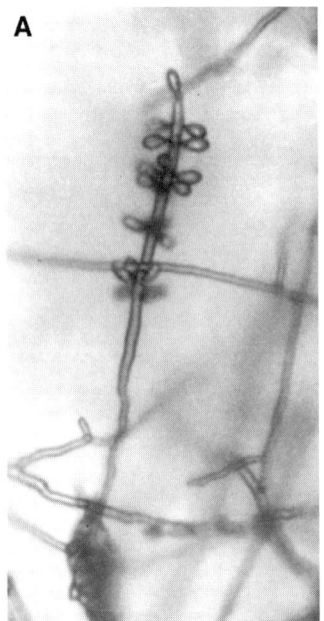
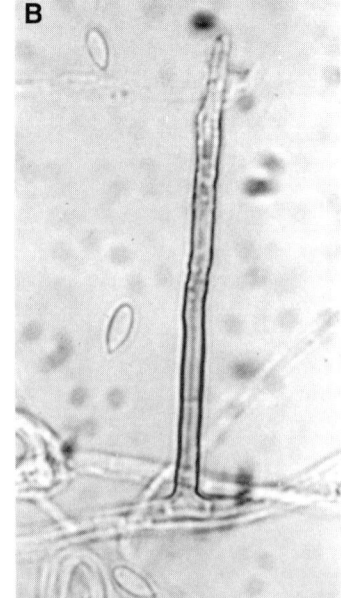
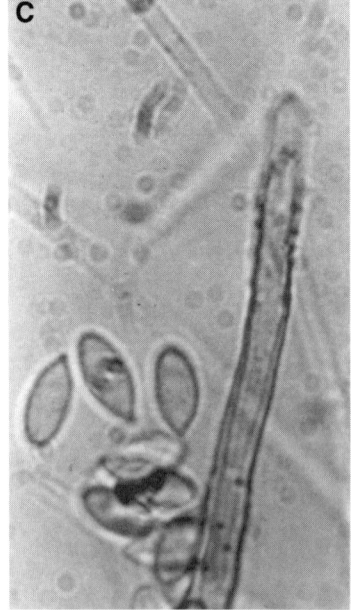

Ramichloridium subulatum. A–C: Conidiophores and conidia. (From Watanabe, T. and Sato, Y. 1988. *Trans. Mycol. Soc. Jpn.*, 29:143–150. With permission.)

Rhizoctonia DC. : Fr.

Syst. Mycol. 2:265, 1823.
Type species: *R. crocorum* (Pers.) DC. : Fr.

Rhizoctonia solani Kühn

Teleomorph: *Thanatephorus cucumeris* (Frank) Donk

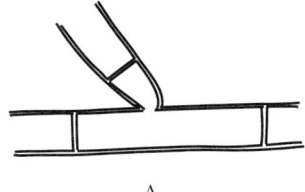
A

References: Parmeter (Ed.) 1970; Sneh et al. 1991.

Morphology: Hyphae pale brown or brown, branched, with nearly right-angled side branches constricted basally, septated closely between main hyphae and side branches. Monilioid cells composed of catenulate cells acropetally developed. Sclerotia brown to dark brown, various in shape.

Dimensions: Hyphae 5–8 (-9) μm wide. Sclerotia 1–3 mm in diameter.

Material: 73-158 (Strawberry root, Shizuoka, Japan); 85-63, 85-65 (Flowering cherry seed, Hachioji, Tokyo, Japan).

C, D

Remarks: Cultures on PDA are brown, and hyphal cells are multinucleate with over 4 nuclei per cell. The hyphal cells of *Rhizoctonia* state of genus *Ceratobasidium* are binucleate. Under subspecies level, *R. solani* isolates are now separated into at least 11 anastomosis groups.

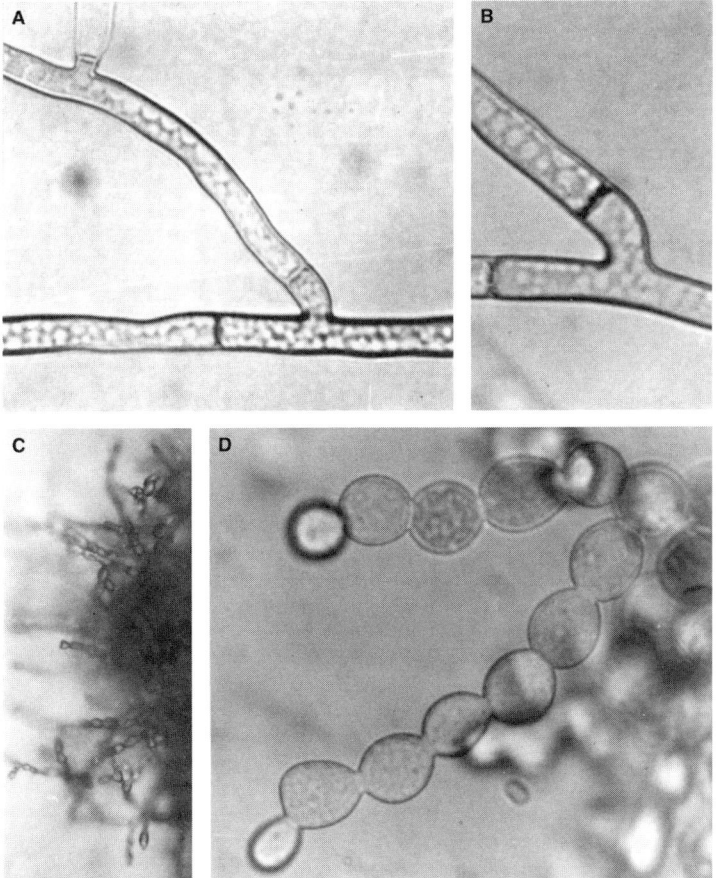

Rhizoctonia solani. A,B: Hyphae. C,D: Monilioid cells.

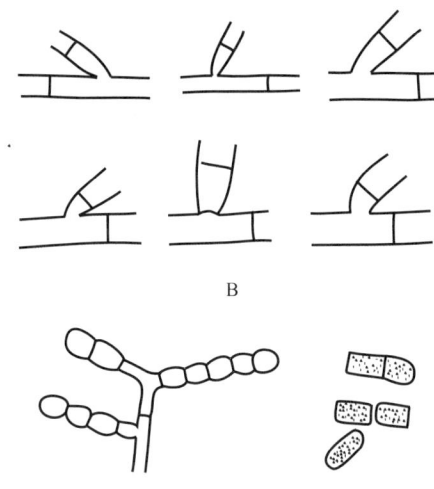

Rhizoctonia spp.

References: Parmeter (Ed.) 1970; Sneh et al. 1991.

Morphology: Conidia not formed. Hyphae pale brown, branched angularly with side branches septated closely near the main hyphae, and constricted basally. Monilioid cells usually formed. Sclerotia discrete or aggregated, pale brown to brown, various in shape and size.

Dimensions: Hyphae 6–10 µm wide. Sclerotia 1 mm long.

Material: 85-60, 85-61, 85-62, 85-63, 85-64, 85-65 (Flowering cherry seed, Hachioji, Tokyo, Japan).

Remarks: Cultures of *Rhizoctonia* spp. on PDA are separated into at least 9 colony types for the cherry seed isolates, showing 6 types among them in Figure A. The sclerotia of Isolate 85-62 are characteristically fragment-like. There are over 20 anastomosis groups in binucleate *Rhizoctonia*.

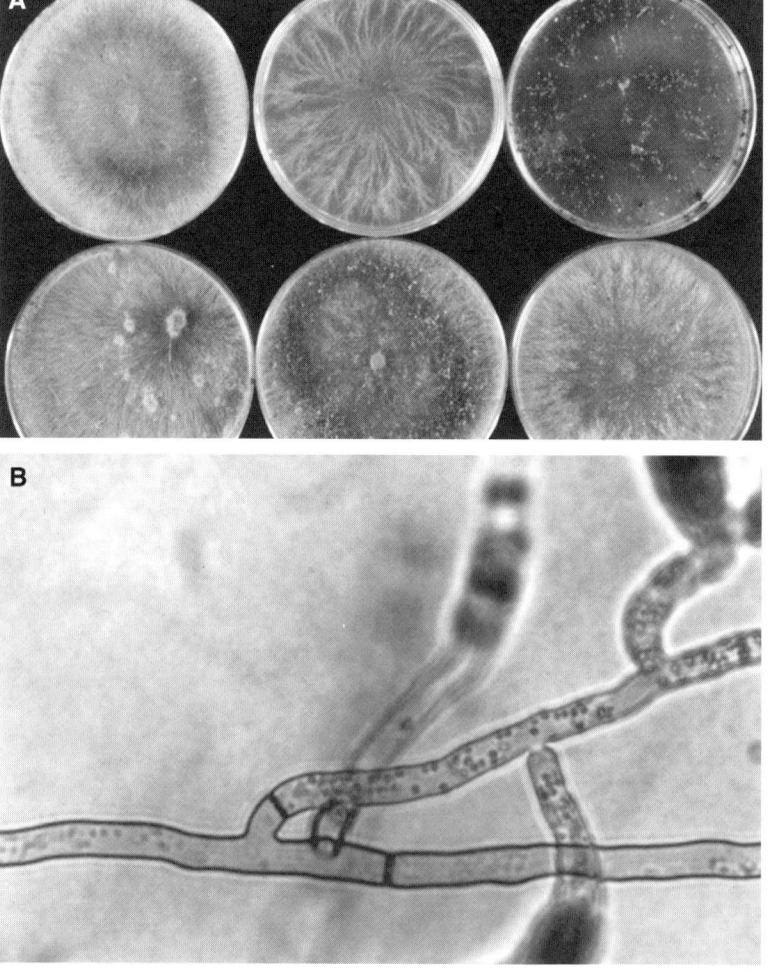

Rhizoctonia spp. obtained from flowering cherry seed. A: Colonies (from upper left to lower right: Isolates 85-60, -61, -62, -63, -64, -65). B: Hypha (Isolate 85-65). C: Monilioid cells. D. Sclerotia (Isolate 85–62).

Robillarda Sacc.

Michelia 2:8, 1880.
Type species: *R. sessilis* Sacc.

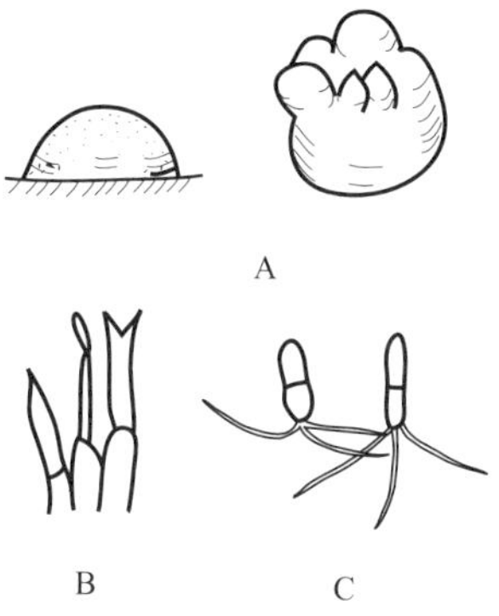

Robillarda agrostidis Sprague

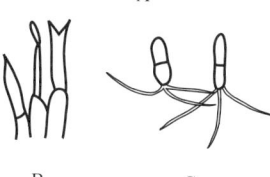

Synonym: *Pseudorobillarda agrostidis* (Sprague) Nag Raj, Morgan-Jones et Kendrick de

References: Cunnell 1958; Nag Raj et al. 1972; Sprague and Cooke 1939.

Morphology: Pycnidia superficial or half-embedded, solitary or aggregated, dark brown, globose or subglobose, irregular on the surface, covered with white hairs. Conidiophores hyaline, simple or branched. Conidia hyaline or pale brown, cylindrical, 2-celled, constricted at or near septa, usually with 3 (2–4) filiform appendages: appendages (setulae) gradually tapering toward apex, hyaline.

Dimensions: Pycnidia 172–494 µm in diameter. Conidiophores 14–27.5 µm tall. Conidia 8–11.5 × 2.2–2.8 µm: appendages 5–26.3 × 0.2–0.5 µm.

Material: 73-222 (Strawberry root, Shizuoka, Japan); 77-69 (Strawberry root, Nishigahara, Tokyo, Japan).

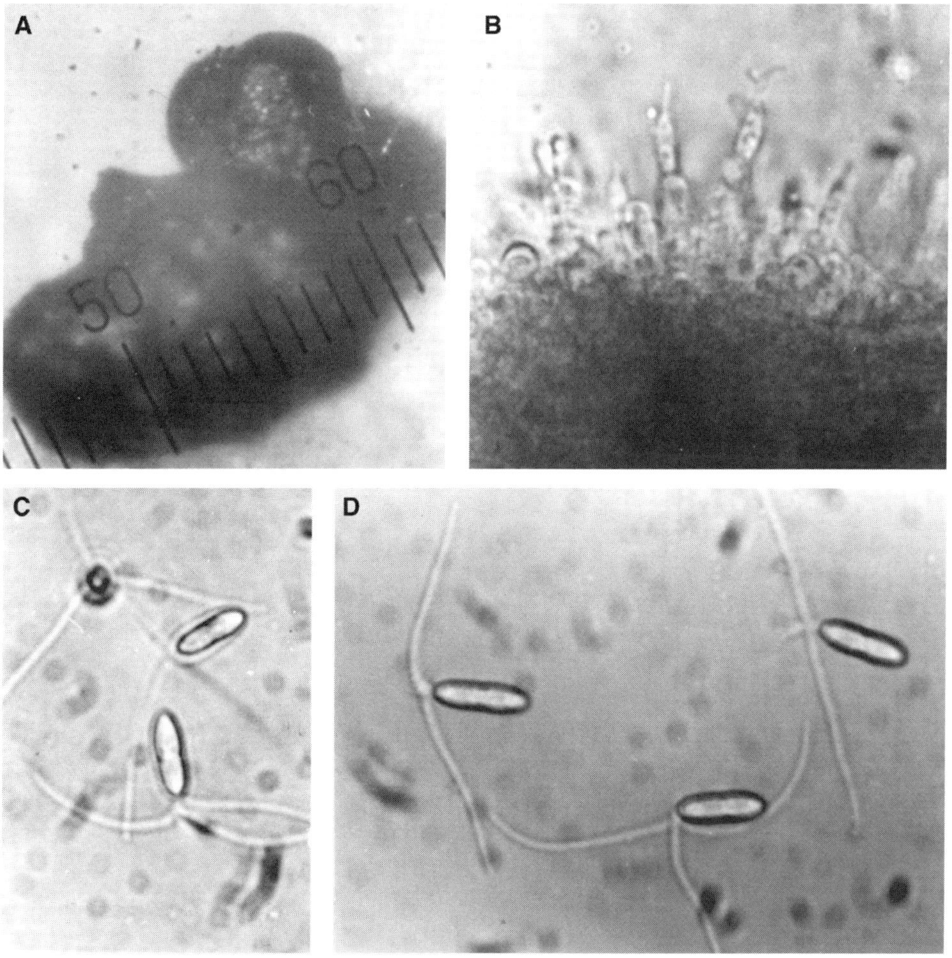

Robillarda agrostidis. A: Pycnidia. B: Conidiophores. C,D: Conidia.

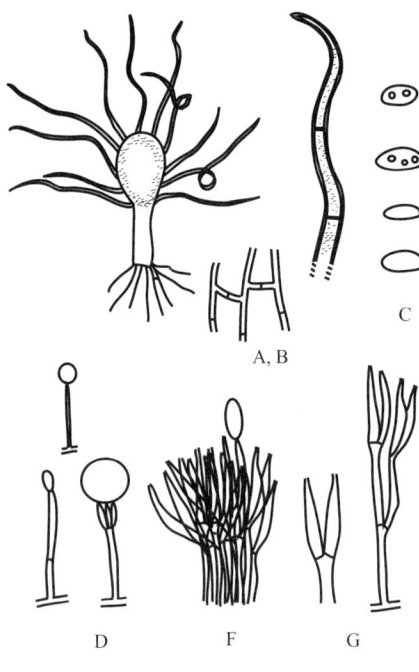

Sarcopodium Ehrenb.

Sylv. Myc. Berol. pp. 12, 23, 1818.
Type species: *S. circinatum* Ehrenb.

Sarcopodium araliae T. Watanabe

References: Ellis 1976; Sutton 1981; Watanabe 1993b.

Morphology: Conidiomata sporodochial, solitary, globose or subglobose, white or pale brown, pinkish basally, furnished with over 30 setae, usually stipitate with pale brown short stipes, rhizoidal basally: setae pale brown or yellowish brown, up to 12-septate, simple, curved or twisted, tapering from base toward tip. Main bodies composed of pallisade layers of conidiophores and spore masses. Conidiophores hyaline, simple or branched, densely aggregated with terminal fasciculate phialides apically. Conidia phialosporous, hyaline, ellipsoidal, with 2–3 oil globules, often with *Acremonium-* or *Gliocladium*-state.

Dimensions: Sporodochia 75–322 μm tall: 35–95 μm in diameter: intact fertile portions 134–322 μm tall: 100–295 μm wide: fertile portions after removal of spore mass 35–113 × 70–263 μm: setae over 500 × 5–8 μm: sporogenous cells 7–21 × 1.5–2.8 μm: rhizoids 1.5–2 μm wide. Conidia 3.5–9 × 2–3.5 μm.

Material: 92-53 (Root of *Araliae elata*, Matsumoto, Nagano, Japan).

Remarks: A key to 11 species of this genus was prepared by Watanabe (1993).

Deuteromycotina (Motosporic Fungi)

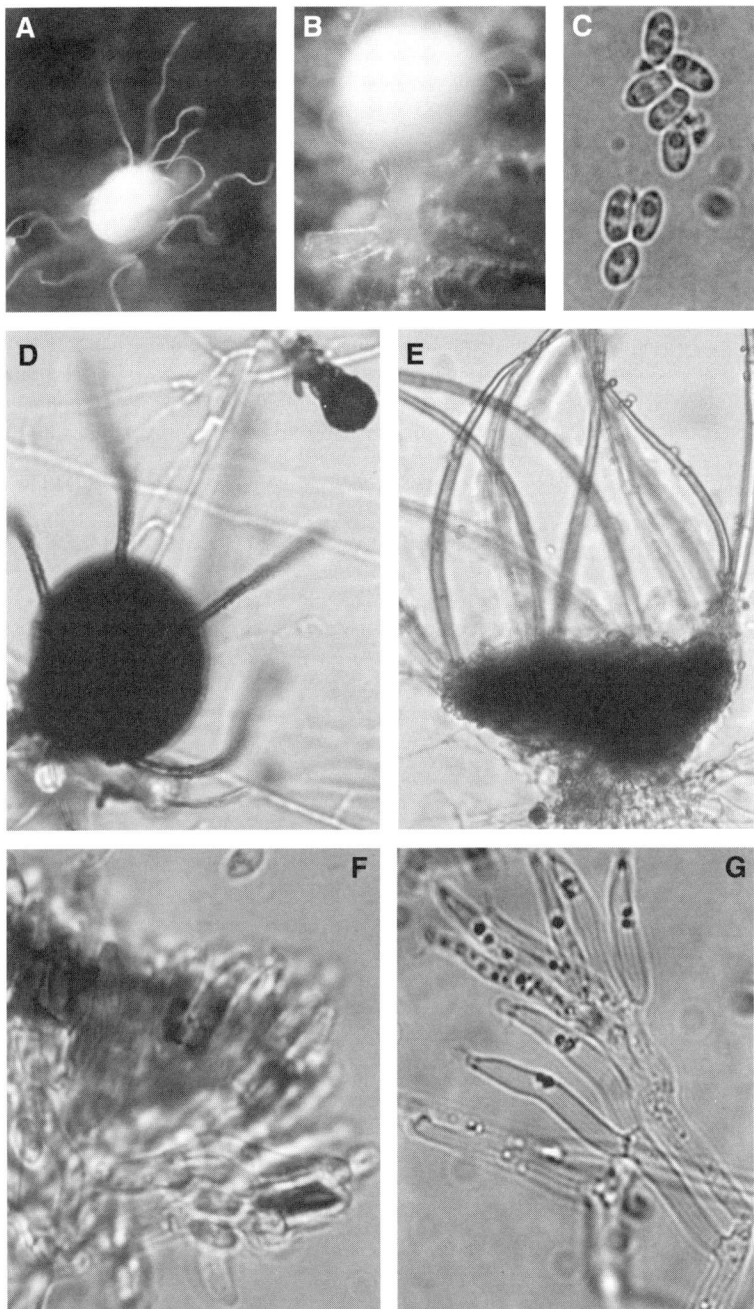

Sarcopodium araliae. A,B: Fruiting structures. C: Conidia. D: Fruiting structure and *Gliocladium*-like sporulation on agar culture. E: Fruiting structure after removal of spore mass. Note setae and rhizoid. F: Palisade layer of conidiophores. G: Conidiophores with phialides. (From Watanabe, T. 1993b. *Mycologia*, 85:520–526. With permission.)

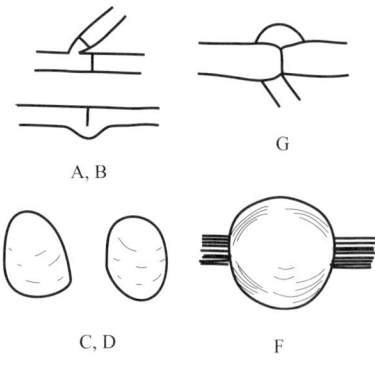

Sclerotium Tode : Fr.

Syst. Mycol. 2:246, 1823.
Type species: *S. complanatum* Tode : Fr.

Sclerotium spp.

References: Barnett and Hunter 1987; Watanabe et al. 1974, 1987b.

Morphology: Conidia not formed. Hyphae pale brown or brown, branched, with side branches septated very closely near the main hyphae and constricted basally, occasionally with clamp connections in some isolates. Sclerotia brown or dark brown, globose or subglobose, smooth on the surface, glossy, compact, composed of well-differentiated rind (outer layer) and medula (inner layer).

Dimensions: Hyphae (3.5-) 7–10 (-12) µm wide. Sclerotia (35-) 73 to over 2000 µm in diameter.

Material: 72-X110 (Sugarcane root, Taiwan, ROC); 73-226 (Strawberry root, Shizuoka, Japan); 85-15 (Flowering cherry seed, Hachioji, Tokyo, Japan).

Remarks: Sclerotia of this fungus are solitary, smooth, globose, whereas those of *Rhizoctonia* are often aggregated, rough marginally, irregular in shape. The teleomorph is Basidiomycetous, if present.

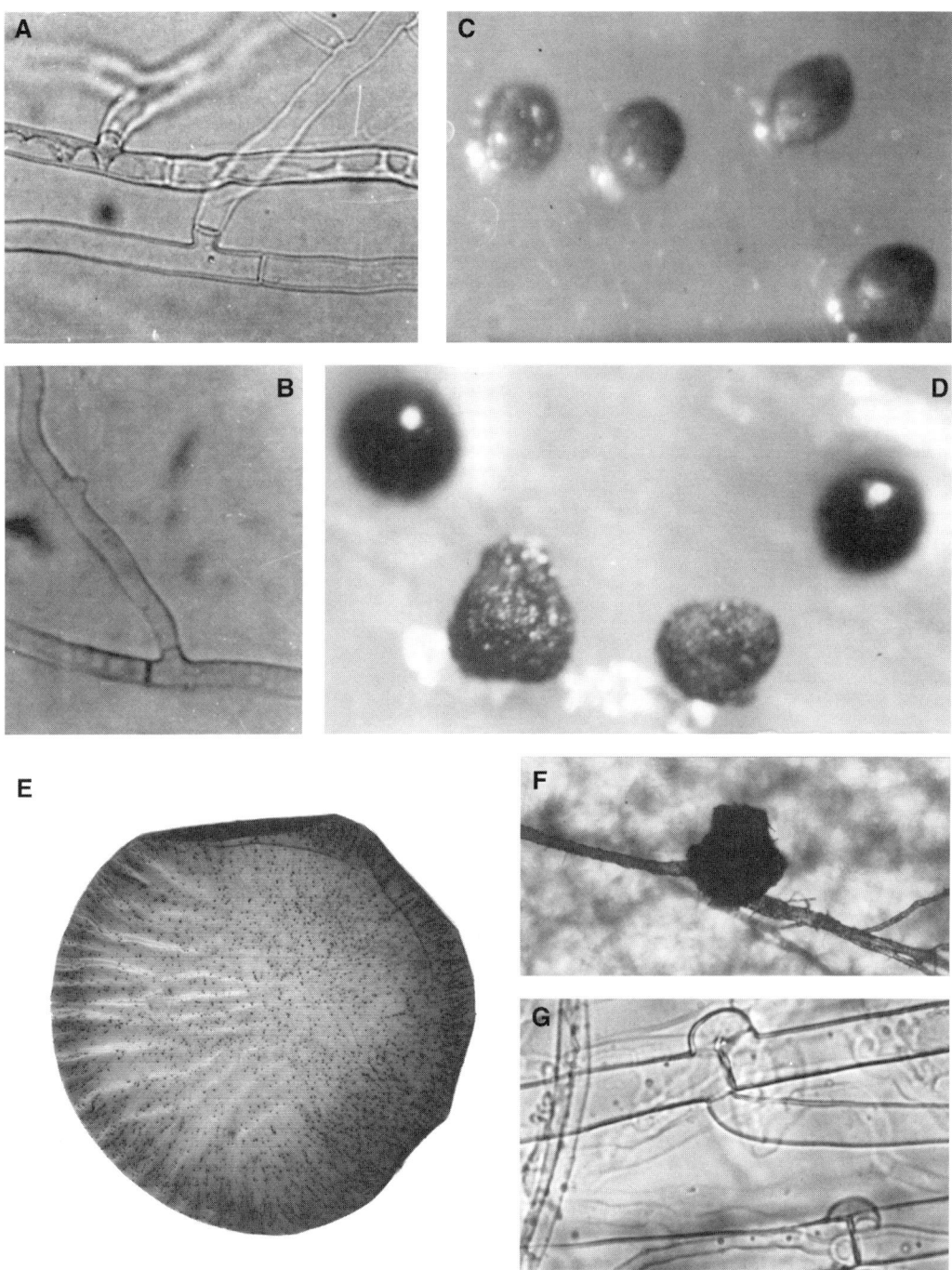

Sclerotium spp. A,B,G: Hyphae with (G) and without clamp connections (A,B). C,D,F: Sclerotia formed in agar (C,D) and on aerial hyphae (F). E: Colony (A,B: 72-X110; C,D: 73-226; E–G: 85-15). (From Watanabe, T. 1987b. *Trans. Mycol. Soc. Jpn.*, 28:475–481. With permission.)

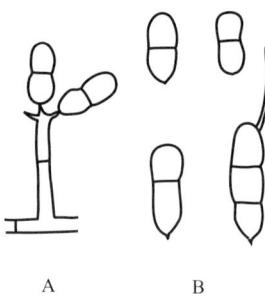

Scolecobasidium Abbott

Mycologia 19:29, 1927.
Type species: *S. terreum* Abbott.

Scolecobasidium constrictum Abbott

Synonym: *Ochroconis constricta* (Abbott) de Hoog et von Arx.

References: Abbott 1927; Barron and Busch 1962; Ellis 1971, 1976.

Morphology: Conidiophores pale brown, erect, simple, bearing over 3 conidia on sterigmata developed on apical fertile portions, constricted basally. Conidia sympodulosporous, pale brown, mainly ellipsoidal, mostly 2-celled, constricted at or near septa, round in the end and apiculate in another end.

Dimensions: Conidiophores 8.7–22.5 (-37.5) × 2.5–3.6 µm. Conidia 6.2–15.3 (-17.5) × 3.7–5.4 µm.

Material: 69-S4-19 (Pineapple root, Okinawa, Japan); 84-581 (Japanese red pine seed, Kiso, Nagano, Japan); 86-52 (Japanese cypress seed, Kasama, Ibaraki, Japan); 86-53 (Japanese cedar seed, Nakashinkawa, Toyama, Japan).

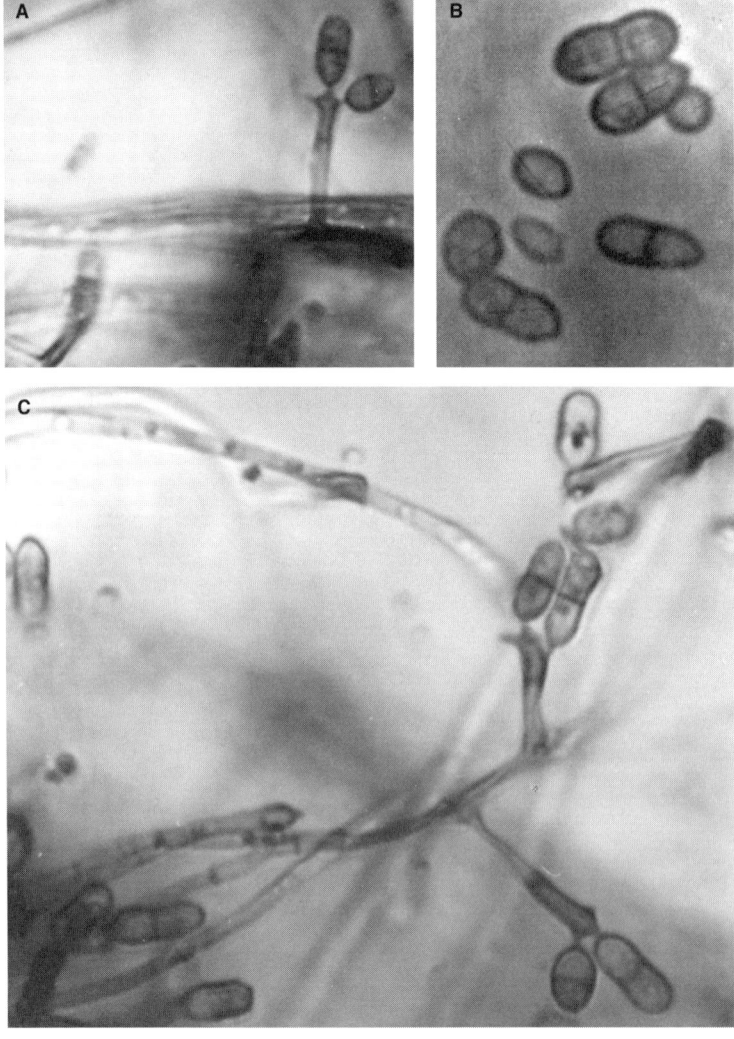

Scolecobasidium constricta. A,C: Conidiophores and conidia. B: Conidia.

Scolecobasidium humicola Barron et Busch

Synonym: *Ochroconis humicola* (Barron et Busch) de Hoog et von Arx

References: Abbott 1927; Barron and Busch 1962; Ellis 1971, 1976.

Morphology: Conidiophores brown, erect, simple, bearing conidia at sterigmata developed on apical fertile portions. Conidia sympodulosporous, brown, cylindrical, mainly 2-celled, round apically, apiculate basally.

Dimensions: Conidiophores (23.5-) 75–105 × 2.2–2.8 µm. Conidia 8.7–15 (-21.3) × (2.7-) 4.2–5.3 µm.

Material: 85-69, 85-70 (Flowering cherry seed, Hachioji, Tokyo, Japan).

Remarks: Conidiophores of this fungus and *S. constrictum* are usually over 75 µm long, and under 37.5 µm long, respectively.

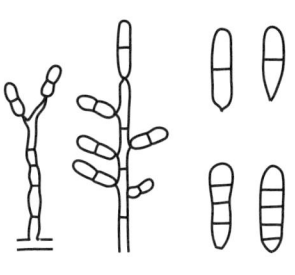

A, B

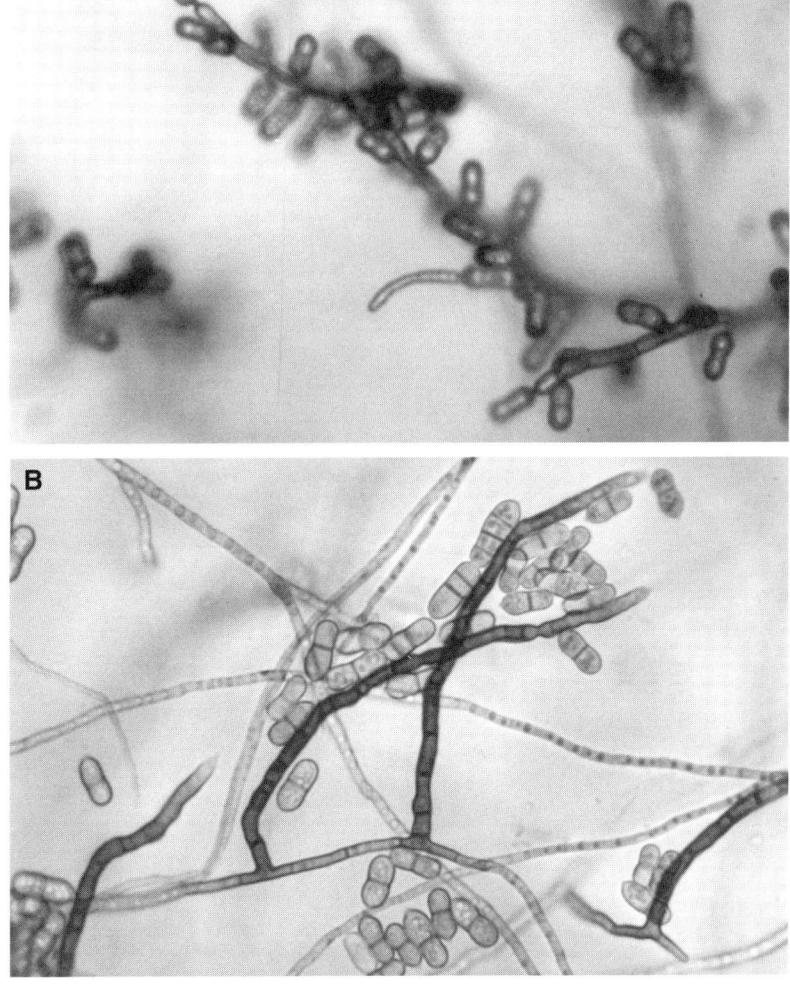

Scolecobasidium humicola.   A,B: Conidiophores and conidia.

Scopulariopsis Bainier

Bull. Soc. Mycol. Fr. 23:98, 1907.
Type species: *S. brevicaulis* (Sacc.) Bainier.

Key to Species

1. Conidia   ovate, smooth .... *S. canadensis*
                       globose, rough on the surface 2

2. Conidia   hyaline or pale brown, globose .... *S. asperula*
                       pale brown, globose, but slightly angled. .... *P. brevicaulis*

Scopulariopsis asperula (Sacc.) Hughes

References: Morton and Smith 1963.

Morphology: Conidiophores hyaline, erect, simple or branched, bearing catenulate conidia basipetally developed on each apical phialide on verticillate branches: phialides hyaline, cylindrical, with characteristic annellations at the apex. Conidia phialosporous, hyaline or pale brown, globose, rough on the surface, truncate at one end.

Dimensions: Conidiophores ca. 60 μm tall: primary branches 5–15 × 2.2–3.3 μm: phialides 5–15 × 2.5–3.8 μm. Conidia 4.7–6.5 μm in diameter.

Material: 74-566 (Strawberry root, Shizuoka, Japan).

Remarks: This fungus is separated from *Penicillium*, based on formation of annellations on apex of phialides.

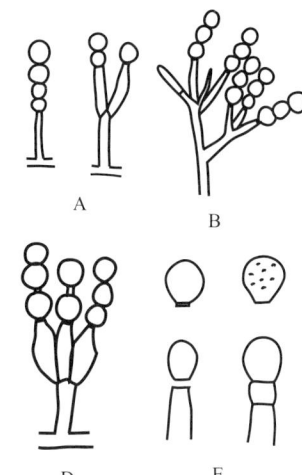

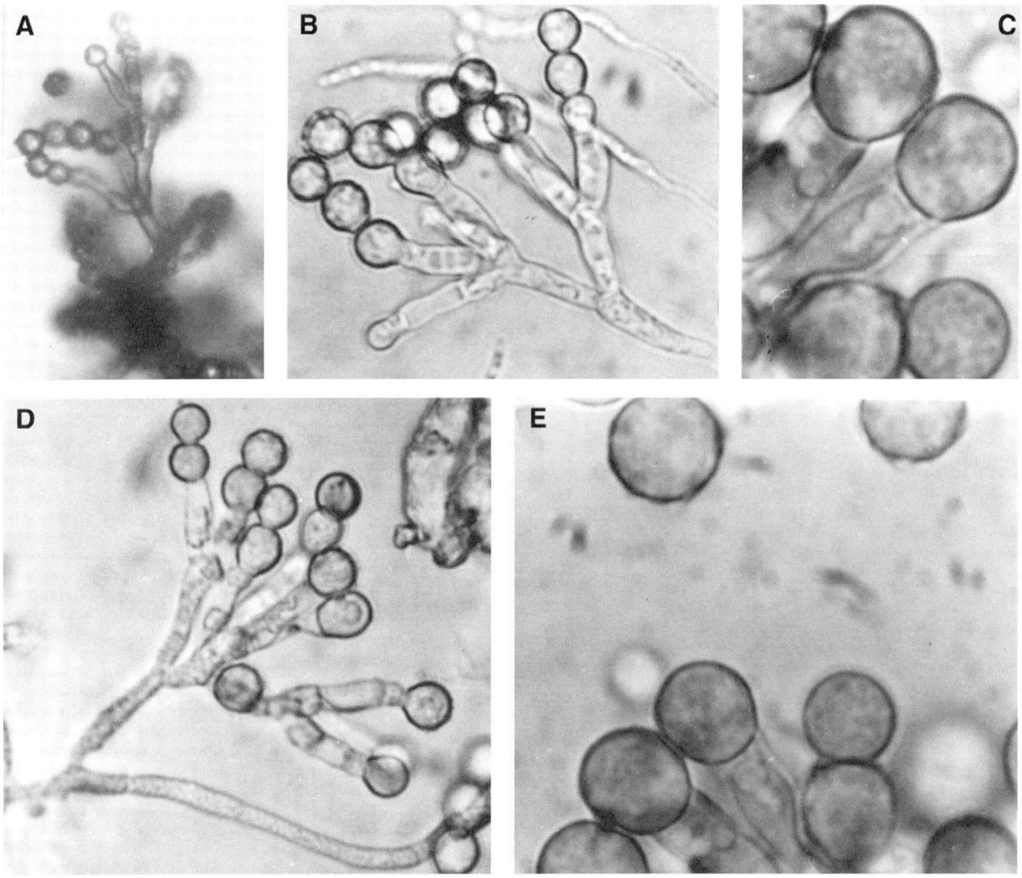

Scopulariopsis asperula. A,B,D: Conidiophores and conidia. C,E: Phialides and conidia.

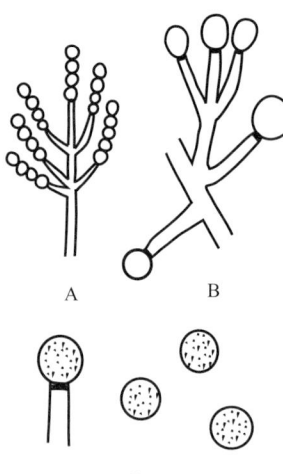

Scopulariopsis brevicaulis (Sacc.) Bainier

References: Morton and Smith 1963; Watanabe et al. 1986a.

Morphology: Conidiophores hyaline, erect, simple or branched, bearing catenulate conidia at the phialides with apical annellations, developed on verticillate or alternate conidiophores: phialides hyaline, cylindrical. Conidia phialosporous, yellowish or pale brown, globose or subglobose, rough on the surface, truncate at one end.

Dimensions: Conidiophore branches 9–16.2 × 2.7–3.6 µm. Conidia 5.4–8.1 µm in diameter.

Material: 84-577 (Japanese red pine seed, Kamiina, Nagano, Japan).

Remarks: Cultures on PDA are pale brown and floury.

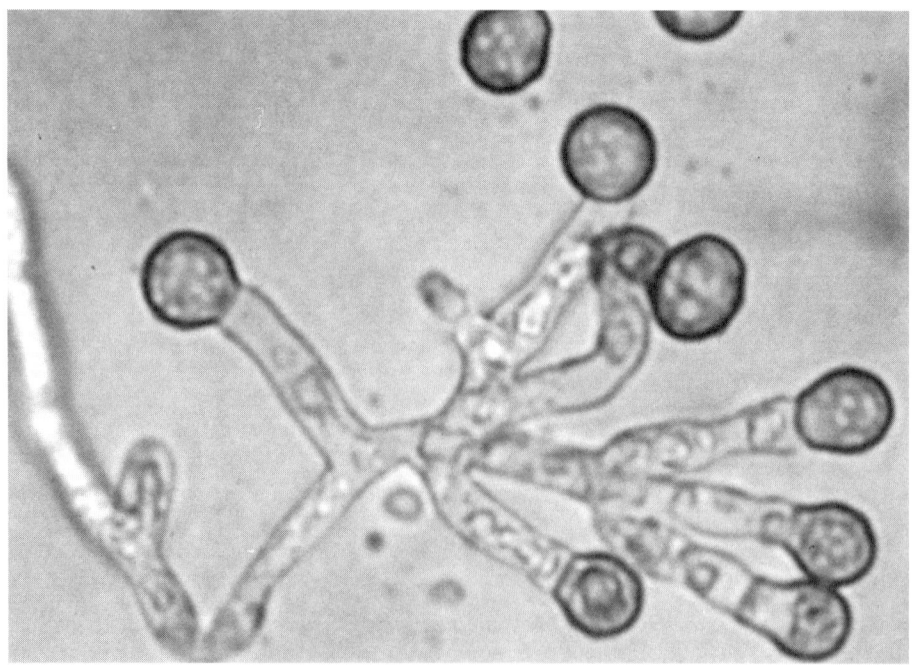

Scopulariopsis brevicaulis. A: Conidiophores, B: phialides, and C: conidia.

Scopulariopsis canadensis Morton & Smith

References: Morton and Smith 1963; Watanabe et al. 1986a.

Morphology: Conidiophores hyaline, erect, simple or branched, bearing catenulate conidia at phialides with apical annellations developed on verticillate or alternate branches: phialides hyaline, cylindrical. Conidia phialosporous, hyaline, smooth, ovate, 1-celled.

Dimensions: Conidiophores ca. 25–67.5 μm long, 2.5–2.8 μm wide: phialides 21.6–42.5 × 2.5–2.8 μm. Conidia 5–8.1 × 3.5–5.4 μm.

Material: 84-514 (Japanese red pine seed, Saihaku, Tottori, Japan).

Remarks: Cultures on PDA are pale brown and floury. Annellations are rather difficult to observe.

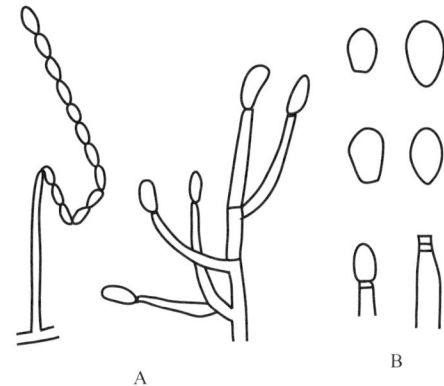

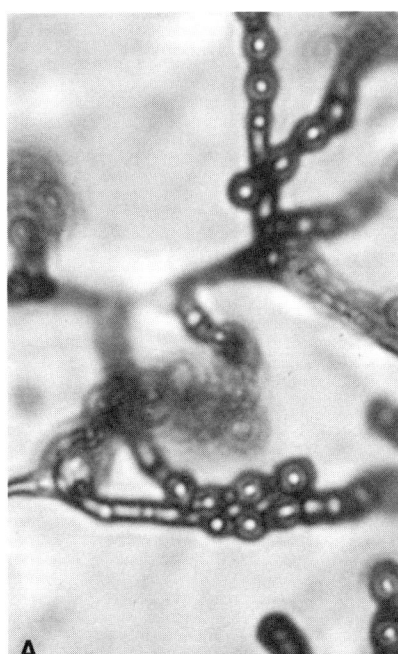
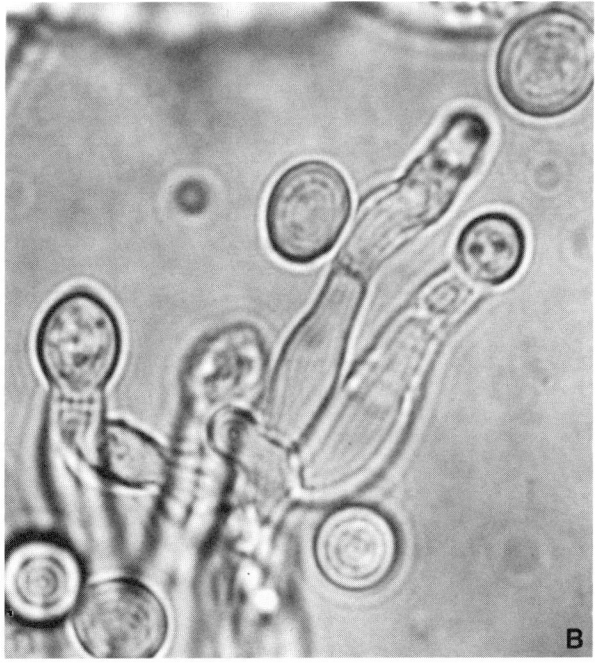

Scopulariopsis canadensis. A: Conidiophores and catenulate conidia. B: Phialides and conidia.

Selenophoma Maire

Bull. Soc. Bot. Fr. 53:87, 1906.
Type species: *S. catananches* Maire

Selenophoma obtusa Sprague et Johnson

References: Sprague 1950; Watanabe 1975b.

Morphology: Pycnidia half-embedded, solitary or aggregated, globose, with black spore masses extruded from indistinct ostioles: peridium black. Conidiophores hyaline, simple, tapering toward apex. Conidia pale brown, cylindrical, ellipsoidal, or various in shape, well curved.

Dimensions: Pycnidia 291–341 µm in diameter. Conidiophores 12.5–20 × 3.7–3.8 µm. Conidia 9.7–15.8 × 2.9–4.7 µm.

Material: 72-X47-801 (= ATCC 32328, Sugarcane root, Taiwan, ROC).

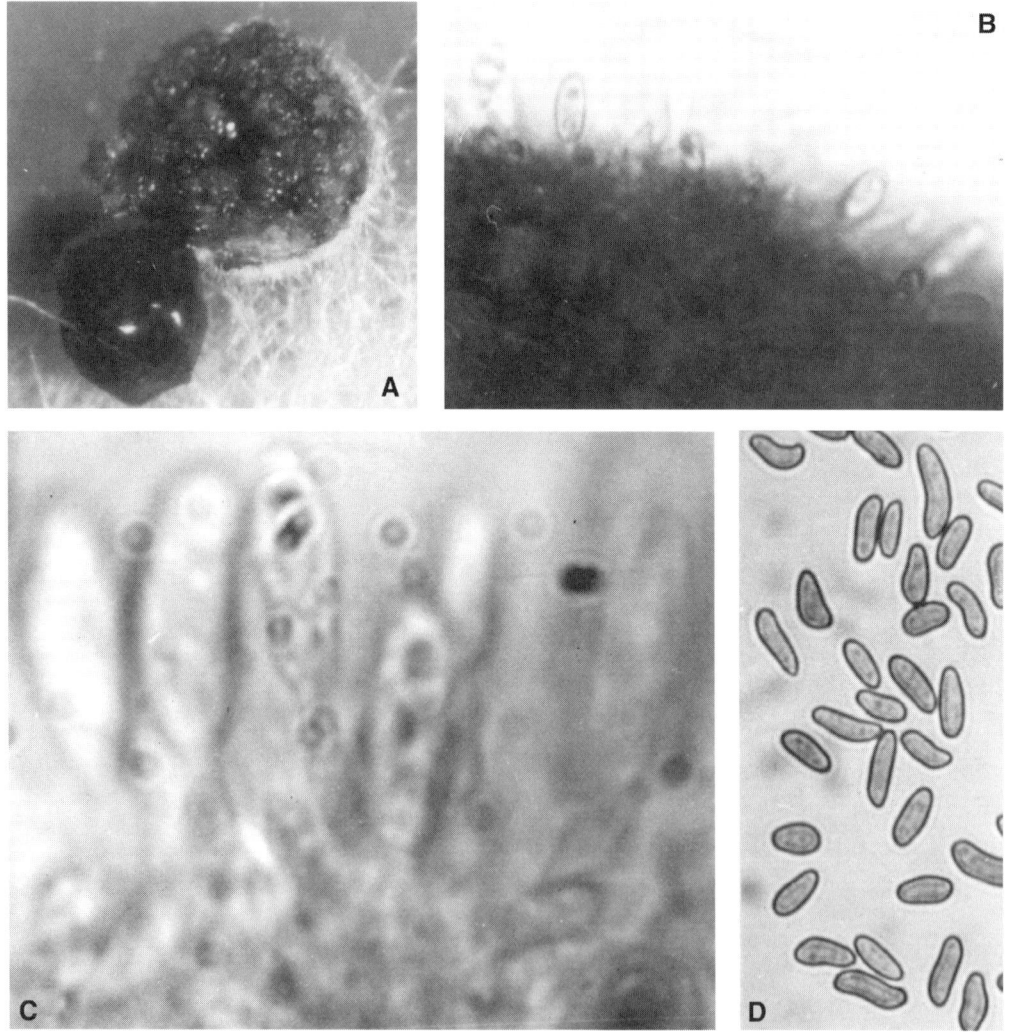

Selenophoma obtusa. A: Pycnidium with extruded spore mass. B,C: Conidiophores and conidia. D: Conidia. (From Watanabe, T. 1975b. *Trans. Mycol. Soc. Jpn.*, 16:28–35. With permission.)

Sepedonium Link : Fr.

Syst. Mycol. 3:428, 1832.
Type species: *S. chrysospermum* (Bull.) Link : Fr.

Sepedonium chrysospermum (Bull.) Link

References: Barron 1968; Howell 1939.

Morphology: Conidiophores short, hyaline, erect, simple or branched, bearing single conidia apically on each conidiophore. Conidia aleuriosporous, hyaline or pale brown, globose, rough superficially and marginally, thick-walled, deciduous in nature.

Dimensions: Conidiophores up to 18 µm long. Conidia 14.5–18 µm in diameter.

Material: 73-248 (Strawberry root, Shizuoka, Japan).

Remarks: Hyaline, large deciduous aleurioconidia are characteristic of this fungus.

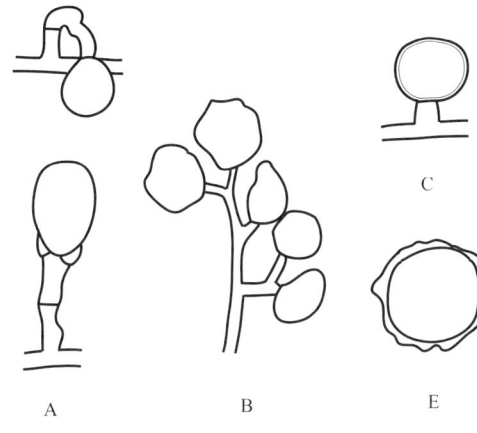

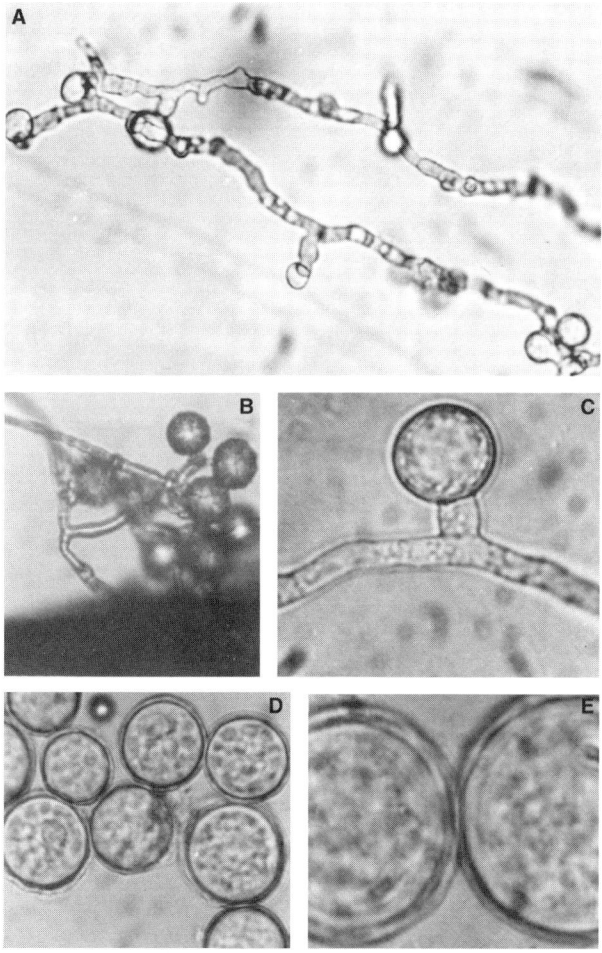

Sepedonium chrysospermum. A–C: Conidiophores and conidia formed on agar (A,C) and on aerial hyphae (B). D,E: Conidia.

Sepedonium sp.

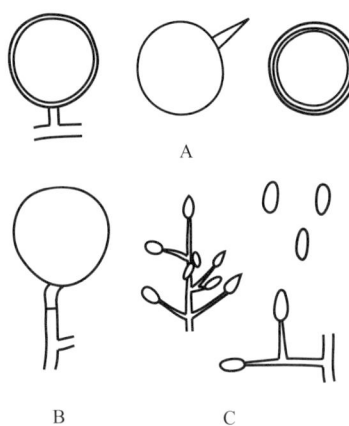

References: Damon 1952; Watanabe et al. 1987a.

Morphology: Spores dimorphic, i.e., aleuriosporous and sympodulosporous. Conidiophores hyaline, erect, simple or branched, mainly simple for aleuriospores, but branched oppositely or verticillately for sympodulospores. Aleurioconidia pale brown, globose, double membranaceous, occasionally deciduous, often pedicellate, occasionally with apically pointed setae. Sympoduloconidia hyaline, ellipsoidal, ovate, 1-celled.

Dimensions: Branches for sympodulospores 17.5–37.5 µm tall. Aleurioconidia 42.5–37.5 µm in diameter. Sympoduloconidia 5–9 × 3.7–5.5 µm.

Material: 85-P106 (Paulownia root, Misiones, Argentina).

Remarks: None of at least 10 described species in this genus fits this fungus morphologically.

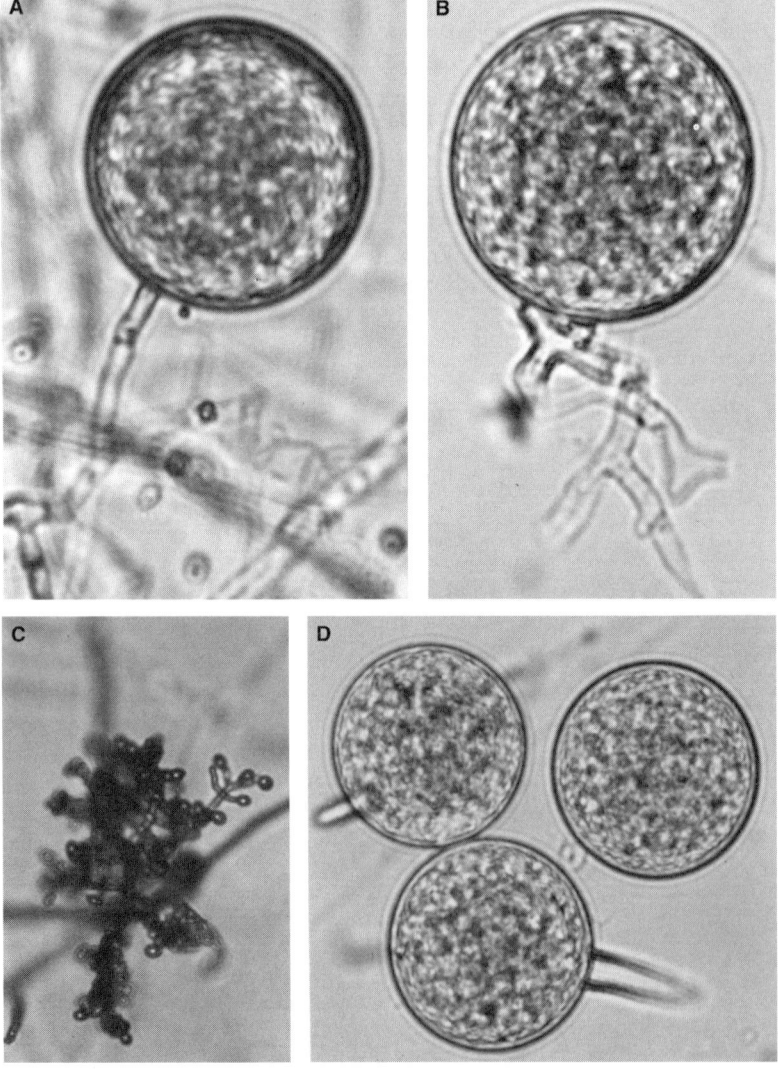

Sepedonium sp. A,B,D: Conidiophores and aleurioconidium. C: Conidiophores and sympoduloconidia.

Septonema Corda

Icon. Fung. 1:9, 1837.
Type species: *S. secedens* Corda

Septonema chaetospira (Grove) Hughes

Synonym: *Heteroconium chaetospira* (Grove) Hughes

References: Ellis 1976; Watanabe and Sato 1988.

Morphology: Conidiophores and conidia not well differentiated. Catenulate conidia formed directly on hyphae: conidial chains acropetally developed, simple or branched. Conidia blastosporous, pale brown, fusiform, occasionally curved, 2- to 4-celled (mainly 2-celled), truncate at both ends.

Dimensions: Hyphae 2–3.5 µm wide. Conidiophores ca. 17.5 × 2.3 µm. Conidia 10–37.5 × 2.2–2.5 µm.

Material: 76-556 (Gentian root, Chino, Nagano, Japan); 86-60 (Japanese cypress seed, Kasama, Ibaraki, Japan).

Remarks: The conidia of this fungus are narrower than those of the original description (3–4 µm).

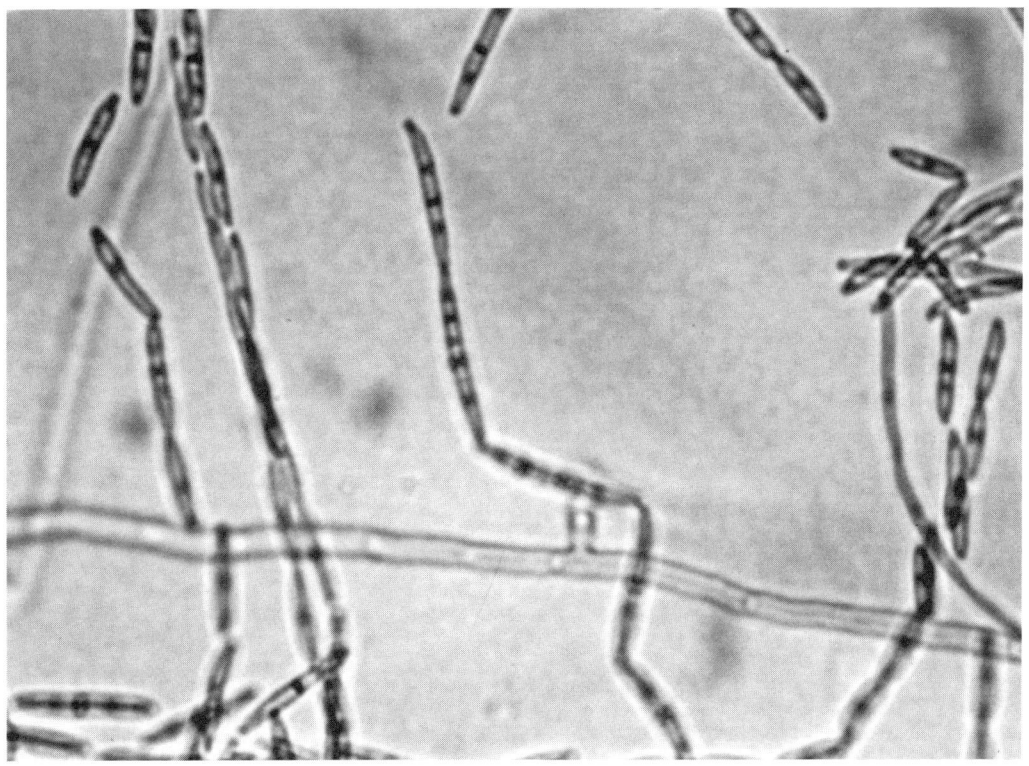

Septonema chaetospira. Hypha and conidia. (From Watanabe, T. and Sato, Y. 1988. *Trans. Mycol. Soc. Jpn.*, 29:143–150. With permission.)

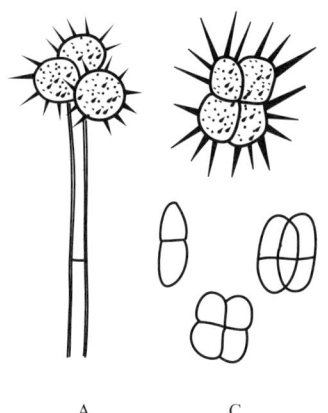

Spegazzinia Sacc.

Michelia 2:37, 1880.
Type species: *S. tessarthra* (Berk. et Curt.) Sacc.

Spegazzinia tessarthra (Berk. et Curt.) Sacc.

References: Damon 1953; Ellis 1971, 1976; Hughes 1953; Watanabe and Sato 1988.

Morphology: Conidiophores brown, erect, various in height, bearing 1 to several conidia apically. Conidia dark brown, at least of two kinds: smooth, square- and disc-shaped, ellipsoidal, 2- to 4-celled; and muriform, composed of cross- and longitudinal septa with setae, constricted at or near septa.

Dimensions: Conidiophores 75–142.5 × 2.2–2.8 µm. Conidia (excluding protuberances) 12.5–16.3 µm in diameter, ca. 7.5 µm thick: protuberances 5–8.8 µm long.

Material: 86-59 (Japanese cypress seed, Kasama, Ibaraki, Japan).

Remarks: Process of sporulation may be interesting to study.

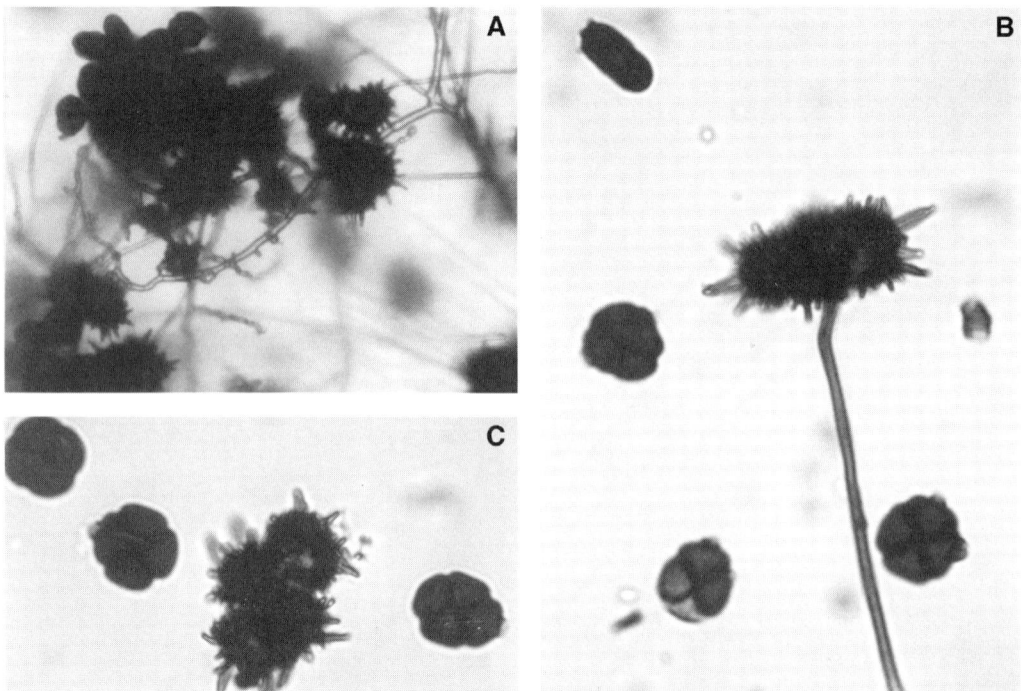

Spegazzinia tessarthra. A: Conidia and sporulation. B: Conidiophore and conidia. C: Conidia. (From Watanabe, T. and Sato, Y. 1988. *Trans. Mycol. Soc. Jpn.*, 29:143–150. With permission.)

Sporidesmium Link : Fr.

Syst. Mycol. 3:492, 1832.
Type species: *S. atrum* Link : Fr.

Sporidesmium bakeri Sydow

References: Ellis 1971; Hughes 1953; Watanabe et al. 1986a.

Morphology: Conidiophores simple, hyaline, straight or curved, bearing single conidia apically. Conidia brown or dark brown, ellipsoidal, ovate or cylindrical, often curved, 2- to 4-celled (usually 3-celled), with round apex, narrowing toward ends, often with frill basally.

Dimensions: Conidiophores 5–27.5 × 1.8–2.2 µm. Conidia 16.2–35 × 7.5–12.5 µm.

Material: 84-527, 84-528 (Japanese black pine seed, Okawa, Kagawa, Japan).

Remarks: Colonies on agar cultures are pale brown, slightly pinkish tinted with zonation.

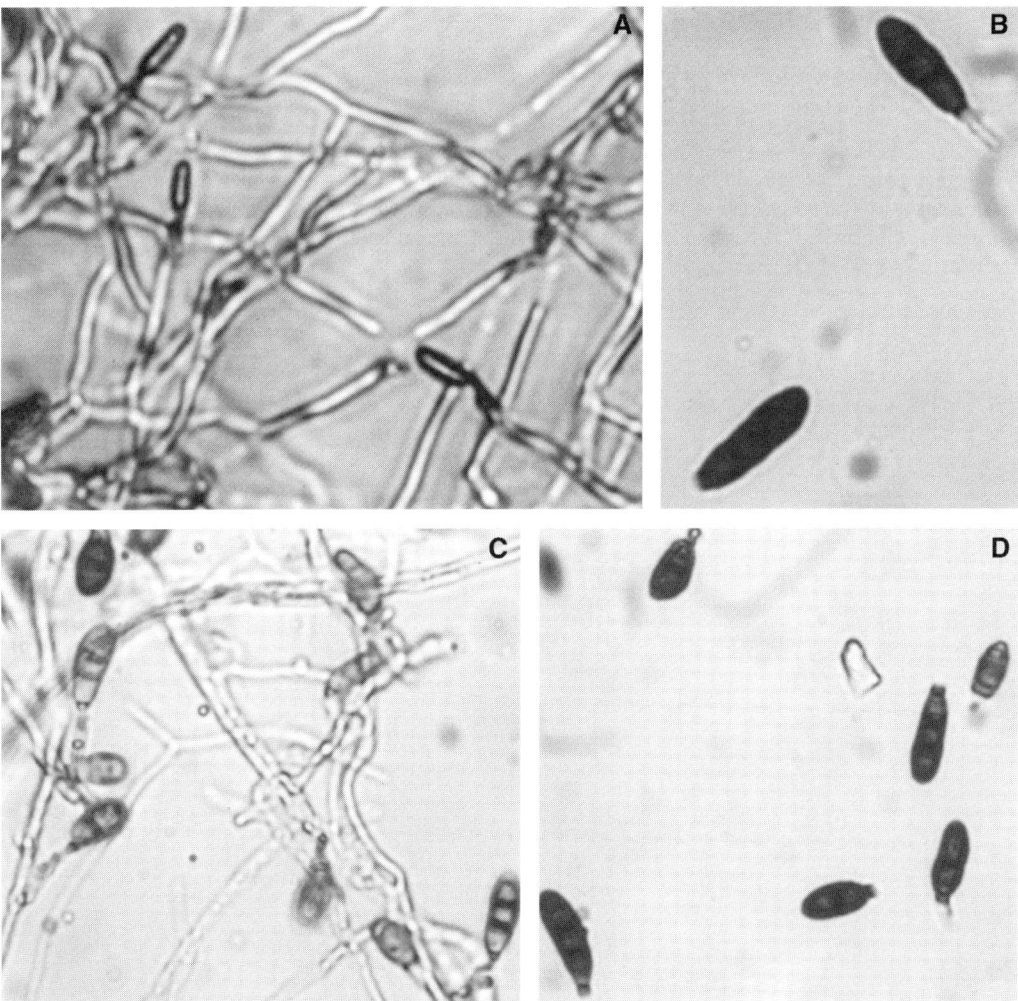

Sporidesmium bakeri.   A: Habit. B,D: Conidia. C: Conidiophores and conidia.

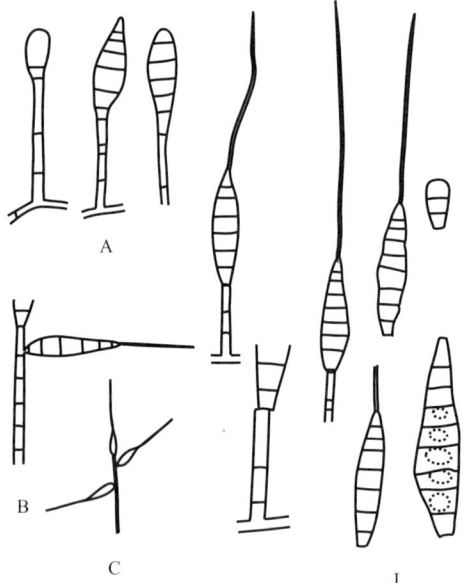

Sporidesmium filiferum Pirozynski

References: Ellis, 1971, 1976; Pirozynski, 1972; Watanabe, 1996.

Morphology: Conidiophores macronematous, mononematous, subhyaline or brown, simple, 1- to 4-septate. Conidiogenous cells integrated, terminal, monoblastic, holoblastic, percurrently or sympodially proliferating, subhyaline or brown, smooth, cylindrical. Conidia solitary, acropleurogenous, ellipsoidal to long-fusiform, rarely curved, mostly 7- to 8-septate, guttulate or eguttulate, brown, hyaline or subhyaline in one to two cells of both ends, smooth, mostly truncate at base, with a filiform, straight, or curved cellular appendage usually from apical (rarely from subapical) cell; appendages hyaline, usually aseptate, rarely 1- to 2-septate basally.

Dimensions: Hyphae 2–3 µm wide. Conidiophores (25-) 32.5–72.5 (-130) µm long, 2–4.5 µm wide. Conidia 27.5–41.3 × 7.5–10 µm, mostly with truncate base, 3–4.5 µm wide: appendages 87.5–137.5 × 0.5–1 µm.

Material: 95-2 (= MAFF 425595, Culture from a fallen leaflet of *Phellodendron amurense* on the soil surface, Tsukuba, Ibaraki, Japan).

Remarks: This fungus is characterized by single phaeophragmospores borne terminally on macronematous, mononematous, simple conidiophores with monoblastic, integrated, terminal, determinate or percurrent conidiogenous cells, and resembles the genus *Phaeotrichoconis* Subraman. morphologically. Conidia of the former are holoblastic, and truncate basally, but those of the latter enteroblastic and tretic. The terminal appendages of mature conidia were often detached.

Deuteromycotina (Motosporic Fungi)

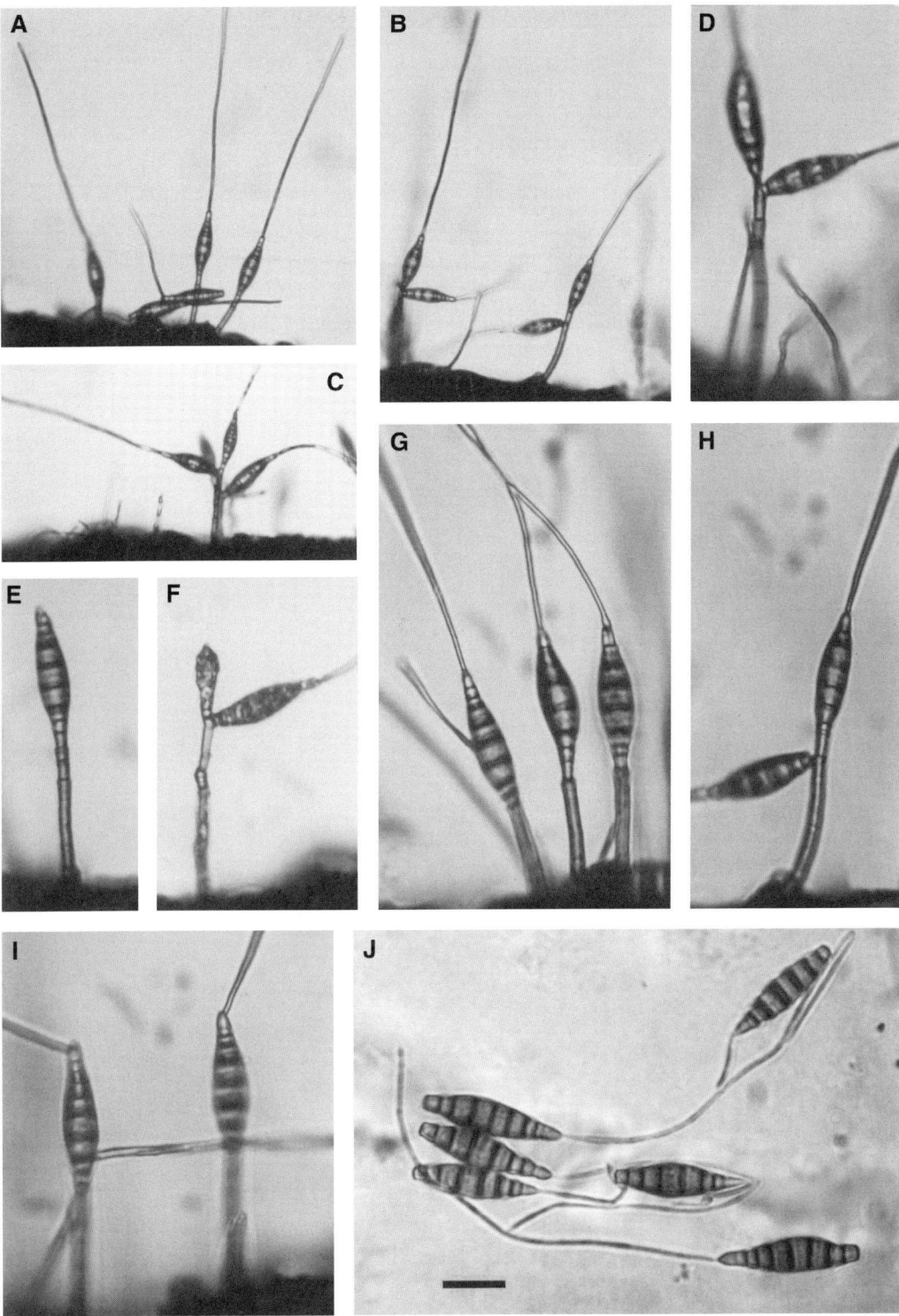

Sporidesmium filiferum. A–I: Conidiophores bearing conidia. J: Detached conidia. Note the truncate bases and apical appendages of conidia. (From Watanabe, T. 1996. *Mycoscience*, 37:367–369. With permission.)

Sporobolomyces Kluyver & van Niel

Zentbl. Bakt. Parasitkde. Abt. 2, 63:19, 1924.
Type species: *S. roseus* Kluyver & van Niel

Sporobolomyces sp.

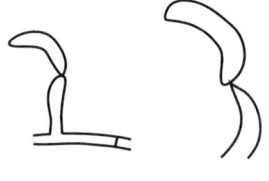

References: Barnett and Hunter 1987; Watanabe et al. 1986a.

Morphology: Conidiophores indistinct. Conidia borne singly on sterigmata developed from hyphae or detached conidia, reproduced by budding, hyaline, cylindrical, lunar- or sickle-shaped, 1-celled.

Dimensions: Hyphae 1.8–2.7 µm wide. Conidia 7.5–12.6 × 1.5–2.5 µm: sterigmata on detached conidia 2.1–2.2 µm long.

Material: 84-578 (Japanese black pine seed, Okawa, Kagawa, Japan).

Remarks: Cultures on PDA are yeast-like, yellowish white in color.

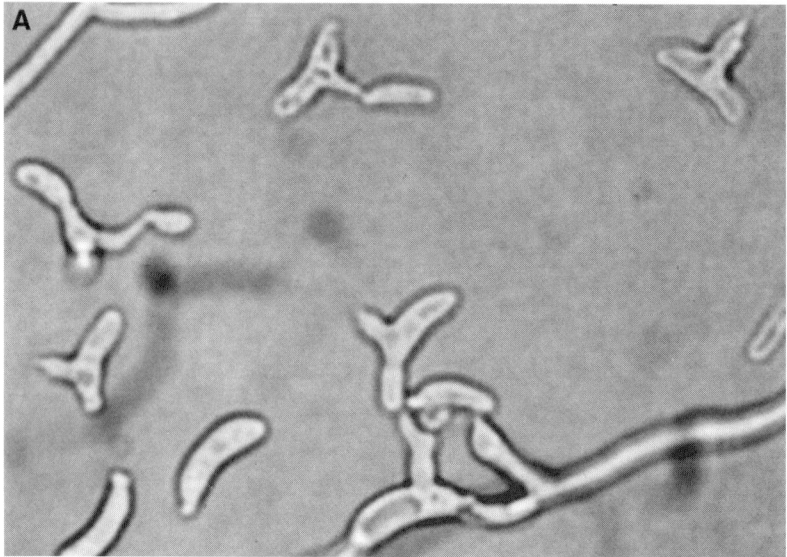

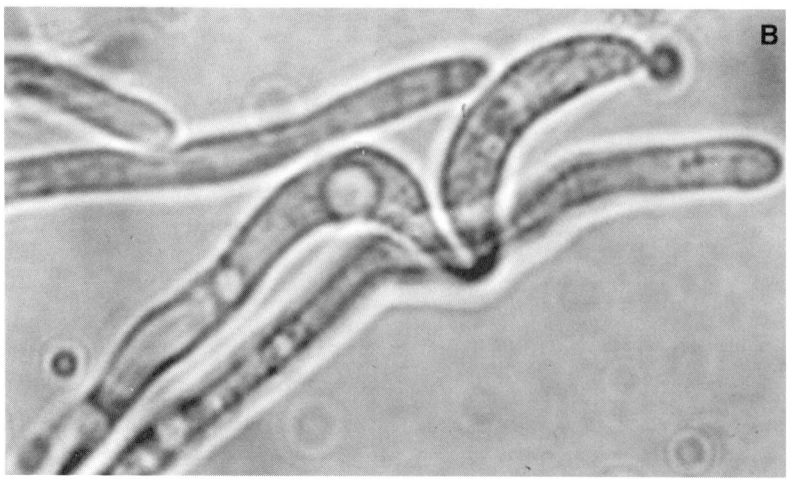

Sporobolomyces sp.   A,B: Hyphae and budding conidia.

Sporoschisma Berk. & Br.

Gnr's Chron. p. 540, 1847.
Type species: *S. mirabile* Berk. and Hughes

Sporoschisma saccardoi Mason & Br.

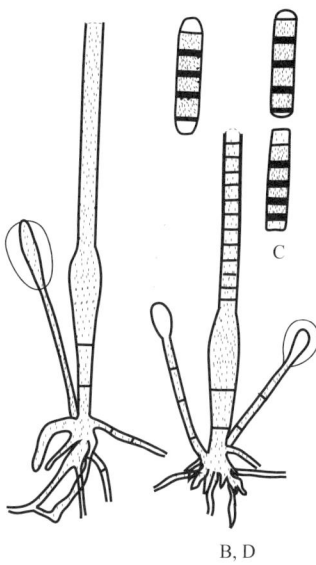

References: Ellis et al. 1951; Gams & Holbova-Jechove 1976; Hughes 1949, Nag Raj and Kendrick 1976.

Morphology: Conidiophores dark, simple, erect, phialidic, inflated below the middle, septate, producing a long chain of catenulate conidia apically, rhizoidal basally. Conidia phialosporous, cylindrical, 5- to 6-septate, brown with thick septum, and subhyaline end cells. Together with conidiophores, the capitate hyphae with mucilaginous envelopes developed.

Dimensions: Conidiophores 240–325 µm long, bulged part 15–16 µm wide, apical cylindrical parts 11–12 µm, basal part 8–9 µm broad. Conidia 35–50 × 9–15 µm: septum 2–2.4 µm broad. Capitate hyphae 100–125 µm long, 6 µm broad apically, 5 µm broad basally.

Material: 01-480 (Kitakou, Hahajima, the Bonin Islands, Tokyo, Japan).

Remarks: The capitate hyphae may be in the process of becoming mature conidiophores. The genus *Sporoschimopsis* is related to this fungus, but its conidiogenous phialidic portion is not obviously bulged and conspicuously percurrent in proliferation.

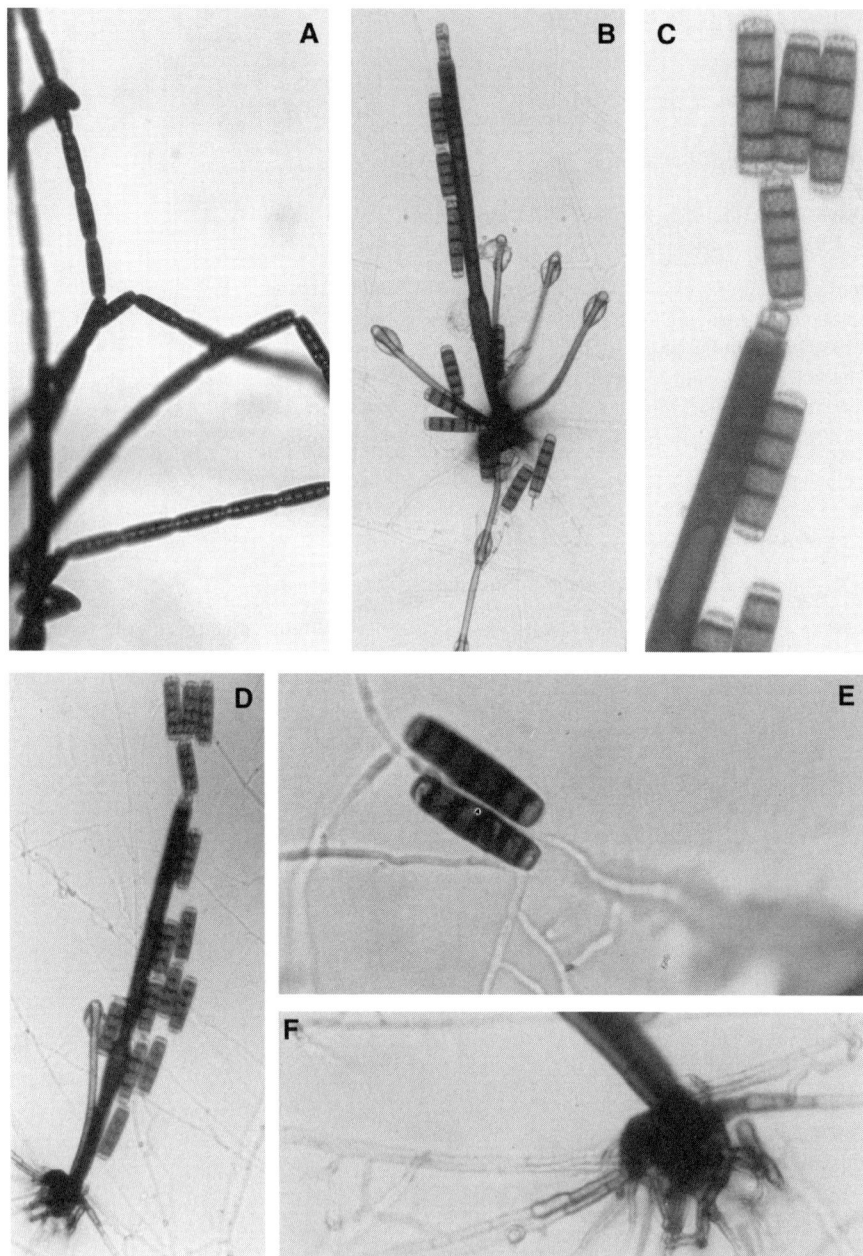

Sporoschisma saccardoi. A: Conidia in a long chain. B,D: Conidiophores, conidia, and capitate hyphae with mucilaginous envelopes. C: Apical portion of conidiophore and conidia. E: Germination of conidium. F: Rhizoid.

Sporotrichum Link : Fr.

Syst. Mycol. 3:415, 1832.
Type species: *S. aureum* Link : Fr.

Sporotrichum aureum Link : Fr.

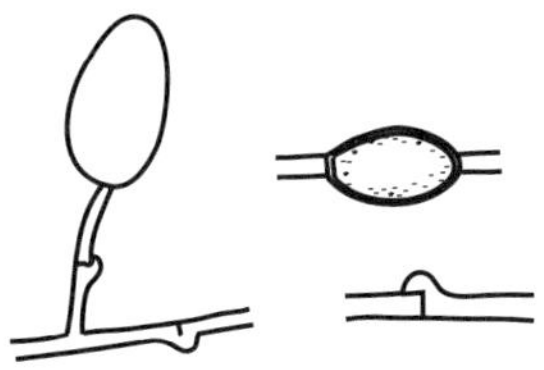

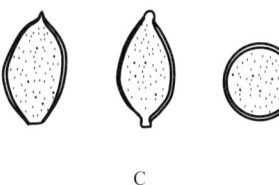

References: Arx 1971; Watanabe 1975d.

Morphology: Conidiophores not well differentiated from hyphae with clamp connections, bearing single conidia apically and laterally on hyphae. Conidia aleuriosporus, yellow, globose or ellipsoidal, 1-celled, thick-walled, apiculate apically, truncate or with hilum basally, readily detached.

Dimensions: Conidia 15–17.5 × 12.1–15.1 µm.

Material: 72-X86-514 (= ATCC 32926, Sugarcane root, Taiwan, ROC).

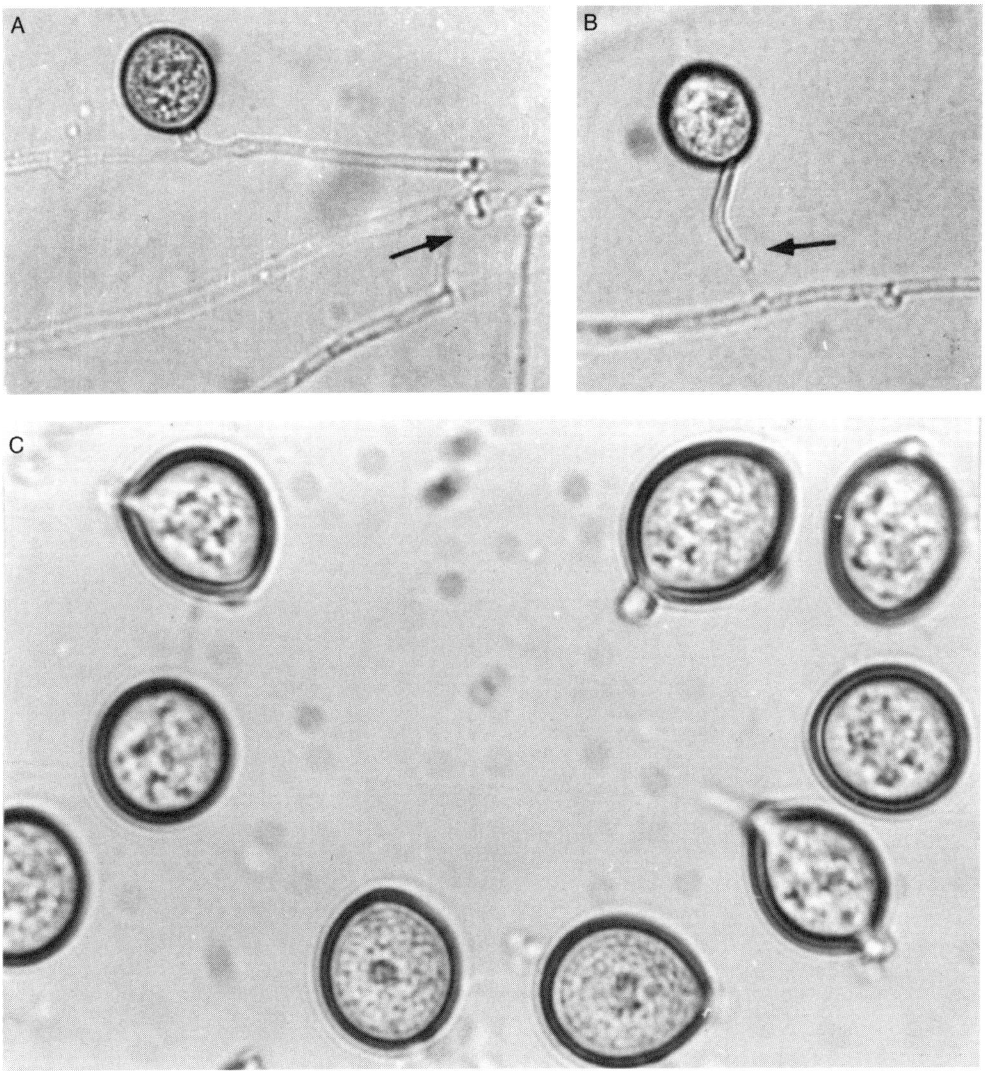

Sporotrichum aureum. A,B: Hyphae, conidiophores, and conidia. Note hyphae with clamp connections (arrow). C: Conidia. (From Watanabe, T. 1975d. *Trans. Mycol. Soc. Jpn.*, 16:264–267. With permission.)

Sporotrichum sp.

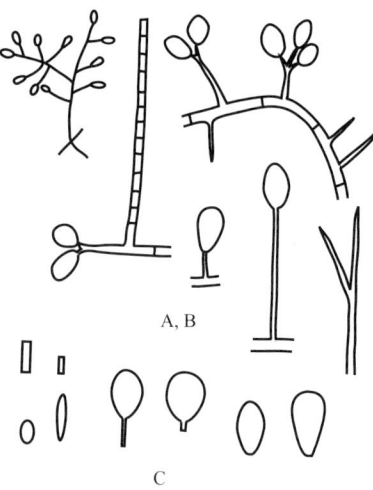

References: Barron 1968.

Morphology: Conidiophores erect, simple or branched, bearing single aleurioconidia apically on alternate or verticillate branches, and catenulate arthroconidia on simple conidiophores. Conidia of two kinds: aleuriosporous or sympodulosporous, hyaline, pale brown, ellipsoidal, spindle-shaped, ovate or subglobose, truncate or pedicellate basally; and arthrosporous, cylindrical, 1-celled.

Dimensions: Conidiophores 3–29 × 0.3–1 μm. Conidia: aleurioconidia (ellipsoidal) (3.5-) 5–8.5 × (1.5-) 3–5.5 μm, and arthroconidia 3–5 × up to 0.5 μm.

Material: 86–76 (Japanese black pine seed, Kyoto, Japan).

Remarks: Colonies on PDA are brownish gray, homogeneous, and floury, the reverse dark gray with slight radiation.

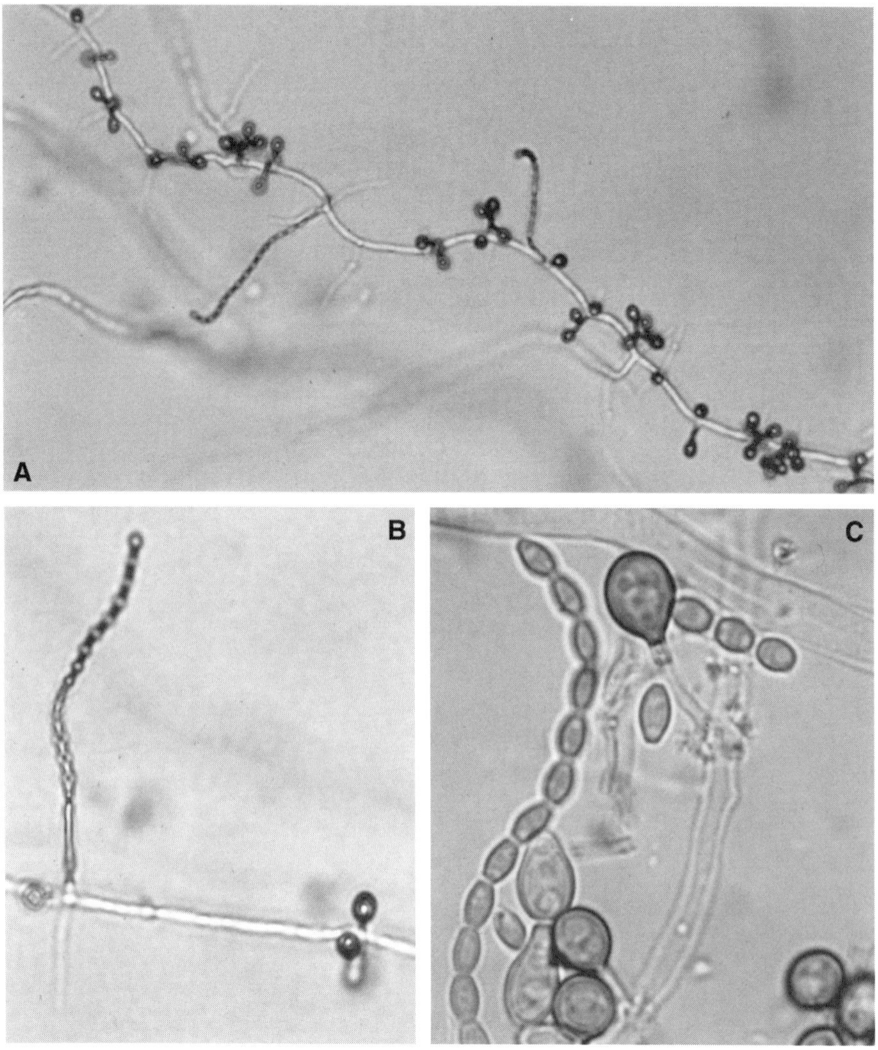

Sporotrichum sp.   A,B: Hyphae and conidia. C: Arthroconidia and aleurioconidia.

Stachybotrys Corda

Icon. Fung. 1:21, 1837.
Type species: *S. chartarum* (Ehrenb. : Link) Hughes

Stachybotrys bisbyi (Srinivasan) Barron

Synonym: *S. sacchari* (Srinivasan) Barron

References: Barron 1964; Ellis 1971; Jong and Davis 1976; Srinivasan 1958; Watanabe 1975d.

Morphology: Conidiophores hyaline or pale brown, simple or branched, erect, occasionally rough on the surface, gradually tapering from base toward apex, septate, bearing mucilaginous spore masses on 3–5 verticillate phialides apically. Conidia phialosporous, hyaline, of two kinds: long-ellipsoidal or spindle-shaped, and subglobose.

Dimensions: Conidiophores 34–138.6 µm tall, 3.5–4.5 µm wide basally, 2.5–2.9 µm wide apically: phialides 9.7–17.5 × 4–5.8 µm. Spore masses 24.3–39 µm in diameter. Conidia: ellipsoidal 7.7–11 × 2.9–4, and globose 7.5–9.5 µm in diameter.

Material: 72-X32-206 (= CBS 268.76), 72-X55 (Sugarcane root, Taiwan, ROC).

Stachybotrys bisbyi. A–C: Conidiophores and spore masses. D: Conidia. E: Conidiophore. (From Watanabe, T. 1975d. *Trans. Mycol. Soc. Jpn.*, 16:264–267. With permission.)

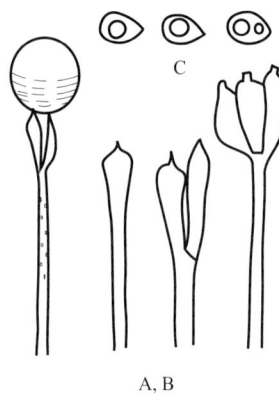

A, B

Stachybotrys elegans (Pidopl.) W. Gams

References: Domsch et al. 1980; Ellis 1971.

Morphology: Conidiophores erect, hyaline or slightly pigmented, rough on the surface, bearing spore masses apically on 1–3 phialides: phialides ellipsoidal, abruptly narrowed toward apex. Conidia phialosporous, hyaline, broadly ellipsoidal, guttulate with 1–2 oil globules.

Dimensions: Conidiophores 50–125 × 3.5–4.5 µm: phialides (10-) 12.5–17.5 (-22.5) × (2.5-) 3.7–5 µm. Conidia 9.5–12.5 × 3.7–7.5 µm.

Material: 74-707 (Strawberry root, Shizuoka, Japan).

Remarks: This fungus is differentiated from *S. bisbyi* on the basis of conidial shape and the number of phialides, but both fungi are treated as a synonym by Domsch et al. (1980).

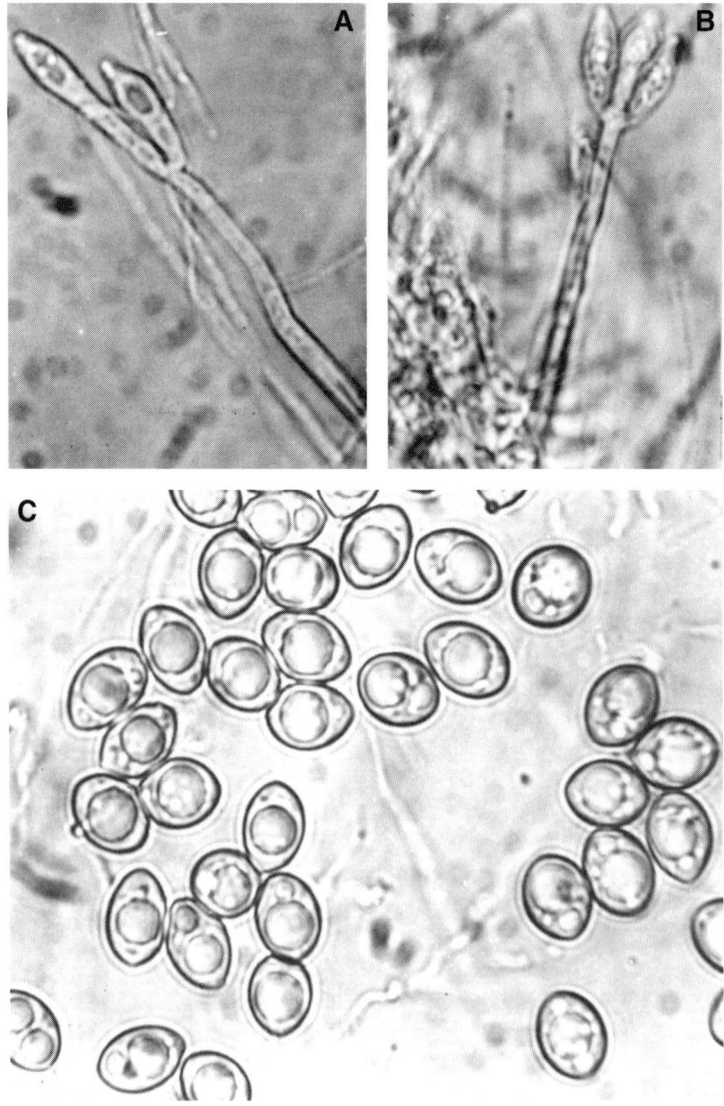

Stachybotrys elegans.   A,B: Conidiophores. C: Conidia.

Stagonospora (Sacc.) Sacc.

Syll. Fung. 3:445, 1884.
Type species: *S. paludosa* (Sacc. & Speg.) Sacc.

Stagonospora subseriata (Desmaz.) Sacc.

References: Sprague 1950; Sutton 1980; Watanabe 1975b.

Morphology: Pycnidia half-embedded, solitary or aggregated, dark brown, subglobose or flask-shaped, with extended ostioles or necks. Conidiophores hyaline, simple or branched. Conidia hyaline, cylindrical, 4-celled, truncate at one end.

Dimensions: Pycnidia 330–430 × 230–350 μm: necks ca. 100 μm wide, 50 μm deep. Conidiophores 20–25 × 3–4.5 μm. Conidia 20–27 × 4.5–5 μm.

Material: 72-X140-1131 (= ATCC 32329, Sugarcane root, Taiwan, ROC).

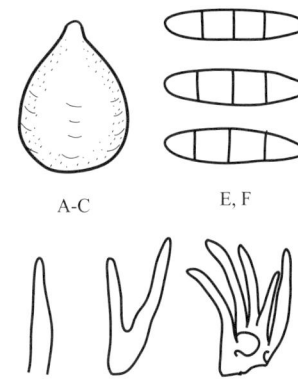

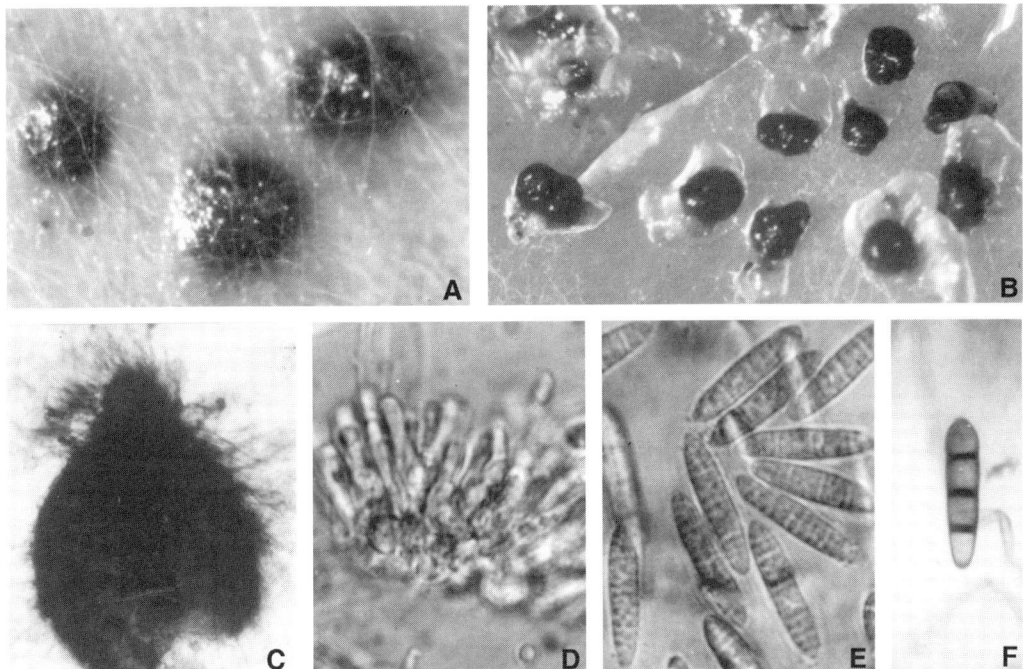

Stagnospora subseriata. A–C: Pycnidia. D: Conidiophores. E,F: Conidia. (From Watanabe, T. 1975d. *Trans. Mycol. Soc. Jpn.*, 16:264–267. With permission.)

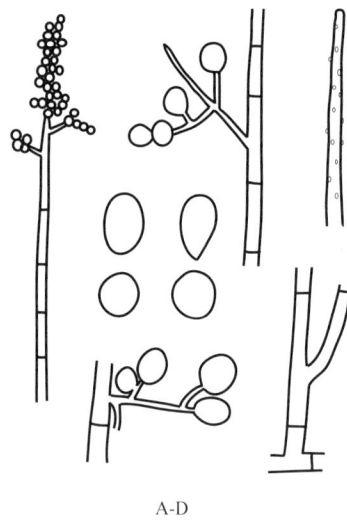

A-D

Staphylotrichum Meyer & Nicot

Bull. Trimest. Soc. Mycol. Fr. 72:322, 1956.
Type species: *S. coccosporum* Meyer & Nicot

Staphylotrichum coccosporum Meyer & Nicot

References: Maciejowska and Williams 1963; Nicot and Meyer 1957.

Morphology: Conidiophores well developed, erect, simple or branched, irregular in upper fertile portions, thick-walled, brown basally, pale brown apically, bearing single or 2–3 conidia in a chain apically, occasionally rhizoidal. Conidia aleuriosporous, hyaline or slightly pigmented, subglobose, ovate, 1-celled, occasionally apiculate apically.

Dimensions: Conidiophores 450–800 × 5–12.5 µm. Conidia (7.5-) 8.7–10.5 (-12.5) µm in diameter.

Material: 70-1161, 70-1362 (Pineapple field soil, Hachijo, Tokyo, Japan).

Remarks: Conidiophores of this fungus are well developed as compared with those of *Botryotrichum piluliferum*, one of the similar fungi.

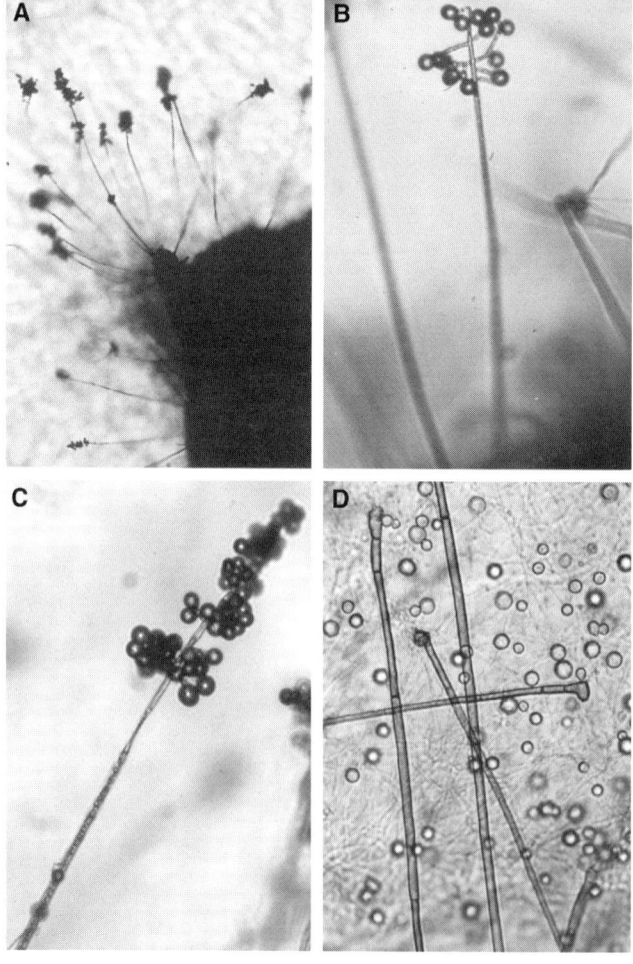

Staphylotrichum coccosporum. A–D: Conidiophores and conidia.

Stemphylium Wallr.

Fl. Crypt. Germ. 2:300, 1833.
Type species: *S. botryosum* Wallr.

Stemphylium botryosum Wallroth

Teleomorph: *Pleospora herbarum* (Pers. : Fr.) Rabenh.

References: Ellis 1971, 1976; Simmons 1967, 1969; Wiltshire 1938.

Morphology: Conidiophores pale brown, erect, simple, often proliferated forming nodes, bearing single conidia on the slightly inflated apex. Conidia porosporous, brown, long-ellipsoidal or cylindrical with rounded end, rough on the surface, muriform composed of over 5 transverse and 1–3 longitudinal septa, constricted at or near median septum.

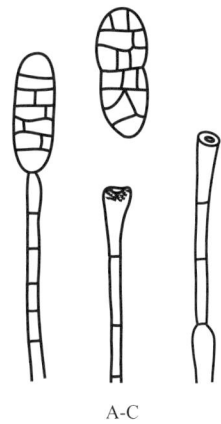
A-C

Dimensions: Conidiophores 75–220 × 2.5–5 µm. Conidia 25–35 × 10.5–15 µm.

Material: 70-1750, 70-1174 (Pineapple root, Hachijo, Tokyo, Japan).

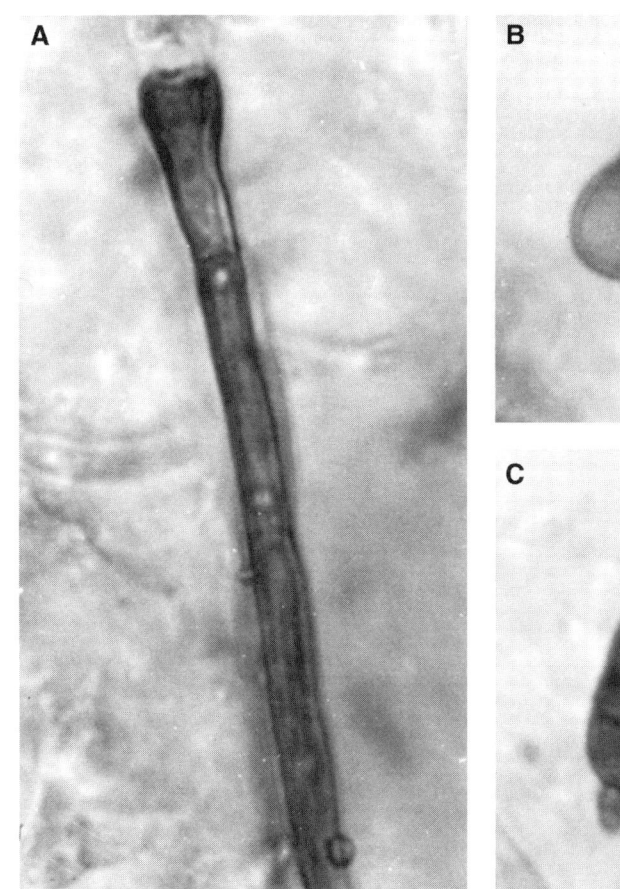
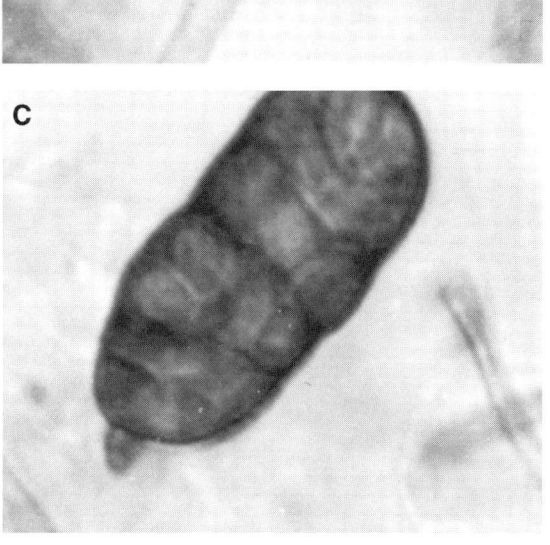

Stemphylium botryosum. A: Conidiophores. B,C: Conidia.

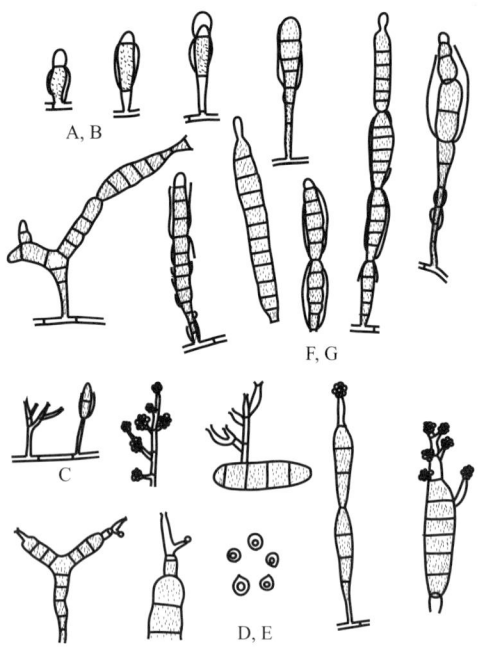

Taeniolella Hughes

Can. J. Bot. 36:816, 1958.
Type species: *T. exilis* (Karst.) Hughes

Taeniolella phialosperma T. Watanabe

References: Ellis 1971, 1976; Watanabe 1992e.

Morphology: Conidiophores pale brown to brown, not well differentiated from conidia, generally simple, short, erect. Conidia dimorphic, i.e., aleuriosporous and phialosporous. Aleurioconidia brown, cylindrical to clavate, reproduced by budding, 2- to 32-transversely septate, very rarely 1-longitudinally septate, covered with hyaline membranous sheath, up to 9 conidia in a chain, catenulate conidia simple or occasionally branched. Phialoconidia, occasionally borne on aleurioconidia in spore masses, hyaline, 1-celled, globose.

Dimensions: Conidiophores up to 375 μm tall: phialides 6.2–17.5 × 2.3–3.8 μm: collarettes 2–2.3 μm wide, 0.2–1.3 μm deep. Aleurioconidia 35–315 × 9–26 μm. Phialoconidia 2–3 μm in diameter.

Material: 70-1070 (Pineapple root, Hachijo, Tokyo, Japan); 73-466 (Strawberry root, Tottori, Japan).

Remarks: This fungus is unique in forming both aleurioconidia and phialoconidia.

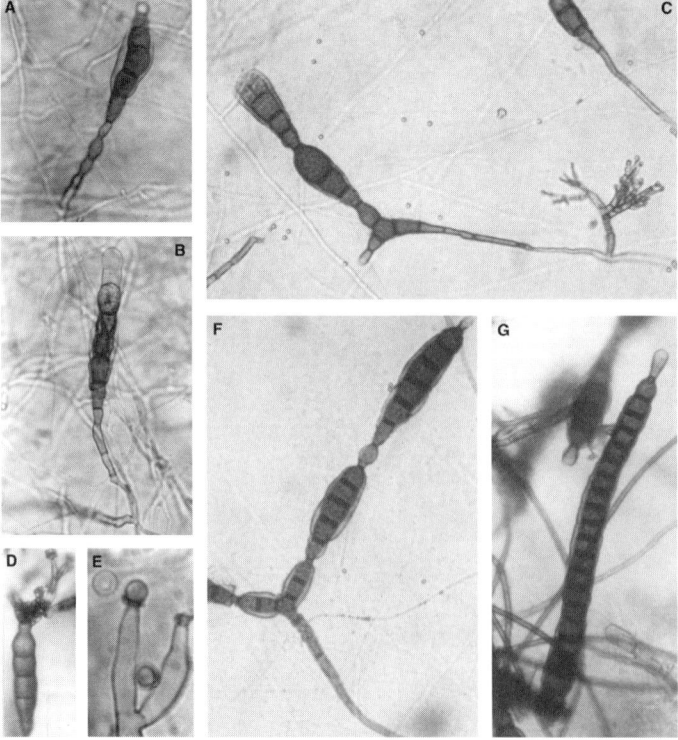

Taeniolella phialosperma. A,B: Conidia and conidiophores in *Taeniolella* state. C: *Taeniolella* and *Phialophora* state on the identical hypha. D,E: *Phialophora* state. F: Catenulate and branched conidia in *Taeniolella* state. G: Aleurioconidia in *Taeniolella* state. (From Watanabe, T. 1992e. *Mycologia*, 84:794–798. With permission.)

Tetracladium de Wild.

Ann. Soc. Belg. Microsc. 17:35, 1899.
Type species: *T. marchalianum* de Wild.

Tetracladium setigerum (Grove) Ingold

B-D

References: Petersen 1962; Watanabe 1975e.

Morphology: Conidiophores hyaline, erect, simple or branched, bearing single conidia or spore masses apically. Conidia aleuriosporous, hyaline, readily detachable, characterized by 3–4 central cells and 2–3 appendicular branches elongated from the central cells: central cells cylindrical, transversely septate: appendicular branches gradually tapering toward apex, transversely septate.

Dimensions: Conidiophores 15–47.5 × 1.3–2.3 µm. Conidia 27.5–47.5 µm tall: central cells 12.5–25 × 2.8–5.3 µm: appendicular branches 10–45 µm long.

Material: 74-875 (= ATCC 34350, Strawberry root, Kuki, Saitama, Japan); 75-137 (= ATCC 34349, Gentian root, Myokou, Niigata, Japan).

Remarks: Known as one of typical aquatic fungi.

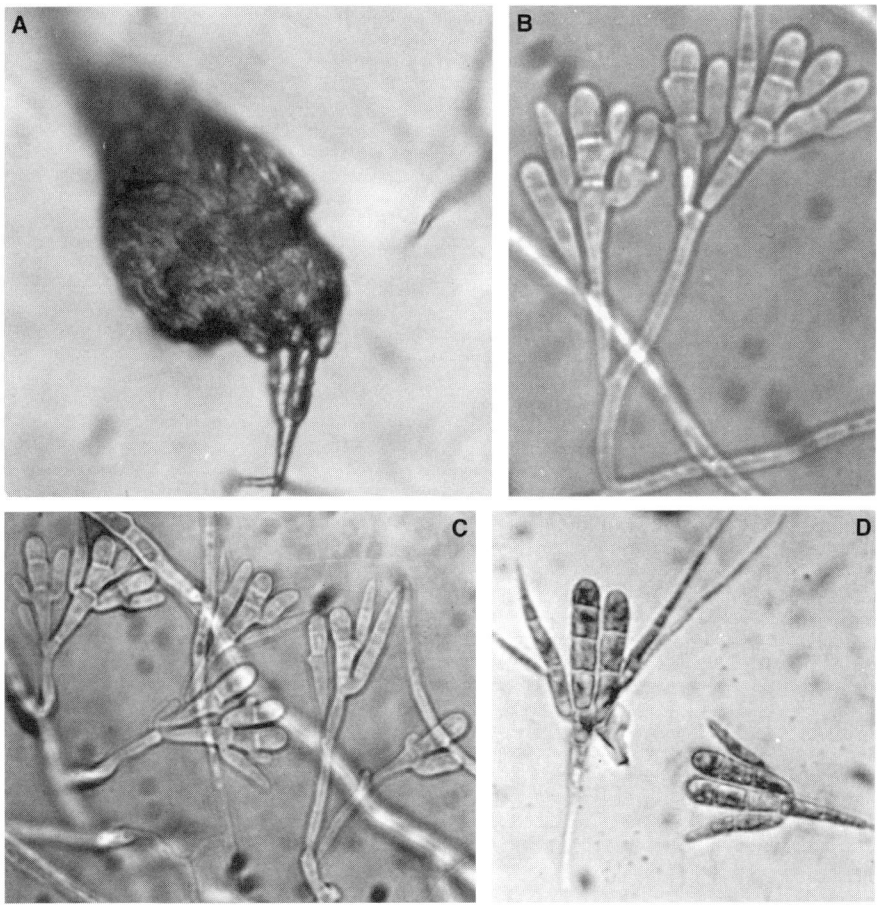

Tetracladium setigerum. A: Conidiophore and spore mass. B,C: Conidiophores and conidia. D: Conidia. (From Watanabe, T. 1975e. *Trans. Mycol. Soc. Jpn.*, 16:348–350. With permission.)

Tetraploa Berk. & Br.

Ann. Mag. Nat. Hist. 2, 5:459, 1850.
Type species: *T. aristata* Berk. & Br.

Tetraploa ellisii Cooke

References: Ellis 1971.

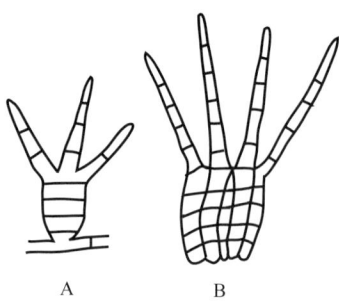

Morphology: Conidiophores lacking. Conidia aleuriosporous, borne directly on hyphae, brown, characterized by single central parts and 4–5 appendicular branches on the central cells: central parts brown, subglobose to cylindrical, longitudinally and transversely septate several times: appendicular branches gradually tapering toward apex, transversely septate numerously.

Dimensions: Conidia: central cells 23.8–30 × 15–20 µm: appendicular branches 42.5–77.5 × 3.8–4.3 µm.

Material: 77-229 (Strawberry root, Shizuoka, Japan).

Remarks: Conidia may be claimed to be aleuriosporus or blastosporous. Colonies on PDA are dark green.

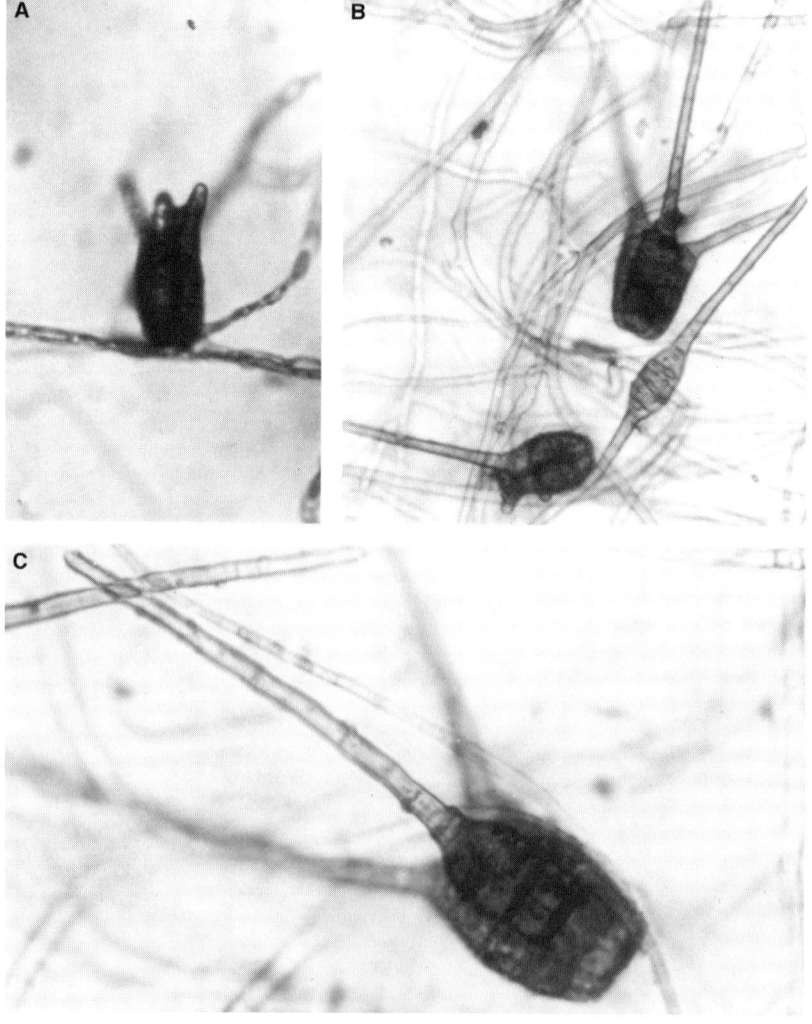

Tetraploa ellisii.   A: Conidium on aerial hypha. B,C: Conidia.

Thielaviopsis Went

Meded. Proefst. West Java 7:4, 1893.
Type species: *T. ethacetica* Went = *T. paradoxa* (de Seynes) Hoehnel

Thielaviopsis adiposa (Butler) C. Moreau

Teleomorph: *Ceratocystis adiposa* (Butler) C. Moreau

References: Butler 1906; Hunt 1956; Sartoris 1927; Watanabe 1975d.

Morphology: Conidiophores (phialides) pale brown, simple, cylindrical, bearing catenulate conidia and chlamydospores apically. Conidia phialosporous, hyaline or pale brown, cylindrical, ellipsoidal or various in shape, 1-celled. Chlamydospores reddish brown, subglobose, polygonal or irregular in shape, thick-walled, minutely echinulate, smooth or rough marginally.

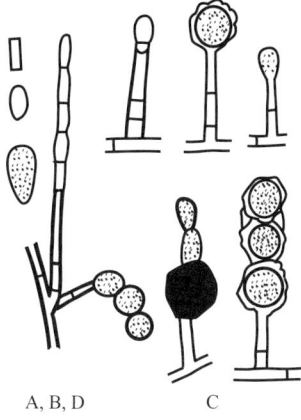

A, B, D   C

Dimensions: Conidiophores 23–34.1 μm long. Conidia 9.4–13.2 × 7.2–10.2 μm. Chlamydospores 18.9–20.7 × 18.2–19.7 μm.

Material: 72-X600-181 (= ATCC 32927, Sugarcane root, Taiwan, ROC).

Remarks: Large globose chlamydospores are very characteristic for this species.

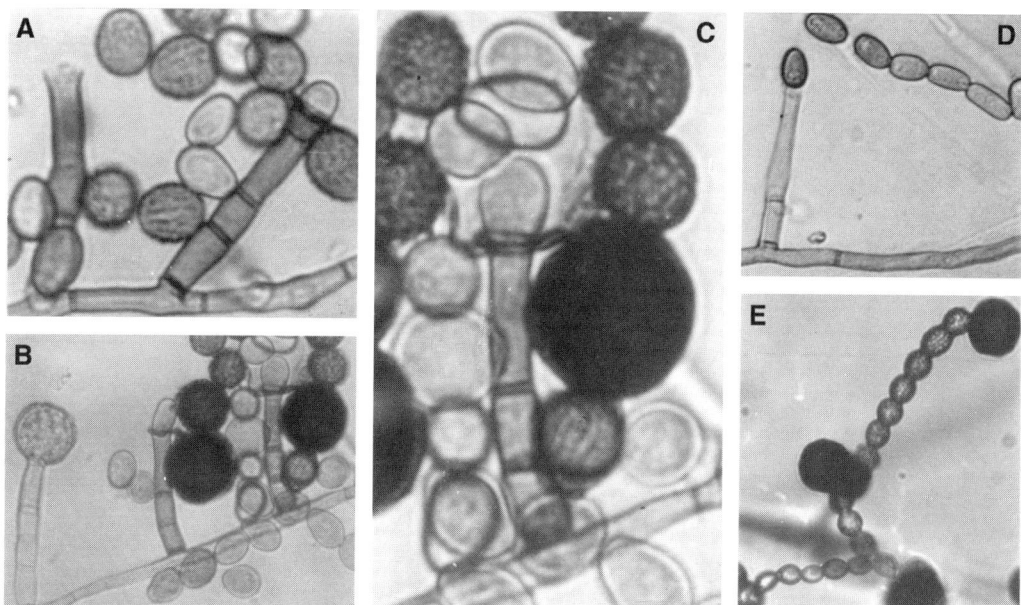

Thielaviopsis adiposa. A–C,E: Conidiophores (phialides), conidia, and chlamydospores. D: Conidiophores and conidia. (From Watanabe, T. 1975d. *Trans. Mycol. Soc. Jpn.*, 16:149–182. With permission.)

Thielaviopsis paradoxa de Seynes

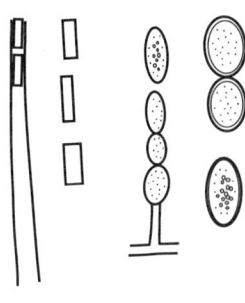

Synonym: *Chalara paradoxa* (de Seyn.) Sacc.

References: Dade 1928; Hunt 1956; Sugiyama 1968; Watanabe 1975d.

Morphology: Conidiophores (phialides) hyaline, simple, cylindrical, bearing conidia and chlamydospores. Conidia phialosporous, 1-celled, mostly cylindrical or spindle-shaped. The cylindrical conidia, hyaline or pale brown, thin-walled, and the spindle-shaped conidia, brown, thick-walled. Chlamydospores dark brown, ellipsoidal, thick-walled, granulate.

A, B   C, D

Dimensions: Conidiophores 121.6–350.3 μm long. Conidia: cylindrical, 8.5–12.7 (-16.3) × 2.5–4.7 μm; and spindle-shaped, 9.2–10 (-12.5) × 5.1–7.5 μm. Chlamydospores 13.3–15.8 (-20.5) × 9.7–11.7 μm.

Material: 69-523 (Pineapple root, Okinawa, Japan); 70-1170 (Pineapple root, Hachijo, Tokyo, Japan); 72-X49-378 (= ATCC 32928, Sugarcane root, Taiwan, ROC).

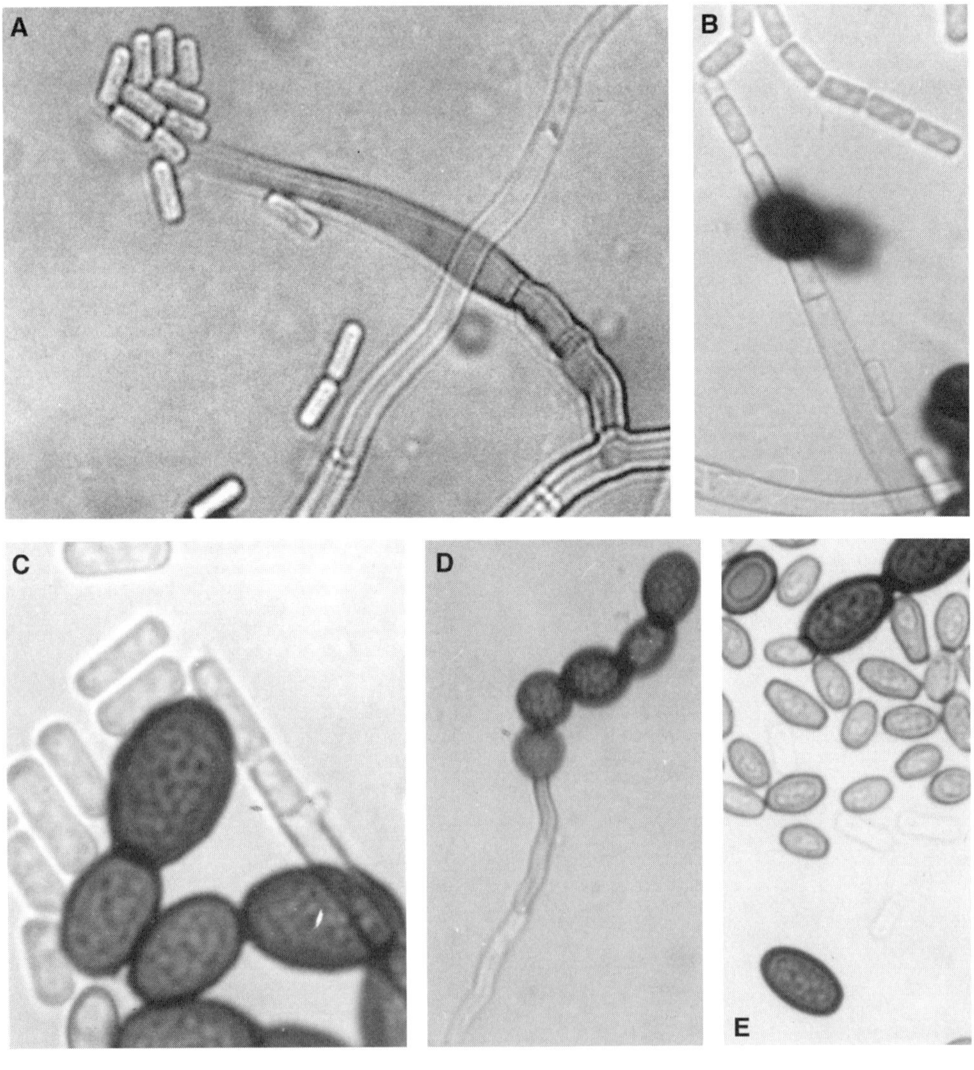

Thielaviopsis paradoxa. A,B: Conidiophores and conidia. C,E: Conidia and chlamydospores. D: Catenulate chlamydospores. (From Watanabe, T. 1975d. *Trans. Mycol. Soc. Jpn.*, 16:149–182. With permission.)

Thysanophora Kendrick

Can. J. Bot. 39:817, 1961.
Type species: *T. penicillioides* (Roum.) Kendrick

Thysanophora penicillioides (Roum.) Kendrick

References: Kendrick 1961; Watanabe et al. 1986a.

Morphology: Conidiophores dark brown, erect, thick-walled, often over 16-septate, bearing spore masses on apical penicillate branches composed of 3–4 primary branches and over 3 verticillate phialides on each primary branch (2-biverticillate). Conidia phialosporous, hyaline or pale brown, ellipsoidal or ovate, 1-celled. Sclerotia brown, globose.

Dimensions: Conidiophores ca. 575 × 7.5 μm: phialides ca. 16 μm long. Conidia 1.8–2.2 μm in diameter. Sclerotia ca. 240 μm in diameter.

Material: 84-531 (Japanese red pine seed, Saihaku, Tottori, Japan).

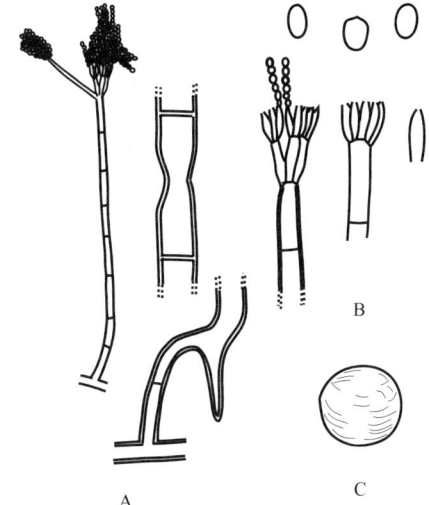

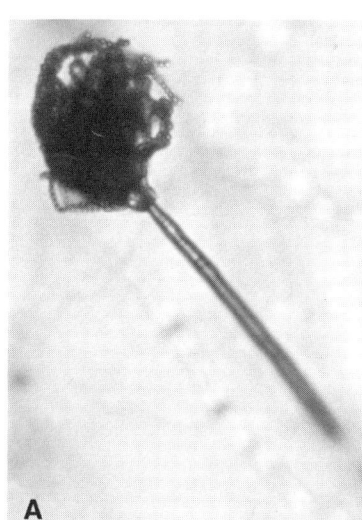
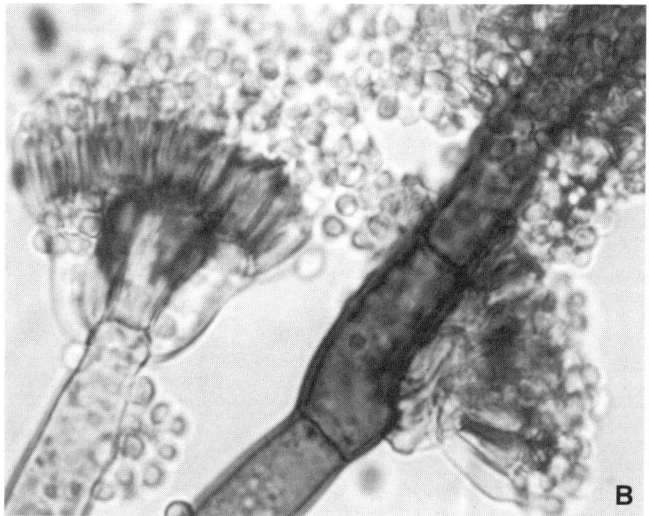

Thysanophora penicillioides. A,B: Conidiophores and conidia. C: Sclerotium.

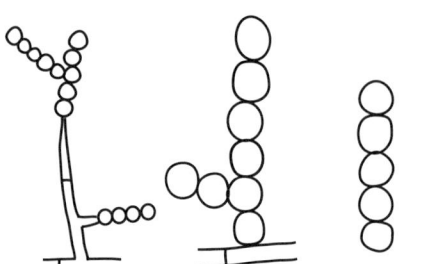

Torula Pers. : Fr.

Syst. Mycol. 3:499, 1932.
Type species: *T. herbarum* (Pers.) Link : S. F. Gray.

Torula herbarum (Persoon) Link : S. F. Gray

References: Crane and Schoknecht 1986; Rao and de Hoog 1975; Watanabe and Sato 1988.

Morphology: Conidiophores lacking. Conidia blastosporous, acropetally developed, forming simple or branched chains borne apically and laterally on hyphae, dark brown, reddish brown, cylindrical, 1- to 8-celled, often detached; component cells globose.

Dimensions: Component cells of conidia 5–7 μm in diameter.

Material: 86-66 (Japanese cedar seed, Kasama, Ibaraki, Japan).

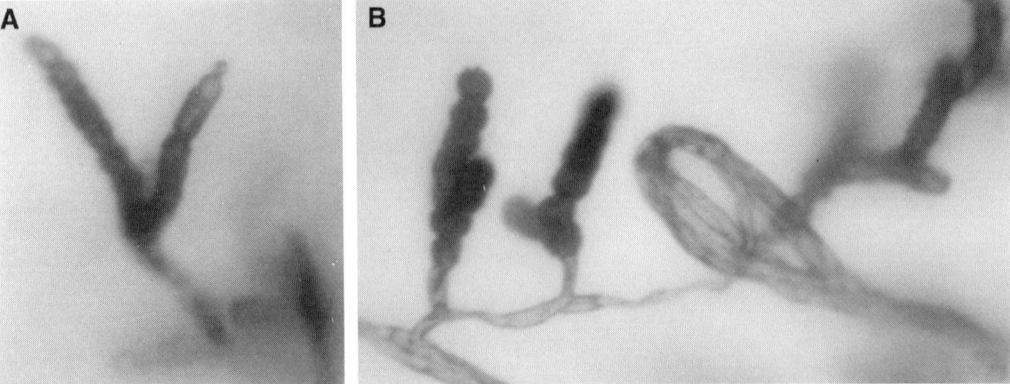

Torula herbarum. Conidiophores and conidia. (From Watanabe, T. and Sato, Y. 1988. *Trans. Mycol. Soc. Jpn.*, 29:143–150. With permission.)

Torula sp.

References: Schoknecht and Crane 1977; Watanabe et al. 1987.

Morphology: Conidiophores lacking. Conidia blastosporous, apically developed, forming simple or branched chains borne apically and laterally on hyphae, often resulting in spore masses, hyaline, pale brown or brown, globose cylindrical, 1-celled. Chlamydospores dark brown, globose, double-membranous.

Dimensions: Conidia 6.2–8.3 × 2.5–3.8 μm. Chlamydospores 8.7–10 μm in diameter.

Material: 85-PP70 (Paulownia root, Itapua, Paraguay); 86-67 (Japanese cypress seed, Higashichikuma, Nagano, Japan).

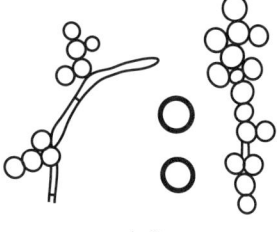

A, B

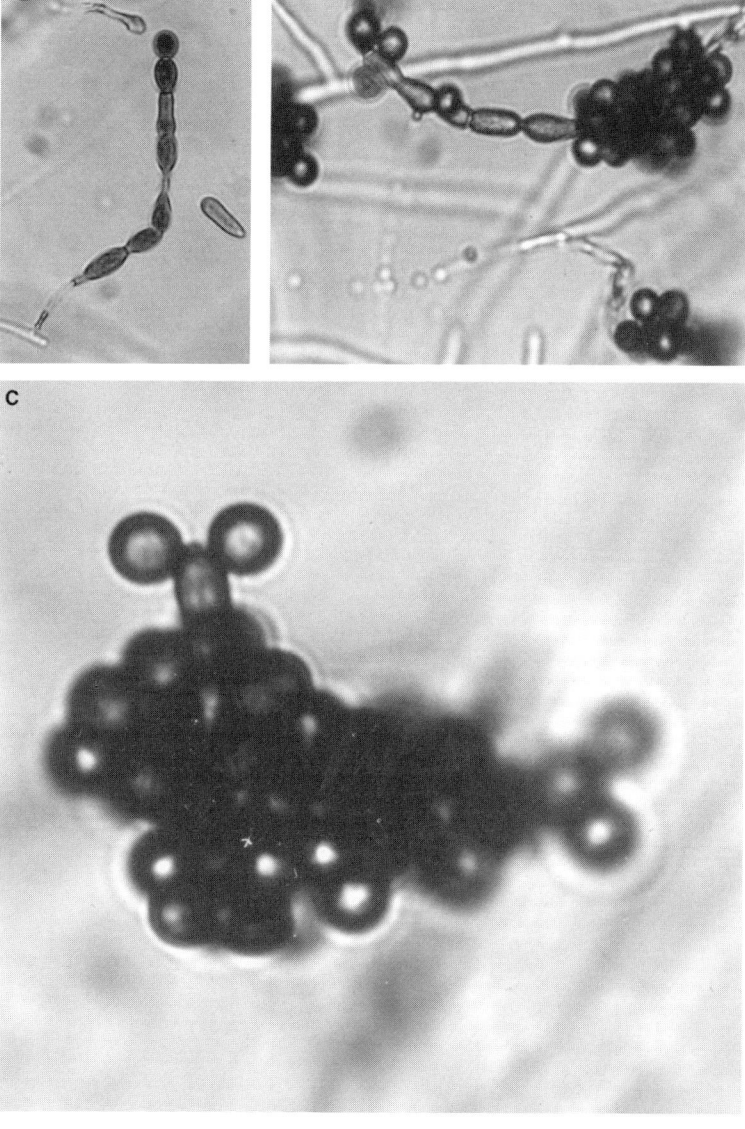

Torula sp. A–C: Conidiophores, conidia, and chlamydospores. (From Watanabe, T. et al. 1987a. *Trans. Mycol. Soc. Jpn.*, 28:453–469. With permission.)

Torulomyces Delitsch

Systematik der Schimmelpilze p. 91, 1943.
Type species: *T. lagena* Delitsch

Torulomyces lagena Delitsch

Synonym: *Monocillium humicola* Barron

References: Barron 1967, 1968; Watanabe and Sato 1988.

Morphology: Conidiophores hyaline, erect, simple or branched, bearing catenulate conidia apically on phialides often with inflated median. Conidia phialosporous, pale brown, globose, often apiculate at one end, minutely echinulate.

Dimensions: Conidiophores 6–13 μm long. Conidia 2.7–4 μm in diameter.

Material: 86-73 (Japanese cedar seed, Kasama, Ibaraki, Japan).

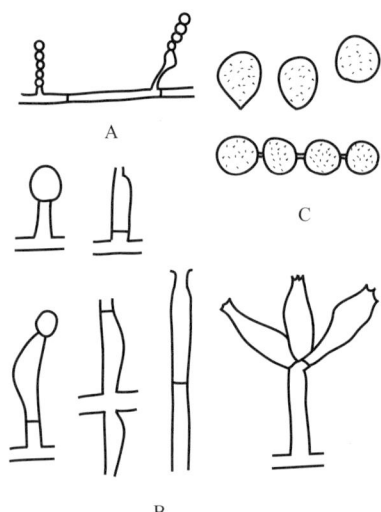

Remarks: Colonies on PDA are yellowish green, floury, with crystals embedded, the reverse yellowish brown.

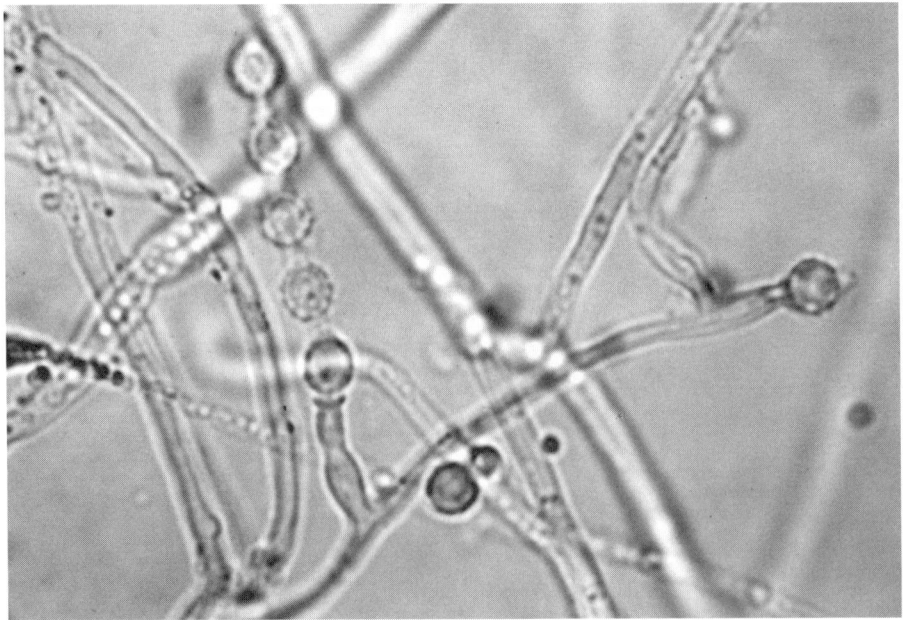

Torulomyces lagena. A: Habit. B: Conidiophores and conidia. C: Conidia. (From Watanabe, T. and Sato, Y. 1988. *Trans. Mycol. Soc. Jpn.*, 29:143–150. With permission.)

Trichocladium Harz

Bull. Soc. Imp. Moscou 44:125, 1871.
Type species: *T. asperum* Harz

Trichocladium canadense Hughes

References: Hughes 1959.

Morphology: Conidiophores not well differentiated from conidia, indistinct or short, cylindrical if present. Conidia aleuriosporous, borne directly and laterally on hyphae, subglobose, ellipsoidal, composed of two to several cells, basal cell often brown, the rest of cells dark brown to black.

Dimensions: Conidiophores 1.2–32.5 × 2–2.5 (-5) µm. Conidia 15–26.4 (-100) × 7.5–11.3 µm.

Material: 99-483 (Forest soil, Agematsu, Nagano, Japan).

Remarks: Multi-septate conidia became more numerous with age.

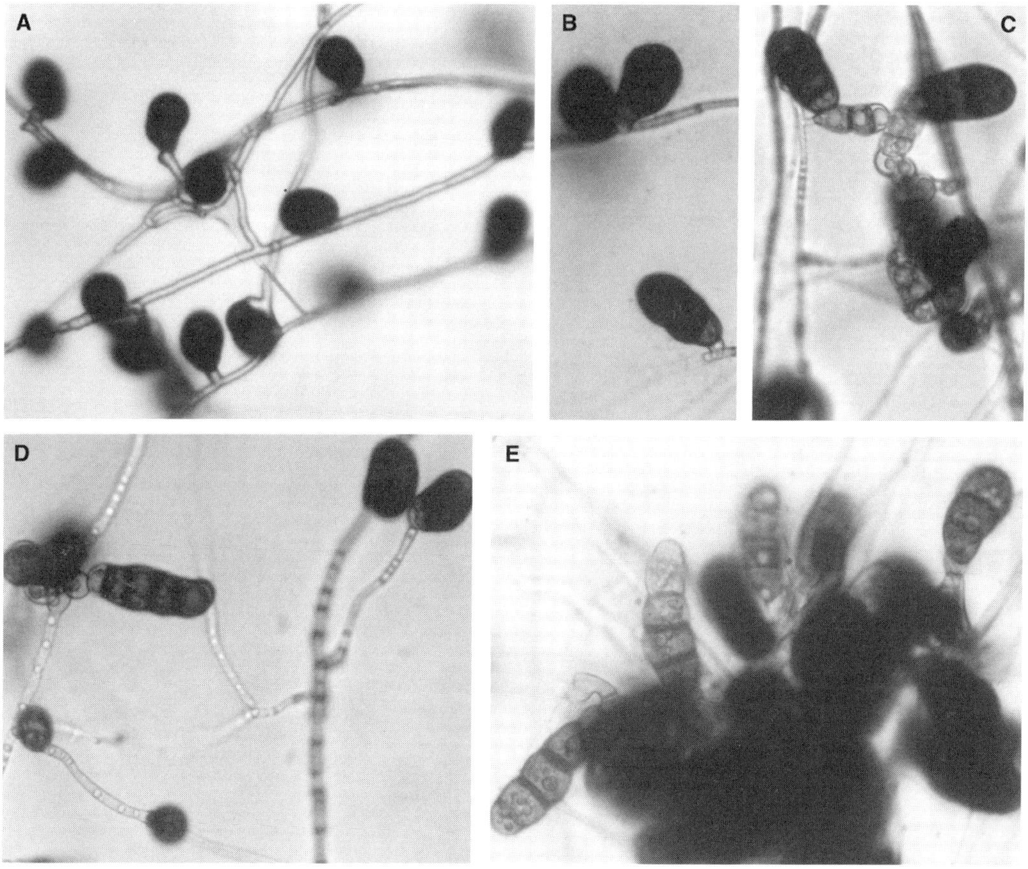

Trichocladium canadense. A. One-celled conidia. B–E: More than 2-celled conidia.

Trichocladium pyriformis Dixon

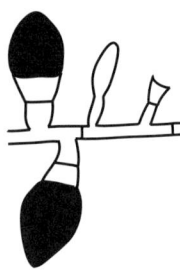

References: Dixon 1968; Ellis 1971; Hughes 1952; Kendrick and Bhatt 1966; Watanabe 1991f.

Morphology: Conidiophores not well differentiated from conidia, short if present, ellipsoidal. Conidia aleuriosporous, borne directly and laterally on hyphae in a series, ellipsoidal or pear-shaped, smooth, 2- to 3-celled, composed of an apical dark brown large cell with germ pore; central and basal, rather smaller pale brown or hyaline cells, constricted at or near the septum.

Dimensions: Conidiophores 3.7–87.5 × 1.2–1.3 µm. Conidia 11.2–15 × 5–6.8 µm.

Material: 85-124 (= IFO 32553, Cucumber seed, Gumma, Japan).

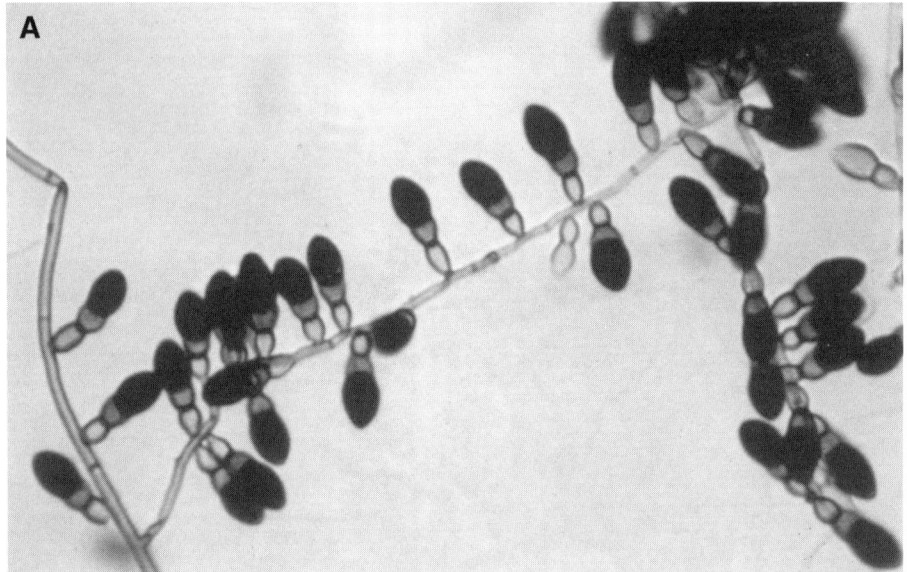

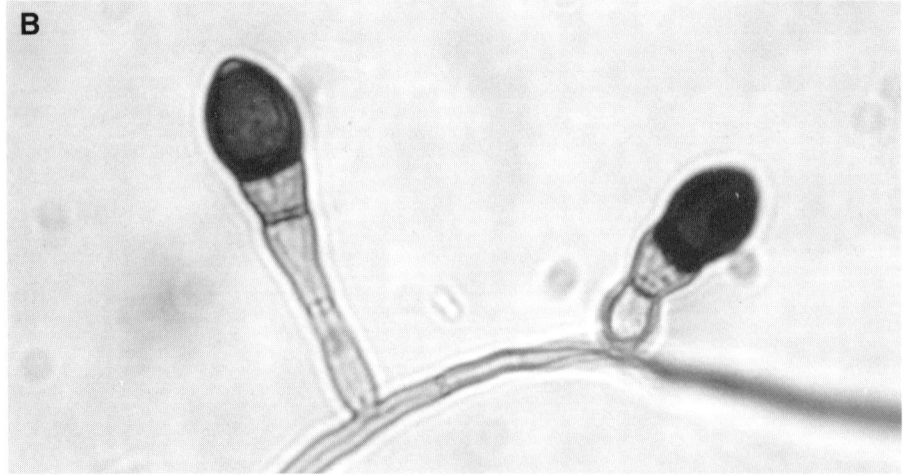

Trichocladium pyriformis. A,B: Hyphae, conidiophores, and conidia. (From Watanabe, T. 1991. *Mycologia*, 83:524–529. With permission.)

Trichoderma Pers. : Fr.

Syst. Mycol. 3:214, 1829.
Type species: *T. viride* Pers. : Fr.

Key to Species

1. Setae-like hyphae present, conidia ovate................................ *T. hamatum*
   lacking, conidia ovate or others.................................. 2

2. Conidia globose.. *T. harzianum*
   not globose... 3

3. Phialides mainly verticillate 4
   irregularly located .... *T. pseudokoningi*

4. Conidia ovate, phialides densely arranged .... *T. paureoviride*
   ellipsoidal, phialides thinly arranged .... *T. koningi*

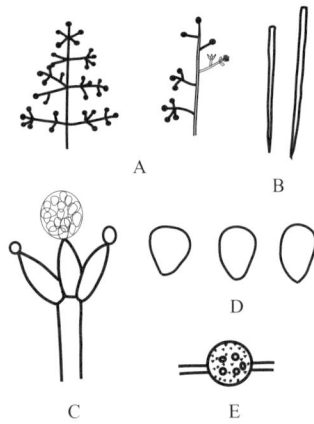

Trichoderma aureoviride Rifai

References: Bissett 1991; Rifai 1969; Watanabe 1975d.

Morphology: Conidiophores branched, bearing spore masses on each of the phialides: phialides often verticillate, short, and thick. Conidia phialosporous, hyaline, ovate, 1-celled. Chlamydospores pale brown, subglobose, granulate.

Dimensions: Phialides 8.5–11 × 2.4–2.7 µm. Conidia 2.4–2.7 × 2.1–2.5 µm. Chlamydospores 5.2–7.5 µm in diameter. Needle-shaped crystals 160 to over 250 × 2.5 µm.

Material: 72-X21-13 (Sugarcane root, Taiwan, ROC).

Remarks: Colonies on PDA are yellowish green, fluffy, the reverse brown. Needle-shaped crystals are produced in agar cultures.

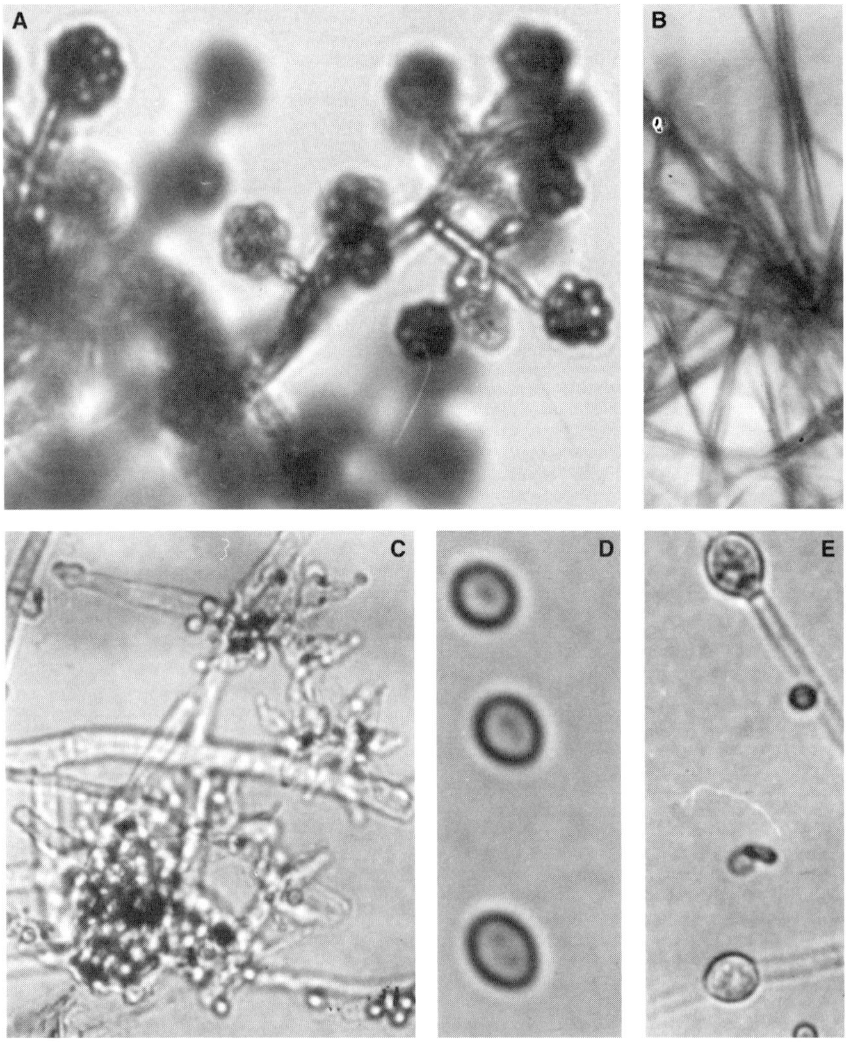

Trichoderma aureoviride. A: Conidiophores and spore masses. B: Needle-shaped crystals. C: Conidiophores and phialides. D: Conidia. E: Chlamydospores. (From Watanabe, T. 1975d. *Trans. Mycol. Soc. Jpn.*, 16:264–267. With permission.)

Trichoderma hamatum (Bonorden) Bainier

References: Bissett 1991; Rifai 1969; Watanabe 1975a.

Morphology: Conidiophores developed on cushion-shaped structures, hyaline erect, branched, bearing spore masses on alternate or verticillate phialides, together with setae-like sterile hyphae elongated: phialides short and thick, densely arranged: setae-like hyphae curved, gradually tapering toward apex, septate. Conidia phialosporous, hyaline, ellipsoidal or ovate, 1-celled. Chlamydospores pale brown, subglobose or ellipsoidal, granulate.

Dimensions: Phialides 4.8–10.2 × 2.4–3 µm. Setae-like hyphae 58.3–100.9 × 2.6–3.9 µm. Conidia 3.4–5 × 2.7–3.8 µm. Chlamydospores 7.5–11.7 µm in diameter.

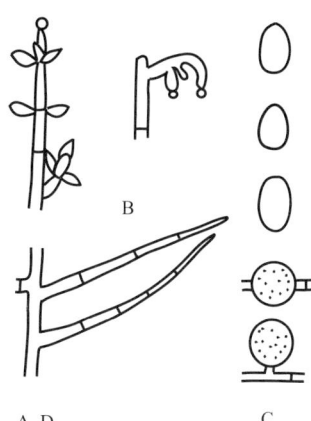

Material: 69-453 (Pineapple root, Ishigaki, Okinawa, Japan); 70-1481 (Pineapple field soil, Hachijo, Tokyo, Japan); 72-X125 (Sugarcane root, Taiwan, ROC); 85-73, 85-107 (Flowering cherry seed, Hachioji, Tokyo, Japan).

Remarks: Colonies on PDA are white initially, greenish, forming cushions distributed with age.

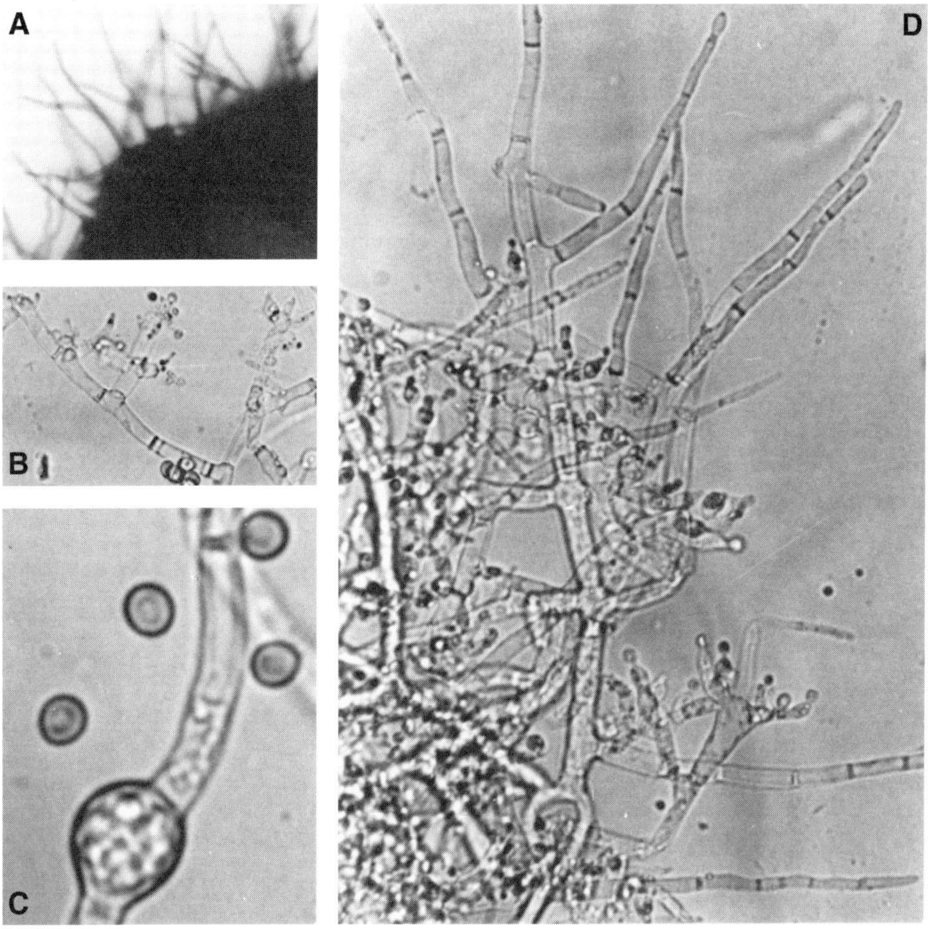

Trichoderma hamatum. A: Setae-like hyphae elongated from cushion-shaped structure. B: Phialides. C: Conidia and chlamydospore. D: Setae-like hyphae and conidiophores. (From Watanabe, T. 1975a. *Trans. Mycol. Soc. Jpn.*, 16:18–27. With permission.)

Trichoderma harzianum Rifai

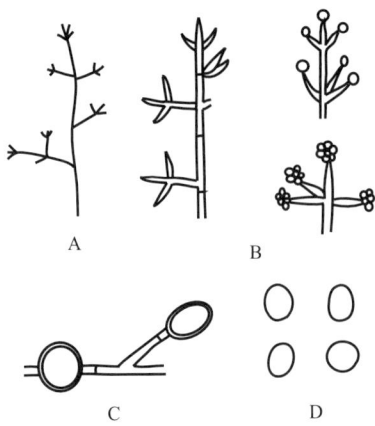

References: Rifai 1969; Watanabe 1975; Watanabe et al. 1986a.

Morphology: Conidiophores hyaline, erect, branched, bearing spore masses apically at verticillate phialides: phialides short and thick. Conidia phialosporous, hyaline, globose, subglobose, or ovate, 1-celled. Chlamydospores brown, subglobose.

Dimensions: Conidiophores ca. 60–110 μm tall: phialides 7.2–9.8 × 2.4–2.7 μm. Conidia: globose 2.1–3 (-4) μm in diameter, and ovate 4–6 × 2.8–4.8 μm. Chlamydospores 6.2–8.8 μm in diameter.

Material: 72-X20-2 (Sugarcane root, Taiwan, ROC); 84-530 (Japanese red pine seed, Saihaku, Tottori, Japan); 85-74 (Flowering cherry seed, Hachioji, Tokyo, Japan).

Remarks: Colonies on PDA are dark green with yellowish tint, with cushion-shaped structures distributed. Needle-shaped crystals were characteristically formed in cultures.

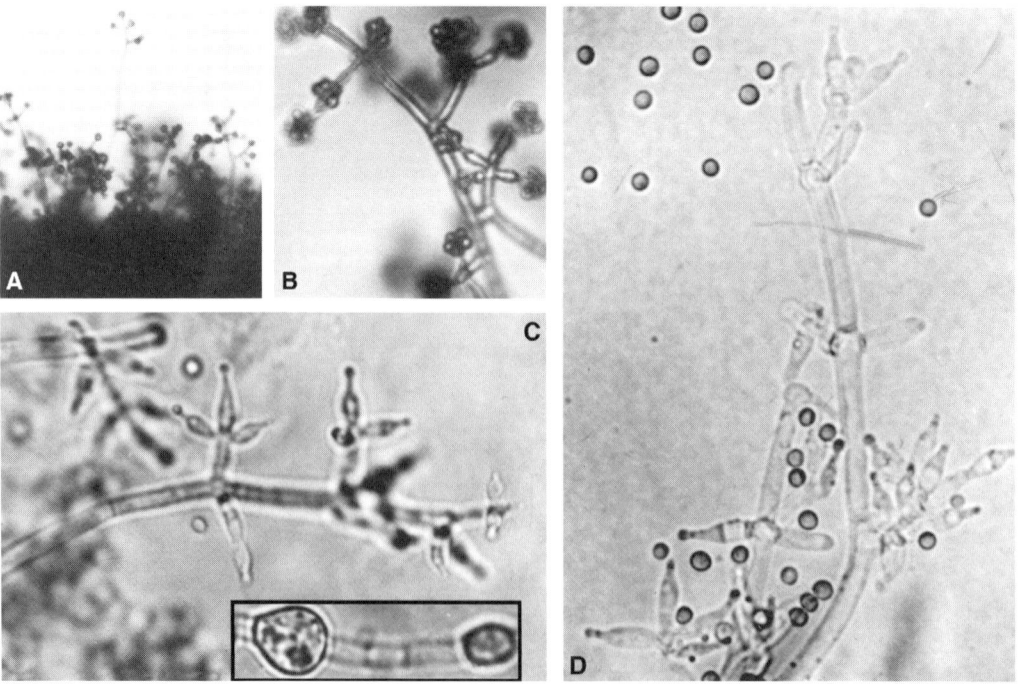

Trichoderma harzianum. A,B: Conidiophores and conidia. C,D: Conidiophores, phialides, and chlamydospores (C) or conidia (D). (From Watanabe, T. 1975c. *Trans. Mycol. Soc. Jpn.*, 16:149–182. With permission.)

Trichoderma koningi Oud.

References: Rifai 1969; Watanabe 1975d.

Morphology: Conidiophores hyaline, erect, branched, bearing spore masses apically at phialides: phialides tapering toward apex. Conidia phialosporous, hyaline, ovate or ellipsoidal, 1-celled. Chlamydospores pale brown, subglobose.

Dimensions: Phialides 7.2–12.2 × 2.1–2.7 µm. Conidia 3–4.5 × 2.4–3 µm. Chlamydospores 9.2–13 µm in diameter.

Material: 72-X70 (Sugarcane root, Taiwan, ROC).

Remarks: Colonies on PDA are white with pale yellowish green tint.

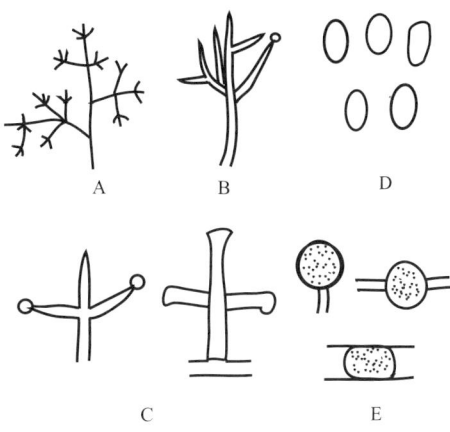

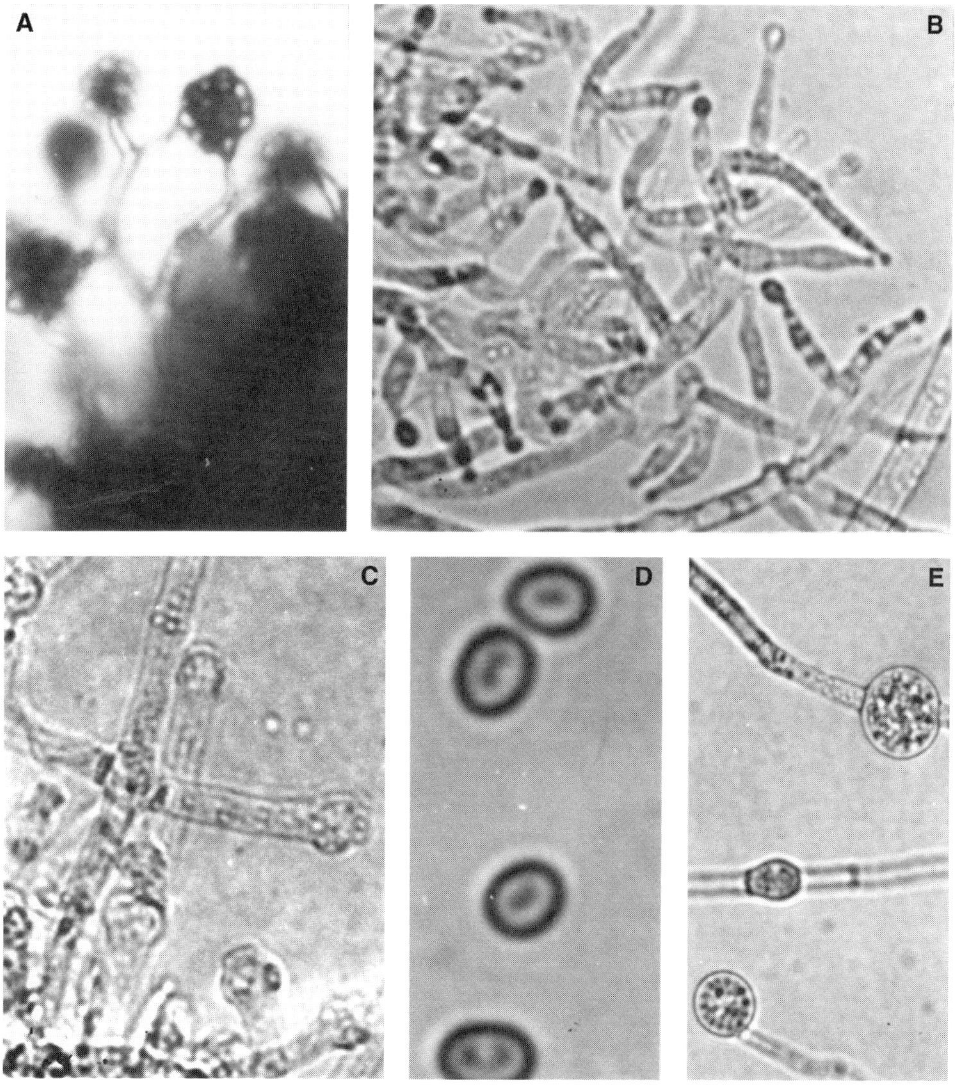

Trichoderma koningi. A: Conidiophores and spore masses. B,C: Conidiophores and phialides. D: Conidia. E: Chlamydospores. (From Watanabe, T. 1975d. *Trans. Mycol. Soc. Jpn.*, 16:264–267. With permission.)

Trichoderma pseudokoningi Rifai

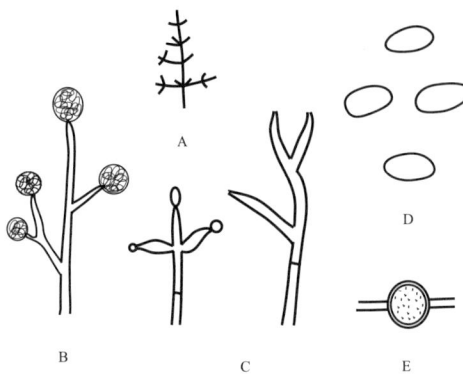

References: Rifai 1969; Watanabe 1975d.

Morphology: Conidiophores hyaline, erect, branched, bearing spore masses apically at irregularly disposed phialides: phialides short and thick. Conidia phialosporous, pale green, ellipsoidal or ovate, 1-celled, apiculate at one end. Chlamydospores brown, subglobose.

Dimensions: Phialides (6-) 9.7–12.2 × 2.1–2.9 µm. Conidia 3.6–4 × 2.2–2.5 µm. Chlamydospores 7.5–12.2 (-15.1) µm in diameter.

Material: 72-X20-1 (Sugarcane root, Taiwan, ROC).

Remarks: Colonies on PDA are white with yellowish green tint.

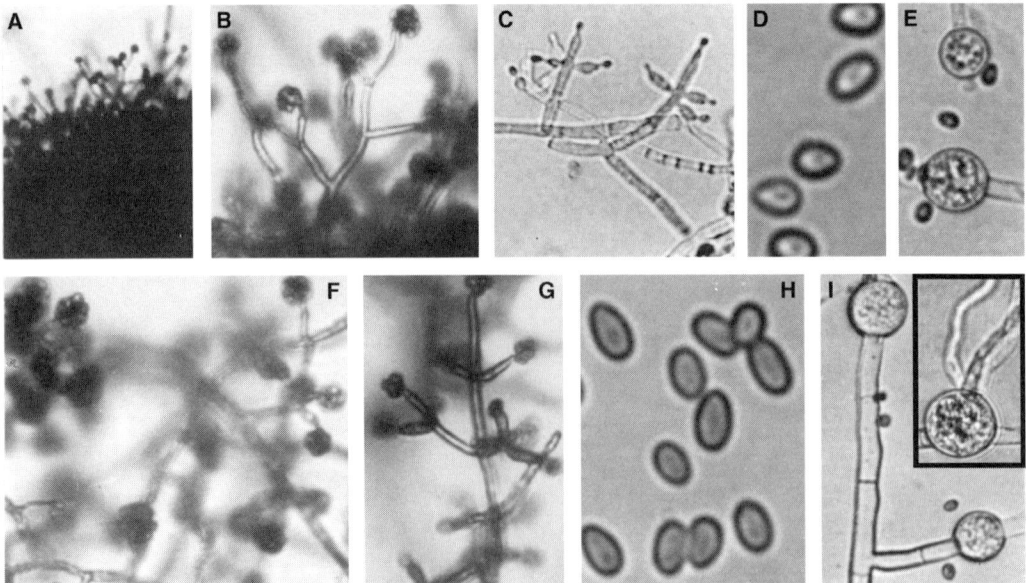

Trichoderma pseudokoningi. A,B: Conidiophores and spore masses. C,G: Conidiophores and phialides. D,H: Conidia. E,I: Chlamydospores and germination. (From Watanabe, T. 1975d. *Trans. Mycol. Soc. Jpn.*, 16:264–267. With permission.)

Trichothecium Link : Fr.

Linn. Spec. Pl. 1:28, 1824.
Type species: *T. roseum* (Pers. : Fr.) Link.

Trichothecium roseum (Pers. : Fr.) Link

References: Ingold 1956; Rifai and Cooke 1966.

Morphology: Conidiophores erect, simple, bearing conidia densely at apical or subapical portions. Conidia meristem arthrosporous, hyaline, ellipsoidal, 2-celled, constricted at or near septum.

Dimensions: Conidiophores 200–300 µm tall. Conidia 15–20 × 7.5–10 µm.

Material: 70-105 (Snap bean seed, Shinsyu-shinmachi, Nagano, Japan).

Remarks: Colonies on PDA are pinkish homogeneously.

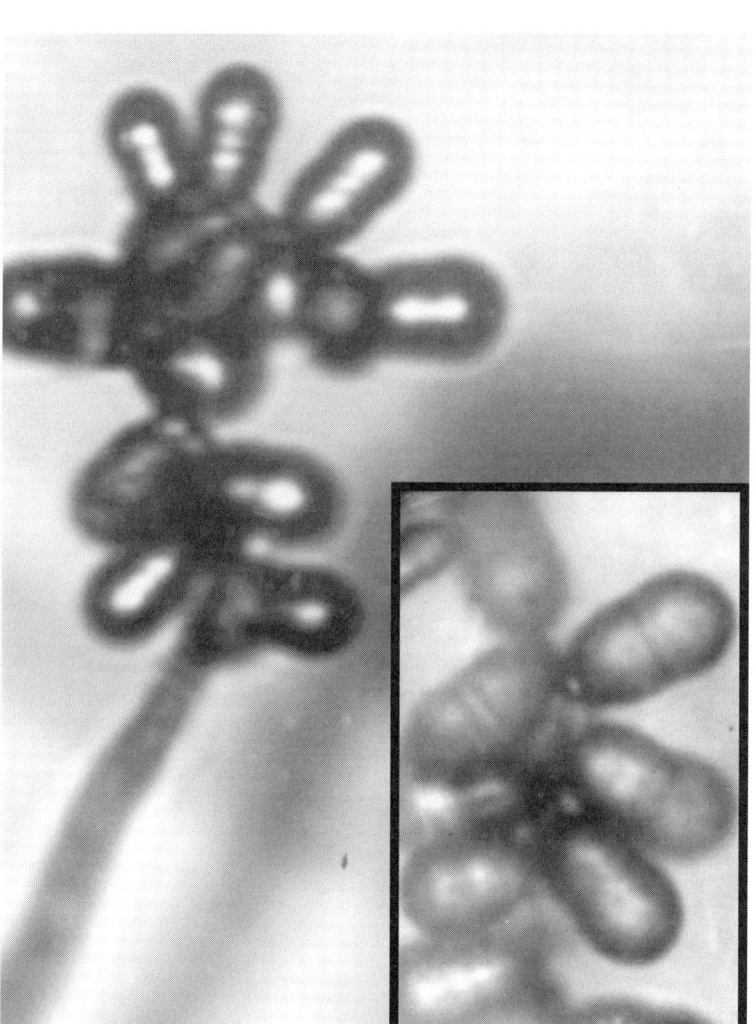

Trichothecium roseum.   Conidiophores and conidia.

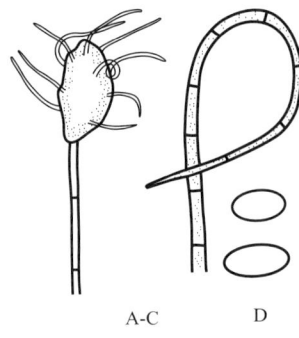

Trichurus Clem. & Shear

Bot. Surv. Nebr. 4:7, 1896.
Type species: *T. cylindricus* Clem. & Shear

Trichurus spiralis Hasselbring

References: Domsch et al. 1980a,b; Ellis 1971.

Morphology: Conidiophores united forming synnemata, erect, brown, simple, bearing spore masses composed of catenulate conidia on apical fertile parts, together with numerous setae: setae brown, over 10-septate, curved, twisted, coiled. Conidia annellosporous, pale brown, ellipsoidal, 1-celled.

Dimensions: Conidiophores up to 3 mm tall. Spore masses 180–340 × 96–100 µm. Conidia 3.8–5.3 × 2–3 µm. Setae ca. 300 µm; ca. 3.5 µm wide basally.

Material: 77-91 (Strawberry root, Nishigahara, Kita, Tokyo, Japan).

Remarks: Annellations of conidium-bearing portion were not observed.

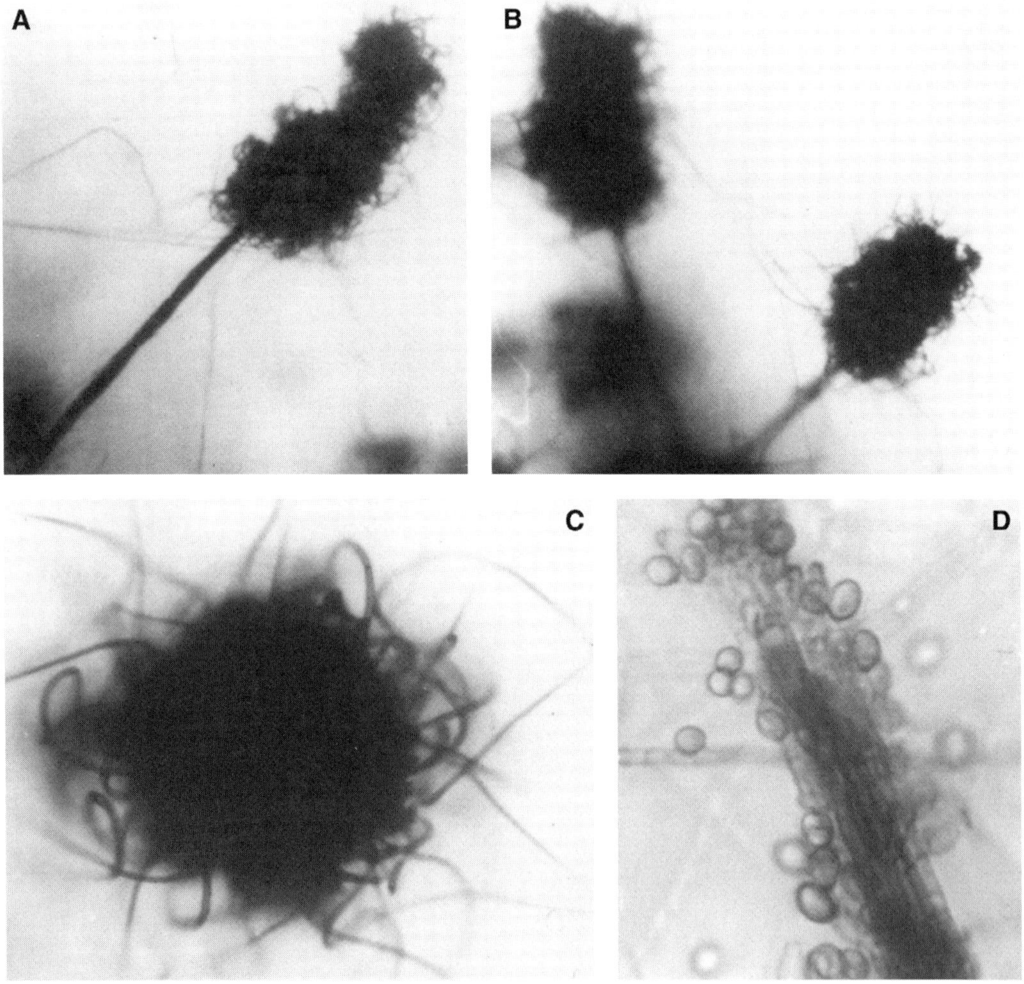

Trichurus spiralis. A,B: Synnema and spore masses (C: top view). D: Part of the synnema and conidia.

Trinacrium Riess

Fres. Beitr. Mykol. 2: 42, 1852.
Type species: *T. subtile* Riess

Trinacrium iridis T. Watanabe

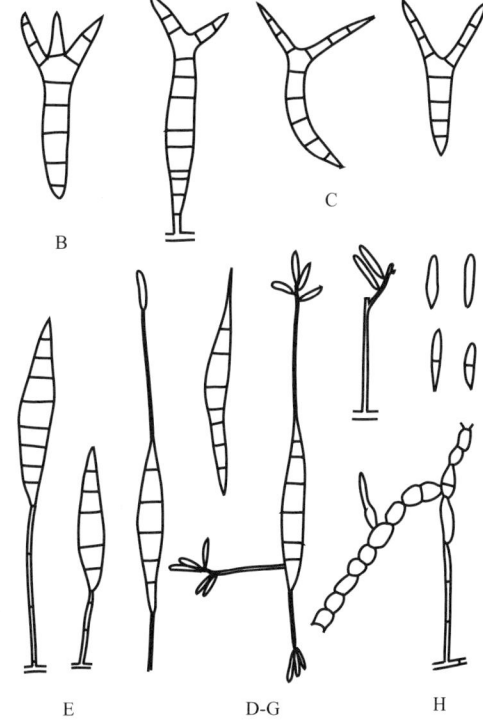

References: Drechsler 1937, 1938; Liu and Qiu 1992; Watanabe 1992g.

Morphology: Conidiophores hyaline, smooth, septate, difficult to differentiate from ordinary hyphae, bearing single macroconidia apically in simple conidiophores and microconidia apically or subapically on conidiophores, developed from hyphae or macroconidia. Conidia hyaline, smooth, of 3 kinds: the first type, simple macroconidia aleuriosporous, solitary, occasionally curved, long-ellipsoidal, inflated in the median, narrowing toward both ends, 5- to 11-celled, occasionally beaked, often proliferated elongating 1–3 conidiophores; the second type, simple macroconidia with 2–3 apical arms, main parts 5- to 9-celled, arms 2- to 6-celled; the third type, microconidia, sympodulosporous, cylindrical, fusiform, obclavate, 1-celled. Chlamydospores globose, catenulate.

Dimensions: Conidiophores for macroconidia (10-) 62.5–117.5 × 2.5–5.5 µm: for microconidia 37.5–225 × 2 µm. Conidia: simple macroconidia 47.5–155 × 7.5–16 µm: main parts of branched macroconidia 55–117.5 × 13.7–17.5 µm: arms 13.7–95 × 5–12.5 µm; microconidia 20–47.5 × 3–5.3 µm. Chlamydospores 10–15 µm in diameter.

Material: 82-567 (= IFO 32554, Iris root, Wakayama, Japan).

Remarks: This fungus resembles species of the genera *Dactylella* Grove, *Vermispora* Deighton & Pirozynski, and *Monacrosporium* Oudem., among others, on the basis of the simple, solitary, terminal, fusiform macroconidia. *Dactylella ramiformis* Liu & Qiu (1992) may be a synonym of this fungus.

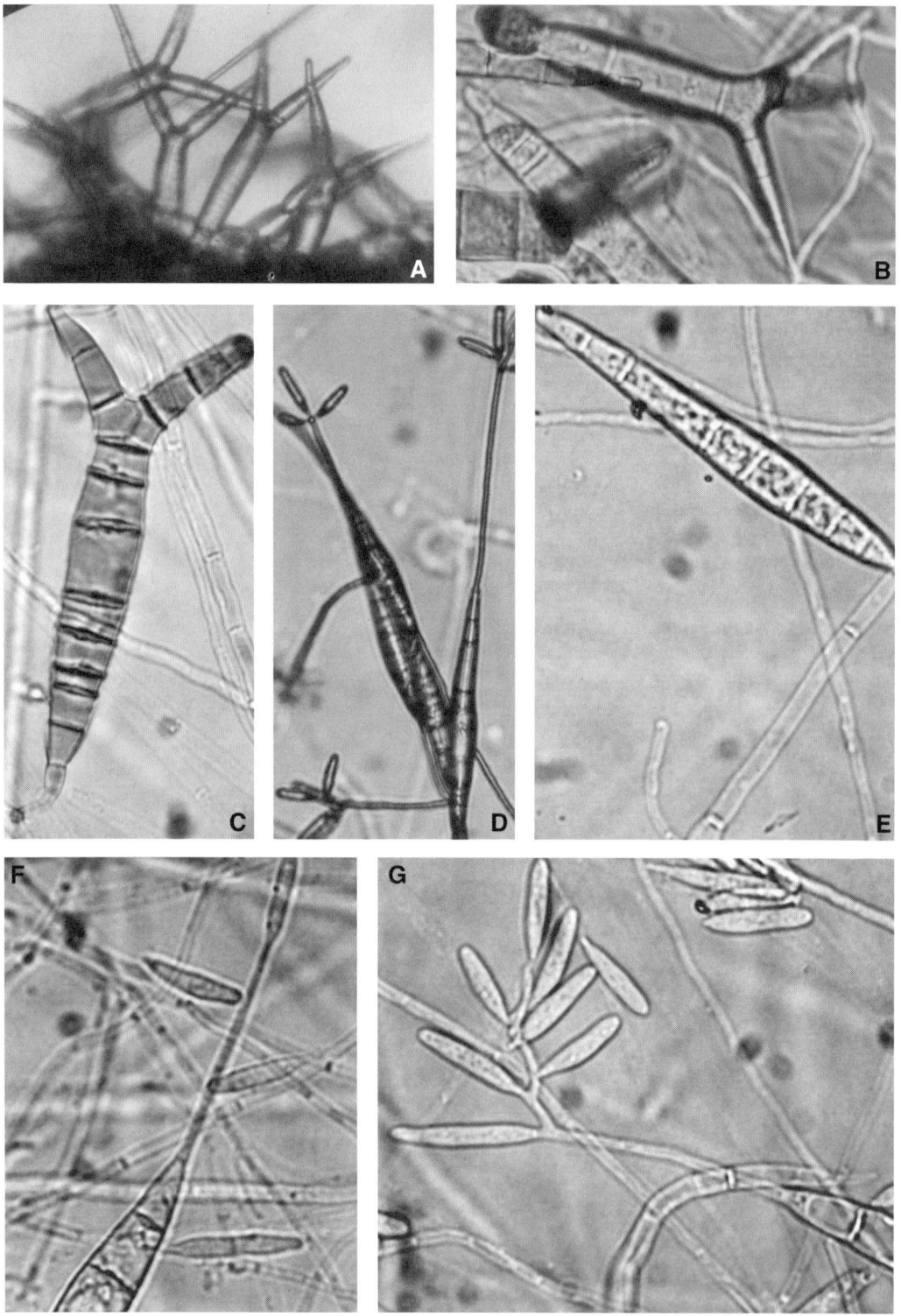

Trinacrium iridis. A: Macroconidia. B,C: Branched macroconidia. D,F,G: Conidiophores elongated from macroconidia and microconidia. E: Macroconidia and conidiophore. H: Chlamydospores. (From Watanabe, T. 1992g. *Mycologia*, 84:794–798. With permission.)

Tripospermum Speg.

Physis B. Aires 4:295, 1918.
Type species: *T. acerinum* (Syd.) Speg.

Tripospermum myrti (Lind) Hughes

References: Ellis 1971; Hughes 1951; Watanabe et al. 1986a.

Morphology: Conidiophores hyaline, simple, short, bearing single conidia apically or directly on hyphae, not well differentiated from conidia. Conidia blastosporous, hyaline or brown, characterized by 1–2 central cells and 1–4 arms elongated from central cells: arms 2- to 3-celled, tapering toward apex.

Dimensions: Conidia: central cells 6.2–10 μm in diameter, and arms 15–26.3 (-42.5) × 5–8.8 μm.

Material: 84-533 (Japanese black pine seed, Okawa, Kagawa, Japan).

Remarks: Colonies on PDA are dark green, very slow in growth, under 20 mm in diameter per 20 days at 25°C, and just like those of *Cladosporium*.

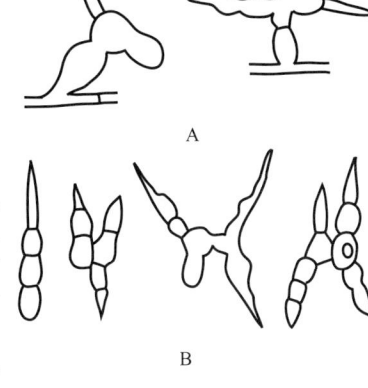

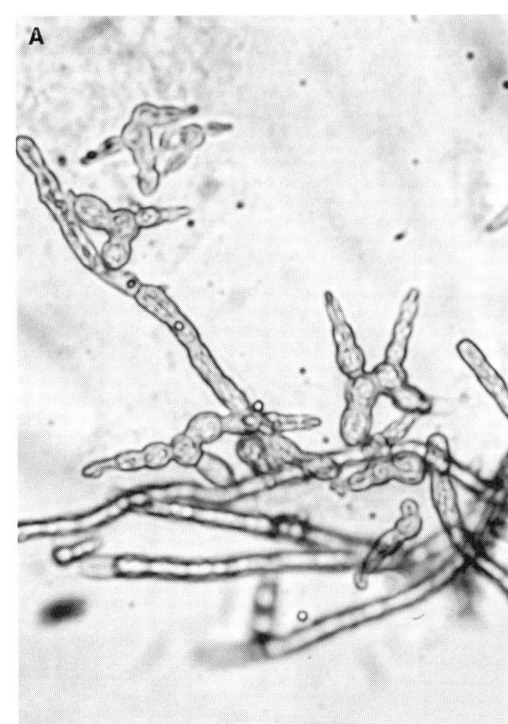
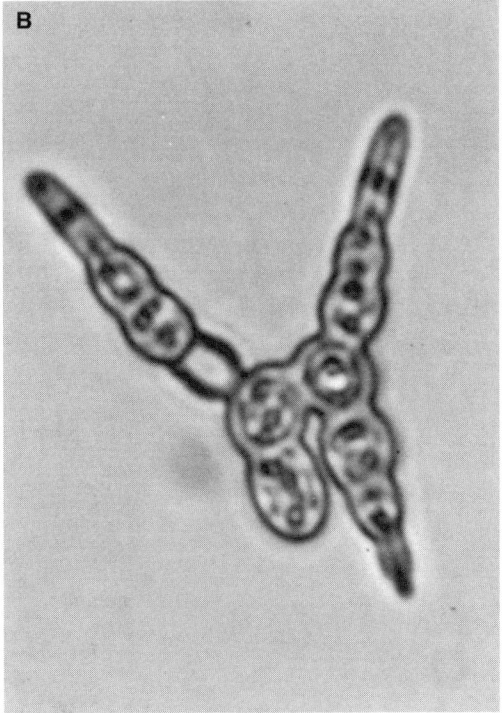

Tripospermum myrti.   A: Conidia on hyphae. B: Conidia.

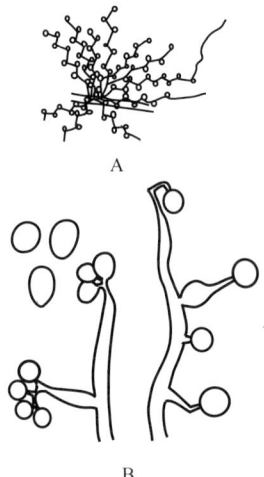

Tritirachium Limber

Mycologia 32:23, 1940.
Type species: *T. dependens* Limber

Tritirachium sp.

References: Hoog 1972; Limber 1940; MacLeod 1954; Watanabe et al. 1986a.

Morphology: Conidiophores hyaline, simple or branched, bearing single conidia on sterigmata or directly on verticillate or zigzag-shaped branchlets, often elongating sterile, twisted, curved branchlets. Conidia sympodulosporous, hyaline, ovate or globose, 1-celled.

Dimensions: Conidiophores 19.8–30.6 × 2.1–2.2 µm. Conidia 2.1–3.8 × 1.8–2.5 µm.

Material: 83-16 (Japanese black pine seed, Kumano, Mie, Japan).

Remarks: This fungus is morphologically close to *Beauveria* species.

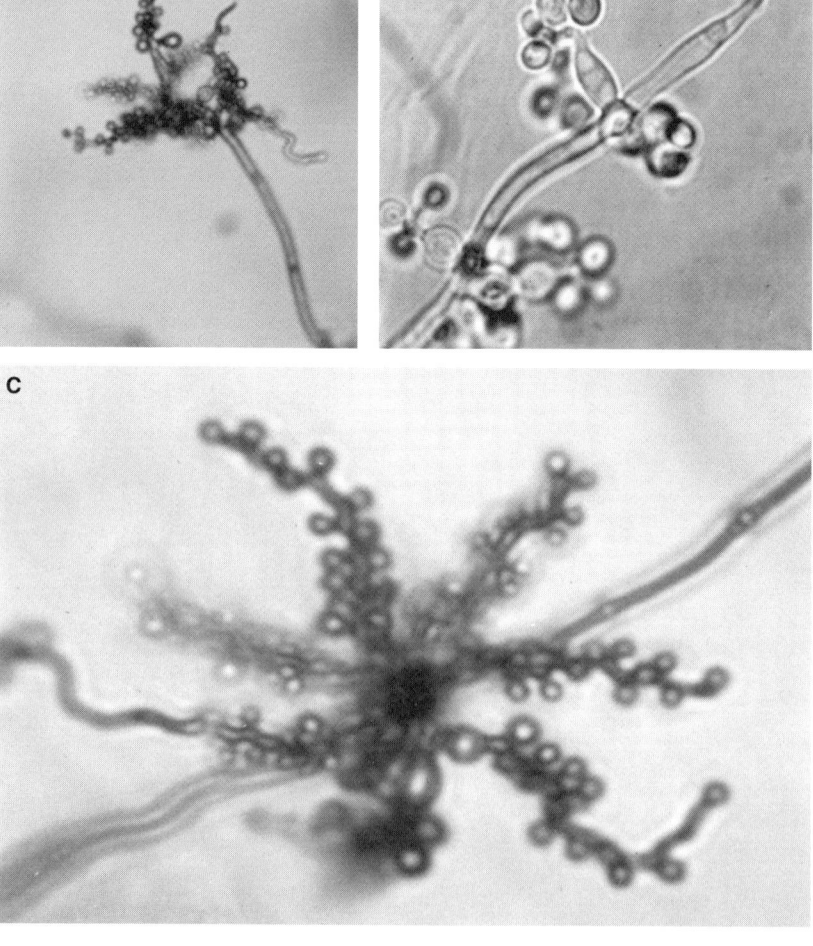

Tritirachium sp.   A,C: Conidiophores and conidia. B: Fertile portions and conidia.

Ulocladium Preuss

Dtschl. Flora, Pilze 3, 3:83, 1851.
Type species: *U. botrytis* Preuss

Ulocladium botrytis Preuss

References: Ellis 1971; Simmons 1967.

Morphology: Conidiophores brown, erect, simple or branched, bearing 1 to several conidia apically or subapically. Conidia porosporous, brown to dark brown, ellipsoidal, ovate, muriform composed of usually 3-transverse septa and 1–3 longitudinal septa, constricted at or near septa, rough marginally.

Dimensions: Conidiophores 2.5–75 × 2.5–4.3 µm. Conidia (17.5-) 19–39 × (7.5-) 15–18 µm.

Material: 70-94 (Snap bean seed, Shinsyu-shinmachi, Nagano, Japan); 74-502 (Strawberry root, Shizuoka, Japan); 77-204 (Strawberry root, Nishigahara, Kita, Tokyo, Japan).

Remarks: Conidia of *Alternaria* and *Ulocladium* spp. are mainly ovate, and broadly ellipsoidal, respectively.

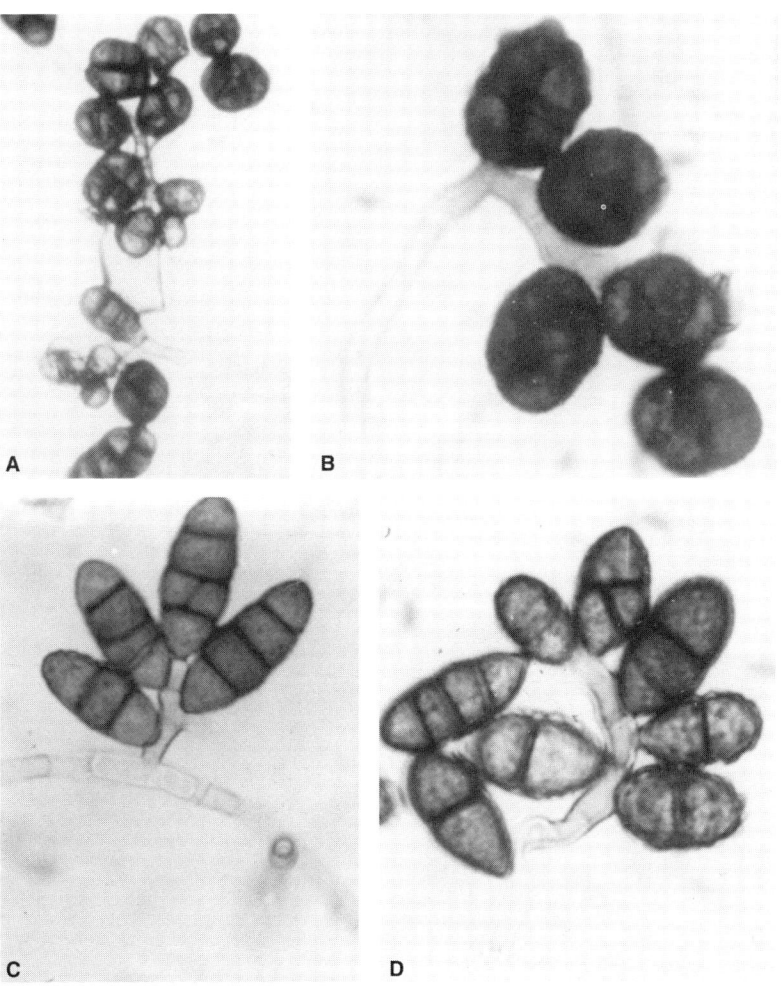

A–D

Ulocladium botrytis.   A–D: Conidiophores and conidia.

Ulocladium chartarum Simmons

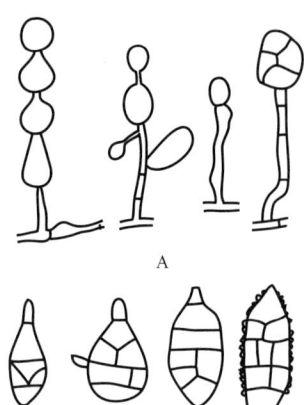

References: Ellis 1971; Simmons 1967; Watanabe et al. 1986a.

Morphology: Conidiophores brown, erect, simple or branched, bearing catenulate conidia apically and laterally in apical fertile portions. Conidia porosporous, brown to dark brown, ellipsoidal, ovate, muriform composed of usually 3–4 transverse septa and 1–2 longitudinal septa, often with 1–2 beaks, rough marginally.

Dimensions: Conidia 21.2–31.3 × 12.5–20 µm: beaks 5–17.5 × 3–3.8 µm.

Material: 84-505 (Japanese red pine seed, Higashichikuma, Nagano, Japan).

Remarks: Conidia of this fungus are mostly 3 to 4 transversely septate, whereas those of *U. botrytis* are 3 transversely septate.

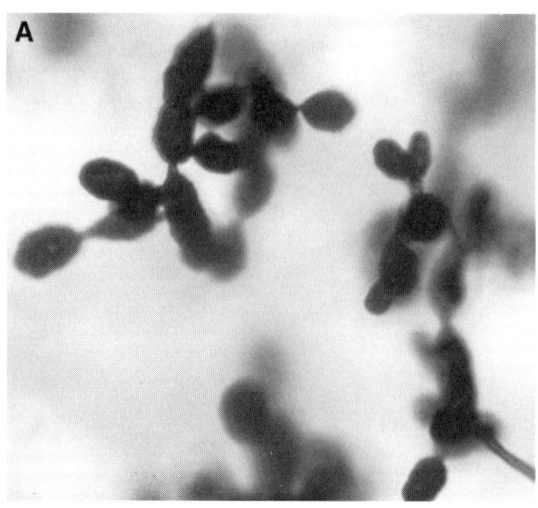
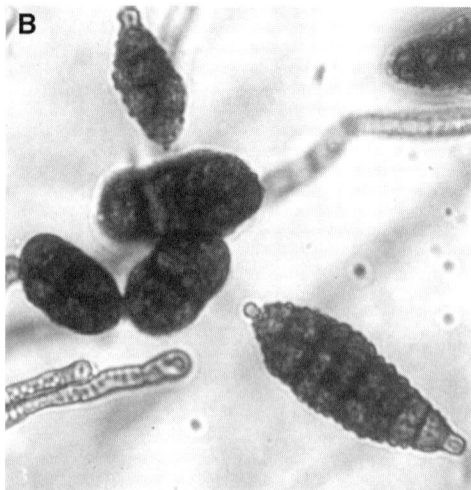

Ulocladium chartarum. A,B: Conidiophores and conidia.

Vermispora Deighton & Piroz.

Mycol. Pap. 128:187, 1972.
Type species: *V. grandispora* Deighton & Pirozynski

Vermispora sp.

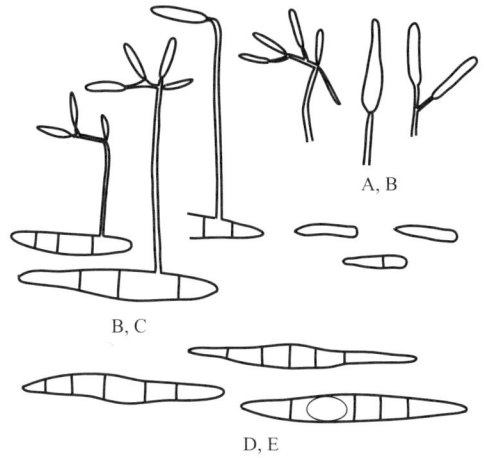

References: Deighton and Pirozynski 1972; Rajashekhar et al. 1991; Vasant & de Hoog. 1986.

Morphology: Conidiophores simple, erect, with terminal single primary macroconidia, conidiogenous cells of secondary microconidia sympodially branched often developing from macroconidia. Conidia hyaline, of two kinds: the long-fusiform, 3- to 5-septate primary macroconidia; and clavate, cylindrical, aseptate or occasionally 1-septate, sharpened at one end, secondary microconidia.

Dimensions: Conidiophores mostly 120–132.5 × 3.7–4 μm. Conidia: primary macroconidia, 70–130 × 12.5–13.8 μm, and secondary microconidia 27.5–36.3 × 5 μm.

Material: 74-632 (Strawberry root, Shizuoka, Japan).

Remarks: This fungus resembles *V. cauveriana* Rajashekar, Bhat et Kaveriappa morphologically, but the macroconidia of the latter are 160–180 μm long, 6- to 9-septate, and the microconidia are truncate at one end.

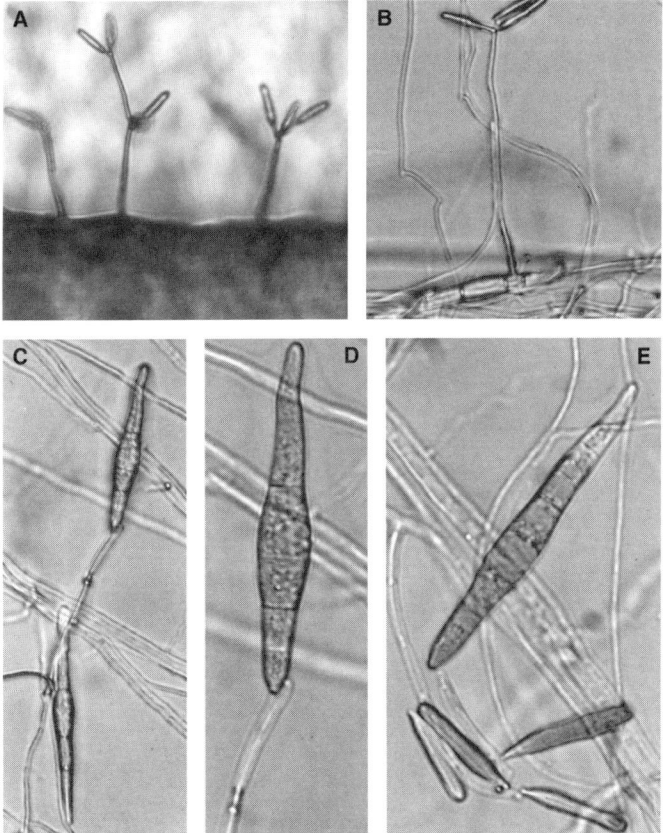

Vermispora sp. A: Habit. B: Conidiophores and microconidia. C,D: Microconidia and conidiophores. E: Macro- and microconidia.

Verticillium Nees : Link

Linn. Spec. Plant. 1:75, 1824.
Type species: *V. lateritium* (Ehrenb. : Link) Rabenh. = *V. tenerum* (Nees : Pers.) Link.

Key to Species

1. Conidia
   - boat-shaped .... *V. fungicola*
   - ellipsoidal or cylindrical 2

2. Chlamydospores or sclerotia
   - not formed 3
   - formed 4

3. Conidia
   - cylindrical .... *V. lecanii*
   - globose, ellipsoidal, or cylindrical,
     endoparasitic to nematodes .... *V. sphaerosporum*

4. Sclerotia
   - formed .... *V. dahliae*
   - not formed 5

5. Chlamydospores
   - solitary .... *Verticillium* sp.
   - catenulate 6

6. Conidia
   - spindle-shaped .... *V. nubilum*
   - cylindrical .... *V. hahajimaense*

Verticillium dahliae Klebahn

References: Domsch et al. 1980; Isaac 1949; Watanabe et al. 1973.

Morphology: Conidiophores hyaline, erect, bearing spore masses apically in each verticillate phialide in 1–3 fertile portions: phialides gradually tapering toward apex. Conidia phialosporous, hyaline, ellipsoidal, 1-celled. Sclerotia composed of catenulate or aggregated dark brown, thick-walled cells.

Dimensions: Conidiophores (85-) 110–275 (-330) × 5 μm: phialides 20–47.5 μm long. Conidia (4.5-) 5.2–8.8 × 2.5–4.5 (-5) μm. Thick-walled resting cells 8.7–17.5 × 7.5–13.8 μm.

Material: 71-315, 71-316 (Chinese cabbage, Nagano, Japan).

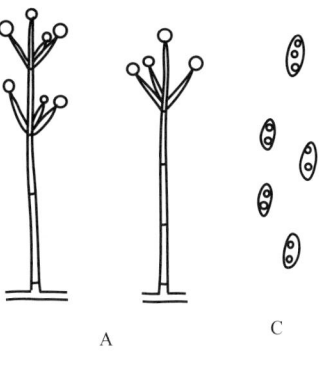

Remarks: *V. dahliae* and *V. albo-atrum* Reinke & Berth. have often been treated synonymously for a long time. The former fungus forms sclerotia and the basal hyaline conidiophores, whereas the latter forms the resting thick-walled cell aggregates and basal dark conidiophores.

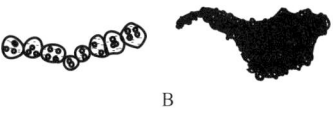

However, interpretation of sclerotia and resting cell aggregates is difficult, because in liquid cultures, the tomato isolate (Isolate 71-314), e.g., forms sclerotia, but the chinese cabbage isolate (Isolate 71-316) forms resting cell aggregates. Pigmentation of basal conidiophores may be a key characteristic to differentiate both species.

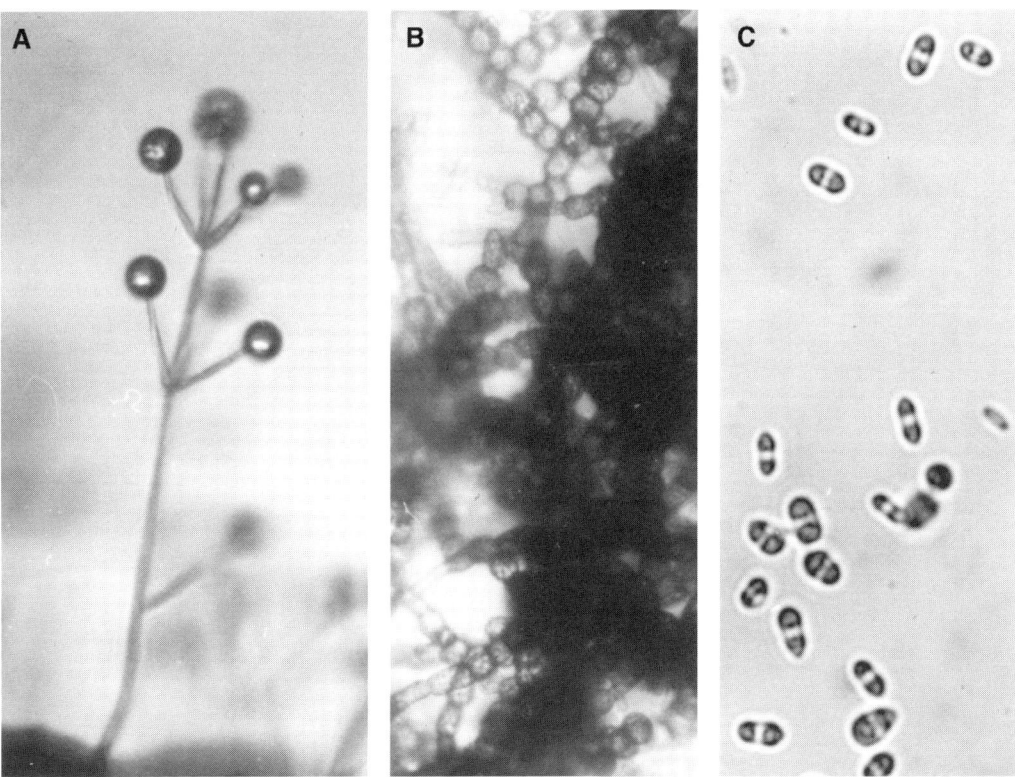

Verticillium dahliae. A: Conidiophore and spore masses. B: Thick-walled resting cells. C: Conidia. (From Watanabe, T. et al. 1973. *Ann. Phytopathol. Soc. Jpn.*, 39:344–350. With permission.)

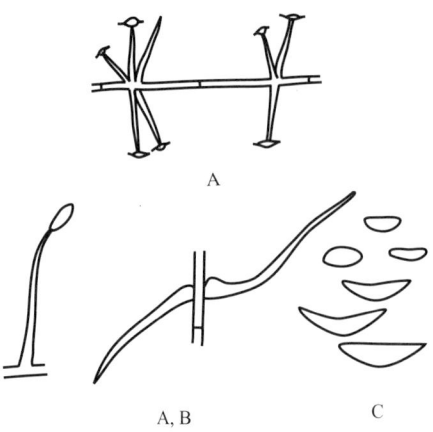

Verticillium fungicola (Preuss) Hassebrauk

Synonym: *V. malthousei* Ware

References: Gams 1971; Ware 1933.

Morphology: Conidiophores (phialides) solitary or verticillate on aerial or creeping hyphae, bearing apparently disc-shaped spore masses apically: phialides constricted and septated basally, tapering toward apex. Conidia phialosporous, hyaline, 1-celled, of two kinds: boat-shaped, and ellipsoidal.

Dimensions: Phialides 15–27.5 × 1–2 µm. Conidia: boat-shaped 5–7.5 (-10) × 1.2–1.8 (-2.5) µm, and ellipsoidal 2.2–3.5 × 0.5–1.0 µm.

Material: 73-81 (Strawberry root, Shizuoka, Japan).

Remarks: Boat-shaped conidia are very characteristic among *Verticillium* spp.

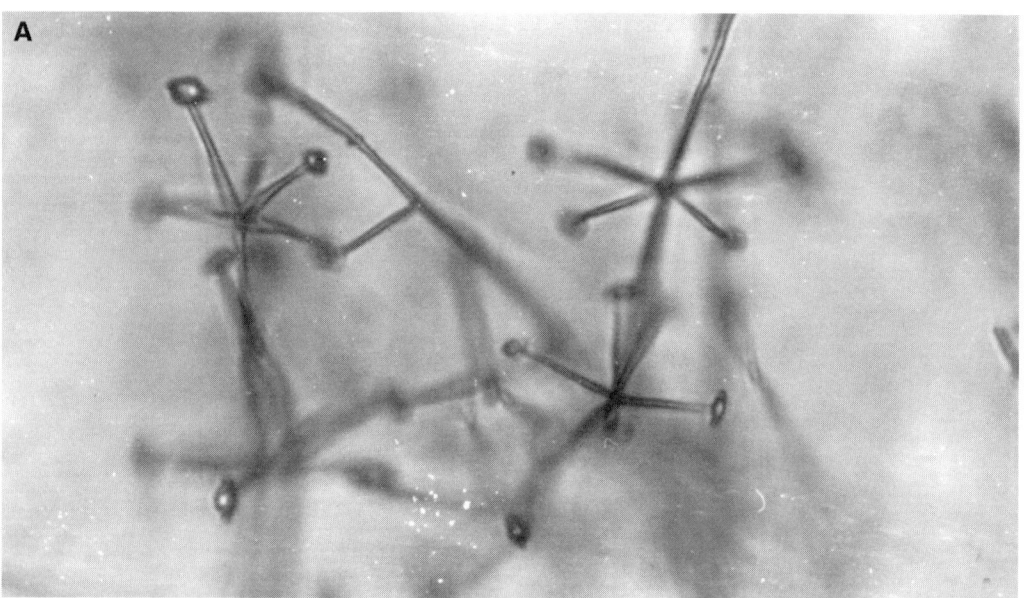
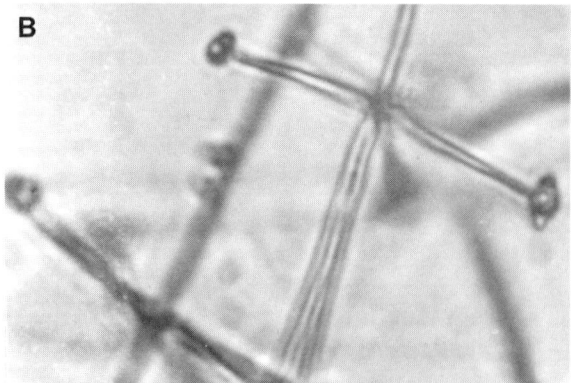
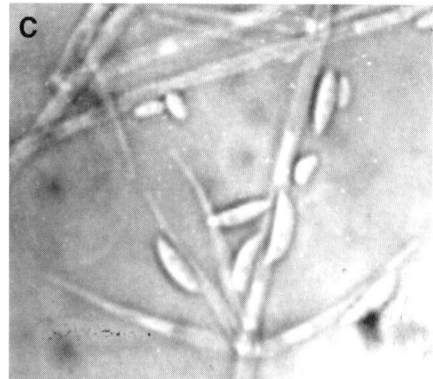

Verticillium fungicola. A–C: Conidiophores (phialides) and conidia.

Verticillium hahajimaense T. Watanabe

References: Gams 1971; Watanabe et al. 2001.

Morphology: Conidiophores erect, hyaline, mostly branched with verticillate phialides bearing terminal spore masses: phialides gradually tapering toward tips, collarette inconspicuous. Conidia hyaline, cylindrical. Chlamydospores brown, granulate, catenulate.

Dimensions: Conidiophores 100–200 × 3 µm: Phialides 20–30 µm long. Spore masses, 10–12 µm in diameter. Conidia 8–10 × 2–2.4 µm. Chlamydospores 15–16 µm in diameter.

Material: 00-65 (= MAFF 238172, Forest soil, Minamizaki, Hahajima, the Bonin Islands, Tokyo, Japan).

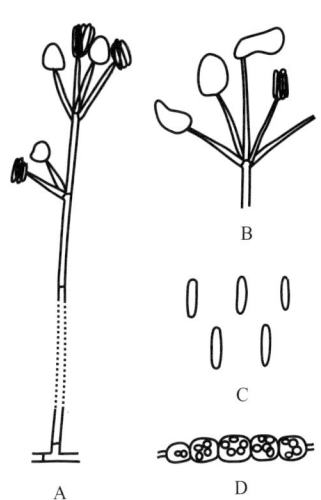

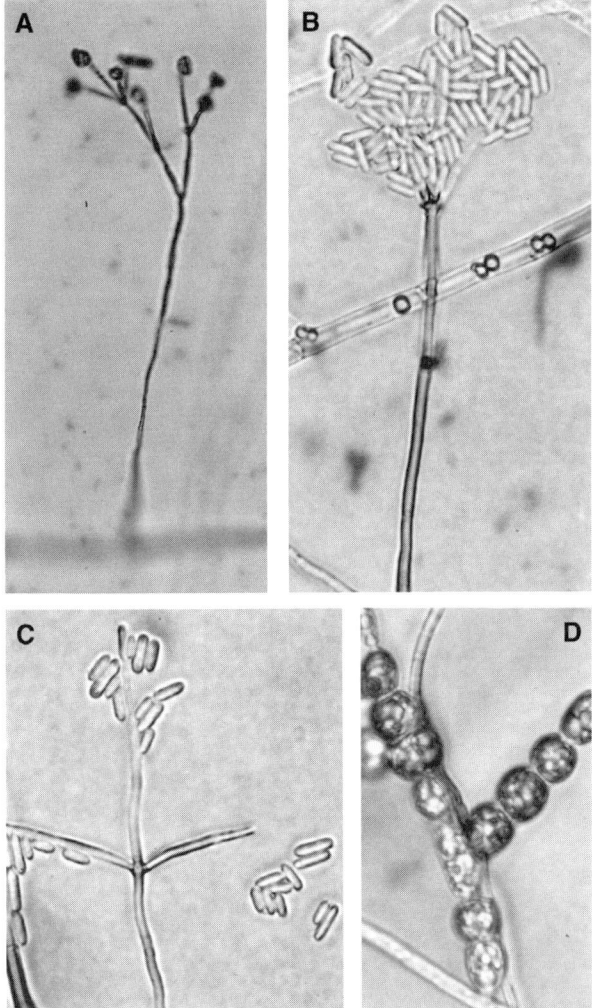

Verticillium hahajimaense. A: Sporulation habit. B,C: Apexes of phialides and conidia. D: Chlamydospores in chains. (From Watanabe, T. et al. 2001b. *Mycoscience*. In press. With permission.)

Verticillium lecanii (Zimmerm.) Viegas

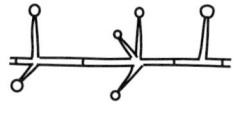

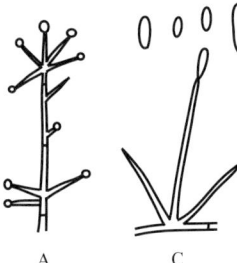

Synonym: *Cephalosporium lecanii* A. Zimmerman

References: Domsch et al. 1980; Gams 1971; Watanabe et al. 1986a.

Morphology: Conidiophores (phialides) solitary or verticillate on aerial or creeping hyphae, bearing spore masses apically. Conidia phialosporous, hyaline, cylindrical with rounded apexes, 1-celled.

Dimensions: Phialides 11.2–30 × 1.7–2.5 μm. Spore masses 6.2–11.3 μm in diameter. Conidia 3.7–10 × 0.7–1.7 μm.

Material: 84-546 (Japanese black pine seed, Okawa, Kagawa, Japan).

Remarks: Colonies on PDA are white, yellowish tinted. This fungus is mostly entomogenous, but often isolated from soils (Domsch et al., 1980a,b).

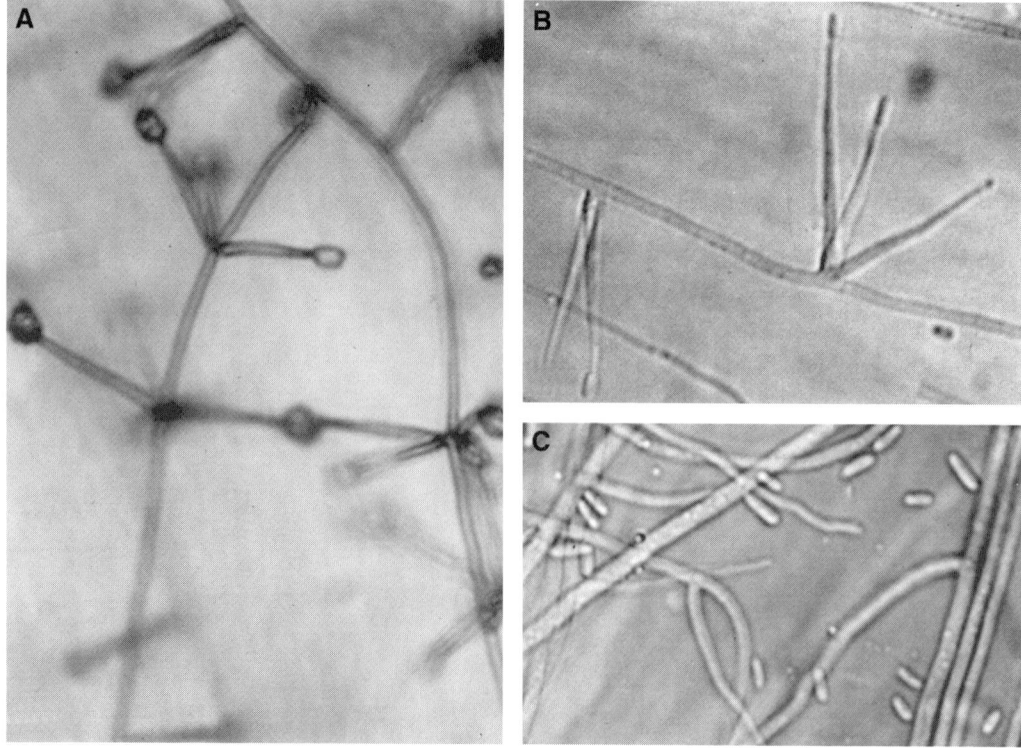

Verticillium lecanii. A–C: Conidiophores and conidia.

Verticillium nubilum Pethybridge

References: Pethybridge 1919.

Morphology: Conidiophores erect, hyaline, bearing spore masses at verticillate or alternate phialides: phialides tapering from base toward apex. Conidia phialosporous, hyaline, spindle-shaped, 1-celled, containing 2 oil globules. Chlamydospores intercalary, globose, solitary or catenulate with up to 12 spores in a chain.

Dimensions: Conidiophores 120 μm tall: phialides 25–30 μm long. Spore masses 7.5–12.5 μm in diameter. Conidia 3.7–5 × 1.5–2 μm. Chlamydospores 5–10 μm in diameter.

Material: 75-251 (Myokou, Niigata, Japan).

Remarks: Catenulate chlamydospores are major characteristics for this fungus.

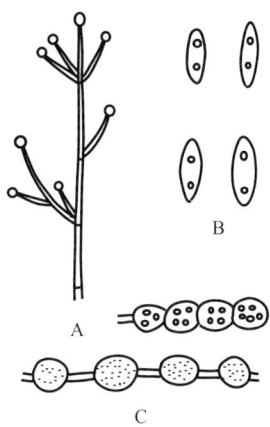

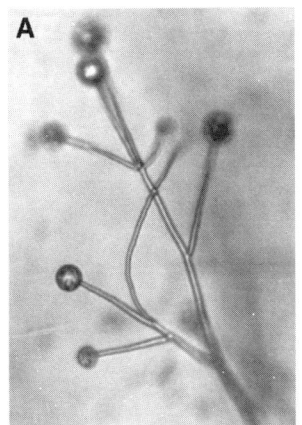 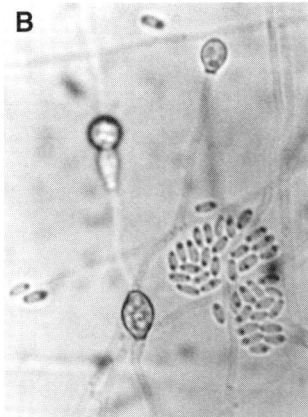 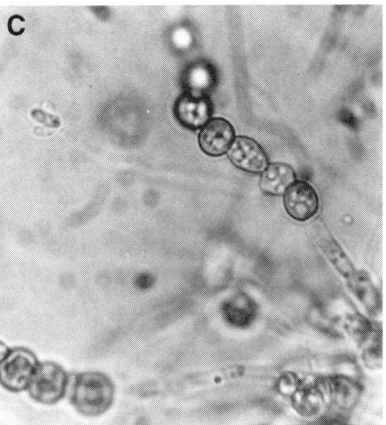

Verticillium nubilum. A: Conidiophores and spore masses. B: Conidia and solitary chlamydospores. C: Catenulate chlamydospores.

Verticillium sphaerosporum Goodey var. bispora T. Watanabe

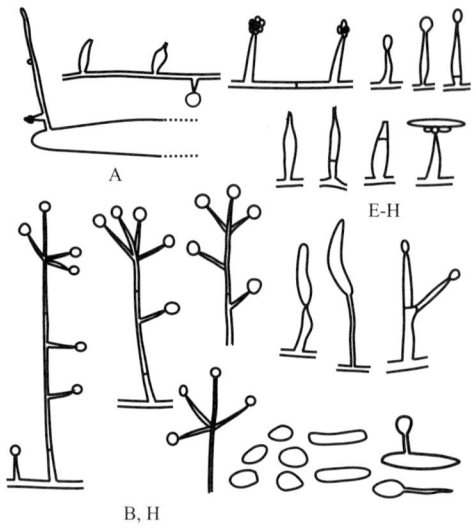

References: Bursnall & Tribe 1974; Cooke & Godfrey 1964; Goodey 1951; Watanabe 1980.

Morphology: Conidiophores erect, hyaline, simple or branched with verticillate phialides bearing terminal spore masses: phialides gradually tapering toward tips, constricted basally, collarette inconspicuous. Conidia hyaline, of two kinds: spherical to ellipsoidal, occasionally apiculate primary conidia; and cylindrical secondary conidia. No chlamydospores observed.

Dimensions: Spore masses 5.5–10 μm in diameter. Phialides 10–30 μm long, 1.7–2.8 μm wide. Conidia, primary 2.5–4.3 × 2–2.8 μm, and secondary 5–13 × 2–2.8 μm.

Material: 75-500 (From nematodes associated with strawberry roots, Tokyo, Japan).

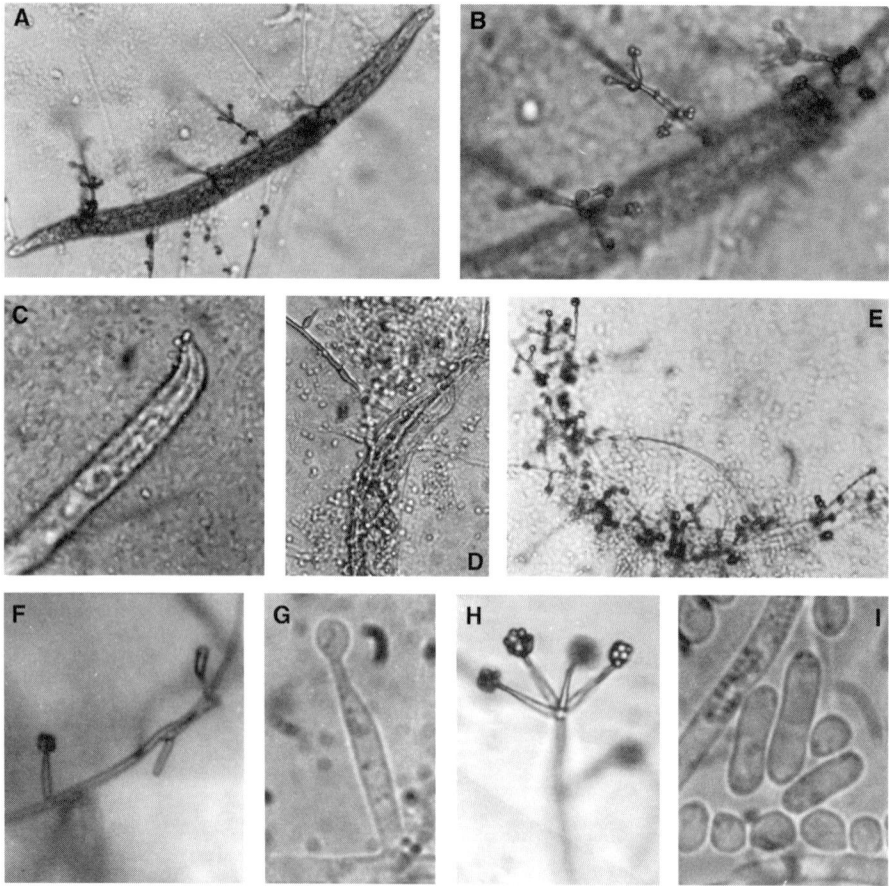

Verticillium sphaerosporum. A,B: Sporulation on dead nematode. C: Conidia attached around the buccal region of a nematode. D: Hyphae inside disintegrated body. E: Sporulation on the disintegrated nematode body. F: Simple conidiophore bearing cylindrical conidia and spore mass composed of globose conidia. G: Phialide and globose conidium. H: Verticillate conidiophore. I: Cylindrical and globose conidia. (From Watanabe, T. 1980. *Ann. Phytopathol. Soc. Jpn.*, 46:598–606. With permission.)

Verticillium sp.

References: Domsch et al. 1980a,b; Watanabe 1975d.

Morphology: Conidiophores erect, hyaline, bearing spore masses at verticillate, alternate, opposite, or simple phialides: phialides tapering from base toward apex. Conidia phialosporous, hyaline, cylindrical, 1-celled. Chlamydospores pale brown, subglobose, minutely echinulate marginally.

Dimensions: Phialides 26.7–53.5 µm long. Conidia 3.6–4.9 × 1.2–2 µm. Chlamydospores 5.6–7.8 µm in diameter.

Material: 72-X25 (Sugarcane root, Taiwan, ROC).

Remarks: Minutely echinulate globose chlamydospores are characteristic for this fungus.

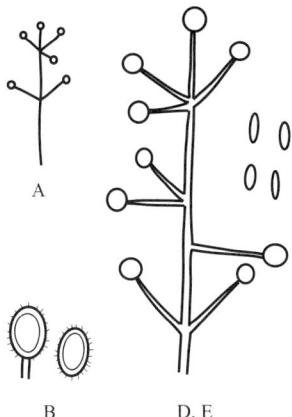

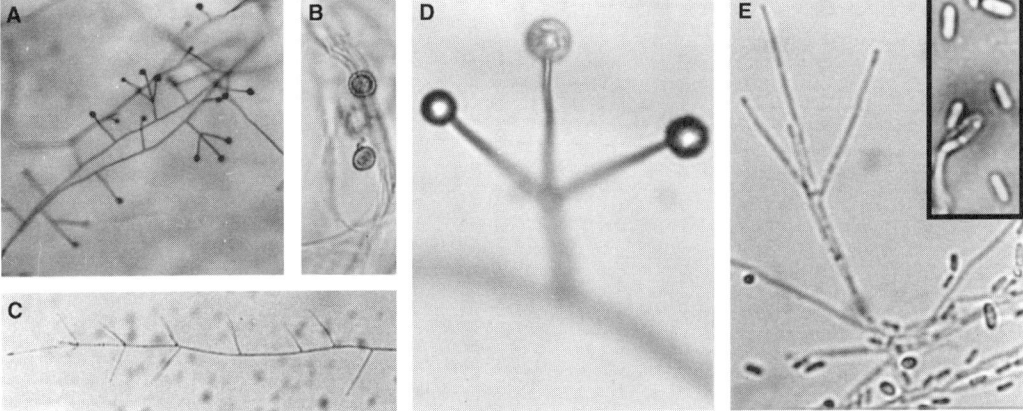

Verticillium sp.   A,D: Conidiophores and spore masses. B: Chlamydospores. C: Apical part of creeping hypha. E: Conidiophores and conidia. (From Watanabe, T. 1975d. *Trans. Mycol. Soc. Jpn.*, 16:264–267. With permission.)

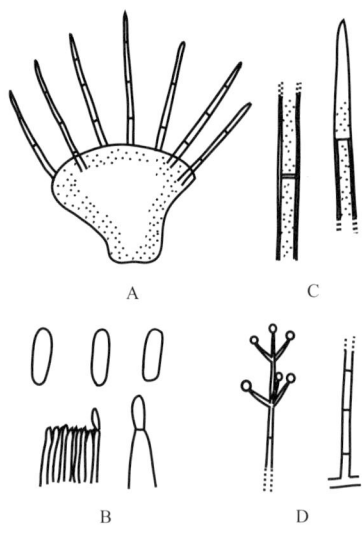

Volutella Tode : Fr.

Syst. Mycol. 3:466, 1832.
Type species: *V. ciliata* (Alb. & Schw.) Fr.

Volutella ciliata (Alb. & Schw.) Fr.

References: Chilton 1954; Samuels 1977.

Morphology: Sporodochia subglobose, usually with over 20 setae around, shortly stipitate basally: setae hyaline, simple, septate tapering from base toward acute apex. Conidiophores phialosporous, hyaline, 1-celled, cylindrical.

Dimensions: Sporodochia 130–440 µm in diameter: setae up to 510 × 5–5.5 µm. Conidia 5–5.5 × 1.7–2 µm.

Material: 73-1093 (Cultivated soil, Nishigahara, Kita, Tokyo, Japan); 75-96 (Gentian root, Kobuchizawa, Yamanashi, Japan); 75-136 (Gentian root, Myokou, Niigata, Japan); 92-52 (Root of *Aralia elata*, Matsumoto, Nagano, Japan).

Remarks: On agar cultures, *Acremonium* and/or *Verticillium* states are observed together with sporodochia.

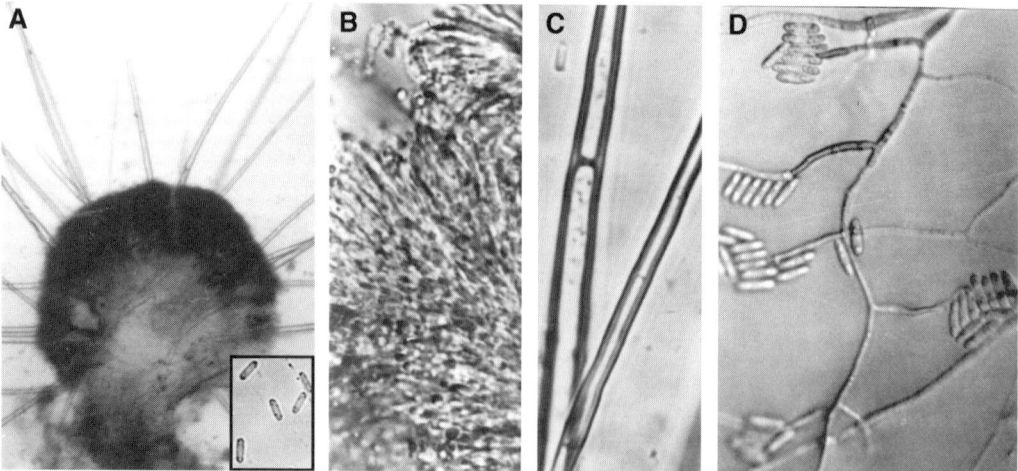

Volutella ciliata. A: Sporodochia and conidia. B: Part of sporodochial tissue. C: Part of the setae. D: Conidia in *Acremonium* or *Verticillium* state.

Wiesneriomyces Koorders

Verh. K. Akad. Wet. Amsterd. 2, 13(4):246, 1907.
Type species: *Wiesneriomyces javanicus* Koorders

Wiesneriomyces javanicus Koorders

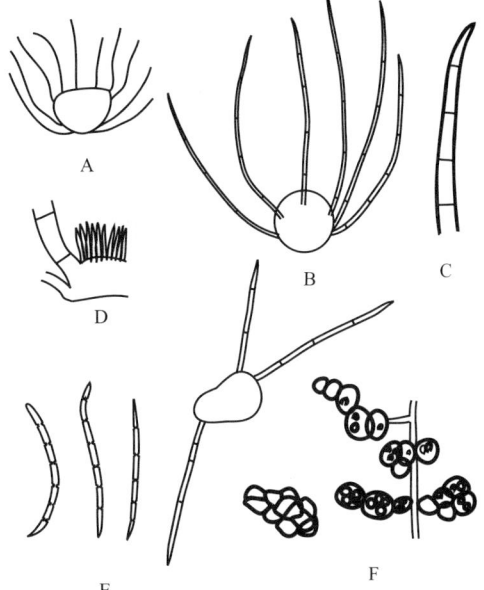

References: Ichinoe 1968; Kirk 1984; Maniotis & Strain 1968; Subramanian 1956.

Morphology: Conidiomata sporodochial, white to dark green, subglobose, hemispherical, or cup-shaped, composed of pallisade layers of single hyaline cylindrical conidiophores and conidial masses, with more than 10 setae: setae simple, usually 4- to 7-septate, brown, thick-walled, slightly curved, tapering toward the tips. Conidia blastosporous, hyaline, cylindrical to elongate-spindle-shaped, slightly curved, mostly 6- to 9-celled, each cell connected with short isthmus. Sclerotia irregular in shape, dark brown, composed of granulate thick-walled chlamydospore-like cells.

Dimensions: Sporodochia nearly 150–170 μm tall, 150–240 μm wide: setae 125–500 × 8–12 μm. Conidiophores nearly 30 μm tall. Conidia 62–76 × 2–2.4 μm. Sclerotia 20–90 μm in diameter: composed cells nearly 10 μm in diameter.

Material: 00-273 (= Forest soil, Mt. Chibusa, Hahajima, the Bonin Islands, Tokyo, Japan).

Remarks: There may be wide variations in the fungus studied (Kirk, 1984). Note the Japanese isolate that lacks sclerotium formation (Ichinoe, 1968).

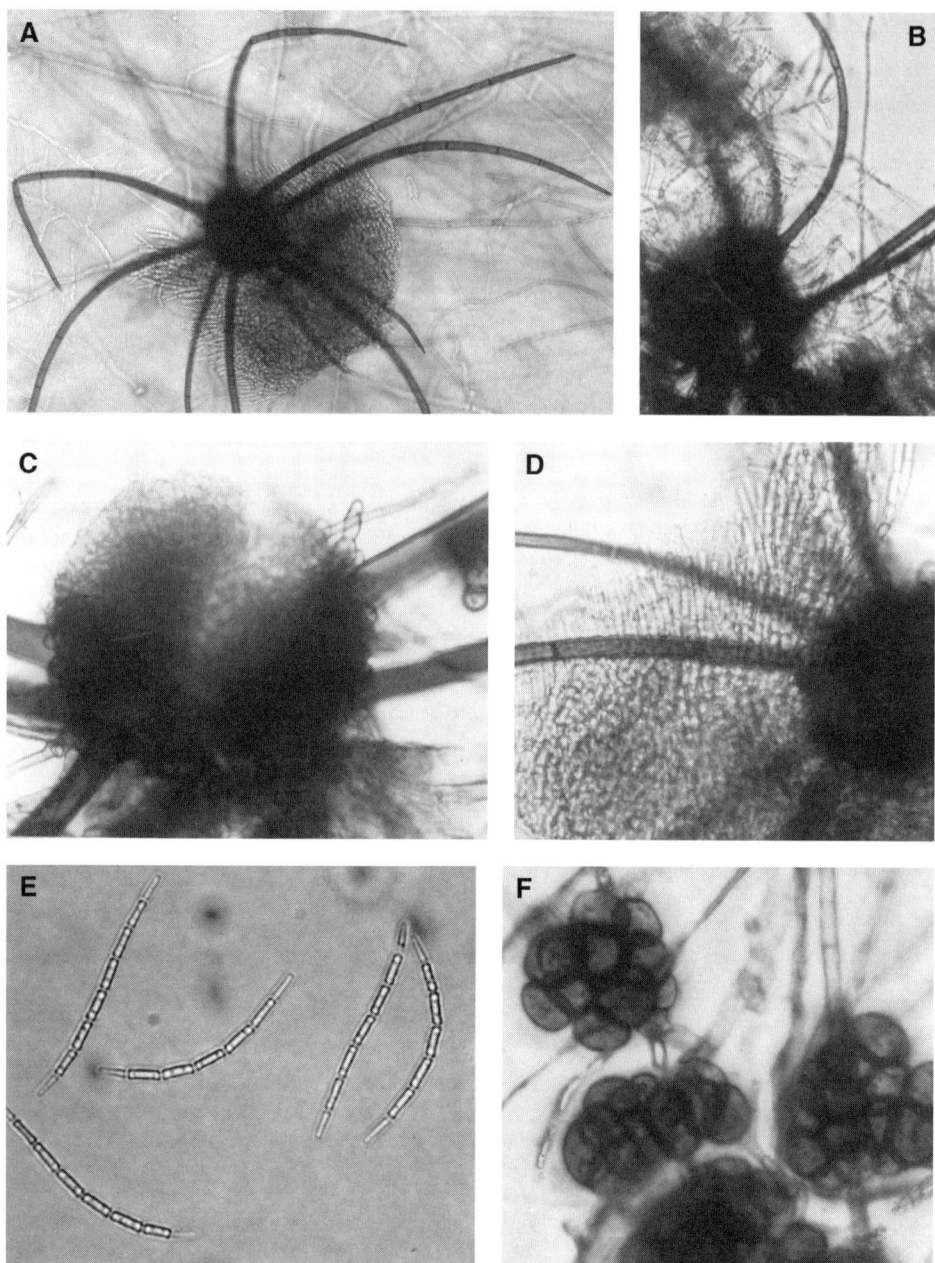

Wiesneriomyces javanicus.   A: Habit. B–D: Part of sporodochia showing fertile portions and setae. E: Conidia. F: Sclerotia.

References

Abbott, E. V. 1927. *Scolecobasidium*, a new genus of soil fungi. *Mycologia* 19:29–31.

Adametz, L. 1886. Untersuchungen über die niederen Pilze der Ackerkrume. *Inaug. Diss.* 1–78. Leipzig.

Agnihothrudu, V. 1960. Rhizosphere microflora of tea (*Camellia sinensis* (L.) O. Kunze) in relation to the root rot caused by *Ustulina zonata* (Lév.) Sacc. *Soil Sci.* 91:133–137.

Ainsworth, G. C., Sparrow, F. K., and Sussman, A. S. (Eds.). 1973. *The Fungi, An Advanced Treatise. IVA. A taxonomic review with keys: Ascomycetes and Fungi Imperfecti. IVB. A taxonomic review with keys: Basidiomycetes and lower fungi.* Academic Press, New York. 621 pp.; 504 pp.

Alcorn, G. D. and Yeager, C. C. 1938. A monograph of the genus *Cunninghamella* with additional descriptions of several common species. *Mycologia* 30:653–658.

Alcorn, J. L. 1994. Appressoria in *Mycoleptodiscus* species. *Aust. Syst. Bot.* 7:591–603.

Al-Doory, Y., Tolba, M. K., and Al-Ani, H. 1959. On the fungal flora of Iraqi soils. II. Central Iraq. *Mycologia* 51:429–439.

Alexopoulos, C. J. and Mims, C. W. 1979. *Introductory Mycology.* 3rd. ed. John Wiley & Sons, New York. 632 pp.

Ali-Shtayeh, M. 1986. Taxonomic notes on three *Pythium* species. *Trans. Br. Mycol. Soc.* 86:659–663.

Ames, L. M. 1949. New cellulose-destroying fungi isolated from military material and equipment. *Mycologia* 41:637–648.

Ames, L. M. 1963. A monograph of the Chaetomiaceae. *U.S. Army Res. Dev.,* Ser. 2. 125 pp.

Amos, R. E. and Barnett, H. L. 1966. *Umbelopsis versiformis*, a new genus and species of the Imperfects. *Mycologia* 58:805–808.

Anderson, P. J. 1919. Rose canker and its control. *Mass. Agric. Exp. Stn. Bull.* 183:11–46. 11 figs.

Aoshima, K., Tubaki, K., and Miura, K. (Eds.). 1983. Mycological research methodology, Kyoritsu, Tokyo, 423 pp., in Japanese.

Aragaki, M. and Uchida, J. Y. 2001. Morphological distinctions between *Phytophthora capsici* and *P. tropicalis* sp. nov. *Mycologia* 93:137–145.

Arx, J. A. von. 1957. Die Arten der Gattung *Colletotrichum* Cda. *Phytopathol. Z.* 29:413–468.

Arx, J. A. von. 1971. Über die Typusart, zwei neue und einige weitere Arten der Gattung *Sporotrichum. Persoonia* 6:179–184.

Arx, J. A. von. 1975. Revision of *Microascus* with the description of a new species. *Persoonia* 8:190–197.

Arx, J. A. von. 1981a. The genera of fungi sporulating in pure culture. *J. Cramer Lehre.* 3rd. ed. 424 pp.

Arx, J. A. von. 1981b. On *Monilia sitophila* and some families of Ascomycetes. *Sydowia* 34:13–29.

Arx, J. A. von. 1982. On Mucoraceae s. str. and other families of the Mucorales. *Sydowia* 35:10–26.

Arx, J. A. von and Gams, W. 1966. Über *Pleurage verruculosa* und die zugehörige *Cladorrhinum*-Konidienfrom. *Nova Hedwigia* 13:199–208.

Arx, J. A. von., Guarro, J., and Figueras, M. J. 1986. The ascomycete genus *Chaetomium. Beih. Nova Hedwigia* 84:1–162.

Ashby, S. F. 1927. *Macrophomina phaseoli* (Maubl.) comb. nov. the pycnidial stage of *Rhizoctonia bataticola* (Taub.) Butl. *Trans. Br. Mycol. Soc.* 12:141–147.

Bain, D. C. and Edgerton, C. W. 1943. The zonate leaf spot, a new disease of sorghum. *Phytopathology* 33:220–226.

Bakshi, B. K. 1950. Fungi associated with ambrosia beetles in Great Britain. *Trans. Br. Mycol. Soc.* 33:111–120.

Bakshi, B. K. 1952. *Oedocephalum lineatum* is a conidial stage of *Fomes annosus. Trans. Br. Mycol. Soc.* 35:195.

Barnett, H. L. and Hunter, B. B. 1987. *Illustrated Genera of Imperfect Fungi.* 4th ed., MacMillan, NY, 218 pp.

Barrasa, J. M., Lundqvist, N., and Moreno, G. 1986. Notes on the genus *Sordaria* in Spain. *Persoonia* 13:83–88.

Barron, G. L. 1961. *Monocillium humicola* sp. nov. and *Paecilomyces variabilis* sp. nov. from soil. *Can. J. Bot.* 39:1573–1578.

Barron, G. L. 1962. New species and new records of *Oidiodendron. Can. J. Bot.* 40:589–607.

Barron, G. L. 1964. A note on the relationship between *Stachybotrys* and *Hyalostachybotrys. Mycologia* 56:313–316.

Barron, G. L. 1967. *Torulomyces* and *Monocillium. Mycologia* 59:716–718.

Barron, G. L. 1968. *The Genera of Hyphomycetes from Soil.* Williams & Wilkins, Baltimore. 364 pp.

Barron, G. L. 1975. Nematophagus fungi: *Helicocephalum. Trans. Br. Mycol. Soc.* 65:309–310.

Barron, G. L., Cain, R. F., and Gilman, J. C. 1961. The genus *Microascus. Can. J. Bot.* 39:1609–1631.

Barron, G. L. and Busch, L. V. 1962. Studies on the soil hyphomycete *Scolecobasidium. Can. J. Bot.* 40:77–84.

Barron, G. L. and Bhatt, G. C. 1967. A new species of *Gonytrichum* from soil. *Mycopathol. Mycol. Appl.* 32:126–128.

Benjamin, R. K. 1966. The merosporangium. *Mycologia* 58:1–42.

Berthet, P. 1964. Formes conidiennes de divers Discomycetes. *Bull. Trimest. Soc. Mycol. Fr.* 80:125–149.

Bessey, E. A. 1961. *Morphology and Taxonomy of Fungi.* Hafner, New York, 791 pp.

Bestagno-Biga, M. L., Ciferri, R., and Bestagno, G. 1958. Ordinamento artificiale delle species del genere *Coniothyrium. Sydowia* 12:258–320.

Bhatt, G. C. and Kendrick, W. B. 1968. The generic concepts of *Diplorhinotrichum* and *Dactylaria*, and a new species of *Dactylaria* from soil. *Can. J. Bot.* 46:1253–1257.

Biesbrock, J. A. and Hendrix, F. F., Jr. 1967. A taxonomic study of *Pythium irregulare* and related species. *Mycologia* 59:943–952.

Bisby, G. R., James, N., and Timonin, M. 1933. Fungi isolated from Manitoba soil by the plate method. *Can. J. Res.* 8:253–275.

Bisby, G. R., Timonin, M. I., and James, N. 1935. Fungi isolated from soil profiles in Manitoba. *Can. J. Res. Sect. C.* 13:47–65.

Bissett, J. 1984. A revision of the genus *Trichoderma*. I. Section *Longibrachiatum* sect. nov. *Can. J. Bot.* 62:924–931.

Bissett, J. 1991. A revision of the genus *Trichoderma*. II. Infrageneric classification; III. Section *Pachybasium*; IV. Additional notes on section *Longibrachiatum*. *Can. J. Bot.* 69: II: 2357–2372; III: 2373–2417; IV: 2418–2420.

Boedijn, K. B. 1962. The Sordariaceae of Indonesia. *Persoonia* 2:305–320.

Boedijn, K. B. and Reitsma, J. 1950. Notes on the genus *Cylindrocladium* (Fungi: Mucedinaceae). *Reinwardtia* 1:51–60.

Boesewinkel, H. J. 1974. *Cylindrocladium floridanum*, a new recording for New Zealand. *Plant Dis. Reptr.* 58:705–707.

Bollard, E. G. 1950. Studies on the genus *Mastigosporium*. I. General account of the species and their host ranges. *Trans. Br. Mycol. Soc.* 33:250–264.

Booth, C. 1959. Studies of Pyrenomycetes: IV. *Nectria* (Part 1). *Mycol. Pap.* 73:1–115.

Booth, C. 1961. Studies of Pyrenomycetes: VI. *Thielavia* with notes on some allied genera. *Mycol. Pap.* 83:1–8.

Booth, C. 1966. The genus *Cylindrocarpon*. *Mycol. Pap.* 104:1–56.

Booth, C. 1971. The genus *Fusarium*. Commonw. Mycol. Inst., Kew. 237 pp.

Booth, C. and Murray, J. S. 1960. *Calonectria hederae* Arnaud and its *Cylindrocladium* conidial state. *Trans. Br. Mycol. Soc.* 43:69–72.

Booth, C. and Shipton, W. A. 1966. *Thielavia pilosa* sp. nov., with key to species of *Thielavia*. *Trans. Br. Mycol. Soc.* 49:665–667.

Bose, S. K. 1961. Studies on *Massarina* Sacc. and related genera. *Phytopathol. Z.* 41:151–213.

Botton, B. and El-Khouri, M. 1978. Synnema and rhizomorph production in *Sphaerostilbe repens* under the influence of other fungi. *Trans. Br. Mycol. Soc.* 70:131–136.

Brasier, C. M. 1971. Induction of sexual reproduction in single A^2 isolates of *Phytophthora* species by *Trichoderma viride*. *Nature (London) New Biol.* 231:283.

Braun, H. 1925. Comparative studies of *Pythium debaryanum* and two relative species from Geranium. *J. Agr. Res.* 30:1043–1062.

Broadbent, P., Baker, K. F., and Waterworth, Y. 1971. Bacteria and Actinomycetes antagonistic to fungal root pathogens in Australian soils. *Aust. J. Biol. Sci.* 24:925–944.

Brown, A. H. S. and Smith, G. 1957. The genus *Paecilomyces* Bainier and its perfect stage *Byssochlamys* Westling. *Trans. Br. Mycol. Soc.* 40:17–89.

Burges, A. 1965. The soil microflora — its nature and biology in Baker, K. F. and Snyder, W. C. (Eds). *Ecology of Soil-Borne Plant Pathogens. Prelude to Biological Control.* University of California Press, Berkeley. pp. 21–32.

Burr, T. J. and Stanghellini, M. E. 1973. Propagules nature and density of *Pythium aphanidermatum* in field soil. *Phytopathology* 63:1499–1501.

Bursnall, L. A. and Tribe, H. T. 1974. Fungal parasitism in cysts of *Heterodera*. II. Egg parasites of *H. schaghtii*. *Trans. Brit. Mycol. Soc.* 62:595–601.

Butler, E. E. 1957. *Rhizoctonia solani* as a parasite of fungi. *Mycologia* 49:354–373.

Butler, E. J. 1906. Fungal diseases of sugar-cane in Bengal. *Mem. Dept. Agr. India, Bot. Ser.* 1:1–53, with 11 plates.

Butler, E. J. 1907. An account of the genus *Pythium* and some Chytridiaceae. *Mem. Dept. Agric. India* 1:1–160, with 10 plates.

Cailleux, R. 1971. Recherches sur la mycoflore coprophile centrafricaine. Les genres *Sordaria, Gelasinospora, Bombardia*. *Bull. Trimest. Soc. Mycol. Fr.* 87:461–626.

Cain, R. F. 1950. Studies of coprophilous Ascomycetes. I. *Gelasinospora*. *Can. J. Bot.* 28, C:566–576.

Cain, R. F. 1952. Studies of Fungi Imperfecti. I. *Phialophora*. *Can. J. Bot.* 30:338–343.

Cain, R. F. 1957. Studies of coprophilous Ascomycetes. VI. Species from the Hudson Bay area. *Can. J. Bot.* 35:255–268.

Cain, R. F. 1961a. Studies of coprophilous Ascomycetes. VII. *Preussia*. *Can. J. Bot.* 39:1633–1666.

Cain, R. F. 1961b. *Anixiella* and *Diplogelasinospora*, two genera with cleistothecia and pitted ascospores. *Can. J. Bot.* 39:1667–1677.

Cain, R. F. and Groves, J. W. 1948. Notes on Seed-borne Fungi. VI. *Sordaria*. *Can. J. Res.*, Sec. C. 26:486–495.

Campbell, W. A. and Hendrix, F. F., Jr. 1967. A new heterothallic *Pythium* from southern United States. *Mycologia* 59:274–278.

Cannon, P. F. 1986. A revision of *Achaetomium, Achaetomiella* and *Subramaniula*, and some similar species of *Chaetomium*. *Trans. Br. Mycol. Soc.* 87:45–76.

Carmichael, J. W. 1957. *Geotrichum candidum*. *Mycologia* 49:820–830.

Carmichael, J. W. 1962. *Chrysosporium* and some other aleuriosporic Hyphomycetes. *Can. J. Bot.* 40:1137–1173.

Carris, L. M., Glawe, D. A., Smyth, C. A., and Edwards, D. I. 1989. Fungi associated with populations of *Heterodera glycines* in two Illinois soy bean fields. *Mycologia* 81:66–75.

Chesters, C. G. C. 1949. Presidential address. Concerning fungi inhabiting soil. *Trans. Br. Mycol. Soc.* 32:197–216.

Chesters, C. G. C. and Hornby, D. 1965. Studies on *Colletotrichum coccodes*. I. The taxonomic significance of variation in isolates from tomato roots. *Trans. Br. Mycol. Soc.* 48:573–581.

Chien, C-Y., Kuhlman, E. G., and Gams, W. 1974. Zygospores in two *Mortierella* species with "stylospores." *Mycologia* 66:114–121.

Chilton, J. E. 1954. *Volutella* species on alfalfa. *Mycologia* 46:800–809.

Christensen, C. M. and Kaufmann, H. H. 1965. Deterioration of stored grains by fungi. *Annu. Rev. Phytopathol.* 3:69–84.

Christensen, M., Whittingham, W. F., and Novak, R. O. 1962. The soil microfungi of wet-mesic forests in Southern Wisconsin. *Mycologia* 54:374–388.

Christensen, M. and Whittingham, W. F. 1965. The soil microfungi of open bogs and conifer swamps in Wisconsin. *Mycologia* 57:882–896.

Cole, G. T. and Kendrick, B. 1973. Taxonomic studies of *Phialophora*. *Mycologia* 65:661–688.

Cooke, R. C. and Godfrey, B. E. S. 1964. A key to the nematode-destroying fungi. *Trans. Br. Mycol. Soc.* 47:61–74.

Cooke, R. C. and Dickinson, C. H. 1965. Nematode-trapping species of *Dactylella* and *Monacrosporium*. *Trans. Br. Mycol. Soc.* 48:621–629.

Cooke, W. B. 1954. The genus *Arthrinium*. *Mycologia* 46:815–822.

Cooke, W. B. 1959. An ecological life history of *Aureobasidium pullulans* (de Bary) Arnaud. *Mycopathologia* 12:1–45.

Cooke, W. B. 1962. A taxonomic study in the genus black yeasts. *Mycopath. Mycol. Appl.* 17:1–43.

Corbaz, R. 1957. Recherches sur le genre *Didymella* Sacc. *Phytopathol. Z.* 28:375–414.

Cormack, M. W. 1937. *Cylindrocarpon ehrenbergi* Wr., and other species, as root parasites of alfalfa and sweet clover in Alberta. *Can. J. Res. Ser. C.* 15:403–424.

Crane, J. L. and Hewings, A. D. 1982. Stilbellaceous fungi. 1. *Didymostibe*. *Mycotaxon* 16:133–140.

Crane, J. L. and Schoknecht, J. D. 1986. Revision of *Torula* and *Hormiscium* species. New names for *Hormiscium undulatum*, *Torula equina*, and *Torula convolvuli*. *Mycologia* 78:86–91.

Christias, C. and Baker, K. F. 1967. Chitinase as a factor in the germination of chlamydospores of *Thielaviopsis basicola*. *Phytopathology* 57:1363–1367.

Crous, R. W. and Wingfield, M. J. 1994. A monograph of *Cylindrocladium*, including anamorphs of *Calonectria*. *Mycotaxon*. 51:341–435.

Cunnell, G. J. 1958. On *Robillarda phragmitis* sp. nov. *Trans. Br. Mycol. Soc.* 41:405–412.

Dade, H. A. 1928. *Ceratostomella paradoxa*, the perfect stage of *Thielaviopsis paradoxa* (de Seynes) von Höhnel. *Trans. Br. Mycol. Soc.* 13:184–194.

Damon, S. C. 1952. Two noteworthy species of *Sepedonium*. *Mycologia* 44:86–96.

Damon, S. C. 1953. Notes on the hyphomycetous genera *Spegazzinia* and *Isthomospora* Stevens. *Bull. Torrey Bot. Club* 80:155–165.

Das, A. C. 1962. New species of *Thielavia* and *Sordaria*. *Trans. Br. Mycol. Soc.* 45:545–548.

Deacon, J. W. 1973. *Phialophora radicicola* and *Gaeumannomyces graminis* on roots of grasses and cereals. *Trans. Br. Mycol. Soc.* 61:471–485.

Deacon, J. W. 1974. Further studies on *Phialophora radicicola* and *Gaeumannomyces graminis* on roots and stem bases of grasses and cereals. *Trans. Br. Mycol. Soc.* 63:307–327.

Deighton, F. C. and Pirozynski, K. A. 1972. Microfungi. V. More hyperparasitic hyphomycetes. *Mycol. Pap.* 128:1–110.

Dennis, R. W. G. 1946. Notes on some British fungi ascribed to *Phoma* and related genera. *Trans. Br. Mycol. Soc.* 29:11–42.

Dennis, R. W. G. 1978. *British Ascomycetes*. 3rd ed. J. Cramer, Vaduz, 585 pp., 44 plates.

Dix, N. J. 1964. Colonization and decay of bean roots. *Trans. Br. Mycol. Soc.* 47:285–292.

Dixon, M. 1968. Notes and brief articles. *Trichocladium pyriformis* sp. nov. *Trans. Br. Mycol. Soc.* 51:160–164.

Dodge, B. O. 1937. The conidial stage of *Peziza pustulata*. *Mycologia* 29:651–655.

Domsch, K. H. and Gams, W. 1972. *Fungi in Agricultural Soils*. Longman, Edinburgh, 290 pp.

Domsch, K. H., Gams, W., and Anderson, T.-H. 1980a. *Compendium of Soil Fungi*. Vol. 1. Academic Press, London, 859 pp.

Domsch, K. H., Gams, W., and Anderson, T.-H. 1980b. *Compendium of Soil Fungi*. Vol. 2. Academic Press, London, 405 pp.

Dorenbosch, M. M. J. 1970. Key to nine ubiquitous soil-borne *Phoma*-like fungi. *Persoonia* 6:1–14.

Downing, M. H. 1953. *Botryotrichum* and *Coccospora*. *Mycologia* 45:934–940.

Drechsler, C. 1923. Some graminicolous species of *Helminthosporium*. *J. Agric. Res.* 24:641–740.

Drechsler, C. 1927. Two water molds causing tomato rootlet injury. *J. Agric. Res.* 34:287–296.

Drechsler, C. 1929. The beet water mold and several related root parasites. *J. Agric. Res.* 38:309–361.

Drechsler, C. 1930. Some new species of *Pythium*. *J. Wash. Acad. Sci.* 20:398–418.

Drechsler, C. 1931. A crown-rot of hollyhocks caused by *Phytophthora megasperma* n. sp. *J. Wash. Acad. Sci.* 21:513–526.

Drechsler, C. 1934. A new species of *Helicocephalum*. *Mycologia* 26:33–37.

Drechsler, C. 1936. *Pythium graminicolum* and *P. arrhenomanes*. *Phytopathology* 26:676–684.

Drechsler, C. 1937. Some hyphomycetes that prey on free-living terricolous nematodes. *Mycologia* 29:447–552.

Drechsler, C. 1938. Two hyphomycetes parasitic on oospores of root-rotting oomycetes. *Phytopathology* 28:81–103.

Drechsler, C. 1939. Several species of *Pythium* causing blossom-end rot of watermelon. *Phytopathology* 29:391–422.

Drechsler, C. 1940. Three species of *Pythium* associated with root rot. *Phytopathology* 30:189–213.

Drechsler, C. 1941. Three species of *Pythium* with proliferous sporangia. *Phytopathology* 31:478–507.

Drechsler, C. 1943. Two species of *Pythium* occurring in southern states. *Phytopathology* 33:261–299.

Drechsler, C. 1946. Several species of *Pythium* peculiar in their sexual development. *Phytopathology* 36:781–864.

Drechsler, C. 1950. Several species of *Dactylella* and *Dactylaria* that capture free-living nematodes. *Mycologia* 42:1–79.

Drechsler, C. 1953. Two new species of *Plectospira* isolated from discolored rootlets. *Bull. Torrey Bot. Club* 80:385–400.

Drechsler, C. 1954. Some hyphomycetes that capture eelworms in southern states. *Mycologia* 46:762–782.

Drechsler, C. 1960. A *Pythium*-causing stem rot of tobacco in Nicaragua and Indonesia. *Sydowia* 14:4–20.

Dube, H. C. and Bilgrami, K. S. 1966. *Pestalotia* or *Pestalotiopsis*? *Mycopathol. Mycol. Appl.* 29:33–54.

Eddins, A. H. 1930. A new *Diplodia* ear rot of corn. *Phytopathology* 20:733–742.

Ellis, M. B. 1940. Some fungi isolated from pinewood soil. *Trans. Br. Mycol. Soc.* 24:87–97.

Ellis, M. B. 1960. Dematiaceous hyphomycetes. I. *Mycol. Pap.* 76:1–36.

Ellis, M. B. 1963. Dematiaceous hyphomycetes. V. *Mycol. Pap.* 93:1–33.

Ellis, M. B. 1965. Dematiaceous hyphomycetes: VI. *Mycol. Pap.* 103:1–46.

Ellis, M. B. 1966. Dematiaceous hyphomycetes. VII. *Curvularia, Brachysporium*, etc. *Mycol. Pap.* 106:1–57.

Ellis, M. B. 1971. Dematiaceous hyphomycetes. *CMI*, Kew. 608 pp.

Ellis, M. B. 1976. More dematiaceous hyphomycetes *CMI*, Kew. 507 pp.

Ellis, M. B., Ellis, E. A., and Ellis, J. P. 1951. British marsh and fen fungi I. *Trans. Br. Mycol. Soc.* 34:147–169.

Emmons, C. W. and Dodge, B. O. 1931. The ascocarpic stage of species of *Scopulariopsis*. *Mycologia* 23:313–331.

Erwin, D. C. 1965. Reclassification of the causal agent of root rot of alfalfa from *Phytophthora cryptogea* to *P. megasperma*. *Phytopathology* 55:1139–1343.

Erwin, D. C., Bartnicki-Garcia, S., and Tsao, P. H. (Eds.). 1983. *Phytophthora, Its Biology, Taxonomy, Ecology and Pathology*. APS Press, St. Paul, MN. 392 pp.

Farr, D. F., Bills, G. F., Chamuris, G. P., and Rossman, A. Y. 1989. *Fungi on Plants and Plant Products in the United States*. APS Press, St. Paul, MN. 1252 pp.

Farrow, W. M. 1954. Tropical soil fungi. *Mycologia* 46:632–646.

Fergus, C. L. 1960. A note on the occurrence of *Peziza ostracoderma*. *Mycologia* 52:959–961.

Ford, E. J., Gold, A. H., and Snyder, W. C. 1970. Induction of chlamydospore formation in *Fusarium solani* by soil bacteria. *Phytopathology* 60:479–484.

Francis, S. M. 1985. *Rosellinia necatrix* — fact or fiction? *Sydowia* 38:75–86.

Friend, R. J. 1965. What is *Fumago vagans*? *Trans. Brit. Mycol. Soc.* 48:371–375.

Furuya, K. and Udagawa, S. 1973. Coprophilous Pyrenomycetes from Japan. III. *Trans. Mycol. Soc. Jpn.* 14:7–30.

Gams, W. 1969. Gliederungsprinzipien in der gattung *Mortierella*. *Nova Hedwigia* 18:30–43.

Gams, W. 1971. *Cephalosporium-artige Shimmelpilze* (Hyphomycetes). G. Fischer, Stuttgart, 126 pp.

Gams, W. 1975. *Cephalosporium*-like hyphomycetes: some tropical species. *Trans. Br. Mycol. Soc.* 64:389–404.

Gams, W. 1976. Some new or noteworthy species of *Mortierella*. *Persoonia* 9:111–140.

Gams, W. 1977. A key to the species of *Mortierella*. *Persoonia* 9:381–391.

Gams, W., Chien, C-Y., and Domsch, K. H. 1972. Zygospore formation by the heterothallic *Mortierella elongata* and a related homothallic species, *M. epigama* sp. nov. *Trans. Br. Mycol. Soc.* 58:5–13.

Gams, W. and Domsch, K. H. 1969. Bemerkungen zu einigen schwer bestimmbaren Bodenpilzen. *Nova Hedwigia* 18:1–29.

Gams W. and Holubová-Jechová, V. 1976. *Chloridium* and some other dematiaceous hyphomycetes growing on decaying wood. *Stud. Mycol.* 13:1–99.

Gams, W. and Hooghiemstra, H. 1976. Notes and brief articles. *Mortierella turficola* Ling Yong. *Persoonia* 9:141–144.

Gerdemann, J. W. 1953. An undescribed fungus causing a root rot of red clover and other leguminosae. *Mycologia* 45:548–554.

Gerdemann, J. W. 1954. Pathogenicity of *Leptodiscus terrestris* on red clover and other leguminosae. *Phytopathology* 44:451–455.

Gerlach, W. 1958. Beiträge zur Kenntnis der Gattung *Cylindrocarpon* Wr. I. *Cylindrocarpon radicicola* Wr. als Krankheitserreger an Alpenveilchen. *Phytopathol. Z.* 26:161–170.

Gerlach, W. 1959. Beiträge zur Kenntnis der Gattung *Cylindrocarpon* Wr. III. *Cylindrocarpon olidum* Wr. und seine phytopathologische Bedeuting. *Phytopathol. Z.* 35:333–346.

Gerlach, W. and Nirenberg, H. 1982. *The Genus Fusarium — a Pictorial Atlas*. Mitt. Biol. Bundesanst. Land U. Fortswirtsh. Berlin-Dahlem, 209:1–406.

Gilman, J. C. 1945. *A Manual of Soil Fungi*. The Collegiate Press, Ames, IA. 392 pp.

Goodey, J. B. 1951. A new species of hyphomycete attacking the stem eelworm *Ditylenchus dipsaci*. *Trans. Brit. Mycol. Soc.* 34:270–272.

Goos, R. D. 1960. Soil fungi from Costa Rica and Panama. *Mycologia* 52:877–883.

Goos, R. D. 1963. Further observations on soil fungi in Honduras. *Mycologia* 55:142–150.

Goos, R. D. 1985. A review of the anamorph genus *Helicomyces*. *Mycologia* 77:606–618.

Goos, R. D. and Timonin, M. I. 1962. Fungi from the rhizosphere of banana in Honduras. *Can. J. Bot.* 40:1371–1377.

Gorenz, A. M., Walker, J. C., and Larson, R. H. 1948. Morphology and taxonomy of the onion pink-root fungus. *Phytopathology* 38:831–840.

Gottlieb, M. and Butler, K. D. 1939. A *Pythium* root rot of cucurbits. *Phytopathology* 29:624–628.

Greathouse, G. A. and Ames, L. M. 1945. Fabric deterioration by thirteen described and three new species of *Chaetomium*. *Mycologia* 37:138–155.

Guarro, J. and Arx, J. A. von. 1987. The Ascomycete genus *Sordaria*. *Persoonia* 13:301–313.

Guba, E. F. 1932. Monograph of the genus *Pestalotia*. Part II. *Mycologia* 24:355–397.

Guba, E. F. 1961. A monograph of *Monochaetia* and *Pestalotia*. Harvard University Press, Cambridge, MA.

Hagem, O. 1908. Untersuchungen über Norwegische Mucorineen Videnskabs–Selskabets Skrifter I. Christiana Math.-Naturw. Klasse, No. 7, 1–50.

Hahn, G. G. 1930. Life history studies of the species of *Phomopsis* occurring conifers. Part I. *Trans. Br. Mycol. Soc.* 15:32–93.

Hammill, T. M. 1970. *Paecilomyces clavisporis* sp. nov., *Trichoderma saturnisporum* sp. nov., and other noteworthy soil from Georgia. *Mycologia* 62:107–122.

Hanlin, R. T. 1961. Studies in the genus *Nectria*. II. Morphology of *N. gliocladioides*. *Am. J. Bot.* 48:900–908.

Hanlin, R. T. 1989. *Illustrated Genera of Ascomycetes.* APS Press, St. Paul, MN. 263 pp.

Hanlin, R. T. 1998. *Illustrated Genera of Ascomycetes.* APS Press, St. Paul, MN. Vol. II, 258 pp.

Hanlin, R. T. 1999. Combined keys to *Illustrated Genera of Ascomycetes.* APS Press, St. Paul, MN. Vols. I & II. 113 pp.

Hansen, H. N. and Snyder, W. C. 1947. Gaseous sterilization of biological materials for use as culture media. *Phytopathology* 37:369–371.

Hasegawa, T. (Ed.) 1984. *Taxonomy and Identification of Microorganisms.* Vol. 1. Japan Scientific Societies Press, Tokyo, Revised, 310 pp., in Japanese.

Hawksworth, D. L. and Ciccarone, A. 1978. Studies on a species of *Monosporascus* isolated from Triticum. *Mycopathologia* 66:147–151.

Hawksworth, D. L., Kirk, P. M., Sutton, B. C., and Pegler, D. N. 1995. *Ainsworth & Bisby's Dictionary of the Fungi.* 8th ed. CAB. International. Surrey, U.K., 616 pp.

Hendrix, F. F., Jr. and Campbell, W. A. 1968a. A new heterothallic *Pythium* from the United States and Canada. *Mycologia* 60:802–805.

Hendrix, F. F., Jr. and Campbell, W. A. 1968b. Pythiaceous fungi isolated from southern forest nursery soils and their pathogenicity to pine seedlings. *For. Sci.* 14:292–297.

Hendrix, F. F., Jr. and Campbell, W. A. 1969. Heterothallism in *Pythium catenulatum. Mycologia* 61:639–641.

Hendrix, F. F. and Campbell, W. A. 1974. Taxonomy of *Pythium sylvaticum* and related fungi. *Mycologia* 66:1049–1053.

Henis, Y. and Inbar, M. 1968. Effect of *Bacillus subtilis* on growth and sclerotium formation by *Rhizoctonia solani. Phytopathology* 58:933–938.

Hennebert, G. L. 1963. Un hyphomycete nouveau *Arachnophora fagicola* gen. nov. spec. nov. *Can. J. Bot.* 41:1165–1169.

Hennebert, G. L. 1968. *Echinobotryum, Wardomyces* and *Mammaria. Trans. Br. Mycol. Soc.* 51:749–762.

Hennebert, G. L. 1973. *Botrytis* and *Botrytis*-like genera. *Persoonia* 7:183–204.

Hermanides-Nijhof, E. J. 1977. *Aureobasidium* and allied genera. *Stud. Mycol.* 15:141–177.

Hesseltine, C. W., Benjamin, C. R., and Mehrotra, B. S. 1959. The genus *Zygorhynchus. Mycologia* 51:173–194.

Hesseltine, C. W. and Ellis, J. J. 1964. The genus *Absidia*: *Gongronella* and cylindrical-spored species of *Absidia. Mycologia* 56:568–601.

Hesseltine, C. W. and Fennell, D. I. 1955. The genus *Circinella. Mycologia* 47:193–212.

Hesseltine, C. W. and Ellis, J. J. 1966. Species of *Absidia* with ovoid sporangiospores. I. *Mycologia* 58:761–785.

Hewings, A. D. and Crane, J. L. 1981. The genus *Codinaea*. Three new species from the Americas. *Mycotaxon* 13:419–427.

Hickman, G. J. 1944. Phycomycetes occurring in Great Britain. 3. *Pythyum aphanidermatum* (Edson) Fitzpatrick. *Trans. Br. Mycol. Soc.* 27:63–67.

Hiura, M. 1967. *Phytopathogenic Fungi.* Yokendo, Tokyo, 342 pp., in Japanese.

Hobbs, T. W., Schmitthenner, A. F., and Kuter, G. A. 1985. A new *Phomopsis* species from soybean. *Mycologia* 77:535–544.

Hodges, C. S. 1962. Fungi isolated from southern forest tree nursery soils. *Mycologia* 54:221–229.

Holliday, P. 1989. *A Dictionary of Plant Pathology.* Cambridge University Press, Cambridge, MA, 369 pp.

Holubová-Jechová, V. 1991. *Helicogoosia*, a new genus of lignicolous Hyphomycetes. *Mycotaxon* 41:445–450.

Hoog, G. S. de. 1972. The genera *Beauveria, Isaria, Tritirachium* and *Acrodontium* gen. nov. *Stud. Mycol.* 1:1–41.

Hoog, G. S. de (Ed.) 1979. The black yeasts, II: *Moniliella* and allied genera. *Stud. Mycol.* 19:1–90.

Hoog, G. S. de (Ed.) 1985. Taxonomy of the *Dactylaria* complex. IV–VI. *Stud. Mycol.* 26:1–122.

Hoog, G. S. de and Hermanides-Nijhof, E. J. 1977. The black yeasts and allied hyphomycetes. *Stud. Mycol.* 15:1–222.

Hoog, G. S. de, Rahman, M. A., and Boekhout, T. 1983. *Ramichloridium, Veronaea* and *Stenella*: generic delimitation, and new combinations and two new species. *Trans. Br. Mycol. Soc.* 81:485–490.

Hotson, H. H. 1942. Some species of *Papulaspora* associated with rots of gladiolus bulbs. *Mycologia* 34:391–399.

Hotson, J. W. 1912. Culture studies of fungi-producing bulbils and similar propagative bodies. *Proc. Am. Acad. Arts Sci.* 48:227–306.

Hotson, J. W. 1917. Notes on bulbiferous fungi with a key to described species. *Bot. Gaz.* 64:265–284.

Howell, A., Jr. 1939. Studies on *Histoplasma capsulatum* and similar form species. I. Morphology and development. *Mycologia* 31:191–216.

Hudson, H. 1963. The perfect state of *Nigrospora oryzae. Trans. Br. Mycol. Soc.* 46:355–360.

Hughes, S. J. 1949. Studies in micro-fungi. II. The genus *Sporoschisma* Berkeley and Broome and a redescription of *Helminthosporium rousselianum* Montagne. *Mycol. Pap.* 31:1–33.

Hughes, S. J. 1951a. Studies on micro-fungi. III. *Mastigosporium, Camposporium* and *Ceratophorum. Mycol. Pap.* 36:1–43.

Hughes, S. J. 1951b. *Stachylidium, Gonytrichum, Mesobotrys, Chaetopsis,* and *Chaetopsella. Trans. Br. Mycol. Soc.* 34:551–576.

Hughes, S. J. 1951c. Studies on micro-fungi. IX. *Calcarisporium, Verticicladium* and *Hansfordia* (gen. nov.). *Mycol. Pap.* 43:1–24.

Hughes, S. J. 1951d. Studies on micro-fungi. XII. *Triposporium, Tripospermum, Ceratosporella,* and *Tetrasporium* (gen. nov.). *Mycol. Pap.* 46:1–35.

Hughes, S. J. 1952. *Trichocladium* Harz. *Trans. Br. Mycol. Soc.* 35:152–157.

Hughes, S. J. 1953a. Conidiophores, conidia, and classification. *Can. J. Bot.* 31:577–659.

Hughes, S. J. 1953b. Fungi from the Gold Coast. 2. *Mycol. Pap. CMI* 50:1–104.

Hughes, S. J. 1958. Revisions Hyphomycetum aliquot cum appendice de nominibus rejiciendis. *Can. J. Bot.* 36:727–836.

Hughes, S. J. 1959. Microfungi. IV. *Trichocladium canadense*. n. sp. *Can. J. Bot.* 37:857–859.
Hughes, S. J. 1976. Sooty moulds. *Mycologia* 68:693–820.
Hughes, S. J. 1983. Five species of *Sarcinella* from North America, with notes on *Questieriella* n. gen., *Mitteriella*, *Endophragmiopsis*, *Schiffnerula*, and *Clypeolella*. *Can. J. Bot.* 61:1727–1767.
Hughes, S. J. and Kendrick, W. B. 1968. New Zealand fungi. 12. *Menispora, Codinaea, Menisporopsis*. *N.Z. J. Bot.* 6:323–375.
Hunt, J. 1956. Taxonomy of the genus *Ceratocystis*. *Lloydia* 19:1–58.
Ichinoe, M. 1968. Japanese Hyphomycete notes II. *Trans. Mycol. Soc. Jpn.* 9:57–64.
Imazeki, R. and Hongo, T. *Colored Illustrations of Mushrooms of Japan*, Hoikusha Publishing, Osaka. Vol. 1. (1987), 325 pp.; Vol. 2 (1989), 315 pp., in Japanese.
Ingold, C. T. 1956. The conidial apparatus of *Trichothecium roseum*. *Trans. Br. Mycol. Soc.* 39:460–464.
Isaac, I. 1949. A comparative study of pathogenic isolates of *Verticillium*. *Trans. Br. Mycol. Soc.* 32:137–157.
Ito, S. 1936. *Mycological Flora of Japan*. Vol. 1. Yokendo, Tokyo. 339 pp., in Japanese.
Ito, S. 1955. *Mycological Flora of Japan*. Vol. 2, No. 4. Yokendo, Tokyo. 450 pp., in Japanese.
Ito, T., Ueda, M., and Yokoyama, T. 1981. Thermophilic and thermotolerant fungi in paddy field soils. *IFO Res. Comm.* 10:20–32.
Jee, H. J., Ho, H. H., and Cho, W. D. 2000. *Pythiogeton zeae* sp. nov. causing root and basal stalk rot of corn in Korea. *Mycologia* 92:522–527.
Jain, B. L. 1962. Two new species of *Curvalaria*. *Trans. Br. Mycol. Soc.* 45:539–544.
Jensen, H. L. 1931. The fungus flora of the soil. *Soil Sci.* 31:123–158.
Joffe, A. Z. 1963. The mycoflora of a continuously cropped soil in Israel, with special reference to effects of manuring and fertilizing. *Mycologia* 55:271–282.
Joffe, A. Z. and Borut, S. Y. 1966. Soil and kernel mycoflora of groundnut fields in Israel. *Mycologia* 58:629–640.
Johnson, T. W., Jr. 1951. An isolate of *Dictyuchus* connecting the false-net and true-net species. *Mycologia* 43:365–372.
Jones, P. M. 1936. A new species of *Microascus* with a *Scopulariopsis* stage. *Mycologia* 28:502–509.
Jong, S. C. and Davis, E. E. 1976. Contribution to the knowledge of *Stachybotrys* and *Memnoniella* in culture. *Mycotaxon* 3:409–485.
Jong, S. C., Dugan, F., and Edwards, M. J. 1996. ATCC Filamentous Fungi. 19th ed., 720 pp.
Katumoto, K. 1986. Two new species of *Eudarluca* hyperparasitic to *Botryosphaeria*. *Trans. Mycol. Soc. Jpn.* 27:11–16.
Katsura, K. 1971. *Phytophthora* diseases of plants. Seibundo-Shinko, Tokyo. 128 pp., in Japanese.
Katsura, K. 1976. Two new species of *Phytophthora* causing damping-off of cucumber and trunk rot of chestnut. *Trans. Mycol. Soc. Jpn.* 17:238–242.
Kendrick, B., Smih, J. E., Neville, J., and Weber, N. S. 1994. A study of morphological variability, character analysis, and taxonomic significance in the Zygomycetous anamorph genus Umbelopsis. *Mycotaxon* 51:15–26.

Kendrick, W. B. 1961. Hyphomycetes of conifer leaf litter. *Thysanophora* gen. nov. *Can. J. Bot.* 39:817–832.
Kendrick, W. B. and Bhatt, G. C. 1966. *Trichocladium opacum*. *Can. J. Bot.* 44:1728–1731.
Kiffer, E. and Morelet, M. 1999. *The Deuteromycetes. Mitosporic Fungi: Classification and Generic Keys.* Science Publishers Inc., Enfield, NH, 273 pp.
Kirk, P. M. 1984. *Volutellaria laurina* Tassi, an earlier name for *Wiesneriomyces javanicus* Koorders. *Trans. Br. Mycol. Soc.* 82:748–749.
Kitz, D. J. and Embree, R. W. 1989. A new species of *Helicocephalum* from Iowa. *Mycologia* 81:164–166.
Kobayashi, T., Katumoto, K., and others (Eds.). 1992. *Illustrated Genera of Plant Pathogonic Fungi in Japan.* Zenkoku Noson Kyoiku Kyokai Publ. Co., Tokyo. 696 pp., in Japanese.
Komada, H. 1972. Selective synthetic medium of *Fusarium oxysporum*. *Ann. Phytopathol. Soc. Jpn.* 38:191. Abstr. in Japanese.
Kuhlman, E. G. 1975. Zygospore formation in *Mortierella alpina* and *M. spinosa*. *Mycologia* 67:678–681.
Kuhlman, E. G. and Hodges, C. S., Jr. 1972. Rediscovery of *Mortierella rostafinskii* and *Mortierella strangulata*. *Mycologia* 64:92–98.
Kurata, H. 1960. Fungal diseases of soybean. *Bull. Natl. Inst. Agric. Sci. Ser.* C:1–153, in Japanese.
Lefebvre, C. L., Johnson, A. G., and Sherwin, H. S. 1949. An undescribed species of *Periconia*. *Mycologia* 41:416–419.
Limber, D. P. 1940. A new form genus of the Moniliaceae. *Mycologia* 32:23–30.
Liu, X. Z. and Qiu, W. F. 1992. A new species of *Dactylella* from China. *Mycol. Res.* 97:359–362.
Liu, X. Z. and Zhang, K. Q. 1994. Nematode-trapping species of *Monacrosporium* with special reference to two new species. *Mycol. Res.* 98:862–868.
Lodder, J. (Ed.). 1970. *The Yeasts. A taxonomic Study.* 2nd ed. North-Holland, Amsterdam. 1385 pp.
Luttrell, E. S. 1963. A *Trichometasphaeria*-perfect stage for a *Helminthosporium* causing leaf blight of *Dactyloctenium*. *Phytopathology* 53:281–285.
MacGarvie, Q. D. 1965. *Diplorhinotrichum funicola* sp. nov., causing a disease of *Juncus effusus*. *Trans. Br. Mycol. Soc.* 48:269–271.
Maciejowska, Z. and Willams, E. B. 1963. Studies on morphological forms of *Staphylotrichum cocosporum*. *Mycologia* 55:221–225.
MacLeod, D. M. 1954. Investigations on the genera *Beauveria* Vuill. and *Tritirachium* Limber. *Can. J. Bot.* 32:818–890.
Malloch, D. and Cain, R. F. 1971. New cleistothecial Sordariaceae and a new family, Coniochaetaceae. *Can. J. Bot.* 49:869–880.
Malloch, D. and Cain, R. F. 1973. The genus *Thielavia*. *Mycologia* 65:1055–1077.
Maniotis, J. and Strain, J. W. 1968. *Wiesneriomyces javanicus* from Panamanian soil. *Mycologia* 60:203–208.
Marasas, W. F. O., Nelson, P. E., Toussoun, T. A., and Wrk, P. S. van. 1986. *Fusarium polyphialidicum*, a new species from South Africa. *Mycologia* 78:678–682.
Martin, J. P., Abbott, E. V., and Hughes, C. G. (Eds.). 1961. *Sugar Cane Diseases of the World.* Vol. 1. Elsevier, Amsterdam. 542 pp.

Marx, D. H. and Haasis, F. A. 1965. Induction of aseptic sporangial formation in *Phytophthora cinnamomi* by metabolic diffusates of soil micro-organisms. *Nature (London)* 206:673–674.

Masago, H., Yoshikawa, M., Fukuda, M., and Nakanishi, N. 1977. Selective inhibition of *Pythium* spp. on a medium for direct isolation of *Phytophthora* spp. from soils and plants. *Phytopathology* 67:425–428.

Mason, E. W. and Ellis, M. B. 1953. British species of *Periconia*. *Mycol. Pap.* 56:1–127.

Matsumoto, T. 1952. *Monograph of Sugar Cane Diseases in Taiwan*, Chinese-American Joint Commission on Rural Development, Taipei, Taiwan, 61 pp., 15 plates.

Matsushima, T. 1975. Icones Microfungorum a Matsushima Lectorum, Kobe, Japan. 209 pp.

Matsushima, T. 1981. *Matsushima Mycological Memoirs*, No.2. Published by the author, Kobe, Japan. 68 pp.

Matsushima, T. 1983. *Matsushima Mycological Memoirs*, No.3. Published by the author, Kobe, Japan. 90 pp.

Matthews, V. D. 1931. *Studies on the genus Pythium*. University of North Carolina Press, Chapel Hill. 136 pp.

Matuo, T., Komada, H., and Matsuda, A. (Eds.). 1980. *Fusarium Disease of Cultivated Plants*. Zenkoku Noson Kyoiku Kyokai Publ. Co., Tokyo. 502 pp., in Japanese.

McKeen, C. D. 1952b. *Aphanomyces cladogamous* Drechs., a cause of damping-off in peppers and certain other vegetables. *Can. J. Bot.* 30:701–709.

McKeen, C. D. and Thorpe, H. J. 1968. A *Pythium* root rot of muskmelon. *Can. J. Bot.* 46:1165–1171.

McKeen, W. E. 1952a. *Phialophora radicicola* Cain, a corn root rot pathogen. *Can. J. Bot.* 30:344–347.

McVey, D. M. and Gerdemann, J. W. 1960. The morphology of *Leptodiscus terrestris*, and the function of setae in spore dispersal. *Mycologia* 52:193–200.

Mehrotra, B. R. 1969. Some new reports of Ascomycetes from Allahabad. *Sydowia* 23:81–91.

Mehrotra, B. S., Baijal, U., and Mehrotra, B. R. 1963. Two new species of *Mortierella* from India. *Mycologia* 55:289–296.

Meurs, A. 1934. Parasitic stemburn of Deli tobacco. *Phytopathol. Z.* 7:169–185.

Middleton, J. T. 1943. The taxonomy, host range, and geographic distribution of the genus *Pythium*. *Mem. Torrey Bot. Club* 20:1–171.

Milanez, A. I. 1978. *Pythium echinulatum* from Michigan soils. *Nova Hedwigia* 29:557–563.

Miller, J. H., Giddens, J. E., and Foster, A. A. 1957. A survey of the fungi of forest and cultivated soils of Georgia. *Mycologia* 49:779–808.

Mills, J. T. and Vlitos, A. J. 1967. Studies on the rhizosphere of sugar-cane in Trinidad. *Trop. Agric., Trin.* 44:151–157.

Mirza, J. H. and Cain, R. F. 1969. Revision of the genus *Podospora*. *Can. J. Bot.* 47:1999–2048.

Misra, P. C. and Talbot, P. H. B. 1964. *Phialomyces*, a new genus of Hyphomycetes. *Can. J. Bot.* 42:1287–1290.

Miyaji, M. and Nishimura, K. (Eds.). 1991. *A Dictionary of Medical Mycology*. Kyowa Kikaku Tsushin, Tokyo, 302 pp., in Japanese.

Moore, R. T. 1955. Index to the *Helicosporae*. *Mycologia* 47:90–103.

Moore, R. T. 1957. Index to the *Helicosporae*: addenda. *Mycologia* 49:580–587.

Morrall, R. A. A. and Vanterpool, T. C. 1968. The soil microfungi of upland boreal forest at Candlelake, Saskatchewan. *Mycologia* 60:642–654.

Morris, E. F. 1956. Tropical Fungi Imperfecti. *Mycologia* 48:728–737.

Morrow, M. B. 1940. The soil fungi of a pine forest. *Mycologia* 24:398–402.

Morton, F. J. and Smith, G. 1963. The genera *Scopulariopsis* Bainier, *Microascus* Zukal, and *Doratomyces* Corda. *Mycol. Pap.* 86:1–96.

Mouchacca, J. and Gams, W. 1993. The hyphomycete genus *Cladorrhinum* and its teleomorph connections. *Mycotaxon* 48:415–440.

Nag Raj, T. R., Morgan-Jones, G., and Kendrick, B. 1972. Genera coelomycetarum. IV. *Pseudorobillarda* gen. nov., a generic segregate of *Robillarda* Sacc. *Can. J. Bot.* 50:861–867.

Nag Raj, T. R. and Kendrick, B. 1975. *A Monograph of Chalara and Allied Genera*. University of Waterloo Press, Waterloo, Ontario. 200 pp.

Nag Raj, T. R. and Kendrick, B. 1976. *A Monograph of Chalara and Allied Genera*. Wilfrid Lanrier University Press, Waterloo, Ontario. 200 pp.

Nagai, M. 1931. Studies on the Japanese Saprolegniaceae. *J. Facul. Agr.*, Hokkaido Imp. Univ., Sapporo 32:1–43, Pls. 1–7.

Nagai, M. 1933. Additional notes on the Japanese Saprolegniaceae. *Bot. Mag. (Tokyo)* 47:136–137.

Nagai, Y., Takeuchi, T., and Watanabe, T. 1978. A stem blight of rose caused by *Phytophthora megasperma*. *Phytopathology* 68:684–688.

Nagai, Y., Takeuchi, T., and Watanabe, T. 1988. Root rot of dasheen caused by *Pythium myriotylum* Drechsler. *Ann. Phytopathol. Soc. Jpn.* 54:529–532, in Japanese with English abstract.

Nash, S. M. and Snyder, W. C. 1965. Quantitative and qualitative comparisons of *Fusarium* populations in cultivated and noncultivated parent soils. *Can. J. Bot.* 43:939–945.

Neergaard, P. and Newhall, A. G. 1951. Notes on the physiology and pathogenicity of *Centrospora acerina* (Hartig) Newhall. *Phytopathology* 41:1021–1033.

Nelson, R. R. and Haasis, F. A. 1964. The perfect stage of *Curvularia lunata*. *Mycologia* 56:316–317.

Nelson, R. R. and Hodges, C. S. 1965. A new species of *Curvularia* with a protuberant conidial hilum. *Mycologia* 57:822–825.

Nelson, P. E., Toussoun, T. A., and Cook, R. J. 1981. *Fusarium, Disease, Biology, and Taxonomy*. Pennsylvania State University Press, University Park. 457 pp.

Nelson, P. E., Toussoun, T. A., and Marasas, W. F. O. 1983. *Fusarium Species. An Illustrated Manual for Identification*. Pennsylvania State University Press, University Park. 193 pp.

Newhall, A. G. 1946. More on the name *Ansatospora acerina*. *Phytopathology* 36:893–896.

Newhook, F. J., Waterhouse, G. M., and Stamps, D. J. 1978. Tabular key to the species of *Phytophthora* de Bary. *Mycol. Pap.* 143:1–20.

Nicot, J. and Meyer, J. 1957. Un hyphomycete nouveau des sols tropicaux: *Staphylotrichum coccosporum* nov. gen., nov. sp. *Bull. Trimest. Soc. Mycol. Fr.* 72:318–323.

Obayashi, N. and Watanabe, T. 1988. A new disease of radish with globose necrotic lesions recently occurred in Miura Peninsula. *Ann. Phytopathol. Soc. Jpn.* 54:68–69. Abstr., in Japanese.

Olive, L. S. and Fantini, A. A. 1961. A new heterothallic species of *Sordaria*. *Am. J. Bot.* 48:124–128.

Omvik, A. 1955. Two new species of *Chaetomium* and one new *Humicola* species. *Mycologia* 47:748–757.

Onions, A. H. S. and Barron, G. L. 1967. Monophialidic species of *Paecilomyces*. *Mycol. Pap.* 107:1–25.

Ostazeski, S. A. 1967. An undescribed fungus associated with a root and crown rot of birdsfoot trefoil (*Lotus corniculatus*). *Mycologia* 59:970–975.

Oudemans, C. A. J. A. and Koning, C. J. 1902. Prodrome d'une flore mycologique obtenue par la culture sur gelatin preparee de la terre humeuse du Spanderwoud, pres de Russum. *Arch. Neerl. Sci. Exactes Nat. Ser.* 2, VII, 266–298.

Padgett, D. E. and Seymour, R. L. 1974. Variability of zoospore discharge in species of the genus *Dictyuchus*. *Mycologia* 66:615–627.

Papa, K. E., Campbell, W. A., and Hendrix, F. F., Jr. 1967. Sexuality in *Pythium sylvaticum*: heterothallism. *Mycologia* 59:589–595.

Papavizas, G. C. 1967. Evaluation of various media and antimicrobial agents for isolation of *Fusarium* from soil. *Phytopathology* 57:848–852.

Park, D. 1973. Germination of the three spore forms of *Mammaria echinobotryoides*. *Trans. Br. Mycol. Soc.* 60:351–354.

Parmeter, J. R., Jr. (Ed.). 1970. *Rhizoctonia solani, Biology and Pathology*. University of California Press, Berkeley. 255 pp.

Peerally, A. 1974. CMI descriptions of pathogenic fungi and bacteria. No. 421. *Calonectria kyotoensis*. Commonw. Mycol. Inst. Kew. U.K.

Peerally, A. 1991. The classification and phytopathology of *Cylindrocladium* species. *Mycotaxon* 40:323–366.

Pentland, G. D. 1967. Ethernol produced by *Aureobasidium pullulans* and its effect on the growth of *Armillaria mellea*. *Can. J. Microbiol.* 13:1631–1639.

Perrin R. 1976. Clef de determination des Nectria d'Europe. *Bull. Soc. Mycol. Fr.* 92:335–347.

Petersen, R. H. 1962. Aquatic Hyphomycetes from North America. I. Aleuriosporae (Part 1), and key to the genera. *Mycologia* 54:117–151.

Pethybridge, G. H. 1919. Notes on some saprophytic species of fungi, associated with diseased potato plants and tubers. *Trans. Br. Mycol. Soc.* 6:104–120.

Pirozynski, K. A. 1963. *Beltrania* and related genera. *Mycol. Pap.* 90:1–37.

Pirozynski, K. A. 1972. Microfungi of Tanzania. *Mycol. Pap.* 129:1–64.

Plaats-Niterink, A. J. van der. 1981. Monograph of the genus *Pythium*. *Stud. Mycol.* 21:1–242.

Plaats-Niterink, A. J. van der, Samson, R. A., Stalpers, J. A., and Weijman. A. C. M. 1976. Some oomycetes and zygomycetes with asexual eghinulate reproductive structures. *Persoonia* 9:85–93.

Pollack, F. G. and Uecker, F. A. 1974. *Monosporascus cannonballus*, an unusual Ascomycete in cantaloupe roots. *Mycologia* 66:346–349.

Pope, S. 1944. A new species of *Metarrhizium* active in decomposing cellulose. *Mycologia* 36:343–350.

Pratt, R. G. and Mitchell, J. E. 1973. A new species of *Pythium* from Wisconsin and Florida isolated from carrots. *Can. J. Bot.* 51:333–339.

Preston, N. C. 1943. Observations on the genus *Myrothecium* I. The three classic species. *Trans. Br. Mycol. Soc.* 26:158–168.

Preston, N. C. 1948. Observations on the genus *Myrothecium* II. *Myrothecium gramineum* Lib. and two new species. *Trans. Br. Mycol. Soc.* 31:271–276.

Rai, J. N., Tewari, J. P., and Mukerji, K. G. 1964. *Achaetomium*, a new genus of Ascomycetes. *Can. J. Bot.* 42:693–697.

Rajashekhar, M., Bhat, D. J., and Kaverippa, K. M. 1991. An undescribed species of *Vermispora* from India. *Mycologia* 83:230–232.

Ramchandra, R. and Baheker, W. S. 1964. Fungi on *Cycas revoluta* Thunb. *Mycopathologia Mycologia Applicata* 23:266–268.

Ranzoni, F. V. 1968. Fungi isolated in culture from soils of the Sonoran desert. *Mycologia* 60:356–371.

Rao, V. and Hoog, G. S. de. 1975. Some notes on *Torula*. *Persoonia* 8:199–206.

Rao, V. and Hoog. G.S. de. 1986. New or critical Hyphomycetes from India. *Stud. Mycol.* 28:1–84.

Raper, K. B. and Fennell, D. I. 1965. *The genus Aspergillus*. Williams & Wilkins, Baltimore. 686 pp.

Raper, K. B. and Thom, C. 1949. *A Manual of the Penicillia*. Williams & Wilkins, Baltimore. 875 pp.

Rattan, S. S., Muhsin, T. M., and Ismail, A. L. S. 1978. Aquatic fungi of Iraq: Species of *Dictuchus* and *Calptyralegnia*. *Sydowia* 31:112–121.

Rifai, M. A. 1969. A revision of the genus *Trichoderma*. *Mycol. Pap.* 116:1–56.

Rifai, M. A. and Cooke, R. C. 1966. Studies on some didymosporous genera of nematode-trapping Hyphomycetes. *Trans. Br. Mycol. Soc.* 49:147–168.

Robison, B. M. 1970. Microfungi of sugar-cane roots and soil in Jamaica. *Trop. Agric., Trin.* 47:23–29.

Rogerson, C. T. and Samuels, G. J. 1985. Species of *Hypomyces* and *Nectria* occurring on Discomycetes. *Mycologia* 77:763–783.

Rossman, A. Y. 1983. The Phragmosporous species of *Nectria* and related genera. *Mycol. Pap.* 150:1–121.

Rossman, A. Y., Howard, R. J., and Valent, B. 1990. *Pyricularia grisea*, the correct name for the rice blast disease fungus. *Mycologia* 82:509–512.

Roxon, J. E. and Jong, S. C. 1974. A new pleomorphic species of *Humicola* from Saskatchewan soil. *Can. J. Bot.* 52:517–520.

Sahni, V. P. 1964. Some filicolous ectoparasites and associated fungi from Jabalpur (M. P.). 1. *Mycopathol. Mycol. Appl.* 23:328–338.

Saksena, S. B. 1953. A new genus of the Mucorales. *Mycologia* 45:426–436.

Samson, R. A. 1974. *Paecilomyces* and some allied Hyphomycetes. *Stud. Mycol.* 6:1–119.

Samson, R. A. and Pitt, J. I. (Eds.) 1985. *Advances in Penicilium and Aspergillus Systematics.* Plenum Press, New York. 483 pp.

Samson, R. A. and Pitt, J. I. (Eds.). 1990. *Modern Concepts in Penicillium and Aspergillus Classification.* Plenum Press, New York. 478 pp.

Samuels, G. J. 1976. Perfect states of *Acremonium.* The genera *Nectria, Actiniopsis, Ijuhya, Neohenningsia, Ophiodictyon,* and *Peristomialis. N.Z. J. Bot.* 14:231–260.

Samuels, G. J. 1977. *Nectria consors* and its *Volutella* state. *Mycologia* 69:255–262.

Samuels, G. J. 1988. Species of *Nectria* (Ascomycetes, Hypocreales) having orange perithecia and colorless, striate ascospores. *Brittonia* 40:306–331.

Samuels, G. J. 1989. *Nectria* and *Penicillifer. Mycologia* 81:347–355.

Sartoris, G. B. 1927. A cytological study of *Ceratostomella adiposum* (Butl.) comb. nov., the black-rot fungus of sugar cane. *J. Agric. Res.* 35:577–585.

Sato, R. and Kitazawa, K. 1980. Occurrence of root rot syndrome of soybean caused by *Corynespora cassiicola.* (Berk. & Curt.) Wei. *Ann. Phytopathol. Soc. Jpn.* 46:193–199, in Japanese.

Schipper, M. A. A. 1973. A study on variability in *Mucor hiemalis* and related species. *Stud. Mycol.* 4:1–40.

Schipper, M. A. A. 1978. 1. On certain species of *Mucor* with a key to all accepted species. 2. On the genera *Rhizomucor* and *Parasitella. Stud. Mycol.* 17:1–70.

Schipper, M. A. A. and Stalpers, J. A. 1984. A revision of the genus *Rhizopus. Stud. Mycologia* 25:1–34.

Schoknecht, J. D. and Crane, J. L. 1977. Revision of *Torula* and *Hormiscium* species. *Torula occulta, T. diversa, T. elasticae, T. bigemina* and *Hormiscium condensatum* reexamined. *Mycologia* 69:533–546.

Schol-Schwarz, M. B. 1959. The genus *Epicoccum* Link. *Trans. Br. Mycol. Soc.* 42:149–173.

Schol-Schwarz, M. B. 1968. *Rhinocladiella,* its synonym *Fonsecaea* and its relation to *Phialophora. Antonie van Leeuwenhoek* 34:119–152.

Schol-Schwarz, M. B. 1970. Revision of the genus *Phialophora* (Moniliales). *Persoonia* 6:59–94.

Scott, P. R. 1970. *Phialophora radicicola,* an avirulent parasite of wheat and grass roots. *Trans. Br. Mycol. Soc.* 55:163–167.

Scott, W. W. 1961. A monograph of the genus *Aphanomyces. Va. Agric. Exp. Stn. Tech. Bull.* 151:1–95.

Seymour, R. L. 1970. The genus *Saprolegnia. Nova Hedwigia* 19:1–124.

Shoemaker, R. A. 1959. Nomenclature of *Drechslera* and *Bipolaris,* grass parasites segregated from *Helminthosporium. Can. J. Bot.* 37:879–887.

Siddiqi, M. A. 1964. Fungus flora of *Coffea arabica* in Nyasaland. *Trans. Br. Mycol. Soc.* 47:281–284.

Sideris, C. P. 1932. Taxonomic studies in the family Pythiacea. 2. *Pythium. Mycologia* 24:14–61.

Simmons, E. G. 1967. Typification of *Alternaria, Stemphilium,* and *Ulocladium. Mycologia* 59:67–92.

Simmons, E. G. 1969. Perfect states of *Stemphilium. Mycologia* 61:1–26.

Sivanesan, A. 1983. Studies on Ascomycetes. *Trans. Br. Mycol. Soc.* 81:313–332.

Sivanesan, A. 1984. The *Bitunicate Ascomycetes and Their Anamorphs.* J. Cramer, Vaduz. 701 pp.

Skolko, A. J. and Groves, J. W. 1948. Notes on seed borne fungi. V. *Chaetomium* species with dichotomously branched hairs. *Can. J. Res. C.* 26:269–280.

Smalley, E. B. and Hansen, H. N. 1957. The perfect stage of *Gliocladium roseum. Mycologia* 49:529–533.

Smith, G. 1962. Some new and interesting species of microfungi. III. *Trans. Br. Mycol. Soc.* 45:387–394.

Sneh, B., Burpee, L., and Ogoshi, A. 1991. *Identification of Rhizoctonia species.* APS Press, St. Paul, MN. 133 p.

Snyder, W. C. and Hansen, H. N. 1940. The species concept in *Fusarium. Am. J. Bot.* 27:64–67.

Snyert, W. C. and Hansen, H. N. 1941. The species concept in *Fusarium* with reference to section Martiella. *Am. J. Bot.* 28:738–742.

Snyder, W. C. and Hansen, H. N. 1945. The species concept in *Fusarium* with reference to Discolor and other sections. *Am. J. Bot.* 32:657–666.

Sobers, E. K. 1972. Morphology and pathogenicity of *Calonectria floridana, Calonectria kyotensis* and *Calonectria uniseptata. Phytopathology* 62:485–487.

Sobers, E. K. and Seymour, C. P. 1967. *Cylindrocladium floridanum* sp. n. associated with decline of peach trees in Florida. *Phytopathology* 57:389–393.

Söderström, B. E. 1975. Vertical distribution of microfungi in a spruce forest in the south of Sweden. *Trans. Br. Mycol. Soc.* 65:419–425.

Sparrow, F. K., Jr. 1931. Two new species of *Pythium* parasitic in green algae. *Ann. Bot.* 45:257–277.

Sparrow, F. K., Jr. 1960. *Aquatic Phycomycetes.* 2nd ed. University of Michigan Press, Ann Arbor. 1187 pp.

Sprague, R. 1948. Some leaf spot fungi on western Gramineae. III. *Mycologia* 40:295–313.

Sprague, R. 1950. *Diseases of Cereals and Grasses in North America* (Fungi, except smuts and rusts). Ronald Press, NY. 538 pp.

Sprague, R. 1951. Some leaf spot fungi on western Gramineae. VI. *Mycologia* 43:549–569.

Sprague, R. and Cooke, W. B. 1939. Some fungi imperfecti from the Pacific Northwest. *Mycologia* 31:43–52.

Srinivasan, K. V. 1958. Fungi of the rhizosphere of sugarcane and allied plants. 1. *Hyalostachybotrys* gen. nov. *J. Ind. Bot. Soc.* 37:334–342.

Srinivasulu, B. V. and Sathe, P. G. 1972. Genus *Massarina* from India. *Sydowia* 26:83–86.

Stalpers, J. A. 1974a. Revision of the genus *Oedocephalum* (Fungi Imperfecti). *Proc. Kon. Ned. Akad. Wet. C* 383–401.

Stalpers, J. A. 1974b. *Spiniger,* a new genus for imperfect states of Basidiomycetes. *Proc. Kon. Ned. Akad. Wet. Ser. C* 77:402–407.

Stanghellini, M. E., White, J. G., Tomlinson, J. A., and Clay, C. 1988. Root rot of hydroponically grown cucumbers caused by zoospore-producing isolates of *Pythium intermedium. Plant Dis.* 72:358–359.

Steyaert, R. L. 1949. Contribution à l'étude monographique de *Pestalotia* de Not. et *Monochaetia* Sacc. (*Truncatella* gen. nov. et *Pestalotiopsis* gen. nov.). *Bull. Jard. Bot. Brux.* 19:285–354.

Steyaert, R. L. 1955. *Pestalotia, Pestalotiopsis,* et *Truncatella. Bull. Jard. Bot. Brux.* 25:191–199.

Steyaert, R. L. 1956. A reply and an appeal to Professor Guba. *Mycologia* 48:767–768.

Stchigel, A. M., Cano J., Guarro, J., and Gugnani, H. C. 2000. A new *Apiosordaria* from Nigeria, with a key to the soil borne species. *Mycologia* 92:1206–1209.

Stolk, A. C. 1963. The genus *Chaetomella* Fuckel. *Trans. Brit. Mycol. Soc.* 46:409–425.

Strong, F. C. and Strong, M. C. 1931. Investigations on the black root of strawberries. *Phytopathology* 21:1041–1060.

Subramanian, C. V. 1956. Hyphomycetes I. *J. Ind. Bot. Sci.* 35:53–91.

Subramanian, C. V. 1956. *Phaeotrichoconis*, a new genus of the Dematiaceae. *Proc. Ind. Acad. Sci. Sect. B.*, 44:1–2.

Subramanian, C.V. 1963. *Dactylella, Monacrosporium* and *Dactylina. J. Ind. Bot. Soc.* 42:291–300.

Sugiyama, J. 1968. Mycoflora in core samples from stratigraphic drillings in Middle Japan. III. The taxonomic status of the genus *Chalaropsis* Peyronel (Hyphomycetes). *J. Fac. Sci. Univ. Tokyo, III* 10:29–48.

Sugiyama, J. 1969. Studies on Himalayan yeasts and moulds. II. *Mammaria echinobotryoides* (Hyphomycetes) and its allies. *Trans. Mycol. Soc. Jpn.* 9:117–124.

Sutton, B. C. 1962. *Colletotrichum dematium* (Pers. ex Fr.) Grove and *C. trichellum* (Fr. ex Fr.) Duke. *Trans. Br. Mycol. Soc.* 45:222–232.

Sutton, B. C. 1964. *Phoma* and related genera. *Trans. Br. Mycol. Soc.* 47:497–509.

Sutton, B. C. 1969. Forest microfungi. III. The hetrogenicity of *Pestalotia* deNot., section Sexloculatae Klebahn sensu Guba. *Can. J. Bot.* 47:2083–2094.

Sutton, B. C. 1973. *Pucciniopsis, Mycoleptodiscus* and *Amerodiscosiella. Trans. Br. Mycol. Soc.* 60:525–536.

Sutton, B. C. 1980. The Coelomycetes. *Commonw. Mycol. Inst.,* Kew, 696 pp.

Sutton, B. C. 1981. *Sarcopodium* and its synonyms. *Trans. Br. Mycol. Soc.* 76:97–102.

Sutton, B. C. and Alcorn, J. L. 1990. New species of *Mycoleptodiscus* (Hyphomycetes). *Mycol. Res.* 94:564–566.

Sutton, B. C., Pirozynski, K. A., and Deighton, F. C. 1972. *Micodochium* Syd. *Can. J. Bot.* 50:1899–1907.

Sutton, B. C. and Sarbhoy, A. K. 1976. Revision of *Chaetomella*, and comments upon *Vermiculariopsis* and *Thyriochaetum. Trans. Brit. Mycol. Soc.* 66:297–303.

Takada, M. and Udagawa, S. 1970. Materials for the fungus flora of Japan (9). *Trans. Mycol. Soc. Jpn.* 11:53–56.

Takahashi, M. 1954. On the morphology and taxonomy of some species of the genus *Pythium* which causes crop diseases. *Ann. Phytopathol. Soc. Jpn.* 18:113–118.

Takahashi, T. 1919. Fungi in soil. *Ann. Phytopathol. Soc. Jpn.* 1:17–22, in Japanese.

Terashita, T. 1968. A new species of *Calonectria* and its conidial state. *Trans. Mycol. Soc. Jpn.* 8:124–129.

Terashita, T. 1969. *Calonectria hederae* and *Cylindrocladium camelliae* in Japan. *Trans. Mycol. Soc. Jpn.* 9:113–116.

Thaxter, R. 1891. On certain new or peculiar North American Hyphomycetes. I. *Oedocephalum, Rhopalomyces* and *Sigmoideomyces* n. g. *Bot. Gaz.* 16:14–26.

Thaxter, R. 1903. New or peculiar North American hyphomycetes. III. *Bot. Gaz.* 35:153–159.

Thirumalachar, M. J., Whitehead, M. D., and Mathur, P. N. 1964. A new genus of Eurotiaceae from soil. *Mycologia* 56:809–815.

Thornton, R. H. 1956. Fungi occurring in mixed oakwood and heath soil profiles. *Trans. Br. Mycol. Soc.* 39:485–494.

Tiffany, L. H. and Gilman, J. C. 1954. Species of *Colletotrichum* from legumes. *Mycologia* 46:52–75.

Timms, A. M. and Hepden, P. M. 1965. British records, 89. *Dactylaria purpurella* (Sacc.) Sacc. *Trans. Br. Mycol. Soc.* 48:666–667.

Timonin, M. I. 1940. The interaction of higher plants and soil micro-organisms. 1. Microbial population of rhizosphere of seedlings of certain cultivated plants. *Can. J. Res. C.* 18:307–317.

Tresner, H. D., Backus, M. P., and Curtis, J. T. 1954. Soil microfungi in relation to the hardwood forest continuum in Southern Wisconsin. *Mycologia* 46:314–333.

Tsao, P. H. 1970. Selective media for isolation of pathogenic fungi. *Annu. Rev. Phytopathol.* 8:157–186.

Tsao, P. H. and Guy, S. O. 1977. Inhibition of *Mortierella* and *Pythium* in a *Phytophthora*-isolation medium containing Hymexazol. *Phytopathology* 67:796–801.

Tsao, P. H. and Ocana, G. 1969. Selective isolation of species of *Phytophthora* from natural soils on an improved antibiotic medium. *Nature* 223:636–638.

Tsuda, M. and Ueyama, A. 1981. *Pseudocochliobolus australiensis*, the ascigerous state of *Biporalis australiensis. Mycologia* 73:88–96.

Tubaki, K. 1956. *Cephaliophora irregularis* newly found in Japan. *J. Jpn. Bot.* 31:161–164.

Tubaki, K. 1963. Notes on the Japanese Hyphomycetes. 1. *Chloridium, Clonostachys, Isthmosperma, Pseudobotrys, Stachybotrys* and *Stephanoma. Trans. Mycol. Soc. Jpn.* 4:83–90.

Tubaki, K. and Ito, T. 1975. Descriptive catalogue of IFO fungus collection IV. 36. *Bispora betulina* (Corda) Hughes. *IFO Res. Comm.* 7:113.

Tubaki, K. and Yokoyama, T. 1971. Notes on the Japanese Hyphomycetes V. *Trans. Mycol. Soc. Jpn.* 12:18–28.

Tucker, C. M. 1931. Taxonomy of the genus *Phytophthora* de Bary. *Mo. Agric. Exp. Stn. Res. Bull.* 153:1–208.

Tulloch, M. 1972. The genus *Myrothecium* Tode ex Fr. *Mycol. Pap.* 130:1–42.

Tulloch, M. 1976. The genus *Metarhizium. Trans. Br. Mycol. Soc.* 66:407–411.

Turner, M. 1963. Studies in the genus *Mortierella*. I. *Mortierella isabellina* and related species. *Trans. Br. Mycol. Soc.* 46:262–272.

Turner, M. and Pugh, G. J. F. 1961. Species of *Mortierella* from a salt marsh. *Trans. Br. Mycol. Soc.* 44:243–252.

Tzean, S. S., Chiu, S. C., Chen, J. L., Hseu, S. H., Lin, G. H., Lion, G. Y., Chen, C. C., and Hsu, W. H. 1994. *Penicillium* and related teleomorphs from Taiwan. CCRC Mycol. Monograph No. 9. Food Industry and Development Institute, Hsinchu, Taiwan, R.O.C. 159 pp.

Udagawa, S. 1960. A taxonomic study on the Japanese species of *Chaetomium. J. Gen. Appl. Microbiol.* 6:223–251.

References

Udagawa, S. 1963. Microascaceae in Japan. *J. Gen. Appl. Microbiol.* 9:137–148.

Udagawa, S. 1965. Notes on some Japanese Ascomycetes II. *Trans. Mycol. Soc. Jpn.* 6:78–90.

Udagawa, S. and Furuya, K. 1972a. Notes on some Japanese Ascomycetes X. *Trans. Mycol. Soc. Jpn.* 13:49–56.

Udagawa, S. and Furuya, K. 1972b. *Zopfiella pilifera*, a new cleistoascomycete from Japanese soil. *Trans. Mycol. Soc. Jpn.* 13:255–259.

Udagawa, S. and Horie, Y. 1974. Notes on some Ascomycetes XII. *Trans. Mycol. Soc. Jpn.* 15:105–112.

Udagawa, S. and Takada, M. 1971. 10. Soil and coprophilous microfungi. *Bull. Nat. Sci. Mus. Tokyo* 14:501–515.

Udagawa, S., et al. 1978. *Illustrated Fungi.* Kodansha, Tokyo, Vols. 1 and 2, 1321 pp., in Japanese.

Vaartaja, O. 1965. New *Pythium* species from South Australia. *Mycologia* 57:417–430.

Van Emden, J. H. 1968. *Penicillifer*, a new genus of Hyphomycetes from soil. *Acta Bot. Neerl.* 17:54–58.

Vasant, R. and de Hoog, G. S. 1986. New or critical hyphomycetes from India. *Stud. Mycol.* 28:1–84.

Varghese, G. 1972. Soil microflora of plantations and natural rain forest of West Malaysia. *Mycopathol. Mycol. Appl.* 48:43–61.

Vries, G. A., de. 1952. Contribution to the knowledge of the genus *Cladosporium* Link ex Fr. Centraalbureau voor Schimmelcultures, Baarn. 121 pp.

Waksman, S. A. 1916. Soil fungi and their activities. *Soil Sci.* 2:103–155, pl. 1–5.

Waksman, S. A. 1917. Is there any soil fungus flora in the soil? *Soil Sci.* 3:565–589.

Walker, J. C. and Minter, D. W. 1981. Taxonomy of *Nematogonum, Gonatobotrys, Gonatobotryum* and *Gonatorrhodiella. Trans. Br. Mycol. Soc.* 77:299–319.

Warcup, J. H. 1951a. The ecology of soil fungi. *Trans. Br. Mycol. Soc.* 34:376–399.

Warcup, J. H. 1951b. Effect of partial sterilization by steam or formalin on the fungus flora of an old forest nursery soil. *Trans. Br. Mycol. Soc.* 34:519–532.

Warcup, J. H. 1957. Studies on the occurrence and activity of fungi in a wheat-field soil. *Trans. Br. Mycol. Soc.* 40:237–262.

Warcup, J. H. 1959. Studies on Basidiomycetes in soil. *Trans. Br. Mycol. Soc.* 42:45–52.

Warcup, J. H. 1976. Studies on soil fumigation. IV. Effects on fungi. *Soil Biol. Biochem.* 8:261–266.

Warcup, J. H. and Talbot, P. H. B. 1962. Ecology and identity of mycelia isolated from soil. *Trans. Br. Mycol. Soc.* 45:495–518.

Warcup, J. H. and Talbot, P. H. B. 1963. Ecology and identity of mycelia isolated from soil. II. *Trans. Brit. Mycol. Soc.* 46:465–472.

Ware, W. M. 1933. A disease of cultivated mushrooms caused by *Verticillium malthousei* sp. nov. *Ann. Bot.* 47:763–785.

Watanabe, T. 1971. Fungi isolated from the rhizosphere soils of wilted pineapple plants in Okinawa. *Trans. Mycol. Soc. Jpn.* 12:35–47.

Watanabe, T. 1972a. Pycnidium formation by fifty different isolates of *Macrophomina phaseoli* originated from soil or kidney bean seed. *Ann. Phytopathol. Soc. Jpn.* 38:106–110.

Watanabe, T. 1972b. Fungi associated with commercial kidney bean seed and their pathogenicity to young seedlings of kidney bean. *Ann. Phytopathol. Soc. Jpn.* 38:111–116.

Watanabe, T. 1974a. Fungi isolated from the underground parts of sugarcane in relation to the poor ratooning in Taiwan. (2) *Pythium* and *Pythiogeton. Trans. Mycol. Soc. Jpn.* 15:343–357.

Watanabe, T. 1974b. Flora of fungi isolated from cultivated plant roots and soils. *Soil Microorganisms* No. 15; 16:39–51, in Japanese.

Watanabe, T. 1975a. Fungi isolated from the underground parts of sugarcane in relation to the poor ratooning in Taiwan, (3) Mucorales. *Trans. Mycol. Soc. Jpn.* 16:18–27.

Watanabe, T. 1975b. Fungi isolated from the underground parts of sugarcane in relation to the poor ratooning in Taiwan, (4) Coelomycetes. *Trans. Mycol. Soc. Jpn.* 16:28–35.

Watanabe, T. 1975c. Fungi isolated from the underground parts of sugarcane in relation to the poor ratooning in Taiwan, (5) Hyphomycetes. *Trans. Mycol. Soc. Jpn.* 16:149–182.

Watanabe, T. 1975d. Fungi isolated from the underground parts of sugarcane in relation to the poor ratooning in Taiwan, (6) *Papulaspora. Trans. Mycol. Soc. Jpn.* 16:264–267.

Watanabe, T. 1975e. *Tetracladium setigerum*, an aquatic hyphomycete associated with gentian and strawberry roots. *Trans. Mycol. Soc. Jpn.* 16:348–350.

Watanabe, T. 1977a. Pathogenicity of *Pythium myriotylum* isolated from strawberry roots in Japan. *Ann. Phytopathol. Soc. Jpn.* 43:306–309.

Watanabe, T. 1977b. A new species of *Umbelopsis* from strawberry roots. *Trans. Mycol. Soc. Jpn.* 18:242–244.

Watanabe, T. 1977c. Fungi associated with strawberry roots in Japan. *Trans. Mycol. Soc. Jpn.* 18:251–256.

Watanabe, T. 1979. *Monosporascus cannonballus*, an ascomycete from wilted melon roots undescribed in Japan. *Trans. Mycol. Soc. Jpn.* 20:312–316.

Watanabe, T. 1980. A new variety of *Verticillium sphaerosporum*, an endoparasite of nematode and its antagonism to soil borne plant pathogens. *Ann. Phytopathol. Soc. Jpn.* 46:598–606.

Watanabe, T. 1981. Detection of *Pythium deliense* in the Ryukyu Islands and its ecological implication. *Ann. Phytopathol. Soc. Jpn.* 47:562–565.

Watanabe, T. 1983a. Distribution of *Pythium aphanidermatum* in Japan: its significance. *Trans. Mycol. Soc. Jpn.* 24:15–23.

Watanabe, T. 1983b. Formation and deciduousness of sporangia of *Pythium intermedium. Trans. Mycol. Soc. Jpn.* 24:25–33.

Watanabe, T. 1984. Identification and taxonomical problems of *Pythium* species in Japan. Shokubutsu Boeki 38:203–211, in Japanese.

Watanabe, T. 1985. *Pythium* species found in the piedomont natural forest at Mt. Fuji. *Trans. Mycol. Soc. Jpn.* 26:41–45.

Watanabe, T. 1986. Rhizomorph production in *Armillaria mellea in vitro* stimulated by *Macrophoma* sp. and several other fungi. *Trans. Mycol. Soc. Jpn.* 27:235–245.

Watanabe, T. 1987. *Plectospira myriandra*, a rediscovered water mold in Japanese soil. *Mycologia* 79:77–81.

Watanabe, T. 1988a. Pathogenic fungi associated with forest seeds including *Pythium* species from cherry seeds. *Trans. Mycol. Soc. Jpn.* 29:197–203.

Watanabe, T. 1988b. Kinds and distribution of *Pythium* species isolated from soils in Shikoku Island. *Ann. Phytopathol. Soc. Jpn.* 54:523–528.

Watanabe, T. 1988c. *Pythium carolinianum* and *P. periplocum* associated with cherry seeds at the cherry tree preservation forest at Asakawa. Int. *Pythium* group, Abstr., 49–50.

Watanabe, T. 1988d. Pathogenic fungi associated with forest seeds including *Pythium* species from cherry seeds. *Trans. Mycol. Soc. Jpn.* 29:197–203.

Watanabe, T. 1989a. Kinds, distribution, and pathogenicity of *Pythium* species isolated from soils of Kyushu Island in Japan. *Ann. Phytopathol. Soc. Jpn.* 55:32–40.

Watanabe, T. 1989b. Further study on *Pythium* species isolated from soils in the Ryukyu Islands. *Ann. Phytopathol. Soc. Jpn.* 55:349–352.

Watanabe, T. 1989c. Kinds and distribution of *Pythium* species isolated from soils in Tohoku district. *Ann. Phytopathol. Soc. Jpn.* 56:88–91.

Watanabe, T. 1989d. Three species of *Sordaria*, and *Eudarluca biconica* from cherry seeds. *Trans. Mycol. Soc. Jpn.* 30:395–400.

Watanabe, T. 1989e. Soil fungal flora in Hachijo-jima island. *Trans. Mycol. Soc. Jpn.* 30:427–435.

Watanabe, T. 1990a. Zygospore induction in *Mortierella chlamydospora* by the soaking-plain-water-agar-culture method. *Mycologia* 82:278–282.

Watanabe, T. 1990b. Kinds and distribution of *Pythium* species isolated from soils in Hokkaido Island. *Ann. Phytopathol. Soc. Jpn.* 56:549–556.

Watanabe, T. 1990c. Three new *Nectria* from Japan. *Trans. Mycol. Soc. Jpn.* 31:227–236.

Watanabe, T. 1990d. Three noteworthy *Pythium* species from Japan. *Jpn. Mycol. Soc., Kanto Div.* p. 4, Abstr., in Japanese.

Watanabe, T. 1991. New species of *Oedocephalum* and *Papulaspora* from Japanese soils. *Mycologia* 83:524–529.

Watanabe, T. 1992a. *Hyphodiscosia radicicola* sp. nov. from Japan. *Mycologia* 84:113–116.

Watanabe, T. 1992b. Sporulation of *Dematophora necatrix in vitro*, and its pathogenicity. *Ann. Phytopathol. Soc. Jpn.* 58:65–71.

Watanabe, T. 1992c. A new species of *Pyrenochaeta* from Japanese black pine seeds. *Trans. Mycol. Soc. Jpn.* 33:21–24.

Watanabe, T. 1992d. Kinds and distribution of *Pythium* species isolated from soils in South Kinki district. *Ann. Phytopathol. Soc. Jpn.* 58:360–365.

Watanabe, T. 1992e. *Taeniolella phialosperma* sp. nov. from Japan. *Mycologia* 84:478–483.

Watanabe, T. 1992f. *Cylindrocarpon olidum, Cylindrocladium parvum* and *Trichocladium pyriformis* not or rarely reported in Japan. *Trans. Mycol. Soc. Jpn.* 33:231–236.

Watanabe, T. 1992g. *Trinacrium iridis* sp. nov. from iris roots in Japan. *Mycologia* 84:794–798.

Watanabe, T. 1993a. *Camposporium laundonii* on root tissue of *Aralia elata* from Japan. *Trans. Mycol. Soc. Jpn.* 34:71–76.

Watanabe, T. 1993b. *Sarcopodium araliae* sp. nov. on root of *Aralia elata* from Japan. *Mycologia* 85:520–526.

Watanabe, T. 1994a. *Cylindrocladium tenue* comb. nov. and two other *Cylindrocladium* species isolated from *Phellodendron amurense* in Japan. *Mycologia* 86:151–156.

Watanabe, T. 1994b. Two new species of homothallic *Mucor* in Japan. *Mycologia* 86:691–695.

Watanabe, T. 1995. *Naranus cryptomeriae* gen. et sp. nov. from Japanese cedar seed. *Mycol. Res.* 99:806–808.

Watanabe, T. 1996. *Sporidesmium filiferum* from Tsukuba, Japan. *Mycoscience* 37:367–369.

Watanabe, T., Hagiwara, S., and Narita, I. 1995. Decline of *Phellodendron amurense* in Tokyo: associated fungi and pathogenicity of associated *Cylindrocladium* spp. *Plant Dis.* 79:1161–1164.

Watanabe, T. and Hashimoto, K. 1978. Recovery of *Gloeocercospora sorghi* from sorghum seed and soil, and its significance in transmission. *Ann. Phytopathol. Soc. Jpn.* 44:633–640.

Watanabe, T., Hashimoto, K., and Sato, M. 1977. *Pythium* species associated with strawberry roots in Japan, and their role in the strawberry stunt disease. *Phytopathology* 67:1324–1332.

Watanabe, T. and Imamura, S. 1977. Brown root rot of gentian and the causal fungi, *Pyrenochaeta* spp. *Ann. Phytopathol. Soc. Jpn.* 43:343. Abstr., in Japanese.

Watanabe, T. and Imamura, S. 1995. Pink root rot, a revised name of brown root rot of gentian and the causal fungi, *Pyrenochaeta gentianicola* sp. nov. and *P. terrestris* in Japan. *Mycoscience* 36:439–445.

Watanabe, T. and Koizumi, S. 1976. Materials for the fungus flora of Japan (20). *Trans. Mycol. Soc. Jpn.* 17:1–3.

Watanabe, T., Moya, J. D., González, J. L., and Matsuda, A. 1996. Fungi associated with roots and fruits of black peppers in the Dominican Republic. *Mycoscience* 37:471–475.

Watanabe, T., Moya, J. D., González, J. L., and Matsuda, A. 1997. *Mycoleptodiscus terrestris* from black pepper roots in the Dominican Republic. *Mycoscience* 38:91–94.

Watanabe, T., Nagai, Y., and Fukami, M. 1986b. Brown-blotted root rot of carrots in Japan. (2) Culture and identification. *Ann. Phytopathol. Soc. Jpn.* 52:287–291.

Watanabe, T., Ozawa, M., and Sakai, R. 1973. A new disease of chinese cabbage caused by *Verticillium albo-atrum* and some factors related to the incidence of the disease. *Ann. Phytopathol. Soc. Jpn.* 39:344–350.

Watanabe, T. and Sato, M. 1995. Root rot of melon caused by *Nodulisporium melonis* in Japan. 2. Identification. *Ann. Phytopathol. Soc. Jpn.* 61:330–333.

Watanabe, T. and Sato, Y. 1988. Fungi isolated from Japanese cedar and cypress seeds in Japan. *Trans. Mycol. Soc. Jpn.* 29:143–150.

Watanabe, T., Shinoda, R. N. Bareiro de, and Ramirez, M. E. 1987a. Fungi isolated from declining paulownia trees in Paraguay and Argentina. *Trans. Mycol. Soc. Jpn.* 28:453–469.

Watanabe, T., Sieber, T. N., and Holdenrieder, O. 1998. *Pythium* in the Swiss Alps. *Mycologia Helvetica* 10:3–13.

Watanabe, T. and Tsudome, K. 1970. Fungi isolated from wilted pineapple plants in Okinawa. *Trans. Mycol. Soc. Jpn.* 11:64–71.

Watanabe, T., Tzean, S. S., and Leu, L. S. 1974. Fungi isolated from the underground parts of sugarcane in relation to the poor ratooning in Taiwan. *Trans. Mycol. Soc. Jpn.* 15:30–41.

Watanabe, T., Uematsu, S., and Sato, Y. 1986a. Fungus isolates from Japanese black and red pine seeds with some taxonomcal notes. *Bull. For. For. Prod. Res. Inst.* 336:1–18.

Watanabe, T., Uematsu, S., and Hayashi, K. 1987b. Fungal isolates from seeds of two cherry species collected at the cherry tree preservation forest at Asakawa. *Trans. Mycol. Soc. Jpn.* 28:475–481.

Watanabe, T., Uematsu, S., and Inoue, Y. 1988. Pathogenicity of twenty-three *Pythium* isolates from soils of Shikoku Island. *Trans. Mycol. Soc. Jpn.* 54:565–570.

Watanabe, T., Watanabe, Y., and Fukatsu, T. 2001b. New species of *Acremonium, Cylindrocarpon* and *Verticillium* from soil in the Bonin (Ogasawara) Islands, Japan. Mycoscience 42, 591–595.

Watanabe, T., Watanabe, Y., and Fukatsu, T. 2001c. *Dactylella chichisimensis* sp. nov. in the Bonin (Ogasawara) Islands, Japan. *Mycoscience* 42, 601–603.

Watanabe, T., Watanabe, Y., and Fukatsu, T. 2001a. Soil fungus flora in the Bonin (Ogasawara) Islands, Japan. *Mycoscience* 42, 503–506.

Watanabe, T., Watanabe, Y., and Fukatsu, T. 2001. Three new *Mortierella* from soil in the Bonin (Ogasawara) Islands, Japan, in preparation.

Watanabe, T., Watanabe, Y., and Fukatsu, T. 2001. *Myrothecium dimorphum* sp. nov. in the Bonin (Ogasawara) Islands, Japan, in preparation.

Watanabe, T., Watanabe, Y., Fukatsu, T., and Kurane, R. 2001. *Mortierella tsukubaensis* sp. Nov. from Japan, with a key to the homothallic species. *Mycol. Res.* 105:506–509.

Waterhouse, G. M. 1963. Key to the species of *Phytophthora* de Bary. *Mycol. Pap.* 92:1–22.

Waterhouse, G. M. 1970. The genus *Phytophthora* de Bary. *Mycol. Pap.* 122:1–59.

Webster, I. 1980. *Introduction to Fungi.* 2nd. ed. Cambridge University Press, Cambridge, MA, 669 pp.

Webster, J. and Lomas, N. 1964. Does *Trichoderma viride* produce gliotoxin and viridin? *Trans. Brit. Mycol. Soc.* 47:535–540.

Webster, J., Rifai, M. A., and El-Abyad, M. S. 1964. Culture observations on some Discomycetes from burnt ground. *Trans. Br. Mycol. Soc.* 47:445–454.

Wehmeyer, L. E. 1946. Studies on some fungi from northwestern Wyoming. II. Fungi Imperfecti. *Mycologia* 38:306–330.

Wei, C. T. 1950. Notes on *Corynespora. Mycol. Pap.*, 34:1–10.

Weindling, R. 1932. *Trichoderma lignorum* as a parasite of other soil fungi. *Phytopathology* 22:837–845.

Weresub, L. K. and LeClair, P. M. 1971. On *Papulaspora* and bulbilliferous basidiomycetes *Burgoa* and *Minimedusa. Can. J. Bot.* 49:2203–2213.

White, J. F., Jr. and Morgan-Jones, G. 1987. Studies in the genus *Phoma.* VII. Concerning *Phoma glomerata. Mycotoxon* 28:437–445.

White, W. L. and Downing, M. H. 1953. *Humicola grisea*, a soil-inhabiting cellulolytic Hyphomycete. *Mycologia* 45:951–963.

Widden, P. and Parkinson, D. 1973. Fungi from Canadian coniferous forest soils. *Can. J. Bot.* 51:2275–2290.

Widden, P. and Parkinson, D. 1979. Populations of fungi in a high arctic ecosystem. *Can. J. Bot.* 57:2408–2417.

Wiltshire, S. P. 1938. The original and modern conceptions of *Stemphylium. Trans. Br. Mycol. Soc.* 21:211–239.

Wolf, F. A. 1949. Two unusual conidial fungi. *Mycologia* 41:561–564.

Yang, B.-Y. and Liu, C.-H. 1972. Preliminary studies on Taiwan Mucorales (1). *Taiwania* 17:293–303.

Yokoyama, T., Asano, I., and Ito, T. 1979. Notes on the filamentous fungi isolated from forest soils in Alaska. *IFO Res. Comm.* 9:46–61.

Zambetakkis, C. 1954. Recherches sur la systematique des Sphaeropsidales. Phaeodidymae. *Bull. Trimest. Soc. Mycol. Fr.* 70:219–350.

Zentmyer, G. A. 1965. Bacterial stimulation of sporangium production in *Phytophthora cinnamomi. Science* 150:1178–1179.

Zycha, H., Siepmann, R., and Linnemann, G. 1969. *Mucorales.* Verlag von J. Cramer. Lehre, 355 pp.

Appendix

List of Living Cultures of Soil Fungi Deposited and Publicized

Fungus	This text (TW)	ATCC	MAFF	Misc.
Acremonium macroclavatum	00-50		238162	
Arthrinium sp.	76-584		425549	
Aureobasidium pullulans	83-50		425042	
Aureobasidium pullulans	83-67		425043	
Aureobasidium pullulans	84-473		425045	
Aureobasidium pullulans	85-6		425047	
Aureobasidium sp.	75-105		425551	
Botryotrichum piluliferum	98-16		237994	
Botryodiplodia	83-64		425049	
Camposporium laudenii	92-51	201310	425341	IFO 32523
Candida sp.	75-4		425552	
Chaetomium cochliodes	70-1026		425212	
Chaetomium cochliodes	70-1836		425215	
Chaetomium funicola	84-484		425066	
Chaetomium funicola	84-573		425058	
Chaetomium fusiforme	98-23		237996	
Chaetomium globosum	84-485		425068	
Chaetomium globosum	84-491		425069	
Chaetomium globosum	86-34		425071	
Chaetomium globosum	70-1385		425217	
Chlomerosporium fulvum	76-573		425572	
Cladorrhinum bulbilosum	72-X10			CBS 267.76
Cladorrhinum bulbilosum	85-20		425074	
Cladorrhinum samala	70-1175		425227	
Cladorrhinum samala	72-X155-1010		425227	CBS 266.76
Cladosporium cladosporioides	83-46		425076	
Cladosporium cladosporioides	85-120		425077	
Cladosporium cladosporioides	70-1035		425230	
Codinaea talbotii	72-X123	32909		
Coprinus sp.	99-1	347998		
Cunninghamella echinulata	72-11-2	32321		
Curvularia affinis	76-545		425573	
Curvularia inaequales	98-102		237999	
Curvularia pallescens	98-101		238000	
Curvularia pallescens	72-X 13-174	32910		
Curvularia protuberata	77-25			IFO 32536
Curvularia senegalensis	72-X12-115	32911		
Cylindrocarpon boninense	00-62		238163	
Cylindrocarpon destructans	92-210		425581	
Cylindrocarpon destructans	98-25		238001	
Cylindrocarpon janthothele	98-14		238002	
Cylindrocarpon olidum	77-96			IFO 32524
Cylindrocarpon olidum	77-127			IFO 32525

List of Living Cultures of Soil Fungi Deposited and Publicized (continued)

Fungus	This text (TW)	ATCC	MAFF	Misc.
Cylindrocarpon olidum	77-149			IFO 32526
Cylindrocarpon olidum	77-185			IFO 32527
Cylindrocladium camelliae	92-202	201118	425363	IFO 32528
Cylindrocladium colhounii	90-211	201116	425360	IFO 32529
Cylindrocladium colhounii	90-260			IFO 32530
Cylindrocladium floridanum	73-231	201117	425364	IFO 32532
Cylindrocladium floridanum	76-77			IFO 32531
Cylindrocladium parvum	87-150	201119	425305	IFO 32534
Cylindrocladium scoparium	92-118			IFO 32535
Cylindrocladium tenue	92-246	201311	425366	ref.: *C. meguroense*, IFO 32533
Cylindrocladium tenue	T 16	200586		
Cytospora sacchari	72-X153-814	32322		
Dactylaria candidula	98-61		238003	
Dactylaria naviculiformis	98-63		238004	
Dactylella chichisimensis	00-315		238165	
Dematophora necatrix	84-373		425314	IFO 32537, ref.: *Rosellinia* necatrix
Dematophora necatrix	84-374		425317	ref.: *Rosellinia* necatrix
Diplodia frumenti	98-105		238005	
Eudarluca biconica	85-100			IFO 32539
Fusarium moniliforme	92-267		425585	
Fusarium roseum	85-30		425094	
Fusarium roseum	70-1176		425242	
Fusarium solani	85-33		425096	
Fusarium solani	70-1239		425245	
Geotrichum sp.	85-39		425098	
Gliocladium catenulatum	72-X92-606	32913		ref.: *G. fimbriatum*
Gliocladium catenulatum	92-227		425587	
Gliocladium deliquesens	31-177	32912		
Gliocladium roseum	85-544		425099	
Gliocladium roseum	84-545		425100	
Gliocladium virens	76-533		425559	
Gongronella butleri	72-X98	32323	425195	
Gongronella butleri	99-453		238014	
Gonytrichum chlamydosporium	72-X61-353	32914		
Gonytrichum chlamydosporium	98-64		238006	
Helicomyces roseus	93-408	201115	425588	
Helicomyces roseus	98-105		238007	
Humicola fuscoatra	72-X68-1	32915		
Humicola fuscoatra	72-X72-1, 341	32916		
Humicola grisea	92-209		425589	
Humicola tainanensis	72-X58	32917		CBS 269. 76 (ref.: *Microdochium tainanense*)
Hyphodiscosia radicicola	77-94	200611	425311	IFO 32540
Monacrosporium bembicodes	98-41		238009	
Mortierella alpina	93-318	200771	425590	
Mortierella alpina	98-67		238010	
Mortierella ambigua	72-X46-40	32324		
Mortierella boninense	01-125		238290	
Mortierella chlamydospora	77-34			IFO 32541

List of Living Cultures of Soil Fungi Deposited and Publicized (continued)

Fungus	This text (TW)	ATCC	MAFF	Misc.
Mortierella chlamydospora	77-647			IFO 32542
Mortierella chlamydospora	80-398			IFO 32543
Mortierella chlamydospora	81-273			IFO 32544
Mortierella elongata	72-X129-854	32325		
Mortierella elongata	92-228		425591	
Mortierella epicladia var. macrochlamydospora	00-21		238166	
Mortierella gemmifera	01-199		238292	
Mortierella isabellina	98-108		238011	
Mortierella tsukubaensis	98-120	204319	237778	
Mortierella uniramosa	00-33		238167	
Mortierella zychae	72-X91-555	32326		
Mucor hachijoensis	70-1179	201000	425203	
Mycoleptodiscus terrestris	SP 22	200587		
Myrothecium cinctum	72-X136-815	32918		ref.: *M. striatisporum*
Myrothecium dimorphum	01-250		238296	
Naranus cryptomeriae	83-83	201312	425384	
Nectria hachijoensis	70-1336			IFO 32545
Nigrospora oryzae	85-45		425111	
Nigrospora oryzae	85-42		425113	
Nigrospora sacchari	84-509		425115	
Nigrospora sacchari	70-1841		425259	ref.: *N. sphaerica*
Nodulisporium melonis	75-355	200606	425439	
Nodulisporium melonis	75-356	200607	425440	
Nodulisporium melonis	95-1	200610	425594	
Ochroconis constricta	86-53		425117	
Ochroconis humicola	85-69		425118	
Ochroconis humicola	85-70		425119	
Oedocephalum nayoroense	81-498	201315	425312	IFO 32546
Oidiodendron cerealis	84-506		425120	
Oidiodendron cerealis	84-507		425121	
Oidiodendron flavum	84-475		425123	
Oidiodendron flavum	84-610		425127	
Oidiodendron flavum	85-117		425128	
Paecilomyces inflatus	72-X 30-142	32919		
Paecilomyces persicinus	72-X 101-489	32920		
Paecilomyces puntoni	70-1870		425262	
Paecilomyces variabilis	72-X 95-540	32921		
Papulaspora irregularis	72-X121-1160	34345		
Papulaspora nishigaharanus	73-1071	200772	425313	IFO 32547
Papulaspora pallidula	72-X 44	34346		
Papulaspora pannosa	72-X 6-88	34347		
Periconia byssoides	86-40		425133	
Periconia macrospinosa	X 39-320	32922		
Periconia macrospinosa	76-509		425576	
Pestalotia sp.	83-1		425134	
Pestalotia sp.	83-3		425136	
Pestalotia sp.	84-512		425139	
Pestalotia sp.	84-513		425140	

List of Living Cultures of Soil Fungi Deposited and Publicized (continued)

Fungus	This text (TW)	ATCC	MAFF	Misc.
Pestalotia sp.	70-1752		425269	ref.: *Pestalotiopsis* sp.
Phialophora atrovirens	84-524		425141	
Phialophora atrovirens	70-1812		425270	
Phialophora cinerecens	86-47		425142	
Phialophora cyclaminis	72-X 142-988	32923		
Phialophora fastigiata	84-574		425143	
Phialophora fastigiata	84-604		425144	
Phialophora malorum	83-37		425145	
Phialophora radicicila	72-X146-1048	32924		
Phialophora richardsiae	72-X 148	32925		
Phialophora sp.	83-38		425146	
Phialophora sp.	70-1226		425271	
Phoma glomerata	83-17		425147	
Phomopsis sp.	83-9		425149	
Phomopsis sp.	84-517		425151	
Pithomyces chartarum	84-585		425152	
Pithomyces maydicus	84-516		425153	
Plectospira myriandra	84-209	64139	425332	IFO 32545
Pyrenochaeta gentianicola	76-501	200774	425519	
Pyrenochaeta gentianicola	75-40		425520	
Pyrenochaeta gentianicola	75-102		425523	
Pyrenochaeta gentianicola	76-589		425529	
Pyrenochaeta globosa	84-523	201314	425154	IFO 32549
Pyrenochaeta terrestris	76-502	200773	425532	IFO 32548
Pyrenochaeta terrestris	72-X 141-897	32327		
Pythium acanthophoron	81-339		425319	
Pythium afertile	74-663	36439		
Pythium angustatum	73-257	36485		
Pythium aphanidermatum	73-54	36431		
Pythium aphanidermatum	79-71	64140	305567	
Pythium aphanidermatum	80-80		305854	
Pythium aphanidermatum	80-265		305857	
Pythium aphanidermatum	80-305		305887	
Pythium aphanidermatum	84-340		305889	
Pythium aphanidermatum	73-54	36431		
Pythium apleroticum	74-841	36441	425515	
Pythium carolinianum	85-54	66260	425155	
Pythium carolinianum	13R-Py	66263		
Pythium carolinianum	73-5	36434	425398	
Pythium carolinianum	80-82		305855	
Pythium catenulatum	72-X99	38892		
Pythium conidiophorum	81-410		425320	
Pythium deliense	72-X108	38893		
Pythium deliense	79-76	64141	305568	
Pythium deliense	79-78	64142	305569	
Pythium deliense	K 6	200559		
Pythium dissotocum	81-708	64149	305576	
Pythium dissotocum	80-571		305858	
Pythium dissotocum	74-377		305891	

List of Living Cultures of Soil Fungi Deposited and Publicized (continued)

Fungus	This text (TW)	ATCC	MAFF	Misc.
Pythium echinulatum	73-29	38891		
Pythium echinulatum	74-822	36437	425516	
Pythium echinulatum	81-219	64153	425517	
Pythium elongatum	79-72	64152	305579	
Pythium elongatum	80-222		305859	
Pythium graminicola	80-214	64150	425413	
Pythium graminicola	80-302	64151	425416	
Pythium cf. *indigoferae*	90-104	200701		
Pythium cf. *indigoferae*	90-113	200702		
Pythium inflatum	72-X117	38894		
Pythium inflatum	73-51	36436		
Pythium inflatum	80-150		305863	
Pythium inflatum	81-359		425322	
Pythium intermedium	90-107	200704		
Pythium intermedium	90-128	200706		
Pythium intermedium	K 5	200569		
Pythium intermedium	73-483-2	36445	425512	
Pythium intermedium	81-395	64143	305570	
Pythium intermedium	80-269		305883	
Pythium irregulare	78-2452	64144	425426	
Pythium irregulare	81-30	64145	305572	
Pythium irregulare	80-243		305864	
Pythium irregulare	84-153		305894	
Pythium irregulare	70-1016		425209	
Pythium myriotylum	74-864-1	36440	305636	
Pythium oedochilum	73-165	36433	425441	
Pythium oedochilum	82-815		425442	
Pythium paroecandrum	74-827	36432		
Pythium paroecandrum	83-400		425513	
Pythium periplocum	85-53	66262	425156	
Pythium periplocum	84-375	66264	305908	
Pythium rostratum	90-124	200703		
Pythium rostratum	84-411		305896	
Pythium rostratum	81-300		425329	
Pythium salpingophorum	01-106		238294	
Pythium spinosum	73-160	36438	425452	
Pythium spinosum	79-468	64146	305573	
Pythium spinosum	80-203		305866	
Pythium spinosum	81-345		425331	
Pythium splendens	K 4	200560		
Pythium splendens	82-17	64147	305574	
Pythium splendens	73-192	36444	425468	
Pythium splendens	80-212		305867	
Pythium splendens	80-311		305868	
Pythium sulcatum	83-161	200997		
Pythium sulcatum	83-160	200998		
Pythium sylvaticum	74-866	36442		
Pythium sylvaticum	90-141	200621		
Pythium sylvaticum	90-142	200622		

List of Living Cultures of Soil Fungi Deposited and Publicized (continued)

Fungus	This text (TW)	ATCC	MAFF	Misc.
Pythium sylvaticum	90-143	200623		homothallic
Pythium sylvaticum	80-164		305869	
Pythium sylvaticum	80-620		305872	
Pythium sylvaticum	84-141		305899	
Pythium torulosum	81-327	64148	425474	
Pythium torulosum	73-256	36484	425471	
Pythium torulosum	90-105	200624		
Pythium torulosum	90-122	200638		
Pythium torulosum	80-149		305873	
Pythium ultimum	74-901	36443	425489	
Pythium ultimum	90-140	200776		
Pythium ultimum	80-186		305875	
Pythium vexans	90-121	200775		
Pythium vexans	80-274		305878	
Pythium vexans	84-152		305904	
Pythium vexans	84-207		305605	
Ramichloridium anceps	84-525		425157	
Ramichloridium sublatum	83-71		425158	
Rhizoctonia solani	74-333		237780	
Rhizoctonia solani	85-63		425159	
Rhizoctonia sp.	SP 24	200568		
Sarcopodium araliae	92-53	201313		
Sclerotium sp.	85-15		425161	
Scopulariopsis brevicaulis	84-577		425163	
Scopulariopsis canadensis	84-514		425164	
Selenophoma obtusa	72-X 47-309	32328		
Septonema chaetospira	76-556		425578	
Septonema sp.	86-60		425165	
Sordaria fimicola	85-84		425166	IFO 32550
Sordaria fimicola	85-85		425167	
Sordaria nodulifera	85-72		425168	IFO 32551
Sordaria tamaensis	85-89		425169	IFO 32552
Spegazzinia tessarthra	86-59		425170	
Sporidesmium bakeri	84-527		425171	
Sporidesmium filiferum	95-2	201120	425595	
Sporotrichum aureum	72-X 86-514	32926		ref.: *S. aurantiacum*
Stachybotrys bisbyi	72-X32-206			CBS 236.82 (ref.: *S. sacchari*)
Stagonospora subseriata	72-X 140-1131	32329		
Staphylotrichum coccosporum	70-1362		425267	
Staphylotrichum coccosporum	70-1698		425268	
Stemphylium botryosum	70-1174		425273	
Syncephalastrum racemosum	72-X 156-764	32330		
Syncephalastrum racemosum	84-529		425174	
Syncephalastrum racemosum	86-62		425175	
Tetracladium setigerum	75-137	34349		
Tetracladium setigerum	74-875	34350	425374	
Thielavia terricola	84-532		435177	
Thielavia terricola	86-64		425178	
Thielaviopsis adiposa	72-X 600-181	32927		

List of Living Cultures of Soil Fungi Deposited and Publicized (continued)

Fungus	This text (TW)	ATCC	MAFF	Misc.
Thielaviopsis paradoxa	72-X 49-378	32928		
Thysanophora penicillioides	84-531		425179	
Torula herbarum	86-66		425180	
Torulomyces lagena	86-73		425182	
Trichocladium pyriformis	85-125	200609	425304	IFO 32553
Trichoderma hamatum	85-107		425183	
Trichoderma harzianum	84-530		425176	
Trichoderma harzianum	85-74		425184	
Trinacrium irides	82-567	200608	425318	IFO 32554
Tripospermum sp.	84-533		425186	
Tritirachium sp.	83-16		425187	
Ulcoladium charrtarum	84-505		425188	
Umbelopsis nana	99-454		238015	
Umbelopsis nana	99-481		238018	
Umbelopsis vinacea	73-723	38089		CBS 236.82 (ref.: *U. multispora*)
Umbelopsis vinacea	99-463		238017	
Verticillium hahajimaense	00-65		238172	
Verticillium lecanii	84-546		425193	
Wiesneriomyces javanicus	00273		238168	
Zygorhynchus moelleri	72-X 11-78	32331		

Note: The fungi described in this text are mostly deposited as the living stock cultures with MAFF, ATCC, IFO, or CBS code number in addition to the list (TW) in this table. Consult the homepage for the MAFF list, at the e-mail address: http://www.gene.affrc.go.jp./micro/T list/index.html.

Index

A

Absidia butleri, See *Gongronella butleri*
Absidia repens, 94
Achaetomium sp., 134–135
Acremonium macroclavatum, 184
Acremonium obclavatum, 184
Acremonium sp., 185, 454
Aleuriosporae, 27–28
Alternaria alternata, 186
Alternaria sp., 376, 443
Anixiella reticulata, 136
Annelosporae, 32
Ansatospora acerina, See *Mycocentrospora acerina*
Aphanomyces cladogamus, 40
Apiocarpella sp., 187
Apiosordaria verruculosa var. *maritima*, 137
Apiospora montagnei, 188
Arthrinium st. (Teleomorph: *Apiospora montagnei*), 188
Arthrobotrys musiformis, See *Candelabrella musiformis*
Arthrobotrys oligospora, 189
Arthrosporae, 28
Ascomycetes, 23–24
Ascomycotina, 35, 134–176
Aspergillus brevipes, 191
Aspergillus fumigatus, 192
Aspergillus niger, 193
Aspergillus parasiticus, 194
Aspergillus sp.
    Sect. *Clavati*, 195
    Sect. *Wentii*, 196
Aureobasidium pullulans, 3, 197
Aureobasidium sp., 198

B

Bacteria
    ammonifying, 1
    description of, 1
    nitrifying, 1
Basidiomycetous fungi, 24, 178–179
Basidiomycotina, 35, 178–181
Basipetospora rubra (Teleomorph: *Monascus ruber*), 199
Beltrania rhombica, 200
Biporalis australiensis (Teleomorph: *Cochliobolus australiensis*), 202
Biporalis holmii, 203
Biporalis sacchari, 204
Bispora betulina, 205
Blastosporae, 28–30
Botryodiplodia sp., 206
Botryosphaeria festucae, 267
Botryosphaeria rhodina, 206
Botryotrichum piluliferum, 207, 418

C

Calcalisporium sp., 326
Calonectria colhouni, 252
Calonectria kyotoensis, 253, 256
Camposporium laundonii, 208–209
Candelabrella musiformis, 210, 211
Candelabrella sp., 211
Candida sp., 212
Centrospora acerina, See *Mycocentrospora acerina*
Cephaliophora irregularis, 213
Cephaliophora tropica, 214
Cephalosporium lecanii, See *Verticillium lecanii*
Cephalosporium sp., 185
Ceratobasidium sp., 389
Ceratocystis adiposa, See *Thielaviopsis adiposa*
Chaetomella sp., 215
Chaetomium africanum, See *Chaetomium funicola*
Chaetomium aureum, 139, 151
Chaetomium brasiliense, 140
Chaetomium cochliodes, 141, 146, 149
Chaetomium cupreum, See *Chaetomium aureum*
Chaetomium dolichotrichum, 142
Chaetomium erectum, 143, 148
Chaetomium funicola, 142, 144
Chaetomium fusiforme, 145
Chaetomium globosum, 141, 146
Chaetomium homopilatum, 147, 150
Chaetomium reflexum, 143, 148
Chaetomium seminudum, 150
Chaetomium spirale, 149
Chaetomium torulosum, 150
Chaetomium virescens, 151
Chaetosphaeria talboti, 223, 224
Chalara paradoxa, See *Thielaviopsis paradoxa*
Chalara thielavioides, 216
Chalaropsis thielavioides, See *Chalara thielavioides*
Chloridium virens var. *chlamydosporum*, 217
Chromelosporium fulvum (Teleomorph: *Peziza ostracoderma*), 218
Chrysosporium keratinophilum, 219
Circinella muscae, 95
Cladorrhinum bulbillosum, 220, 221
Cladorrhinum samala, 221
Cladosporium cladosporioides, 222
Cladosporium sp., 441
Classification, 21–33
Cochliobolus australiensis, 202
Cochliobolus lunatus, 237

Cochliobolus pallescens, 238
Cochliobolus tuberculatus, 242
Codinea parva, 223
Codinea st. (Teleomorph: *Chaetosphaeria talboti*), 224
Colletotrichum coccodes, 226
Colletotrichum dematium, 229
Colletotrichum destructivum, 154
Colletotrichum falcatum (Teleomorph: *Glomerella tucumanensis*), 227
Colletotrichum lindemuthianum (Teleomorph: *Glomerella lindemuthiana*), 228
Colletotrichum truncatum, 229
Coniothyrium fuckelii (Teleomorph: *Diapleella coniothyrium*), 230
Coprinus sp., 180–181
Cordyceps memorabilis, 335
Corynespora cassiicola, 231, 232
Corynespora citricola, 232
Cunninghamella echinulata, 96, 97
Cunninghamella elegans, 96, 97
Curvularia affinis, 234
Curvularia brachyspora, 235
Curvularia clavata, 236
Curvularia lunata (Teleomorph: *Cochliobolus lunatus*), 237
Curvularia pallescens (Teleomorph: *Cochliobolus pallescens*), 238
Curvularia prasadii, 239
Curvularia protuberata, 240
Curvularia senegalensis, 241
Curvularia tuberculata (Teleomorph: *Cochliobolus tuberculatus*), 242
Cylindrocarpon boninense, 244
Cylindrocarpon destructans (Teleomorph: *Nectria radicicola*), 245
Cylindrocarpon janthothele, 246
Cylindrocarpon obtusisporum, 247
Cylindrocarpon olidum, 248
Cylindrocladium camelliae, 250–251
Cylindrocladium colhounii (Teleomorph: *Calonectria colhouni*), 252
Cylindrocladium floridanum (Teleomorph: *Calonectria kyotoensis*), 253
Cylindrocladium parvum, 254–255
Cylindrocladium pteridis, 254
Cylindrocladium scoparium, 253, 256
Cylindrocladium sp., 258
Cylindrocladium tenue, 257
Cytospora sacchari, 259–260

D

Dactylaria candidula, 261
Dactylaria naviculiformis, 262
Dactylaria sp., 323
Dactylella bemicodes, See *Monacrosporium bembicodes*
Dactylella chichisimensis, 263
Dactylella sp., 439
Dematophora necatrix (Teleomorph: *Rosellinia necatrix*), 264–265
Deuteromycetes fungi, 25
Deuteromycotina, 36–37, 184–456

Diapleella coniothyrium, 230
Dictyuchus sp., 41
Didymella effusa, 152
Didymostilbe eichleriana, 266
Didymostilbe sp., 266
Diplodia frumenti (Teleomorph: *Botryosphaeria festucae*), 267
Discohainesia oenotherae, 290
Drechslera australiensis, See *Biporalis australiensis*

E

Epicoccum nigrum, See *Epicoccum purpurescens*
Epicoccum purpurescens, 268
Eudarluca biconica, 153
Experimental materials and methods
    agar cultures, 16
    collection sites and samples, 13
    culture preservation, 15
    isolation method
        description of, 13–14
        fungi from soil, 14–15
        general principles, 14
    morphogenesis, 16
    sporulation, 16

F

Fumago sp., 269
Fungi
    airborne, 1
    historical studies, 1–2
    identification
        basal knowledge for, 17–18
        binomial selection, 18
        experimentation for, 18
        morphological characteristics for, 18–20
        physiological characteristics for, 20
    isolation methods for, 14–15
    microorganisms and, 3–4
    research problems, 4–5
    root-inhabiting, 5–6
    seed fungi vs., 6
    studies of, 7–12
    types of, 1–3
Fusarium ciliatum, 271
Fusarium moniliforme (Teleomorph: *Gibberella fujikuroi*), 272
Fusarium oxysporum, 273
Fusarium roseum, 274
Fusarium solani (Teleomorph: *Nectria haematococca*), 275
Fusarium ventricosum, 276–277

G

Gaeumannomyces graminis, 371
Gelasinospora reticulata, See *Anixiella reticulata*
Geotrichum candidum, 278
Gibberella fujikuroi, 272
Gliocladium catenulatum, 280

Index

Gliocladium deliquescens, See *Gliocladium viride*
Gliocladium fimbriatum, 280
Gliocladium penicilloides, 283
Gliocladium roseum (Teleomorph: *Nectria gliocladioides*), 163, 281
Gliocladium virens, 282
Gliocladium viride, 283–284
Gloeocercospora sorghi, 285
Glomerella glycines, 154
Glomerella lindemuthiana, 228
Glomerella tucumanensis, 227
Gonatobotrys simplex, 286
Gonatobotrys sp., 286
Gongronella butleri, 98
Gonytrichum chlamydosporium, 287–288
Gonytrichum macrocladum, 289

H

Hainesia lythri (Teleomorph: *Pezizella lythri*), 290
Hansfordia biophila, 291
Hansfordia sp., 326
Helicocephalum oligosporum, 99
Helicogoosia sp., 292
Helicomyces roseus, 292
Helminthosporium holmii, See *Biporalis holmii*
Helminthosporium sacchari, See *Biporalis sacchari*
Helminthosporium solani, 293
Heteroconium chaetospira, See *Septonema chaetospira*
Humicola dimorphospora, 295
Humicola fuscoatra, 296, 297
Humicola grisea, 296, 297
Humicola sp., 298
Humicola st., 175
Humicola tainanensis, 299
Hyalodendron sp., 300
Hyphodiscosia radicicola, 301–302
Hypoxylon sp., 326

K

Khuskia oryzae, 324

L

Leptodiscus terrestris, See *Mycoleptodiscus terrestris*
Leptosphaeria coniothyrium, 230

M

Macrophomina phaseoli, See *Macrophomina phaseolina*
Macrophomina phaseolina, 303–304
Mammaria echinobotryoides, 305
Mammaria sp., 305
Massarina sp. 1, 155
Massarina sp. 2, 156
Mastigomyectous fungi
 description of, 5
 genera, 35
 key to, 22
 morphology of, 40–92
Metarhizium anisopliae, 306
Microascus longirostris, 157
Microdochium tainanense, See *Humicola tainanensis*
Microsphaeropsis sp., 307
Mitosporic fungi, 25
Monacrosporium bembicodes, 309
Monacrosporium ellipsosporum, 310
Monacrosporium phymatopagum, 310
Monacrosporium sclerohyphum, 311
Monascus ruber, 199
Monilia pruinosa, 312
Monilia sp., 313
Monocillium humicola, See *Torulomyces lagena*
Monodictys sp., 323
Monosporascus cannonballus, 158
Monosporascus monosporus, 158
Mortierella alpina, 102
Mortierella ambigua, 103
Mortierella bisporalis, 110
Mortierella boninense, 104
Mortierella chlamydospora, 105
Mortierella echisphaera, 105
Mortierella elongata, 106
Mortierella epicladia var. *chlamydospora*, 107
Mortierella exigua, 108
Mortierella gemmifera, 109
Mortierella humilis, 110
Mortierella hyalina, 111
Mortierella hygrophia, See *Mortierella hyalina*
Mortierella isabellina, 112
Mortierella marburgenesis, See *Mortierella verticillata*
Mortierella minutissima, 113
Mortierella nana, See *Umbelopsis nana*
Mortierella tsukubaensis, 114–115
Mortierella uniramosa, 116
Mortierella verticillata, 117
Mortierella zychae, 118
Mucor circinelloides, 120
Mucor hachijoensis, 121
Mucor hiemalis f. *luteus*, 122
Mucor luteus, See *Mucor hiemalis* f. *luteus*
Mucor microsporus, 123
Mucor plumbeus, 124
Mycocentrospora acerina, 314
Mycoleptodiscus terrestris, 315–316
Myrioconium sp., 317
Myrothecium cinctum, 319
Myrothecium dimorphum, 320
Myrothecium striatisporum, See *Myrothecium cinctum*
Myrothecium verrucaria, 321

N

Naranus cryptomeriae, 322
Nectria asakawaensis, 160
Nectria fragariae, 161–162
Nectria gliocladioides, 163, 281
Nectria hachijoensis, 164–165
Nectria haematococca, 275
Nectria radicicola, 245

Neta quadriguttata, 323
Nigrospora oryzae (Teleomorph: *Khuskia oryzae*), 324
Nigrospora sacchari, 325
Nigrospora sp., 324
Nigrospora sphaerica, 325
Nodulisporiuim melonis, 326–327

O

Ochroconis constricta, See Scolecobasidium constrictum
Ochroconis humicola, See Scolecobasidium humicola
Oedocephalum nayoroense, 328–329
Oidiodendron cerealis, 331
Oidiodendron citrinum, 332
Oidiodendron flavum, 333
Ostracoderma sp., *See Chromelosporium fulvum*

P

Paecilomyces farinosus (Teleomorph: *Cordyceps memorabilis),* 335
Paecilomyces inflatus, 336
Paecilomyces javanicus, 337
Paecilomyces persicinus, 338
Paecilomyces puntoni, 339
Paecilomyces roseolus, 340
Paecilomyces variabilis, 341
Paecilomyces victoriae, 342
Papulaspora irregularis, 344
Papulaspora nishigaharanus, 345–346
Papulaspora pallidula, 347
Papulaspora pannosa, 348
Papulaspora sp. 1, 349
Papulaspora sp. 2, 350
Penicillifer fragariae, 161
Penicillium corylophilum, 352
Penicillium janthinellum, 353
Penicillium lanosum, 354
Penicillium nigricans, 355
Penicillium resticulosum, 356
Periconia byssoides, 358
Periconia macrospinosa, 359
Periconia saraswatipurensis, 360
Pestalotia sp., 361–363
Pestalotiopsis sp., *See Pestalotia* sp.
Peziza ostracoderma, 218
Pezizella lythri, 290
Pezizella oenotherae, 290
Phaeotrichosphaeria sp., 166
Phialomyces macrosporus, 364
Phialophora atrovirens, 366
Phialophora cinerescens, 367
Phialophora cyclaminis, 368, 374
Phialophora fastigiata, 369
Phialophora malorum, 370
Phialophora radicicola, 371
Phialophora richardsiae, 372
Phialophora sp. 1, 373
Phialophora sp. 2, 374
Phialophora verrucosa, 373
Phialosporae, 30–31

Phoma glomerata, 376
Phoma medicaginis var. *pinodella,* 377
Phoma sp., 378
Phoma terrestris, See Pyrenochaeta terrestris
Phomopsis sp., 379
Physalospora tucumanensis, 227
Phytophthora capsici, 43
Phytophthora cinnamomi, 3
Phytophthora cryptogea, 44
Phytophthora erythroseptica, 45
Phytophthora megasperma, 46
Phytophthora melonis, 47
Phytophthora nicotianae var. *parasitica* (heterothallic), 49
Phytophthora nicotianae var. *parasitica* (homothallic), 48
Phytophthora parasitica, See Phytophthora nicotianae var. *parasitica* (heterothallic); *Phytophthora nicotianae* var. *parasitica* (homothallic)
Pithomyces chartarum, 380, 381
Pithomyces maydicus, 381
Plectospira myriandra, 50–51
Pleospora herbarum, 419
Porosporae, 31–32
Preussia terricola, 167
Pseudobillarda agrostidis, See Robillarda agrostidis
Pycnidium-forming fungi, 25–26
Pyrenochaeta gentianicola, 383
Pyrenochaeta globosa, 384–385
Pyrenochaeta terrestris, 386
Pythiogeton ramosum, 52
Pythium acanthicum, 56
Pythium acanthophoron, 57
Pythium afertile, 58
Pythium angustatum, 59
Pythium aphanidermatum, 60–61
Pythium apleroticum, 62
Pythium butleri, See Pythium aphanidermatum
Pythium carolinianum, 63
Pythium catenulatum, 64
Pythium conidiophorum, 65
Pythium deliense, 66
Pythium dissimile, 67
Pythium dissotocum, 59, 68
Pythium echinulatum, 69
Pythium elongatum, 70
Pythium graminicola, 71
Pythium helicoides, 78
Pythium cf. *indigoferae,* 72
Pythium inflatum, 73
Pythium intermedium, 74
Pythium irregulare, 75
Pythium myriotylum, 76, 91
Pythium nayoroense, 77
Pythium oedochilum, 78
Pythium ostracodes, 78
Pythium palingenes, 78
Pythium paroecandrum, 79
Pythium periplocum, 80–81
Pythium polytylum, 78
Pythium rostratum, 82
Pythium salpingophorum, 83
Pythium spinosum, 84

Pythium splendens, 85
Pythium sulcatum, 86
Pythium sylvaticum, 87
Pythium torulosum, 88
Pythium ultimum, 89
Pythium vexans, 90
Pythium volutum, 91
Pythium zingiberis, 91
Pythium zinjeberum, 91

R

Ramichloridium anceps, 387
Ramichloridium subulatum, 388
Rhinocladiella anceps, See *Ramichloridium anceps*
Rhizoctonia baticola, See *Macrophomina phaseolina*
Rhizoctonia solani (Teleomorph: *Thanatephorus cucumeris*), 389
Rhizoctonia sp., 312, 389, 390
Rhizopus arrhizus, See *Rhizopus oryzae*
Rhizopus oryzae, 125
Robillarda agrostidis, 391
Rosellinia necatrix, 264–265

S

Saksenae vasiformis, 126
Saprolegnia anisospora, 92
Sarcinella sp., 269
Sarcopodium araliae, 392–393
Sclerotinia folicola, 317
Sclerotium bataticola, See *Macrophomina phaseolina*
Sclerotium sp., 394–395
Scolecobasidium constrictum, 396, 397
Scolecobasidium humicola, 397
Scopulariopsis asperula, 399
Scopulariopsis brevicaulis, 400
Scopulariopsis canadensis, 401
Scopulariopsis st., 157
Seed fungi, 6
Selenophoma obtusa, 402
Sepedonium chrysospermum, 403
Sepedonium sp., 404
Septonema chaetospira, 405
Soil fungi
    airborne, 1
    historical studies, 1–2
    identification
        basal knowledge for, 17–18
        binomial selection, 18
        experimentation for, 18
        morphological characteristics for, 18–20
        physiological characteristics for, 20
    isolation methods for, 14–15
    living cultures of, 471–477
    microorganisms and, 3–4
    research problems, 4–5
    root-inhabiting, 5–6
    seed fungi vs., 6
    studies of, 7–12
    types of, 1–3

Sordaria fimicola, 169
Sordaria nodulifera, 170
Sordaria tamaensis, 171
Spegazzinia tessarthra, 406
Sporidesmium bakeri, 407
Sporidesmium filiferum, 408–409
Sporobolomyces sp., 410
Sporodochium-forming fungi, 26–27
Sporoschisma sacchari, 411–412
Sporotrichum aureum, 413
Sporotrichum sp., 414
Stachybotryna hachijoensis, 164
Stachybotrys bisbyi, 415, 416
Stachybotrys elegans, 416
Stachybotrys sacchari, See *Stachybotrys bisbyi*
Stagonospora subseriata, 417
Staphylotrichum cocosporum, 418
Stemphylium botryosum (Teleomorph: *Pleospora herbarum*), 419
Sterile fungi, 33
Sympodulosporae, 32
Syncephalastrum racemosum, 127
Syncephalastrum verruculosum, 127
Synnema-forming fungi, 27

T

Taeniolella phialosperma, 420
Tetracladium setigerum, 421
Tetraploa ellisii, 422
Thanatephorus cucumeris, 389
Thielavia terricola, 172
Thielaviopsis adiposa, 423
Thielaviopsis paradoxa, 424
Thysanophora penicillioides, 425
Torula herbarum, 426
Torula sp., 427
Torulomyces lagena, 428
Trichocladium canadense, 429
Trichocladium pyriformis, 430
Trichoderma aureoviride, 432
Trichoderma hamatum, 433
Trichoderma harzianum, 434
Trichoderma koningi, 435
Trichoderma pseudokoningi, 436
Trichothecium roseum, 437
Trichurus spiralis, 438
Trinacrium iridis, 439–440
Tripospermum myrti, 441
Tritirachium sp., 442

U

Ulocladium botrytis, 443, 444
Ulocladium chartarum, 444
Ulocladium sp., 443
Umbelopsis multispora, See *Umbelopsis vinacea*
Umbelopsis nana, 128–129
Umbelopsis versiformis, See *Umbelopsis nana*
Umbelopsis vinacea, 130

V

Vermispora sp., 439, 445
Verticillium albo-atrum, 447
Verticillium cauveriana, 445
Verticillium dahliae, 447
Verticillium fungicola, 448
Verticillium hahajimaense, 449
Verticillium lecanii, 450
Verticillium malthousei, See *Verticillium fungicola*
Verticillium nubilum, 451
Verticillium sp., 453
Verticillium sphaerosporum var. *bispora*, 452
Volutella ciliata, 454

W

Wiesneriomyces javanicus, 455–456

X

Xylaria sp., 326

Z

Zopfiella curvata, 174
Zopfiella latipes, 175
Zopfiella pilifera, 176
Zygomycetes, 23
Zygomycotina, 23, 93–131
Zygorhynchus moelleri, 131

Afterword to the Second Edition

About 800 species of fungi are being catalogued in the *Index of Fungi* every year according to Hawksworth et al. (1995). If 10% of them are assumed to be soilborne, more than 500 species must be added to the number of soil fungi for the 7 years since the publication of the English edition of this book in 1994. In the revised second edition, some new species include such recent knowledge on soil fungi.

Biodiversity of soil fungi and their knowledge are significant in relation to environmental problems practically and scientifically. Most soil fungi in any location may be identified if we know nearly 300 fungal species because any soil fungus flora may not include more than 200 species (see Supplement, Chapter 1). Therefore, the revised second edition, which includes more than 350 species belonging to at least 153 genera, may meet the demand for identification with quantitative and qualitative improvement.

Tsuneo Watanabe
September 20, 2001

Afterword

About 600 species of soil fungi were described by Gilman in 1945, but no one knows how many species have been described as soil fungi since then. Total fungal species have increased by 1.7 times from 1950 to 1983. Based on records on the number of fungi in the 3rd and 7th edition of *Ainsworth & Bisby's Dictionary of the Fungi* published in 1950 and 1983, respectively, at least 1200 species of soil fungi have been described. Among 308 species described in this book, at least 20 are newly described as soil fungi.

Although there are many mycological taxonomists worldwide, all of them have specialties in particular fields, and they may be amateurs in other fields. Therefore, it is almost impossible to describe various fungi covering the whole fungal fields alone, and would probably result in misinterpretation or misidentification. However, bearing this in mind, this book was summarized on the basis of the author's own data. Therefore, all data and observations are not systematic and purposive, including both detailed observations and data, and unsatisfactory observations and poor data. However, all pictures and illustrations are based on the author's own work, collected from various sources during the past 25 years, irrespective of time, collection sites, hosts, types of works, etc. Among materials, some are common, but others are different in having compared or summarized all the data together, even if originally they were thought to be the same.

Although the results obtained are not always satisfactory, the author is happy to contribute to influencing mycological interest in soil fungi. By summarizing this book, some new species could be described, including the fresh data recently submitted for printing.

The author hopes finally for a prosperous and successful future in soil mycology, including soil fungus taxonomy.

Tsuneo Watanabe